SPACECRAFT ATTITUDE DYNAMICS

Peter C. Hughes
Professor
Institute for Aerospace Studies
University of Toronto

John Wiley & Sons
New York Chichester Brisbane Toronto Singapore

***Library of Congress Cataloging in Publication Data*:**

Hughes, Peter C. (Peter Carlisle), 1940–
Spacecraft attitude dynamics.

Bibliography
Includes index.
1. Astrodynamics. 2. Space vehicles—
Attitude control systems. I. Title.
TL1050.H84 1986 629.4'11 85-6556
ISBN 0-471-81842-9

Printed in the United States of America

10 9 8 7 6 5 4 3 2 1

For Professor Bernard Etkin

PREFACE

When the first artificial satellites were inserted into Earth orbit a quarter-century ago, the subject of spacecraft attitude stabilization barely existed. *Sputnik I* just tumbled slowly in orbit; and *Explorer I*, after a brave but ill-fated attempt at spin stabilization, ended up doing much the same thing. Notwithstanding these shaky beginnings, a complex and successful body of engineering practice has arisen: three-axis control of satellite orientation to within a small fraction of a degree is now commonplace. Indeed, no modern space vehicle, whether a low-orbit resources satellite, a geostationary communications platform, or an interplanetary probe, can accomplish mission objectives without a properly functioning attitude stabilization system.

All techniques of spacecraft attitude stabilization rest on the twin disciplines of dynamics and control, as adapted to the special problem of regulating the orientation of a relatively small but very expensive hardware package that must function for long periods of time, faultlessly, far beyond reach. This book is about *attitude dynamics*. The contiguous topics of *orbital dynamics* and "active" attitude *control* are also addressed, but only to the extent that they interact with attitude dynamics.

WHO CAN PROFIT FROM READING THIS BOOK?

I have written this book, in the first instance, for students. Surprising as it may seem considering the vastness of the relevant technical literature, no textbook with adequate coverage of this subject has been available; this book is therefore partly a response to the urgings of my own students. To understand most of the material presented, student readers will require only the usual undergraduate courses in vector dynamics and matrix algebra, and they will undoubtedly benefit from the more than 250 figures that illustrate the text. Those committed to more specialized study will want to work out some of the problems at the end of each chapter. In so doing, they will find that the problems are not simply exercises; on the contrary, they usually contain important extensions to the analysis. The student who conquers all 175 problems will have become a mature analyst in the subject.

During the preparation of this book I have also incorporated the interests of a group who, based on their proven record, deserve great credit for making spacecraft attitude dynamics more than just a collection of dynamical analyses—practicing aerospace engineers. For them, this book will provide for ready reference a coherent exposition of the fundamentals of spacecraft attitude dynamics in a unified notation. Readers interested in the operational application of basic attitude stabilization ideas will especially appreciate their application to actual spacecraft in Chapters 9, 10, and 11. In some cases, typical on-orbit flight data are included.

I am also aware of the needs of a third group of individuals—those who, through their research endeavors, continue to extend the theoretical foundations of the discipline. Many such persons work in universities, as I do; others make their contributions within a government or industrial setting. Such persons will, I hope, find this book to be an attractive reference work. I hope they will also agree with most, if not all, of the subjective comments occasionally inserted. And the advanced topics interspersed from time to time will likely lead to ideas for further research. Moreover, 350 references have been cited, placing the researcher in contact with the extensive analysis available in the technical literature. Finally, teachers should find the extensive problem sets at the end of each chapter a handy resource. (Solutions to the problems can be obtained by writing to the author.)

A QUICK LOOK AT CONTENTS

After an introductory chapter to set the context, Chapters 2 and 3 provide kinematical and dynamical fundamentals. Chapter 4 then examines a classical subject—the torque-free motion of a rigid body—from a modern viewpoint, and Chapter 5 treats the all-important subject of how energy dissipation affects the attitude stability of spinning bodies. Chapters 6 and 7, the last of the "torque-free motion" chapters, do for dual-spin systems (systems that include one or more spinning rotors) what Chapters 4 and 5 do for monospinners.

Among the most characteristic features of satellite attitude dynamics are the subtle effects produced by small environmental torques encountered in space. Chapter 8 is set aside for a treatment of this topic; it includes a detailed examination of gravitational, aerodynamic, and solar-radiation pressure torques.

The ultimate objective of this book—to explicate the dynamics underlying modern spacecraft attitude stabilization systems—is achieved in Chapters 9, 10, and 11. These final chapters are concerned, respectively, with gravity gradient stabilization, spin stabilization, and dual-spin stabilization. Within the last category, both external-rotor gyrostats and bias momentum satellites are extensively discussed.

FUNDING ACKNOWLEDGMENTS

Financial support for this book was provided by the University of Toronto Institute for Aerospace Studies (UTIAS) and by the Natural Sciences and Engineering Research Council of Canada (NSERC).

PERSONAL ACKNOWLEDGMENTS

Finally, and with pleasure, my personal acknowledgments. For brevity I will confine my public gratitude to individuals whose contributions have been crucial. My thanks to Dr. Jaap de Leeuw, Director of UTIAS, for his confidence in me; to my students, for suffering through (and helpfully criticizing) early manuscripts; to Dr. Glen Sincarsin of Dynacon Enterprises Ltd., for his technical assistance on innumerable occasions; to Sam Altman and Howard Reynaud of the Communications Research Centre, Ottawa, for their support; to Ida Krauze, for preparing the figures with care; and to Rose Yager for her expert copy editing.

Last, and most, my abiding thanks to my wife, Joanne, whose expert typing of a large and difficult manuscript was the most tangible expression of her unflagging support during the preparation of this book.

Maple Lake, Ontario **Peter C. Hughes**

CONTENTS

CHAPTER 1 INTRODUCTION 1

CHAPTER 2 ROTATIONAL KINEMATICS 6

2.1 Reference Frames and Rotations 6
2.2 Angular Displacement Parameters 15
2.3 Angular Velocity 22
2.4 Comments on Parameter Alternatives 29
2.5 Problems 31

CHAPTER 3 ATTITUDE MOTION EQUATIONS 39

3.1 Motion Equations for a Point Mass, $\mathscr{P}$ 40
3.2 Motion Equations for a System of Point Masses, $\Sigma\mathscr{P}_n$ 42
3.3 Motion Equations for a Rigid Body, $\mathscr{R}$ 55
3.4 A System with Damping, $\mathscr{R} + \mathscr{P}$ 61
3.5 A Dual-Spin System, $\mathscr{R} + \mathscr{W}$ 65
3.6 A Simple Multi-Rigid-Body System, $\mathscr{R}_1 + \mathscr{R}_2$ 70
3.7 Dynamics of a System of Rigid Bodies 76
3.8 Problems 83

CHAPTER 4 ATTITUDE DYNAMICS OF A RIGID BODY 93

4.1 Basic Motion Equations 93
4.2 Torque-Free Motion; $\mathscr{R}$ Inertially Axisymmetrical 96
4.3 Torque-Free Motion; $\mathscr{R}$ Tri-inertial 104
4.4 Stability of Motion for $\mathscr{R}$ 114
4.5 Motion of a Rigid Body Under Torque 124
4.6 Problems 129

CHAPTER 5 EFFECT OF INTERNAL ENERGY DISSIPATION ON THE DIRECTIONAL STABILITY OF SPINNING BODIES 139

5.1 Quasi-Rigid Body with an Energy Sink, $\mathscr{Q}$ 140
5.2 Rigid Body with a Point Mass Damper, $\mathscr{R} + \mathscr{P}$ 146
5.3 Problems 152

CHAPTER 6 DIRECTIONAL STABILITY OF MULTISPIN VEHICLES 156

6.1 The $\mathscr{R} + \mathscr{W}$ Gyrostat 156
6.2 Gyrostat with Nonspinning Carrier 161
6.3 The Zero Momentum Gyrostat 164
6.4 The General Case 165
6.5 System of Coaxial Wheels 178
6.6 Problems 184

CHAPTER 7 EFFECT OF INTERNAL ENERGY DISSIPATION ON THE DIRECTIONAL STABILITY OF GYROSTATS 192

7.1 Energy Sink Analyses 193
7.2 Gyrostats with Discrete Dampers 217
7.3 Problems 225

CHAPTER 8 SPACECRAFT TORQUES 232

8.1 Gravitational Torque 233
8.2 Aerodynamic Torque 248
8.3 Radiation Torques 260
8.4 Other Environmental Torques 264
8.5 Nonenvironmental Torques 269
8.6 Closing Remarks 271
8.7 Problems 272

CHAPTER 9 GRAVITATIONAL STABILIZATION 281

9.1 Context 282
9.2 Equilibria for a Rigid Body in a Circular Orbit 293
9.3 Design of Gravitationally Stabilized Satellites 313
9.4 Flight Experience 335
9.5 Problems 346

CHAPTER 10 SPIN STABILIZATION IN ORBIT 354

10.1 Spinning Rigid Body in Orbit 356
10.2 Design of Spin-Stabilized Satellites 381
10.3 Long-Term Effects of Environmental Torques, and Flight Data 400
10.4 Problems 416

CHAPTER 11 DUAL-STABILIZATION IN ORBIT: GYROSTATS AND BIAS MOMENTUM SATELLITES 423

11.1 The Gyrostat in Orbit 424
11.2 Gyrostats with External Rotors 444
11.3 Bias Momentum Satellites 455
11.4 Problems 470

APPENDIX A ELEMENTS OF STABILITY THEORY 480

A.1 Stability Definitions 481

A.2 Stability of the Origin 492

A.3 The Linear Approximation 493

A.4 Nonlinear Inferences from Infinitesimal Stability Properties 502

A.5 Liapunov's Method 504

A.6 Stability of Linear Stationary Mechanical Systems 510

A.7 Stability Ideas Specialized to Attitude Dynamics 520

APPENDIX B VECTRICES 522

B.1 Remarks on Terminology 523

B.2 Vectrices 523

B.3 Several Reference Frames 527

B.4 Kinematics of Vectrices 530

B.5 Derivative with Respect to a Vector 532

APPENDIX C LIST OF SYMBOLS 535

C.1 Lowercase Symbols 535

C.2 Uppercase Symbols 536

C.3 Lowercase Greek Symbols 538

C.4 Uppercase Greek Symbols 539

C.5 Other Notational Conventions 539

REFERENCES 541

INDEX 559

CHAPTER 1

INTRODUCTION

A spacecraft must point in the right direction. Many satellites are intended to be Earth oriented; others are intended to face the sun or certain stars of interest; still others are designed to point first at one object, then at another. Often one part of a spacecraft (a communications antenna perhaps) must point toward Earth, while another part (a solar panel) must face the sun. To achieve such mission objectives, it is evident that an attitude stabilization and control system must be an important part of spacecraft design.

The subject of this book is *spacecraft attitude dynamics*—the applied science whose aim is to understand and predict how spacecraft orientation evolves. As a part of the larger science of dynamics, the book deals with the special problems associated with rotational motion, including how to describe the motion (see Chapter 2); how to formulate differential equations that govern the motion (Chapter 3); and how to make physically meaningful inferences from these equations (Chapters 4 through 7). And as part of spacecraft technology, it seeks to predict the torques acting on specific spacecraft (Chapter 8) and what attitude motion these torques will cause, including how to minimize undesirable motions through the application of suitable design strategies (Chapters 9, 10, and 11).

Spacecraft attitude dynamics does not, of course, exist in splendid isolation—it interacts with many other sister disciplines. Indeed, the closer one comes to applying the principles of spacecraft attitude dynamics to actual spacecraft, the more these interrelationships manifest themselves. Some of the most important interfaces are shown in Fig. 1.1.

Orbit Interface

In reality, attitude dynamics (rotational dynamics) and orbital dynamics (translational dynamics) are mutually coupled. (This tends to be less true, of

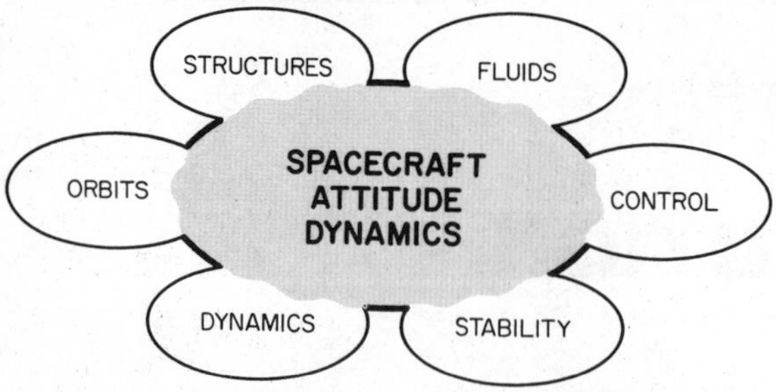

FIGURE 1.1 Disciplines related to spacecraft attitude dynamics.

course, for those occasional spacecraft whose missions are interplanetary in nature.) Even if gravity is the only force field considered (the "classical" problem), orbit dynamics affects attitude dynamics, and vice versa. Because gravity is a conservative force field, the system energy, momentum, and angular momentum are conserved; but note that this "system" includes the translation and rotation of both the satellite and its gravitational primary (usually Earth). According to such a theory, the translational and rotational motions of both Earth and the satellite are mutually coupled by their gravitational interaction, not to mention other celestial bodies in the solar system.

Now it is certainly reasonable to assume that Earth's motion is negligibly affected by the presence of a small artificial satellite. Nevertheless, there remains the coupling between satellite translation (the orbital variables) and spacecraft rotation (the attitude variables); this coupling will be discussed in Section 9.1. But there are many nongravitational sources of coupling as well. As an example of orbit affecting attitude, the torques experienced by a satellite often depend on altitude (Fig. 8.15). As an example of attitude affecting orbit, interplanetary solar sailing, using solar radiation pressure, is based on adjusting attitude to produce the desired trajectory. As a more mundane example, a vernier thruster affects both attitude and orbit—unless its line of action is through the mass center, in which case it affects only orbit, or unless two equal and opposite thrusters are fired as a pair, in which case they affect only attitude.

In spite of these various coupling effects (indicated by the intersection of the "spacecraft attitude dynamics" and the "orbit" ellipses in Fig. 1.1), much of attitude dynamics analysis proceeds by largely ignoring orbital effects. There are two justifications for this policy. First, many spacecraft are either spinning or contain rotors or wheels that spin; to a good approximation, these can be analyzed as though they were isolated from all external influences. Chapters 4 through 7 are based on this type of analysis. The immediate goal is to ascertain whether a given basic design is asymptotically attitude stable; if so, external torques, whether orbitally induced or otherwise, can be admitted subsequently as disturbances. Second, for those spacecraft which rely on an environmental torque for their stabilization, and to the extent that this environmental torque depends on orbital variables, the effect of orbit on attitude is included in the analyses. Other torques are again treated as disturbances.

Structures Interface

Although real bodies are not rigid, much of spacecraft attitude dynamics relies on "rigid body" dynamics, and some aspects of the observed behavior of spacecraft orientation can be explained on this basis. The advantage a "rigid body" brings to the analysis is that it contributes, at most, six degrees of freedom, and only three of these are rotational. Unfortunately, a rigid body is essentially a geometrical abstraction; very little physics is represented. The extra internal degrees of freedom possessed by a real body contribute important coupling effects, among the most important of which is energy dissipation.

Rigid-body dynamics applies the Second Law of Motion precisely, while ignoring the Second Law of Thermodynamics entirely. Perhaps it should not be surprising that analysis from which perpetual-motion machines can be deduced is not always a reliable guide in the design of real spacecraft. This is the lesson of Explorer I. On the other hand, accurate structural analysis implies an enormous increase in math model complexity; indeed, for some spacecraft, especially those of recent vintage and those soon to be constructed in space, complex dynamical models are unavoidable. Nevertheless, an understanding of the basic principles of passive stabilization requires that only two general properties of structural dynamics be dealt with: the effect of structural energy dissipation on the attitude stability of spinning (or partly spinning) spacecraft, and the adverse effects of internal degrees of freedom on the Major-Axis Rule for spinners and dual spinners. Moreover, the discussion of these topics will be handled in such a manner that the reader will find it relatively easy to extrapolate to more general cases. Detailed structural analysis will be reluctantly excluded on the grounds that an adequate treatment would require a much larger book.

Fluids Interface

The dynamics of fluids also bears on the subject of spacecraft attitude dynamics in important ways. The sloshing of thruster fuel and the transport of cooling liquid for thermal control are examples that come to mind in this connection. Indeed, the dynamical problem of a slug of fuel that is sloshing about, in free fall, inside a fuel tank, with a free surface (or, worse still, separated into discrete globlets) has not yet been solved in a form suitable for incorporation into attitude dynamics modeling. Fluid dynamics is outside the scope of this book, although one of its chief influences, energy dissipation, will be represented heuristically by the Energy Sink Hypothesis in Chapters 5 and 7.

Dynamics Interface

Many formulations of dynamics exist, and each has its benefits and limitations. Conclusions about suitability can be made even more sharply for attitude dynamics in particular [Likins, 4], although for linear analyses many of the distinctions disappear. In any case, the competent analyst will be familiar with all methods and their idiosyncrasies. Here, a vectorial mechanics approach will be taken (Chapter 3) to analyze a sequence of fundamental dynamical models each of which illustrates an important principle. These basic models will then be used through the remainder of the book. There does remain, however, a substantial element of personal taste in the choice of dynamical procedures, and the motion equations used in Chapters 4 through 7 and Chapters 9 through 11 can also be derived using other equally legitimate formulations.

Stability Interface

Exact differential equations for spacecraft attitude motion are nonlinear and cannot generally be solved in closed form. The first logical step is to locate equilibria, that is, to identify conditions under which the system is either at rest or in uniform motion. Each such equilibrium suggests a potential attitude stabilization scheme—especially if the equilibrium is stable. Much of this book is taken up with a discussion of such equilibria and their stability properties.

The subject of stability is itself a subtle one, especially when applied to mechanical systems, where many specialized results are available. In the absence of an equivalent reference source, the most important elements of stability theory, as related especially to spacecraft attitude dynamics, are summarized in Appendix A. Several original results are also given. This appendix will be called upon repeatedly in the text.

Control Interface

In practice, there is no clear-cut interface between spacecraft attitude dynamics and spacecraft attitude control. Most modern spacecraft are stabilized through a combination of "passive" control (meaning dynamics only) and "active" control (meaning at least one sensor, one actuator, one energy source, and a control law). However, we shall explore in this book only passive stabilization techniques in detail, including the interface between passive and active control. Strictly "active" control is a subject left to other sources.

Analysis vis-à-vis Simulation

Many of the differential equations governing spacecraft motion cannot be solved analytically in their "exact" form (Fig. 1.2). The general kinematical equations of Chapter 2 and the general motion equations of Chapter 3 are inherently nonlinear. Exact solutions can usually be found only in a numerical sense from some type of computer simulation. Such simulation may be carried out either on an analog computer, with the motion equations "patched" onto the appropriate analog circuitry, or on a digital computer using established methods for the numerical solution of differential equations. In either case, the output is essentially numerical—sheets filled with numbers, or wavy time plots. It is difficult to learn anything of a fundamental nature about the subject from these sheets and plots (although precise information is thereby provided about a very

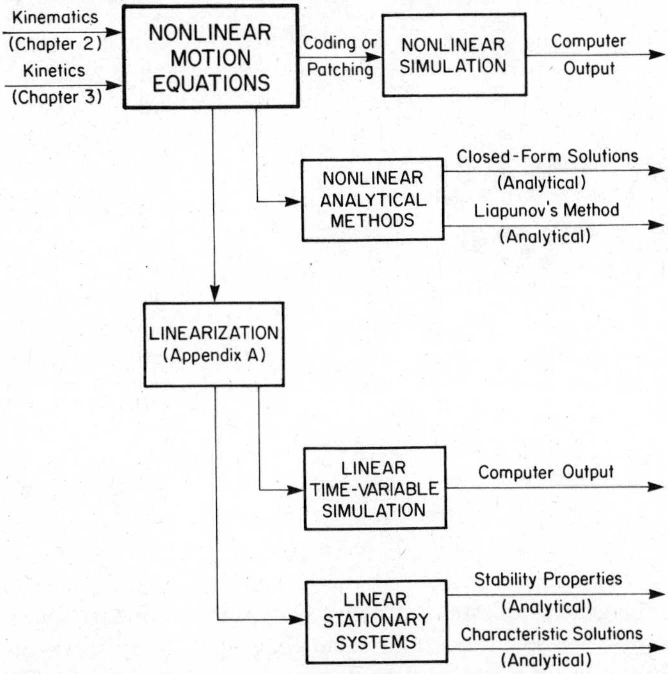

FIGURE 1.2 Analysis vis-à-vis simulation.

specific configuration and situation). Therefore, as with most text/reference books, we shall only rarely give simulation results, preferring instead to build a framework of analytical approaches to specialized and approximate cases. This framework, taken as a whole, generates a deep and fundamental understanding that cannot be had from any amount of numerical computer output.

Certain opportunities do arise, however, for closed-form nonlinear results. Examples include the analytical solution to Euler's motion equations for a rigid body (Section 4.2) and the bounding of motions on the unit sphere (e.g., Fig. 9.10) in the spirit of Liapunov's method (Section A.5). Notwithstanding such opportunities, most analytical progress must be made via the process of linearization.

Even linearization does not permit analytical solutions if the resulting motion equations have *time-variable* coefficients; once again one must resort to computer simulation. For linear *stationary* systems, however, a wide range of analytical techniques are available, including closed-form expressions for the stability boundaries in parameter space. This will therefore prove to be the primary methodology used to derive the analytical results in this book.

CHAPTER 2

ROTATIONAL KINEMATICS

Before confronting the general question of how to describe the orientation of all or part of a spacecraft, we require certain basic concepts, among them the techniques for describing the orientation of *vectors* and *reference frames*. (See Appendix B for the meaning of "vector" as used in this book.) Figure 2.1 depicts some examples: we may wish to describe the direction of the sun, Earth's center, Polaris, or the local magnetic field; we may also wish to describe the orientation of the vehicle structure, a solar-cell array, an internal spinning wheel, or Earth. The orientation of a vector can be given only with respect to other *reference* vectors. As is well known, the minimum number of (noncoplanar) reference vectors is three, and the orientation of any vector $\underrightarrow{\mathbf{v}}$ is uniquely specified by the three direction cosines between $\underrightarrow{\mathbf{v}}$ and these reference vectors. The latter constitute a *reference frame*. Thus, in Fig. 2.1, the orientation of any vector of interest can be specified with respect to any reference frame of interest.

This idea has an important extension. We speak of the orientation of one frame $\mathscr{F}_a$ with respect to a second frame $\mathscr{F}_b$, meaning that the orientation of the three constituent vectors in $\mathscr{F}_a$ is known with respect to $\mathscr{F}_b$. Furthermore, it is known—read [Goldstein], for example—that the orientation of a *rigid* body can be defined as the orientation of a reference frame that is fixed, or *imbedded* within it. Using Fig. 2.1 again as an example, the orientations of Earth, the main structure, the wheel, and the solar-cell array can be specified, each with respect to the other, assuming them to be rigid bodies.

2.1 REFERENCE FRAMES AND ROTATIONS

The most convenient set of reference vectors is a *dextral* (i.e., right-handed), *orthonormal* (i.e., mutually perpendicular and of unit length) *triad*, and we shall

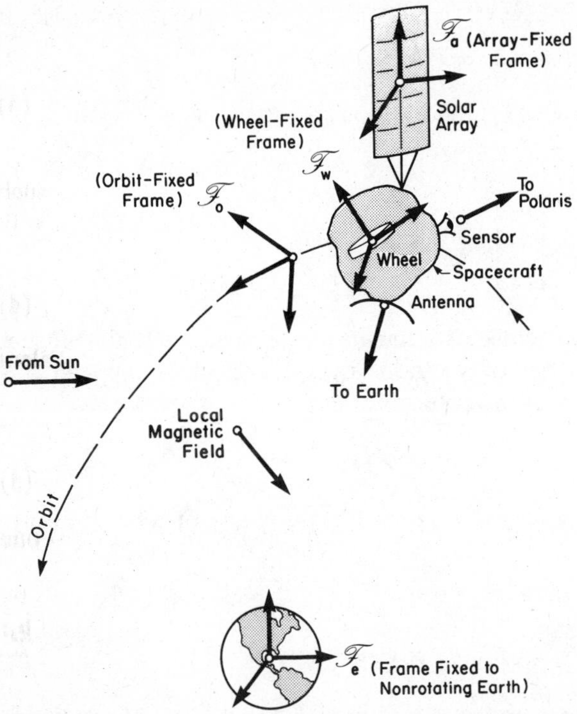

FIGURE 2.1 Sampling of the many reference directions and reference frames that may be required in a typical attitude dynamics analysis.

use these exclusively. Such a triad is given the descriptive term *reference frame*. Denote by $\mathscr{F}_a$ a reference frame whose three constituent vectors are $\underset{\rightarrow}{\hat{\mathbf{a}}}_1$, $\underset{\rightarrow}{\hat{\mathbf{a}}}_2$, and $\underset{\rightarrow}{\hat{\mathbf{a}}}_3$ (Fig. 2.2), and denote by c_1, c_2, and c_3 the direction cosines of $\underset{\rightarrow}{\mathbf{v}}$ with respect to $\mathscr{F}_a$. Then, we write

$$\underset{\rightarrow}{\mathbf{v}} = v(c_1\underset{\rightarrow}{\hat{\mathbf{a}}}_1 + c_2\underset{\rightarrow}{\hat{\mathbf{a}}}_2 + c_3\underset{\rightarrow}{\hat{\mathbf{a}}}_3) \tag{1}$$

where v is the length of $\mathbf{v}$. Defining $v_i \triangleq vc_i$ $(i = 1, 2, 3)$, we have the alternate

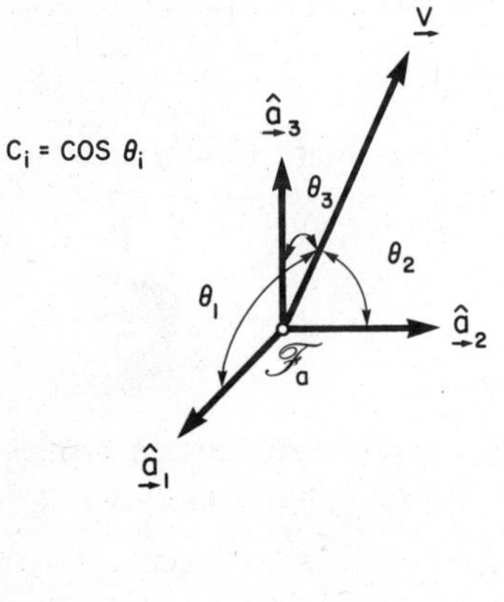

FIGURE 2.2 Direction cosines between a vector $\underset{\rightarrow}{\mathbf{v}}$ and a frame $\mathscr{F}_a$.

form

$$\underset{\rightarrow}{\mathbf{v}} = v_1\hat{\underset{\rightarrow}{\mathbf{a}}}_1 + v_2\hat{\underset{\rightarrow}{\mathbf{a}}}_2 + v_3\hat{\underset{\rightarrow}{\mathbf{a}}}_3 \tag{2}$$

Some writers call v_i (others call $v_i\hat{\underset{\rightarrow}{\mathbf{a}}}_i$) the *components* of $\underset{\rightarrow}{\mathbf{v}}$ in $\mathscr{F}_a$.

Now define two column matrices thus:

$$\mathbf{v} \triangleq \begin{bmatrix} v_1 \\ v_2 \\ v_3 \end{bmatrix}; \qquad \underset{\rightarrow}{\mathscr{F}}_a \triangleq \begin{bmatrix} \hat{\underset{\rightarrow}{\mathbf{a}}}_1 \\ \hat{\underset{\rightarrow}{\mathbf{a}}}_2 \\ \hat{\underset{\rightarrow}{\mathbf{a}}}_3 \end{bmatrix} \tag{3}$$

The second is not conventional, since its elements are vectors. We shall call such matrices *vectrices*; the properties of vectrices are the subject of Appendix B. Definitions (3) now facilitate a compact form for (2):

$$\underset{\rightarrow}{\mathbf{v}} = \mathbf{v}^T\underset{\rightarrow}{\mathscr{F}}_a \equiv \underset{\rightarrow}{\mathscr{F}}_a^T\mathbf{v} \tag{4}$$

Conversely, we may write $\mathbf{v}$ in terms of $\underset{\rightarrow}{\mathbf{v}}$ and $\underset{\rightarrow}{\mathscr{F}}_a$ using any of four equivalent expressions:

$$\mathbf{v} = \underset{\rightarrow}{\mathscr{F}}_a \cdot \underset{\rightarrow}{\mathbf{v}} \equiv \underset{\rightarrow}{\mathbf{v}} \cdot \underset{\rightarrow}{\mathscr{F}}_a$$

$$\mathbf{v}^T = \underset{\rightarrow}{\mathscr{F}}_a^T \cdot \underset{\rightarrow}{\mathbf{v}} \equiv \underset{\rightarrow}{\mathbf{v}} \cdot \underset{\rightarrow}{\mathscr{F}}_a^T \tag{5}$$

In summary, each frame $\mathscr{F}$ has an associated vectrix $\underset{\rightarrow}{\mathscr{F}}$ in terms of which one can compactly write the orientation, with respect to $\mathscr{F}$, of any vector.

Consider next two reference frames, $\mathscr{F}_a$ and $\mathscr{F}_b$, and denote their vectrices by $\underset{\rightarrow}{\mathscr{F}}_a$ and $\underset{\rightarrow}{\mathscr{F}}_b$. A relation of the form (1) can be written for each of $\hat{\underset{\rightarrow}{\mathbf{b}}}_1$, $\hat{\underset{\rightarrow}{\mathbf{b}}}_2$, and $\hat{\underset{\rightarrow}{\mathbf{b}}}_3$:

$$\begin{aligned} \hat{\underset{\rightarrow}{\mathbf{b}}}_1 &= c_{11}\hat{\underset{\rightarrow}{\mathbf{a}}}_1 + c_{12}\hat{\underset{\rightarrow}{\mathbf{a}}}_2 + c_{13}\hat{\underset{\rightarrow}{\mathbf{a}}}_3 \\ \hat{\underset{\rightarrow}{\mathbf{b}}}_2 &= c_{21}\hat{\underset{\rightarrow}{\mathbf{a}}}_1 + c_{22}\hat{\underset{\rightarrow}{\mathbf{a}}}_2 + c_{23}\hat{\underset{\rightarrow}{\mathbf{a}}}_3 \\ \hat{\underset{\rightarrow}{\mathbf{b}}}_3 &= c_{31}\hat{\underset{\rightarrow}{\mathbf{a}}}_1 + c_{32}\hat{\underset{\rightarrow}{\mathbf{a}}}_2 + c_{33}\hat{\underset{\rightarrow}{\mathbf{a}}}_3 \end{aligned} \tag{6}$$

where c_{ij} is the direction cosine between $\hat{\underset{\rightarrow}{\mathbf{b}}}_i$ and $\hat{\underset{\rightarrow}{\mathbf{a}}}_j$. Just as (2) was compactly written as (4) with the aid of vectrices, (6) can be expressed as follows:

$$\underset{\rightarrow}{\mathscr{F}}_b = \mathbf{C}_{ba}\underset{\rightarrow}{\mathscr{F}}_a \tag{7}$$

where $\mathbf{C}_{ab} = \{c_{ij}\}$. The direction cosine matrix $\mathbf{C}_{ab}$ is a rotation matrix, whose properties will shortly be derived. For brevity, we may write simply $\mathbf{C}$ for $\mathbf{C}_{ab}$ if only two reference frames are involved.

Orthonormal Matrices

Because $\mathbf{C}$ relates two orthonormal dextral triads, it possesses certain properties. To state these properties, we introduce some needed definitions. An $n \times n$

real matrix $\mathbf{A}$ is *row-orthogonal* if the inner products between its rows are zero. Denote the rows of $\mathbf{A}$ by $[\mathbf{r}_1, \ldots, \mathbf{r}_n] \triangleq \mathbf{A}^T$. Then, for row-orthogonality, $\mathbf{r}_i^T\mathbf{r}_j = 0$ $(i \neq j)$, leading to the conclusion that $\mathbf{A}\mathbf{A}^T$ is a diagonal matrix. Similarly, denoting the columns of $\mathbf{A}$ by $[\mathbf{c}_1, \ldots, \mathbf{c}_n] \triangleq \mathbf{A}$, we say that $\mathbf{A}$ is *column-orthogonal* if the inner products between its columns are zero, that is, $\mathbf{c}_i^T\mathbf{c}_j = 0$ $(i \neq j)$. This implies that $\mathbf{A}^T\mathbf{A}$ is a diagonal matrix. If $\mathbf{A}$ is both row- and column-orthogonal, it is said to be *orthogonal*. For example, by these definitions, the matrix

$$\begin{bmatrix} 1 & 0 \\ 0 & 2 \end{bmatrix}$$

is orthogonal. If, in addition, the rows of $\mathbf{A}$ have a (Euclidean) norm of unity ($\mathbf{r}_i^T\mathbf{r}_i = 1$), then $\mathbf{A}$ is row-normal; similarly, if the columns of $\mathbf{A}$ have a norm of unit magnitude, $\mathbf{A}$ is column-normal, and $\mathbf{c}_i^T\mathbf{c}_i = 1$. When the properties of orthogonality and normality are both present, we speak of *orthonormality*. It is straightforward to show that if the rows of $\mathbf{A}$ are orthonormal, that is, if

$$\mathbf{A}\mathbf{A}^T = \mathbf{1} \tag{8}$$

then the columns of $\mathbf{A}$ are orthonormal also:

$$\mathbf{A}^T\mathbf{A} = \mathbf{1} \tag{9}$$

To derive (9) from (8), pre- and postmultiply (8) by $\mathbf{A}^{-1}$ and $\mathbf{A}$, respectively. Likewise, (8) follows from (9), and so we say simply that $\mathbf{A}$ is *orthonormal*. Orthonormal matrices have the convenient property that

$$\mathbf{A}^{-1} = \mathbf{A}^T \tag{10}$$

We conclude our discussion of orthonormal matrices by observing that their determinant is always ± 1. To demonstrate this, form the determinant of (9):

$$1 = \det \mathbf{1} = \det \mathbf{A}^T\mathbf{A} = \det \mathbf{A}^T \det \mathbf{A} = \det{}^2\mathbf{A}$$

and therefore, for an orthonormal matrix $\mathbf{A}$,

$$\det \mathbf{A} = \pm 1 \tag{11}$$

Rotation Matrices

Let us now show that the transformation matrix in (7), $\mathbf{C}$, is orthonormal. We have

$$\mathcal{F}_b \cdot \mathcal{F}_b^T = \mathbf{C}\mathcal{F}_a \cdot \mathcal{F}_a^T\mathbf{C}^T$$

Because $\mathcal{F}_b \cdot \mathcal{F}_b^T$ and $\mathcal{F}_a \cdot \mathcal{F}_a^T$ reduce to unit matrices (see Appendix B),

$$\mathbf{1} = \mathbf{C}\mathbf{C}^T \tag{12}$$

whence **C** is orthonormal. Orthonormality implies 12 scalar relationships involving the coefficients c_{ij}. These are given in Problem 2.24.

We are also able to resolve the sign ambiguity in (11). Since $\mathscr{F}_a$ and $\mathscr{F}_b$ are both right-handed,

$$\underset{\rightarrow}{\hat{\mathbf{a}}}_1 \times \underset{\rightarrow}{\hat{\mathbf{a}}}_2 = \underset{\rightarrow}{\hat{\mathbf{a}}}_3 \qquad \underset{\rightarrow}{\hat{\mathbf{a}}}_2 \times \underset{\rightarrow}{\hat{\mathbf{a}}}_3 = \underset{\rightarrow}{\hat{\mathbf{a}}}_1 \qquad \underset{\rightarrow}{\hat{\mathbf{a}}}_3 \times \underset{\rightarrow}{\hat{\mathbf{a}}}_1 = \underset{\rightarrow}{\hat{\mathbf{a}}}_2$$

$$\underset{\rightarrow}{\hat{\mathbf{b}}}_1 \times \underset{\rightarrow}{\hat{\mathbf{b}}}_2 = \underset{\rightarrow}{\hat{\mathbf{b}}}_3 \qquad \underset{\rightarrow}{\hat{\mathbf{b}}}_2 \times \underset{\rightarrow}{\hat{\mathbf{b}}}_3 = \underset{\rightarrow}{\hat{\mathbf{b}}}_1 \qquad \underset{\rightarrow}{\hat{\mathbf{b}}}_3 \times \underset{\rightarrow}{\hat{\mathbf{b}}}_1 = \underset{\rightarrow}{\hat{\mathbf{b}}}_2$$

it follows from (6) that

$$\begin{aligned} c_{11} &= c_{22}c_{33} - c_{23}c_{32} \\ c_{21} &= c_{32}c_{13} - c_{33}c_{12} \\ c_{31} &= c_{12}c_{23} - c_{13}c_{22} \end{aligned} \tag{13a}$$

$$\begin{aligned} c_{12} &= c_{23}c_{31} - c_{21}c_{33} \\ c_{22} &= c_{33}c_{11} - c_{31}c_{13} \\ c_{32} &= c_{13}c_{21} - c_{11}c_{23} \end{aligned} \tag{13b}$$

$$\begin{aligned} c_{13} &= c_{21}c_{32} - c_{22}c_{31} \\ c_{23} &= c_{31}c_{12} - c_{32}c_{11} \\ c_{33} &= c_{11}c_{22} - c_{12}c_{21} \end{aligned} \tag{13c}$$

These nine relationships among the elements of **C**, which are not independent, may be neatly summarized in terms of the adjoint matrix adj **C**:

$$\mathbf{C}^T = \operatorname{adj} \mathbf{C} \tag{14}$$

We next recall the general result for the inverse of any square matrix **M**:

$$\mathbf{M}^{-1} = \frac{\operatorname{adj} \mathbf{M}}{\det \mathbf{M}}$$

Applying this formula to **C** and noting that $\mathbf{C}^{-1} = \mathbf{C}^T$, we conclude that

$$\det \mathbf{C} = +1 \tag{15}$$

We summarize these developments by defining a rotation matrix as a 3×3 orthonormal matrix whose determinant is $+1$. For a discussion of rotation matrices as a *mathematical group*, see [Mayer].

Euler's Theorem

Euler observed that the general displacement of a rigid body with one fixed point is a rotation about an axis through that point (Euler's theorem). In the present context, Euler's theorem requires that the displacement of $\mathscr{F}_b$ with

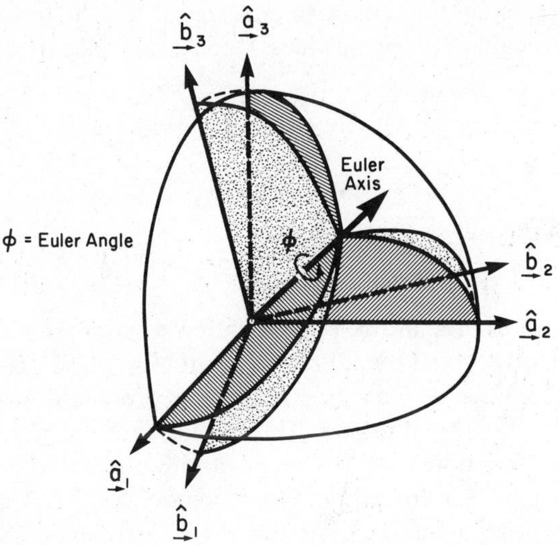

FIGURE 2.3 Geometry pertaining to Euler's theorem.

respect to $\mathscr{F}_a$ be a rotation about some axis through their common origin (Fig. 2.3). We can determine the location of this axis, and the angular rotation about it, by examining the eigenvalues and eigencolumns of the rotation matrix **C**.

Let **e** be an eigencolumn of **C**, and λ the corresponding eigenvalue; then,

$$\mathbf{Ce} = \lambda \mathbf{e} \tag{16}$$

Note that **e** and λ may be complex. We first show that λ must have unit magnitude; premultiply (16) by its Hermitian conjugate to deduce that $(\bar{\lambda}\lambda - 1)\mathbf{e}^H\mathbf{e} = 0$. Since **e** is nontrivial, it follows that $\bar{\lambda}\lambda = 1$, whence λ has unit magnitude. Now at least one of the three eigenvalues of **C**, say λ_1, must be real, and therefore the other two are, in general, complex: $\lambda_{2,3} = \exp(\pm j\phi)$. The determinant of any matrix, including **C**, is equal to the product of its eigenvalues: $\det \mathbf{C} = \lambda_1\lambda_2\lambda_3 = +1$. Therefore, the real eigenvalue is

$$\lambda_1 = +1 \tag{17}$$

The corresponding eigencolumn, which we denote by **a**, has from (16) the property that

$$\mathbf{Ca} = \mathbf{a} \tag{18}$$

and the normalization

$$\mathbf{a}^T\mathbf{a} = 1 \tag{19}$$

is used. The symbol **a** is employed because the vector $\mathscr{F}_a^T\mathbf{a} = \mathscr{F}_b^T\mathbf{a}$ is precisely the *axis of rotation* referred to in Euler's theorem; the angle ϕ, furthermore, is the *angle of rotation* mentioned in Euler's theorem.

To justify these last statements, we continue our examination of **C**. The trace of any matrix is known to equal the sum of its eigenvalues. Denote the trace by σ; then,

$$\sigma \triangleq \operatorname{trace} \mathbf{C} = c_{11} + c_{22} + c_{33} = \lambda_1 + \lambda_2 + \lambda_3 = 1 + e^{j\phi} + e^{-j\phi} = 1 + 2\cos\phi \tag{20}$$

We can now show that the matrix

$$\mathbf{M} \triangleq \cos\phi\mathbf{1} + (1 - \cos\phi)\mathbf{a}\mathbf{a}^T - \sin\phi\mathbf{a}^\times \tag{21}$$

is, in fact, an expression for **C** in terms of **a** and ϕ. (The notation $\mathbf{a}^\times$ is explained in Appendix B.) **M** is shown to be orthogonal by a direct computation of $\mathbf{M}^T\mathbf{M}$; and the identities $\mathbf{a}^\times\mathbf{a} = \mathbf{0}$, $\mathbf{a}^T\mathbf{a} = 1$, and $\mathbf{a}^\times\mathbf{a}^\times = \mathbf{a}\mathbf{a}^T - \mathbf{1}$ reduce the product down to $\mathbf{M}^T\mathbf{M} = \mathbf{1}$. Therefore, **M** is an orthonormal matrix. This implies that $\det \mathbf{M} = \pm 1$. Since $\det \mathbf{M}$ is a continuous function of ϕ, however, and since $\det \mathbf{M}$ is evidently $+1$ when $\phi = 0$, it must be $+1$ for all ϕ. We conclude that **M** is a rotation matrix. To complete the proof that **M** is identical to **C**, it remains to demonstrate that $\mathbf{M}\mathbf{a} = \mathbf{a}$ and that $\operatorname{trace} \mathbf{M} = \sigma$, both of which follow immediately from (20) and (21). In summary, we have confirmed that

$$\mathbf{C} = \cos\phi\mathbf{1} + (1 - \cos\phi)\mathbf{a}\mathbf{a}^T - \sin\phi\mathbf{a}^\times \tag{22}$$

This result is an important one and will be the basis of the kinematical formulations described in the remainder of the chapter. Written out in full, (22) becomes

$$\begin{aligned}
c_{11} &= (1 - \cos\phi)a_1^2 + \cos\phi \\
c_{22} &= (1 - \cos\phi)a_2^2 + \cos\phi \\
c_{33} &= (1 - \cos\phi)a_3^2 + \cos\phi \\
c_{12} &= (1 - \cos\phi)a_1a_2 + a_3\sin\phi \\
c_{21} &= (1 - \cos\phi)a_2a_1 - a_3\sin\phi \\
c_{23} &= (1 - \cos\phi)a_2a_3 + a_1\sin\phi \\
c_{32} &= (1 - \cos\phi)a_3a_2 - a_1\sin\phi \\
c_{31} &= (1 - \cos\phi)a_3a_1 + a_2\sin\phi \\
c_{13} &= (1 - \cos\phi)a_1a_3 - a_2\sin\phi
\end{aligned}$$

Lastly, we confirm the alleged interpretation of $\underrightarrow{a}$ and ϕ as the axis and angle in Euler's theorem. Let $\underrightarrow{v}$ be an arbitrary vector; as shown in Fig. 2.4, $\underrightarrow{v}$ can be written as the sum of a vector parallel to $\underrightarrow{a}$ and a vector perpendicular to $\underrightarrow{a}$. Furthermore, let the components of $\underrightarrow{v}$ in $\mathscr{F}_a$ by $\mathbf{v}_a = \underrightarrow{\mathscr{F}}_a \cdot \underrightarrow{v}$. As $\mathscr{F}_a$ rotates about $\underrightarrow{a}$ through an angle ϕ, it will appear to an observer fixed in $\mathscr{F}_a$ that $\underrightarrow{v}$ is rotating about $\underrightarrow{a}$ through an angle $-\phi$; to this observer, then, the rotation corresponds to $\underrightarrow{v} \to \underrightarrow{v}'$, where

$$\underrightarrow{v}' = (\underrightarrow{a} \cdot \underrightarrow{v})\underrightarrow{a} - \underrightarrow{a} \times (\underrightarrow{a} \times \underrightarrow{v})\cos\phi - \underrightarrow{a} \times \underrightarrow{v}\sin\phi$$

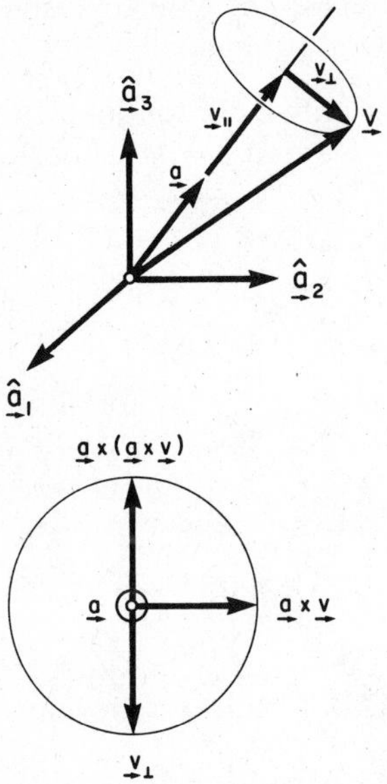

FIGURE 2.4 Geometrical interpretation of the rotation matrix **C**.

So the components of $\underset{\rightarrow}{v}'$ in $\mathscr{F}_b$ should be

$$\mathbf{v}_b = \left[\mathbf{a}\mathbf{a}^T + (\mathbf{1} - \mathbf{a}\mathbf{a}^T)\cos\phi - \mathbf{a}^\times \sin\phi\right]\mathbf{v}_a$$

But this is precisely $\mathbf{v}_b = \mathbf{C}\mathbf{v}_a$, with $\mathbf{C}$ given by (22). This shows the validity of the interpretation of $\mathbf{a}$ as the axis of rotation (containing the components of $\underset{\rightarrow}{a}$ in both $\mathscr{F}_b$ and $\mathscr{F}_a$) and of ϕ as the angle of (right-handed) rotation about $\underset{\rightarrow}{a}$.

Finding a and ϕ from C

Based on the preceding results, an algorithm is now given for calculating $\mathbf{a}$ and ϕ when $\mathbf{C}$ is known. Equation (20) provides the first two steps:

(i) $\sigma = c_{11} + c_{22} + c_{33}$

(ii) $\cos\phi = \frac{1}{2}(\sigma - 1)$ (23)

The calculation of $\mathbf{a}$ is based on the observation, from (22), that $-\sin\phi\mathbf{a}^\times$ is the skew-symmetric portion of $\mathbf{C}$. We recall that any square matrix $\mathbf{M}$ can be written as the sum of a symmetric matrix, $\frac{1}{2}(\mathbf{M} + \mathbf{M}^T)$, and a skew-symmetric matrix, $\frac{1}{2}(\mathbf{M} - \mathbf{M}^T)$. We deduce that

$$\mathbf{a}^\times = \frac{1}{2\sin\phi}(\mathbf{C}^T - \mathbf{C}) \tag{24}$$

provided $\phi \neq 0, \pm\pi, \pm 2\pi, \ldots$, that is, provided $\sigma \neq +3, -1$. When $\sigma = 3$, then $\phi = 0, \pm 2\pi, \pm 4\pi, \ldots$, and $\mathbf{C} = \mathbf{1}$; this means that there has been either no rotation or a number of complete rotations through 2π. In either case, the axis of rotation cannot be determined. When $\sigma = -1$, however, $\phi = \pm\pi, \pm 3\pi, \ldots$, and

$$\mathbf{C} = 2\mathbf{a}\mathbf{a}^T - \mathbf{1}$$

In the light of these observations, the algorithm begun with (23) continues as follows:

$$\text{(iii)} \quad \left.\begin{aligned} a_1 &= \frac{1}{2}\frac{c_{23} - c_{32}}{\sin\phi} \\ a_2 &= \frac{1}{2}\frac{c_{31} - c_{13}}{\sin\phi} \\ a_3 &= \frac{1}{2}\frac{c_{12} - c_{21}}{\sin\phi} \end{aligned}\right\} \quad (\sigma \neq 3, -1) \tag{25}$$

$$\text{(iv)} \quad \mathbf{a} \text{ undefined if } \sigma = +3 \tag{26}$$

$$\text{(v)} \quad \left.\begin{aligned} a_1 &= \pm\left(\frac{1 + c_{11}}{2}\right)^{1/2} \\ a_2 &= \pm\left(\frac{1 + c_{22}}{2}\right)^{1/2} \\ a_3 &= \pm\left(\frac{1 + c_{33}}{2}\right)^{1/2} \\ a_1 a_2 &= \frac{c_{12}}{2} \\ a_2 a_3 &= \frac{c_{23}}{2} \\ a_3 a_1 &= \frac{c_{31}}{2} \end{aligned}\right\} \quad (\sigma = -1) \tag{27}$$

As noted by [Meyer], the last three equations in (27) serve to resolve the internal sign ambiguities in the first three.

The only remaining point to clarify is the uniqueness of $\mathbf{a}$ and ϕ, given $\mathbf{C}$. It is plain that there is a one-to-one correspondence between $\mathbf{C}$ and the orientation of $\mathscr{F}_b$ with respect to $\mathscr{F}_a$. It is equally clear from (22) that there is a unique rotation matrix $\mathbf{C}$ for a given $(\mathbf{a}, \phi)$ pair. The nonuniqueness is in the other direction: given $\mathbf{C}$, an infinitude of $(\mathbf{a}, \phi)$ pairs can be deduced, unless we wish to artificially introduce uniqueness. Equation (23ii) reflects this. If ϕ satisfies $\cos\phi = (\sigma - 1)/2$, then $\phi' \triangleq 2\pi - \phi$ does so also. From (24), the axis of rotation has the opposite sense: a rotation ϕ about $\mathbf{a}$ brings $\mathscr{F}_b$ to the same position as a rotation $2\pi - \phi$ about $-\mathbf{a}$. One can also add any integral multiple of 2π to ϕ without changing $\mathbf{C}$. Uniqueness may be achieved, however, simply by restricting ϕ to the range $0 \leq \phi \leq \pi$, when this is desirable. It may not always be desirable; for example, a body spinning about a fixed axis is more naturally thought of as having a fixed axis of rotation and ϕ increasing to arbitrarily large values than as having an axis whose sense reverses every π radians with ϕ' alternately positive and negative. Be that as it may, a general attitude displacement has three degrees of freedom, represented here by ϕ and the three elements of $\mathbf{a}$ under the constraint $\mathbf{a}^T\mathbf{a} = 1$.

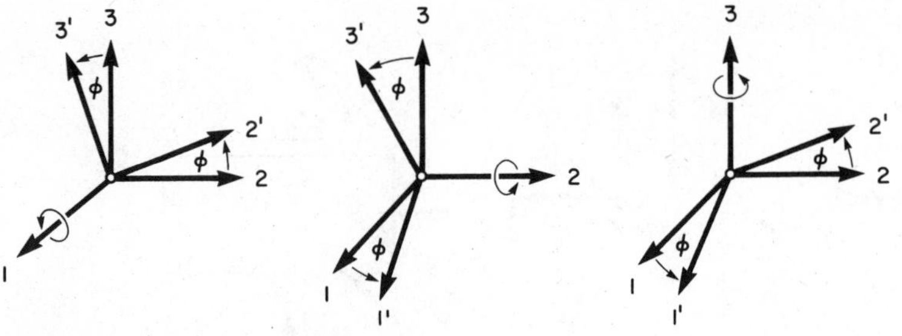

FIGURE 2.5 Principal rotations.

Principal Rotations

There are three particular rotation matrices whose importance justifies a special notation. These *principal rotations* are about the 1-, 2-, or 3-axis and are denoted by $\mathbf{C}_1$, $\mathbf{C}_2$, and $\mathbf{C}_3$, respectively (Fig. 2.5). They can be derived by setting $\mathbf{a} = \mathbf{1}_1, \mathbf{1}_2$, and $\mathbf{1}_3$, respectively, in the general formula for $\mathbf{C}$, namely (22). Thus,

$$\mathbf{C}_1(\phi) \triangleq \begin{bmatrix} 1 & 0 & 0 \\ 0 & \cos\phi & \sin\phi \\ 0 & -\sin\phi & \cos\phi \end{bmatrix}; \qquad \mathbf{C}_2(\phi) \triangleq \begin{bmatrix} \cos\phi & 0 & -\sin\phi \\ 0 & 1 & 0 \\ \sin\phi & 0 & \cos\phi \end{bmatrix};$$

$$\mathbf{C}_3(\phi) \triangleq \begin{bmatrix} \cos\phi & \sin\phi & 0 \\ -\sin\phi & \cos\phi & 0 \\ 0 & 0 & 1 \end{bmatrix} \tag{28}$$

are the three principal rotation matrices.

2.2 ANGULAR DISPLACEMENT PARAMETERS

It is a relatively simple matter to record the displacement of a point $\mathscr{P}_2$ relative to another point $\mathscr{P}_1$; this is done by using a *displacement vector* $\underrightarrow{\mathbf{r}}$. It would be pleasing if the general angular displacement of a frame $\mathscr{F}_b$ relative to another frame $\mathscr{F}_a$ could be dealt with as simply. Unfortunately, this is not the case. At first glance, it may appear that an angular displacement vector is readily available in terms of the Euler axis and angle, $\underrightarrow{\mathbf{a}}$ and ϕ. Based on the discussion in the last section, it is tempting to consider

$$\underrightarrow{\boldsymbol{\phi}} \triangleq \phi \underrightarrow{\mathbf{a}} \tag{1}$$

as a suitable candidate vector. It has a straightforward geometrical interpretation. It even meets one of the criteria that a candidate vector must satisfy, namely, multiplication by a scalar (for, if the rotation about $\underrightarrow{\mathbf{a}}$ is $\kappa\phi$ instead of ϕ, one gets $\kappa\underrightarrow{\boldsymbol{\phi}}$ instead of $\underrightarrow{\boldsymbol{\phi}}$). These promising properties of $\underrightarrow{\boldsymbol{\phi}}$ unfortunately founder on the shoals of vector addition.

We return to the displacement vector $\underrightarrow{\mathbf{r}}$ for a moment. If a point $\mathscr{P}_1$ undergoes first a displacement $\underrightarrow{\mathbf{r}}_{21}$ to $\mathscr{P}_2$, and then a displacement $\underrightarrow{\mathbf{r}}_{32}$ to $\mathscr{P}_3$, its total

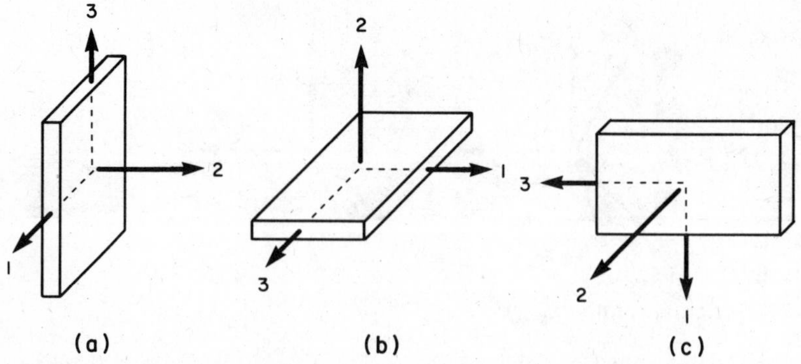

FIGURE 2.6 Finite rotations do not commute in multiplication.

displacement is written $\underset{\rightarrow}{r}_{21} + \underset{\rightarrow}{r}_{32}$. Because $\underset{\rightarrow}{r}_{21} + \underset{\rightarrow}{r}_{32} = \underset{\rightarrow}{r}_{32} + \underset{\rightarrow}{r}_{21}$, the point could have undergone the displacements in the reverse order and still have arrived at the same point $\mathscr{P}_3$. Does the proposed angular displacement vector, given by (1), behave in the same manner for angular displacements? Denote rotation matrices corresponding to $\vec{\phi}_{ba}$ and $\vec{\phi}_{cb}$ by $\mathbf{C}_{ba}$ and $\mathbf{C}_{cb}$. The total angular displacement corresponding to the sequence {first $\vec{\phi}_{ba}$, then $\vec{\phi}_{cb}$} corresponds to the rotation matrix $\mathbf{C}_{cb}\mathbf{C}_{ba}$. On the other hand, the angular displacement corresponding to the sequence {first $\vec{\phi}_{cb}$, then $\vec{\phi}_{ba}$} corresponds to the rotation matrix $\mathbf{C}_{ba}\mathbf{C}_{cb}$. Since rotation matrices do not, in general, commute in multiplication, we conclude that there is no meaning to the equation $\vec{\phi}_{ba} + \vec{\phi}_{cb} = \vec{\phi}_{cb} + \vec{\phi}_{ba}$ and that $\vec{\phi}$ is not, in this sense, an angular displacement vector. Because rotation matrices do not commute in multiplication, the search for an angular displacement "vector" is futile. The standard demonstration of this idea is shown in Fig. 2.6.

Two Sequential Angular Displacements

Consider a reference frame that undergoes an angular displacement from $\mathscr{F}_a$ to $\mathscr{F}_b$ represented by $(\mathbf{a}_1, \phi_1)$ followed by an angular displacement from $\mathscr{F}_b$ to $\mathscr{F}_c$ represented by $(\mathbf{a}_2, \phi_2)$. From Euler's theorem, $\mathscr{F}_c$ can also be obtained directly from $\mathscr{F}_a$ by a single angular displacement $(\mathbf{a}_3, \phi_3)$. We now derive expressions for $\mathbf{a}_3$ and ϕ_3 in terms of $(\mathbf{a}_1, \phi_1)$ and $(\mathbf{a}_2, \phi_2)$. From (22) in Section 2.1,

$$\mathbf{C}_{ca} = \left[c_2\mathbf{1} + (1 - c_2)\mathbf{a}_2\mathbf{a}_2^T - s_2\mathbf{a}_2^{\times}\right]\left[c_1\mathbf{1} + (1 - c_1)\mathbf{a}_1\mathbf{a}_1^T - s_1\mathbf{a}_1^{\times}\right] \tag{2}$$

where, for brevity, $s_i = \sin\phi_i$, $c_i = \cos\phi_i$.

Multiplying out (2) and extracting the trace of $\mathbf{C}_{ca}$, we find that

$$\operatorname{trace}\mathbf{C}_{ca} = c_1 + c_2 + c_1c_2 + (1 - c_1)(1 - c_2)\cos^2\gamma - 2s_1s_2\cos\gamma \tag{3}$$

where $\cos\gamma$ is the direction cosine between the two constituent axes of rotation:

$$\cos\gamma \triangleq \mathbf{a}_1^T\mathbf{a}_2 \tag{4}$$

According to (23) of Section 2.1,

$$\cos\phi_3 = \tfrac{1}{2}(\operatorname{trace}\mathbf{C}_{ca} - 1) \tag{5}$$

The final result for ϕ_3 is conveniently expressed in terms of half-angles:

$$\cos\frac{\phi_3}{2} = \cos\frac{\phi_1}{2}\cos\frac{\phi_2}{2} - \sin\frac{\phi_1}{2}\sin\frac{\phi_2}{2}\cos\gamma \tag{6}$$

(There is a sign ambiguity encountered en route from (3) and (5) to (6); it is resolved on the basis that $\phi_3 = 0$ when $\phi_1 = \phi_2 = 0$.) Observe that the two constituent rotations could have had their order reversed with no consequent change in ϕ_3.

The equivalent axis of rotation, $\mathbf{a}_3$, is found from (25) in Section 2.1, provided $\sin\phi_3 \neq 0$. The pathological case, $\sin\phi_3 = 0$, is left to Problem 2.10; otherwise, we may use (24) of Section 2.1 to arrive at the result

$$\begin{aligned} 2\sin\phi_3\,\mathbf{a}_3 = &\left[s_1(1+c_2) - s_2(1-c_1)\cos\gamma\right]\mathbf{a}_1 \\ &+\left[s_2(1+c_1) - s_1(1-c_2)\cos\gamma\right]\mathbf{a}_2 \\ &+\left[s_1 s_2 - (1-c_1)(1-c_2)\cos\gamma\right]\mathbf{a}_1^{\times}\mathbf{a}_2 \end{aligned} \tag{7}$$

This may again be rewritten in terms of half-angles:

$$\mathbf{a}_3 = \left(\mathbf{a}_1\sin\frac{\phi_1}{2}\cos\frac{\phi_2}{2} + \mathbf{a}_2\cos\frac{\phi_1}{2}\sin\frac{\phi_2}{2} + \mathbf{a}_1^{\times}\mathbf{a}_2\sin\frac{\phi_1}{2}\sin\frac{\phi_2}{2}\right)\Big/\sin\frac{\phi_3}{2} \tag{8}$$

This result, with (6), specifies the single angular displacement $(\mathbf{a}_3, \phi_3)$ that is equivalent to $(\mathbf{a}_1, \phi_1)$ followed by $(\mathbf{a}_2, \phi_2)$. The set $(\mathbf{a}, \phi)$ will be called *Euler axis/angle variables*, as suggested by [Grubin, 2, 3].

Euler Parameters

The results just obtained, (6) and (8), suggest that the product $\mathbf{a}\sin\frac{1}{2}\phi$ is a useful combination of variables. Accordingly, we define the three parameters

$$[\varepsilon_1 \quad \varepsilon_2 \quad \varepsilon_3]^T \equiv \boldsymbol{\varepsilon} \triangleq \mathbf{a}\sin\frac{\phi}{2} \tag{9}$$

and the auxiliary parameter

$$\eta \triangleq \cos\frac{\phi}{2} \tag{10}$$

The parameters $(\varepsilon_1, \varepsilon_2, \varepsilon_3, \eta)$ are called *Euler parameters*. An angular displacement is then specified by $(\boldsymbol{\varepsilon}, \eta)$. Note that

$$\boldsymbol{\varepsilon}^T\boldsymbol{\varepsilon} + \eta^2 \equiv \varepsilon_1^2 + \varepsilon_2^2 + \varepsilon_3^2 + \eta^2 = 1 \tag{11}$$

Returning to the problem posed in the last section, we ask: What single angular displacement $(\boldsymbol{\varepsilon}_3, \eta_3)$ corresponds to the two successive displacements, first $(\boldsymbol{\varepsilon}_1, \eta_1)$, then $(\boldsymbol{\varepsilon}_2, \eta_2)$? According to (6) and (8), the answer is

$$\begin{aligned} \boldsymbol{\varepsilon}_3 &= \eta_2\boldsymbol{\varepsilon}_1 + \eta_1\boldsymbol{\varepsilon}_2 + \boldsymbol{\varepsilon}_1^{\times}\boldsymbol{\varepsilon}_2 \\ \eta_3 &= \eta_1\eta_2 - \boldsymbol{\varepsilon}_1^T\boldsymbol{\varepsilon}_2 \end{aligned} \tag{12}$$

The partial symmetry in these expressions has led some writers to call $(\boldsymbol{\varepsilon}, \eta)$ Euler's "symmetrical" parameters. Note that trigonometric functions have been avoided in (12) and that the $\boldsymbol{\varepsilon}_1^{\times}\boldsymbol{\varepsilon}_2$ term leads to the noncommutativity of rotational sequences. The parameters $(\boldsymbol{\varepsilon}, \eta)$ are also sometimes called *quaternions*, a term invented by Hamilton for four-element hypercomplex numbers whose law of multiplication is equivalent to (12). Hamilton's notation will not be introduced because matrix notation is entirely adequate for dealing with $(\boldsymbol{\varepsilon}, \eta)$ displacements. Since $(\boldsymbol{\varepsilon}, \eta)$ are not, strictly speaking, quaternions unless Hamilton's quaternion notation is used, we shall continue to refer to them as Euler's parameters. They appear in yet another disguise as *Cayley–Klein parameters*. According to [Goldstein], Cayley–Klein parameters are beneficial in connection with quantum mechanics; they have not, however, been shown to have any advantages in the treatment of spacecraft attitude dynamics.

The rotation matrix associated with the angular displacement $(\boldsymbol{\varepsilon}, \eta)$ is now noted. From (22) in Section 2.1,

$$\mathbf{C} = \left(\eta^2 - \boldsymbol{\varepsilon}^T\boldsymbol{\varepsilon}\right)\mathbf{1} + 2\boldsymbol{\varepsilon}\boldsymbol{\varepsilon}^T - 2\eta\boldsymbol{\varepsilon}^{\times} \tag{13}$$

or, displayed in full,

$$\mathbf{C} = \begin{bmatrix} 1 - 2\left(\varepsilon_2^2 + \varepsilon_3^2\right) & 2(\varepsilon_1\varepsilon_2 + \varepsilon_3\eta) & 2(\varepsilon_1\varepsilon_3 - \varepsilon_2\eta) \\ 2(\varepsilon_2\varepsilon_1 - \varepsilon_3\eta) & 1 - 2\left(\varepsilon_3^2 + \varepsilon_1^2\right) & 2(\varepsilon_2\varepsilon_3 + \varepsilon_1\eta) \\ 2(\varepsilon_3\varepsilon_1 + \varepsilon_2\eta) & 2(\varepsilon_3\varepsilon_2 - \varepsilon_1\eta) & 1 - 2\left(\varepsilon_1^2 + \varepsilon_2^2\right) \end{bmatrix} \tag{14}$$

Euler's parameters are normally treated as a four-parameter set, subject to a constraint, namely, $\boldsymbol{\varepsilon}^T\boldsymbol{\varepsilon} + \eta^2 = 1$, because a general angular displacement has only three degrees of freedom.

The inverse of (13) is also desirable: given $\mathbf{C}$, what are $(\boldsymbol{\varepsilon}, \eta)$? From (23) in Section 2.1,

$$\eta = \pm\tfrac{1}{2}(1 + c_{11} + c_{22} + c_{33})^{1/2} \tag{15}$$

The plus sign is chosen if it is advantageous to have a unique η; this corresponds to $0 \le \phi \le \pi$ (refer to the discussion on page 14). Next, from (25) in Section 2.1,

$$\boldsymbol{\varepsilon} = \frac{1}{4\eta}\begin{bmatrix} c_{23} - c_{32} \\ c_{31} - c_{13} \\ c_{12} - c_{21} \end{bmatrix} \tag{16}$$

provided $\eta \neq 0$. If $\eta = 0$, then $\boldsymbol{\varepsilon}$ is simply $\mathbf{a}$, and (27) in Section 2.1 is used to determine $\boldsymbol{\varepsilon}$.

Euler Angles

Unlike the previous representations, which are based largely on the $(\mathbf{a}, \phi)$ interpretation of angular displacement, Euler conceived of his *Euler angles* from a different viewpoint. The product of any number of rotation matrices is itself a rotation matrix (Problem 2.1). Therefore, any product of the so-called principal

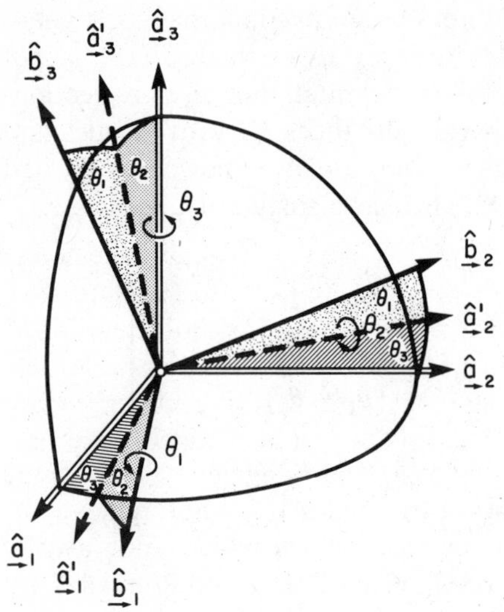

FIGURE 2.7 The 3 – 2 – 1 sequence of Euler angles.

rotation matrices given by (28) in Section 2.1 is also a rotation matrix. Since a general angular displacement is known to have three degrees of freedom, and since each principal rotation matrix has but one degree of freedom, a minimum of three principal rotations must be combined to represent a general angular displacement. The sequence used classically, and considered in Problem 2.4, is as follows:

$$\mathbf{C}(\boldsymbol{\theta}) \equiv \mathbf{C}(\theta_1, \theta_2, \theta_3) \triangleq \mathbf{C}_3(\theta_1)\mathbf{C}_1(\theta_2)\mathbf{C}_3(\theta_3) \tag{17}$$

In aerospace flight vehicle dynamics, however, one frequency encounters situations in which the angular displacement ϕ is small. With frame $\mathscr{F}_b$ imbedded in the vehicle and $\mathscr{F}_a$ representing the vehicle orientation desired, an effective attitude stabilization scheme will ensure that $\mathscr{F}_b \doteq \mathscr{F}_a$, $\mathbf{C}_{ba} \doteq \mathbf{1}$, and $\phi \ll 1$. We shall return to this theme below, but it suffices for now to note that the Euler angle scheme of (17) is not well suited to this important class of problems: when $\theta_2 = 0$, we have $\mathbf{C}(\theta_1, 0, \theta_3) = \mathbf{C}_3(\theta_1)\,\mathbf{1}\mathbf{C}_3(\theta_3) = \mathbf{C}_3(\theta_1 + \theta_3)$ (see Problem 2.4). Therefore, at $\theta_2 = 0$, θ_1 and θ_3 are associated with the same degree of freedom; this condition produces a *singularity*. Occurring as it does right at the reference flight condition ($\boldsymbol{\theta} = \mathbf{0}$), this singularity is most unwelcome. Fortunately, this difficulty can be circumvented by selecting a different sequence of principal rotations and different meanings for θ_1, θ_2, and θ_3. In place of (17), we instead define (see Fig. 2.7)

$$\mathbf{C}(\boldsymbol{\theta}) \triangleq \mathbf{C}_1(\theta_1)\mathbf{C}_2(\theta_2)\mathbf{C}_3(\theta_3) \tag{18}$$

$$\mathbf{C}(\boldsymbol{\theta}) = \begin{bmatrix} c_2c_3 & c_2s_3 & -s_2 \\ s_1s_2c_3 - c_1s_3 & s_1s_2s_3 + c_1c_3 & s_1c_2 \\ c_1s_2c_3 + s_1s_3 & c_1s_2s_3 - s_1c_3 & c_1c_2 \end{bmatrix} \tag{19}$$

The shorthand $c_i \triangleq \cos\theta_i$, $s_i \triangleq \sin\theta_i$ should not obscure the fact that six trigonometric functions have to be calculated to form **C**. Even though (18) is not precisely the same sequence used by Euler, this representation is also conventionally called *Euler angles*, although [Wittenburg] calls them "Bryant" angles and [Roberson, 3] calls them "Tait–Bryan" angles. (They are also known as "Cardan angles.") The 3–2–1 set of Euler angles, (18), also has a singularity, at $\theta_2 = \pi/2$, where

$$\mathbf{C}\left(\theta_1, \frac{\pi}{2}, \theta_3\right) = \begin{bmatrix} 0 & 0 & -1 \\ \sin(\theta_1 - \theta_3) & \cos(\theta_1 - \theta_3) & 0 \\ \cos(\theta_1 - \theta_3) & -\sin(\theta_1 - \theta_3) & 0 \end{bmatrix}$$

and the two degrees of freedom corresponding to θ_1 and θ_3 again coalesce to one degree of freedom. The second choice of Euler angles does not remove the singularity entirely; instead, it moves it to a location where it is not usually troublesome. The determination of θ_1, θ_2, and θ_3, given **C**, is left to Problem 2.13. In general, there are 12 possible ways to define Euler angles, each possibility resulting in a different form for the rotation matrix **C**. These are summarized in Table 2.1. Numerical aspects of forming **C** according to (17) or (18), or one of the ten other similar constructions, are discussed by [Cupit], [Ohkami], and [Wilkes].

TABLE 2.1
Rotation Matrices Based on Euler Angles

$$\mathbf{C}_1(\theta_1)\mathbf{C}_2(\theta_2)\mathbf{C}_1(\theta_3) = \begin{bmatrix} c_2 & s_2 s_3 & -s_2 c_3 \\ s_1 s_2 & c_1 c_3 - s_1 c_2 s_3 & c_1 s_3 + s_1 c_2 c_3 \\ c_1 s_2 & -s_1 c_3 - c_1 c_2 s_3 & -s_1 s_3 + c_1 c_2 c_3 \end{bmatrix}$$

$$\mathbf{C}_1(\theta_1)\mathbf{C}_3(\theta_2)\mathbf{C}_1(\theta_3) = \begin{bmatrix} c_2 & s_2 c_3 & s_2 s_3 \\ -c_1 s_2 & c_1 c_2 c_3 - s_1 s_3 & c_1 c_2 s_3 + s_1 c_3 \\ s_1 s_2 & -s_1 c_2 c_3 - c_1 s_3 & -s_1 c_2 s_3 + c_1 c_3 \end{bmatrix}$$

$$\mathbf{C}_1(\theta_1)\mathbf{C}_3(\theta_2)\mathbf{C}_2(\theta_3) = \begin{bmatrix} c_2 c_3 & s_2 & -c_2 s_3 \\ -c_1 s_2 c_3 + s_1 s_3 & c_1 c_2 & c_1 s_2 s_3 + s_1 c_3 \\ s_1 s_2 c_3 + c_1 s_3 & -s_1 c_2 & -s_1 s_2 s_3 + c_1 c_3 \end{bmatrix}$$

$$\mathbf{C}_1(\theta_1)\mathbf{C}_2(\theta_2)\mathbf{C}_3(\theta_3) = \begin{bmatrix} c_2 c_3 & c_2 s_3 & -s_2 \\ -c_1 s_3 + s_1 s_2 c_3 & c_1 c_3 + s_1 s_2 s_3 & s_1 c_2 \\ s_1 s_3 + c_1 s_2 c_3 & -s_1 c_3 + c_1 s_2 s_3 & c_1 c_2 \end{bmatrix}$$

$$\mathbf{C}_2(\theta_1)\mathbf{C}_1(\theta_2)\mathbf{C}_2(\theta_3) = \begin{bmatrix} c_1 c_3 - s_1 c_2 s_3 & s_1 s_2 & -c_1 s_3 - s_1 c_2 c_3 \\ s_2 s_3 & c_2 & s_2 c_3 \\ s_1 c_3 + c_1 c_2 s_3 & -c_1 s_2 & -s_1 s_3 + c_1 c_2 c_3 \end{bmatrix}$$

$$\mathbf{C}_3(\theta_1)\mathbf{C}_2(\theta_2)\mathbf{C}_1(\theta_3) = \begin{bmatrix} c_1 c_2 & c_1 s_2 s_3 + s_1 c_3 & -c_1 s_2 c_3 + s_1 s_3 \\ -s_1 c_2 & -s_1 s_2 s_3 + c_1 c_3 & s_1 s_2 c_3 + c_1 s_3 \\ s_2 & -c_2 s_3 & c_2 c_3 \end{bmatrix}$$

Infinitesimal Angular Displacements

A rotation matrix $\mathbf{C}$ often corresponds to the angular displacement of a reference frame from its desired orientation. When an infinitesimal analysis is employed—for example, for stability determination (see Appendix A)—$\mathbf{C}$ is considered to differ from $\mathbf{1}$ by an infinitesimal amount. Thus motivated, we expand the basic expression for $\mathbf{C}$, given by (22) in Section 2.1, as a Taylor series in ϕ about $\phi = 0$:

$$\mathbf{C} = \mathbf{1} - \phi\mathbf{a}^{\times} + \tfrac{1}{2}\phi^2(\mathbf{a}\mathbf{a}^T - \mathbf{1}) + O(\phi^3) \tag{20}$$

Note that $\mathbf{a}$ itself is not "small" in any sense. It must retain its unit magnitude. It is the rotation about $\mathbf{a}$ that is small. It is noteworthy that *infinitesimal rotation matrices commute in multiplication*. To be specific, let $\mathbf{C}_{ba} = \mathbf{C}(\mathbf{a}_1, \phi_1)$ and $\mathbf{C}_{cb} = \mathbf{C}(\mathbf{a}_2, \phi_2)$. Then,

$$\mathbf{C}_{cb}\mathbf{C}_{ba} \doteq \mathbf{C}_{ba}\mathbf{C}_{cb} \doteq \mathbf{1} - \phi_1\mathbf{a}_1^{\times} - \phi_2\mathbf{a}_2^{\times} \tag{21}$$

to first order in ϕ_1 and ϕ_2. Although rotation sequence is important for finite rotations, for first-order infinitesimal rotations it is not.

Another characteristic of infinitesimal rotations is that the distinction between the parameter sets tends to vanish. If $(\mathbf{a}, \phi)$ is treated as a three-parameter set,

$$\mathbf{C}_2(\theta_1)\mathbf{C}_3(\theta_2)\mathbf{C}_2(\theta_3) = \begin{bmatrix} c_1c_2c_3 - s_1s_3 & c_1s_2 & -c_1c_2s_3 - s_1c_3 \\ -s_2c_3 & c_2 & s_2s_3 \\ s_1c_2c_3 + c_1s_3 & s_1s_2 & -s_1c_2s_3 + c_1c_3 \end{bmatrix}$$

$$\mathbf{C}_2(\theta_1)\mathbf{C}_3(\theta_2)\mathbf{C}_1(\theta_3) = \begin{bmatrix} c_1c_2 & c_1s_2c_3 + s_1s_3 & c_1s_2s_3 - s_1c_3 \\ -s_2 & c_2c_3 & c_2s_3 \\ s_1c_2 & s_1s_2c_3 - c_1s_3 & s_1s_2s_3 + c_1c_3 \end{bmatrix}$$

$$\mathbf{C}_2(\theta_1)\mathbf{C}_1(\theta_2)\mathbf{C}_3(\theta_3) = \begin{bmatrix} c_1c_3 - s_1s_2s_3 & c_1s_3 + s_1s_2c_3 & -s_1c_2 \\ -c_2s_3 & c_2c_3 & s_2 \\ s_1c_3 + c_1s_2s_3 & s_1s_3 - c_1s_2c_3 & c_1c_2 \end{bmatrix}$$

$$\mathbf{C}_3(\theta_1)\mathbf{C}_1(\theta_2)\mathbf{C}_3(\theta_3) = \begin{bmatrix} c_1c_3 - s_1c_2s_3 & c_1s_3 + s_1c_2c_3 & s_1s_2 \\ -s_1c_3 - c_1c_2s_3 & -s_1s_3 + c_1c_2c_3 & c_1s_2 \\ s_2s_3 & -s_2c_3 & c_2 \end{bmatrix}$$

$$\mathbf{C}_3(\theta_1)\mathbf{C}_2(\theta_2)\mathbf{C}_3(\theta_3) = \begin{bmatrix} c_1c_2c_3 - s_1s_3 & c_1c_2s_3 + s_1c_3 & -c_1s_2 \\ -s_1c_2c_3 - c_1s_3 & -s_1c_2s_3 + c_1c_3 & s_1s_2 \\ s_2c_3 & s_2s_3 & c_2 \end{bmatrix}$$

$$\mathbf{C}_3(\theta_1)\mathbf{C}_1(\theta_2)\mathbf{C}_2(\theta_3) = \begin{bmatrix} c_1c_3 + s_1s_2s_3 & s_1c_2 & -c_1s_3 + s_1s_2c_3 \\ -s_1c_3 + c_1s_2s_3 & c_1c_2 & s_1s_3 + c_1s_2c_3 \\ c_2s_3 & -s_2 & c_2c_3 \end{bmatrix}$$

$\boldsymbol{\phi} \triangleq \phi \mathbf{a}$, then

$$\mathbf{C} \doteq \mathbf{1} - \boldsymbol{\phi}^{\times} \tag{22}$$

to first order. Similarly, (9) and (10) show that the first three Euler parameters are first order, while the fourth is approximately unity. Thus, from (13),

$$\mathbf{C} \doteq \mathbf{1} - 2\boldsymbol{\varepsilon}^{\times} \tag{23}$$

to first order. Finally, we look at the Euler angles of (18). When these angles are small, from (19),

$$\mathbf{C} \doteq \mathbf{1} - \boldsymbol{\theta}^{\times} \tag{24}$$

to first order in $\boldsymbol{\theta}$, where

$$\boldsymbol{\theta} \triangleq [\theta_1 \quad \theta_2 \quad \theta_3]^T \tag{25}$$

It is evident from (22) through (24) that (apart from the factor of 2) axis/angle parameters, Euler parameters, and Euler angles are identical for infinitesimal rotations.

2.3 ANGULAR VELOCITY

Thus far, we have examined the general properties of angular displacements and have identified several possibilities for representing them parametrically. We turn now to *kinematics*, in which we encounter angular *displacements that evolve with time*. When the relationship between two reference frames is time dependent, the associated rotation matrix is time dependent also, $\mathbf{C} = \mathbf{C}(t)$, as are the several parameterizations: Euler axis/angle variables $\mathbf{a}(t)$ and $\phi(t)$; Euler parameters $\boldsymbol{\varepsilon}(t)$ and $\eta(t)$; and Euler angles $\boldsymbol{\theta}(t)$. To treat all these possible representations in a unified manner, the logical starting point is, once again, the rotation matrix $\mathbf{C}$.

Differential Equation for C

Let $\mathcal{F}_a$ and $\mathcal{F}_b$ be any two reference frames and $\underset{\rightarrow}{\mathcal{F}}_a$ and $\underset{\rightarrow}{\mathcal{F}}_b$ their associated vectrices (see Appendix B). Then,

$$\underset{\rightarrow}{\mathcal{F}}_a^T = \underset{\rightarrow}{\mathcal{F}}_b^T \mathbf{C}_{ba} \tag{1}$$

Denoting time derivatives in $\mathcal{F}_a$ by an overdot, it follows that $\dot{\underset{\rightarrow}{\mathcal{F}}}_a$ vanishes and that, from (3) in Section B.4,

$$\dot{\underset{\rightarrow}{\mathcal{F}}}_b^T = \underset{\rightarrow}{\omega}_{ba} \times \underset{\rightarrow}{\mathcal{F}}_b^T \tag{2}$$

where $\underset{\rightarrow}{\omega}_{ba}$ is the angular velocity of $\mathcal{F}_b$ with respect to $\mathcal{F}_a$. Now express $\underset{\rightarrow}{\omega}_{ba}$ in $\mathcal{F}_b$:

$$\underset{\rightarrow}{\omega}_{ba} = \boldsymbol{\omega}_{ba}^T \underset{\rightarrow}{\mathcal{F}}_b \tag{3}$$

Taking $\mathcal{F}_a$-derivatives of (1),

$$\begin{aligned} \underset{\rightarrow}{\mathbf{O}} &= \boldsymbol{\omega}_{ba}^T \underset{\rightarrow}{\mathcal{F}}_b \times \underset{\rightarrow}{\mathcal{F}}_b^T \mathbf{C}_{ba} + \underset{\rightarrow}{\mathcal{F}}_b^T \dot{\mathbf{C}}_{ba} \\ &= \underset{\rightarrow}{\mathcal{F}}_b^T \left(\boldsymbol{\omega}_{ba}^{\times} \mathbf{C}_{ba} + \dot{\mathbf{C}}_{ba} \right) \end{aligned} \tag{4}$$

We conclude that

$$\dot{\mathbf{C}}_{ba} + \boldsymbol{\omega}_{ba}^{\times}\mathbf{C}_{ba} = \mathbf{O} \tag{5}$$

This important equation is the basis for most of what follows in this section.

Equation (5) is a differential equation for the rotation matrix $\mathbf{C}$. It implies that if we know the history of the angular velocity $\boldsymbol{\omega}_{ba}(t)$, we may integrate and find $\mathbf{C}_{ba}(t)$, that is, the attitude history of $\mathscr{F}_b$ with respect to $\mathscr{F}_a$. Although (5) is a linear differential equation, it has the time-dependent coefficient $\boldsymbol{\omega}_{ba}(t)$: unfortunately, this means it cannot be solved in closed form except in special cases. Two general applications of (5) will be mentioned. The first is strapdown inertial navigation, in which attitude rate sensors measure $\boldsymbol{\omega}_{ba}(t)$ because they are "strapped down" to $\mathscr{F}_b$. In this case, $\mathscr{F}_a$ is an inertial reference frame. The scalar differential equations corresponding to (5), or to one of its descendants below, are then integrated numerically using on-board software. The second general application of (5) is to dynamical analysis and simulation, where $\boldsymbol{\omega}_{ba}$ is found by solving the attitude motion equations (see Chapter 3).

An alternative view of (5) is that is provides a formula for $\boldsymbol{\omega}_{ba}$ if the attitude history $\mathbf{C}_{ba}(t)$ is known:

$$\boldsymbol{\omega}_{ba}^{\times} = -\dot{\mathbf{C}}_{ba}\mathbf{C}_{ab} \equiv \mathbf{C}_{ba}\dot{\mathbf{C}}_{ab} \tag{6}$$

The last expression follows from the orthonormality condition $\mathbf{C}_{ba}\mathbf{C}_{ab} = \mathbf{1}$.

Equation (6) is now used to prove that angular velocity vectors satisfy the law of vector addition. At least three frames are needed for this discussion: $\mathscr{F}_a$, $\mathscr{F}_b$, and $\mathscr{F}_c$. From (6),

$$\boldsymbol{\omega}_{ba}^{\times} = -\dot{\mathbf{C}}_{ba}\mathbf{C}_{ab}; \qquad \boldsymbol{\omega}_{cb}^{\times} = -\dot{\mathbf{C}}_{cb}\mathbf{C}_{bc}; \qquad \boldsymbol{\omega}_{ca}^{\times} = -\dot{\mathbf{C}}_{ca}\mathbf{C}_{ac} \tag{7}$$

Note that by convention

$$\boldsymbol{\omega}_{ba} = \mathscr{F}_b \cdot \underset{\rightarrow}{\omega}_{ba}; \qquad \boldsymbol{\omega}_{cb} = \mathscr{F}_c \cdot \underset{\rightarrow}{\omega}_{cb}; \qquad \boldsymbol{\omega}_{ca} = \mathscr{F}_c \cdot \underset{\rightarrow}{\omega}_{ca} \tag{8}$$

Since $\mathbf{C}_{ca} = \mathbf{C}_{cb}\mathbf{C}_{ba}$ [see (5) in Section B.3],

$$\begin{aligned}\boldsymbol{\omega}_{ca}^{\times} &= -(\dot{\mathbf{C}}_{cb}\mathbf{C}_{ba} + \mathbf{C}_{cb}\dot{\mathbf{C}}_{ba})\mathbf{C}_{ac}\\ &= -\dot{\mathbf{C}}_{cb}\mathbf{C}_{bc} - \mathbf{C}_{cb}\dot{\mathbf{C}}_{ba}\mathbf{C}_{ab}\mathbf{C}_{bc}\\ &= \boldsymbol{\omega}_{cb}^{\times} + \mathbf{C}_{cb}\boldsymbol{\omega}_{ba}^{\times}\mathbf{C}_{bc}\\ &= \boldsymbol{\omega}_{cb}^{\times} + (\mathbf{C}_{cb}\boldsymbol{\omega}_{ba})^{\times}\end{aligned}$$

The last step follows from the identity (10) in Section B.3. It follows that

$$\boldsymbol{\omega}_{ca} = \boldsymbol{\omega}_{cb} + \mathbf{C}_{cb}\boldsymbol{\omega}_{ba} \tag{9}$$

Multiply through by $\mathscr{F}_c^T$, converting this matrix equation to a vector equation:

$$\mathscr{F}_c^T\boldsymbol{\omega}_{ca} = \mathscr{F}_c^T\boldsymbol{\omega}_{cb} + \mathscr{F}_c^T\mathbf{C}_{cb}\boldsymbol{\omega}_{ba} = \mathscr{F}_c^T\boldsymbol{\omega}_{cb} + \mathscr{F}_b^T\boldsymbol{\omega}_{ba}$$

whence

$$\underset{\rightarrow}{\omega}_{ca} = \underset{\rightarrow}{\omega}_{cb} + \underset{\rightarrow}{\omega}_{ba} \tag{10}$$

showing that the angular velocity of $\mathscr{F}_c$ with respect to $\mathscr{F}_a$ is equal to the angular velocity of $\mathscr{F}_c$ with respect to $\mathscr{F}_b$ plus the angular velocity of $\mathscr{F}_b$ with respect to $\mathscr{F}_a$. Although (10) is valid for vectors, the presence of $\mathbf{C}_{cb}$ in (9) shows that caution is needed when the addition of angular velocities is done in terms of their components.

Direction Cosines as Attitude Variables

In the search for variables to represent angular displacements, we should not overlook the rotation matrix elements themselves. These elements c_{ij} are the direction cosines between the two sets of basis vectors; see (6) and (7) in Section 2.1. If the c_{ij} are used in this manner as attitude variables, then the differential equations for attitude, given the angular velocity, are the elements of (5):

$$\begin{aligned} \dot{c}_{11} &= \omega_3 c_{21} - \omega_2 c_{31} \\ \dot{c}_{21} &= \omega_1 c_{31} - \omega_3 c_{11} \\ \dot{c}_{31} &= \omega_2 c_{11} - \omega_1 c_{21} \end{aligned} \tag{10a}$$

$$\begin{aligned} \dot{c}_{12} &= \omega_3 c_{22} - \omega_2 c_{32} \\ \dot{c}_{22} &= \omega_1 c_{32} - \omega_3 c_{12} \\ \dot{c}_{32} &= \omega_2 c_{12} - \omega_1 c_{22} \end{aligned} \tag{10b}$$

$$\begin{aligned} \dot{c}_{13} &= \omega_3 c_{23} - \omega_2 c_{33} \\ \dot{c}_{23} &= \omega_1 c_{33} - \omega_3 c_{13} \\ \dot{c}_{33} &= \omega_2 c_{13} - \omega_1 c_{23} \end{aligned} \tag{10c}$$

These are the kinematical equations of Poisson. In practice, it is not necessary to integrate all three sets of equations. Any one set (a, b, or c) can be replaced by the corresponding set of algebraic relations in (13) of Section 2.1. For example, (10c) can be replaced by (13c) of Section 2.1. Geometrically, if we know the orientation of two of the $\mathscr{F}_b$ basis vectors, we know the orientation of the third by the right-hand rule.

The Poisson equations are seen to be linear but with variable coefficients. In general, they cannot be solved in closed form and must be integrated with the aid of a computer. The orthonormality condition $\mathbf{C}^T\mathbf{C} = \mathbf{1}$ should be examined at regular intervals as a partial check on the accuracy of the integration.

Axis / Angle Parameters

The rotation matrix $\mathbf{C}$ can always be expressed in terms of the $(\mathbf{a}, \phi)$ implied by Euler's theorem:

$$\mathbf{C} = \cos\phi\mathbf{1} + (1 - \cos\phi)\mathbf{a}\mathbf{a}^T - \sin\phi\mathbf{a}^\times \tag{11}$$

From (11) and the definition

$$\boldsymbol{\omega}^\times \triangleq -\dot{\mathbf{C}}\mathbf{C}^T \tag{12}$$

we proceed to obtain $\boldsymbol{\omega}$ in terms of $\dot{\mathbf{a}}$ and $\dot{\phi}$. The identities

$$\mathbf{a}^T\mathbf{a} = 1; \qquad \mathbf{a}^T\dot{\mathbf{a}} = 0$$

$$\mathbf{a}^\times\mathbf{a} = \mathbf{0}; \qquad \dot{\mathbf{a}}^\times\mathbf{a} = -\mathbf{a}^\times\dot{\mathbf{a}}; \qquad \dot{\mathbf{a}}^\times\mathbf{a}^\times = \mathbf{a}\dot{\mathbf{a}}^T$$

$$\mathbf{a}\dot{\mathbf{a}}^T\mathbf{a}^\times + \mathbf{a}^\times\dot{\mathbf{a}}\mathbf{a}^T = -\dot{\mathbf{a}}^\times; \qquad \dot{\mathbf{a}}\mathbf{a}^T - \mathbf{a}\dot{\mathbf{a}}^T = (\mathbf{a}^\times\dot{\mathbf{a}})^\times$$

are useful in reducing the resulting expression to its desired form:

$$\boldsymbol{\omega}^\times = \dot{\phi}\mathbf{a}^\times - (1 - \cos\phi)(\mathbf{a}^\times\dot{\mathbf{a}})^\times + \sin\phi\dot{\mathbf{a}}^\times \tag{13}$$

It follows immediately that

$$\boldsymbol{\omega} = \dot{\phi}\mathbf{a} - (1 - \cos\phi)\mathbf{a}^\times\dot{\mathbf{a}} + \sin\phi\dot{\mathbf{a}} \tag{14}$$

An alternate derivation has been given by [Gelman, 3], who also provides a geometrical interpretation. When the rotation is about a fixed axis, (14) reduces to

$$\boldsymbol{\omega} = \dot{\phi}\mathbf{a} \tag{15}$$

as expected. Three examples of rotation about a fixed axis are the "principal" rotations of Section 2.1 for which

$$\begin{aligned} \mathbf{a} = \mathbf{1}_1 &\to \boldsymbol{\omega} = [\dot{\phi} \quad 0 \quad 0]^T \\ \mathbf{a} = \mathbf{1}_2 &\to \boldsymbol{\omega} = [0 \quad \dot{\phi} \quad 0]^T \\ \mathbf{a} = \mathbf{1}_3 &\to \boldsymbol{\omega} = [0 \quad 0 \quad \dot{\phi}]^T \end{aligned} \tag{16}$$

as we should expect.

We next find the relation inverse to (14); this is an important step because it will produce differential equations for $\mathbf{a}$ and ϕ that can be solved to find attitude if $\boldsymbol{\omega}(t)$ is known. Premultiply (14) by $\mathbf{a}^T$ and use the usual identities for $\mathbf{a}$ to reduce the result to

$$\dot{\phi} = \mathbf{a}^T\boldsymbol{\omega} \tag{17}$$

To complete the derivation, we need $\dot{\mathbf{a}}$ in terms of $\boldsymbol{\omega}$. Take (14) again and premultiply by $\mathbf{a}^\times$:

$$\mathbf{a}^\times\boldsymbol{\omega} = (1 - \cos\phi)\dot{\mathbf{a}} + \sin\phi\mathbf{a}^\times\dot{\mathbf{a}} \tag{18}$$

Next, premultiply (18) by $\mathbf{a}^\times$:

$$\mathbf{a}^\times\mathbf{a}^\times\boldsymbol{\omega} = -\sin\phi\dot{\mathbf{a}} + (1 - \cos\phi)\mathbf{a}^\times\dot{\mathbf{a}} \tag{19}$$

The last two equations are viewed as two algebraic equations in the two unknowns $\dot{\mathbf{a}}$ and $\mathbf{a}^\times\dot{\mathbf{a}}$. Solving for $\dot{\mathbf{a}}$,

$$\dot{\mathbf{a}} = \tfrac{1}{2}\left[\mathbf{a}^\times - \cot\frac{\phi}{2}\mathbf{a}^\times\mathbf{a}^\times\right]\boldsymbol{\omega} \tag{20}$$

This equation and (17) are the desired result. If one wishes to use $(\mathbf{a}, \phi)$ as the variables representing angular displacements, then (17) and (20) are available for integration. The constraint $\mathbf{a}^T\mathbf{a} = 1$ should be checked at suitable intervals.

Euler Parameters

The rate of change of the Euler parameters, defined by

$$\boldsymbol{\varepsilon} \triangleq \mathbf{a} \sin\frac{\phi}{2}; \qquad \eta \triangleq \cos\frac{\phi}{2} \tag{21}$$

is found by differentiating (21) and inserting $\dot{\mathbf{a}}$ and $\dot{\phi}$, available from (17) and (20):

$$\dot{\boldsymbol{\varepsilon}} = \tfrac{1}{2}(\boldsymbol{\varepsilon}^{\times} + \eta\mathbf{1})\boldsymbol{\omega}$$

$$\dot{\eta} = -\tfrac{1}{2}\boldsymbol{\varepsilon}^{T}\boldsymbol{\omega} \tag{22}$$

The form of the differential equations (22) is once again linear, although the variable coefficients preclude closed-form solutions. The constraint $\boldsymbol{\varepsilon}^T\boldsymbol{\varepsilon} + \eta^2 = 1$ can be used to advantage during a computer-aided solution; alternatively, we may integrate only for $\boldsymbol{\varepsilon}$ in (22), using $\eta = (1 - \boldsymbol{\varepsilon}^T\boldsymbol{\varepsilon})^{1/2}$.

By explicitly inverting (22a), or otherwise, the inverse relationship—$\boldsymbol{\omega}$ in terms of $\dot{\boldsymbol{\varepsilon}}$—is obtained:

$$\boldsymbol{\omega} = 2\left[\frac{\eta^2\mathbf{1} - \eta\boldsymbol{\varepsilon}^{\times} + \boldsymbol{\varepsilon}\boldsymbol{\varepsilon}^T}{\eta}\right]\dot{\boldsymbol{\varepsilon}} \tag{23}$$

This form is useful when it is desired to calculate the angular velocity from a known attitude history.

Euler Angles

This series of results is completed by obtaining the relationships between angular velocity components and the time derivatives of the 3–2–1 Euler angles of (18) in Section 2.2. The method used is identical to that used in the derivation of the addition law for angular velocities, which resulted in (9). With $\mathbf{C}$ given by (18) in Section 2.2, we calculate

$$\begin{aligned}
\boldsymbol{\omega}^{\times} &= -\dot{\mathbf{C}}\mathbf{C}^T \\
&= -(\dot{\mathbf{C}}_1\mathbf{C}_2\mathbf{C}_3 + \mathbf{C}_1\dot{\mathbf{C}}_2\mathbf{C}_3 + \mathbf{C}_1\mathbf{C}_2\dot{\mathbf{C}}_3)\mathbf{C}_3^T\mathbf{C}_2^T\mathbf{C}_1^T \\
&= -\dot{\mathbf{C}}_1\mathbf{C}_1^T - \mathbf{C}_1\dot{\mathbf{C}}_2\mathbf{C}_2^T\mathbf{C}_1^T - \mathbf{C}_1\mathbf{C}_2\dot{\mathbf{C}}_3\mathbf{C}_3^T\mathbf{C}_2^T\mathbf{C}_1^T
\end{aligned}$$

Referring to (16), however, we recognize that

$$-\dot{\mathbf{C}}_1\mathbf{C}_1^T = (\mathbf{1}_1\dot{\theta}_1)^{\times}; \qquad -\dot{\mathbf{C}}_2\mathbf{C}_2^T = (\mathbf{1}_2\dot{\theta}_2)^{\times}; \qquad -\mathbf{C}_3\mathbf{C}_3^T = (\mathbf{1}_3\dot{\theta}_3)^{\times}$$

Hence,

$$\begin{aligned}
\boldsymbol{\omega}^{\times} &= (\mathbf{1}_1\dot{\theta}_1)^{\times} + \mathbf{C}_1(\mathbf{1}_2\dot{\theta}_2)^{\times}\mathbf{C}_1^T + \mathbf{C}_1\mathbf{C}_2(\mathbf{1}_3\dot{\theta}_3)^{\times}(\mathbf{C}_1\mathbf{C}_2)^T \\
&= (\mathbf{1}_1\dot{\theta}_1)^{\times} + (\mathbf{C}_1\mathbf{1}_2\dot{\theta}_2)^{\times} + (\mathbf{C}_1\mathbf{C}_2\mathbf{1}_3\dot{\theta}_3)^{\times}
\end{aligned}$$

The last step follows from identity (10) in Section B.3. The desired result is therefore

$$\boldsymbol{\omega} = \mathbf{1}_1\dot{\theta}_1 + \mathbf{C}_1(\theta_1)\mathbf{1}_2\dot{\theta}_2 + \mathbf{C}_1(\theta_1)\mathbf{C}_2(\theta_2)\mathbf{1}_3\dot{\theta}_3$$
$$= \mathbf{S}(\theta_1, \theta_2)\dot{\boldsymbol{\theta}} \quad (24)$$

where

$$\mathbf{S}(\theta_1, \theta_2) = \begin{bmatrix} 1 & 0 & -\sin\theta_2 \\ 0 & \cos\theta_1 & \sin\theta_1\cos\theta_2 \\ 0 & -\sin\theta_1 & \cos\theta_1\cos\theta_2 \end{bmatrix}; \qquad \boldsymbol{\theta} \triangleq \begin{bmatrix} \theta_1 \\ \theta_2 \\ \theta_3 \end{bmatrix} \quad (25)$$

The inverse relationship is found by inverting the 3×3 matrix $\mathbf{S}$:

$$\dot{\boldsymbol{\theta}} = \mathbf{S}^{-1}(\theta_1, \theta_2)\boldsymbol{\omega} \quad (26)$$

where

$$\mathbf{S}^{-1} = \begin{bmatrix} 1 & \sin\theta_1\tan\theta_2 & \cos\theta_1\tan\theta_2 \\ 0 & \cos\theta_1 & -\sin\theta_1 \\ 0 & \sin\theta_1\sec\theta_2 & \cos\theta_1\sec\theta_2 \end{bmatrix} \quad (27)$$

The singularity at $\theta_2 = \pi/2$, noted earlier, causes $\mathbf{S}^{-1}$ to be undefined at $\theta_2 = \pi/2$. The form of $\mathbf{S}^{-1}$ for each of the 12 possible sequences of Euler angles (Table 2.1) is shown in Table 2.2.

Infinitesimal Angular Displacements

When $\mathbf{C}$ represents an infinitesimal angular displacement, as assumed in (22) through (24) in Section 2.2, the preceding results become especially simple. We regard $\boldsymbol{\omega}(t)$, $\boldsymbol{\phi}(t)$, $\boldsymbol{\varepsilon}(t)$, and $\boldsymbol{\theta}(t)$ as first-order infinitesimals. Then, for direction cosines, (10) becomes, to first order,

$$c_{11} \equiv 1; \qquad c_{22} \equiv 1; \qquad c_{33} \equiv 1$$
$$\dot{c}_{12} = -\dot{c}_{21} = \omega_3(t); \quad \dot{c}_{31} = -\dot{c}_{13} = \omega_2(t); \quad \dot{c}_{23} = -\dot{c}_{32} = \omega_1(t) \quad (28)$$

The use of four Euler axis/angle variables $(\mathbf{a}, \phi)$ is not attractive in this context owing to the singularity of $\phi = 0$; see (20). This difficulty is overcome by defining $\boldsymbol{\phi} \triangleq \phi\mathbf{a}$, as in (22) in Section 2.2; to first order, from (6) or otherwise:

$$\dot{\boldsymbol{\phi}} = \boldsymbol{\omega}(t) \quad (29)$$

As for Euler parameters, (22) shows that, to first order,

$$\dot{\boldsymbol{\varepsilon}} = \tfrac{1}{2}\boldsymbol{\omega}(t); \qquad \eta \equiv 1 \quad (30)$$

Finally, (26) in Section 2.2 implies that 3–2–1 Euler angles may, to first order, be integrated from

$$\dot{\boldsymbol{\theta}} = \boldsymbol{\omega}(t) \quad (31)$$

TABLE 2.2
$S^{-1}(\theta_1, \theta_2)$, Matrix Relating Euler Angle Rates to Angular Velocity Components [See Eq. (26) in Section 2.3]

$$(1,2,1):\begin{bmatrix} 1 & -s_1c_2/s_2 & -c_1c_2/s_2 \\ 0 & c_1 & -s_1 \\ 0 & s_1/s_2 & c_1/s_2 \end{bmatrix} \qquad (2,3,2):\begin{bmatrix} -c_1c_2/s_2 & 1 & -s_1c_2/s_2 \\ -s_1 & 0 & c_1 \\ c_1/s_2 & 0 & s_1/s_2 \end{bmatrix}$$

$$(1,3,1):\begin{bmatrix} 1 & c_1c_2/s_2 & -s_1c_2/s_2 \\ 0 & s_1 & c_1 \\ 0 & -c_1/s_2 & s_1/s_2 \end{bmatrix} \qquad (2,3,1):\begin{bmatrix} c_1s_2/c_2 & 1 & s_1s_2/c_2 \\ -s_1 & 0 & c_1 \\ c_1/c_2 & 0 & s_1/c_2 \end{bmatrix}$$

$$(1,3,2):\begin{bmatrix} 1 & -c_1s_2/c_2 & s_1s_2/c_2 \\ 0 & s_1 & c_1 \\ 0 & c_1/c_2 & -s_1/c_2 \end{bmatrix} \qquad (2,1,3):\begin{bmatrix} s_1s_2/c_2 & 1 & -c_1s_2/c_2 \\ c_1 & 0 & s_1 \\ -s_1/c_2 & 0 & c_1/c_2 \end{bmatrix}$$

$$(1,2,3):\begin{bmatrix} 1 & s_1s_2/c_2 & c_1s_2/c_2 \\ 0 & c_1 & -s_1 \\ 0 & s_1/c_2 & c_1/c_2 \end{bmatrix} \qquad (3,1,3):\begin{bmatrix} -s_1c_2/s_2 & -c_1c_2/s_2 & 1 \\ c_1 & -s_1 & 0 \\ s_1/s_2 & c_1/s_2 & 0 \end{bmatrix}$$

$$(2,1,2):\begin{bmatrix} -s_1c_2/s_2 & 1 & c_1c_2/s_2 \\ c_1 & 0 & s_1 \\ s_1/s_2 & 0 & -c_1/s_2 \end{bmatrix} \qquad (3,2,3):\begin{bmatrix} c_1c_2/s_2 & -s_1c_2/s_2 & 1 \\ s_1 & c_1 & 0 \\ -c_1/s_2 & s_1/s_2 & 0 \end{bmatrix}$$

$$(3,2,1):\begin{bmatrix} -s_2c_1/c_2 & s_1s_2/c_2 & 1 \\ s_1 & c_1 & 0 \\ c_1/c_2 & -s_1/c_2 & 0 \end{bmatrix} \qquad (3,1,2):\begin{bmatrix} s_1s_2/c_2 & c_1s_2/c_2 & 1 \\ c_1 & -s_1 & 0 \\ s_1/c_2 & c_1/c_2 & 0 \end{bmatrix}$$

NOTE: Each entry in this table corresponds to the Euler angle sequence in the same position in Table 2.1.

It is clear from (28) through (31) that there is no real distinction between these four parameter sets in a linearized analysis.

A more subtle situation occurs when the orientation of a reference frame deviates slightly from a frame rotating at constant angular velocity. To deal with this case, we need three frames: $\mathscr{F}_a$; $\mathscr{F}_b$, which rotates at constant angular velocity $\underset{\rightarrow}{\boldsymbol{\nu}}$ with respect to $\mathscr{F}_a$; and $\mathscr{F}_c$, which is only infinitesimally different from $\mathscr{F}_b$. Denote by $\boldsymbol{\nu}$ the components in $\mathscr{F}_b$ of $\underset{\rightarrow}{\boldsymbol{\nu}}$. From (24) in Section 2.2, we have, to first order,

$$\mathbf{C}_{cb} = \mathbf{1} - \boldsymbol{\alpha}^{\times} \tag{32}$$

(There is no need to distinguish between the several parameters sets.) According to (9), the angular velocity of $\mathscr{F}_c$ with respect to $\mathscr{F}_a$, expressed in $\mathscr{F}_c$, is

$$\boldsymbol{\omega}_{ca} = \dot{\boldsymbol{\alpha}} + (\mathbf{1} - \boldsymbol{\alpha}^{\times})\boldsymbol{\nu} \tag{33}$$

for $\|\boldsymbol{\alpha}\| \ll 1$. The presence of the term $-\boldsymbol{\alpha}^{\times}\boldsymbol{\nu}$ should be noted.

2.4 COMMENTS ON PARAMETER ALTERNATIVES

Many candidate sets of orientation variables have been discussed in this chapter, and the question of their relative merits naturally arises. The alternatives are summarized in Table 2.3, which shows how to calculate **C** and $\boldsymbol{\omega}$ for each set of variables.

We have seen that in linear analyses (for which infinitesimal rotations are appropriate) no useful distinction can be made between parameter sets. For finite rotations, however, the choice of variables should be made with care. A three-parameter set is attractive because it has as many parameters as there are degrees of freedom, although one must face the fact that *no three-parameter set can be both global and nonsingular* ([Stuelpnagel]). To treat all orientations uniformly, at least *four* variables are required. On the other hand, if it is anticipated that a certain parameter region will not be entered, then a three-parameter set is suitable, provided that its inevitable singularity is in the unentered region.

A second basic consideration is computational efficiency. With finite displacements, we have seen that the kinematical differential equations cannot be solved in closed form because of either time-varying coefficients, nonlinearities, or both. Equations whose right sides include trigonometric functions (e.g., Euler angles) or square roots (e.g., Euler parameters with η implicit) are unattractive in this respect. Euler parameters (the four-parameter set, with explicit η) and direction cosines are well suited for computation, and they have consistently been well spoken of in comparative studies reported in the literature. In fact, [A. C. Robinson], [Sabroff et al.], [Fang and Zimmerman], [Wilcox], [Mortensen], [Ickes], and [Grubin, 3], all conclude that Euler parameters are computationally superior. Owing to this superiority, Euler parameters were used on Skylab ([Duty and

TABLE 2.3
Comparison of Parameter Alternatives

Name	n	Parameters, v	Definition	$\mathbf{C(v)}$
Direction cosines	9	c_{ij}	(2.1, 6)	$\mathbf{C} = \{c_{ij}\}$
Direction cosines	6	c_{ij}	(2.1, 6)	$\mathbf{C} = \{c_{ij}\}$
Axis / angle variables	4	a_1, a_2, a_3, ϕ	Euler's theorem	$\mathbf{C} = \mathbf{1}\cos\phi + (1 - \cos\phi)\mathbf{a}\mathbf{a}^T - \sin\phi\,\mathbf{a}^\times$
Axis / angle variables	3	ϕ_1, ϕ_2, ϕ_3	$\boldsymbol{\phi} = \phi\mathbf{a}$	$\mathbf{C} = \mathbf{1} + b_1(\phi)\boldsymbol{\phi}^\times + b_2(\phi)\boldsymbol{\phi}^\times\boldsymbol{\phi}^\times$; $b_1 \triangleq -\phi^{-1}\sin\phi$; $b_2 \triangleq 2\phi^{-2}\sin^2\frac{1}{2}\phi$
Euler–Rodriguez (see Probs. 2.12, 19)	3	p_1, p_2, p_3	$\mathbf{p} = \mathbf{a}\tan\frac{1}{2}\phi$	$\mathbf{C} = \mathbf{1} + \dfrac{2}{1 + \mathbf{p}^T\mathbf{p}}(\mathbf{p}^\times\mathbf{p}^\times - \mathbf{p}^\times)$
Euler parameters	4	$\varepsilon_1, \varepsilon_2, \varepsilon_3, \eta$	$\boldsymbol{\varepsilon} = \mathbf{a}\sin\frac{1}{2}\phi$; $\eta = \cos\frac{1}{2}\phi$	$\mathbf{C} = (\eta^2 - \boldsymbol{\varepsilon}^T\boldsymbol{\varepsilon})\mathbf{1} + 2\boldsymbol{\varepsilon}\boldsymbol{\varepsilon}^T - 2\eta\boldsymbol{\varepsilon}^\times$
Euler parameters	3	$\varepsilon_1, \varepsilon_2, \varepsilon_3$	$\boldsymbol{\varepsilon} = \mathbf{a}\sin\frac{1}{2}\phi$	$\mathbf{C} = \mathbf{1} + 2\boldsymbol{\varepsilon}^\times\boldsymbol{\varepsilon}^\times - 2(1 - \boldsymbol{\varepsilon}^T\boldsymbol{\varepsilon})^{1/2}\boldsymbol{\varepsilon}^\times$
Euler angles	3	$\theta_1, \theta_2, \theta_3$	Fig. 2.2	$\mathbf{C} = \mathbf{C}_i(\theta_1)\mathbf{C}_j(\theta_2)\mathbf{C}_k(\theta_3)$; $i \neq j$; $j \neq k$ (see Table 2.1)

Bean]) and on the NASA Space Shuttle ([Klumpp, 1]). For readers interested in the computational aspects of Euler parameters, the sequence of notes by [Klumpp, 1], [Spurrier], [Klumpp, 2], [Shepperd], and [Grubin, 3] is recommended.

Euler angles have probably been overemphasized in textbooks. For infinitesimal analyses, they are equivalent to other parametrizations; for arbitrary displacements, computation is made difficult by both the trigonometric functions and the singularity. There is one instance, however, where Euler angles are inescapable on physical grounds: whenever a reference frame corresponds to a body suspended in gimbals, the gimbal angles form a natural set of Euler angles. The singularity, known as *gimbal lock*, will normally not be reached; to avoid it, a four-gimbal suspension is sometimes used.

Although it is dangerous to generalize, the best overall choice appears to be Euler parameters. They are computationally efficient and can be visualized (with practice), and the extra parameter is a price willingly paid to avoid singularities. Nevertheless, the expert analyst will be familiar with all the alternatives and will be able to select the one most appropriate to his specific need. It should also be realized that most of the differential equations in this chapter—the nonlinear ones and those that are linear but have periodic coefficients—cannot be solved analytically. Although such equations are essential for general simulation, we shall rely in this book primarily on the linear relationships (24) in Section 2.2 and (31) and (33) in Section 2.3 for analytical results.

Name	$\boldsymbol{\omega} = \mathbf{S}(\mathbf{v})\dot{\mathbf{v}}$	$\dot{\mathbf{v}} = \mathbf{S}^{-1}(\mathbf{v})\boldsymbol{\omega}$
Direction cosines	$\boldsymbol{\omega}^{\times} = -\dot{\mathbf{C}}\mathbf{C}^T \equiv \mathbf{C}\dot{\mathbf{C}}^T$	$[\mathbf{c}_1 \quad \mathbf{c}_2 \quad \mathbf{c}_3] \triangleq \mathbf{C}$ $\dot{\mathbf{c}}_1 = \mathbf{c}_1^{\times}\boldsymbol{\omega};\ \dot{\mathbf{c}}_2 = \mathbf{c}_2^{\times}\boldsymbol{\omega};\ \dot{\mathbf{c}}_3 = \mathbf{c}_3^{\times}\boldsymbol{\omega}$
Direction cosines	—	$\dot{\mathbf{c}}_1 = \mathbf{c}_1^{\times}\boldsymbol{\omega};\ \dot{\mathbf{c}}_2 = \mathbf{c}_2^{\times}\boldsymbol{\omega};\ \mathbf{c}_3 = \mathbf{c}_1^{\times}\mathbf{c}_2$
Axis / angle variables	$\boldsymbol{\omega} = \dot{\phi}\mathbf{a} - (1 - \cos\phi)\mathbf{a}^{\times}\dot{\mathbf{a}} + \dot{\mathbf{a}}\sin\phi$	$\dot{\mathbf{a}} = \frac{1}{2}(\mathbf{a}^{\times} - \cot\frac{1}{2}\phi\,\mathbf{a}^{\times}\mathbf{a}^{\times})\boldsymbol{\omega}$ $\dot{\phi} = \mathbf{a}^T\boldsymbol{\omega}$
Axis / angle variables	$\boldsymbol{\omega} = (\mathbf{1} + c_1(\phi)\boldsymbol{\phi}^{\times} + c_2(\phi)\boldsymbol{\phi}^{\times}\boldsymbol{\phi}^{\times})\dot{\boldsymbol{\phi}}$ $c_1 \triangleq -2\phi^{-2}\sin^2\frac{1}{2}\phi;$ $c_2 \triangleq \phi^{-3}(\phi - \sin\phi)$	$\dot{\boldsymbol{\phi}} = (\mathbf{1} + \frac{1}{2}\boldsymbol{\phi}^{\times} + a_1(\phi)\boldsymbol{\phi}^{\times}\boldsymbol{\phi}^{\times})\boldsymbol{\omega}$ $\phi \triangleq (\boldsymbol{\phi}^T\boldsymbol{\phi})^{1/2};\ a_1 \triangleq \phi^{-2}(1 - \frac{1}{2}\phi\cot\frac{1}{2}\phi)$
Euler – Rodriguez (see Probs. 2.12, 19)	$\boldsymbol{\omega} = \dfrac{2}{1 + \mathbf{p}^T\mathbf{p}}(\mathbf{1} - \mathbf{p}^{\times})\dot{\mathbf{p}}$	$\dot{\mathbf{p}} = \frac{1}{2}(\mathbf{p}\mathbf{p}^T + \mathbf{1} + \mathbf{p}^{\times})\boldsymbol{\omega}$
Euler parameters	$\boldsymbol{\omega} = 2(\eta\dot{\boldsymbol{\varepsilon}} - \dot{\eta}\boldsymbol{\varepsilon}) - 2\boldsymbol{\varepsilon}^{\times}\dot{\boldsymbol{\varepsilon}}$	$\dot{\boldsymbol{\varepsilon}} = \frac{1}{2}(\boldsymbol{\varepsilon}^{\times} + \eta\mathbf{1})\boldsymbol{\omega}$ $\dot{\eta} = -\frac{1}{2}\boldsymbol{\varepsilon}^T\boldsymbol{\omega}$
Euler parameters	$\boldsymbol{\omega} = 2\left[\dfrac{\mathbf{1} + \boldsymbol{\varepsilon}^{\times}\boldsymbol{\varepsilon}^{\times}}{(1 - \boldsymbol{\varepsilon}^T\boldsymbol{\varepsilon})^{1/2}} - \boldsymbol{\varepsilon}^{\times}\right]\dot{\boldsymbol{\varepsilon}}$	$\dot{\boldsymbol{\varepsilon}} = \frac{1}{2}\{\boldsymbol{\varepsilon}^{\times} + (1 - \boldsymbol{\varepsilon}^T\boldsymbol{\varepsilon})^{1/2}\mathbf{1}\}\boldsymbol{\omega}$
Euler angles	$\boldsymbol{\omega} = [\mathbf{1}_i \quad \mathbf{C}_i\mathbf{1}_j \quad \mathbf{C}_i\mathbf{C}_j\mathbf{1}_k]\dot{\boldsymbol{\theta}}$	$\dot{\boldsymbol{\theta}} = \mathbf{S}^{-1}(\theta_1, \theta_2)\boldsymbol{\omega}$ (see Table 2.2)

2.5 PROBLEMS

2.1 Show by induction that the product of any number or rotation matrices is itself a rotation matrix.

2.2 Show that the eigenvalues λ_1, λ_2, and λ_3 of a rotation matrix $\mathbf{C}$ satisfy the characteristic equation

$$\lambda^3 - \sigma\lambda^2 + \sigma\lambda - 1 = 0 \tag{1}$$

where $\sigma \triangleq \operatorname{trace}\mathbf{C}$. Based on (16) in Section 2.1 and the ensuing discussion, explain why $\det[\mathbf{C} - \mathbf{1}]$ and $\det[\mathbf{C} - \exp(\pm j\phi)\mathbf{1}]$ vanish.

2.3 Consider the following rotation matrix:

$$\mathbf{C}(\theta_2, \theta_3) = \mathbf{C}_2(\theta_2)\,\mathbf{C}_2(\theta_3) \tag{2}$$

where $\mathbf{C}_2$ and $\mathbf{C}_3$ are principal rotations defined in (28) in Section 2.1. Show that the axis and angle referred to in Euler's theorem are given by

$$\cos\frac{\phi}{2} = \cos\frac{\theta_2}{2}\cos\frac{\theta_3}{2}$$

$$\mathbf{a}\sin\frac{\phi}{2} = \begin{bmatrix} -\sin\dfrac{\theta_2}{2}\sin\dfrac{\theta_3}{2} \\ \sin\dfrac{\theta_2}{2}\cos\dfrac{\theta_3}{2} \\ \cos\dfrac{\theta_2}{2}\sin\dfrac{\theta_3}{2} \end{bmatrix} \tag{3}$$

2.4 Consider the rotation matrix formed by the classical 3–1–3 Euler angle sequence

$$\mathbf{C}(\theta_1, \theta_2, \theta_3) = \mathbf{C}_3(\theta_1)\mathbf{C}_1(\theta_2)\mathbf{C}_3(\theta_3) \tag{4}$$

(*a*) Show that

$$\mathbf{C} = \begin{bmatrix} c_1c_3 - s_1c_2s_3 & c_1s_3 + s_1c_2c_3 & s_1s_2 \\ -s_1c_3 - c_1c_2s_3 & -s_1s_3 + c_1c_2c_3 & c_1s_2 \\ s_2s_3 & -s_2c_3 & c_2 \end{bmatrix} \tag{5}$$

(*b*) Show that $\mathbf{C}(\theta_1, 0, \theta_3) = \mathbf{C}_3(\theta_1 + \theta_3)$, indicating that, when $\theta_2 = 0$, the two variables θ_1 and θ_3 are both associated with the same degree of freedom.

2.5 Show that

$$\mathbf{C}(\mathbf{a}, \phi) = \exp(-\phi\mathbf{a}^\times) \tag{6}$$

as noted by [Meyer] and [Gelman, 1], where $\mathbf{a}$ and ϕ are, as usual, the axis and angle in Euler's theorem.

2.6 Given are two vectrices $\vec{\mathscr{F}}_a$ and $\vec{\mathscr{F}}_b$. Denote by ψ_i the angle between $\hat{\vec{a}}_i$ and $\hat{\vec{b}}_i$ $(i = 1, 2, 3)$. Show that

$$\phi \geq \psi_i \qquad (i = 1, 2, 3) \tag{7}$$

as noted by [Meyer], where all angles are restricted to the range $[0, \pi]$.

2.7 Consider the 3×3 matrix $\mathbf{U}$ defined by

$$\mathbf{U} \triangleq \mathbf{1} - 2\mathbf{n}\mathbf{n}^T \tag{8}$$

where $\mathbf{n}$ is a unit column, $\mathbf{n}^T\mathbf{n} = 1$. Show that $\mathbf{U}$ is an orthonormal matrix but that $\det\mathbf{U} = -1$. The matrix $\mathbf{U}$ is called a *reflection matrix* because, as noted by [Gelman, 2], it reflects any vector in a plane whose normal vector has components $\mathbf{n}$; verify this assertion. A matrix of the form $\mathbf{CU}$ or $\mathbf{UC}$ is called an *improper rotation*. If $\vec{\mathscr{F}}$ is a vectrix of right-handed orthogonal unit vectors (as usual), show that $\mathbf{CU}\vec{\mathscr{F}}$ and $\mathbf{UC}\vec{\mathscr{F}}$ are vectrices of *left-handed* orthogonal unit vectors.

2.8 The rotation matrix $\mathbf{C}(\mathbf{a}, \phi)$ is given by (22) in Section 2.1. Show that (i) $\mathbf{C}^T\mathbf{C} = \mathbf{1}$; (ii) $\det\mathbf{C} = +1$; and (iii) $\mathbf{Ca} = \mathbf{a}$.

2.9 Show that the two rotation matrices $\mathbf{C}(\mathbf{a}_1, \phi_1)$ and $\mathbf{C}(\mathbf{a}_2, \phi_2)$ commute in multiplication if and only if

$$\mathbf{a}_1^\times\mathbf{a}_2 = \mathbf{0} \tag{9}$$

2.10 Consider the rotation matrix corresponding to two angular displacements in succession,

$$\mathbf{C}(\mathbf{a}_3, \phi_3) = \mathbf{C}(\mathbf{a}_2, \phi_2)\mathbf{C}(\mathbf{a}_1, \phi_1) \tag{10}$$

The aim is to obtain $\mathbf{a}_3$ and ϕ_3 in terms of $\mathbf{a}_1, \mathbf{a}_2, \phi_1$, and ϕ_2.

(*a*) Derive the results mentioned in the text, namely, (6) and (8) in Section 2.2. The latter assumes that $\sin\phi_3 \neq 0$.

(*b*) When $\sin\phi_3 = 0$, either $\phi_3 = 0, \pm 2\pi, \ldots$, in which case $\mathbf{a}_3$ is undefined; or else $\phi_3 = \pm\pi, \pm 3\pi, \ldots$, in which case show that

$$\begin{aligned}\sin^2\gamma\mathbf{a}_3 = &\left(\cos^2\frac{\phi_1}{2} + \cos^2\gamma\sin^2\frac{\phi_1}{2}\right)(\mathbf{a}_2 - \cos\gamma\mathbf{a}_1)\\ &+\left(\cos^2\frac{\phi_2}{2} + \cos^2\gamma\sin^2\frac{\phi_2}{2}\right)(\mathbf{a}_1 - \cos\gamma\mathbf{a}_2)\\ &+\left(1 - \cos^2\frac{\phi_1}{2} - \cos^2\frac{\phi_2}{2}\right)\sin\gamma\mathbf{a}_2^{\times}\mathbf{a}_1 \qquad (11)\end{aligned}$$

This formula yields $\mathbf{a}_3$ when $\sin\gamma \neq 0$. (If $\sin\gamma = 0$, then $\mathbf{a}_1$ and $\mathbf{a}_2$ must be parallel or antiparallel; hence, $\mathbf{a}_3 = \mathbf{a}_1$.)

2.11 Assume that a rotation matrix is specified in terms of 3–2–1 Euler angles, as given by (18) in Section 2.2. Based on (15) and (16) in the same section (or otherwise), shows that the Euler parameters for this rotation matrix are

$$\begin{aligned}\varepsilon_1 &= S_1C_2C_3 - C_1S_2S_3\\ \varepsilon_2 &= C_1S_2C_3 + S_1C_2S_3\\ \varepsilon_3 &= C_1C_2S_3 - S_1S_2C_3\\ \eta &= C_1C_2C_3 + S_1S_2S_3 \qquad (12)\end{aligned}$$

where $C_i \triangleq \cos\theta_i/2$, $S_i \triangleq \sin\theta_i/2$, for $i = 1, 2, 3$. An alternative derivation of (12) has been given by [A. C. Robinson].

2.12 Consider the following three-parameter for representing angular displacements:

$$\mathbf{p} \triangleq \tan\frac{\phi}{2}\mathbf{a} \qquad (13)$$

(*a*) Show, using (12) in Section 2.2, or otherwise, that the displacement $\mathbf{p}_3$ equivalent to {first $\mathbf{p}_1$, then $\mathbf{p}_2$} is

$$\mathbf{p}_3 = (\mathbf{p}_1 + \mathbf{p}_2 + \mathbf{p}_1^{\times}\mathbf{p}_2)/(1 - \mathbf{p}_1^T\mathbf{p}_2) \qquad (14)$$

(*b*) Show that the rotation matrix associated with $\mathbf{p}$ is

$$\mathbf{C} = \frac{(1 - \mathbf{p}^T\mathbf{p})\mathbf{1} + 2\mathbf{p}\mathbf{p}^T - 2\mathbf{p}^{\times}}{1 + \mathbf{p}^T\mathbf{p}} \equiv (\mathbf{1} - \mathbf{p}^{\times})(\mathbf{1} + \mathbf{p}^{\times})^{-1} \qquad (15)$$

(*c*) Given a rotation matrix $\mathbf{C}$, show that the associated $\mathbf{p}$ is

$$\mathbf{p} = (1 + c_{11} + c_{22} + c_{33})^{-1}\begin{bmatrix} c_{23} - c_{32}\\ c_{31} - c_{13}\\ c_{12} - c_{21}\end{bmatrix} \qquad (16)$$

Where is the singularity for these three parameters? Based on historical evidence, [Roberson, 3] has called these *Euler–Rodrigues parameters*. An exposition of essentially the same parameters has been given by [Gibbs], and so they are also referred to in the literature as *Gibbs parameters*.

2.13 Given a rotation matrix $\mathbf{C}$, show that the associated 3–2–1 Euler angles are given by

$$\left.\begin{aligned} \theta_2 &= -\sin^{-1} c_{13} \\ \theta_1 &= \tan^{-1}\left(\frac{c_{23}}{\cos\theta_2}, \frac{c_{33}}{\cos\theta_2}\right) \\ \theta_3 &= \tan^{-1}\left(\frac{c_{12}}{\cos\theta_2}, \frac{c_{11}}{\cos\theta_2}\right) \end{aligned}\right\} \quad \left(\theta_2 \neq \frac{\pi}{2}\right) \tag{17}$$

What does one do when $\theta_2 = \pi/2$?

2.14 Given four Euler parameters $(\boldsymbol{\varepsilon}, \eta)$, define two 4×4 matrices as follows:

$$\mathbf{E}_+(\boldsymbol{\varepsilon}, \eta) \triangleq \begin{bmatrix} \eta & -\boldsymbol{\varepsilon}^T \\ \boldsymbol{\varepsilon} & \eta\mathbf{1} + \boldsymbol{\varepsilon}^\times \end{bmatrix} \tag{18}$$

$$\mathbf{E}_-(\boldsymbol{\varepsilon}, \eta) \triangleq \begin{bmatrix} \eta & -\boldsymbol{\varepsilon}^T \\ \boldsymbol{\varepsilon} & \eta\mathbf{1} - \boldsymbol{\varepsilon}^\times \end{bmatrix} \tag{19}$$

(a) Show that (12) in Section 2.2, which states the Euler parameters $(\boldsymbol{\varepsilon}_3, \eta_3)$ corresponding to the two consecutive angular displacements, first $(\boldsymbol{\varepsilon}_1, \eta_1)$, then $(\boldsymbol{\varepsilon}_2, \eta_2)$, may be written in either of the alternate forms:

$$\begin{aligned} \begin{bmatrix} \eta_3 \\ \boldsymbol{\varepsilon}_3 \end{bmatrix} &= \mathbf{E}_+(\boldsymbol{\varepsilon}_1, \eta_1)\begin{bmatrix} \eta_2 \\ \boldsymbol{\varepsilon}_2 \end{bmatrix} \\ \begin{bmatrix} \eta_3 \\ \boldsymbol{\varepsilon}_3 \end{bmatrix} &= \mathbf{E}_-(\boldsymbol{\varepsilon}_2, \eta_2)\begin{bmatrix} \eta_1 \\ \boldsymbol{\varepsilon}_1 \end{bmatrix} \end{aligned} \tag{20}$$

According to [Whittaker], the scalar formulas equivalent to (20) were discovered independently by Gauss, Rodrigues, Hamilton, and Cayley. See also [Ickes].

(b) Show that $\mathbf{E}_+$ and $\mathbf{E}_-$ are orthonormal matrices.

(c) Define the 4×4 matrix $\boldsymbol{\Omega}$ as follows:

$$\boldsymbol{\Omega}(\boldsymbol{\omega}) = \begin{bmatrix} 0 & -\boldsymbol{\omega}^T \\ \boldsymbol{\omega} & -\boldsymbol{\omega}^\times \end{bmatrix} \tag{21}$$

Show that $\boldsymbol{\Omega}$ is skew-symmetric and that (22) in Section 2.3 can be written

$$\begin{bmatrix} \dot{\eta} \\ \dot{\boldsymbol{\varepsilon}} \end{bmatrix} = \tfrac{1}{2}\boldsymbol{\Omega}(\boldsymbol{\omega})\begin{bmatrix} \eta \\ \boldsymbol{\varepsilon} \end{bmatrix} \tag{22}$$

As noted by [Kane, 3], show that

$$\exp\left(\tfrac{1}{2}\boldsymbol{\Omega}t\right) = \mathbf{1}\cos\left(\tfrac{1}{2}\omega t\right) + \omega^{-1}\boldsymbol{\Omega}\sin\left(\tfrac{1}{2}\omega t\right)$$

and hence solve (22) for $\eta(t)$ and $\boldsymbol{\varepsilon}(t)$, given $\eta(0)$ and $\boldsymbol{\varepsilon}(0)$, for the special case of a steady rotation at rate ω about a fixed axis.

(d) Show that the relation inverse to (22), that is, the equivalent of (23) in Section 2.3, can be written

$$\begin{bmatrix} 0 \\ \boldsymbol{\omega} \end{bmatrix} = 2\mathbf{E}_-(\boldsymbol{\varepsilon}, \eta)\begin{bmatrix} -\dot{\eta} \\ \dot{\boldsymbol{\varepsilon}} \end{bmatrix} \tag{23}$$

2.15 A rotation matrix $\mathbf{C} \equiv \{c_{ij}\}$ is given, and it is desired to find the equivalent Euler parameters. Show that

$$\begin{aligned} \varepsilon_1\varepsilon_2 &= \tfrac{1}{4}(c_{12} + c_{21}) \\ \varepsilon_2\varepsilon_3 &= \tfrac{1}{4}(c_{23} + c_{32}) \\ \varepsilon_3\varepsilon_1 &= \tfrac{1}{4}(c_{31} + c_{13}) \end{aligned} \tag{24}$$

These formulas, together with (15) and (16) in Section 2.2, are used in modern on-board computer algorithms ([Shepperd]).

2.16 Let $\boldsymbol{\omega}_{ij}$ represent the components, in $\mathscr{F}_i$, of the angular velocity of $\mathscr{F}_i$ with respect to $\mathscr{F}_j$, where i and j can be a, b, c, or d. Show that

$$\boldsymbol{\omega}_{da} = \boldsymbol{\omega}_{db} + \mathbf{C}_{db}\boldsymbol{\omega}_{bc} + \mathbf{C}_{dc}\boldsymbol{\omega}_{ca}$$

2.17 In connection with (20) in Section 2.3, explain geometrically why $\dot{\mathbf{a}}$ is unbounded as ϕ passes through zero.

2.18 Perform the scalar inversion of (22a) in Section 2.3 to obtain (23) in that section.

2.19 (a) Referring to Problem 2.12 above, show that the kinematical equations for the Euler–Rodrigues parameters are

$$\dot{\mathbf{p}} = \tfrac{1}{2}(\mathbf{1} + \mathbf{p}\mathbf{p}^T + \mathbf{p}^\times)\boldsymbol{\omega} \tag{25}$$

These parameters are not often used because the term $\mathbf{p}\mathbf{p}^T\boldsymbol{\omega}$ leads to instabilities during numerical integration ([Churchyard]).

(b) Perform a scalar inversion of (25) to show that

$$\boldsymbol{\omega} = \frac{2(\mathbf{1} - \mathbf{p}^\times)\dot{\mathbf{p}}}{1 + \mathbf{p}^T\mathbf{p}} \tag{26}$$

(c) For infinitesimal rotations, show that $\dot{\mathbf{p}} = \frac{1}{2}\boldsymbol{\omega}(t)$. This indicates that Euler–Rodrigues parameters are indistinguishable from other parameter sets when the context is limited to infinitesimal rotations.

2.20 Two constant vectrices $\mathscr{F}_a$ and $\mathscr{F}_b$ are given, and $\mathbf{C}_{ba}$ is the (constant) rotation matrix relating them. There are an infinite number of continuous paths from $\mathscr{F}_a$ to $\mathscr{F}_b$, and for each of these paths there is a continuous matrix function $\mathbf{C}(t)$, $0 \le t \le T$, T fixed, such that $\mathbf{C}(0) = \mathbf{1}$ and $\mathbf{C}(T) = \mathbf{C}_{ba}$. For each matrix function, define $\boldsymbol{\omega}(t)$ by $\boldsymbol{\omega}^\times \triangleq -\dot{\mathbf{C}}\mathbf{C}^T$. Let Ω be the set of all such functions $\boldsymbol{\omega}$. Show that

$$\phi = \min_{\Omega} \int_0^T \|\boldsymbol{\omega}(t)\|\, dt \tag{27}$$

where $\boldsymbol{\omega}(t) \in \Omega$, and ϕ is the Euler theorem angle associated with $\mathbf{C}_{ba}$. Explain how it may be interpreted that ϕ is the "minimum angular distance" between $\mathscr{F}_a$ and $\mathscr{F}_b$. This problem is based on the work of [Meyer].

2.21 The reference frame $\mathscr{F}_b$ is imbedded in a rigid body $\mathscr{R}$. Three noncolinear points in $\mathscr{R}$ are located at $\underset{\rightarrow}{\mathbf{r}}_1$, $\underset{\rightarrow}{\mathbf{r}}_2$, and $\underset{\rightarrow}{\mathbf{r}}_3$, and the velocities of these points, as seen in a second frame, $\mathscr{F}_r$, are $\underset{\rightarrow}{\mathbf{v}}_1$, $\underset{\rightarrow}{\mathbf{v}}_2$, and $\underset{\rightarrow}{\mathbf{v}}_3$, respectively Show that the angular velocity of $\mathscr{F}_b$ (and therefore of $\mathscr{R}$) with respect to $\mathscr{F}_r$ is given by

$$\omega = \frac{\underset{\rightarrow}{\mathbf{v}}_1 \times \underset{\rightarrow}{\mathbf{v}}_2 + \underset{\rightarrow}{\mathbf{v}}_2 \times \underset{\rightarrow}{\mathbf{v}}_3 + \underset{\rightarrow}{\mathbf{v}}_3 \times \underset{\rightarrow}{\mathbf{v}}_1}{\underset{\rightarrow}{\mathbf{v}}_1 \cdot (\underset{\rightarrow}{\mathbf{r}}_2 - \underset{\rightarrow}{\mathbf{r}}_3) + \underset{\rightarrow}{\mathbf{v}}_2 \cdot (\underset{\rightarrow}{\mathbf{r}}_3 - \underset{\rightarrow}{\mathbf{r}}_1) + \underset{\rightarrow}{\mathbf{v}}_3 \cdot (\underset{\rightarrow}{\mathbf{r}}_1 - \underset{\rightarrow}{\mathbf{r}}_2)} \tag{28}$$

This problem is taken from [Wittenburg], who attributes it to [Charlamov].

2.22 The kinematical equation for infinitesimal rotations with respect to a constant angular velocity, (33) in Section 2.3 may be written

$$\dot{\boldsymbol{\alpha}} + \boldsymbol{\nu}^{\times}\boldsymbol{\alpha} = \boldsymbol{\omega}_{ca}(t) - \boldsymbol{\nu} \tag{29}$$

where $\boldsymbol{\alpha}(t)$ represents the infinitesimal rotation of $\mathscr{F}_c$ with respect to $\mathscr{F}_b$, and $\boldsymbol{\nu}$ is the (constant) angular velocity of $\mathscr{F}_b$ with respect to $\mathscr{F}_a$. It is convenient to introduce the unit column $\hat{\boldsymbol{\nu}} = \boldsymbol{\nu}/\nu$ and the angle $\psi = \nu t$, where ν is the magnitude of $\boldsymbol{\nu}$. Show that the closed-form solution for $\boldsymbol{\alpha}(t)$ is

$$\boldsymbol{\alpha}(t) = \mathbf{C}(\hat{\boldsymbol{\nu}}, \psi)\left[\boldsymbol{\alpha}(0) + \int_0^{\psi} \mathbf{C}^T(\hat{\boldsymbol{\nu}}, \psi')\mathbf{u}(\psi')\, d\psi'\right] \tag{30}$$

where

$$\mathbf{C}(\hat{\boldsymbol{\nu}}, \psi) \triangleq \cos\psi \mathbf{1} + (1 - \cos\psi)\hat{\boldsymbol{\nu}}\hat{\boldsymbol{\nu}}^T - \sin\psi\hat{\boldsymbol{\nu}}^{\times}$$

$$\mathbf{u}(\psi) \triangleq \frac{\boldsymbol{\omega}_{ca}(\psi)}{\nu} - \hat{\boldsymbol{\nu}} \tag{31}$$

2.23 The concept of *instantaneous axis* is sometimes helpful in the kinematics of rotation (Fig. 2.8).

(*a*) First consider motion in the 1–2-plane. The velocity is the superposition of the translational velocity of a reference point O and a "rigid" angular velocity about O. Thus, for a point P at $\mathbf{r}$ with respect to O in the 1–2-plane,

$$\mathbf{v}(\mathbf{r}, t) = [v_{o1} - \omega r_2 \quad v_{o2} + \omega r_1 \quad 0]^T \tag{32}$$

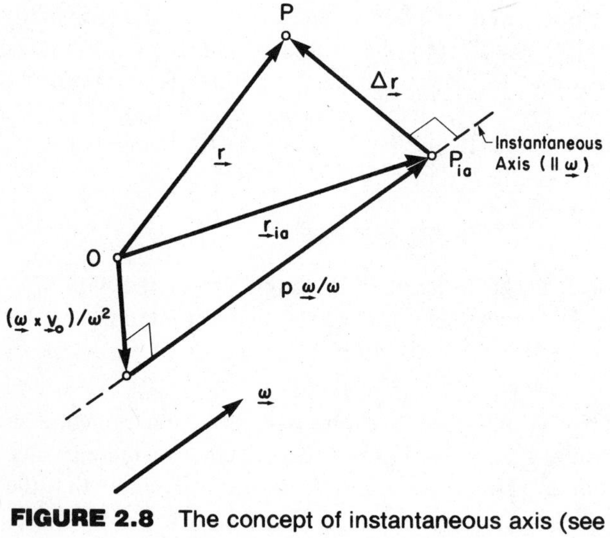

FIGURE 2.8 The concept of instantaneous axis (see Problem 2.23).

where v_{o1} and v_{o2} are the components of the velocity of O, and ω is the angular speed about the 3-axis at O. Show that the point P_{ia}, defined by

$$\mathbf{r}_{ia} \triangleq \begin{bmatrix} \frac{-v_{o2}}{\omega} & \frac{v_{o1}}{\omega} & 0 \end{bmatrix}^T \tag{33}$$

has zero velocity. The instantaneous axis, parallel to the 3-axis, thus passes through this point.

(*b*) Show that the velocity field (32) may be written in the equivalent form

$$\mathbf{v}(\Delta\mathbf{r}, t) = [-\omega\Delta r_2 \quad \omega\Delta r_1 \quad 0]^T \tag{34}$$

where $\Delta\mathbf{r} \triangleq \mathbf{r} - \mathbf{r}_{ia}$ is the location of P with respect to P_{ia}. In other words, show that the velocity field can be represented as a "rigid" angular velocity about P_{ia}.

(*c*) Now consider the more general three-dimensional instantaneous axis. The velocity field is now a superposition of the translational velocity of the reference point O and a rigid angular velocity about O. Thus, for a point P at $\mathbf{r}$ with respect to O,

$$\mathbf{v}(r, t) = \mathbf{v}_o(t) + \boldsymbol{\omega}^{\times}(t)\mathbf{r} \tag{35}$$

Show that the line defined by

$$\mathbf{r} = \mathbf{r}_{ia}(p) \triangleq \frac{\boldsymbol{\omega}^{\times}\mathbf{v}_o}{\omega^2} + \frac{p\boldsymbol{\omega}}{\omega} \tag{36}$$

where p is a parameter, $-\infty < p < \infty$, consists of points having the common velocity

$$\mathbf{v}(\mathbf{r}_{ia}, t) = \mathbf{v}_{ia} \triangleq \frac{\boldsymbol{\omega}\boldsymbol{\omega}^T v_o}{\omega^2} \tag{37}$$

(*d*) Show further that the velocity of an arbitrary point P, at $\mathbf{r}$ with respect to O, can be written as

$$\mathbf{v}(\Delta\mathbf{r}, t) = \mathbf{v}_{ia} + \boldsymbol{\omega}^{\times}\Delta\mathbf{r} \tag{28}$$

where $\Delta\mathbf{r}$ is the displacement of P from the instantaneous axis.

(*e*) Explain why any point P not on the instantaneous axis has a speed greater than v_{ia}, where

$$v_{ia} \triangleq \|\mathbf{v}_{ia}\| \triangleq v_o \cos(\boldsymbol{\omega}, \mathbf{v}_o) \tag{39}$$

2.24 Let $\mathbf{C} \equiv \{c_{ij}\}$ be any rotation matrix.

(*a*) Explain why the orthonormality condition, when expressed as $\mathbf{C}\mathbf{C}^T = \mathbf{1}$, implies that the rows of $\mathbf{C}$ are orthonormal. Show that the following scalar conditions follow from row-orthonormality:

$$\begin{aligned} c_{11}^2 + c_{12}^2 + c_{13}^2 &= 1 \qquad & c_{11}c_{21} + c_{12}c_{22} + c_{13}c_{23} &= 0 \\ c_{21}^2 + c_{22}^2 + c_{23}^2 &= 1 \qquad & c_{21}c_{31} + c_{22}c_{32} + c_{23}c_{33} &= 0 \\ c_{31}^2 + c_{32}^2 + c_{33}^2 &= 1 \qquad & c_{31}c_{11} + c_{32}c_{12} + c_{33}c_{13} &= 0 \end{aligned} \tag{40}$$

Are these six equations independent?

(*b*) Explain why the orthonormality condition, when expressed as $\mathbf{C}^T\mathbf{C} = \mathbf{1}$, implies that the columns of $\mathbf{C}$ are orthonormal. Show that the following scalar equations follow from column-orthonormality:

$$\begin{aligned} c_{11}^2 + c_{21}^2 + c_{31}^2 &= 1 \qquad & c_{11}c_{12} + c_{21}c_{22} + c_{31}c_{32} &= 0 \\ c_{12}^2 + c_{22}^2 + c_{32}^2 &= 1 \qquad & c_{12}c_{13} + c_{22}c_{23} + c_{32}c_{33} &= 0 \\ c_{13}^2 + c_{23}^2 + c_{33}^2 &= 1 \qquad & c_{13}c_{11} + c_{23}c_{21} + c_{33}c_{31} &= 0 \end{aligned} \tag{41}$$

Are these six equations independent? Are they independent of (40)?

2.25 The three-parameter set of orientation variables $[\phi_1 \quad \phi_2 \quad \phi_3]^T \equiv \boldsymbol{\phi} \triangleq \mathbf{a}\phi$ is not considered suitable for numerical computation because of the trigonometric functions associated with these variables; see, for example, the fourth entry in Table 2.3. For first-order analyses, however, (22) in Section 2.2 and (29) in Section 2.3 indicate that

$$\mathbf{C} \doteq \mathbf{1} - \boldsymbol{\phi}^\times; \qquad \boldsymbol{\omega} \doteq \dot{\boldsymbol{\phi}} \tag{42}$$

and the objection no longer applies. Show from (22) in Section 2.1 and (14) in Section 2.3 that, to second order in ϕ ($\equiv \|\boldsymbol{\phi}\|$),

$$\mathbf{C} \doteq \mathbf{1} - \boldsymbol{\phi}^\times + \tfrac{1}{2}(\boldsymbol{\phi}\boldsymbol{\phi}^T - \phi^2\mathbf{1}) \tag{43}$$

$$\boldsymbol{\omega} \doteq \dot{\boldsymbol{\phi}} - \tfrac{1}{2}\boldsymbol{\phi}^\times\dot{\boldsymbol{\phi}} \tag{44}$$

and comment.

CHAPTER 3

ATTITUDE MOTION EQUATIONS

With rare exceptions, the dynamics of spacecraft can be adequately treated in terms of *classical* mechanics. Indeed, the modern need to predict the orbital and attitude motions of man-made satellites has helped to infuse new vigor into this discipline. It is not our intention here, however, to embark on a general exposition of classical mechanics. For one thing, there is no need to do so; many excellent textbooks on the subject are available including those by [Goldstein], [Greenwood], [Halfman], [Symon], [Kane, 2], [Likins, 3], and [Meirovitch, 2]; the last three are of special interest because the research interests of the authors include spacecraft attitude dynamics, and this is reflected in their selection of material. A second reason for limiting the discussion is that we have in mind a single well-defined subject of study—spacecraft attitude dynamics. It is therefore not necessary for us to approach dynamics from the general viewpoint needed to provide a wide range of application. In fact, we shall require in this book the dynamics of only three mathematical abstractions, alone or in combination: a *point mass*, a *rigid body*, and a *quasi-rigid body*.

Our objectives are limited in yet another respect. There are many dynamical formulations available for deriving motion equations. In the limited space available, however, we shall be content with but one derivation in each situation. The formulation principally employed below is *vectorial mechanics*, so called because its basic equations [$\dot{\underrightarrow{\mathbf{p}}} = \underrightarrow{\mathbf{f}}$, $\dot{\underrightarrow{\mathbf{h}}} = \underrightarrow{\mathbf{g}}$; see (4) in Section 3.1 and (22) in Section 3.2] are expressed in terms of vectors. These fundamental motion equations are respectively associated with Newton and Euler, and so the approach is also referred to as the *Newton–Euler* formulation; we shall in addition emphasize the importance of the system kinetic energy. Vectorial mechanics is the best-known and most physically direct method and forms the mainstream of published attitude dynamics analyses. Moreover, [Likins, 4] has examined eight dynamical formulations and compared their use in the generation of motion equations for several models of interest (including a single rigid body and a topological tree of point-connected rigid bodies). Without suggesting that one method is always best, he concludes (in part) that "there appears to be no demonstrable advantage to any of the methods of analytical dynamics over the Newton–Euler methods... ."

Vectorial mechanics is applied to a single point mass in Section 3.1, and then to an arbitrary number of point masses (and, by extension, to a continuum) in Section 3.2. Motion equations are then derived (Sections 3.3 through 3.6) for a rigid body, first alone, then with a point mass damper, next with a spinning wheel, and finally with a second (hinge-connected) rigid body. In each case, the common *structure* of the equations of motion is emphasized, a structure that persists when an arbitrary number of bodies is considered briefly in Section 3.7. In this chapter, motion equations are formulated but not solved; properties of the motion are studied in the subsequent four chapters.

3.1 MOTION EQUATIONS FOR A POINT MASS, $\mathscr{P}$

A *point mass* is a material body of vanishingly small size. Strictly speaking, it is a mathematical abstraction; there are no point masses in nature. The concept is nevertheless useful in situations where the size of the approximated physical body is sufficiently small compared to the other characteristic dimensions of the system. One of the first bodies approximated (by Newton) as a point mass is more than 100 million meters in diameter—the planet Jupiter. This was a useful model because the other characteristic dimension of his model, Jupiter's distance from the sun, was 10,000 times larger. This example also serves to illustrate why "point mass" is preferred here to the older term "particle" which, according to the Oxford English Dictionary, is a "minute portion of matter." There is in physics an abundance of particles that more nearly comply with this general definition, including the particles in Brownian motion and the fundamental particles of matter. The term "point mass" also seems more descriptive.

Vectorial Motion Equation

Figure 3.1 shows an inertial (or Newtonian) reference frame $\mathscr{F}_i$ and a point mass $\mathscr{P}$ of mass m. If no force is acting ($\underrightarrow{\mathbf{f}} = \underrightarrow{\mathbf{0}}$), Newton's First Law asserts that an inertial observer sees $\mathscr{P}$ move with constant velocity $\underrightarrow{\mathbf{v}}$. In other words, $\mathscr{P}$ moves with respect to $\mathscr{F}_i$ in a straight line with constant speed $v = \|\underrightarrow{\mathbf{v}}\|$. When $\underrightarrow{\mathbf{f}} \neq \underrightarrow{\mathbf{0}}$, $\mathscr{P}$ moves in accordance with Newton's Second Law:

$$m\dot{\underrightarrow{\mathbf{v}}} = \underrightarrow{\mathbf{f}} \tag{1}$$

It is essential that the temporal derivative in (1) be measured in $\mathscr{F}_i$. A solid dot over a vector, as in (1), is used in this book exclusively for such inertial derivatives. Moreover, the velocity $\underrightarrow{\mathbf{v}}$ itself must be measured with respect to O_i; it is called the *absolute velocity*. The position of $\mathscr{P}$, denoted by $\underrightarrow{\mathbf{R}}$, can be computed

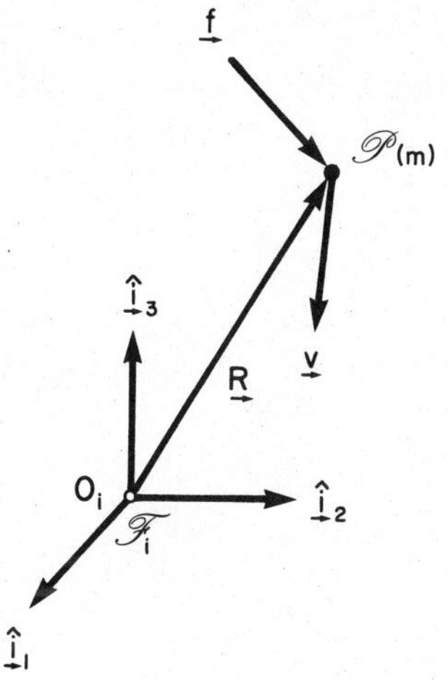

FIGURE 3.1 A point mass $\mathscr{P}$ of mass m.

by integrating the kinematical equation

$$\dot{\underset{\rightarrow}{\mathbf{R}}} = \underset{\rightarrow}{\mathbf{v}}(t) \tag{2}$$

The *absolute momentum* of $\mathscr{P}$ is defined as

$$\underset{\rightarrow}{\mathbf{p}} \triangleq m\underset{\rightarrow}{\mathbf{v}} \tag{3}$$

In terms of $\underset{\rightarrow}{\mathbf{p}}$, the motion equation is slightly simplified:

$$\dot{\underset{\rightarrow}{\mathbf{p}}} = \underset{\rightarrow}{\mathbf{f}} \tag{4}$$

The momentum of $\mathscr{P}$ bears so simple a relationship to its velocity that it seems scarcely worthwhile to make the distinction. The momentum concept is extremely important in mechanics, however, as will become evident in the next section when the dynamical system comprises many point masses.

Unlike momentum—which is a vector quantity—the other dynamical quantity of paramount importance is a scalar—the kinetic energy T. For $\mathscr{P}$,

$$T = \tfrac{1}{2}m\underset{\rightarrow}{\mathbf{v}} \cdot \underset{\rightarrow}{\mathbf{v}} \tag{5}$$

Its importance stems partly from the fact that its time derivative during the motion is

$$\dot{T} = m\dot{\underset{\rightarrow}{\mathbf{v}}} \cdot \underset{\rightarrow}{\mathbf{v}} = \underset{\rightarrow}{\mathbf{f}} \cdot \underset{\rightarrow}{\mathbf{v}} \tag{6}$$

If $\underset{\rightarrow}{\mathbf{f}} \equiv \underset{\rightarrow}{\mathbf{0}}$, or if $\underset{\rightarrow}{\mathbf{f}} \perp \underset{\rightarrow}{\mathbf{v}}$ always, the kinetic energy is constant throughout the motion. The significance of the kinetic energy is more evident when complex systems are considered. As a final observation, there is an important relationship between the two dynamical quantities $\underset{\rightarrow}{\mathbf{p}}$ and T. From (5) and (3),

$$T = \frac{1}{2}\underset{\rightarrow}{\mathbf{p}} \cdot \underset{\rightarrow}{\mathbf{v}}; \qquad \frac{dT}{d\underset{\rightarrow}{\mathbf{v}}} = \underset{\rightarrow}{\mathbf{p}} \tag{7a, b}$$

The notation on the left side of (7*b*) is explained in Section B.5 of Appendix B.

Scalar Motion Equations

The scalar equivalents of the above results are most succinctly expressed in matrix notation. We select a specific inertial reference frame $\mathcal{F}_i$ whose vectrix is $\underset{\rightarrow}{\mathcal{F}}_i$ and express all vectors in $\mathcal{F}_i$:

$$[\mathbf{R} \quad \mathbf{v} \quad \mathbf{p} \quad \mathbf{f}] \triangleq \underset{\rightarrow}{\mathcal{F}}_i \cdot \left[\underset{\rightarrow}{\mathbf{R}} \quad \underset{\rightarrow}{\mathbf{v}} \quad \underset{\rightarrow}{\mathbf{p}} \quad \underset{\rightarrow}{\mathbf{f}}\right] \tag{8}$$

The results obtained above now become

$$m\dot{\mathbf{v}} = \mathbf{f} \tag{9}$$

$$\dot{\mathbf{R}} = \mathbf{v} \tag{10}$$

$$\mathbf{p} = m\mathbf{v} \tag{11}$$

$$\dot{\mathbf{p}} = \mathbf{f} \tag{12}$$

$$T = \tfrac{1}{2}m\mathbf{v}^T\mathbf{v} \tag{13}$$

$$\dot{T} = \mathbf{f}^T\mathbf{v} \tag{14}$$

$$T = \frac{1}{2}\mathbf{p}^T\mathbf{v}; \qquad \frac{\partial T}{\partial \mathbf{v}} = \mathbf{p} \tag{15}$$

We note from (12) and (15) that another (somewhat unusual) version of the motion equation is

$$\frac{d}{dt}\left(\frac{\partial T}{\partial \mathbf{v}}\right) = \mathbf{f} \tag{16}$$

This is herein termed a *quasi-Lagrangian* equation of motion, owing to its similarity to the true Lagrangian equation of motion, which is, for $\mathcal{P}$,

$$\frac{d}{dt}\left(\frac{\partial T}{\partial \dot{\mathbf{R}}}\right) = \mathbf{f} \tag{17}$$

The distinction between (16) and (17) may appear trivial, but for the rotating systems shortly to be considered, it is a distinction well made.

3.2 MOTION EQUATIONS FOR A SYSTEM OF POINT MASSES, $\Sigma\mathcal{P}_n$

The last section treated the simplest dynamical system: a single point mass $\mathcal{P}$. It served mainly to introduce basic ideas and notation in their simplest forms. We now consider a system of N point masses, $\mathcal{P}_1, \ldots, \mathcal{P}_N$, which mutually interact. The mass of $\mathcal{P}_n$ is m_n, and the system is illustrated in Fig. 3.2. An important contrast with Fig. 3.1 is that the position of $\mathcal{P}_n$ with respect to O_i is now expressed as a vector sum of two displacements, $\underset{\rightarrow}{\mathbf{R}}_n = \underset{\rightarrow}{\mathbf{R}}_o + \underset{\rightarrow}{\mathbf{r}}_n$, where $\underset{\rightarrow}{\mathbf{R}}_o$ is the position with respect to O_i of O, a point of interest to be chosen precisely later, and $\underset{\rightarrow}{\mathbf{r}}_n$ locates $\mathcal{P}_n$ with respect to O.

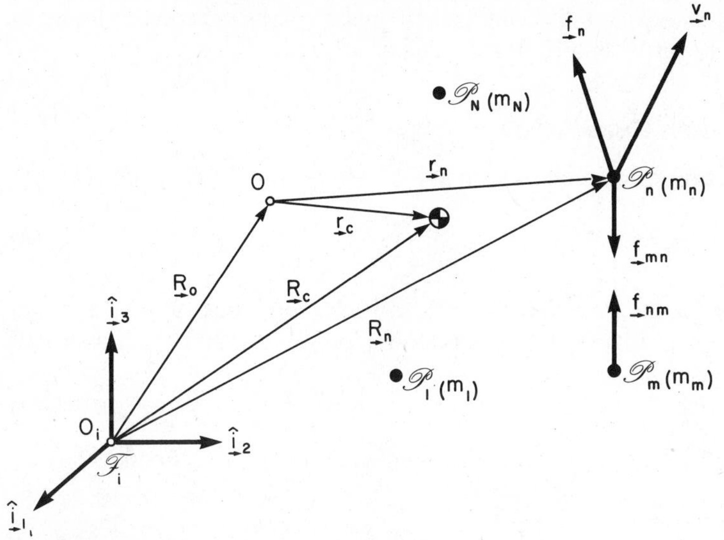

FIGURE 3.2 System of point masses, $\Sigma\mathscr{P}_n$.

The total mass of the system is m:

$$m \triangleq \sum_{n=1}^{N} m_n \tag{1}$$

The *first and second moments of inertia* with respect to O are likewise defined as follows:

$$\underset{\rightarrow}{\mathbf{c}} \triangleq \sum_{n=1}^{N} m_n \underset{\rightarrow}{\mathbf{r}}_n \tag{2}$$

$$\underset{\rightarrow}{\mathbf{J}} \triangleq \sum_{n=1}^{N} m_n \left(r_n^2 \underset{\rightarrow}{\mathbf{1}} - \underset{\rightarrow}{\mathbf{r}}_n \underset{\rightarrow}{\mathbf{r}}_n \right) \tag{3}$$

The motivation for these definitions will be clear presently. Normally, the second-moment-of-inertia dyadic is referred to simply as the *moment of inertia*. It has two important properties: symmetry and positive definiteness. To demonstrate symmetry, one shows that

$$\underset{\rightarrow}{\mathbf{u}}_1 \cdot \underset{\rightarrow}{\mathbf{J}} \cdot \underset{\rightarrow}{\mathbf{u}}_2 - \underset{\rightarrow}{\mathbf{u}}_2 \cdot \underset{\rightarrow}{\mathbf{J}} \cdot \underset{\rightarrow}{\mathbf{u}}_1 = 0 \tag{4}$$

for arbitrary vectors $\underset{\rightarrow}{\mathbf{u}}_1$ and $\underset{\rightarrow}{\mathbf{u}}_2$. As for positive definiteness,

$$\begin{aligned} \underset{\rightarrow}{\mathbf{u}} \cdot \underset{\rightarrow}{\mathbf{J}} \cdot \underset{\rightarrow}{\mathbf{u}} &= \sum_{n=1}^{N} m_n \left[r_n^2 u^2 - \left(\underset{\rightarrow}{\mathbf{r}}_n \cdot \underset{\rightarrow}{\mathbf{u}} \right)^2 \right] \\ &= \sum_{n=1}^{N} m_n r_n^2 u^2 \left[1 - \cos^2 \left(\underset{\rightarrow}{\mathbf{r}}_n, \underset{\rightarrow}{\mathbf{u}} \right) \right] \\ &= u^2 \sum_{n=1}^{N} m_n r_n^2 \sin^2 \left(\underset{\rightarrow}{\mathbf{r}}_n, \underset{\rightarrow}{\mathbf{u}} \right) \end{aligned} \tag{5}$$

which is always positive ($\mathbf{u} \neq \mathbf{0}$) unless all the mass is concentrated in a line in the $\mathbf{u}$-direction. The latter should be regarded as a limiting case.

Vectorial Motion Equations

The motion equation for each point mass is of the form (1) in Section 3.1:

$$m_n \dot{\mathbf{v}}_n = \mathbf{f}_n + \sum_{m=1}^{N} \mathbf{f}_{mn} \qquad (n = 1, \ldots, N) \tag{6}$$

Here, $\mathbf{f}_n$ is no longer the *total* force on $\mathscr{P}_n$; it is the force due only to agencies *external* to the system. Likewise, $\mathbf{f}_{mn}$ is the force on $\mathscr{P}_n$ due to $\mathscr{P}_m$. A glance at Fig. 3.2 shows that

$$\mathbf{v}_n = \mathbf{v}_o + \dot{\mathbf{r}}_n; \qquad \mathbf{v}_o \triangleq \dot{\mathbf{R}}_o \tag{7}$$

By analogy with (3) in Section 3.1, the momentum of $\mathscr{P}_n$ is defined to be

$$\mathbf{p}_n \triangleq m_n \mathbf{v}_n \tag{8}$$

a quantity whose importance rests on its contribution to the total momentum:

$$\mathbf{p} \triangleq \sum_{n=1}^{N} \mathbf{p}_n \tag{9}$$

From these definitions, it follows that

$$\mathbf{p} = \sum_{n=1}^{N} m_n(\mathbf{v}_o + \dot{\mathbf{r}}_n) = m\mathbf{v}_o + \dot{\mathbf{c}} \tag{10}$$

The motion equations (6) are now rewritten in terms of momenta:

$$\dot{\mathbf{p}}_n = \mathbf{f}_n + \sum_{m=1}^{N} \mathbf{f}_{mn} \qquad (n = 1, \ldots, N) \tag{11}$$

We sum these N equations to obtain

$$\dot{\mathbf{p}} = \mathbf{f} \tag{12}$$

where

$$\mathbf{f} \triangleq \sum_{n=1}^{N} \mathbf{f}_n \tag{13}$$

is the total external force on the system. (The sum of internal forces is

$$\sum_{n=1}^{N} \sum_{m=1}^{N} \mathbf{f}_{mn} = \mathbf{0}$$

by virtue of Newton's Third Law, which affirms that $\mathbf{f}_{nm} + \mathbf{f}_{mn} = \mathbf{0}$.) The importance of momentum is now evident from (12): using $\mathbf{p}$ yields for N particles a motion equation of the same form as (4) in Section 3.1 for one particle.

It should be observed that there is always a point, which we denote by $\oplus$, about which the first moment of inertia vanishes. Its locations with respect to O

and O_i are given (Fig. 3.2) by

$$\underset{\rightarrow}{\mathbf{r}}_c \triangleq \frac{\underset{\rightarrow}{\mathbf{c}}}{m}; \qquad \underset{\rightarrow}{\mathbf{R}}_c \triangleq \underset{\rightarrow}{\mathbf{R}}_o + \underset{\rightarrow}{\mathbf{r}}_c \tag{14}$$

For, by definition,

$$\sum_{n=1}^{N} m_n(\underset{\rightarrow}{\mathbf{r}}_n - \underset{\rightarrow}{\mathbf{r}}_c) = \underset{\rightarrow}{\mathbf{c}} - m\underset{\rightarrow}{\mathbf{r}}_c = \underset{\rightarrow}{\mathbf{0}}$$

as asserted. The point $\oplus$ is called the *mass center*, or *centroid*, of the system. Now consider (10): we know $\mathbf{p}(t)$ by integrating (12); if we are free to choose $\oplus$ as our point O, then $\underset{\rightarrow}{\mathbf{c}} = \underset{\rightarrow}{\mathbf{0}}$, and the motion of the mass center is found by integrating

$$\dot{\underset{\rightarrow}{\mathbf{r}}}_c = \frac{1}{m}\underset{\rightarrow}{\mathbf{p}}(t) \tag{15}$$

As a second special case, we may choose O as an inertially fixed point ($\underset{\rightarrow}{\mathbf{v}}_o \equiv \underset{\rightarrow}{\mathbf{0}}$), leading again to an equation of the form (15). In the attitude dynamics of spacecraft, however, it is not normally appropriate to regard a point in the vehicle as inertially fixed, and the mass center, while always of interest, is frequently not a convenient point about which to build a dynamical analysis (Section 3.7). In such cases, the distinction between O and $\oplus$ should be retained.

The moment of the momentum of $\mathscr{P}_n$ about O is $\underset{\rightarrow}{\mathbf{r}}_n \times \underset{\rightarrow}{\mathbf{p}}_n$. The total (system) moment of momentum is

$$\underset{\rightarrow}{\mathbf{h}}_o \triangleq \sum_{n=1}^{N} \underset{\rightarrow}{\mathbf{r}}_n \times \underset{\rightarrow}{\mathbf{p}}_n \tag{16}$$

From (8), it is seen that

$$\underset{\rightarrow}{\mathbf{h}}_o = \underset{\rightarrow}{\mathbf{c}} \times \underset{\rightarrow}{\mathbf{v}}_o + \sum_{n=1}^{N} m_n \underset{\rightarrow}{\mathbf{r}}_n \times \dot{\underset{\rightarrow}{\mathbf{r}}}_n \tag{17}$$

The nomenclature used by dynamicists in connection with (17) is not uniform. There are two names commonly used for both $\mathbf{h}_o$ and the sum in (17): "moment of momentum" and "angular momentum." These four possibilities can be a source of confusion when differing analyses are compared. The convention we shall use is the following: denoting by $\underset{\rightarrow}{\mathbf{h}}_\Sigma$ the sum in (17), $\mathbf{h}_\Sigma$ is the *angular momentum* and $\mathbf{h}_o$ is the *absolute angular momentum* ([Wittenburg]). Note also from (17) that the distinction vanishes if the reference point O is either at the mass center or inertially fixed.

The temporal derivative of $\mathbf{h}_o$ is

$$\begin{aligned}
\dot{\underset{\rightarrow}{\mathbf{h}}}_o &= \sum_{n=1}^{N} \left(\dot{\underset{\rightarrow}{\mathbf{r}}}_n \times \underset{\rightarrow}{\mathbf{p}}_n + \underset{\rightarrow}{\mathbf{r}}_n \times \dot{\underset{\rightarrow}{\mathbf{p}}}_n\right) \\
&= \sum_{n=1}^{N} \left[(\underset{\rightarrow}{\mathbf{v}}_n - \underset{\rightarrow}{\mathbf{v}}_o) \times m_n \underset{\rightarrow}{\mathbf{v}}_n + \underset{\rightarrow}{\mathbf{r}}_n \times \underset{\rightarrow}{\mathbf{f}}_n\right]
\end{aligned}$$

where (7), (8), and (11) have been used. The double sum is

$$\sum_{n=1}^{N} \sum_{m=1}^{N} \underset{\rightarrow}{\mathbf{r}}_n \times \underset{\rightarrow}{\mathbf{f}}_{mn} = \underset{\rightarrow}{\mathbf{0}} \tag{18}$$

assuming that the equal and opposite forces between $\mathscr{P}_n$ and $\mathscr{P}_m$ act along the line joining them. (For the material bodies we have in mind, this assumption is valid.) Continuing the derivation,

$$\dot{\underset{\rightarrow}{\mathbf{h}}}_o + \underset{\rightarrow}{\mathbf{v}}_o \times \underset{\rightarrow}{\mathbf{p}} = \underset{\rightarrow}{\mathbf{g}}_o \tag{19}$$

where

$$\underset{\rightarrow}{\mathbf{g}}_o \triangleq \sum_{n=1}^{N} \underset{\rightarrow}{\mathbf{r}}_n \times \underset{\rightarrow}{\mathbf{f}}_n \tag{20}$$

is the moment of the external forces (the torque) about the point O.

Equations (19) and (12) are the two basic motion equations derived in this section. We again see the simplification that attends the choice of an inertially fixed point for O ($\underset{\rightarrow}{\mathbf{v}}_o = \underset{\rightarrow}{\mathbf{0}}$); we get simply

$$\dot{\underset{\rightarrow}{\mathbf{h}}}_o = \underset{\rightarrow}{\mathbf{g}}_o \tag{21}$$

and the distinction between angular momentum and absolute angular momentum also vanishes. Or, if the mass center is chosen as the reference point, the distinction again disappears, and (19) reduces to

$$\dot{\underset{\rightarrow}{\mathbf{h}}}_c = \underset{\rightarrow}{\mathbf{g}}_c \tag{22}$$

(since $\underset{\rightarrow}{\mathbf{v}}_o \times \underset{\rightarrow}{\mathbf{p}} \rightarrow \underset{\rightarrow}{\mathbf{v}}_c \times m\underset{\rightarrow}{\mathbf{v}}_c = \underset{\rightarrow}{\mathbf{0}}$). The forms of (21) and (22) are so similar that the precise meaning of the symbols should be studied to avoid confusion.

For multibody systems, the reference point is, in general, neither inertially fixed nor the system mass center, and the forms (12) and (19) must be used. To completely specify the motion of the system in Fig. 3.2, a further $(N-2)$ vector motion equations are needed; these involve the "internal" forces $\underset{\rightarrow}{\mathbf{f}}_{mn}$ and therefore depend on the internal physics of each particular system. The virtue of (12) and (19) is that they hold independently of the details of the $\underset{\rightarrow}{\mathbf{f}}_{mn}$.

Kinetic Energy

By analogy to (5) in Section 3.1 for a point mass, the system kinetic energy is

$$T \triangleq \frac{1}{2} \sum_{n=1}^{N} m_n \underset{\rightarrow}{\mathbf{v}}_n \cdot \underset{\rightarrow}{\mathbf{v}}_n \tag{23}$$

The importance of T is due, in part, to its rate of change during the motion:

$$\dot{T} = \sum_{n=1}^{N} m_n \dot{\underset{\rightarrow}{\mathbf{v}}}_n \cdot \underset{\rightarrow}{\mathbf{v}}_n = \sum_{n=1}^{N} \left(\underset{\rightarrow}{\mathbf{f}}_n + \sum_{m=1}^{N} \underset{\rightarrow}{\mathbf{f}}_{mn} \right) \cdot \underset{\rightarrow}{\mathbf{v}}_n \tag{24}$$

from (6), showing that the kinetic energy can increase due to work done by both internal and external forces. Inserting (7) in (23) allows the expansion

$$T = \frac{1}{2} m \underset{\to}{\mathbf{v}}_o \cdot \underset{\to}{\mathbf{v}}_o + \underset{\to}{\mathbf{v}}_o \cdot \dot{\underset{\to}{\mathbf{c}}} + \frac{1}{2} \sum_{n=1}^{N} m_n \dot{\underset{\to}{\mathbf{r}}}_n \cdot \dot{\underset{\to}{\mathbf{r}}}_n \tag{25}$$

If O is the mass center, this simplifies to

$$T = \frac{1}{2} m \underset{\to}{\mathbf{v}}_c \cdot \underset{\to}{\mathbf{v}}_c + \frac{1}{2} \sum_{n=1}^{N} m_n \dot{\underset{\to}{\mathbf{r}}}_n \cdot \dot{\underset{\to}{\mathbf{r}}}_n \tag{26}$$

The first term is loosely referred to as the kinetic energy of the mass center, while the second can be described as the kinetic energy of the system with respect to the mass center.

Returning to the general form, (25), we observe that

$$\frac{\partial T}{\partial \underset{\to}{\mathbf{v}}_o} = \underset{\to}{\mathbf{p}} \tag{27}$$

where (1) has been employed. This relationship between the two important dynamical quantities, kinetic energy and momentum, is analogous to (7) in Section 3.1 for a single point mass and will shortly be augmented by a complementary relationship.

Rotating Reference Frame

We have had the foresight to retain freedom in picking our reference point O. For similar reasons, we wish to provide freedom of choice in selecting a reference frame $\mathscr{F}_b$, centered at O, in which eventually to express our scalar motion equations. To cite the simplest and most important example: all the mass elements in a "rigid" body are fixed with respect to a frame "imbedded" in the body, and so it is advantageous to use such frames in rigid-body dynamics. Although, in general, the best choice for $\mathscr{F}_b$ is less obvious, all that matters for the present analysis is that $\mathscr{F}_b$ has an angular velocity $\underset{\to}{\boldsymbol{\omega}}$ with respect to $\mathscr{F}_i$, as shown in Fig. 3.3.

Time derivatives measured in $\mathscr{F}_i$ and $\mathscr{F}_b$ will be denoted by overdots and overcircles, respectively. For any vector $\underset{\to}{\mathbf{u}}$, then,

$$\dot{\underset{\to}{\mathbf{u}}} = \mathring{\underset{\to}{\mathbf{u}}} + \underset{\to}{\boldsymbol{\omega}} \times \underset{\to}{\mathbf{u}} \tag{28}$$

In particular, the motion equations (12) and (19) are rewritten thus:

$$\mathring{\underset{\to}{\mathbf{p}}} + \underset{\to}{\boldsymbol{\omega}} \times \underset{\to}{\mathbf{p}} = \underset{\to}{\mathbf{f}} \tag{29}$$

$$\mathring{\underset{\to}{\mathbf{h}}}_o + \underset{\to}{\boldsymbol{\omega}} \times \underset{\to}{\mathbf{h}}_o + \underset{\to}{\mathbf{v}}_o \times \underset{\to}{\mathbf{p}} = \underset{\to}{\mathbf{g}}_o \tag{30}$$

The expressions for $\underset{\to}{\mathbf{p}}$ and $\underset{\to}{\mathbf{h}}$ themselves undergo expansion: from (10),

$$\underset{\to}{\mathbf{p}} = m \underset{\to}{\mathbf{v}}_o - \underset{\to}{\mathbf{c}} \times \underset{\to}{\boldsymbol{\omega}} + \mathring{\underset{\to}{\mathbf{c}}} \tag{31}$$

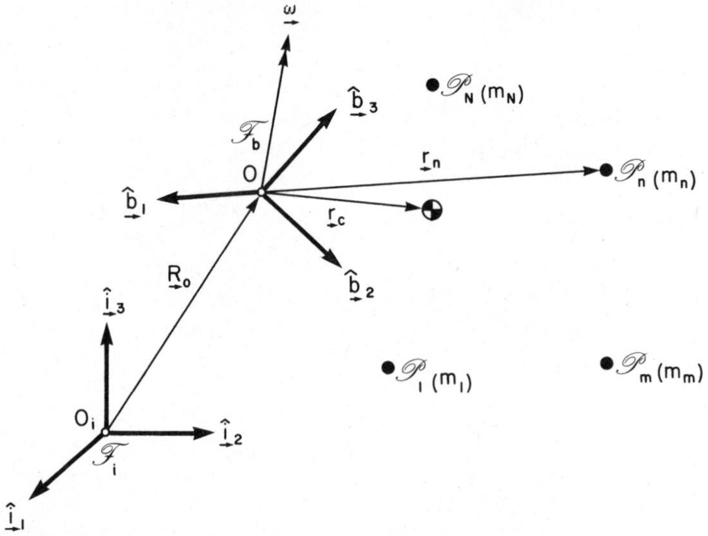

FIGURE 3.3 Introduction of the rotating reference frame $\mathscr{F}_b$.

while from (17),

$$\underset{\to}{\mathbf{h}}_o = \underset{\to}{\mathbf{c}} \times \underset{\to}{\mathbf{v}}_o + \sum_{n=1}^{N} m_n \underset{\to}{\mathbf{r}}_n \times (\underset{\to}{\boldsymbol{\omega}} \times \underset{\to}{\mathbf{r}}_n) + \sum_{n=1}^{N} m_n \underset{\to}{\mathbf{r}}_n \times \underset{\to}{\mathring{\mathbf{r}}}_n$$

The middle sum, with the aid of standard vector identities, reduces to $\underset{\to}{\mathbf{J}} \cdot \underset{\to}{\boldsymbol{\omega}}$, $\underset{\to}{\mathbf{J}}$ being the moment-of-inertia dyadic introduced earlier in (3). Thus,

$$\underset{\to}{\mathbf{h}}_o = \underset{\to}{\mathbf{c}} \times \underset{\to}{\mathbf{v}}_o + \underset{\to}{\mathbf{J}} \cdot \boldsymbol{\omega} + \sum_{n=1}^{N} m_n \underset{\to}{\mathbf{r}}_n \times \underset{\to}{\mathring{\mathbf{r}}}_n \tag{32}$$

In general, $\underset{\to}{\mathbf{J}}$ depends on the configuration of the point masses and is therefore time varying.

The expression for kinetic energy, (25), is expanded further by using $\mathscr{F}_b$ derivatives:

$$T = \frac{1}{2} m \underset{\to}{\mathbf{v}}_o \cdot \underset{\to}{\mathbf{v}}_o + \underset{\to}{\mathbf{v}}_o \cdot (\underset{\to}{\mathring{\mathbf{c}}} + \underset{\to}{\boldsymbol{\omega}} \times \underset{\to}{\mathbf{c}}) + \frac{1}{2} \sum_{n=1}^{N} m_n (\underset{\to}{\mathring{\mathbf{r}}}_n + \underset{\to}{\boldsymbol{\omega}} \times \underset{\to}{\mathbf{r}}_n) \cdot (\underset{\to}{\mathring{\mathbf{r}}}_n + \underset{\to}{\boldsymbol{\omega}} \times \underset{\to}{\mathbf{r}}_n)$$

With liberal use of vector identities and previous definitions, this reduces to

$$T = \frac{1}{2} m \underset{\to}{\mathbf{v}}_o \cdot \underset{\to}{\mathbf{v}}_o + \frac{1}{2} \underset{\to}{\boldsymbol{\omega}} \cdot \underset{\to}{\mathbf{J}} \cdot \underset{\to}{\boldsymbol{\omega}} + \frac{1}{2} \sum_{n=1}^{N} m_n \underset{\to}{\mathring{\mathbf{r}}}_n \cdot \underset{\to}{\mathring{\mathbf{r}}}_n + \underset{\to}{\mathbf{v}}_o \cdot \underset{\to}{\mathring{\mathbf{c}}}$$

$$+ \underset{\to}{\boldsymbol{\omega}} \cdot \underset{\to}{\mathbf{c}} \times \underset{\to}{\mathbf{v}}_o + \underset{\to}{\boldsymbol{\omega}} \cdot \sum_{n=1}^{N} m_n \underset{\to}{\mathbf{r}}_n \times \underset{\to}{\mathring{\mathbf{r}}}_n \tag{33}$$

We may observe that

$$\frac{\partial T}{\partial \underset{\rightarrow}{\boldsymbol{\omega}}} = \underset{\rightarrow}{\mathbf{h}}_o \tag{34}$$

using (32). This result is complementary to (27).

Scalar Motion Equations

Thus far in this section, we have avoided the expressions of our vector relations in terms of their components in *specific* reference frames. Experience shows that such components are best not taken so long as manipulations are possible in vector terms. This strategy—deferring the choice of reference frame in which to express a given vector—is simply a matter of keeping one's options open as long as possible. At the same time, detailed solution for the system motion generally cannot proceed without reducing the motion equations to scalar form. These scalar equations are most conveniently expressed in matrix notation, whose compactness promotes a direct understanding of their structure.

We have at present two reference frames at our disposal, $\mathscr{F}_i$ and $\mathscr{F}_b$. Each vector and dyadic can be expressed in either frame; furthermore, the vector equations in which these vectors and dyadics appear can also be expressed in either frame. There is no "best" set of choices, especially in the present general context; in specific situations, the analyst must choose as wisely as possible.

To illustrate the process, denote by $\underset{\rightarrow}{\mathscr{F}}_i$ and $\underset{\rightarrow}{\mathscr{F}}_b$ the vectrices associated with $\mathscr{F}_i$ and $\mathscr{F}_b$, and let us express the vectors as follows:

$$[\mathbf{R}_o \quad \mathbf{v}_o \quad \mathbf{p} \quad \mathbf{f}] \triangleq \underset{\rightarrow}{\mathscr{F}}_i \cdot \left[\underset{\rightarrow}{\mathbf{R}}_o \quad \underset{\rightarrow}{\mathbf{v}}_o \quad \underset{\rightarrow}{\mathbf{p}} \quad \underset{\rightarrow}{\mathbf{f}}\right] \tag{35a}$$

$$[\mathbf{r}_c \quad \boldsymbol{\omega} \quad \mathbf{h}_o \quad \mathbf{g}_o \quad \mathbf{c} \quad \mathbf{p}_n \quad \mathbf{f}_n \quad \mathbf{f}_{mn}]$$
$$\triangleq \underset{\rightarrow}{\mathscr{F}}_b \cdot \left[\underset{\rightarrow}{\mathbf{r}}_c \quad \underset{\rightarrow}{\boldsymbol{\omega}} \quad \underset{\rightarrow}{\mathbf{h}}_o \quad \underset{\rightarrow}{\mathbf{g}}_o \quad \underset{\rightarrow}{\mathbf{c}} \quad \underset{\rightarrow}{\mathbf{p}}_n \quad \underset{\rightarrow}{\mathbf{f}}_n \quad \underset{\rightarrow}{\mathbf{f}}_{mn}\right] \tag{35b}$$

$$\mathbf{J} \triangleq \underset{\rightarrow}{\mathscr{F}}_b \cdot \underset{\rightarrow}{\mathbf{J}} \cdot \underset{\rightarrow}{\mathscr{F}}_b^T \tag{35c}$$

Using the vectrix formalism of Appendix B, we may derive for each of the preceding vector equations an equivalent matrix equation. Because we are using two frames, the rotation matrix relating them,

$$\mathbf{C} \triangleq \underset{\rightarrow}{\mathscr{F}}_b \cdot \underset{\rightarrow}{\mathscr{F}}_i^T \tag{36}$$

occurs in the expressions below. The more important vector results are now expressed in matrix form. (The antecedent vector equation is given in the left margin.) The first and second moments of inertia are

$$(2): \qquad \mathbf{c} = \sum_{n=1}^{N} m_n \mathbf{r}_n \tag{37}$$

$$(3): \qquad \mathbf{J} = \sum_{n=1}^{N} m_n \left(\mathbf{r}_n^T \mathbf{r}_n \mathbf{1} - \mathbf{r}_n \mathbf{r}_n^T\right) \tag{38}$$

where **1** is the 3×3 unit matrix. Proceeding toward the motion equations, the momentum-related quantities are

$$(31):\qquad \mathbf{p} = m\mathbf{v}_o - \mathbf{C}^T\mathbf{c}^{\times}\boldsymbol{\omega} + \mathbf{C}^T\sum_{n=1}^{N} m_n\dot{\mathbf{r}}_n \tag{39}$$

$$(32):\qquad \mathbf{h}_o = \mathbf{c}^{\times}\mathbf{C}\mathbf{v}_o + \mathbf{J}\boldsymbol{\omega} + \sum_{n=1}^{N} m_n\mathbf{r}_n^{\times}\dot{\mathbf{r}}_n \tag{40}$$

$$(8):\qquad \mathbf{p}_n = m_n(\mathbf{C}\mathbf{v}_o - \mathbf{r}_n^{\times}\boldsymbol{\omega} + \dot{\mathbf{r}}_n) \tag{41}$$

Based on these quantities, the motion equations take the form

$$(12):\quad \dot{\mathbf{p}} = \mathbf{f} \tag{42}$$

$$(30):\quad \dot{\mathbf{h}}_o = -\boldsymbol{\omega}^{\times}\mathbf{h}_o - \mathbf{C}\mathbf{v}_o^{\times}\mathbf{p} + \mathbf{g}_o \tag{43}$$

$$(6,8):\quad \dot{\mathbf{p}}_n = \mathbf{f}_n + \sum_{m=1}^{N}\mathbf{f}_{mn} \qquad (n = 1,\ldots,N) \tag{44}$$

This set of motion equations is, of course, redundant because $\mathbf{p}$ and $\mathbf{h}_o$ are known when all the $\mathbf{p}_n$ are known. This redundancy could be removed by dropping two of equations (44). However, the purpose here is not to solve these equations but to examine their *structure*. In the limit as $N \to \infty$, one arrives at a continuum (see remarks at the end of this section). Any collection of rigid or flexible bodies (or both) can be regarded as a limiting case of foregoing system with $N \to \infty$. It is for this reason that the structure of these N-body equations is of great interest.

The System Inertia Matrix $\mathcal{M}$

To examine this structure, we assemble the following matrices:

$$\boldsymbol{v} \triangleq \begin{bmatrix} \mathbf{v}_o \\ \boldsymbol{\omega} \\ \dot{\mathbf{r}}_1 \\ \vdots \\ \dot{\mathbf{r}}_N \end{bmatrix}; \qquad \boldsymbol{p} \triangleq \begin{bmatrix} \mathbf{p} \\ \mathbf{h}_o \\ \mathbf{p}_1 \\ \vdots \\ \mathbf{p}_N \end{bmatrix}; \qquad \boldsymbol{f} \triangleq \begin{bmatrix} \mathbf{f} \\ \mathbf{g}_o \\ \mathbf{f}_1 + \Sigma\mathbf{f}_{m,1} \\ \vdots \\ \mathbf{f}_N + \Sigma\mathbf{f}_{m,N} \end{bmatrix} \tag{45}$$

$$\mathcal{M} \triangleq \begin{bmatrix} m\mathbf{1} & -\mathbf{C}^T\mathbf{c}^{\times} & m_1\mathbf{C}^T & \cdots & m_N\mathbf{C}^T \\ \mathbf{c}^{\times}\mathbf{C} & \mathbf{J} & m_1\mathbf{r}_1^{\times} & \cdots & m_N\mathbf{r}_N^{\times} \\ m_1\mathbf{C} & -m_1\mathbf{r}_1^{\times} & m_1\mathbf{1} & \cdots & \mathbf{O} \\ \vdots & \vdots & \vdots & & \vdots \\ m_N\mathbf{C} & -m_N\mathbf{r}_N^{\times} & \mathbf{O} & \cdots & m_N\mathbf{1} \end{bmatrix} \tag{46}$$

Note that $\mathcal{M}$ is symmetric. We shall call it the *system inertia matrix*. From (39–41), the momenta are related to the velocities by

$$\boldsymbol{p} = \mathcal{M}\boldsymbol{v} \tag{47}$$

Another important dynamical quantity—the kinetic energy—is also concisely expressed in terms of $\mathscr{M}$. From (33),

$$T = \frac{1}{2}m\mathbf{v}_o^T\mathbf{v}_o + \frac{1}{2}\boldsymbol{\omega}^T\mathbf{J}\boldsymbol{\omega} + \frac{1}{2}\sum_{n=1}^{N} m_n\dot{\mathbf{r}}_n^T\dot{\mathbf{r}}_n + \mathbf{v}_o^T\mathbf{C}^T\sum_{n=1}^{N} m_n\dot{\mathbf{r}}_n$$

$$+\boldsymbol{\omega}^T\mathbf{c}^{\times}\mathbf{C}\mathbf{v}_o + \boldsymbol{\omega}_n^T\sum_{n=1}^{N} m_n\mathbf{r}_n^{\times}\dot{\mathbf{r}}_n \tag{48}$$

With the definitions of $\boldsymbol{v}$ and $\mathscr{M}$ above,

$$T = \tfrac{1}{2}\boldsymbol{v}^T\mathscr{M}\boldsymbol{v} \tag{49}$$

Note that $\mathscr{M}$ is positive-*semi*definite; $\mathscr{M}$ would be positive-*definite* were it not for the redundancy mentioned above. (*Proof*: A set of points motionless with respect to $\mathscr{F}_i$ will have $T = 0$ although a translating and rotating $\mathscr{F}_b$ will record nonzero $\mathbf{v}_o, \boldsymbol{\omega}, \dot{\mathbf{r}}_n, n = 1,\ldots, N$.) Because of relationships such as (48), it may be argued that the absolute angular momentum $\mathbf{h}_o$ is a more fundamental quantity than the angular momentum $\mathbf{h}_\Sigma$.

Remark on the Newton – Euler Formulation

The above development also shows why Lagrange's equations of motion, although eminently useful in other areas of dynamics, are not ideal for formulating the motion equations of vehicles undergoing large, three-dimensional, angular displacements. This is because Lagrange's equations require that some particular set of displacement variables be chosen (one needs to specify the "q_j"). As we have seen in Chapter 2, the choice for rotational variables is often a difficult one; many of the common parameter sets are redundant, yet Lagrange's equations are best suited for use when the generalized coordinates (the q_j) are independent. It is, after all, the elegant elimination of constraint forces for which Lagrange's equations are justly celebrated, but this elimination is not possible unless the q_j are independent. Lagrange's equations are inherently *second*-order motion equations, with the dynamics and kinematics interwoven in a single set of equations. Momentum-based motion equations such as (42) through (44), on the other hand, are *first* order. After integrating for the momenta, one solves algebraic equations, like (39) through (41), for the velocities, and only then does one make kinematical choices for q_j. Thus, in the Newton–Euler formulation, the dynamical and kinematical differential equations can be considered separately. For these reasons, many dynamicists prefer momentum-related methods to Lagrangian-related methods for treating vehicle attitude dynamics.

In an attempt to garner in a single formulation the benefits from both these approaches, the use of *quasi-coordinates* has been suggested; see [Whittaker], for example. Unfortunately, even for relatively simple problems, the details become long and tedious; for example, [Likins, 4] has sketched the operations needed to arrive at motion equations for a single rigid body and showed that the complete solution requires many hand-written pages. We shall not dwell on this method here, however, since the same result can be derived more economically from our momentum-based motion equations (42) through (44), provided that the general relationship between energy and momentum is utilized. To derive this relationship, we note from (47) and (39) that

$$\frac{\partial T}{\partial \boldsymbol{\upsilon}} = \mathcal{M}\boldsymbol{\upsilon} = \boldsymbol{\not{p}} \tag{50}$$

To display this result in greater detail, we use the definition (45) and find

$$\frac{\partial T}{\partial \mathbf{v}_o} = \mathbf{p}; \qquad \frac{\partial T}{\partial \boldsymbol{\omega}} = \mathbf{h}_o; \qquad \frac{\partial T}{\partial \dot{\mathbf{r}}_n} = \mathbf{p}_n \qquad (n = 1, \ldots, N) \tag{51}$$

The first two relationships were noted earlier, in vector form, as (27) and (34).

We are now in a position to combine (51) with the motion equations derived earlier, (42) through (44), to obtain

$$\frac{d}{dt}\left(\frac{\partial T}{\partial \mathbf{v}_o}\right) = \mathbf{f} \tag{52}$$

$$\frac{d}{dt}\left(\frac{\partial T}{\partial \boldsymbol{\omega}}\right) + \boldsymbol{\omega}^{\times}\left(\frac{\partial T}{\partial \boldsymbol{\omega}}\right) + \mathbf{C}\mathbf{v}_o^{\times}\left(\frac{\partial T}{\partial \mathbf{v}_o}\right) = \mathbf{g}_o \tag{53}$$

$$\frac{d}{dt}\left(\frac{\partial T}{\partial \dot{\mathbf{r}}_n}\right) = \mathbf{f}_n + \sum_{m=1}^{N} \mathbf{f}_{mn} \qquad (n = 1, \ldots, N) \tag{54}$$

These *quasi-Lagrangian* equations are identical to the motion equations arrived at by using quasi-coordinates. It is clear, however, that they are just another way of writing (43) through (45).

Continua as Limiting Cases

Our study of a system of N point masses is a means to an end, not an end in itself. Continua can be viewed as a system of point masses in which the number of point masses has increased indefinitely ($N \to \infty$) and the individual masses have become vanishingly small ($m_n \to 0$). The limit is carried out in such a manner that

$$0 < \lim \sum_{n=1}^{N} m_n < \infty$$

This limiting process is not without mathematical interest and, in view of what is known about the structure of matter, evidently requires additional physical assumptions. We shall not dwell on these matters, however; the limiting process is intuitively plausible and, perhaps more importantly, leads to the "continuum view" of material bodies, a view that has had enduring practical utility for material bodies of the size that is typical of space vehicles.

The systems that directly interest us—rigid bodies, flexible bodies, and their combinations—can be regarded as special cases of continua. Instead of a configuration of point masses m_n at $\underset{\rightarrow}{\mathbf{R}}_n(t)$, we have for continua a mass distribution

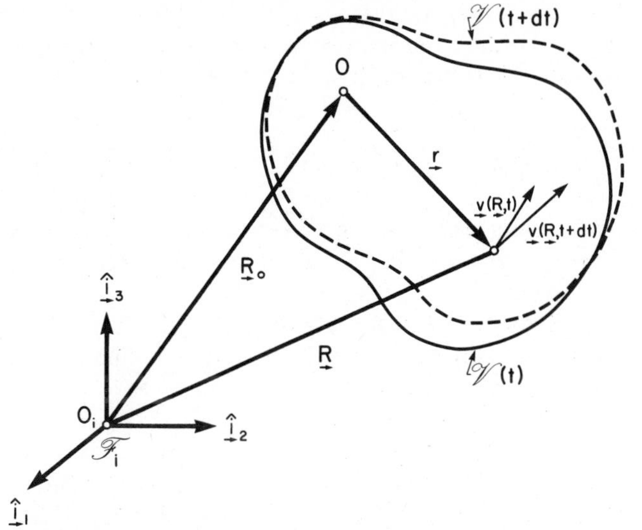

FIGURE 3.4 A nonrigid continuum.

or *mass density* function $\sigma(\underset{\rightarrow}{\mathbf{R}}, t)$. Note that $\underset{\rightarrow}{\mathbf{R}}$ is now an *independent* vector variable, the position of a geometrical point with respect to an inertial origin O_i. Physical quantities of interest thus depend on position $\underset{\rightarrow}{\mathbf{R}}$ and on time t. An infinitesimal element of volume at $\underset{\rightarrow}{\mathbf{R}}$ is denoted dV for brevity; in Cartesian coordinates, for example, dV is shorthand for $dR_1\,dR_2\,dR_3$. Similarly, $\sigma(\underset{\rightarrow}{\mathbf{R}}, t)\,dV$ is often denoted dm and may be interpreted as the mass contained in dV, although, of course, m is not an independent variable.

Consider now a spacecraft modeled as a continuum (Fig. 3.4) and assume that the absolute material velocity is $\underset{\rightarrow}{\mathbf{v}}(\underset{\rightarrow}{\mathbf{R}}, t)$. According to Reynolds' transport theorem,

$$\frac{d}{dt}\int_{\mathscr{V}(t)} s(\underset{\rightarrow}{\mathbf{R}}, t)\,dV = \int_{\mathscr{V}(t)}\left[\frac{\partial s}{\partial t} + \underset{\rightarrow}{\nabla}\cdot(s\underset{\rightarrow}{\mathbf{v}})\right]dV \tag{55}$$

where $s(\underset{\rightarrow}{\mathbf{R}}, t)$ is any scalar function. The need for this theorem arises because the domain of integration, $\mathscr{V}(t)$, is itself time dependent. The gradient vector $\underset{\rightarrow}{\nabla}$ represents differentiation with respect to the vector $\underset{\rightarrow}{\mathbf{R}}$, and the expression of $\underset{\rightarrow}{\nabla}$ in various reference frames is treated in Section B.5 of Appendix B. If in (55) we suppose that the scalar $s(\underset{\rightarrow}{\mathbf{R}}, t)$ is the product of the density function $\sigma(\underset{\rightarrow}{\mathbf{R}}, t)$ and another scalar function $\phi(\underset{\rightarrow}{\mathbf{R}}, t)$, we find that

$$\frac{d}{dt}\int_{\mathscr{V}(t)} \phi\sigma\,dV = \int_{\mathscr{V}(t)}\left[\frac{\partial \sigma}{\partial t} + \underset{\rightarrow}{\nabla}\cdot(\sigma\underset{\rightarrow}{\mathbf{v}})\right]\phi\,dV + \int_{\mathscr{V}(t)}\frac{D\phi}{Dt}\sigma\,dV \tag{56}$$

where D/Dt, defined by

$$\frac{D}{Dt} \triangleq \frac{\partial}{\partial t} + \underset{\rightarrow}{\mathbf{v}}\cdot\underset{\rightarrow}{\nabla} \tag{57}$$

is the *convective*, or *material-fixed*, time derivative. However, the quantity in brackets is known to vanish based on the continuity equation, which ensures that mass is neither created nor destroyed anywhere in $\mathscr{V}$. Therefore, it has been proved that

$$\frac{d}{dt}\int_{\mathscr{V}(t)} \phi \, dm = \int_{\mathscr{V}(t)} \frac{D\phi}{Dt} \, dm \tag{58}$$

Several significant dynamical results can now be discerned corresponding to special choices for $\phi(\underset{\rightarrow}{\mathbf{R}}, t)$. These have been pointed out in the spacecraft context by [Jankovic].

By imposing (58) consecutively to $\phi = v_1$, $\phi = v_2$, and $\phi = v_3$, where $[v_1 \quad v_2 \quad v_3] \triangleq \mathscr{F}_i^T \cdot \underset{\rightarrow}{\mathbf{v}}$ and $\mathscr{F}_i$ is an inertial reference frame, we find

$$\dot{\underset{\rightarrow}{\mathbf{p}}} = \int_{\mathscr{V}(t)} \frac{D\underset{\rightarrow}{\mathbf{v}}}{Dt} \, dm \tag{59}$$

where $\underset{\rightarrow}{\mathbf{p}}$ is the total momentum

$$\underset{\rightarrow}{\mathbf{p}} \triangleq \int_{\mathscr{V}(t)} \underset{\rightarrow}{\mathbf{v}} \, dm \tag{60}$$

and the overdot designates time differentiation measured in $\mathscr{F}_i$. Realizing that $D\underset{\rightarrow}{\mathbf{v}}/Dt$ is the material acceleration at $(\underset{\rightarrow}{\mathbf{R}}, t)$, we may denote

$$d\underset{\rightarrow}{\mathbf{f}} \triangleq \frac{D\underset{\rightarrow}{\mathbf{v}}}{Dt} \, dm \tag{61}$$

a notation in the same spirit as dm, so that (59) becomes

$$\dot{\underset{\rightarrow}{\mathbf{p}}} = \int_{\mathscr{V}(t)} d\underset{\rightarrow}{\mathbf{f}} \tag{62}$$

which corresponds to (12) derived for a system of point masses. In fact, this correspondence is so clear that the subtle distinction may easily be overlooked. Here, although $\underset{\rightarrow}{\mathbf{v}}(\underset{\rightarrow}{\mathbf{R}}, t)$ is the velocity of the material at $(\underset{\rightarrow}{\mathbf{R}}, t)$, it remains "attached to" $(\underset{\rightarrow}{\mathbf{R}}, t)$ and does not move with the material, as $\underset{\rightarrow}{\mathbf{v}}_n(t)$ did with m_n.

In the same manner, one can define the absolute angular momentum about O_i:

$$\underset{\rightarrow}{\mathbf{h}}_i \triangleq \int_{\mathscr{V}(t)} \underset{\rightarrow}{\mathbf{R}} \times \underset{\rightarrow}{\mathbf{v}} \, dm \tag{63}$$

Based on (58), one then finds

$$\dot{\underset{\rightarrow}{\mathbf{h}}}_i = \int_{\mathscr{V}(t)} \underset{\rightarrow}{\mathbf{R}} \times d\underset{\rightarrow}{\mathbf{f}} \equiv \int_{\mathscr{V}(t)} d\underset{\rightarrow}{\mathbf{g}}_i \tag{64}$$

where $d\underset{\rightarrow}{\mathbf{g}}_i \triangleq \underset{\rightarrow}{\mathbf{R}} \times d\underset{\rightarrow}{\mathbf{f}}$ is another "differential" of the same type as dm and $d\underset{\rightarrow}{\mathbf{f}}$. If it is desired to find the equivalent of (64) about a noninertial point O of interest, $\underset{\rightarrow}{\mathbf{R}} = \underset{\rightarrow}{\mathbf{R}}_o + \underset{\rightarrow}{\mathbf{r}}$, one defines

$$\underset{\rightarrow}{\mathbf{h}}_o \triangleq \int_{\mathscr{V}(t)} \underset{\rightarrow}{\mathbf{r}} \times \underset{\rightarrow}{\mathbf{v}} \, dm = \underset{\rightarrow}{\mathbf{h}}_i - \underset{\rightarrow}{\mathbf{R}}_o \times \underset{\rightarrow}{\mathbf{p}} \tag{65}$$

and finds, from (65),

$$\dot{\underset{\rightarrow}{\mathbf{h}}}_o + \underset{\rightarrow}{\mathbf{v}}_o \times \underset{\rightarrow}{\mathbf{p}} = \int_{\mathscr{V}(t)} d\underset{\rightarrow}{\mathbf{g}}_o \tag{66}$$

where $d\underset{\rightarrow}{\mathbf{g}}_o = \underset{\rightarrow}{\mathbf{r}} \times d\underset{\rightarrow}{\mathbf{f}}$ and (62) has been inserted.

Lastly, if one sets $\phi = \frac{1}{2}\underset{\rightarrow}{\mathbf{v}} \cdot \underset{\rightarrow}{\mathbf{v}}$ in (58), one finds, after a few lines,

$$\dot{T} = \int_{\mathscr{V}(t)} \underset{\rightarrow}{\mathbf{v}} \cdot d\underset{\rightarrow}{\mathbf{f}} \tag{67}$$

where the kinetic energy of the continuum has been defined as

$$T \triangleq \frac{1}{2}\int_{\mathscr{V}(t)} \underset{\rightarrow}{\mathbf{v}} \cdot \underset{\rightarrow}{\mathbf{v}}\, dm \tag{68}$$

in analogy with (23) for N point masses.

3.3 MOTION EQUATIONS FOR A RIGID BODY, $\mathscr{R}$

In common with many other engineering applications, dynamical models for spacecraft attitude motion can be analyzed in terms of a number of elemental submodels. For our purposes, three submodels suffice. These are the point mass $\mathscr{P}$ discussed earlier in Section 3.1; the rigid body $\mathscr{R}$, which forms the subject of the present section; and the quasi-rigid body $\mathscr{Q}$, which will be introduced in Chapter 5. All spacecraft models in this book will consist of these submodels and combinations of them. This should not be taken to imply that other models are not on occasion important; a fuel tank partially filled with liquid propellant in a free-fall environment is a challenging exception, and an elastic continuum is often a good model for flexible structures.

A point mass has at most three degrees of freedom, and these specify its position. Because of its zero size, one cannot speak of the attitude or orientation of a point mass. A rigid body, on the other hand, occupies a finite volume and furthermore has the property that the distance between any two elements of mass within it remains fixed. This property is not satisfied precisely for any real material body, and the "rigid body" idea is therefore a mathematical abstraction, albeit a more sophisticated one than "point mass." A "rigid" body is the simplest possible model available for spacecraft attitude dynamics. Although it has some severe restrictions on its use (see Chapter 5), it can often be safely used by an experienced dynamicist to gain useful information in special situations.

Euler showed that the most general displacement of a rigid body $\mathscr{R}$ with one point fixed is equivalent to a single rotation about some axis through that point. Since this is exactly the property of reference frames (Chapter 2), we can speak of *imbedding* a reference frame in $\mathscr{R}$; the orientation and rotational kinematics of $\mathscr{R}$ are identical to the orientation and kinematics of the imbedded reference frame. The material in Chapter 2 is therefore directly applicable to rigid bodies. In addition to the three (rotational) degrees of freedom attributed by Euler to a rigid body with a fixed point, there are the additional three (translational) degrees of freedom created by releasing the "fixed" point. The most general motion of a rigid body is therefore a six-degree-of-freedom motion, three associated with rotation and three associated with translation. We are primarily interested in the rotational three, although dynamical coupling frequently compels us to reckon with translational motion also.

Fundamental to dynamics is the notion of *inertia*. Unlike the point mass abstraction, where the mass is assumed to be concentrated at a geometrical point, the mass in a rigid body is distributed continuously through the body. We symbolize this distribution by the scalar function $\sigma(\underrightarrow{\mathbf{r}})$, which denotes the mass density at the point $\underrightarrow{\mathbf{r}}$ in the body. Although the notion of a *continuum* is in conflict with the known physical nature of matter, the mathematical benefits attending the continuum assumption are great: the theory of continuous functions of a real variable is immediately applicable. Through the continuum assumption, the properties of the 10^{30} molecules in a typical spacecraft are treated in terms of local averages, and the more tractable model that results is entirely adequate for engineering analysis.

Vectorial Motion Equations

Figure 3.5 shows a rigid body $\mathscr{R}$. Its motion with respect to an inertial frame $\mathscr{F}_i$ is specified in terms of (i) the location $\underrightarrow{\mathbf{R}}_o$ of a body-fixed point O with respect to O_i, and (ii) the orientation of a body-fixed frame $\mathscr{F}_b$ with respect to $\mathscr{F}_i$. This orientation is conveniently expressed in terms of a rotation matrix $\mathbf{C}$, which represents any set of orientation parameters (see Chapter 2).

The motion equations from the preceding section are immediately applicable once we pass from discrete mass concentrations to a continuous mass distribution and provided that we enforce the rigid-body condition that there be no relative motion within $\mathscr{R}$. The former is accomplished by passing from finite sums to integrals of continuous functions, while the latter requires that time derivatives of $\underrightarrow{\mathbf{r}}$ measured in $\mathscr{F}_b$ (denoted by overcircles) be set to zero. With respect to O, the first and second moments of inertia are, by extension from (2) and (3) in Section 3.2,

$$\underrightarrow{\mathbf{c}} \triangleq \int_{\mathscr{R}} \underrightarrow{\mathbf{r}}\sigma(\underrightarrow{\mathbf{r}})\, dV \tag{1}$$

$$\underrightarrow{\mathbf{J}} \triangleq \int_{\mathscr{R}} \left(r^2 \underrightarrow{\mathbf{1}} - \underrightarrow{\mathbf{r}}\underrightarrow{\mathbf{r}}\right) \sigma(\underrightarrow{\mathbf{r}})\, dV \tag{2}$$

where $\sigma(\underrightarrow{\mathbf{r}})$ is the mass density at $\underrightarrow{\mathbf{r}}$ and dV is an element of volume at $\underrightarrow{\mathbf{r}}$. The total mass of $\mathscr{R}$ is

$$m \triangleq \int_{\mathscr{R}} \sigma(\underrightarrow{\mathbf{r}})\, dV \tag{3}$$

Where no confusion can arise, we shall for brevity use the symbol dm for $\sigma(\underrightarrow{\mathbf{r}})\, dV$.

It may be recalled from the last section that simplifications ensue if the reference point O is either (i) inertially fixed ($\underrightarrow{\mathbf{v}}_o \equiv \underrightarrow{\mathbf{0}}$) or (ii) the mass center of the system. In the absence of a fixed point, we should like to choose O to be the mass center ⊕ of $\mathscr{R}$. If $\mathscr{R}$ is a model for the entire spacecraft, we might very well

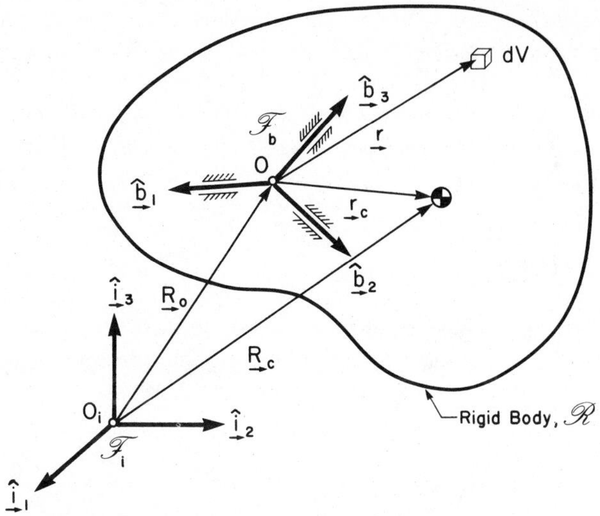

FIGURE 3.5 Rigid body $\mathscr{R}$.

make this choice. If $\mathscr{R}$ represents only part of the spacecraft, on the other hand, we may wish to choose O as some other point of importance—a point of attachment to the rest of the vehicle, for example. Thus, we intend our study of $\mathscr{R}$ to have two related applications: a model for the whole vehicle and a model for part of the vehicle. Owing to the latter, we shall retain the generality of distinguishing between O and $\oplus$. The results appropriate to the special case $O \equiv \oplus$ are readily deduced by setting the first moment of inertia to zero, $\mathbf{c} \equiv \mathbf{0}$, and by the notational adjustment $\mathbf{J} \rightarrow \mathbf{I}$.

Based on (12) and (19) in Section 3.2, we can assemble the motion equations:

$$\dot{\mathbf{p}} = \mathbf{f} \tag{4}$$

$$\dot{\mathbf{h}}_o + \mathbf{v}_o \times \mathbf{p} = \mathbf{g}_o \tag{5}$$

with $\mathbf{f}$ and $\mathbf{g}_o$ being continuum extensions to the earlier discrete definitions, (13) and (20) in Section 3.2. In a general case, one might have distributions of force per unit volume, $\mathbf{f}_v(\mathbf{r})$, and per unit area, $\mathbf{f}_s(\mathbf{r})$. One might in addition have forces best modeled as being discrete (a force from a jet thruster, for example): $\mathbf{f}_j$ acting at $\mathbf{r}_j$. Thus,

$$\mathbf{f}(t) = \int_{\mathscr{R}} \mathbf{f}_v(\mathbf{r}, t)\, dV + \oint_{\mathscr{R}} \mathbf{f}_s(\mathbf{r}, t)\, dS + \sum_j \mathbf{f}_j(t) \tag{6}$$

$$\begin{aligned} \mathbf{g}_o(t) = {} & \int_{\mathscr{R}} \mathbf{r} \times \mathbf{f}_v(\mathbf{r}, t)\, dV + \oint_{\mathscr{R}} \mathbf{r} \times \mathbf{f}_s(\mathbf{r}, t)\, dS + \sum_j \mathbf{r}_j \times \mathbf{f}_j(t) \\ & + \int_{\mathscr{R}} \mathbf{g}_v(\mathbf{r}, t)\, dV + \oint_{\mathscr{R}} \mathbf{g}_s(\mathbf{r}, t)\, dS + \sum_j \mathbf{g}_j(t) \end{aligned} \tag{7}$$

Included in the definition of $\mathbf{g}_o$ are terms representing volume and surface distributions of torque and discrete couples. In specific analyses, only a few of the terms in (6) and (7) are usually needed. The origins of these terms are discussed more fully in Chapter 8.

Having discussed the right sides of the motion equations (4) and (5), we now clarify the left sides. The momentum and (absolute) angular momentum are given by (31) and (32) in Section 3.2:

$$\underset{\rightarrow}{\mathbf{p}} = m\underset{\rightarrow}{\mathbf{v}}_o + \underset{\rightarrow}{\mathbf{c}} \times \underset{\rightarrow}{\boldsymbol{\omega}} \tag{8}$$

$$\underset{\rightarrow}{\mathbf{h}}_o = \underset{\rightarrow}{\mathbf{c}} \times \underset{\rightarrow}{\mathbf{v}}_o + \underset{\rightarrow}{\mathbf{J}} \cdot \underset{\rightarrow}{\boldsymbol{\omega}} \tag{9}$$

The integral $\int \underset{\rightarrow}{\mathbf{r}} \times \overset{\circ}{\underset{\rightarrow}{\mathbf{r}}}\, dm$ implied by (32) in Section 3.2 is zero, because $\overset{\circ}{\underset{\rightarrow}{\mathbf{r}}} \equiv \underset{\rightarrow}{\mathbf{0}}$ (the rigidity assumption). Lastly, we record the expression for kinetic energy: from (33) in Section 3.2,

$$T = \tfrac{1}{2}m\underset{\rightarrow}{\mathbf{v}}_o \cdot \underset{\rightarrow}{\mathbf{v}}_o + \underset{\rightarrow}{\boldsymbol{\omega}} \cdot \underset{\rightarrow}{\mathbf{c}} \times \underset{\rightarrow}{\mathbf{v}}_o + \tfrac{1}{2}\underset{\rightarrow}{\boldsymbol{\omega}} \cdot \underset{\rightarrow}{\mathbf{J}} \cdot \underset{\rightarrow}{\boldsymbol{\omega}} \tag{10}$$

under present assumptions.

Scalar Motion Equations

To obtain scalar equations, the vectors and vector equations above are expressed in terms of their components in appropriate reference frames. We choose to express all vectors in the body-fixed frame $\mathscr{F}_b$:

$$\begin{aligned}&[\mathbf{p} \quad \mathbf{R}_o \quad \mathbf{R}_c \quad \mathbf{v}_o \quad \mathbf{f} \quad \mathbf{r} \quad \mathbf{h}_o \quad \mathbf{g}_o \quad \mathbf{c} \quad \boldsymbol{\omega}] \\ &\quad \triangleq \mathscr{F}_b \cdot [\underset{\rightarrow}{\mathbf{p}} \quad \underset{\rightarrow}{\mathbf{R}}_o \quad \underset{\rightarrow}{\mathbf{R}}_c \quad \underset{\rightarrow}{\mathbf{v}}_o \quad \underset{\rightarrow}{\mathbf{f}} \quad \underset{\rightarrow}{\mathbf{r}} \quad \underset{\rightarrow}{\mathbf{h}}_o \quad \underset{\rightarrow}{\mathbf{g}}_o \quad \underset{\rightarrow}{\mathbf{c}} \quad \underset{\rightarrow}{\boldsymbol{\omega}}]\end{aligned} \tag{11a}$$

$$\mathbf{J} = \mathscr{F}_b \cdot \underset{\rightarrow}{\mathbf{J}} \cdot \mathscr{F}_b^T \tag{11b}$$

For another alternative, see Problem 3.28. With these conventions, the motion equations are (consult Appendix B for the methodology used in the intervening steps):

$$\begin{aligned}\dot{\mathbf{p}} &= -\boldsymbol{\omega}^{\times}\mathbf{p} + \mathbf{f} \\ \dot{\mathbf{h}}_o &= -\boldsymbol{\omega}^{\times}\mathbf{h}_o - \mathbf{v}_o^{\times}\mathbf{p} + \mathbf{g}_o\end{aligned} \tag{12}$$

where

$$\mathbf{p} = m\mathbf{v}_o - \mathbf{c}^{\times}\boldsymbol{\omega}; \qquad \mathbf{h}_o = \mathbf{c}^{\times}\mathbf{v}_o + \mathbf{J}\boldsymbol{\omega} \tag{13}$$

$$\mathbf{J} = \int_{\mathscr{R}} (r^2\mathbf{1} - \mathbf{r}\mathbf{r}^T)\, dm \tag{14}$$

The inertia matrix $\mathbf{J}$ is constant, symmetrical, and positive-definite. Its properties are examined in Problem 3.8.

We recall once again the two principal contexts in which these equations are developed. In one of these, $\mathscr{R}$ is one of several dynamical elements in the spacecraft model, and $\mathbf{f}$ and $\mathbf{g}_o$ will include the forces and torques of interaction with the other elements. Equations (12) and (13) are then the building blocks from which a comprehensive system of motion equations can be built (Section 3.7). In the other context, $\mathscr{R}$ represents the entire spacecraft, and we can be more specific in our inferences. We note immediately that (12) and (13) comprise two matrix differential equations and two matrix algebraic equations, and that they are coupled. If $\mathscr{R}$ represents the whole spacecraft, however, we might wish to choose the mass center $\mathscr{R}$ as our reference point. These equations then simplify to

$$\dot{\mathbf{p}} = -\boldsymbol{\omega}^{\times}\mathbf{p} + \mathbf{f}; \qquad \dot{\mathbf{h}}_c = -\boldsymbol{\omega}^{\times}\mathbf{h}_c + \mathbf{g}_c \tag{15}$$

$$\mathbf{p} = m\mathbf{v}_c; \qquad \mathbf{h}_c = \mathbf{I}\boldsymbol{\omega} \tag{16}$$

For this relatively simple model, we make the considered substitution of (16) into (15), which eliminates the momentum variables:

$$m\dot{\mathbf{v}}_c = -m\boldsymbol{\omega}^{\times}\mathbf{v}_c + \mathbf{f}; \qquad \mathbf{I}\dot{\boldsymbol{\omega}} = -\boldsymbol{\omega}^{\times}\mathbf{I}\boldsymbol{\omega} + \mathbf{g}_c \tag{17a, b}$$

This substitution should in more complex situations be made only after careful thought. One now integrates for $\mathbf{v}_c(t)$ and $\boldsymbol{\omega}(t)$ and then subsequently integrates the appropriate kinematical equations for the displacement variables:

$$\dot{\mathbf{R}}_c + \boldsymbol{\omega}^{\times}\mathbf{R}_c = \mathbf{v}_c; \qquad \dot{\mathbf{C}} = -\boldsymbol{\omega}^{\times}\mathbf{C} \tag{18}$$

(It should be understood that $\mathbf{C}$ is used here as a generic symbol to represent whichever attitude variables, discussed in Chapter 2, are being used.) At first glance, it may appear that at least the rotational equation, (17*b*), is uncoupled. Unfortunately, (17) and (18) are in general mutually coupled through the explicit dependence of the external torque on the displacement and velocity:

$$\mathbf{f} = \mathbf{f}(\mathbf{R}_c, \mathbf{C}, \mathbf{v}_c, \boldsymbol{\omega}, t); \qquad \mathbf{g} = \mathbf{g}_c(\mathbf{R}_c, \mathbf{C}, \mathbf{v}_c, \boldsymbol{\omega}, t) \tag{19}$$

This poses the integration of a coupled nonlinear system of differential equations whose order is 12 or higher, depending on the attitude variables chosen. This system is discussed more fully in Chapter 9.

Equation (17*b*) is the quintessential attitude motion equation and is due to Euler. Provided $\mathscr{F}_b$ is chosen to be a *principal-axis* frame (see Problem 3.8*c*), it is equivalent to the following three scalar equations:

$$\begin{aligned} I_1\dot{\omega}_1 &= (I_2 - I_3)\omega_2\omega_3 + g_1 \\ I_2\dot{\omega}_2 &= (I_3 - I_1)\omega_3\omega_1 + g_2 \\ I_3\dot{\omega}_3 &= (I_1 - I_2)\omega_1\omega_2 + g_3 \end{aligned} \tag{20}$$

The solution to these equations is discussed in Chapter 4.

Kinetic Energy and Quasi-Lagrangian Equations

The kinetic energy, given vectorially by (10), is expressed in terms of vector components as follows:

$$T = \tfrac{1}{2}m\mathbf{v}_o^T\mathbf{v}_o + \boldsymbol{\omega}^T\mathbf{c}^{\times}\mathbf{v}_o + \tfrac{1}{2}\boldsymbol{\omega}^T\mathbf{J}\boldsymbol{\omega} \tag{21}$$

The rate of change of T during the motion is

$$\begin{aligned}\dot{T} &= \left(m\dot{\mathbf{v}}_o^T + \dot{\boldsymbol{\omega}}^T\mathbf{c}^\times\right)\mathbf{v}_o + \boldsymbol{\omega}^T\left(\mathbf{c}^\times\dot{\mathbf{v}}_o + \mathbf{J}\dot{\boldsymbol{\omega}}\right)\\ &= \mathbf{f}^T\mathbf{v}_o + \mathbf{g}_o^T\boldsymbol{\omega}\end{aligned} \tag{22}$$

(Equations (12) and (13) have been inserted.) With no external force and torque, the kinetic energy is an "integral" of (constant during) the motion. When O is ⊕, the kinetic energy simplifies to two terms: a "translational" term and a "rotational" term:

$$T = T_{\text{trans}} + T_{\text{rot}}; \qquad T_{\text{trans}} \triangleq \tfrac{1}{2}m\mathbf{v}_c^T\mathbf{v}_c; \qquad T_{\text{rot}} \triangleq \tfrac{1}{2}\boldsymbol{\omega}^T\mathbf{I}\boldsymbol{\omega} \tag{23}$$

When $\mathbf{f} \equiv \mathbf{0}$, T_{trans} is constant; when $\mathbf{g}_c \equiv \mathbf{0}$, T_{rot} is constant.

To establish a quasi-Lagrangian equation for rotational motion, we note from (21) that

$$\begin{aligned}\frac{\partial T}{\partial \boldsymbol{\omega}} &= \mathbf{c}^\times\mathbf{v}_o + \mathbf{J}\boldsymbol{\omega} \equiv \mathbf{h}_o\\ \frac{\partial T}{\partial \mathbf{v}_o} &= m\mathbf{v}_o - \mathbf{c}^\times\boldsymbol{\omega} \equiv \mathbf{p}\end{aligned} \tag{24}$$

and therefore, noting (12), we have

$$\begin{aligned}\frac{d}{dt}\left(\frac{\partial T}{\partial \mathbf{v}_o}\right) + \boldsymbol{\omega}^\times\left(\frac{\partial T}{\partial \mathbf{v}_o}\right) &= \mathbf{f}\\ \frac{d}{dt}\left(\frac{\partial T}{\partial \boldsymbol{\omega}}\right) + \boldsymbol{\omega}^\times\left(\frac{\partial T}{\partial \boldsymbol{\omega}}\right) + \mathbf{v}_o^\times\left(\frac{\partial T}{\partial \mathbf{v}_o}\right) &= \mathbf{g}_o\end{aligned} \tag{25}$$

These quasi-Lagrangian equations can also be derived (ardously) using quasi-coordinates, but they are just a fancy way of writing equation (12).

The System Inertia Matrix $\mathcal{M}$

The structure of the relationship between momenta, kinetic energy, and velocities is most succinctly expressed using the system inertia matrix $\mathcal{M}$, defined as follows

$$\mathcal{M} \triangleq \begin{bmatrix} m\mathbf{1} & -\mathbf{c}^\times \\ \mathbf{c}^\times & \mathbf{J} \end{bmatrix} \tag{26}$$

In terms of $\mathcal{M}$, (13) and (21) are written

$$\boldsymbol{\mathscr{p}} = \mathcal{M}\boldsymbol{\mathscr{v}}; \qquad T = \tfrac{1}{2}\boldsymbol{\mathscr{v}}^T\mathcal{M}\boldsymbol{\mathscr{v}} \tag{27a, b}$$

where

$$\boldsymbol{\mathscr{p}} \triangleq \begin{bmatrix}\mathbf{p}\\ \mathbf{h}_o\end{bmatrix}; \qquad \boldsymbol{\mathscr{v}} \triangleq \begin{bmatrix}\mathbf{v}_o\\ \boldsymbol{\omega}\end{bmatrix}; \qquad \boldsymbol{\mathscr{f}} \triangleq \begin{bmatrix}\mathbf{f}\\ \mathbf{g}_o\end{bmatrix} \tag{28}$$

These relations should be compared with (45) through (49) in Section 3.2, the latter applying to a system of point masses.

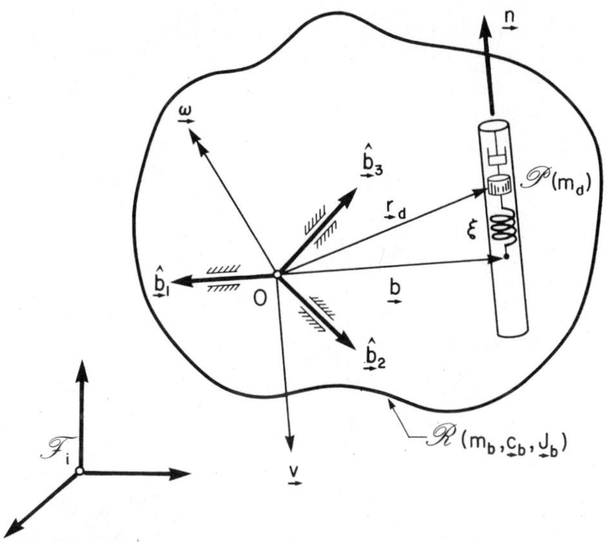

FIGURE 3.6 Spinning rigid body $\mathscr{R}$, with an internal point mass damper $\mathscr{P}$.

We can also show that $\mathscr{M}$ is positive-definite by exploiting its association with the kinetic energy, (27*b*). For $\mathscr{R}$,

$$T = \tfrac{1}{2}\int_{\mathscr{R}} \underset{\rightarrow}{\mathbf{v}} \cdot \underset{\rightarrow}{\mathbf{v}}\, dm \tag{29}$$

Therefore, $T > 0$, unless $\underset{\rightarrow}{\mathbf{v}} \equiv \underset{\rightarrow}{\mathbf{0}}$ everywhere in $\mathscr{R}$. But $\underset{\rightarrow}{\mathbf{v}} = \underset{\rightarrow}{\mathbf{v}}_o + \underset{\rightarrow}{\boldsymbol{\omega}} \times \underset{\rightarrow}{\mathbf{r}}$. Therefore, $\underset{\rightarrow}{\mathbf{v}} \equiv \underset{\rightarrow}{\mathbf{0}}$ everywhere in $\mathscr{R}$ only if $\underset{\rightarrow}{\mathbf{v}}_o$ and $\underset{\rightarrow}{\boldsymbol{\omega}}$ are both zero. This demonstrates the positive definiteness of $\mathscr{M}$.

3.4 A SYSTEM WITH DAMPING, $\mathscr{R}+\mathscr{P}$

We embark now on a series of three examples intended to illustrate the process of formulating motion equations for rotating systems. They are more than didactic examples, however; each can serve as a model for actual spacecraft. They may be viewed as extensions to a single-rigid-body model $\mathscr{R}$. In this and the following two sections, $\mathscr{R}$ is augmented to include either a point mass damper (this section), a rotor (Section 3.5), or an appended rigid body (Section 3.6).

We turn first to the situation shown in Fig. 3.6. The system comprises a rigid body $\mathscr{R}$ and a point mass $\mathscr{P}$. The latter is constrained to move in a rectilinear slot with respect to $\mathscr{R}$. This motion is damped by a linear viscous damper (damping constant c_d), and $\mathscr{P}$ is also connected to $\mathscr{R}$ by a linear spring (spring constant k_d). The interaction between $\mathscr{P}$ and $\mathscr{R}$ provides a mechanism for dissipating the energy of $\mathscr{R}$. As we shall see in Chapters 5 and 7, the effects of energy dissipation are often crucial to the character of the motion, and this model will help to illustrate this point.

Referring to Fig. 3.6, we note that the position of $\mathscr{P}$ with respect to O is

$$\underset{\rightarrow}{\mathbf{r}}_d = \underset{\rightarrow}{\mathbf{b}} + \xi \underset{\rightarrow}{\mathbf{n}} \qquad (\underset{\rightarrow}{\mathbf{n}} \cdot \underset{\rightarrow}{\mathbf{n}} = 1) \tag{1}$$

where $\xi = 0$ corresponds to a relaxed spring, and the unit vector $\underset{\rightarrow}{\mathbf{n}}$ defines the direction of $\mathscr{P}$'s travel. The masses of $\mathscr{R}$ and $\mathscr{P}$ are, respectively, m_b and m_d, and therefore the total system mass is

$$m = m_b + m_d \tag{2}$$

The first and second moments of inertia of the system with respect to O are

$$\underset{\rightarrow}{\mathbf{c}} = \underset{\rightarrow}{\mathbf{c}}_b + m_d \underset{\rightarrow}{\mathbf{r}}_d \tag{3}$$

$$\underset{\rightarrow}{\mathbf{J}} = \underset{\rightarrow}{\mathbf{J}}_b + m_d \left(r_d^2 \underset{\rightarrow}{\mathbf{1}} - \underset{\rightarrow}{\mathbf{r}}_d \underset{\rightarrow}{\mathbf{r}}_d \right) \tag{4}$$

where $\underset{\rightarrow}{\mathbf{c}}_b$ and $\underset{\rightarrow}{\mathbf{J}}_b$ are the first and second moments of inertia of $\mathscr{R}$ about O.

Vectorial Motion Equations

We denote by $\underset{\rightarrow}{\mathbf{v}}$ and $\underset{\rightarrow}{\boldsymbol{\omega}}$ the absolute velocity of O and the absolute angular velocity of $\mathscr{R}$. Then, the velocity at a point $\underset{\rightarrow}{\mathbf{r}}$ in $\mathscr{R}$ is $\underset{\rightarrow}{\mathbf{v}} + \underset{\rightarrow}{\boldsymbol{\omega}} \times \underset{\rightarrow}{\mathbf{r}}$, and the velocity of $\mathscr{P}$ is $\underset{\rightarrow}{\mathbf{v}} + \underset{\rightarrow}{\boldsymbol{\omega}} \times \underset{\rightarrow}{\mathbf{r}}_d + \dot{\xi} \underset{\rightarrow}{\mathbf{n}}$. It follows that the momenta of $\mathscr{R}$ and $\mathscr{P}$ are

$$\underset{\rightarrow}{\mathbf{p}}_b \triangleq \int_{\mathscr{R}} (\underset{\rightarrow}{\mathbf{v}} + \underset{\rightarrow}{\boldsymbol{\omega}} \times \underset{\rightarrow}{\mathbf{r}}) \, dm = m_b \underset{\rightarrow}{\mathbf{v}} + \underset{\rightarrow}{\boldsymbol{\omega}} \times \underset{\rightarrow}{\mathbf{c}}_b \tag{5}$$

$$\underset{\rightarrow}{\mathbf{p}}_d \triangleq m_d (\underset{\rightarrow}{\mathbf{v}} + \underset{\rightarrow}{\boldsymbol{\omega}} \times \underset{\rightarrow}{\mathbf{r}}_d + \dot{\xi} \underset{\rightarrow}{\mathbf{n}}) \tag{6}$$

The total momentum of the system is

$$\underset{\rightarrow}{\mathbf{p}} \triangleq \underset{\rightarrow}{\mathbf{p}}_b + \underset{\rightarrow}{\mathbf{p}}_d = m \underset{\rightarrow}{\mathbf{v}} - \underset{\rightarrow}{\mathbf{c}} \times \underset{\rightarrow}{\boldsymbol{\omega}} + m_d \dot{\xi} \underset{\rightarrow}{\mathbf{n}} \tag{7}$$

We are particularly interested in the component of $\underset{\rightarrow}{\mathbf{p}}_d$ along $\underset{\rightarrow}{\mathbf{n}}$, which we denote by p_n:

$$p_n \triangleq \underset{\rightarrow}{\mathbf{n}} \cdot \underset{\rightarrow}{\mathbf{p}}_d = m_d (\underset{\rightarrow}{\mathbf{n}} \cdot \underset{\rightarrow}{\mathbf{v}} - \underset{\rightarrow}{\mathbf{n}} \times \underset{\rightarrow}{\mathbf{b}} \cdot \underset{\rightarrow}{\boldsymbol{\omega}} + \dot{\xi}) \tag{8}$$

Similarly, the absolute angular momentum of $\mathscr{R}$ about O is

$$\underset{\rightarrow}{\mathbf{h}}_b \triangleq \int_{\mathscr{R}} \underset{\rightarrow}{\mathbf{r}} \times (\underset{\rightarrow}{\mathbf{v}} + \underset{\rightarrow}{\boldsymbol{\omega}} \times \underset{\rightarrow}{\mathbf{r}}) \, dm = \underset{\rightarrow}{\mathbf{c}}_b \times \underset{\rightarrow}{\mathbf{v}} + \underset{\rightarrow}{\mathbf{J}}_b \cdot \underset{\rightarrow}{\boldsymbol{\omega}} \tag{9}$$

and the total absolute angular momentum of $\mathscr{R} + \mathscr{P}$ about O is

$$\underset{\rightarrow}{\mathbf{h}} \triangleq \underset{\rightarrow}{\mathbf{h}}_b + \underset{\rightarrow}{\mathbf{r}}_d \times \underset{\rightarrow}{\mathbf{p}}_d = \underset{\rightarrow}{\mathbf{c}} \times \underset{\rightarrow}{\mathbf{v}} + \underset{\rightarrow}{\mathbf{J}} \cdot \underset{\rightarrow}{\boldsymbol{\omega}} + m_d \dot{\xi} \underset{\rightarrow}{\mathbf{b}} \times \underset{\rightarrow}{\mathbf{n}} \tag{10}$$

The other dynamical quantity of interest is the kinetic energy:

$$\begin{aligned} T \triangleq & \frac{1}{2} \int_{\mathscr{R}} (\underset{\rightarrow}{\mathbf{v}} + \underset{\rightarrow}{\boldsymbol{\omega}} \times \underset{\rightarrow}{\mathbf{r}}) \cdot (\underset{\rightarrow}{\mathbf{v}} + \underset{\rightarrow}{\boldsymbol{\omega}} \times \underset{\rightarrow}{\mathbf{r}}) \, dm \\ & + \frac{1}{2} m_d (\underset{\rightarrow}{\mathbf{v}} + \underset{\rightarrow}{\boldsymbol{\omega}} \times \underset{\rightarrow}{\mathbf{r}}_d + \dot{\xi} \underset{\rightarrow}{\mathbf{n}}) \cdot (\underset{\rightarrow}{\mathbf{v}} + \underset{\rightarrow}{\boldsymbol{\omega}} \times \underset{\rightarrow}{\mathbf{r}}_d + \dot{\xi} \underset{\rightarrow}{\mathbf{n}}) \\ & = \frac{1}{2} m \underset{\rightarrow}{\mathbf{v}} \cdot \underset{\rightarrow}{\mathbf{v}} + \frac{1}{2} \underset{\rightarrow}{\boldsymbol{\omega}} \cdot J \cdot \underset{\rightarrow}{\boldsymbol{\omega}} + \frac{1}{2} m_d \dot{\xi}^2 - \underset{\rightarrow}{\mathbf{v}} \cdot (\underset{\rightarrow}{\mathbf{c}} \times \underset{\rightarrow}{\boldsymbol{\omega}}) \\ & - m_d \dot{\xi} (\underset{\rightarrow}{\mathbf{n}} \times \underset{\rightarrow}{\mathbf{b}}) \cdot \underset{\rightarrow}{\boldsymbol{\omega}} + m_d \dot{\xi} \underset{\rightarrow}{\mathbf{v}} \cdot \underset{\rightarrow}{\mathbf{n}} \end{aligned} \tag{11}$$

after reduction utilizing earlier definitions.

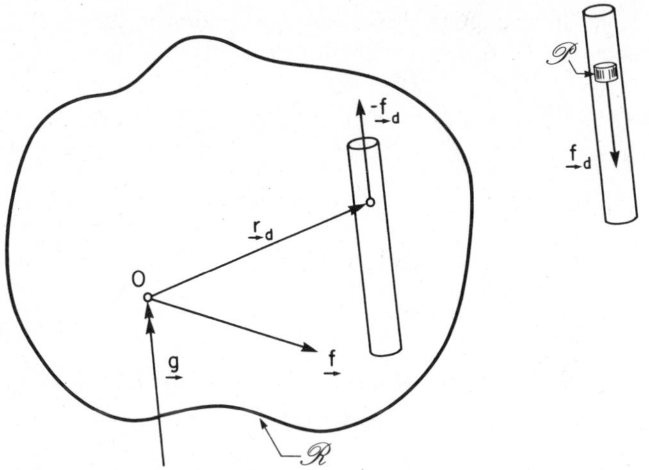

FIGURE 3.7 Free-body diagram for $\mathscr{R} + \mathscr{P}$.

To clarify the forces and torques acting on the system, consider Fig. 3.7. On $\mathscr{R}$, the external force and torque are $\underset{\rightarrow}{\mathbf{f}}$ and $\underset{\rightarrow}{\mathbf{g}}$, and $\mathscr{P}$ exerts a force $-\underset{\rightarrow}{\mathbf{f}}_d$. By Newton's Third Law, $\mathscr{R}$ exerts a force $\underset{\rightarrow}{\mathbf{f}}_d$ on $\mathscr{P}$ (external forces on $\mathscr{P}$ can also be included, if desired). We are now in a position to write motion equations for $\mathscr{R}$ and $\mathscr{P}$ based on the results of Sections 3.1 and 3.3:

$$(3.1,4): \quad \dot{\underset{\rightarrow}{\mathbf{p}}}_d = \underset{\rightarrow}{\mathbf{f}}_d = -\underset{\rightarrow}{\mathbf{n}}(c_d\dot{\xi} + k_d\xi) + \underset{\rightarrow}{\mathbf{f}}_{\text{con}} \qquad (\underset{\rightarrow}{\mathbf{n}} \cdot \underset{\rightarrow}{\mathbf{f}}_{\text{con}} = 0) \tag{12}$$

$$(3.3,4): \quad \dot{\underset{\rightarrow}{\mathbf{p}}}_b = \underset{\rightarrow}{\mathbf{f}} - \underset{\rightarrow}{\mathbf{f}}_d \tag{13}$$

$$(3.3,5): \quad \dot{\underset{\rightarrow}{\mathbf{h}}}_b + \underset{\rightarrow}{\mathbf{v}} \times \underset{\rightarrow}{\mathbf{p}}_b = \underset{\rightarrow}{\mathbf{g}} - \underset{\rightarrow}{\mathbf{r}}_d \times \underset{\rightarrow}{\mathbf{f}}_d \tag{14}$$

where $\underset{\rightarrow}{\mathbf{f}}_{\text{con}}$ is the constraining force normal to the direction of $\mathscr{P}$'s travel. Adding (12) and (13) gives

$$\dot{\underset{\rightarrow}{\mathbf{p}}} = \underset{\rightarrow}{\mathbf{f}} \tag{15}$$

Substitution of (14) into (10) yields (after reduction using earlier definitions)

$$\dot{\underset{\rightarrow}{\mathbf{h}}} + \underset{\rightarrow}{\mathbf{v}} \times \underset{\rightarrow}{\mathbf{p}} = \underset{\rightarrow}{\mathbf{g}} \tag{16}$$

The last motion equation is found from (8) and (12):

$$\dot{p}_n = \dot{\underset{\rightarrow}{\mathbf{n}}} \cdot \underset{\rightarrow}{\mathbf{p}}_d + \underset{\rightarrow}{\mathbf{n}} \cdot \dot{\underset{\rightarrow}{\mathbf{p}}}_d = (\underset{\rightarrow}{\boldsymbol{\omega}} \times \underset{\rightarrow}{\mathbf{n}}) \cdot \underset{\rightarrow}{\mathbf{p}}_d - c_d\dot{\xi} - k_d\xi$$

Or, from (6),

$$\dot{p}_n = m_d\underset{\rightarrow}{\boldsymbol{\omega}} \cdot \underset{\rightarrow}{\mathbf{n}} \times (\underset{\rightarrow}{\mathbf{v}} - \underset{\rightarrow}{\mathbf{r}}_d \times \underset{\rightarrow}{\boldsymbol{\omega}}) - c_d\dot{\xi} - k_d\xi \tag{17}$$

We are also interested in the change in the kinetic energy during the motion. By analogy with (68) in Section 3.2 for a contimuum, we calculate $\dot{T}$ from

$$\begin{aligned}\dot{T} &= \int_{\mathscr{R}} (\underset{\rightarrow}{\mathbf{v}} + \underset{\rightarrow}{\boldsymbol{\omega}} \times \underset{\rightarrow}{\mathbf{r}}) \cdot d\underset{\rightarrow}{\mathbf{f}} + (\underset{\rightarrow}{\mathbf{v}} + \underset{\rightarrow}{\boldsymbol{\omega}} \times \underset{\rightarrow}{\mathbf{r}}_d + \dot{\xi}\underset{\rightarrow}{\mathbf{n}}) \cdot \underset{\rightarrow}{\mathbf{f}}_d \\ &= \underset{\rightarrow}{\mathbf{v}} \cdot \int_{\mathscr{R}} d\underset{\rightarrow}{\mathbf{f}} + \underset{\rightarrow}{\boldsymbol{\omega}} \cdot \int_{\mathscr{R}} \underset{\rightarrow}{\mathbf{r}} \times d\underset{\rightarrow}{\mathbf{f}} + \underset{\rightarrow}{\mathbf{v}} \cdot \underset{\rightarrow}{\mathbf{f}}_d + \underset{\rightarrow}{\boldsymbol{\omega}} \cdot \underset{\rightarrow}{\mathbf{r}}_d \times \underset{\rightarrow}{\mathbf{f}}_d + \dot{\xi}\underset{\rightarrow}{\mathbf{n}} \cdot \underset{\rightarrow}{\mathbf{f}}_d \\ &= \underset{\rightarrow}{\mathbf{v}} \cdot (\underset{\rightarrow}{\mathbf{f}} - \underset{\rightarrow}{\mathbf{f}}_d) + \underset{\rightarrow}{\boldsymbol{\omega}} \cdot (\underset{\rightarrow}{\mathbf{g}} - \underset{\rightarrow}{\mathbf{r}}_d \times \underset{\rightarrow}{\mathbf{f}}_d) + \underset{\rightarrow}{\mathbf{v}} \cdot \underset{\rightarrow}{\mathbf{f}}_d + \underset{\rightarrow}{\boldsymbol{\omega}} \cdot \underset{\rightarrow}{\mathbf{r}}_d \times \underset{\rightarrow}{\mathbf{f}}_d + \dot{\xi}\underset{\rightarrow}{\mathbf{n}} \cdot \underset{\rightarrow}{\mathbf{f}}_d \\ &= \underset{\rightarrow}{\mathbf{f}} \cdot \underset{\rightarrow}{\mathbf{v}} + \underset{\rightarrow}{\mathbf{g}} \cdot \underset{\rightarrow}{\boldsymbol{\omega}} - c_d\dot{\xi}^2 - k_d\xi\dot{\xi} \end{aligned} \tag{18}$$

The term $k_d\xi\dot{\xi}$ is the rate of change of potential energy stored in the spring, $V \triangleq 1/2k_d\xi^2$. Therefore, the change in the total energy $E = T + V$ is

$$\dot{E} = \underset{\rightarrow}{\mathbf{f}} \cdot \underset{\rightarrow}{\mathbf{v}} + \underset{\rightarrow}{\mathbf{g}} \cdot \underset{\rightarrow}{\boldsymbol{\omega}} - c_d\dot{\xi}^2 \tag{19}$$

showing that the energy increases as work is done by $\underset{\rightarrow}{\mathbf{f}}$ and $\underset{\rightarrow}{\mathbf{g}}$ and decreases through the dissipation in the damper.

Scalar Motion Equations

To deduce scalar equations of motion, we must express vectors in terms of their components. We choose to express all vectors in $\mathscr{F}_b$. Note that $\mathbf{n}$, $\mathbf{b}$, $\mathbf{c}_b$, and $\mathbf{J}_b$ are constant when expressed in $\mathscr{F}_b$. The momenta and kinetic energy, given by equations (7), (10), (8), (11), and (19) in this section, are restated, respectively, as follows:

$$\mathbf{p} = m\mathbf{v} - \mathbf{c}^{\times}\boldsymbol{\omega} + m_d\dot{\xi}\mathbf{n} \tag{20}$$

$$\mathbf{h} = \mathbf{c}^{\times}\mathbf{v} + \mathbf{J}\boldsymbol{\omega} + m_d\dot{\xi}\mathbf{b}^{\times}\mathbf{n} \tag{21}$$

$$p_n = m_d(\mathbf{n}^T\mathbf{v} - \mathbf{n}^T\mathbf{b}^{\times}\boldsymbol{\omega} + \dot{\xi}) \tag{22}$$

$$T = \tfrac{1}{2}m\mathbf{v}^T\mathbf{v} + \tfrac{1}{2}\boldsymbol{\omega}^T\mathbf{J}\boldsymbol{\omega} + \tfrac{1}{2}m_d\dot{\xi}^2 - \mathbf{v}^T\mathbf{c}^{\times}\boldsymbol{\omega} - m_d\dot{\xi}\mathbf{n}^T\mathbf{b}^{\times}\boldsymbol{\omega} + m_d\dot{\xi}\mathbf{v}^T\mathbf{n} \tag{23}$$

$$\dot{E} = \mathbf{f}^T\mathbf{v} + \mathbf{g}^T\boldsymbol{\omega} - c_d\dot{\xi}^2 \tag{24}$$

The structure of these equations is succinctly stated in terms of the system inertia matrix $\mathscr{M}$, defined as follows:

$$\mathscr{M} \triangleq \begin{bmatrix} m\mathbf{1} & -\mathbf{c}^{\times} & m_d\mathbf{n} \\ \mathbf{c}^{\times} & \mathbf{J} & m_d\mathbf{b}^{\times}\mathbf{n} \\ m_d\mathbf{n}^T & -m_d\mathbf{n}^T\mathbf{b}^{\times} & m_d \end{bmatrix} \tag{25}$$

The momenta and energy can now be expressed concisely in terms of $\mathscr{M}$:

$$\mathscr{p} = \mathscr{M}\mathscr{v} \tag{26}$$

$$T = \tfrac{1}{2}\mathscr{v}^T\mathscr{M}\mathscr{v} \tag{27}$$

where

$$\not{p} \triangleq \begin{bmatrix} \mathbf{p} \\ \mathbf{h} \\ p_n \end{bmatrix}; \qquad v \triangleq \begin{bmatrix} \mathbf{v} \\ \boldsymbol{\omega} \\ \dot{\xi} \end{bmatrix}; \qquad \not{f} \triangleq \begin{bmatrix} \mathbf{f} \\ \mathbf{g} \\ 0 \end{bmatrix} \tag{28}$$

The matrix $\mathscr{M}$ is evidently symmetric. It is also positive-definite; it would be tedious to prove this directly from the definition (25), but it follows directly from (27) and (11). If $T = 0$, then

$$\underset{\rightarrow}{\mathbf{v}} + \underset{\rightarrow}{\boldsymbol{\omega}} \times \underset{\rightarrow}{\mathbf{r}} = \underset{\rightarrow}{\mathbf{0}} \qquad (\text{for all } \underset{\rightarrow}{\mathbf{r}} \in \mathscr{R})$$

$$\underset{\rightarrow}{\mathbf{v}} + \underset{\rightarrow}{\boldsymbol{\omega}} \times \underset{\rightarrow}{\mathbf{r}}_d + \dot{\xi}\underset{\rightarrow}{\mathbf{n}} = 0 \tag{29}$$

This is only possible if $\underset{\rightarrow}{\mathbf{v}}$, $\underset{\rightarrow}{\boldsymbol{\omega}}$, and $\dot{\xi}$ all vanish; hence, $\mathscr{M}$ is positive definite. Unfortunately, $\mathscr{M}$ is not a constant matrix because $\mathbf{c}$ and $\mathbf{J}$ are time dependent. It is readily shown that

$$\frac{\partial T}{\partial \mathbf{v}} = \mathbf{p}; \qquad \frac{\partial T}{\partial \boldsymbol{\omega}} = \mathbf{h}; \qquad \frac{\partial T}{\partial \dot{\xi}} = p_n \tag{30}$$

which relate the important dynamical quantities.

As for the motion equations themselves, the scalar equivalents of (15) through (17) are

$$\dot{\mathbf{p}} = -\boldsymbol{\omega}^{\times}\mathbf{p} + \mathbf{f} \tag{31}$$

$$\dot{\mathbf{h}} = -\boldsymbol{\omega}^{\times}\mathbf{h} - \mathbf{v}^{\times}\mathbf{p} + \mathbf{g} \tag{32}$$

$$\dot{p}_n = m_d\boldsymbol{\omega}^T\mathbf{n}^{\times}(\mathbf{v} - \mathbf{r}_d^{\times}\boldsymbol{\omega}) - c_d\dot{\xi} - k_d\xi \tag{33}$$

It is tempting to substitute for $\mathbf{p}$, $\mathbf{h}$, and p_n in the motion equations (30) through (32) using (20) through (22). The resulting differential equations for $\mathbf{v}$, $\boldsymbol{\omega}$, and $\dot{\xi}$ would no longer contain momenta explicitly. It is not clear, however, that this substitution is advantageous. It may be preferable to integrate (31) through (33) for $\mathbf{p}$, $\mathbf{h}$, and p_n while simultaneously solving the algebraic equations (20) through (22) for $\{\mathbf{v}, \boldsymbol{\omega}, \dot{\xi}\}$. In terms of $\mathscr{M}$, this solution takes the form

$$v = \mathscr{M}^{-1}\not{p} \tag{34}$$

3.5 A DUAL-SPIN SYSTEM, $\mathscr{R} + \mathscr{W}$

The last section dealt with a rigid body $\mathscr{R}$ which contained a point mass $\mathscr{P}$ having linear motion with respect to $\mathscr{R}$. The chief aim of that model was to set up a simple situation which could be used to demonstrate the important influence of energy dissipation on attitude stability (as is done in Chapter 5). A different model is now studied whose aim is to illustrate gyric effects. Consider two rigid bodies $\mathscr{R}$ and $\mathscr{W}$. A special case of a rigid body, $\mathscr{W}$ is a spinning *wheel*, or *rotor*, having an axis of inertial symmetry (unit vector $\underset{\rightarrow}{\mathbf{a}}$) and whose mass center is on this axis. As shown in Fig. 3.8, the symmetry axis $\underset{\rightarrow}{\mathbf{a}}$ is fixed with respect to $\mathscr{R}$ and (using bearings) is allowed to rotate about $\underset{\rightarrow}{\mathbf{a}}$. In applications, $\mathscr{W}$ may either be inside or outside $\mathscr{R}$.

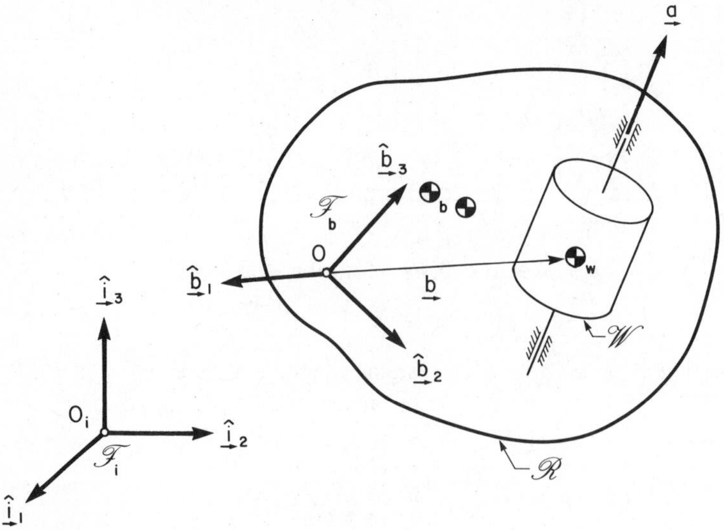

FIGURE 3.8 Rigid body $\mathscr{R}$ with rotor or wheel $\mathscr{W}$.

This model $\mathscr{R} + \mathscr{W}$ is sometimes called a *gyrostat*, a term whose meaning has undergone some evolution. According to [Gray], the term originated when a box, concealing a flywheel, was set up on its edge and resisted attempts by gravity or men to lay it flat; the system did not move toward its statically stable equilibrium. A more modern view is to define a "gyrostat" as a mechanical system that consists of a rigid body $\mathscr{R}$ and one or more (symmetrical) rotors, $\mathscr{W}_1, \mathscr{W}_2, \ldots,$ whose axes are fixed in $\mathscr{R}$ and which are allowed to rotate with respect to $\mathscr{R}$ about their axes of symmetry. This arrangement has the interesting consequence that the inertia matrix of the system, expressed in an $\mathscr{R}$-fixed frame, is constant (Problem 3.17). The simplest such model, $\mathscr{R} + \mathscr{W}$, provides an opportunity to study the implications of the gyric torques from $\mathscr{W}$ on $\mathscr{R}$.

The location of $\mathscr{W}$'s mass center $\oplus_w$ is located by $\underrightarrow{\mathbf{b}}$ with respect to O. The mass of the system is

$$m \triangleq m_b + m_w \tag{1}$$

where m_b and m_w are the masses of $\mathscr{R}$ and $\mathscr{W}$. The first and second moments of inertia of $\mathscr{R} + \mathscr{W}$ about O are

$$\underrightarrow{\mathbf{c}} \triangleq \underrightarrow{\mathbf{c}}_b + m_w \underrightarrow{\mathbf{b}} \tag{2}$$

$$\underrightarrow{\mathbf{J}} \triangleq \underrightarrow{\mathbf{J}}_b + \underrightarrow{\mathbf{I}}_w + m_w\left(b^2 \underrightarrow{\mathbf{1}} - \underrightarrow{\mathbf{b}}\underrightarrow{\mathbf{b}}\right) \tag{3}$$

where $\mathbf{J}_b$ and $\boldsymbol{I}_w$ are the moments of inertia of $\mathscr{P}$ and $\mathscr{W}$, respectively, about O and $\oplus_w$. According to Problem 3.7, the axial symmetry of $\mathscr{W}$ permits us to write its inertia dyadic in the special form:

$$\underrightarrow{\boldsymbol{I}}_w = I_t \underrightarrow{\mathbf{1}} + (I_s - I_t)\underrightarrow{\mathbf{a}}\underrightarrow{\mathbf{a}} \tag{4}$$

where I_s is the moment of inertia about the symmetry axis $\underset{\rightarrow}{\mathbf{a}}$, and I_t is the moment of inertia about (any) transverse axis.

Vectorial Motion Equations

We denote by $\underset{\rightarrow}{\mathbf{v}}$ and $\underset{\rightarrow}{\boldsymbol{\omega}}$ the absolute velocity of O and the absolute angular velocity of $\mathscr{R}$. The velocity at a point $\underset{\rightarrow}{\mathbf{r}}$ in $\mathscr{R}$ is $\underset{\rightarrow}{\mathbf{v}} + \underset{\rightarrow}{\boldsymbol{\omega}} \times \underset{\rightarrow}{\mathbf{r}}$, and the velocity at a point $\underset{\rightarrow}{\mathbf{r}}_w$ in $\mathscr{W}$ is $\underset{\rightarrow}{\mathbf{v}} + \underset{\rightarrow}{\boldsymbol{\omega}} \times \underset{\rightarrow}{\mathbf{b}} + \underset{\rightarrow}{\boldsymbol{\omega}}_w \times \underset{\rightarrow}{\mathbf{r}}_w$, where $\underset{\rightarrow}{\boldsymbol{\omega}}_w$ is the *absolute* angular velocity of $\mathscr{W}$ and $\underset{\rightarrow}{\mathbf{r}}_w \triangleq \underset{\rightarrow}{\mathbf{r}} - \underset{\rightarrow}{\mathbf{b}}$. The momenta of $\mathscr{R}$ and $\mathscr{W}$ are therefore

$$\underset{\rightarrow}{\mathbf{p}}_b \triangleq \int_{\mathscr{R}} (\underset{\rightarrow}{\mathbf{v}} + \underset{\rightarrow}{\boldsymbol{\omega}} \times \underset{\rightarrow}{\mathbf{r}})\, dm = m_b \underset{\rightarrow}{\mathbf{v}} + \underset{\rightarrow}{\boldsymbol{\omega}} \times \underset{\rightarrow}{\mathbf{c}}_b \tag{5}$$

$$\underset{\rightarrow}{\mathbf{p}}_w \triangleq \int_{\mathscr{W}} (\underset{\rightarrow}{\mathbf{v}} + \underset{\rightarrow}{\boldsymbol{\omega}} \times \underset{\rightarrow}{\mathbf{b}} + \underset{\rightarrow}{\boldsymbol{\omega}}_w \times \underset{\rightarrow}{\mathbf{r}}_w)\, dm = m_w \underset{\rightarrow}{\mathbf{v}} + m_w \underset{\rightarrow}{\boldsymbol{\omega}} \times \underset{\rightarrow}{\mathbf{b}} \tag{6}$$

The total momentum of the system is

$$\underset{\rightarrow}{\mathbf{p}} = \underset{\rightarrow}{\mathbf{p}}_b + \underset{\rightarrow}{\mathbf{p}}_w = m\underset{\rightarrow}{\mathbf{v}} - \underset{\rightarrow}{\mathbf{c}} \times \underset{\rightarrow}{\boldsymbol{\omega}} \tag{7}$$

where (1) and (2) have been used. In a similar manner, the (absolute) angular momenta (about O and $\oplus_w$) are calculated:

$$\underset{\rightarrow}{\mathbf{h}}_b \triangleq \int_{\mathscr{R}} \underset{\rightarrow}{\mathbf{r}} \times (\underset{\rightarrow}{\mathbf{v}} + \underset{\rightarrow}{\boldsymbol{\omega}} \times \underset{\rightarrow}{\mathbf{r}})\, dm = \underset{\rightarrow}{\mathbf{c}}_b \times \underset{\rightarrow}{\mathbf{v}} + \underset{\rightarrow}{\mathbf{J}}_b \cdot \underset{\rightarrow}{\boldsymbol{\omega}} \tag{8}$$

$$\underset{\rightarrow}{\mathbf{h}}_w \triangleq \int_{\mathscr{W}} \underset{\rightarrow}{\mathbf{r}}_w \times (\underset{\rightarrow}{\mathbf{v}} + \underset{\rightarrow}{\boldsymbol{\omega}} \times \underset{\rightarrow}{\mathbf{b}} + \underset{\rightarrow}{\boldsymbol{\omega}}_w \times \underset{\rightarrow}{\mathbf{r}}_w)\, dm = \underset{\rightarrow}{\mathbf{I}}_w \cdot \underset{\rightarrow}{\boldsymbol{\omega}}_w \tag{9}$$

The total absolute angular momentum (about O) is

$$\underset{\rightarrow}{\mathbf{h}} \triangleq \underset{\rightarrow}{\mathbf{h}}_b + \underset{\rightarrow}{\mathbf{h}}_w + \underset{\rightarrow}{\mathbf{b}} \times \underset{\rightarrow}{\mathbf{p}}_w = \underset{\rightarrow}{\mathbf{c}} \times \underset{\rightarrow}{\mathbf{v}} + \underset{\rightarrow}{\mathbf{J}} \cdot \underset{\rightarrow}{\boldsymbol{\omega}} + \underset{\rightarrow}{\mathbf{I}}_w \cdot \underset{\rightarrow}{\boldsymbol{\omega}}_s \tag{10}$$

In arriving at (10), we have used (2) and (3) and have defined the *relative* angular velocity of $\mathscr{W}$ with respect to $\mathscr{R}$ by $\underset{\rightarrow}{\boldsymbol{\omega}}_s$:

$$\underset{\rightarrow}{\boldsymbol{\omega}}_s \triangleq \underset{\rightarrow}{\boldsymbol{\omega}}_w - \underset{\rightarrow}{\boldsymbol{\omega}} = \omega_s \underset{\rightarrow}{\mathbf{a}} \tag{11}$$

We may refer to ω_s as the *rotor spin rate*. A superior form for (10) develops by recognizing (4) and (11):

$$\underset{\rightarrow}{\mathbf{h}} = \underset{\rightarrow}{\mathbf{c}} \times \underset{\rightarrow}{\mathbf{v}} + \underset{\rightarrow}{\mathbf{J}} \cdot \underset{\rightarrow}{\boldsymbol{\omega}} + \underset{\rightarrow}{\mathbf{a}} I_s \omega_s \tag{12}$$

Of particular interest is the component of $\underset{\rightarrow}{\mathbf{h}}_w$ along the rotor spin axis, $\underset{\rightarrow}{\mathbf{a}}$:

$$h_a \triangleq \underset{\rightarrow}{\mathbf{a}} \cdot \underset{\rightarrow}{\mathbf{h}}_w = I_s \underset{\rightarrow}{\mathbf{a}} \cdot \underset{\rightarrow}{\boldsymbol{\omega}} + I_s \omega_s \tag{13}$$

The other dynamical quantity of interest is the kinetic energy:

$$T = \frac{1}{2} \int_{\mathscr{R}} (\underset{\rightarrow}{\mathbf{v}} + \underset{\rightarrow}{\boldsymbol{\omega}} \times \underset{\rightarrow}{\mathbf{r}}) \cdot (\underset{\rightarrow}{\mathbf{v}} + \underset{\rightarrow}{\boldsymbol{\omega}} \times \underset{\rightarrow}{\mathbf{r}})\, dm$$

$$+ \frac{1}{2} \int_{\mathscr{W}} (\underset{\rightarrow}{\mathbf{v}}_{cw} + \underset{\rightarrow}{\boldsymbol{\omega}}_w \times \underset{\rightarrow}{\mathbf{r}}_w) \cdot (\underset{\rightarrow}{\mathbf{v}}_{cw} + \underset{\rightarrow}{\boldsymbol{\omega}}_w \times \underset{\rightarrow}{\mathbf{r}}_w)\, dm \tag{14}$$

with $\underset{\rightarrow}{\mathbf{v}}_{cw}$ introduced as a contraction for $\underset{\rightarrow}{\mathbf{v}} + \underset{\rightarrow}{\boldsymbol{\omega}} \times \underset{\rightarrow}{\mathbf{b}}$. Working out the details of (14) in terms of earlier definitions, we obtain

$$T = \tfrac{1}{2}m\underset{\rightarrow}{\mathbf{v}} \cdot \underset{\rightarrow}{\mathbf{v}} + \tfrac{1}{2}\underset{\rightarrow}{\boldsymbol{\omega}} \cdot \mathbf{J} \cdot \underset{\rightarrow}{\boldsymbol{\omega}} + \tfrac{1}{2}I_s\omega_s^2 - \underset{\rightarrow}{\mathbf{v}} \cdot \underset{\rightarrow}{\mathbf{c}} \times \underset{\rightarrow}{\boldsymbol{\omega}} + \underset{\rightarrow}{\boldsymbol{\omega}} \cdot \underset{\rightarrow}{\mathbf{a}} I_s\omega_s \tag{15}$$

Let the force and torque acting on $\mathscr{W}$ due to $\mathscr{R}$ be $\underset{\rightarrow}{\mathbf{f}}_{bw}$ and $\underset{\rightarrow}{\mathbf{g}}_{bw}$. An equal and opposite force and torque act on $\mathscr{R}$ due to $\mathscr{W}$; and, in addition, we include an external force $\underset{\rightarrow}{\mathbf{f}}$ and torque $\underset{\rightarrow}{\mathbf{g}}$ on $\mathscr{R}$. The motion equations for $\mathscr{R}$ and $\mathscr{W}$ are therefore as follows: from (4) and (5) in Section 3.3,

$$\dot{\underset{\rightarrow}{\mathbf{p}}}_b = \underset{\rightarrow}{\mathbf{f}} - \underset{\rightarrow}{\mathbf{f}}_{bw} \tag{16}$$

$$\dot{\underset{\rightarrow}{\mathbf{p}}}_w = \underset{\rightarrow}{\mathbf{f}}_{bw} \tag{17}$$

$$\dot{\underset{\rightarrow}{\mathbf{h}}} - \underset{\rightarrow}{\mathbf{v}} \times (\underset{\rightarrow}{\mathbf{c}}_b \times \underset{\rightarrow}{\boldsymbol{\omega}}) = \underset{\rightarrow}{\mathbf{g}} - \underset{\rightarrow}{\mathbf{g}}_{bw} - \underset{\rightarrow}{\mathbf{b}} \times \underset{\rightarrow}{\mathbf{f}}_{bw} \tag{18}$$

$$\dot{\underset{\rightarrow}{\mathbf{h}}}_w = \underset{\rightarrow}{\mathbf{g}}_{bw} \tag{19}$$

Summing the first two of these, and observing (7), we obtain

$$\dot{\underset{\rightarrow}{\mathbf{p}}} = \underset{\rightarrow}{\mathbf{f}} \tag{20}$$

Similarly, adding (18) and (19) and substituting (17) for $\underset{\rightarrow}{\mathbf{f}}_{bw}$ yields the intermediate result

$$\dot{\underset{\rightarrow}{\mathbf{h}}} - \underset{\rightarrow}{\mathbf{b}} \times \dot{\underset{\rightarrow}{\mathbf{p}}}_w - \underset{\rightarrow}{\mathbf{v}} \times (\underset{\rightarrow}{\mathbf{c}}_b \times \underset{\rightarrow}{\boldsymbol{\omega}}) = \underset{\rightarrow}{\mathbf{g}}$$

With $\dot{\underset{\rightarrow}{\mathbf{b}}} = \underset{\rightarrow}{\boldsymbol{\omega}} \times \underset{\rightarrow}{\mathbf{b}}$, and $\underset{\rightarrow}{\mathbf{p}}_w$ from (6), we arrive at the form we want:

$$\dot{\underset{\rightarrow}{\mathbf{h}}} + \underset{\rightarrow}{\mathbf{v}} \times \underset{\rightarrow}{\mathbf{p}} = \underset{\rightarrow}{\mathbf{g}} \tag{21}$$

These motion equations, (20) and (21), could have been anticipated from the earlier derivation of (12) and (19) in Section 3.2 for a system of point masses.

Our current system, $\mathscr{R} + \mathscr{W}$, has seven degrees of freedom, corresponding to the position of O, the rotation of $\mathscr{R}$ about O, and the rotation of $\mathscr{W}$ about $\underset{\rightarrow}{\mathbf{a}}$. We need an additional equation to complete (20) and (21), an equation governing wheel speed. To this end, we define

$$g_a \triangleq \underset{\rightarrow}{\mathbf{a}} \cdot \underset{\rightarrow}{\mathbf{g}}_{bw} \tag{22}$$

This is the axial torque on the rotor. In a passive system, g_a would be a frictional torque. With a motor, g_a could be used to control ω_s. From (13),

$$\begin{aligned} \dot{h}_a &= \dot{\underset{\rightarrow}{\mathbf{a}}} \cdot \underset{\rightarrow}{\mathbf{h}}_w + \underset{\rightarrow}{\mathbf{a}} \cdot \dot{\underset{\rightarrow}{\mathbf{h}}}_w \\ &= (\underset{\rightarrow}{\boldsymbol{\omega}} \times \underset{\rightarrow}{\mathbf{a}}) \cdot \underset{\rightarrow}{\mathbf{h}}_w + \underset{\rightarrow}{\mathbf{a}} \cdot \underset{\rightarrow}{\mathbf{g}}_{bw} \\ &= \underset{\rightarrow}{\boldsymbol{\omega}} \cdot (\underset{\rightarrow}{\mathbf{a}} \times \underset{\rightarrow}{\mathbf{h}}_w) + g_a \\ &= \underset{\rightarrow}{\boldsymbol{\omega}} \cdot \underset{\rightarrow}{\mathbf{a}} \times \underset{\rightarrow}{\mathbf{I}}_w \cdot \underset{\rightarrow}{\boldsymbol{\omega}}_w + g_a \end{aligned}$$

However, from (4), $\underset{\rightarrow}{\mathbf{a}} \times \underset{\rightarrow}{\mathbf{I}}_w = I_t \underset{\rightarrow}{\mathbf{a}} \times \underset{\rightarrow}{\mathbf{1}}$, and from (11), $\underset{\rightarrow}{\boldsymbol{\omega}} \cdot \underset{\rightarrow}{\mathbf{a}} \times \underset{\rightarrow}{\boldsymbol{\omega}}_w = 0$. Therefore, we have the simple motion equation for the wheel

$$\dot{h}_a = g_a \tag{23}$$

We are also interested in the variation in kinetic energy throughout the motion. By analogy with (24) in Section 3.2, we calculate

$$\begin{aligned}\dot{T} &= \int_{\mathscr{R}} (\underset{\rightarrow}{\mathbf{v}} + \underset{\rightarrow}{\boldsymbol{\omega}} \times \underset{\rightarrow}{\mathbf{r}}) \cdot d\underset{\rightarrow}{\mathbf{f}} + \int_{\mathscr{W}} (\underset{\rightarrow}{\mathbf{v}} + \underset{\rightarrow}{\boldsymbol{\omega}} \times \underset{\rightarrow}{\mathbf{b}} + \underset{\rightarrow}{\boldsymbol{\omega}}_w \times \underset{\rightarrow}{\mathbf{r}}_w) \cdot d\underset{\rightarrow}{\mathbf{f}} \\ &= \underset{\rightarrow}{\mathbf{v}} \cdot (\underset{\rightarrow}{\mathbf{f}} - \underset{\rightarrow}{\mathbf{f}}_{bw}) + \underset{\rightarrow}{\boldsymbol{\omega}} \cdot (\underset{\rightarrow}{\mathbf{g}} - \underset{\rightarrow}{\mathbf{g}}_{bw} - \underset{\rightarrow}{\mathbf{b}} \times \underset{\rightarrow}{\mathbf{f}}_{bw}) + \underset{\rightarrow}{\mathbf{v}} \cdot \underset{\rightarrow}{\mathbf{f}}_{bw} + \underset{\rightarrow}{\boldsymbol{\omega}} \cdot \underset{\rightarrow}{\mathbf{b}} \times \underset{\rightarrow}{\mathbf{f}}_{bw} + \underset{\rightarrow}{\boldsymbol{\omega}}_w \cdot \underset{\rightarrow}{\mathbf{g}}_{bw} \\ &= \underset{\rightarrow}{\mathbf{f}} \cdot \underset{\rightarrow}{\mathbf{v}} + \underset{\rightarrow}{\mathbf{g}} \cdot \underset{\rightarrow}{\boldsymbol{\omega}} + g_a \omega_s \end{aligned} \tag{24}$$

As expected, the increase in kinetic energy is equal to the work done by the external force and torque and by the friction (and motor) torque on the wheel. Equation (24) can be obtained also by differentiating (15) and inserting the motion equations (Problem 3.18), but this derivation is more arduous than the one sketched above. It provides a useful check, however.

Scalar Motion Equations

To proceed further, we now express all vectors in $\mathscr{F}_b$. Note that $\underset{\rightarrow}{\mathbf{b}}$, $\underset{\rightarrow}{\mathbf{a}}$, $\underset{\rightarrow}{\mathbf{c}}$, and $\mathbf{J}$ are constants when expressed in $\mathscr{F}_b$. Note that due to the axial symmetry of $\mathscr{W}$ we have not needed to utilize a wheel-fixed reference frame (Problem 3.17).

In terms of these components, the momenta and kinetic energy, equations (7), (12), (13), (15), and (24), can be expressed, respectively, as follows:

$$\mathbf{p} = m\mathbf{v} - \mathbf{c}^{\times}\boldsymbol{\omega} \tag{25}$$

$$\mathbf{h} = \mathbf{c}^{\times}\mathbf{v} + \mathbf{J}\boldsymbol{\omega} + \mathbf{a}I_s\omega_s \tag{26}$$

$$h_a = I_s\mathbf{a}^T\boldsymbol{\omega} + I_s\omega_s \tag{27}$$

$$T = \tfrac{1}{2}m\mathbf{v}^T\mathbf{v} + \tfrac{1}{2}\boldsymbol{\omega}^T\mathbf{J}\boldsymbol{\omega} + \tfrac{1}{2}I_s\omega_s^2 - \mathbf{v}^T\mathbf{c}^{\times}\boldsymbol{\omega} + I_s\omega_s\mathbf{a}^T\boldsymbol{\omega} \tag{28}$$

$$\dot{T} = \mathbf{f}^T\mathbf{v} + \mathbf{g}^T\boldsymbol{\omega} + g_a\omega_s \tag{29}$$

In parallel with the other models in this series, there is a "system inertia matrix" that crystallizes the content of the above dynamical quantities:

$$\mathscr{M} \triangleq \begin{bmatrix} m\mathbf{1} & -\mathbf{c}^{\times} & \mathbf{0} \\ \mathbf{c}^{\times} & \mathbf{J} & I_s\mathbf{a} \\ \mathbf{0}^T & I_s\mathbf{a}^T & I_s \end{bmatrix} \tag{30}$$

If we also define

$$\mathscr{p} \triangleq \begin{bmatrix} \mathbf{p} \\ \mathbf{h} \\ h_a \end{bmatrix}; \qquad \mathscr{v} \triangleq \begin{bmatrix} \mathbf{v} \\ \boldsymbol{\omega} \\ \omega_s \end{bmatrix}; \qquad \mathscr{f} \triangleq \begin{bmatrix} \mathbf{f} \\ \mathbf{g} \\ g_a \end{bmatrix} \tag{31}$$

we can express these relationships as follows:

$$\mathscr{p} = \mathscr{M}\mathscr{v} \tag{32}$$

$$T = \tfrac{1}{2}\mathscr{v}^T\mathscr{M}\mathscr{v} \tag{33}$$

Note that $\mathscr{M}$ is a constant, symmetric, positive-definite matrix. [Proof of the latter is similar to the proof following (28) in section 3.4.] In fact, it is plain from (30) that we can choose the mass center of the system as the reference point O, fixed in $\mathscr{R}$. This simplification, $\mathbf{c} = \mathbf{0}$, is possible because $\mathbf{c}$ is constant:

$$\mathscr{M} = \begin{bmatrix} m\mathbf{1} & \mathbf{O} & \mathbf{0} \\ \mathbf{O} & \mathbf{I} & I_s\mathbf{a} \\ \mathbf{0}^T & I_s\mathbf{a}^T & I_s \end{bmatrix} \tag{34}$$

The translational and rotational motions are now uncoupled, at least as far as $\mathscr{M}$ is concerned. Whether or not $\mathbf{c} = \mathbf{0}$, it is clear that

$$\frac{\partial T}{\partial \mathbf{v}} = \mathbf{p}; \qquad \frac{\partial T}{\partial \boldsymbol{\omega}} = \mathbf{h}; \qquad \frac{\partial T}{\partial \omega_s} = h_a \tag{35}$$

The motion equations, (20), (21), and (23), in matrix notation are respectively

$$\dot{\mathbf{p}} = -\boldsymbol{\omega}^{\times}\mathbf{p} + \mathbf{f} \tag{36}$$

$$\dot{\mathbf{h}} = -\boldsymbol{\omega}^{\times}\mathbf{h} - \mathbf{v}^{\times}\mathbf{p} + \mathbf{g} \tag{37}$$

$$\dot{h}_a = g_a \tag{38}$$

It is possible, of course, now to substitute (25), (26), and (27) and thereby obtain motion equations in terms of $\mathbf{v}$, $\boldsymbol{\omega}$, and ω_s. A reasonable alternative is to integrate (36), (37), and (38) for $\mathbf{p}$, $\mathbf{h}$, and h_a and to solve collaterally the algebraic equations for $\mathbf{v}$, $\boldsymbol{\omega}$, and ω_s:

$$\mathscr{v} = \mathscr{M}^{-1}\mathscr{p} \tag{39}$$

Because $\mathscr{M}$ is constant, its inverse need by found only once.

3.6 A SIMPLE MULTI-RIGID-BODY SYSTEM, $\mathscr{R}_1 + \mathscr{R}_2$

Our third example concerns two rigid bodies $\mathscr{R}_1$ and $\mathscr{R}_2$, touching at a common point P. This example has a substantial range of application to practical problems in spacecraft attitude dynamics. Many space vehicles assume their final configuration after an initial deployment phase. During this phase, one part of the vehicle moves with respect to another part as the configuration evolves from its initial to its final form. Or, to cite another class of applications, a scientific satellite may have an instrument package (e.g., a telescope) that must be aimed at celestial points of interest; such an attached body must be attitude controlled with respect to the main body during active pointing, and it must furthermore be *slewed* from one aiming point to another. A final case in point is a satellite whose function (e.g., communications) requires it to be Earth-pointing but which includes a solar-cell array for generating energy from the sun; thus, the array must be sun-pointing.

Referring to Fig. 3.9, we choose O as the reference point fixed in $\mathscr{R}_1$ and P as the reference point fixed in $\mathscr{R}_2$. Point P is located at $\underrightarrow{\mathbf{b}}$ with respect to O. The mass of the system is

$$m \triangleq m_1 + m_2 \tag{1}$$

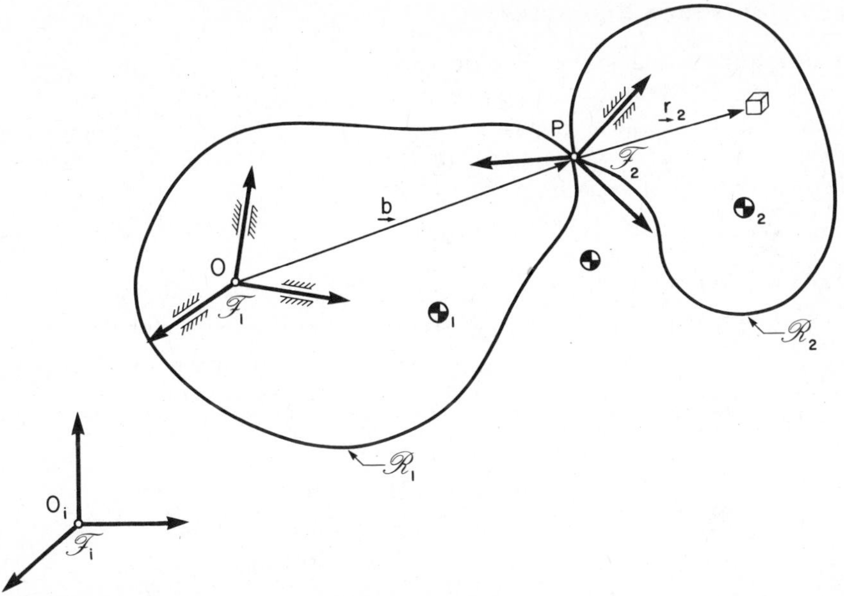

FIGURE 3.9 System of two connected rigid bodies, $\mathcal{R}_1 + \mathcal{R}_2$.

where m_j is the mass of $\mathcal{R}_j$. The first and second moments of inertia of $\mathcal{R}_1$ and $\mathcal{R}_2$ about their respective reference points, O and P, are

$$\underrightarrow{\mathbf{c}}_1 \triangleq \int_{\mathcal{R}_1} \underrightarrow{\mathbf{r}}\, dm; \qquad \underrightarrow{\mathbf{c}}_2 \triangleq \int_{\mathcal{R}_2} \underrightarrow{\mathbf{r}}_2\, dm \qquad \left(\underrightarrow{\mathbf{r}}_2 \triangleq \underrightarrow{\mathbf{r}} - \underrightarrow{\mathbf{b}}\right) \tag{2}$$

$$\underrightarrow{\mathbf{J}}_1 \triangleq \int_{\mathcal{R}_1} \left(r^2 \underrightarrow{\mathbf{1}} - \underrightarrow{\mathbf{r}}\underrightarrow{\mathbf{r}}\right) dm; \qquad \underrightarrow{\mathbf{J}}_2 \triangleq \int_{\mathcal{R}_2} \left(r_1^2 \underrightarrow{\mathbf{1}} - \underrightarrow{\mathbf{r}}_2\underrightarrow{\mathbf{r}}_2\right) dm \tag{3}$$

The first and second moments of inertia for $\mathcal{R}_1 + \mathcal{R}_2$ about O are therefore

$$\underrightarrow{\mathbf{c}} \triangleq \underrightarrow{\mathbf{c}}_1 + \underrightarrow{\mathbf{c}}_2 + m_2 \underrightarrow{\mathbf{b}} \tag{4}$$

$$\underrightarrow{\mathbf{J}} \triangleq \underrightarrow{\mathbf{J}}_1 + \int_{\mathcal{R}_2} \left[(\underrightarrow{\mathbf{b}} + \underrightarrow{\mathbf{r}}_2)\cdot(\underrightarrow{\mathbf{b}} + \underrightarrow{\mathbf{r}}_2)\underrightarrow{\mathbf{1}} - (\underrightarrow{\mathbf{b}} + \underrightarrow{\mathbf{r}}_2)(\underrightarrow{\mathbf{b}} + \underrightarrow{\mathbf{r}}_2)\right] dm \tag{5}$$

It is straightforward to show from these definitions that

$$\underrightarrow{\mathbf{J}} = \underrightarrow{\mathbf{J}}_1 + \underrightarrow{\mathbf{J}}_2 + m_2\left(b^2 \underrightarrow{\mathbf{1}} - \underrightarrow{\mathbf{b}}\underrightarrow{\mathbf{b}}\right) + \left(2\underrightarrow{\mathbf{b}} \cdot \underrightarrow{\mathbf{c}}_2 \underrightarrow{\mathbf{1}} - \underrightarrow{\mathbf{b}}\underrightarrow{\mathbf{c}}_2 - \underrightarrow{\mathbf{c}}_2\underrightarrow{\mathbf{b}}\right) \tag{6}$$

Anticipating our needs somewhat, we also define

$$\underrightarrow{\mathbf{J}}_{12} \triangleq \int_{\mathcal{R}_2} \left[\underrightarrow{\mathbf{r}}_2 \cdot(\underrightarrow{\mathbf{b}} + \underrightarrow{\mathbf{r}}_2)\underrightarrow{\mathbf{1}} - \underrightarrow{\mathbf{r}}_2(\underrightarrow{\mathbf{b}} + \underrightarrow{\mathbf{r}}_2)\right] dm \tag{7}$$

$$\underrightarrow{\mathbf{J}}_{21} \triangleq \int_{\mathcal{R}_2} \left[(\underrightarrow{\mathbf{b}} + \underrightarrow{\mathbf{r}}_2) \cdot \underrightarrow{\mathbf{r}}_2 \underrightarrow{\mathbf{1}} - (\underrightarrow{\mathbf{b}} + \underrightarrow{\mathbf{r}}_2)\underrightarrow{\mathbf{r}}_2\right] dm \tag{8}$$

These inertia dyadics are *mixed*: one of the two constituent position vectors refers to $\mathcal{R}_1$, while the other refers to $\mathcal{R}_2$. We shall call $\underrightarrow{\mathbf{J}}_{12}$ and $\underrightarrow{\mathbf{J}}_{21}$ *mixed moments of inertia*.

Vectorial Motion Equations

We denote by $\mathbf{v}$ and $\boldsymbol{\omega}$ the absolute velocity of O and the absolute angular velocity of $\mathscr{R}_1$. The velocity at a point $\mathbf{r}$ in $\mathscr{R}_1$ is therefore $\mathbf{v} + \boldsymbol{\omega} \times \mathbf{r}$. Similarly, denoting by $\boldsymbol{\omega}_2$ the *absolute* angular velocity of $\mathscr{R}_2$, the velocity at a point $\mathbf{r}_2$ in $\mathscr{R}_2$ is $\mathbf{v} + \boldsymbol{\omega} \times \mathbf{b} + \boldsymbol{\omega}_2 \times \mathbf{r}_2 = \mathbf{v} + \boldsymbol{\omega} \times (\mathbf{b} + \mathbf{r}_2) + \boldsymbol{\omega}_p \times \mathbf{r}_2$, where $\boldsymbol{\omega}_p$ is the *relative* angular velocity of $\mathscr{R}_2$ with respect to $\mathscr{R}_1$:

$$\boldsymbol{\omega}_p \triangleq \boldsymbol{\omega}_2 - \boldsymbol{\omega} \tag{9}$$

We shall assume a "universal joint" allows arbitrary angular motion at P.

Knowing the velocity fields, we can compute the momenta and the kinetic energy:

$$\mathbf{p}_1 \triangleq \int_{\mathscr{R}_1} (\mathbf{v} + \boldsymbol{\omega} \times \mathbf{r})\, dm = m_1 \mathbf{v} - \mathbf{c}_1 \times \boldsymbol{\omega} \tag{10}$$

$$\mathbf{p}_2 \triangleq \int_{\mathscr{R}_2} \left[\mathbf{v} + \boldsymbol{\omega} \times (\mathbf{b} + \mathbf{r}_2) + \boldsymbol{\omega}_p \times \mathbf{r}_2\right] dm$$

$$= m_2 \mathbf{v} - (m_2 \mathbf{b} + \mathbf{c}_2) \times \boldsymbol{\omega} - \mathbf{c}_2 \times \boldsymbol{\omega}_p \tag{11}$$

The total system momentum is

$$\mathbf{p} = \mathbf{p}_1 + \mathbf{p}_2 = m\mathbf{v} - \mathbf{c} \times \boldsymbol{\omega} - \mathbf{c}_2 \times \boldsymbol{\omega}_p \tag{12}$$

Similarly, the (absolute) angular momenta (respectively about O and P) are

$$\mathbf{h}_1 \triangleq \int_{\mathscr{R}_1} \mathbf{r} \times (\mathbf{v} + \boldsymbol{\omega} \times \mathbf{r})\, dm = \mathbf{c}_1 \times \mathbf{v} + \mathbf{J}_1 \cdot \boldsymbol{\omega} \tag{13}$$

$$\mathbf{h}_2 \triangleq \int_{\mathscr{R}_2} \mathbf{r}_2 \times \left[\mathbf{v} + \boldsymbol{\omega} \times (\mathbf{b} + \mathbf{r}_2) + \boldsymbol{\omega}_p \cdot \mathbf{r}_2\right] dm$$

$$= \mathbf{c}_2 \times \mathbf{v} + \mathbf{J}_{21} \cdot \boldsymbol{\omega} + \mathbf{J}_2 \cdot \boldsymbol{\omega}_p \tag{14}$$

The total moment of momentum about O therefore is

$$\mathbf{h} \triangleq \mathbf{h}_1 + \mathbf{h}_2 + \mathbf{b} \times \mathbf{p}_2 = \mathbf{c} \times \mathbf{v} + \mathbf{J} \cdot \boldsymbol{\omega} + \mathbf{J}_{12} \cdot \boldsymbol{\omega}_p \tag{15}$$

The last quantity of dynamical significance is the kinetic energy:

$$T = \frac{1}{2} \int_{\mathscr{R}_1} (\mathbf{v} + \boldsymbol{\omega} \times \mathbf{r}) \cdot (\mathbf{v} + \boldsymbol{\omega} \times \mathbf{r})\, dm$$

$$+ \frac{1}{2} \int_{\mathscr{R}_2} \left[\mathbf{v} + \boldsymbol{\omega} \times (\mathbf{b} + \mathbf{r}_2) + \boldsymbol{\omega}_p \times \mathbf{r}_2\right] \cdot \left[\mathbf{v} + \boldsymbol{\omega} \times (\mathbf{b} \times \mathbf{r}_2) + \boldsymbol{\omega}_p \times \mathbf{r}_2\right] dm$$

$$= \frac{1}{2} m \mathbf{v} \cdot \mathbf{v} + \frac{1}{2} \boldsymbol{\omega} \cdot \mathbf{J} \cdot \boldsymbol{\omega} + \frac{1}{2} \boldsymbol{\omega}_p \cdot \mathbf{J}_2 \cdot \boldsymbol{\omega}_p$$

$$- \mathbf{v} \cdot \mathbf{c} \times \boldsymbol{\omega} - \mathbf{v} \cdot \mathbf{c}_2 \times \boldsymbol{\omega}_p + \boldsymbol{\omega} \cdot \mathbf{J}_{12} \cdot \boldsymbol{\omega}_p \tag{16}$$

in terms of our earlier definitions.

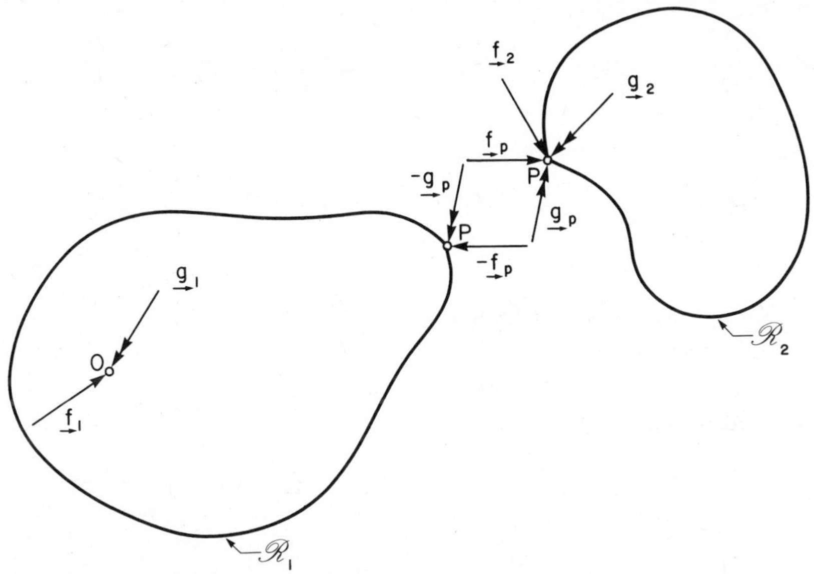

FIGURE 3.10 Free-body diagram for $\mathscr{R}_1 + \mathscr{R}_2$.

The notations $\underset{\rightarrow}{\mathbf{f}}_1$, $\underset{\rightarrow}{\mathbf{f}}_2$, $\underset{\rightarrow}{\mathbf{g}}_1$, and $\underset{\rightarrow}{\mathbf{g}}_2$ are now used for the external forces and torques on $\mathscr{R}_1$ and $\mathscr{R}_2$. Note that $\underset{\rightarrow}{\mathbf{g}}_1$ is about O, while $\underset{\rightarrow}{\mathbf{g}}_2$ is about P (see Fig. 3.10). Furthermore, $\underset{\rightarrow}{\mathbf{f}}_p$ and $\underset{\rightarrow}{\mathbf{g}}_p$ are the force and torque acting at P on $\mathscr{R}_2$ due to $\mathscr{R}_1$. Motion equations for $\mathscr{R}_1$ and $\mathscr{R}_2$ can be constructed by using the appropriate equations from Section 3.3, namely, (4) and (5):

$$\dot{\underset{\rightarrow}{\mathbf{p}}}_1 = \underset{\rightarrow}{\mathbf{f}}_1 - \underset{\rightarrow}{\mathbf{f}}_p \tag{17}$$

$$\dot{\underset{\rightarrow}{\mathbf{p}}}_2 = \underset{\rightarrow}{\mathbf{f}}_2 + \underset{\rightarrow}{\mathbf{f}}_p \tag{18}$$

$$\dot{\underset{\rightarrow}{\mathbf{h}}}_1 + \underset{\rightarrow}{\mathbf{v}} \times \underset{\rightarrow}{\mathbf{p}}_1 = \underset{\rightarrow}{\mathbf{g}}_1 - \underset{\rightarrow}{\mathbf{g}}_p - \underset{\rightarrow}{\mathbf{b}} \times \underset{\rightarrow}{\mathbf{f}}_p \tag{19}$$

$$\dot{\underset{\rightarrow}{\mathbf{h}}}_2 + (\underset{\rightarrow}{\mathbf{v}} + \underset{\rightarrow}{\boldsymbol{\omega}} \times \underset{\rightarrow}{\mathbf{b}}) \times \underset{\rightarrow}{\mathbf{p}}_2 = \underset{\rightarrow}{\mathbf{g}}_2 + \underset{\rightarrow}{\mathbf{g}}_p \tag{20}$$

Summing (17) and (18), we find

$$\dot{\underset{\rightarrow}{\mathbf{p}}} = \underset{\rightarrow}{\mathbf{f}} \tag{21}$$

where

$$\underset{\rightarrow}{\mathbf{f}} \triangleq \underset{\rightarrow}{\mathbf{f}}_1 + \underset{\rightarrow}{\mathbf{f}}_2 \tag{22}$$

is the total external force on the system. Similarly, adding (19) and (20), and substituting (18) for $\underset{\rightarrow}{\mathbf{f}}_p$, we find

$$\dot{\underset{\rightarrow}{\mathbf{h}}} + \underset{\rightarrow}{\mathbf{v}} \times \underset{\rightarrow}{\mathbf{p}} = \underset{\rightarrow}{\mathbf{g}} \tag{23}$$

where

$$\underset{\rightarrow}{\mathbf{g}} \triangleq \underset{\rightarrow}{\mathbf{g}}_1 + \underset{\rightarrow}{\mathbf{g}}_2 + \underset{\rightarrow}{\mathbf{b}} \times \underset{\rightarrow}{\mathbf{f}}_2 \tag{24}$$

it the total external torque on the system.

The kinetic energy changes at the following rate:

$$\dot{T} = \int_{\mathscr{R}_1} (\underrightarrow{\mathbf{v}} + \underrightarrow{\boldsymbol{\omega}} \times \underrightarrow{\mathbf{r}}) \cdot d\underrightarrow{\mathbf{f}} + \int_{\mathscr{R}_2} \left[\underrightarrow{\mathbf{v}} + \underrightarrow{\boldsymbol{\omega}} \times (\underrightarrow{\mathbf{b}} \times \underrightarrow{\mathbf{r}}_2) + \underrightarrow{\boldsymbol{\omega}}_p \times \underrightarrow{\mathbf{r}}_2\right] \cdot d\underrightarrow{\mathbf{f}}$$

$$= \underrightarrow{\mathbf{v}} \cdot (\underrightarrow{\mathbf{f}}_1 - \underrightarrow{\mathbf{f}}_p) + \underrightarrow{\boldsymbol{\omega}} \cdot (\underrightarrow{\mathbf{g}}_1 - \underrightarrow{\mathbf{g}}_p - \underrightarrow{\mathbf{b}} \times \underrightarrow{\mathbf{f}}_p)$$

$$+ \underrightarrow{\mathbf{v}} \cdot (\underrightarrow{\mathbf{f}}_2 + \underrightarrow{\mathbf{f}}_p) + \underrightarrow{\boldsymbol{\omega}} \cdot \underrightarrow{\mathbf{b}} \times (\underrightarrow{\mathbf{f}}_2 + \underrightarrow{\mathbf{f}}_p) + (\underrightarrow{\boldsymbol{\omega}} + \underrightarrow{\boldsymbol{\omega}}_p) \cdot (\underrightarrow{\mathbf{g}}_2 + \underrightarrow{\mathbf{g}}_p)$$

$$= \underrightarrow{\mathbf{f}} \cdot \underrightarrow{\mathbf{v}} + \underrightarrow{\mathbf{g}} \cdot \underrightarrow{\boldsymbol{\omega}} + (\underrightarrow{\mathbf{g}}_2 + \underrightarrow{\mathbf{g}}_p) \cdot \underrightarrow{\boldsymbol{\omega}}_p \tag{25}$$

This result shows the power contributed by external forces and torques, and by the control torque at the joint, $\underrightarrow{\mathbf{g}}_p$.

Scalar Motion Equations

At this stage in each of our spacecraft models, each vector is expressed in a suitable reference frame. As shown in Fig. 3.9, we have three reference frames to choose from: $\mathscr{F}_i$, $\mathscr{F}_1$, and $\mathscr{F}_2$. The latter two are fixed in $\mathscr{R}_1$ and $\mathscr{R}_2$, respectively. Their advantage is that the moments of inertia, $\mathbf{c}_j$ and $\mathbf{J}_j$, are constant when expressed In $\mathscr{F}_j$ $(j = 1, 2)$. We express the vectors and dyadics accordingly as follows:

$$[\mathbf{p} \quad \mathbf{f} \quad \mathbf{v}] \triangleq \mathscr{F}_1 \cdot [\underrightarrow{\mathbf{p}} \quad \underrightarrow{\mathbf{f}} \quad \underrightarrow{\mathbf{v}}]$$

$$[\mathbf{c} \quad \mathbf{c}_1 \quad \boldsymbol{\omega} \quad \mathbf{h} \quad \mathbf{b} \quad \mathbf{g}] \triangleq \mathscr{F}_1 \cdot [\underrightarrow{\mathbf{c}} \quad \underrightarrow{\mathbf{c}}_1 \quad \underrightarrow{\boldsymbol{\omega}} \quad \underrightarrow{\mathbf{h}} \quad \underrightarrow{\mathbf{b}} \quad \underrightarrow{\mathbf{g}}]$$

$$[\mathbf{J}_1 \quad \mathbf{J}] \triangleq \mathscr{F}_1 \cdot [\underrightarrow{\mathbf{J}}_1 \quad \underrightarrow{\mathbf{J}}] \cdot \mathscr{F}_1^T$$

$$[\mathbf{c}_2 \quad \boldsymbol{\omega}_p \quad \mathbf{h}_2 \quad \mathbf{g}_2 \quad \mathbf{g}_p] \triangleq \mathscr{F}_2 \cdot [\underrightarrow{\mathbf{c}}_2 \quad \underrightarrow{\boldsymbol{\omega}}_p \quad \underrightarrow{\mathbf{h}}_2 \quad \underrightarrow{\mathbf{g}}_2 \quad \underrightarrow{\mathbf{g}}_p]$$

$$\mathbf{J}_2 \triangleq \mathscr{F}_2 \cdot \underrightarrow{\mathbf{J}}_2 \cdot \mathscr{F}_2^T$$

Note that $\underrightarrow{\mathbf{J}}_{12}$ and $\underrightarrow{\mathbf{J}}_{21}$ have not been mentioned yet. To see why, let us begin to form scalar equations and use the vectrix procedures of Appendix B. From (12),

$$\mathbf{p} = \mathscr{F}_1 \cdot \underrightarrow{\mathbf{p}} = \mathscr{F}_1 \cdot (m\underrightarrow{\mathbf{v}} - \underrightarrow{\mathbf{c}} \times \underrightarrow{\boldsymbol{\omega}} - \underrightarrow{\mathbf{c}}_2 \times \underrightarrow{\boldsymbol{\omega}}_p)$$

$$= m\mathbf{v} - \mathbf{c}^{\times}\boldsymbol{\omega} - \mathscr{F}_1 \cdot \mathbf{c}_2^T \mathscr{F}_2 \times \mathscr{F}_2^T \boldsymbol{\omega}_p$$

$$= m\mathbf{v} - \mathbf{c}^{\times}\boldsymbol{\omega} - \mathscr{F}_1 \cdot \mathscr{F}_2^T \mathbf{c}_2^{\times} \boldsymbol{\omega}_p$$

$$= m\mathbf{v} - \mathbf{c}^{\times}\boldsymbol{\omega} - \mathbf{C}_{12}\mathbf{c}_2^{\times}\boldsymbol{\omega}_p \tag{26}$$

Next, from (15),

$$\mathbf{h} = \mathscr{F}_1 \cdot \underrightarrow{\mathbf{h}} = \mathscr{F}_1 \cdot (\underrightarrow{\mathbf{c}} \times \underrightarrow{\mathbf{v}} + \underrightarrow{\mathbf{J}} \cdot \underrightarrow{\boldsymbol{\omega}} + \underrightarrow{\mathbf{J}}_{12} \cdot \underrightarrow{\boldsymbol{\omega}}_p)$$

$$= \mathbf{c}^{\times}\mathbf{v} + \mathbf{J}\boldsymbol{\omega} + \mathscr{F}_1 \cdot \underrightarrow{\mathbf{J}}_{12} \cdot \mathscr{F}_2^T \boldsymbol{\omega}_p \tag{27}$$

We now realize that *the most convenient way to express* $\underrightarrow{\mathbf{J}}_{12}$ *is partially in* $\mathscr{F}_1$ *and partially in* $\mathscr{F}_2$, as follows:

$$\mathbf{J}_{12} \triangleq \mathscr{F}_1 \cdot \underset{\rightarrow}{\mathbf{J}}_{12} \cdot \mathscr{F}_2^T \tag{28}$$

Similarly, for $\mathbf{J}_{21}$,

$$\mathbf{J}_{21} \triangleq \mathscr{F}_2 \cdot \underset{\rightarrow}{\mathbf{J}}_{21} \cdot \mathscr{F}_1^T \tag{30}$$

It is not difficult to show that $\mathbf{J}_{21}$ is the transpose of $\mathbf{J}_{12}$:

$$\mathbf{J}_{12}^T = \mathbf{J}_{21} \tag{31}$$

With this definition for $\mathbf{J}_{21}$, it can be further shown that, from (14),

$$\mathbf{h}_2 = \mathbf{c}_2^{\times}\mathbf{C}_{21}\mathbf{v} + \mathbf{J}_{21}\boldsymbol{\omega} + \mathbf{J}_2\boldsymbol{\omega}_p \tag{32}$$

Lastly, the kinetic energy is given in terms of vector components from (16):

$$T = \tfrac{1}{2}m\mathbf{v}^T\mathbf{v} + \tfrac{1}{2}\boldsymbol{\omega}^T\mathbf{J}\boldsymbol{\omega} + \tfrac{1}{2}\boldsymbol{\omega}_p^T\mathbf{J}_2\boldsymbol{\omega}_p$$
$$-\mathbf{v}^T\mathbf{c}^{\times}\boldsymbol{\omega} - \mathbf{v}^T\mathbf{C}_{12}\mathbf{c}_2^{\times}\boldsymbol{\omega}_p + \boldsymbol{\omega}^T\mathbf{J}_{12}\boldsymbol{\omega}_p \tag{33}$$

and its rate of change, from (25), is given by

$$\dot{T} = \mathbf{f}^T\mathbf{v} + \mathbf{g}^T\boldsymbol{\omega} + (\mathbf{g}_2 + \mathbf{g}_p)^T\boldsymbol{\omega}_p \tag{34}$$

The system inertia matrix for this system is

$$\mathscr{M} \triangleq \begin{bmatrix} m\mathbf{1} & -\mathbf{c}^{\times} & -\mathbf{C}_{12}\mathbf{c}_2^{\times} \\ \mathbf{c}^{\times} & \mathbf{J} & \mathbf{J}_{12} \\ \mathbf{c}_2^{\times}\mathbf{C}_{21} & \mathbf{J}_{21} & \mathbf{J}_2 \end{bmatrix} \tag{35}$$

in terms of which we have

$$\mathscr{p} = \mathscr{M}\mathscr{v} \tag{36}$$
$$T = \tfrac{1}{2}\mathscr{v}^T\mathscr{M}\mathscr{v} \tag{37}$$

where

$$\mathscr{p} \triangleq \begin{bmatrix} \mathbf{p} \\ \mathbf{h} \\ \mathbf{h}_2 \end{bmatrix}; \quad \mathscr{v} \triangleq \begin{bmatrix} \mathbf{v} \\ \boldsymbol{\omega} \\ \boldsymbol{\omega}_p \end{bmatrix}; \quad \mathscr{f} \triangleq \begin{bmatrix} \mathbf{f} \\ \mathbf{g} \\ \mathbf{g}_2 + \mathbf{g}_p \end{bmatrix} \tag{38}$$

The system inertia matrix is symmetrical [recall (31)] and positive-definite (Problem 3.21). It is unfortunately time dependent. The rotation matrix $\mathbf{C}_{12}$ varies with time, as do the inertia matrices $\mathbf{J}$, $\mathbf{J}_{12}$, and $\mathbf{J}_{21}$ (see Problem 3.26). It may appear that some simplification can be provided by choosing the reference point in $\mathscr{R}_1$ to be ⊕, the mass center of the whole system; then, $\mathbf{c} = \mathbf{0}$. Unfortunately, ⊕ is not fixed in $\mathscr{R}_1$. For, from (4),

$$\mathbf{c} = \mathbf{c}_1 + \mathbf{C}_{12}\mathbf{c}_2 + m_2\mathbf{b} \tag{39}$$

Because of $\mathbf{C}_{12}$, $\mathbf{c}$ is also time dependent. There is no net advantage—indeed, matters may get worse—trying to choose ⊕ as the reference point. It is probably better to have the reference point tied down to the "home" body $\mathscr{R}_1$ than to have it "floating" in an attempt to uncouple the translational motion equation. We also note from (37) that

$$\frac{\partial T}{\partial \mathbf{v}} = \mathbf{p}; \qquad \frac{\partial T}{\partial \boldsymbol{\omega}} = \mathbf{h}; \qquad \frac{\partial T}{\partial \boldsymbol{\omega}_p} = \mathbf{h}_2 \tag{40}$$

thereby increasing our catalog of momentum–energy relationships.

The scalar motion equations (21), (23), and (20) (in matrix form for compactness and clarity, as usual), are respectively as follows:

$$\dot{\mathbf{p}} = -\boldsymbol{\omega}^{\times}\mathbf{p} + \mathbf{f} \tag{41}$$

$$\dot{\mathbf{h}} = -\boldsymbol{\omega}^{\times}\mathbf{h} - \mathbf{v}^{\times}\mathbf{p} + \mathbf{g} \tag{42}$$

$$\dot{\mathbf{h}}_2 = \left[\mathbf{h}_2^{\times} + \mathbf{C}_{21}(\mathbf{v} + \boldsymbol{\omega}^{\times}\mathbf{b})^{\times}\mathbf{C}_{12}\mathbf{c}_2^{\times}\right]\left(\mathbf{C}_{21}\boldsymbol{\omega} + \boldsymbol{\omega}_p\right) + \mathbf{g}_2 + \mathbf{g}_p \tag{43}$$

after substitution of (11) and further reduction. The last equation can be expressed in several equivalent forms, some of which may be more computationally efficient than others, but the issue here is to express the right side in terms of $\mathscr{p}$, $\mathscr{v}$, and $\mathscr{f}$. In the general form shown, these differential equations of motion have to be integrated numerically for $\{\mathbf{p}, \mathbf{h}, \mathbf{h}_2\}$ concurrently with solving the linear algebraic system (36) for $\{\mathbf{v}, \boldsymbol{\omega}, \boldsymbol{\omega}_p\}$ and integrating the appropriate rotational kinematical equations for $\mathbf{C}_{1i}$ and $\mathbf{C}_{21}$. The latter give the orientations of $\mathscr{R}_1$ and $\mathscr{R}_2$ as functions of time. Alternatively, (36) can be substituted into (41), (42), and (43), producing motion equations in terms of $\{\mathbf{v}, \boldsymbol{\omega}, \boldsymbol{\omega}_p\}$ directly. Bearing in mind the time dependence of $\mathscr{M}$, these equations are quite complex; it is not clear that the elimination of the linear algebraic equations (36) confers a net benefit.

3.7 DYNAMICS OF A SYSTEM OF RIGID BODIES

The preceding sequence of spacecraft models can be extended to accommodate N rigid bodies. This more comprehensive model could allow some "bodies" to be point masses as limiting cases, the implied loss of three rotational degrees of freedom being reflected in a corresponding reduction in system order. Rotors could also be accommodated by appropriate simplifications in the inertial properties. Such an N-body model would clearly be useful as a general-purpose instrument in the analysis and simulation of spacecraft attitude dynamics. In spite of the fact that the underlying physics is classical and undisputed, the detailed generation of such general models is more art than science and has become a subspecialty in the field. This can be traced at least in part to the large number of decisions faced by the analyst, some of which are discussed in the following paragraphs.

Modeling Decisions

The basic and most important question is how to model the spacecraft. The first step is usually to partition the vehicle into dynamically interacting substructures (antennas, rotors, solar-cell arrays, dampers, etc.). For each substructure, it

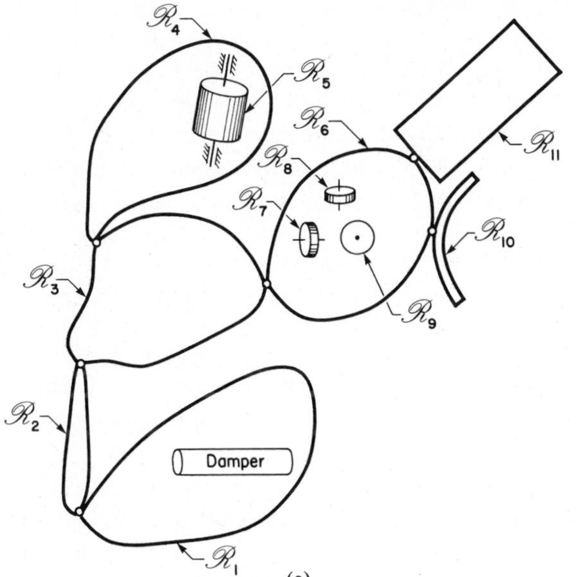

(a)

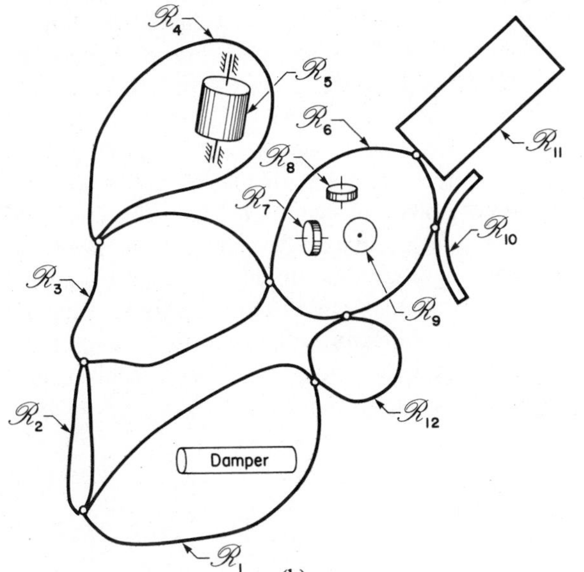

(b)

FIGURE 3.11 A system of rigid bodies. (*a*) Topological tree structure. (*b*) A closed path.

must be decided what mathematical abstraction to use (point mass, rigid body, elastic body, viscoelastic body, fluid, etc.). In this connection, it has occasionally been suggested in the literature that an elastic body might be modeled as a system of N rigid bodies connected to each other by springs. Not usually accompanying these suggestions is a recipe for choosing the inertial properties of the constituent bodies, the number of such bodies, or the strength of the mysterious interbody springs. Fortunately, this information is not much needed, for although this approximate representation of an elastic body may in some sense converge to the correct motion as $N \to \infty$, other methods are much to be preferred, at least for small deformations. The present discussion is intended to apply primarily to spacecraft that can be modeled *naturally* as a collection of point masses and rigid bodies. Many of the remarks and concepts, however, remain applicable to more sophisticated models (e.g., with elastic bodies included).

An important concept in multibody dynamical modeling is the *topological tree*. Between any two contiguous bodies there is a connection which is, in general, termed a *joint*. Each joint may impart one, two, or three *rotational* degrees of freedom (in which case it is often called a *hinge*); and, depending on the relative motions permitted between the bodies, it may also include one, two, or three *translational* degrees of freedom. At each joint, translational and rotational springs and dampers or control forces and torques are permitted in the appropriate coordinate directions. Now consider Fig. 3.11, which shows schematically two nearly identical multibody systems of rigid bodies connected by hinges. Between any body and any other body, a path can be found through hinges and bodies. The important distinction made by Fig. 3.11 is this: while in Fig. 3.11*a* there is a *unique* path between body i and body j, in Fig. 3.11*b* there are *two* distinct paths from body i to body j. Stated another way, there is a closed loop in Fig. 3.11*b*. By analogy with a botanical tree, which has smaller branches proceeding from larger branches but no closed loops, the system in Fig. 3.11*a* is said to have a *tree* topology. This concept is important in assessing the number of system degrees of freedom. Whereas in a tree topology this number is equal to six plus the number of degrees of freedom at each hinge, closed paths imply kinematical constraints and add a further measure of complexity to the analysis.

Also needful of modeling are the external force and torque on each body, and this subject is addressed for spacecraft in Chapter 8.

Formulational Decisions

It is plain from the literature on the dynamics of multibody systems that the formulation of appropriate motion equations for a system of N rigid bodies is far from straightforward. This may at first appear astonishing since motion equations for each constituent body have been known and well understood for over a century. The explanation for this paradox lies in the multiplicity of options open to the dynamicist in the management of the equations. These options include, but are not limited to, the following:

(i) Symbol selection (the analysis is sufficiently complex that notation must be chosen with care).

(ii) Coordinates for the three translational degrees of freedom (the two leading candidates are the coordinates of the system mass center and the coordinates of a reference material point in one of the bodies, normally labeled body 1).

(iii) Momenta or velocities as dynamical variables.

(iv) Absolute angular velocities vis-à-vis relative angular velocities.

(v) Reference point when writing rotational motion equations for a body or a body cluster (intuition and practice suggest that either a hinge point or a mass center be used).

(vi) Principle of mechanics used in the initial statement of motion equations.

(vii) Strategy for combining motion equations for individual bodies to form system motion equations.

(viii) Closely related to (vii), the method for eliminating unwanted constraint forces and torques at hinges.

(ix) Choice of rotational displacement parameters (see Chapter 2).

When these alternatives are concatenated, it is not surprising that the challenge of multibody dynamics has endured for three decades. It is also noteworthy that this subject did not achieve its present maturity until long after the basic motion equations were postulated. The urgent necessity of application to modern space vehicles and the promise of machine solutions were apparently required to motivate the refinement of this classical discipline—a development that subsequently found many other engineering applications.

A full exposition of multibody dynamics would require a book in itself (indeed, it already has; see [Wittenburg]). In the limited space available, we can offer only some general comments and encourage the interested reader to peruse the cited references to the aerospace literature. An early modern treatment was offered by [Abzug], who presented the vector equivalent of $6N$ scalar equations of N rigid bodies, acknowledging in a footnote that appropriate constraint equations were also necessary. Slightly later, [Hooker and Margulies] made important discoveries about the geometrical and inertial implications of a conjoined system of bodies. Their most celebrated finding calls for the temporary replacement (on paper) of each body by an *augmented body*.

An illustrative construction of an augmented body is shown in Fig. 3.12. Referring back to the topological tree of rigid bodies depicted in Fig. 3.11a, Fig. 3.12 shows augmented body 3; at each joint of body 3 is added a point mass equal to the total mass of all bodies outward from that joint. After forming N such augmented bodies, *connection barycenter* n is then defined to be the mass center of augmented body ($n = 1, \ldots, N$). Furthermore, denote by $\mathbf{I}_n^{(\Delta)}$ the inertia dyadic for augmented body n, computed about connection barycenter n.

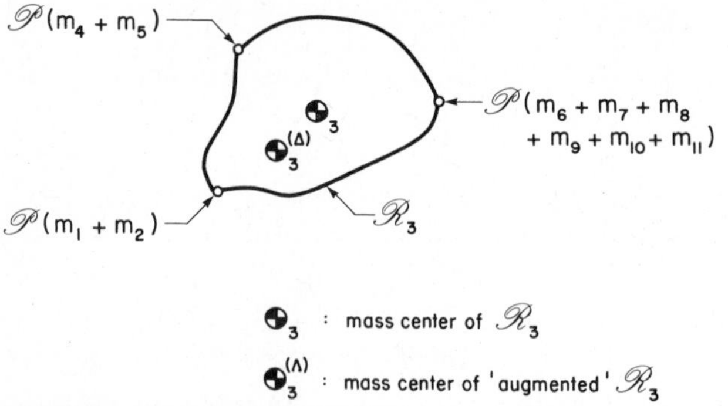

FIGURE 3.12 Augmented body $\mathscr{R}_3$ (see Fig. 3.11), with its connection barycenter.

Hooker and Margulies observed that when rotational motion equations written for each body about its mass center in terms of absolute angular velocities are appropriately summed to eliminate interbody forces at the hinges, the inertia dyadics appearing in the resulting equations are precisely $\mathbf{I}_n^{(\Delta)}$ $(n = 1, \ldots, N)$. This pattern was observed also by [Roberson and Wittenburg], who were studying this problem at about the same time. Interestingly, they used graph theory to describe the interconnectional configuration of the bodies. Their work has since been extended by [Roberson, 10] to encompass translational motion at the joints, and [Wittenburg] recently has published a complete treatment of the Roberson–Wittenburg formalism, applied to complex multibody systems, including closed loops, and with applications to spacecraft and other mechanical systems of engineering interest. The closed loops are handled with the aid of Lagrangian multipliers.

Meanwhile, [Hooker, 1] was making an essential extension to the Hooker–Margulies formulation. He showed how to eliminate constraint torques at hinges that have fewer than three rotational degrees of freedom (the usual case), whereupon the total number of Hooker–Margulies equations is reduced to the total number of degrees of freedom. A comprehensive review of this formulation has also been given by [Likins and Fleischer]; they show in detail how to apply it to spacecraft-motivated problems and extend the method by identifying modes of motion. Certain of these modes may be judiciously neglected thereby facilitating a reduction in model size.

The formulations in the last two paragraphs share several common characteristics: they all use the Newton–Euler motion equations for individual bodies as their starting point; they all use coordinates of the system mass center as the three translational degrees of freedom; and they all use the augmented-body and connection barycenter conceptions. In spite of the indisputable elegance of these formulations, however, attention is now drawn to an alternative which appears to be gaining the ascendancy. (This second methodology also happens to be preferred by the author, and Sections 3.3 through 3.6 can be regarded as simple examples of its use.) The occurrence of "augmented bodies" is closely related to the choice of system mass center to represent the translational degrees of freedom. This latter choice is presumably motivated by a desire to produce a simple translational motion equation ($m\mathbf{a}_c = \mathbf{f}$) which can, in many cases, be uncoupled from the rotational equations. This simplification is not however without cost: the remaining (rotational) equations are more complicated in their coefficients, and despite the elegant augmented-body interpretation this complexity is arguably undesirable. The alternative is to pick one body of the system as *home body*, normally labeled body 1 or body 0, with the coordinates of a material point of interest fixed in the home body representing the translational degrees of freedom. The material point of interest might be the mass center or a hinge point. The translational and rotational motion equations are now coupled, but the calculation and interpretation of the inertial coefficients are straightforward: these coefficients are the self-inertias and mixed inertias of the bodies. Expositions have been given by [Frisch, 3] and [Ho] of this approach, which Ho has called the *direct-path method*. Another aspect to the direct-path strategy is initially to write rotational motion equations (for each body) about the hing point (instead of about the body mass center) to eliminate *a priori* the interbody constraint forces. Another derivation is found in Hooker's last paper; significantly, he affirms that "the complicated topological manipulations that characterized our earlier work are no longer necessary."

The discussion thus far has emphasized the choice between augmented-body and direct-path ideas. Mention should also be made of other contemporaneous contributions. [Velman] was among the first to employ *relative* angular velocities between contiguous bodies and to select successive subgroups of bodies in building the final system equations. Constraints are eliminated by projection operators, and it is interesting to compare Velman's work in this respect with that of [Turner]. The methodology used by [Russell] is in the same vein; he isolates subsets of bodies, writing motion equations for each "cluster" in terms of the angular momenta. The use of momenta is also recommended by [Williams]. It should also be noted that while vectorial mechanics is evidently the dynamical principle chosen by most analysts, the equations produces by other principles are often identical (they must always, of course, by *equivalent*). For example, [Likins, 5] has demonstrated that the augmented-body motion equations of Hooker and Margulies also emerge from three of the principles of analytical mechanics, including the quasi-coordinate formulation of [Kane and Wang].

Structure of Motion Equations

Multibody system motion equations tend to have a structure that is independent of the method of formulation. This structure has been emphasized in Sections 3.2 through 3.6 and is typically of the following form:

$$\dot{\mathscr{p}} = \mathscr{f} + \mathscr{n}_p \tag{1}$$

$$\mathscr{p} = \mathscr{M}\mathscr{v} \tag{2}$$

$$T = \tfrac{1}{2}\mathscr{v}^T\mathscr{M}\mathscr{v} \tag{3}$$

In (1), the rate of change of momentum-like quantities is equated to the sum of the two terms; the first, $\mathscr{f}$ represents internal or external forces and torques; the second, $\mathscr{n}_p$, is usually highly nonlinear and comprises dynamical coupling terms. Examples of (1) in this chapter are (42) through (44) in Section 3.2 for a system of N point masses, (12) in Section 3.3 for a rigid body $\mathscr{R}$, (31) through (33) in Section 3.4 for $\mathscr{R}+\mathscr{P}$, (36) through (38) in Section 3.5 for $\mathscr{R}+\mathscr{W}$, and (41) through (43) in Section 3.6 for $\mathscr{R}_1+\mathscr{R}_2$. Furthermore, (2) and (3) above show how the dynamical variables $\mathscr{p}$ and T are related to the kinematical variables $\mathscr{v}$ via the system inertia matrix $\mathscr{M}$. To assess the complexity of these equations, the functional dependences of $\mathscr{f}$, $\mathscr{n}_p$, and $\mathscr{M}$ should be mentioned. If, for example, relative rotations (but not relative translations) are permitted at the hinges, the functional dependences may include the following:

$$\mathscr{f}=\mathscr{f}(\mathbf{R}_o,\mathbf{C}_n,\mathscr{v},t) \tag{4}$$

$$\mathscr{n}_p=\mathscr{n}_p(\mathbf{C}_n,\mathscr{v},\mathscr{p}) \tag{5}$$

$$\mathscr{M}=\mathscr{M}(\mathbf{C}_n) \tag{6}$$

where $\mathbf{R}_o$ is the position of the home point in the home body and the $\mathbf{C}_n$ $(n=1,\ldots,N)$ represent generically the absolute or relative rotation parameters of the N bodies. These displacement variables themselves satisfy appropriate kinematical differential equations whose generic form is

$$\dot{\mathbf{R}}_o=\boldsymbol{\phi}_o(\mathbf{R}_o,\mathscr{v}) \tag{7}$$

$$\dot{\mathbf{C}}_n=\boldsymbol{\phi}_n(\mathbf{C}_m,\mathscr{v}) \qquad (m,n=1,\ldots,N) \tag{8}$$

where $\boldsymbol{\phi}_o$ and $\boldsymbol{\phi}_n$ $(n=1,\ldots,N)$ are known functions. It is reiterated that not all the dependences listed above are present in every case. To summarize, the system to be integrated is the following:

$$\dot{\mathscr{p}}=\mathscr{f}(\mathbf{R}_o,\mathbf{C}_n,\mathscr{v},t)+\mathscr{n}_p(\mathbf{C}_n,\mathscr{v},\mathscr{p}) \tag{9}$$

$$\mathscr{p}=\mathscr{M}(\mathbf{C}_n)\mathscr{v} \tag{10}$$

$$\dot{\mathbf{R}}_o=\boldsymbol{\phi}_o(\mathbf{R}_o,\mathscr{v}) \tag{11}$$

$$\dot{\mathbf{C}}_n=\boldsymbol{\phi}_n(\mathbf{C}_m,\mathscr{v}) \tag{12}$$

The linear algebraic equation (10) forms a bridge between the dynamical and kinematical variables. In the process of solving (9) through (12) the evaluation of $\mathscr{M}^{-1}$, or its equivalent, cannot be avoided. In this evaluation, the properties of symmetry and positive definiteness should be exploited. It may also be that literal inverses are possible for important subclasses of problems.

It is customary, but perhaps not always advisable, to eliminate $\mathscr{p}$ by substitution of (10) into (9). The ensuing system of equations then reduces (in number) to the following:

$$\dot{\boldsymbol{v}} = \mathscr{M}^{-1}(\mathbf{C}_n)\boldsymbol{f}(\mathbf{R}_o, \mathbf{C}_n, \boldsymbol{v}, t) + \boldsymbol{n}_v(\mathbf{C}_n, \boldsymbol{v}) \tag{13}$$

$$\dot{\mathbf{R}}_o = \boldsymbol{\phi}_o(\mathbf{R}_o, \boldsymbol{v}) \tag{14}$$

$$\dot{\mathbf{C}}_n = \boldsymbol{\phi}_n(\mathbf{C}_m, \boldsymbol{v}) \tag{15}$$

where

$$\boldsymbol{n}_v(\mathbf{C}_n, \boldsymbol{v}) \triangleq \mathscr{M}^{-1}\left(\boldsymbol{n}_p - \dot{\mathscr{M}}\boldsymbol{v}\right) \tag{16}$$

These equations can be further reduced in number (and increased in complexity) by eliminating the translational coordinates $\mathbf{R}_o$. This is possible provided that $\boldsymbol{f}$ in the rotational members of (13) does not depend on $\mathbf{R}_o$.

For readers wishing to delve further into the structure of multibody equations, the presentation by [Jerkovsky] is recommended. Most of the remarks in this section apply also to multi-elastic-body systems.

3.8 PROBLEMS

3.1 A certain scalar, s, is defined as the inner product of two vectors:

$$s(t) \triangleq \underrightarrow{\mathbf{u}}(t) \cdot \underrightarrow{\mathbf{v}}(t) \tag{1}$$

The derivative $\dot{s}(t)$ is measured in two reference frames that are rotating with respect to each other. Using vector identities, show that both frames record the same $\dot{s}$.

3.2 A point mass $\mathscr{P}$ of mass m moves (Fig. 3.1) under the influence of a potential function $V(\underrightarrow{\mathbf{R}})$; that is, $\underrightarrow{\mathbf{f}}(\underrightarrow{\mathbf{R}}) = -dV/d\underrightarrow{\mathbf{R}}$. (The notation $d/d\underrightarrow{\mathbf{R}}$ is explained in Section B.5 of Appendix B.) Show from (6) in Section 3.1 that $T + V$ is constant for this motion.

3.3 Consider again the motion in Problem 3.2, but as observed (Fig. 3.13) in the frame $\mathscr{F}_b$ whose origin has a known motion, $\underrightarrow{\mathbf{R}}_o(t)$, and whose orientation with respect to $\mathscr{F}_i$ is given by the known rotation matrix $\mathbf{C}(t)$. Define $\underrightarrow{\mathbf{v}}_o \triangleq \dot{\underrightarrow{\mathbf{R}}}_o$ and $\boldsymbol{\omega}^\times \triangleq -\dot{\mathbf{C}}\mathbf{C}^T$, and note that $\underrightarrow{\mathbf{r}}$ is the position of $\mathscr{P}$ with respect to O. Furthermore, define components of $\underrightarrow{\mathbf{v}}_o$, $\underrightarrow{\mathbf{f}}$, and $\underrightarrow{\mathbf{r}}$ as follows:

$$\mathbf{v}_o \triangleq \mathscr{F}_i \cdot \underrightarrow{\mathbf{v}}_o; \qquad \mathbf{f} \triangleq \mathscr{F}_b \cdot \underrightarrow{\mathbf{f}}; \qquad \mathbf{r} \triangleq \mathscr{F}_b \cdot \underrightarrow{\mathbf{r}} \tag{2}$$

(*a*) Show that the kinetic energy of $\mathscr{P}$ is

$$T = T_2 + T_1 + T_0 \tag{3}$$

where

$$\begin{aligned} T_2 &\triangleq \tfrac{1}{2}m\dot{\mathbf{r}}^T\dot{\mathbf{r}} \\ T_1 &\triangleq m\left(\mathbf{v}_o^T\mathbf{C}^T - \mathbf{r}^T\boldsymbol{\omega}^\times\right)\dot{\mathbf{r}} \\ T_0 &\triangleq \tfrac{1}{2}m\left(\mathbf{v}_o^T\mathbf{v}_o - \mathbf{r}^T\boldsymbol{\omega}^\times\boldsymbol{\omega}^\times\mathbf{r}\right) + m\mathbf{v}_o^T\mathbf{C}^T\boldsymbol{\omega}^\times\mathbf{r} \end{aligned} \tag{4}$$

Note that T_2 is the kinetic energy apparent in $\mathscr{F}_b$.

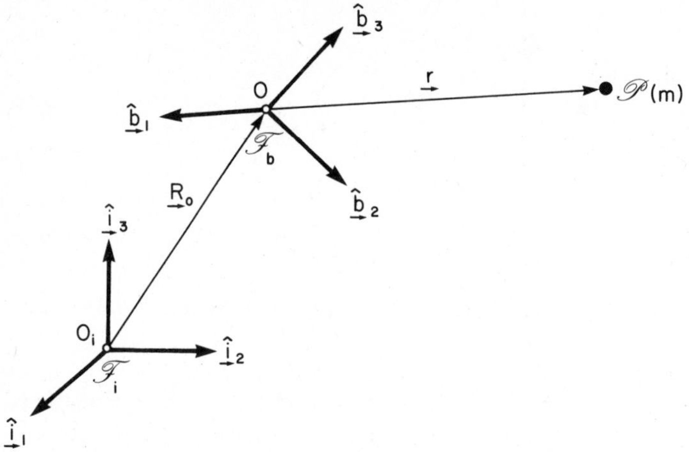

FIGURE 3.13 Point mass $\mathscr{P}$ observed from a rotating reference frame $\mathscr{F}_b$.

(*b*) Define an "apparent" potential function as follows:

$$V_a \triangleq V - T_1 - T_0 \tag{5}$$

Note that the "apparent Lagrangian," $L_a \triangleq T_2 - V_a$, is equal to $T - V$. Calculate the "apparent forces" on $\mathscr{P}$ from

$$\mathbf{f}_a \triangleq -\frac{\partial V_a}{\partial \mathbf{r}} + \frac{d}{dt}\left(\frac{\partial V_a}{\partial \dot{\mathbf{r}}}\right) \tag{6}$$

to show that

$$\mathbf{f}_a = -\frac{\partial V}{\partial \mathbf{r}} - m\left[\mathbf{C}\dot{\mathbf{v}}_o + 2\boldsymbol{\omega}^\times\dot{\mathbf{r}} + (\dot{\boldsymbol{\omega}}^\times + \boldsymbol{\omega}^\times\boldsymbol{\omega}^\times)\mathbf{r}\right] \tag{7}$$

(*c*) Is $T_2 + V_a$ constant? Define the "apparent Hamiltonian" as follows:

$$H_a \triangleq \dot{\mathbf{r}}^T\left[\frac{\partial L}{\partial \dot{\mathbf{r}}}\right] - L \tag{8}$$

where $L = L_a = T - V$. Show that

$$H_a = T_2 + (V - T_0) \tag{9}$$

and that H_a is a constant of the motion.

3.4 Generalize the results of Problem 3.3 to a system of N point masses.

3.5 (*a*) Verify that (32) in Section 3.2 follows from (17) in that section.
(*b*) Verify that (33) in Section 3.2 follows from (25) in that section.
(*c*) Verify (49) in Section 3.2.

3.6 For a system of N point masses, show that

$$T = \frac{1}{2}\left(\mathbf{p}^T\mathbf{v}_o + \mathbf{h}_o^T\boldsymbol{\omega} + \sum_{n=1}^{N}\mathbf{p}_n^T\dot{\mathbf{r}}_n\right) \tag{10}$$

The meaning of the symbols is given in Section 3.2.

3.7 Referring to Fig. 3.14, consider a rigid body $\mathscr{R}$ with inertia dyadic $\underset{\to}{\mathbf{J}}$ with respect to the point O, and a unit vector at O, $\underset{\to}{\mathbf{a}}$.

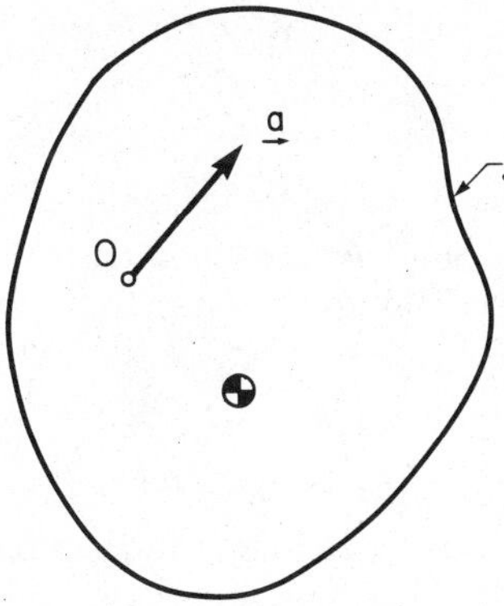

FIGURE 3.14 Rigid body $\mathscr{R}$ with direction of interest $\underrightarrow{\mathbf{a}}$ and point of interest O.

(*a*) Show that the moment of inertia of $\mathscr{R}$ about $\underrightarrow{\mathbf{a}}$ at O is $\underrightarrow{\mathbf{a}} \cdot \underrightarrow{\mathbf{J}} \cdot \underrightarrow{\mathbf{a}}$.

(*b*) Suppose further that O is the mass center of $\mathscr{R}$, $O \equiv$ ⊕, and $\underrightarrow{\mathbf{a}}$ is an axis of inertial symmetry. Denote by I_a the moment of inertia about $\underrightarrow{\mathbf{a}}$ at ⊕, and denote by I_t the moment of inertia about any axis transverse to $\underrightarrow{\mathbf{a}}$ at ⊕. Show that

$$\underrightarrow{\mathbf{I}} = I_t \underrightarrow{\mathbf{1}} + (I_a - I_t)\underrightarrow{\mathbf{a}}\underrightarrow{\mathbf{a}} \tag{11}$$

3.8 This problem studies the inertial matrix $\mathbf{J}$ for a rigid body $\mathscr{R}$ at point O, which need not coincide with the mass center (Fig. 3.15).

(*a*) Show that

$$\mathbf{J} = -\int_{\mathscr{R}} \mathbf{r}^{\times}\mathbf{r}^{\times}\, dm \tag{12}$$

(*b*) Show that $\mathbf{J}$ is symmetric and positive-definite.

(*c*) With the properties shown in (*b*), it is known from linear algebra that $\mathbf{J}$ has three real positive eigenvalues, which we shall denote J_1, J_2, and J_3, and three associated real orthogonal eigencolumns. The latter define the *principal axes* for $\mathbf{J}$. Denote by $\mathscr{F}_p$ a reference frame fixed in $\mathscr{R}$,

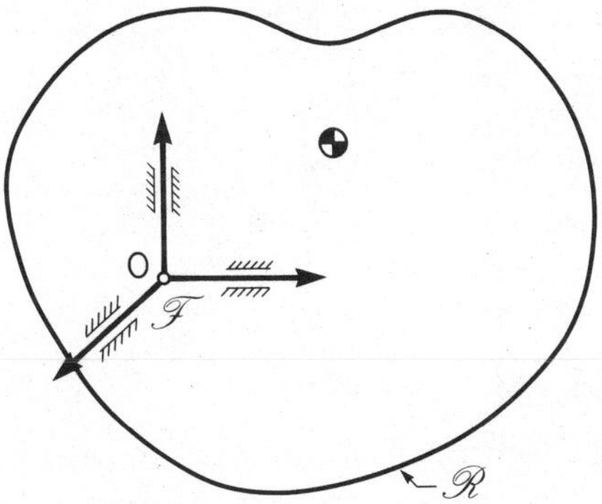

FIGURE 3.15 Rigid body $\mathscr{R}$ with point of interest O and frame of interest $\mathscr{F}$, both fixed in $\mathscr{R}$.

with origin O, and coincident with the principal axes at O. Show that, expressed in $\mathscr{F}_p$,

$$\mathbf{J} = \mathbf{J}_p \triangleq \text{diag}\{J_1, J_2, J_3\} \tag{13}$$

(*d*) For a general (not necessarily principal) body-fixed reference frame $\mathscr{F}_b$ at O, denote the elements of $\mathbf{J}$ by $\{J_{ij}\}$. Since $\mathbf{J}$ is symmetric and positive-definite, deduce from Sylvester's inequalities that

$$\begin{gathered} J_{11} > 0; \qquad J_{22} > 0; \qquad J_{33} > 0 \\ J_{11}J_{22} - J_{12}^2 > 0; \qquad J_{22}J_{33} - J_{23}^2 > 0; \qquad J_{33}J_{11} - J_{31}^2 > 0 \\ J_{11}J_{22}J_{33} + 2J_{12}J_{23}J_{31} - J_{11}J_{23}^2 - J_{22}J_{31}^2 - J_{33}J_{12}^2 > 0 \end{gathered} \tag{14}$$

(*e*) The eigenvalues of $\mathbf{J}$, and therefore the characteristic equation for $\mathbf{J}$, must be independent of the orientation of $\mathscr{F}_p$. Use this observation to show that

$$\begin{gathered} J_{11} + J_{22} + J_{33} = J_1 + J_2 + J_3 \\ J_{11}J_{22} + J_{22}J_{33} + J_{33}J_{11} - J_{12}^2 - J_{23}^2 - J_{31}^2 = J_1J_2 + J_2J_3 + J_3J_1 \\ J_{11}J_{22}J_{33} + 2J_{12}J_{23}J_{31} - J_{11}J_{23}^2 - J_{22}J_{31}^2 - J_{33}J_{12}^2 = J_1J_2J_3 \end{gathered} \tag{15}$$

(*f*) Show that

$$J_{11} + J_{22} > J_{33}; \qquad J_{22} + J_{33} > J_{11}; \qquad J_{33} + J_{11} > J_{22} \tag{16}$$

These are the so-called *triangle inequalities*.

(*g*) Show that

$$J_{11}^2 > 4J_{23}; \qquad J_{22}^2 > 4J_{31}; \qquad J_{33}^2 > 4J_{12} \tag{17}$$

3.9 Figure 3.16 shows three body-fixed reference frames: $\mathscr{F}_1$, $\mathscr{F}_2$, and $\mathscr{F}_3$. $\mathscr{F}_1$ and $\mathscr{F}_2$ are parallel but have different origins; $\mathscr{F}_2$ and $\mathscr{F}_3$ have the same origin, but there is an angular displacement between them. The vector $\underrightarrow{\mathbf{b}}$ is also shown. Denote the inertia matrices expressed in $\mathscr{F}_i$ by $\mathbf{J}_i$.

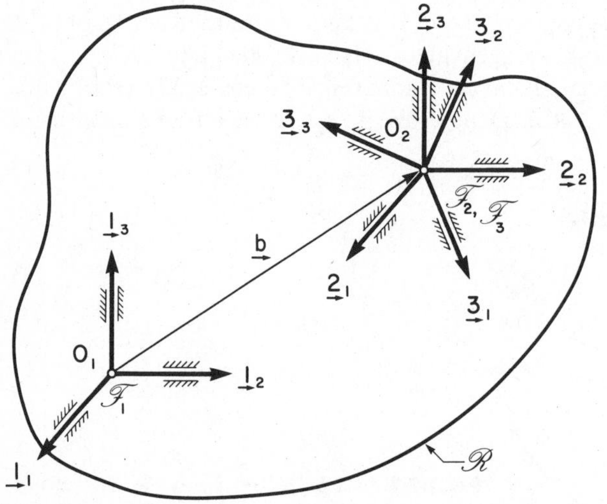

FIGURE 3.16 Rigid body with three body-fixed reference frames: $\mathscr{F}_1, \mathscr{F}_2, \mathscr{F}_3$. (See Problem 3.9.)

(*a*) Prove the *parallel-axis theorem*:

$$\mathbf{J}_1 = \mathbf{J}_2 + m(b^2\mathbf{1} - \mathbf{b}\mathbf{b}^T) + 2\mathbf{c}^T\mathbf{b}\mathbf{1} - \mathbf{b}\mathbf{c}^T - \mathbf{c}\mathbf{b}^T \tag{18}$$

where $\mathbf{b} \triangleq \mathscr{F}_1 \cdot \underset{\rightarrow}{\mathbf{b}} \equiv \mathscr{F}_2 \cdot \underset{\rightarrow}{\mathbf{b}}$, $\mathbf{c} \triangleq \mathscr{F}_1 \cdot \underset{\rightarrow}{\mathbf{c}} \equiv \mathscr{F}_2 \cdot \underset{\rightarrow}{\mathbf{c}}$, and $\underset{\rightarrow}{\mathbf{c}}$ is the first moment of inertia of $\mathscr{R}$ with respect to O_2. In particular, if O_2 is the mass center of $\mathscr{R}$, show that

$$\mathbf{J}_1 = \mathbf{I} + m(b^2\mathbf{1} - \mathbf{b}\mathbf{b}^T) \tag{19}$$

(*b*) Show, using the vectrix notation of Appendix B (or otherwise), that

$$\mathbf{J}_3 = \mathbf{C}_{32}\mathbf{J}_2\mathbf{C}_{23} \tag{20}$$

where $\mathbf{C}_{32} = \mathscr{F}_3 \cdot \mathscr{F}_2^T$.

(*c*) Given $\mathbf{J}_3$, $\mathbf{b}$, $\mathbf{C}_{32}$, and $\mathbf{c}$, calculate $\mathbf{J}_1$.

3.10 Denote by $\mathbf{I}$ the centroidal inertia matrix for $\mathscr{R}$ with respect to the principal-axis reference frame $\mathscr{F}_p$:

$$\mathbf{I} \triangleq \text{diag}\{I_1, I_2, I_3\} \tag{21}$$

(*a*) Show that the components of the angular momentum in $\mathscr{F}_p$ are [see (16) in Section 3.3]

$$\mathbf{h} = [I_1\omega_1 \quad I_2\omega_2 \quad I_3\omega_3]^T \tag{22}$$

(*b*) Show that the rotational kinetic energy is [see (23*c*) in Section 3.2]

$$T_{\text{rot}} = \frac{1}{2}\left(I_1\omega_1^2 + I_2\omega_2^2 + I_3\omega_3^2\right) \tag{23}$$

(*c*) If $\mathbf{I}$ is expressed instead in a frame whose axes do not coincide with principal axes, what is the expression for $\mathbf{h}$? For T_{rot}? How must the motion equations (20) in Section 3.3 be modified?

3.11 Show directly from (20) in Section 3.3 that in the absence of external torques the angular momentum and energy integrals for $\mathscr{R}$ are

$$h^2 = I_1^2\omega_1^2 + I_2^2\omega_2^2 + I_3^2\omega_3^2 = \text{constant}$$
$$2T_{\text{rot}} = I_1\omega_1^2 + I_2\omega_2^2 + I_3\omega_3^2 = \text{constant} \tag{24}$$

3.12 Consider the general expression for the kinetic energy of a rigid body $\mathscr{R}$, given by (10) in Section 3.3:

$$T = \tfrac{1}{2}m\underset{\rightarrow}{\mathbf{v}}_o \cdot \underset{\rightarrow}{\mathbf{v}}_o + \underset{\rightarrow}{\boldsymbol{\omega}} \cdot \underset{\rightarrow}{\mathbf{c}} \times \underset{\rightarrow}{\mathbf{v}}_o + \tfrac{1}{2}\underset{\rightarrow}{\boldsymbol{\omega}} \cdot \underset{\rightarrow}{\mathbf{J}} \cdot \underset{\rightarrow}{\boldsymbol{\omega}} \tag{25}$$

(*a*) By writing (25) for two reference points in $\mathscr{R}$, O_1, and O_2, show that T is independent of O.

(*b*) Show from (25) and the motion equations that

$$\dot{T} = \underset{\rightarrow}{\mathbf{f}} \cdot \underset{\rightarrow}{\mathbf{v}}_o + \underset{\rightarrow}{\mathbf{g}}_o \cdot \underset{\rightarrow}{\boldsymbol{\omega}} \tag{26}$$

(*c*) Show from (26) that $\dot{T}$ is independent of the reference point O.

3.13 A rigid body $\mathscr{R}$ is subjected to a conservative "body" force in which the potential per unit volume is $\phi(\underset{\rightarrow}{\mathbf{r}})$ and the total potential is

$$V \triangleq \int_{\mathscr{R}} \phi(\underset{\rightarrow}{\mathbf{r}})\, dv \tag{27}$$

The resulting force on a volume element dv is

$$d\underset{\rightarrow}{\mathbf{f}} = -\underset{\rightarrow}{\nabla}\phi\, dv \tag{28}$$

(The symbol $\underset{\rightarrow}{\nabla}$ stands for $d/d\underset{\rightarrow}{\mathbf{r}}$, a notation explained in Section B.5, Appendix B.)

(*a*) Show that

$$\underset{\rightarrow}{\mathbf{f}} \triangleq \int_{\mathscr{R}} d\underset{\rightarrow}{\mathbf{f}} = -\oint \phi \, d\underset{\rightarrow}{\mathbf{S}} \tag{29}$$

$$\underset{\rightarrow}{\mathbf{g}}_o \triangleq \int_{\mathscr{R}} \underset{\rightarrow}{\mathbf{r}} \times d\underset{\rightarrow}{\mathbf{f}} = -\oint \phi \underset{\rightarrow}{\mathbf{r}} \times d\underset{\rightarrow}{\mathbf{S}} \tag{30}$$

where $d\underset{\rightarrow}{\mathbf{S}}$ is on outwardly directed normal vector at the surface of $\mathscr{R}$, with magnitude equal to that of the element of surface area dS.

(*b*) Show further that

$$\underset{\rightarrow}{\mathbf{f}} \cdot \underset{\rightarrow}{\mathbf{v}}_o + \underset{\rightarrow}{\mathbf{g}}_o \cdot \underset{\rightarrow}{\boldsymbol{\omega}} = -\dot{V}$$

$$Hint\text{: } \oint \phi \underset{\rightarrow}{\mathbf{v}} \cdot d\underset{\rightarrow}{\mathbf{S}} = \frac{d}{dt}\int_{\mathscr{R}} \phi \, dv \tag{31}$$

(*c*) Compare (31) and (26). What do you conclude?

3.14 (*a*) Show from (12) and (13) in Section 3.3 that the general motion equations for a rigid body may be written

$$m\dot{\mathbf{v}}_o - \mathbf{c}^\times \dot{\boldsymbol{\omega}} = \mathbf{f} - \boldsymbol{\omega}^\times \boldsymbol{\omega}^\times \mathbf{c}$$
$$\mathbf{c}^\times \dot{\mathbf{v}}_o + \mathbf{J}\dot{\boldsymbol{\omega}} = \mathbf{g}_o - \boldsymbol{\omega}^\times \mathbf{J}\boldsymbol{\omega} \tag{32}$$

(*b*) Using the notation of (25) through (27) in Section 3.3, show that (32) above can be rewritten in the compact form

$$\dot{\mathscr{v}} = \mathscr{M}^{-1}\mathscr{f} + \mathscr{n}_v \tag{33}$$

(*c*) Verify that the 6×6 matrix $\mathscr{M}$, given by (25) in Section 3.3, has the explicit inverse

$$\mathscr{M}^{-1} = \begin{bmatrix} m^{-1}\mathbf{1} - \mathbf{r}_c^\times \mathbf{I}^{-1}\mathbf{r}_c^\times & \mathbf{r}_c^\times \mathbf{I}^{-1} \\ -\mathbf{I}^{-1}\mathbf{r}_c^\times & \mathbf{I}^{-1} \end{bmatrix} \tag{34}$$

where $\mathbf{r}_c = \mathbf{c}/m$ and $\mathbf{I}$ is the inertia matrix for axes parallel to $\mathscr{F}_b$ but with origin at ⊕; thus, $\mathbf{I} = \mathbf{J} + m\mathbf{r}_c^\times \mathbf{r}_c^\times \equiv \mathbf{J} + \mathbf{c}^\times \mathbf{c}^\times/m$. (*Harder problem*: *Derive* (34).)

(*d*) Show that $\mathscr{n}_v$ in (33) is given by

$$\mathscr{n}_v = \begin{bmatrix} \boldsymbol{\omega}^\times \mathbf{r}_c^\times \boldsymbol{\omega} - \mathbf{r}_c^\times \mathbf{I}^{-1}\boldsymbol{\omega}^\times \mathbf{I}\boldsymbol{\omega} \\ -\mathbf{I}^{-1}\boldsymbol{\omega}^\times \mathbf{I}\boldsymbol{\omega} \end{bmatrix} \tag{35}$$

(*e*) Explain how (32), (34), and (35) simplify when $\mathbf{c} = \mathbf{0}$, that is, when the mass center is chosen as origin.

(*f*) Try to show that $\mathscr{M}$, given by (25) in Section 3.3, is positive-definite, by working with the elements of $\mathscr{M}$ themselves, that is, without using the kinetic-energy argument centering around (28) in Section 3.3.

3.15 For a rigid body with a point mass damper, the $\mathscr{R} + \mathscr{P}$ system of Section 3.4, fill in the missing steps in the derivation of the kinetic-energy expression (11) in Section 3.4.

3.16 The system inertia matrix for $\mathscr{R} + \mathscr{P}$ is given by (25) in Section 3.4. Can you find a literal inverse for this 7×7 matrix? Compare with (34) above.

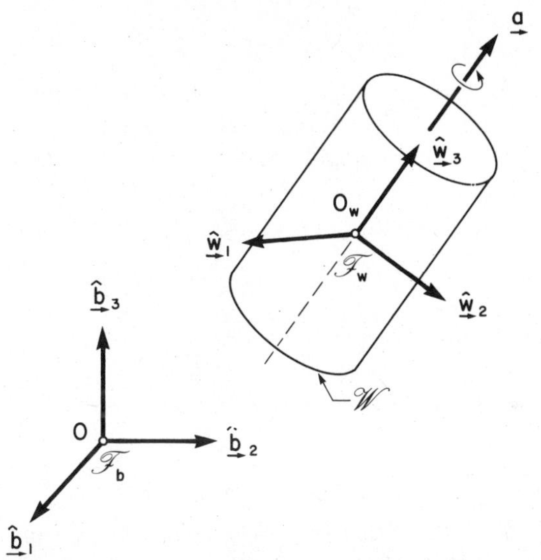

FIGURE 3.17 Inertially symmetrical wheel (or rotor). (See Problem 3.17.)

3.17 Consider a rotor $\mathscr{W}$ rotating about an axis (unit vector $\underset{\rightarrow}{a}$) that is fixed with respect to $\mathscr{F}_b$ (Fig. 3.17). Let $\mathscr{F}_w$ be a $\mathscr{W}$-fixed frame aligned with $\underset{\rightarrow}{a}$ as shown. The inertia matrix for $\mathscr{W}$ is given in $\mathscr{F}_w$ by

$$\mathscr{F}_w \cdot \underset{\rightarrow}{\mathbf{I}}_w \cdot \mathscr{F}_w^T = \begin{bmatrix} I_t & 0 & 0 \\ 0 & I_t & 0 \\ 0 & 0 & I_s \end{bmatrix} \tag{36}$$

Show (the result of Problem 3.7*b* may be useful) that $\underset{\rightarrow}{\mathbf{I}}_w$ is expressed in $\mathscr{F}_b$ by

$$\mathscr{F}_b \cdot \underset{\rightarrow}{\mathbf{I}}_w \cdot \mathscr{F}_b^T \triangleq \mathbf{I}_w = I_t\mathbf{1} + (I_s - I_t)\mathbf{a}\mathbf{a}^T \tag{37}$$

where $\mathbf{a} \triangleq \mathscr{F}_b \cdot \underset{\rightarrow}{a}$. On physical grounds, do you expect $\mathbf{I}_w$ to depend on the rotation of $\mathscr{F}_w$ about $\underset{\rightarrow}{a}$? Does this expectation concur with (37)?

3.18 For the $\mathscr{R} + \mathscr{W}$ system of Section 3.5, show that the rate of change of kinetic energy, $\dot{T}$, is given by (24) in Section 3.5 by differentiating the expression for T, (15) in Section 3.5, and inserting the motion equations.

3.19 Still with the $\mathscr{R} + \mathscr{W}$ system, consider the system inertia matrix given by (30) in Section 3.5.
(*a*) Show that $\mathscr{M}$ is positive definite.
(*b*) Can you find a literal inverse for $\mathscr{M}$?

3.20 For the $\mathscr{R}_1 + \mathscr{R}_2$ system of Section 3.6, show that the rate of change of kinetic energy, $\dot{T}$, is given by (34) in Section 3.6 by differentiating the expression for T, (33) in Section 3.6, and inserting the motion equations.

3.21 Still with the $\mathscr{R}_1 + \mathscr{R}_2$ system, consider the system inertia matrix $\mathscr{M}$ given by (35) in Section 3.6.

(*a*) Show that $\mathcal{M}$ is positive-definite.
(*b*) Can you find a literal inverse for $\mathcal{M}$?

3.22 Suppose in the development of motion equations for $\mathcal{R}_1 + \mathcal{R}_2$ (Section 3.6) that the absolute angular velocity of $\mathcal{R}_2$, namely, $\boldsymbol{\omega}_2$, were used instead of the relative angular velocity $\underrightarrow{\omega}_p$. What would the new motion equations be? What would the new system inertia matrix $\mathcal{M}$ be?

3.23 Fill in the missing details in the derivation of (16) in Section 3.6.

3.24 Show that $\underrightarrow{\mathbf{J}}$, given by (5) in Section 3.6, may be expressed as follows:

$$\underrightarrow{\mathbf{J}} = \underrightarrow{\mathbf{J}}_1 + \left[\underrightarrow{\mathbf{J}}_2 - \frac{c_2^2 \underrightarrow{\mathbf{1}} - \underrightarrow{\mathbf{c}}_2 \underrightarrow{\mathbf{c}}_2}{m_2} \right] + \left[\frac{(m_2 \underrightarrow{\mathbf{b}} + \underrightarrow{\mathbf{c}}_2) \cdot (m_2 \underrightarrow{\mathbf{b}} + \underrightarrow{\mathbf{c}}_2) \underrightarrow{\mathbf{1}} - (m_2 \underrightarrow{\mathbf{b}} + \underrightarrow{\mathbf{c}}_2)(m_2 \underrightarrow{\mathbf{b}} + \underrightarrow{\mathbf{c}}_2)}{m_2} \right] \tag{38}$$

Interpret (38) in terms of the parallel-axis theorem of Problem 3.9*a*.

3.25 Consider further the system $\mathcal{R}_1 + \mathcal{R}_2$ of Section 3.6, and suppose now that $\mathcal{R}_2$ is constrained to rotate with respect to $\mathcal{R}_1$ about a single-axis hinge. The axis of the hinge is defined by the unit vector $\underrightarrow{\mathbf{a}}$, and $\mathbf{a} = \mathcal{F}_2 \cdot \underrightarrow{\mathbf{a}}$. Furthermore, define the following 7×9 matrix:

$$\mathbf{U} \triangleq \begin{bmatrix} \mathbf{1}_{3\times 3} & \mathbf{O}_{3\times 3} & \mathbf{O}_{3\times 3} \\ \mathbf{O}_{3\times 3} & \mathbf{1}_{3\times 3} & \mathbf{O}_{3\times 3} \\ \mathbf{0}^T & \mathbf{0}^T & \mathbf{a}^T \end{bmatrix} \tag{39}$$

(*a*) Show that $\mathbf{U}\mathbf{U}^T$ is the 7×7 unit matrix.
(*b*) The new velocity and momentum variables are

$$\boldsymbol{\mathcal{v}} \triangleq \begin{bmatrix} \mathbf{v} \\ \boldsymbol{\omega} \\ \omega_a \end{bmatrix}; \qquad \boldsymbol{\mathcal{p}} \triangleq \begin{bmatrix} \mathbf{p} \\ \mathbf{h} \\ h_a \end{bmatrix} \tag{40}$$

where $\omega_a = \mathbf{a}^T \boldsymbol{\omega}_p$ is the angular speed of $\mathcal{R}_2$ with respect to $\mathcal{R}_1$ and $h_a = \mathbf{a}^T \mathbf{h}_2$. In other symbols, $\boldsymbol{\mathcal{v}}_a = \mathbf{U}\boldsymbol{\mathcal{v}}$ and $\boldsymbol{\mathcal{p}}_a = \mathbf{U}\boldsymbol{\mathcal{p}}$. Show that

$$\boldsymbol{\mathcal{p}}_a = \mathcal{M}_a \boldsymbol{\mathcal{v}}_a \tag{41}$$

$$T = \tfrac{1}{2} \boldsymbol{\mathcal{v}}_a^T \mathcal{M}_a \boldsymbol{\mathcal{v}}_a \tag{42}$$

where

$$\mathcal{M}_a \triangleq \mathbf{U}\mathcal{M}\mathbf{U}^T = \begin{bmatrix} m\mathbf{1} & -\mathbf{c}^{\times} & -\mathbf{C}_{12}\mathbf{c}_2^{\times}\mathbf{a} \\ \mathbf{c}^{\times} & \mathbf{J} & \mathbf{J}_{12}\mathbf{a} \\ \mathbf{a}^T\mathbf{c}_2^{\times}\mathbf{C}_{21} & \mathbf{a}^T\mathbf{J}_{21} & J_a \end{bmatrix} \tag{43}$$

and $J_a \triangleq \mathbf{a}^T \mathbf{J}_2 \mathbf{a}$.

(*c*) Write the motion equations for this single-axis-hinge system and compare with (41) through (43) in Section 3.6.

3.26 For the $\mathcal{R}_1 + \mathcal{R}_2$ system, the relevant inertia dyadics $\{\underrightarrow{\mathbf{J}}_1, \underrightarrow{\mathbf{J}}_2, \underrightarrow{\mathbf{J}}, \underrightarrow{\mathbf{J}}_{12}, \underrightarrow{\mathbf{J}}_{12}\}$ were given by (3) through (8) in Section 3.6. Expressing these dyadics in

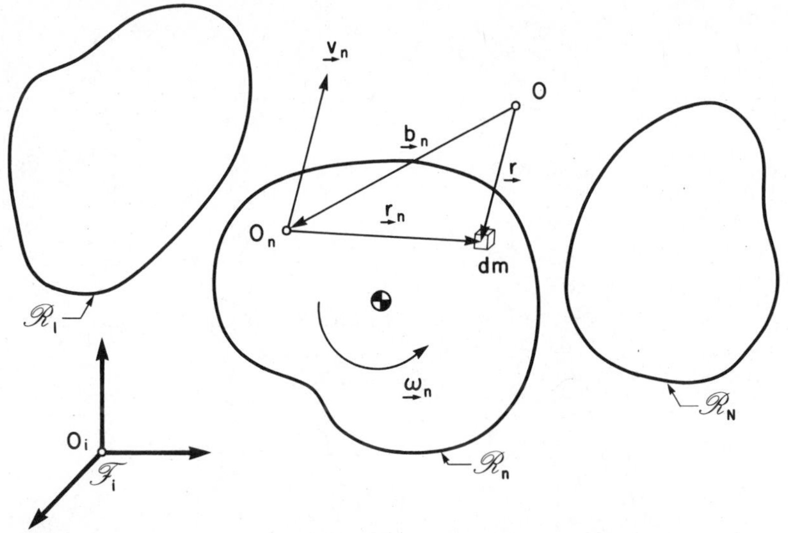

FIGURE 3.18 System of N rigid bodies. (See Problem 3.27.)

frames $\mathscr{F}_1$ and $\mathscr{F}_2$ as indicated in Section 3.6, show that

$$\mathbf{J}_1 = \int_{\mathscr{R}_1} (r^2\mathbf{1} - \mathbf{r}\mathbf{r}^T)\, dm; \qquad \mathbf{J}_2 = \int_{\mathscr{R}_2} \left(r_2^2\mathbf{1} - \mathbf{r}_2\mathbf{r}_2^T\right) dm \tag{44}$$

$$\begin{aligned} \mathbf{J} = \mathbf{J}_1 &+ \mathbf{C}_{12}\mathbf{J}_2\mathbf{C}_{21} + m_2(b^2\mathbf{1} - \mathbf{b}\mathbf{b}^T) \\ &+ \left(2\mathbf{b}^T\mathbf{C}_{12}\mathbf{c}_2 - \mathbf{b}\mathbf{c}_2^T\mathbf{C}_{21} - \mathbf{C}_{12}\mathbf{c}_2\mathbf{b}^T\right) \end{aligned} \tag{45}$$

$$\begin{aligned} \mathbf{J}_{12} &= \mathbf{C}_{12}\mathbf{J}_2 + \mathbf{C}_{12}\mathbf{c}_2^T\mathbf{C}_{21}\mathbf{b} - \mathbf{C}_{12}\mathbf{c}_2\mathbf{b}^T\mathbf{C}_{12} \\ \mathbf{J}_{21} &= \mathbf{J}_2\mathbf{C}_{21} + \mathbf{b}^T\mathbf{C}_{12}\mathbf{c}_2\mathbf{C}_{21} - \mathbf{C}_{21}\mathbf{b}\mathbf{c}_2^T\mathbf{C}_{21} \end{aligned} \tag{46}$$

Note in particular that $\mathbf{J}$, $\mathbf{J}_{12}$, and $\mathbf{J}_{21}$ are time dependent and that $\mathbf{J}_{12}^{\mathrm{T}} = \mathbf{J}_{21}$.

3.27 A collection of N rigid bodies, $\mathscr{R}_1, \ldots, \mathscr{R}_N$ is shown in Fig. 3.18. A reference point O_n is chosen in $\mathscr{R}_n$, fixed in $\mathscr{R}_n$. Denote by $\underrightarrow{\mathbf{r}}_n$ the location with respect to O_n of a mass element dm, and by $\underrightarrow{\mathbf{r}}$ the location of the same mass element with respect to O, a general reference point. The location of O_n with respect to O is $\underrightarrow{\mathbf{b}}_n$. The point O_n has a velocity $\underrightarrow{\mathbf{v}}_n$ with respect to O_i, and $\mathscr{R}_n$ has angular velocity $\underrightarrow{\boldsymbol{\omega}}_n$ with respect to $\mathscr{F}_i$. Therefore, the velocity distribution is

$$\underrightarrow{\mathbf{v}}(\underrightarrow{\mathbf{r}}) = \underrightarrow{\mathbf{v}}_n + \underrightarrow{\boldsymbol{\omega}}_n \times \underrightarrow{\mathbf{r}}_n \qquad (\underrightarrow{\mathbf{r}}, \underrightarrow{\mathbf{r}}_n \in \mathscr{R}_n)$$

Define the following momenta and (absolute) angular momenta:

$$\underrightarrow{\mathbf{p}}_n \triangleq \int_{\mathscr{R}_n} \underrightarrow{\mathbf{v}}\, dm; \qquad \underrightarrow{\mathbf{p}} \triangleq \int_{\Sigma\mathscr{R}_n} \underrightarrow{\mathbf{v}}\, dm$$

$$\underrightarrow{\mathbf{h}}_n \triangleq \int_{\mathscr{R}_n} \underrightarrow{\mathbf{r}}_n \times \underrightarrow{\mathbf{v}}\, dm; \qquad \underrightarrow{\mathbf{h}} \triangleq \int_{\Sigma\mathscr{R}_n} \underrightarrow{\mathbf{r}} \times \underrightarrow{\mathbf{v}}\, dm$$

where $\Sigma\mathscr{R}_n$ stands for $\mathscr{R}_1 + \cdots + \mathscr{R}_N$.

(*a*) Show that

(i) $$\underset{\rightarrow}{\mathbf{p}} = \sum_{n=1}^{N} \underset{\rightarrow}{\mathbf{p}}_n \tag{47}$$

(ii) $$\underset{\rightarrow}{\mathbf{h}} = \sum_{n=1}^{N} (\underset{\rightarrow}{\mathbf{h}}_n + \underset{\rightarrow}{\mathbf{b}}_n \times \underset{\rightarrow}{\mathbf{p}}_n) \tag{48}$$

(*b*) Show that

(i) $$\underset{\rightarrow}{\mathbf{p}}_n = m_n \underset{\rightarrow}{\mathbf{v}}_n + \underset{\rightarrow}{\boldsymbol{\omega}}_n \times \underset{\rightarrow}{\mathbf{c}}_n \tag{49}$$

(ii) $$\underset{\rightarrow}{\mathbf{h}}_n = \underset{\rightarrow}{\mathbf{c}}_n \times \underset{\rightarrow}{\mathbf{v}}_n + \underset{\rightarrow}{\mathbf{J}}_n \cdot \underset{\rightarrow}{\boldsymbol{\omega}}_n \tag{50}$$

where m_n is the mass of $\mathscr{R}_n$, and $\underset{\rightarrow}{\mathbf{c}}_n$ and $\underset{\rightarrow}{\mathbf{J}}_n$ are the first and second moments of inertia of $\mathscr{R}_n$ about O_n.

(*c*) Show that the kinetic energy of the system is

$$T = \sum_{n=1}^{N} \left[\frac{1}{2} m_n \underset{\rightarrow}{\mathbf{v}}_n \cdot \underset{\rightarrow}{\mathbf{v}}_n + \underset{\rightarrow}{\mathbf{v}}_n \cdot \underset{\rightarrow}{\boldsymbol{\omega}}_n \times \underset{\rightarrow}{\mathbf{c}}_n + \frac{1}{2} \underset{\rightarrow}{\boldsymbol{\omega}}_n \cdot \underset{\rightarrow}{\mathbf{J}}_n \cdot \underset{\rightarrow}{\boldsymbol{\omega}}_n \right]$$
$$= \frac{1}{2} \sum_{n=1}^{N} \left(\underset{\rightarrow}{\mathbf{v}}_n \cdot \underset{\rightarrow}{\mathbf{p}}_n + \underset{\rightarrow}{\boldsymbol{\omega}}_n \cdot \underset{\rightarrow}{\mathbf{h}}_n \right) \tag{51}$$

3.28 For the motion of a single rigid body, suppose the external force $\underset{\rightarrow}{\mathbf{f}}$ depends only on the position and velocity of the mass center (and possibly also on time):

$$\underset{\rightarrow}{\mathbf{f}} = \underset{\rightarrow}{\mathbf{f}}(\underset{\rightarrow}{\mathbf{R}}_c, \underset{\rightarrow}{\mathbf{v}}_c, t)$$

Show that the coupling of the translational motion equation, (17*a*) in Section 3.3, to the rotational motion equation, (17*b*) in Section 3.3, can be eliminated by choosing the reference point O to be the mass center ⊕, and expressing the translation-related vectors ($\mathbf{p}, \underset{\rightarrow}{\mathbf{R}}_c, \underset{\rightarrow}{\mathbf{v}}_c, \underset{\rightarrow}{\mathbf{f}}$) in $\mathscr{F}_i$ instead of $\mathscr{F}_b$. Specifically, with this new interpretation for $\mathbf{v}_c$ and $\mathbf{R}_c$, show that (17*a*) and (18*a*) in Section 3.3 are respectively replaced by

$$m\dot{\mathbf{v}}_c = \mathbf{f}(\mathbf{R}_c, \mathbf{v}_c, t)$$
$$\dot{\mathbf{R}}_c = \mathbf{v}_c \tag{52}$$

so that the translational motion can now be solved for independently.

CHAPTER 4

ATTITUDE DYNAMICS OF A RIGID BODY

In approaching the attitude dynamics of spacecraft, the importance of understanding the motion of a single rigid body cannot be overemphasized. This is true not because it is generally advisable to model spacecraft as strictly rigid bodies, but because rigid-body motion is the reference behavior to which more general motion can usually be compared. To make an analogy, one normally starts a study of music with the scale of C major not because most musical compositions are written in C major but because C major is the logical place to begin.

Most of this chapter is concerned with *free* (i.e., torque-free) attitude motion. The results obtained are immediately applicable to spacecraft in situations where the relatively weak torques in space (Chapter 8) are small compared to the other dynamical terms in the motion equations. In fact, it is often the object of stabilization schemes to bring about this condition. The analysis strategy followed under these circumstances, to be described in more detail in Chapters 10 and 11, is to regard external torques as weak disturbances causing perturbations to a torque-free motion that itself has desirable stability properties. These stability properties are the chief subject of this and the next three chapters.

4.1 BASIC MOTION EQUATIONS

The motion equations for a rigid body $\mathscr{R}$ were derived in Section 3.3, and the required kinematical relations were supplied in Chapter 2. Because the system comprises $\mathscr{R}$ only (Fig. 4.1), we are free to choose the mass center as the origin of a body-fixed reference frame $\mathscr{F}_b$. The vectrix associated with $\mathscr{F}_b$ is denoted $\mathscr{F}_b$.

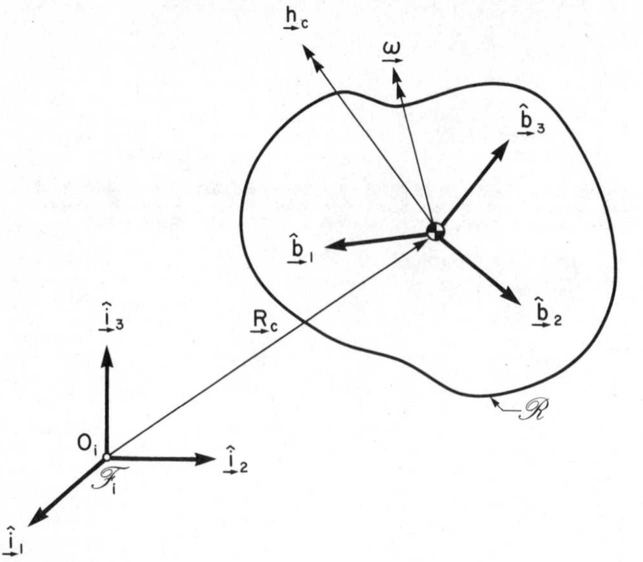

FIGURE 4.1 Rigid body $\mathscr{R}$ with body-fixed frame $\mathscr{F}_b$ and inertial frame $\mathscr{F}_i$.

According to (3.3, 15 and 16), the appropriate motion equations are

$$\begin{aligned} \dot{\mathbf{p}} &= -\boldsymbol{\omega}^{\times}\mathbf{p} + \mathbf{f} \\ \mathbf{p} &= m\mathbf{v}_c \\ \dot{\mathbf{R}}_c &= \mathbf{v}_c \end{aligned} \tag{1}$$

$$\begin{aligned} \dot{\mathbf{h}}_c &= -\boldsymbol{\omega}^{\times}\mathbf{h}_c + \mathbf{g}_c \\ \mathbf{h}_c &= \mathbf{I}\boldsymbol{\omega} \\ \dot{\mathbf{C}} &= -\boldsymbol{\omega}^{\times}\mathbf{C} \end{aligned} \tag{2}$$

The first trio are the translational equations; the rotational equations comprise the second trio. In each case an appropriate kinematical equation has been added to complete the system.

For the sake of clarity, the meaning of the symbols in (1) and (2) is reiterated: the position and velocity of the mass center ⊕ are denoted by the vectors $\underrightarrow{\mathbf{R}}_c$ and $\underrightarrow{\mathbf{v}}_c$; the (total) external force is denoted $\underrightarrow{\mathbf{f}}$; and $\underrightarrow{\mathbf{p}}$ is the momentum of $\mathscr{R}$. Then,

$$[\mathbf{R}_c \quad \mathbf{v}_c \quad \mathbf{p} \quad \mathbf{f}] \triangleq \mathscr{F}_b \cdot [\underrightarrow{\mathbf{R}}_c \quad \underrightarrow{\mathbf{v}}_c \quad \underrightarrow{\mathbf{p}} \quad \underrightarrow{\mathbf{f}}] \tag{3}$$

In like manner, $\mathbf{C} \triangleq \mathscr{F}_b \cdot \mathscr{F}_i^T$; $\underrightarrow{\boldsymbol{\omega}}$ is the angular velocity of $\mathscr{F}_b$ (i.e., of $\mathscr{R}$) with respect to $\mathscr{F}_i$, and $\boldsymbol{\omega} \triangleq \mathscr{F}_b \cdot \underrightarrow{\boldsymbol{\omega}}$; $\underrightarrow{\mathbf{h}}_c$ is the angular momentum (identical to the absolute angular momentum in this case) of $\mathscr{R}$ about ⊕; $\underrightarrow{\mathbf{g}}_c$ is the (total) external torque about ⊕; $\underrightarrow{\mathbf{I}}$ is the centroidal inertia dyadic $\mathbf{I} \triangleq \mathscr{F}_b \cdot \underrightarrow{\mathbf{I}} \cdot \mathscr{F}_b^T$; and both $\underrightarrow{\mathbf{h}}_c$ and $\underrightarrow{\mathbf{g}}_c$ are expressed in $\mathscr{F}_b$.

Provided the external torque $\mathbf{g}_c$ is independent of body position and velocity, $\mathbf{g}_c \neq \mathbf{g}_c(\mathbf{R}_c, \mathbf{v}_c)$, the rotational system (2) can be solved independently of the translational system (1). We choose for the present to make this assumption and to drop (1) from explicit consideration. In the general case, discussed further in Chapter 9, there is no alternative but to treat (1) and (2) as a coupled system.

We shall in this section drop the "c" subscript on $\mathbf{h}$ and $\mathbf{g}$. The first of three equations in (2) is so important that it is now written out in detail:

$$\begin{aligned}\dot{h}_1 &= \omega_3 h_2 - \omega_2 h_3 + g_1\\ \dot{h}_2 &= \omega_1 h_3 - \omega_3 h_1 + g_2\\ \dot{h}_3 &= \omega_2 h_1 - \omega_1 h_2 + g_3\end{aligned} \tag{4}$$

We shall also assume that $\mathscr{F}_b$ is aligned with the principal axes, $\mathscr{F}_b \to \mathscr{F}_p$, whereupon

$$\mathbf{I} = \operatorname{diag}\{I_1, I_2, I_3\} \tag{5}$$

The more general case $\mathscr{F}_b \neq \mathscr{F}_p$ can be handled via a transformation (a constant rotation matrix $\mathbf{C}_{bp}$) of the results obtained in this section. With principal axes, the second of equations (2) becomes

$$h_1 = I_1\omega_1; \qquad h_2 = I_2\omega_2; \qquad h_3 = I_3\omega_3 \tag{6}$$

Furthermore, because the motion equations of a single rigid body are relatively simple, considerable progress can be made by substituting (6) into (4), thereby arriving at Euler's motion equations for $\mathscr{R}$:

$$\begin{aligned}I_1\dot{\omega}_1 &= (I_2 - I_3)\omega_2\omega_3 + g_1\\ I_2\dot{\omega}_2 &= (I_3 - I_1)\omega_3\omega_1 + g_2\\ I_3\dot{\omega}_3 &= (I_1 - I_2)\omega_1\omega_2 + g_3\end{aligned} \tag{7}$$

These equations pose an analytical challenge because of the nonlinear terms on the right-hand side and because of the functional dependence of $\mathbf{g}$ (see Section 4.5). They have attracted the interest of mathematicians since Euler, but closed-form solutions are available only in special cases.

Our plan of attack on the important equations (7) in this chapter is as follows: First, we shall solve for the motion when $\mathbf{g} \equiv \mathbf{0}$ and two of the principal moments of inertia are equal. Then, we shall allow the principal moments of inertia to be distinct (still with $\mathbf{g} \equiv \mathbf{0}$). Finally we shall comment briefly on the case of excited motion, $\mathbf{g} \not\equiv \mathbf{0}$. Central to the discussion will be the dynamical quantities T and $\mathbf{h}$. The rotational kinetic energy, from (3.3, 23), is given by

$$T = \tfrac{1}{2}\left(I_1\omega_1^2 + I_2\omega_2^2 + I_3\omega_3^2\right) \tag{8}$$

and it changes at the rate

$$\dot{T} = \omega_1 g_1 + \omega_2 g_2 + \omega_3 g_3 \tag{9}$$

according to (3.3, 22). The change in angular momentum $\underrightarrow{\mathbf{h}}$ is the subject of the motion equations (4); the magnitude of $\underrightarrow{\mathbf{h}}$, given by

$$h = \left(I_1^2\omega_1^2 + I_2^2\omega_2^2 + I_3^2\omega_3^2\right)^{1/2} \tag{10}$$

changes at the rate

$$\dot{h} = \frac{h_1 g_1 + h_2 g_2 + h_3 g_3}{h} \tag{11}$$

When the external torque is identically zero, T and h are valuable constants (integrals) of the motion.

4.2 TORQUE-FREE MOTION; $\mathscr{R}$ INERTIALLY AXISYMMETRICAL

Consider first the torque-free motion of a rigid body. The mass center may follow an arbitrary path in response to external *forces* so long as the *torque* remains zero. For spacecraft, the general attitude motion in response to torques can normally be treated as a deviation from some reference torque-free motion, and it is therefore important to study this reference motion. Although it is tempting to simply integrate the motion equations (4.1, 7) numerically after modeling the torque (see Chapter 8), important insights are obtained from a familiarity with the torque-free motion, which has an analytical solution and a geometrical interpretation.

Under our present assumptions, $\mathscr{R}$ is completely characterized by the three positive constants $\{I_1, I_2, I_3\}$. The case where all three are equal (the *isoinertial* case) is easily disposed of: according to (4.1, 7), if $I_1 = I_2 = I_3$ and $\mathbf{g} \equiv \mathbf{0}$, the motion is $\boldsymbol{\omega} =$ constant, and $\mathscr{R}$ continues to spin at a constant rate about an axis fixed in itself. Moreover, this axis is also fixed in $\mathscr{F}_i$, since $\dot{\vec{\omega}} = \overset{\circ}{\vec{\omega}}$.

Consider now the less trivial case where only two of $\{I_1, I_2, I_3\}$ are equal. To be specific, let

$$I_t \triangleq I_1 = I_2; \qquad I_a \triangleq I_3 \tag{1}$$

We say that $\mathscr{R}$ is *inertially axisymmetrical*. [Note that $\mathscr{R}$ need not be a body of revolution; for example, a rectangular block of uniform density and with sides a, a, and b is inertially axisymmetrical (Fig. 4.2), the axis of symmetry being the $\vec{\mathbf{p}}_3$ axis.] Euler's motion equations can be readily solved in this case. Substituting (1) and $\mathbf{g} \equiv \mathbf{0}$ into (4.1, 7), we find

$$\begin{aligned} I_t \dot{\omega}_1 &= (I_t - I_a)\omega_2\omega_3 \\ I_t \dot{\omega}_2 &= (I_a - I_t)\omega_3\omega_1 \\ I_a \dot{\omega}_3 &= 0 \end{aligned} \tag{2}$$

Solution for $\omega(t)$

From the third of the three equations in (2),

$$\omega_3 \equiv \omega_{3o} \triangleq \nu \tag{3}$$

We may with no loss in generality take $\nu > 0$. This can be thought of as specifying the polarity of $\vec{\mathbf{p}}_3$ in $\mathscr{R}$. The first two equations have the solution

$$\begin{aligned} \omega_1(t) &= \omega_{1o}\cos\Omega t + \omega_{2o}\sin\Omega t \\ \omega_2(t) &= \omega_{2o}\cos\Omega t - \omega_{1o}\sin\Omega t \end{aligned} \tag{4}$$

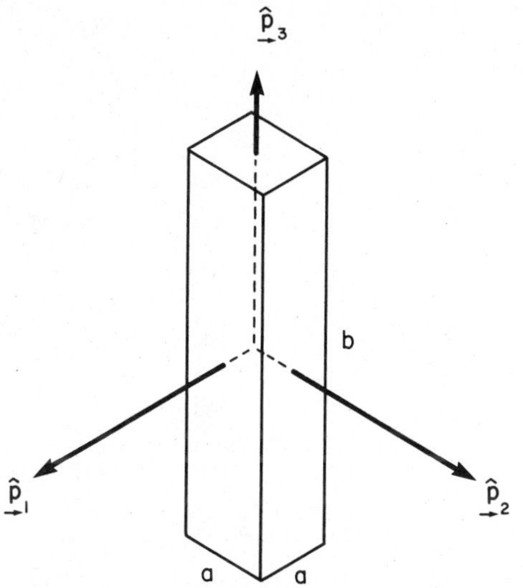

FIGURE 4.2 An inertially symmetrical rigid body. (Assume uniform mass distribution.)

where the *relative spin rate* Ω is defined by

$$\Omega \triangleq \left(\frac{I_t - I_a}{I_t}\right)\nu \tag{5}$$

Note that ω_{1o}, ω_{2o}, and Ω may be of either sign. It follows also from (4) that if a "transverse" angular velocity is defined by

$$\omega_t \triangleq \left(\omega_1^2 + \omega_2^2\right)^{1/2} \equiv \left(\omega_{1o}^2 + \omega_{2o}^2\right)^{1/2} \geq 0 \tag{6}$$

then ω_t is constant. The value of $\boldsymbol{\omega}$ at $t = 0$, namely, $[\omega_{1o} \quad \omega_{2o} \quad \omega_{3o}]^T$, can therefore be replaced by the three constants ν, ω_t, and t_o, where $\omega_1 = 0$ and $\omega_2 = \omega_t$ at $t = t_o$:

$$t_o \triangleq \Omega^{-1}\tan^{-1}(-\omega_{1o}, \omega_{2o}) \tag{7}$$

The double argument in (7) ensures that $\tan^{-1}$ is defined over the range $0 \leq \tan^{-1} < 2\pi$. To summarize, the solution for the angular velocity is

$$\omega_1 = \omega_t \sin\mu; \qquad \omega_2 = \omega_t \cos\mu; \qquad \omega_3 = \nu \tag{8}$$

where

$$\mu(t) \triangleq \Omega(t - t_o) \tag{9}$$

Furthermore, since

$$\omega^2 \equiv \omega_1^2 + \omega_2^2 + \omega_3^2 = \nu^2 + \omega_t^2 \tag{10}$$

it is recognized that the magnitude of $\boldsymbol{\omega}$ is constant (see Fig. 4.3).

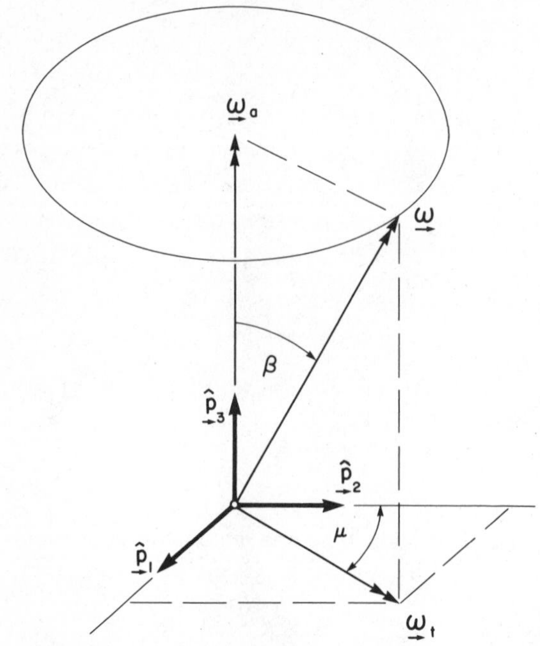

FIGURE 4.3 Torque-free motion of $\underset{\rightarrow}{\omega}$, as seen in principal axes.

Solution for Attitude History C(*t*)

Although we have derived the motion of $\boldsymbol{\omega}$ with respect to $\mathscr{F}_p$, our ultimate objective is to describe the motion of $\mathscr{R}$ with respect to $\mathscr{F}_i$. From (4.1, 6) and (1) and (8) above, the components of $\underset{\rightarrow}{\mathbf{h}}$ in $\mathscr{F}_p$ are

$$h_1 = I_t\omega_t \sin\mu; \qquad h_2 = I_t\omega_t\cos\mu; \qquad h_3 = I_a\nu \tag{11}$$

This leads naturally to the definition of "axial" and "transverse" momentum components:

$$h_a \triangleq I_a\nu; \qquad h_t = I_t\omega_t \tag{12}$$

whereupon

$$h_1 = h_t\sin\mu; \qquad h_2 = h_t\cos\mu; \qquad h_3 = h_a \tag{13}$$

and

$$h = \left(h_t^2 + h_a^2\right)^{1/2} \tag{14}$$

Note that h_t, h_a, and h are all constant; also $h_t \geq 0$, and $h_a \geq 0$ by convention (see Fig. 4.4). Moreover, we denote by γ the angle between $\underset{\rightarrow}{\mathbf{h}}$ and the symmetry axis $\underset{\rightarrow}{\hat{\mathbf{p}}}_3$, with $0 \leq \gamma < \pi/2$:

$$h_a = h\cos\gamma; \qquad h_t = h\sin\gamma \tag{15}$$

Because $\underset{\rightarrow}{\mathbf{h}}$ is fixed in $\mathscr{F}_i$, the angles γ and μ are directly related to the attitude of $\mathscr{R}$.

To complete our solution for the attitude motion of $\mathscr{R}$, a natural choice is an appropriate set of Euler angles, because for this relatively simple problem one of

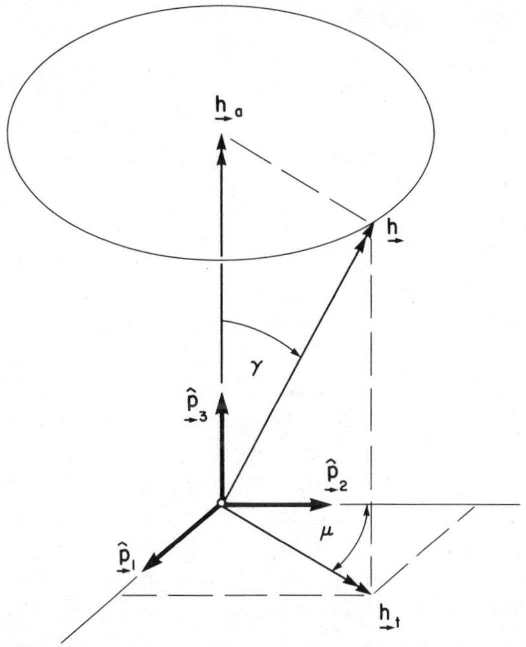

FIGURE 4.4 Torque-free motion of $\underset{\rightarrow}{\mathbf{h}}$, as seen in principal axes.

the three Euler angles is constant. Indeed, it will now be shown that γ and μ are two such Euler angles. Denoting the direction cosine matrix $\mathscr{F}_p \cdot \mathscr{F}_i^T$ by $\mathbf{C}$, we choose

$$\mathbf{C} \triangleq \mathbf{C}_3(\mu)\mathbf{C}_1(\gamma)\mathbf{C}_3(\theta_3) \tag{16}$$

although [Greenwood] and others have given an interesting alternative in terms of $\mathbf{C}_1(\phi)\mathbf{C}_2(\theta)\mathbf{C}_3(\psi)$. A great simplification, at no loss in essential generality, results from choosing the inertial axis $\hat{\underset{\rightarrow}{\mathbf{i}}}_3$ to coincide with $\hat{\underset{\rightarrow}{\mathbf{h}}}$:

$$\hat{\underset{\rightarrow}{\mathbf{i}}}_3 = \hat{\underset{\rightarrow}{\mathbf{h}}} \triangleq \frac{\underset{\rightarrow}{\mathbf{h}}}{h} \tag{17}$$

From (16), we note (e.g., in Table 2.1 in Chapter 2) that the third column of $\mathbf{C}$ is $[\sin\mu\sin\gamma \quad \cos\mu\sin\gamma \quad \cos\gamma]^T$. Thus,

$$h_1 = h\sin\mu\sin\gamma; \qquad h_2 = h\cos\mu\sin\gamma; \qquad h_3 = h\cos\gamma \tag{18}$$

in accordance with (13) and (15). We recall that

$$\gamma = \text{constant}; \qquad \mu(t) = \Omega(t - t_o) \tag{19}$$

In classical mechanics literature, γ is called the *nutation angle*. As for the third angle, θ_3, a derivation analogous to the one used in producing (2.3, 24) reveals that (or use Table 2.2)

$$\boldsymbol{\omega} = \mathbf{1}_3\dot{\mu} + \mathbf{C}_3(\mu)\mathbf{1}_1\dot{\gamma} + \mathbf{C}_3(\mu)\mathbf{C}_1(\gamma)\mathbf{1}_3\dot{\theta}_3 \tag{20}$$

and that, in particular, for the *precession angle* θ_3,

$$\dot{\theta}_3 = \frac{\omega_3 - \dot{\mu}}{\cos\gamma} = \frac{\nu - \Omega}{\cos\gamma} = \frac{h}{I_t} \triangleq \Omega_p \tag{21}$$

where Ω_p is called the *precession rate*, a constant shortly to be interpreted geometrically. Evidently,

$$\theta_3 = \Omega_p t + \theta_{3o} \tag{22}$$

where θ_{3o} is a constant to be evaluated at $t = 0$.

Prior to establishing the geometrical interpretation of these results, it is desirable to pause and reflect on the several constants that have been defined and on the solution of the initial condition problem: given $\boldsymbol{\omega}$ and $\mathbf{C}$ at $t = 0$, find $\boldsymbol{\omega}(t)$ and $\mathbf{C}(t)$ for $t > 0$. The following algorithm indicates a possible sequence. Given the body $\{I_a, I_t\}$ and the initial conditions $\{\omega_{1o}, \omega_{2o}, \omega_{3o}, \theta_{3o}\}$, the algorithm proceeds as follows: calculate $\{\nu, \Omega, \omega_t, t_o, \omega, h, h_a, h_t, \gamma, \mu(t), \Omega_p, \theta_3(t)\}$, in that order, respectively from equations {(3), (5), (6), (7), (10), (14), (12), (12), (15), (9), (21), and (22)} of this section. It appears that the solution is less than completely general inasmuch as arbitrary initial conditions γ_o, μ_o are not admitted. This generality was sacrificed when $\underset{\rightarrow}{\hat{\mathbf{i}}}_3$ was chosen to be in the $\underset{\rightarrow}{\hat{\mathbf{h}}}$ direction; see (17). This preordination corresponds to

$$\gamma_o = \gamma; \qquad \mu_o = -\Omega t_o \tag{23}$$

as is clear from (19). A completely general solution, which does not make the constraining assumption $\underset{\rightarrow}{\hat{\mathbf{i}}}_3 = \underset{\rightarrow}{\hat{\mathbf{h}}}$, is not difficult to derive by applying a two-angle transformation on the above results.

Geometrical Interpretation

The preceding analytical solution has an elegant geometrical interpretation, a display piece for most textbooks on classical mechanics. Let us denote by α and β the respective angles between $\underset{\rightarrow}{\omega}$ and $\underset{\rightarrow}{\hat{\mathbf{i}}}_3$ and between $\underset{\rightarrow}{\omega}$ and $\underset{\rightarrow}{\hat{\mathbf{p}}}_3$. Then,

$$\omega\cos\alpha \triangleq \underset{\rightarrow}{\omega} \cdot \underset{\rightarrow}{\hat{\mathbf{i}}}_3 = \boldsymbol{\omega}^T\mathscr{F}_p \cdot \mathscr{F}_i^T\mathbf{1}_3 = \boldsymbol{\omega}^T\mathscr{F}_p \cdot \mathscr{F}_p^T\mathbf{C}\mathbf{1}_3 = \boldsymbol{\omega}^T\mathbf{C}\mathbf{1}_3$$

$$\cos\alpha = \frac{\boldsymbol{\omega}^T\mathbf{h}}{\omega h} = \frac{2T}{\omega h} \tag{24}$$

after inserting (18). Note that the rotational kinetic energy is

$$T = \tfrac{1}{2}\left(I_a\nu^2 + I_t\omega_t^2\right) \tag{25}$$

The second angle, β, is found from

$$\cos\beta \triangleq \frac{\underset{\rightarrow}{\omega} \cdot \underset{\rightarrow}{\hat{\mathbf{p}}}_3}{\omega} = \frac{\nu}{\omega} \tag{26}$$

Note that α and β are constants of the motion, both lying in the range $[0, \pi/2]$. We can calculate $\cos(\alpha + \beta)$ from (24) and (26), noting that $\sin\alpha \geq 0$ and $\sin\beta \geq 0$. Thus,

$$\cos(\alpha + \beta) = \frac{(I_a\nu^2 + I_t\omega_t^2)\nu - |I_a - I_t|\nu\omega_t^2}{h\omega^2} \tag{27}$$

which is not particularly helpful unless it is known that $I_a \leq I_t$, in which case

$$\cos(\alpha + \beta) = \frac{I_a\nu}{h} = \cos\gamma \tag{28}$$

or unless it is known that $I_a \geq I_t$, in which case

$$\cos(\alpha - \beta) = \cos\gamma \tag{29}$$

To sort out the possible cases, recall that $\{\alpha, \beta, \gamma\}$ all lie in the range $[0, \pi/2]$. From (28) and (29),

$$I_a \leq I_t: \quad \alpha + \beta = \gamma \qquad \text{(case A)}$$

$$I_a \geq I_t: \quad \begin{cases} \beta - \alpha = \gamma & \text{(case B)} \\ \alpha - \beta = \gamma & \text{(case C)} \end{cases}$$

These results are the key to the geometrical construction used in Fig. 4.5. Consider first case A in Fig. 4.5, which shows a circular cone $\mathscr{C}_i$ of half-angle α and fixed in $\mathscr{F}_i$. Its axis is $\hat{\underset{\rightarrow}{\mathbf{i}}}_3$. Also shown is a second circular cone $\mathscr{C}_p$ of half-angle β. Its axis is $\hat{\mathbf{p}}_3$. Suppose now that $\mathscr{C}_p$ rolls without slipping on $\mathscr{C}_i$. The angular velocity of $\mathscr{C}_p$, denoted $\underset{\rightarrow}{\omega}'$, also called the *instantaneous axis of rotation* (see Problem 2.23), must then lie on the line of contact between $\mathscr{C}_i$ and $\mathscr{C}_p$. This implies that $\underset{\rightarrow}{\omega}' \cdot \hat{\underset{\rightarrow}{\mathbf{i}}} = \omega' \cos\alpha$; $\underset{\rightarrow}{\omega}' \cdot \hat{\mathbf{p}}_3 = \omega' \cos\beta$; and $\underset{\rightarrow}{\omega}' \cdot (\hat{\mathbf{p}}_3 \times \hat{\underset{\rightarrow}{\mathbf{i}}}_3) = 0$. These are precisely the components of $\underset{\rightarrow}{\omega}' \equiv \underset{\rightarrow}{\omega}$. Evidently, $\mathscr{C}_p$ is fixed in $\mathscr{R}$. In summary, to visualize the attitude motion of $\mathscr{R}$ when $I_a \leq I_t$, one constructs a cone $\mathscr{C}_p$, fixed in $\mathscr{R}$, whose axis coincides with the axis of inertial symmetry for $\mathscr{R}$; $\mathscr{C}_p$ then rolls without slipping and at fixed ω along a second cone $\mathscr{C}_i$, which is fixed in $\mathscr{F}_i$ and whose axis coincides with $\underset{\rightarrow}{\mathbf{h}}$.

In a similar manner, case B is portrayed in Fig. 4.5. One again constructs an inertially fixed cone $\mathscr{C}_i$ and a body-fixed cone $\mathscr{C}_p$, but now $\mathscr{C}_i$ lies inside $\mathscr{C}_p$. However, $\mathscr{C}_p$ still rolls without slipping at fixed ω on $\mathscr{C}_i$. Case A is called *direct*, or *prograde*, precession because $\Omega > 0$; case B is called *retrograde* precession because $\Omega < 0$. Case C is shown in Problem 4.19 to be physically impossible.

The only remaining geometrical point to be settled is the interpretation of the precession rate Ω_p: it is the angular speed at which the symmetry axis $\hat{\mathbf{p}}_3$ rotates (precesses) about $\underset{\rightarrow}{\mathbf{h}}$. This can be demonstrated in various ways. It follows, for example, from the definition of θ_3 and (22). Alternatively, from Fig. 4.6, the rate

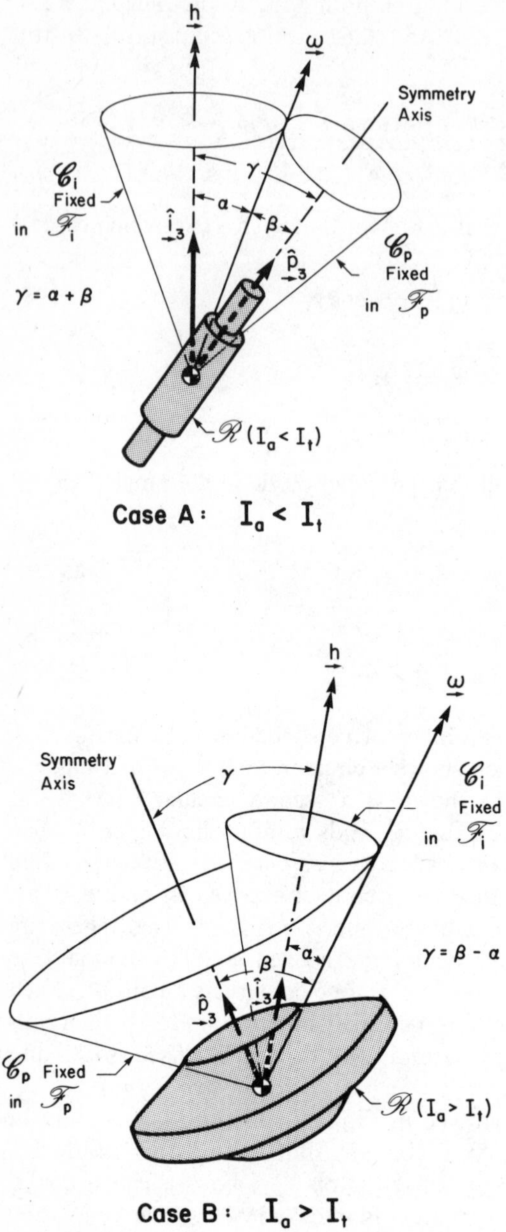

FIGURE 4.5 Torque-free motion of an inertially axisymmetrical rigid body.

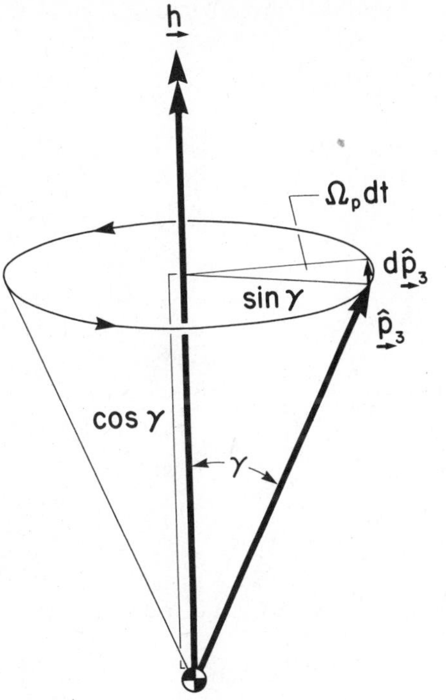

FIGURE 4.6 Precession of the symmetry axis about the angular momentum vector. The nutation angle γ is constant.

of rotation of $\hat{\underset{\rightarrow}{p}}_3$ about $\underset{\rightarrow}{h}$ is

$$\frac{|\dot{\hat{\underset{\rightarrow}{p}}}_3|}{\sin\gamma} = \frac{|\underset{\rightarrow}{\omega} \times \hat{\underset{\rightarrow}{p}}_3|}{\sin\gamma} = \frac{\left(\omega_1^2 + \omega_2^2\right)^{1/2}}{\sin\gamma} = \frac{\omega_t}{\sin\gamma}$$

By the definition (15) of γ, however, the last expression is

$$\frac{\omega_t}{\sin\gamma} = \frac{h\omega_t}{h_t} = \frac{h\omega_t}{I_t\omega_t} = \frac{h}{I_t} = \Omega_p \tag{30}$$

as stated earlier. This coning motion, or precession, described here in detail for a spinning axisymmetrical body, is often also typical of the motion in more general circumstances—for example, a nonsymmetrical body, or a body containing spinning wheels. Some modifications have to be made in the geometrical interpretation for these more complex cases, but the basic coning picture remains qualitatively valid.

Euler Parameters

To conclude this section, the solution for the attitude motion of ℛ will be expressed in terms of Euler parameters. Retaining the convention that $\hat{\underset{\rightarrow}{i}}_3$ is aligned with $\hat{\underset{\rightarrow}{h}}$, we note the third column of $\mathbf{C}$, expressed in terms of $\{\varepsilon_1, \varepsilon_2, \varepsilon_3, \eta\}$

in (2.2, 14), and compare element by element with the third column of **C** indicated by (16):

$$2(\varepsilon_1\varepsilon_3 - \varepsilon_2\eta) = \sin\gamma\sin\mu$$

$$2(\varepsilon_2\varepsilon_3 + \varepsilon_1\eta) = \sin\gamma\cos\mu$$

$$1 - 2(\varepsilon_1^2 + \varepsilon_2^2) = \cos\gamma \tag{31}$$

It is concluded that

$$\varepsilon_1 = \sin\frac{\gamma}{2}\cos\lambda$$

$$\varepsilon_2 = \sin\frac{\gamma}{2}\sin\lambda$$

$$\varepsilon_3 = \cos\frac{\gamma}{2}\sin(\mu + \lambda)$$

$$\eta = \cos\frac{\gamma}{2}\cos(\mu + \lambda) \tag{32}$$

where $\lambda(t)$ is yet to be specified. To determine $\lambda(t)$, (2.3, 22) is used:

$$\dot{\eta} = -\tfrac{1}{2}(\varepsilon_1\omega_1 + \varepsilon_2\omega_2 + \varepsilon_3\omega_3) \tag{33}$$

which, upon substitution of (32) and (8), produces the intermediate result

$$2(\dot{\mu} + \dot{\lambda})\sin\gamma = \omega_t(1 - \cos\gamma) + \nu\sin\gamma \tag{34}$$

Noting the definitions of $\{\gamma, h_a, h_t, \Omega_p, \Omega\}$ given respectively by equations {(15), (12), (12), (21), and (5)} of this section, the simpler form

$$2(\dot{\mu} + \dot{\lambda}) = \Omega_p + \Omega \tag{35}$$

is obtained. Since $\dot{\mu} = \Omega$, the variable λ is evidently given by

$$\lambda(t) = \tfrac{1}{2}(\Omega_p - \Omega)(t - t_o) + \lambda_o \tag{36}$$

where λ_o is the value of λ at $t = t_o$. Once again it is observed that the choice of $\vec{\hat{\mathbf{i}}}_3 = \vec{\hat{\mathbf{h}}}$ limits the initial conditions on attitude to a single constant—in this case λ_o.

4.3 TORQUE-FREE MOTION; $\mathscr{R}$ TRI-INERTIAL

When I_1, I_2, and I_3 are distinct, the rigid body $\mathscr{R}$ is said to be *tri-inertial*, or *triaxial*, or *asymmetrical*, and the analytical solution is more difficult. In brief, the circular functions of time that characterize the axisymmetrical case must, for asymmetrical bodies, be generalized to become elliptic functions.

We return to Euler's motion equations (4.1, 7), with $g_1 = g_2 = g_3 = 0$:

$$I_1\dot{\omega}_1 = (I_2 - I_3)\omega_2\omega_3$$
$$I_2\dot{\omega}_2 = (I_3 - I_1)\omega_3\omega_1$$
$$I_3\dot{\omega}_3 = (I_1 - I_2)\omega_1\omega_2 \tag{1}$$

It is clear from the cyclic symmetry in (1) that we may, without loss in generality, label the principal axes so that

$$I_1 > I_2 > I_3 \tag{2}$$

although when, as later in Section 4.4, the motion is essentially a spin about a principal axis, we shall drop this convention. The two motion integrals

$$2T = I_1\omega_1^2 + I_2\omega_2^2 + I_3\omega_3^2$$
$$h^2 = I_1^2\omega_1^2 + I_2^2\omega_2^2 + I_3^2\omega_3^2 \tag{3}$$

will again be exploited. They represent two of the three quadratures required by (1); the third requires substantially more effort. The following definition is also useful:

$$I \triangleq \frac{h^2}{2T} \tag{4}$$

Note that I has the dimensions of a moment of inertia. According to (3), if h^2 is given, T is maximal when $\omega_1 = \omega_2 = 0$ and is minimal when $\omega_2 = \omega_3 = 0$:

$$\frac{h^2}{2I_1} \le T \le \frac{h^2}{2I_3} \tag{5}$$

These limits are in accordance with (4.6, 1) in Problem 4.1 for an inertially symmetrical body. In terms of I, the corresponding bounds are

$$I_1 \ge I \ge I_3 \tag{6}$$

Note that I depends on the constants of the motion, and therefore on initial conditions. The value $I = I_2$ will prove to be a bifurcation value.

Analytical Solution for $\boldsymbol{\omega}(t)$

There is, as mentioned above, a closed-form solution for $\boldsymbol{\omega}(t)$ in terms of elliptic functions, dating back to Jacobi. The opinion held of these solutions by dynamicists is remarkably nonuniform. Although treated as expedient in older treatises, for example, [Whittaker] and [MacMillan], more modern texts tend to avoid it (that of [Synge and Griffith] is a notable exception). Mild concern can be raised concerning the desirability of its inclusion on two grounds: the relative unfamiliarity of elliptic functions, and their suitability for numerical calculations.

The unfamiliarity is largely subjective; one naturally encounters the circular functions earlier and more often. On the other hand, the elliptic functions are, after all, relatively simple generalizations of the more familiar circular functions, and the enterprising analyst will not withdraw from them. As to their appropriateness for digital computation, much depends on the availability of these functions as library subroutines. Efficient algorithms have been developed for their evaluation, so there is no intrinsic reason why elliptic functions are not as readily available as the circular functions, "sin" and "cos." It is also apparent that whether a solution is called "closed form" depends largely on giving that solution a name, establishing its analytical properties, and developing efficient numerical methods for its calculation. By these standards, the solution of Euler's equations in terms of elliptic functions is a welcome result and may often be preferrable to the blind numerical integration of Euler's equations, (1). [Thomson, 1] was among the first to bring the analytical solution of Euler's equations to the attention of spacecraft attitude dynamicists. It might also be mentioned parenthetically that the introduction of the parameter I, given by (4), does not seem to appeal to modern dynamicists, perhaps for reasons of computational efficiency. Since our goals are slightly different, the diagnostic parameter I will play an important role in the developments to follow.

The amplitudes of ω_1, ω_2, and ω_3 can be determined directly from the integrals T and h^2; see (3). These amplitudes also are required in the solution for $\boldsymbol{\omega}(t)$. Thus, for example, to maximize ω_1, one has a standard maximization problem with constraints: Find

$$\max\left[\omega_1 + \mu(2T - \boldsymbol{\omega}^T\mathbf{I}\boldsymbol{\omega}) + \lambda(h^2 - \boldsymbol{\omega}^T\mathbf{I}^2\boldsymbol{\omega})\right]$$

where μ and λ are Lagrangian multipliers. The result of this maximization (Problem 4.5) is

$$\omega_{1m} \triangleq \omega_{1,\max} = h\left[\frac{I - I_3}{II_1(I_1 - I_3)}\right]^{1/2} \tag{7}$$

Similarly,

$$\omega_{3m} \triangleq \omega_{3,\max} = h\left[\frac{I_1 - I}{II_3(I_1 - I_3)}\right]^{1/2} \tag{8}$$

As for $\omega_2(t)$, the expression for its maximum depends on the size of I relative to I_2:

$$\omega_{2m} = \begin{cases} h\left[\dfrac{I_1 - I}{II_2(I_1 - I_2)}\right]^{1/2} & (I_1 \geq I > I_2) \\ \dfrac{h}{I_2} & (I = I_2) \\ h\left[\dfrac{I - I_3}{II_2(I_2 - I_3)}\right]^{1/2} & (I_2 > I \geq I_3) \end{cases} \tag{9}$$

In terms of these constants, the solution to Euler's equations (1) is as follows:

$$\left.\begin{aligned}
&\left.\begin{aligned}
\omega_1 &= s_1 \omega_{1m}\, dn(\tau; k)\\
\omega_2 &= s_2 \omega_{2m}\, sn(\tau; k)\\
\omega_3 &= s_3 \omega_{3m}\, cn(\tau; k)
\end{aligned}\right\} \quad (s_1 s_2 s_3 = -1)\\
&\tau = h\left[\frac{(I_1 - I_2)(I - I_3)}{I I_1 I_2 I_3}\right]^{1/2}(t - t_o)\\
&k = \left[\frac{(I_2 - I_3)(I_1 - I)}{(I_1 - I_2)(I - I_3)}\right]^{1/2}
\end{aligned}\right\} \quad (I_1 \geq I > I_2) \qquad (10)$$

$$\left.\begin{aligned}
&\left.\begin{aligned}
\omega_1 &= s_1 \omega_{1m}\, cn(\tau; k)\\
\omega_2 &= s_2 \omega_{2m}\, sn(\tau; k)\\
\omega_3 &= s_3 \omega_{3m}\, dn(\tau; k)
\end{aligned}\right\} \quad (s_1 s_2 s_2 = -1)\\
&\tau = h\left[\frac{(I_2 - I_3)(I_1 - I)}{I I_1 I_2 I_3}\right]^{1/2}(t - t_o)\\
&k = \left[\frac{(I_1 - I_2)(I - I_3)}{(I_2 - I_3)(I_1 - I)}\right]^{1/2}
\end{aligned}\right\} \quad (I_2 > I \geq I_3) \qquad (11)$$

Here, $\{s_1, s_2, s_3\}$ are all equal to ± 1; they determine signs. Since $s_1 s_2 s_3 = -1$, four combinations of signs are allowed. The functions $\{sn, cn, dn\}$ are Jacobian elliptic functions. It is helpful to think of $\{sn, cn, dn\}$ as generalizations of $\{\sin, \cos, 1\}$, respectively (see Fig. 4.7). Each of $\{sn, cn, dn\}$ is really a family of functions. They vary with τ, of course, but depend also on the parameter $k, 0 \leq k < 1$. When $k \to 0$, $\{sn, cn, dn\} \to \{\sin, \cos, 1\}$. For details of the derivation of (10) and (11) as the solution to (1), see [Wittenburg], for example. The

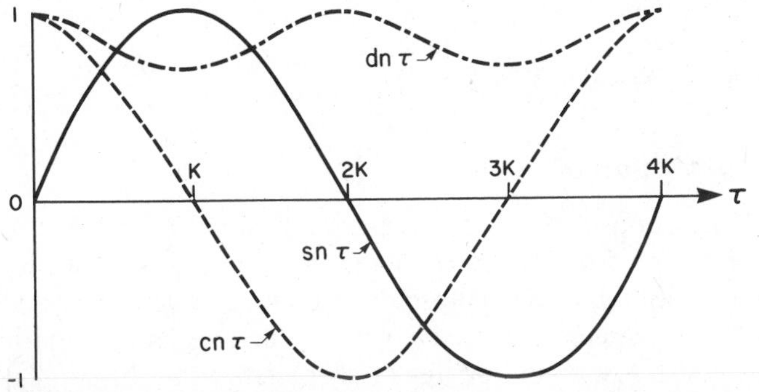

FIGURE 4.7 The Jacobian elliptic functions $\{sn, cn, dn\}$ for $k^2 = \frac{1}{2}$ (after [Korn and Korn]). Here, $K(k)$ is Legendre's complete elliptic integral of the first kind, and $K(1/\sqrt{2}) \doteq 1.854$.

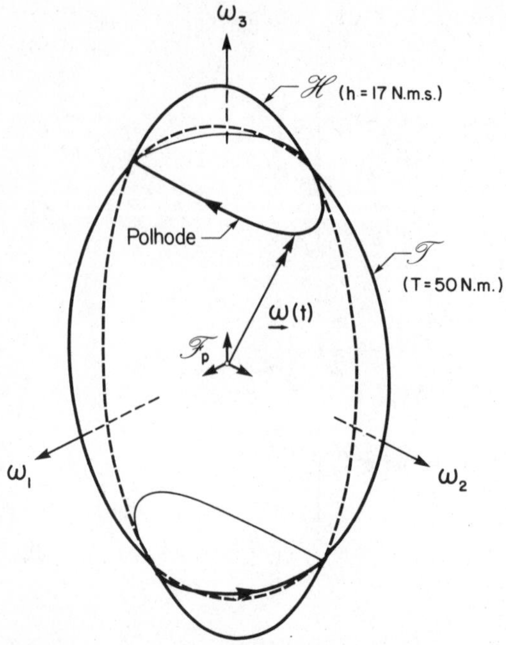

FIGURE 4.8 Intersection of the energy ellipsoid $\mathscr{T}$ and angular-momentum ellipsoid $\mathscr{H}$ for a rigid body with $\{I_1, I_2, I_3\} = \{6, 4, 2\}$ kg · m². For $\mathscr{T}$, $T = 50$ N · m; and for $\mathscr{H}$, $h = 17$ N · m · s.

missing case, $I = I_2$, is now recorded (see also Problem 4.6):

$$\left.\begin{array}{l}\left.\begin{array}{l}\omega_1 = s_1\omega_{1m}\operatorname{sech}\tau \\ \omega_2 = s_2\omega_{2m}\tanh\tau \\ \omega_3 = s_3\omega_{3m}\operatorname{sech}\tau\end{array}\right\} \quad (s_1s_2s_3 = -1) \\ \tau = \dfrac{h}{I_2}\left[\dfrac{(I_1 - I_2)(I_2 - I_3)}{I_1 I_3}\right]^{1/2}(t - t_o)\end{array}\right\} \quad (I = I_2) \tag{12}$$

This completes the analytical solution for $\boldsymbol{\omega}(t)$.

Geometrical Interpretation of $\omega(t)$

The geometrical interpretation of these results begins by recognizing that in $\boldsymbol{\omega}$-space the motion is represented by the tip of $\vec{\omega}$ traveling on a curve, called a *polhode*, defined by the intersection of two ellipsoids: the *energy ellipsoid* $\mathscr{T}$ and the *momentum ellipsoid* $\mathscr{H}$, represented respectively by the first and second parts of (3). For example, Fig. 4.8 shows $\mathscr{T}$ and $\mathscr{H}$ for $\mathscr{R}$ characterized by $\{I_1, I_2, I_3\} = \{6, 4, 2\}$ kg · m². Here, $\mathscr{T}$ corresponds to $T = 50$ N · m; and, for $\mathscr{H}$, $h = 17$ N · m · s. Thus, for the case shown, $I = \frac{1}{2}h^2/T = 2.89$ kg · m². Evidently, the curve of intersection, at least for this case, consists of two separate closed

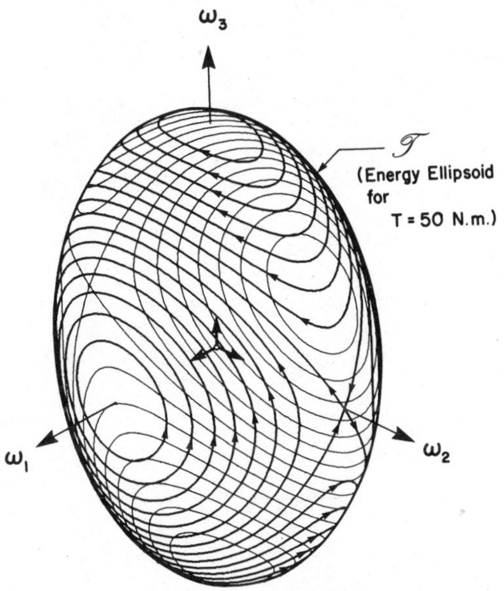

FIGURE 4.9 Family of polhodes, for varying angular momentum h, on the energy ellipsoid $\mathscr{T}$ corresponding to $T = 50$ N · m. Here, $\{I_1, I_2, I_3\} = \{6, 4, 2\}$ kg · m^2.

polhodes. Moreover, since the motion must be continuous, $\underset{\rightarrow}{\omega}$ cannot jump from one polhode to the other. This picture is a generalization for tri-inertial bodies of what Fig. 4.3 shows for inertially axisymmetrical bodies.

For a given energy T, and thus a fixed ellipsoid $\mathscr{T}$, one may vary I from its minimum value, $I = I_3$, for which $h = (2I_3T)^{1/2}$, to its maximum, $I = I_1$, for which $h = (2I_1T)^{1/2}$, and for each I find a polhode on $\mathscr{T}$. This is illustrated in Fig. 4.9, where I is varied from 2.25 to 5.75, in steps of 0.25; $\mathscr{T}$ is subdivided into four regions by the boundary polhode associated with $I = I_2$. The projections of these polhodes on the coordinate planes are shown in Figs. 4.10*a*, and the corresponding regions are labeled in Figs. 4.10*b*. The boundary polhodes ($I = I_2$) are given analytically by (12). (Problem 4.10 studies these projections further.) All of these conclusions can of course be deduced directly from the energy and momentum integrals (3). The added information provided by the third and final quadrature is the time dependence of $\boldsymbol{\omega}(t)$ as it proceeds along the polhodes—see (10) through (12). If $\mathscr{H}$ lies entirely within $\mathscr{T}$, or vice versa, no real motion is possible. The ellipsoid $\mathscr{T}$ (Figs. 4.4 through 4.9) has an important property, considered below, in connection with the attitude motion of $\mathscr{R}$ with respect to $\mathscr{F}_i$.

Instead of holding T fixed and varying h, one may instead do the converse, which results in a family of polhodes on $\mathscr{H}$. This construction, which is illustrated in Fig. 4.11, has an important application to be exploited in the next chapter. Note that the "boundary" polhodes, the ones corresponding to $I = I_2$, have been arranged to be identical in Figs. 4.9 and 4.11. For these, $T = 50$ N · m and $h = 20$ N · m · s.

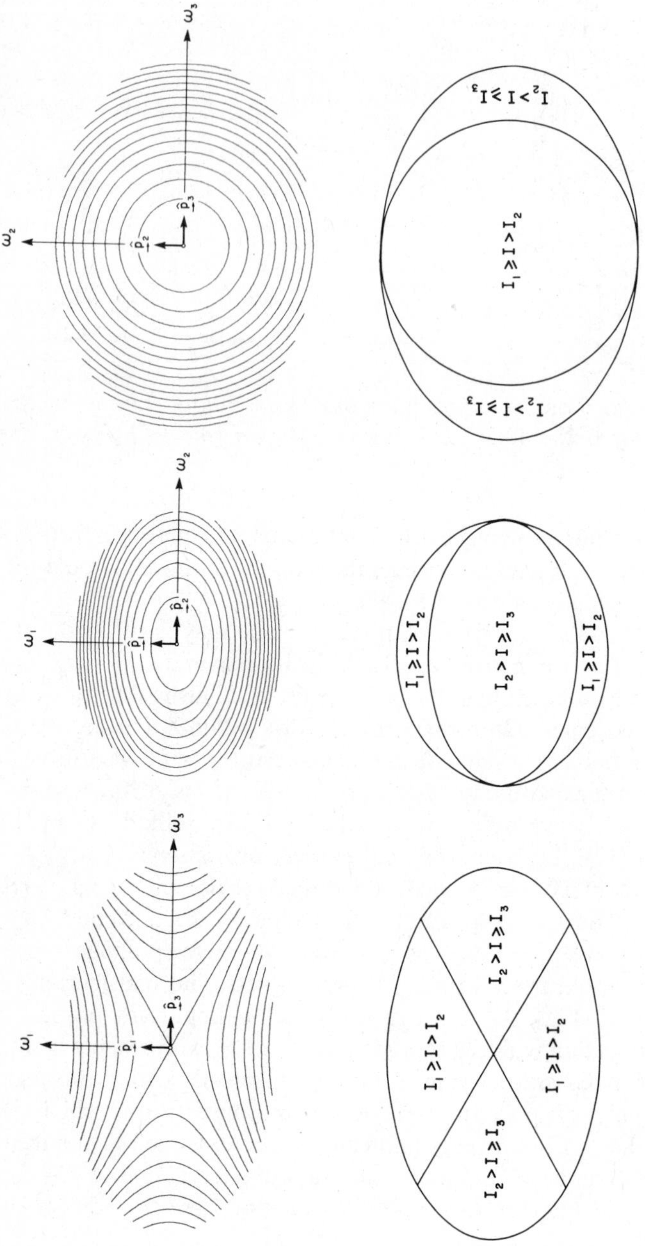

FIGURE 4.10 Projections of $\mathscr{T}$-polhodes onto principal planes.

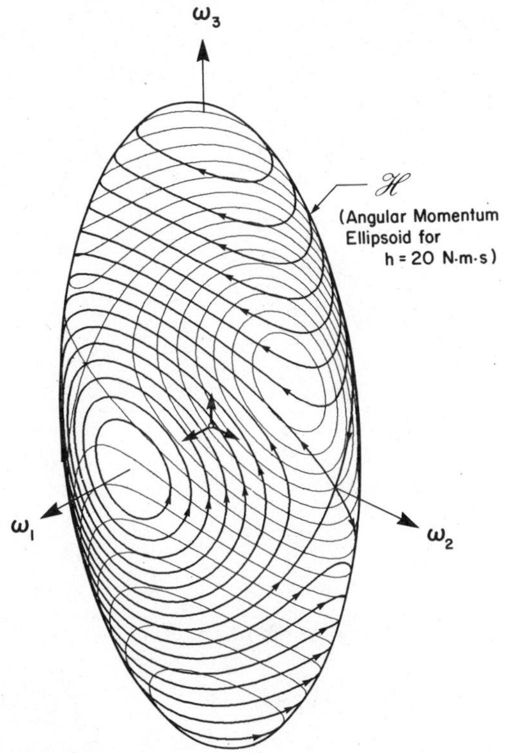

FIGURE 4.11 Family of polhodes, for varying energy T, on the angular-momentum ellipsoid $\mathscr{H}$ corresponding to $h = 20$ N · m · s. Here, $\{I_1, I_2, I_3\} = \{6, 4, 2\}$ kg · m^2.

Solution for Attitude History C(t)

Having obtained and interpreted $\boldsymbol{\omega}(t)$, we are near to our final goal, the attitude motion of $\mathscr{R}$ with respect to $\mathscr{F}_i$. Consider first Euler angles. Again choosing, as in (4.2, 17), to align $\underset{\rightarrow}{\hat{\mathbf{i}}}_3$ with $\underset{\rightarrow}{\hat{\mathbf{h}}}$, we combine (4.2, 18) with (4.1, 6):

$$h \sin\mu \sin\gamma = I_1\omega_1$$

$$h \cos\mu \sin\gamma = I_2\omega_2$$

$$h \cos\gamma = I_3\omega_3 \tag{13}$$

The solution for γ and μ is therefore

$$\gamma(t) = \cos^{-1}\left(\frac{I_3\omega_3}{h}\right) \qquad (0 \le \gamma \le \pi) \tag{14}$$

$$\mu(t) = \tan^{-1}(I_1\omega_1, I_2\omega_2) \qquad (0 \le \mu < 2\pi) \tag{15}$$

with (10) through (12) to be inserted. The double argument in (15) provides the

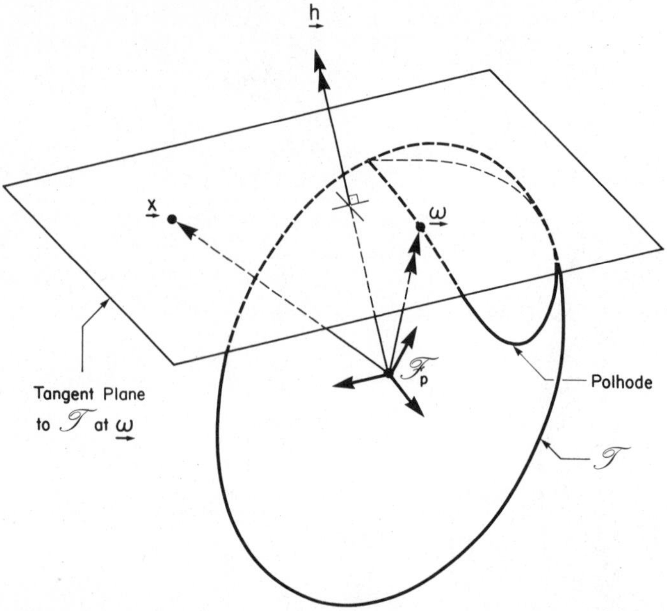

FIGURE 4.12 Poinsot's geometrical interpretation of rigid-body motion.

range $0 \leq \mu < 2\pi$. The third angle, θ_3, is found from (4.2, 21):

$$\dot{\theta}_3 = \frac{\omega_3 - \dot{\mu}}{\cos\gamma} \tag{16}$$

After substitution of μ, the needed integration has the form of an elliptic integral of the third kind (see [Wittenburg]). This development for θ_3 is not presented here in detail because our interest centers primarily on the nutation angle γ. For an analytical solution in terms of Euler parameters, the reader is referred to [Morton, Junkins, and Blanton].

Geometrical Interpretation of C(t)

The key to geometrically interpreting the attitude motion lies in recognizing that $\underset{\rightarrow}{h}$, as viewed in $\underset{\rightarrow}{\mathscr{F}}_i$, is a constant vector. (This is essentially the same technique used in deriving the rolling-cone analogy for inertially axisymmetrical bodies; see Fig. 4.5.) With Poinsot, we consider the energy ellipsoid $\mathscr{T}$ (Figs. 4.8 and 4.9). Let $\mathbf{x}$ be any point on the plane tangent to this ellipsoid (Fig. 4.12) at the point $\boldsymbol{\omega}$. From analytical geometry, the equation for this tangent plane is

$$I_1\omega_1 x_1 + I_2\omega_2 x_2 + I_3\omega_3 x_3 = 2T \tag{17}$$

where $\boldsymbol{\omega}$ is regarded as temporarily fixed, and $\mathbf{x}$ is free to move in the plane. In preparation for the key step to follow, (17) is rewritten as follows:

$$\mathbf{h}^T\mathbf{x} = 2T \tag{18}$$

representing the tangent plane to $\mathscr{T}$, at one instant in time, for a motion whose integrals are h^2 and T. A complementary observation is that, from the definitions of $\mathbf{h}$, $\boldsymbol{\omega}$, and T,

$$\mathbf{h}^T\boldsymbol{\omega} = 2T \tag{19}$$

for the entire motion ($\mathbf{h}$ and T given). Originating from this equation is the notion of an invariable plane—a plane that, for constant $\underrightarrow{\mathbf{h}}$ and T, is fixed (invariable) in an inertial (nonrotating) frame $\mathscr{F}_i$. Apparently, once $\mathbf{h}$ and T are given, the tip of $\boldsymbol{\omega}$ must remain in this invariable plane. The geometrical interpretation of Poinsot rests on the simple but elegant inference drawn from comparing (18) with (19): the plane instantaneously tangent to $\mathscr{T}$ is in fact a plane fixed in $\mathscr{F}_i$. Since $\mathscr{T}$ is fixed in $\mathscr{F}_b$, we conclude (Fig. 4.12) that *the torque-free motion of a rigid body consists of the energy ellipsoid rolling without slipping on the invariable plane*. The path traced out on the invariable plane by the moving point of contact is called the *herpolhode*. This interpretation reduces to the rolling-cone analogy, as it should, when $I_2 = I_1$ or $I_2 = I_3$. Other geometrical interpretations of rigid-body motion have been contributed by MacCullagh and others; many further references are found in [Leimanis]. A recent analogy in terms of nonlinear oscillators has also been given by [Junkins, Jacobson, and Blanton].

Another (related) geometrical interpretation is to examine the motion of the unit vector $\underrightarrow{\hat{\mathbf{i}}}_3 \equiv \underrightarrow{\hat{\mathbf{h}}}$ within $\mathscr{R}$. Let the direction cosines of $\underrightarrow{\hat{\mathbf{h}}}$ in $\mathscr{F}_p$ be $\{c_1, c_2, c_3\}$. In other words, $[c_1 \quad c_2 \quad c_3]^T$ is the third column in the rotation matrix $\mathbf{C}_{pi}(t)$, which relates $\mathscr{F}_p$ to $\mathscr{F}_i$ via $\mathbf{C}_{pi} \triangleq \mathscr{F}_p \cdot \mathscr{F}_i^T$. Thus, the components of $\underrightarrow{\mathbf{h}}$ in $\mathscr{F}_p$ are $[hc_1 \quad hc_2 \quad hc_3]^T$. However, from (4.1, 6), these same components are known to be $[I_1\omega_1 \quad I_2\omega_2 \quad I_3\omega_3]^T$. Therefore, the following correspondence

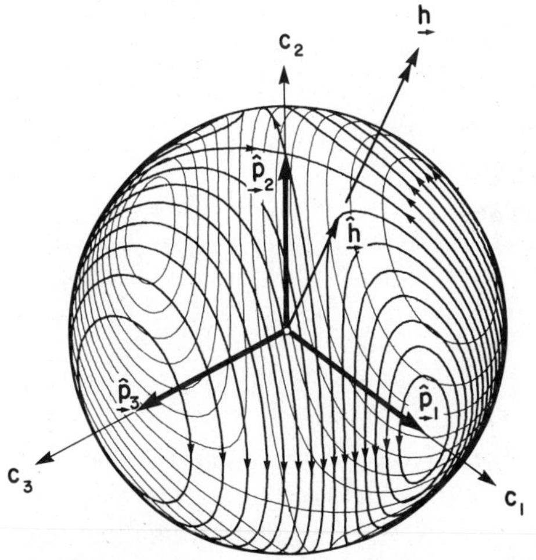

FIGURE 4.13 Attitude motion of a rigid body $\mathscr{R}$. The unit sphere is fixed in $\mathscr{R}$, and $\underrightarrow{\mathbf{h}}$ is inertially fixed. Each curve corresponds to a fixed energy. Here, $\{I_1, I_2, I_3\} = \{6, 4, 2\}$ kg · m^2.

has been established

$$hc_1 = I_1\omega_1; \qquad hc_2 = I_2\omega_2; \qquad hc_3 = I_3\omega_3 \tag{20}$$

In place of the momentum ellipsoid $\mathscr{H}$, we now have the unit sphere:

$$c_1^2 + c_2^2 + c_3^2 = 1 \tag{21}$$

Similarly, the energy ellipsoid, now expressed in terms of $\{c_1, c_2, c_3\}$ is defined by

$$\mathscr{T}: \quad \frac{c_1^2}{I_1} + \frac{c_2^2}{I_2} + \frac{c_3^2}{I_3} = \frac{2T}{h^2} = \frac{1}{I} \tag{22}$$

and the intersection of the two (Fig. 4.13) defines the attitude motion of the inertially fixed direction $\underset{\rightarrow}{\mathbf{h}}$ as seen in $\mathscr{R}$. In viewing Fig. 4.13, one should maintain $\underset{\rightarrow}{\mathbf{h}}$ fixed and move the frame $\mathscr{F}_p$ so that $\underset{\rightarrow}{\mathbf{h}}$ stays in the "slot" defined by the curve of intersection. Once again, if $\mathscr{T}$ lies entirely with the unit sphere, or vice versa, no corresponding physical motion exists.

4.4 STABILITY OF MOTION FOR $\mathscr{R}$

Implicit in the foregoing are three elementary solutions to Euler's equations (4.3, 1). These are the "pure spins":

$$\omega_1 \equiv \nu = \text{constant}; \qquad \omega_2 = \omega_3 \equiv 0 \tag{1}$$

$$\omega_2 \equiv \nu = \text{constant}; \qquad \omega_3 = \omega_1 \equiv 0 \tag{2}$$

$$\omega_3 \equiv \nu = \text{constant}; \qquad \omega_1 = \omega_2 \equiv 0 \tag{3}$$

They satisfy the motion equations for any value of ν. For the present, we keep the convention that $I_1 > I_2 > I_3$ and refer to (1), (2), and (3) as a "major-axis spin," an "intermediate-axis spin," and a "minor-axis spin," respectively. These solutions correspond, respectively to $I = I_1, I_2, I_3$ and may be readily discerned as the pivotal solutions in the diagrams shown in Figs. 4.8 and 4.9. The stability of these solutions is a central issue in spacecraft attitude dynamics.

Liapunov's Second Method; ω-Stability

Before a discussion of stability is initiated, it is mandatory that the meaning of the word "stability" be clearly understood. As used in this book, "stable" means *Liapunov stable*. Appendix A is devoted to the needed stability definitions and theorems. Briefly, a reference motion $\mathbf{x}(t)$ is Liapunov stable if a neighboring motion $\mathbf{x}(t) + \Delta\mathbf{x}(t)$ can be made to lie arbitrarily close to $\mathbf{x}(t)$ (in other words, $\|\Delta\mathbf{x}\| < \varepsilon$ for any $\varepsilon > 0$) for all $t > t_o$ by sufficiently reducing their disparity at $t = t_o$ $[0 < \|\mathbf{x}(t_o)\| < \delta(\varepsilon, t_o)]$. In the present context, the state vector $\mathbf{x}$ consists of the three angular velocity components $\{\omega_1, \omega_2, \omega_3\}$ and the attitude variables (e.g., Euler angles or Euler parameters). Several techniques will now be brought to bear on this problem, including infinitesimal (linear) stability analysis (see Sections A.3 and A.4) and Liapunov's indirect method (Section A.5).

Among the three simple-spin solutions under scrutiny, one can be treated with dispatch: the intermediate-axis spin is unstable. This instability is strongly suggested in geometrical terms by the polhodes of Fig. 4.8. We can show that the solution ($\omega_2 \equiv \nu$; $\omega_1 = \omega_3 \equiv 0$) is unstable with respect to variations in $\boldsymbol{\omega}$, designated $\boldsymbol{\xi}$. The attitude variables need not even be considered. To confirm this assertion, we call upon Theorem A.22 of Appendix A and, following [Chetayev], select the Liapunov function:

$$v \triangleq \xi_1 \xi_3 \tag{4}$$

From the motion equations (4.3, 1), in which we now use $\omega_1 \equiv \xi_1$ and $\omega_3 \equiv \xi_3$,

$$\dot{v} = (\nu + \xi_2) \frac{I_1(I_1 - I_2)\xi_1^2 + I_3(I_2 - I_3)\xi_3^2}{I_1 I_3} \tag{5}$$

The neighborhood $\mathscr{N}$ and region $\mathscr{R}$ (in $\boldsymbol{\omega}$-space) referred to in the theorem are now identified:

$$\mathscr{N}: \qquad \xi_2 > -\nu \tag{6}$$

$$\mathscr{R}: \qquad \xi_1 \xi_3 > 0 \tag{7}$$

The boundary of $\mathscr{R}$ is $\xi_1 \xi_3 = 0$, which clearly includes the point $\boldsymbol{\xi} = \mathbf{0}$. Since all the conditions of the theorem are met (remember $I_1 > I_2 > I_3$), it is concluded that the intermediate-axis spin is unstable. The elegant feature of this proof is that no linearization is required and an explicit solution for $\boldsymbol{\omega}(t)$ is not needed.

The stability of the major- and minor-axis spins can also be investigated by this method, whose strength, to repeat, is that no restriction is placed on the size of $\|\boldsymbol{\xi}\|$ (linearization is not performed) and that it is not necessary to solve the motion equations. Again confining attention to perturbations $\boldsymbol{\xi}$, and ignoring attitude variables for the present, consider the reference motion consisting of a major-axis spin ($\omega_1 \equiv \nu$; $\omega_2 = 0 = \omega_3$). We might select as a Liapunov function

$$v \triangleq 2I_1 T - h^2 \equiv 2(I_1 - I)T \qquad \text{(tentative)}$$

where I and $\mathbf{h}$ are the kinetic energy and angular momentum of the new (neighboring) motion. According to (4.3, 3), with $\omega_1 = \nu + \xi_1$, $\omega_2 \equiv \xi_2$, and $\omega_3 \equiv \xi_3$,

$$v \triangleq I_2(I_1 - I_2)\xi_2^2 + I_3(I_1 - I_3)\xi_3^2 \qquad \text{(tentative)}$$

Because $I_1 > I_2$ and $I_1 > I_3$, this v is positive-definite with respect to ω_2 and ω_3. However, it is not positive-definite with respect to ξ_1 because v can be zero without ξ_1 being zero. To repair this defect, we instead define

$$v = 2(I_1 - I)T + 4(T - T_\nu)^2 \tag{8}$$

where $T_\nu \triangleq \frac{1}{2} I_1 \nu^2$. Even when $\omega_2 = \omega_3 = 0$, this new $v = (2I_1 \nu \xi_1 + I_1 \xi_1^2)^2$. That is, v is now positive-definite with respect to $\boldsymbol{\xi}$. It is therefore an appropriate Liapunov function in $\boldsymbol{\xi}$. Moreover, since all the entries in (8) are constants of the motion,

$$\dot{v} = 0 \tag{9}$$

evaluated for any motion. Therefore, by the authority of Theorem A.18 in Appendix A, a major-axis spin is $\boldsymbol{\omega}$-stable with respect to perturbations $\boldsymbol{\xi}$. Similarly, by choosing

$$v = 2(I - I_3)T + 4(T - T_\nu)^2 \tag{10}$$

where $T_\nu \triangleq \frac{1}{2}I_3\nu^2$, we conclude that a minor-axis spin is $\boldsymbol{\omega}$-stable with respect to perturbations $\boldsymbol{\xi}$.

To summarize the stability results thus far: The intermediate-axis spin of a rigid body is unstable and the major- and minor-axis spins are $\boldsymbol{\omega}$-stable with respect to perturbations $\boldsymbol{\xi}$. A major question left unanswered at this point is whether they are also *attitude* stable, and whether they are stable also with respect to initial changes in attitude. The answer, as we shall see, must be a carefully qualified one.

Attitude Stability: Infinitesimal Analysis

Of great interest from a spacecraft stabilization standpoint is *attitude* stability. We shall first study *infinitesimal* deviations from the reference motion, using a variational stability analysis whose procedures, strengths, and limitations are discussed in Appendix A (Section A.4). In particular, an apprehended instability will imply instability for the "exact" nonlinear system of equations. Using this technique, we shall show further that none of the three relative equilibria (pure-spin solutions) for $\mathscr{R}$ is *attitude* stable. A lesser form of attitude stability, *directional* stability, will be defined and demonstrated to exist for the variational (linear) equations under certain circumstances. This finding will then be confirmed to persist for *finite* perturbations using Liapunov functions.

To discuss attitude stability, it is of course necessary to introduce attitude variables. The most convenient attitude variables for an infinitesimal analysis of deviations from a pure-spin motion are the quantities $\{\alpha_1, \alpha_2, \alpha_3\}$ described in Chapter 2, above equation (2.3, 33). They measure the attitude deviation of $\mathscr{R}$ from its reference constant-$\boldsymbol{\omega}$ motion. At this point, we also *abandon the convention $I_1 > I_2 > I_3$ used thus far in this chapter*, preferring now to choose $\vec{\hat{\mathbf{p}}}_2$ as the axis about which the reference pure-spin motion occurs. Thus, I_2 is no longer necessarily the intermediate moment of inertia, and all relative magnitudes of $\{I_1, I_2, I_3\}$ are allowed. The advantage of this new convention is that a single derivation of variational equations suffices for all three reference motions. The choice of the 2-axis for the reference spin, instead of the 1-axis or the 3-axis, anticipates the once-per-orbit spin (with respect to $\mathscr{F}_i$) of Earth-pointing spacecraft, considered in Chapter 9. This reference rotation is about the orbit normal, which is conventionally called the 2-axis (*pitch* axis) in aerospace flight mechanics. Thus, we select $\boldsymbol{\nu} = \mathbf{1}_2\nu$ and substitute this selection in (2.3, 33). Written out in full,

$$\begin{aligned} \omega_1 &= \dot{\alpha}_1 + \nu\alpha_3 \\ \omega_2 &= \dot{\alpha}_2 + \nu \\ \omega_3 &= \dot{\alpha}_3 - \nu\alpha_1 \end{aligned} \tag{11}$$

These are accurate to first order in $\boldsymbol{\alpha}$ and $\dot{\boldsymbol{\alpha}}$. Note that *all six* state variables, $\{\alpha_1, \alpha_2, \alpha_3, \dot{\alpha}_1, \dot{\alpha}_2, \dot{\alpha}_3\}$, are now taken into consideration.

The state variable α_2 proves to be a source of instability. Substituting (11) into Euler's motion equations (4.3, 1) and neglecting terms higher than first order, one arrives at the linearized motion equations

$$I_2\ddot{\alpha}_2 = 0 \tag{12}$$

$$\left.\begin{aligned} I_1\ddot{\alpha}_1 + (I_3 + I_1 - I_2)\nu\dot{\alpha}_3 + (I_2 - I_3)\nu^2\alpha_1 &= 0 \\ I_3\ddot{\alpha}_3 - (I_3 + I_1 - I_2)\nu\dot{\alpha}_1 + (I_2 - I_1)\nu^2\alpha_3 &= 0 \end{aligned}\right\} \tag{13}$$

Clearly, the motion is unstable by virtue of the fact that perturbations $\dot{\alpha}_2(0)$, however small, cause unbounded growth in $\alpha_2(t)$.

More promising, however, is the motion of α_1 and α_3. The structure of (13) can be recognized as being of the canonical matrix second-order form discussed in Section A.6 of Appendix A:

$$\mathscr{M}\ddot{\mathbf{q}} + \mathscr{G}\dot{\mathbf{q}} + \mathscr{K}\mathbf{q} = \mathbf{0} \tag{14}$$

where

$$\mathbf{q} \triangleq [\alpha_1 \quad \alpha_3]^T; \qquad \mathscr{M} \triangleq \operatorname{diag}\{I_1, I_3\}$$

$$\mathscr{G} \triangleq (I_3 + I_1 - I_2)\nu\begin{bmatrix} 0 & 1 \\ -1 & 0 \end{bmatrix}; \qquad \mathscr{K} \triangleq \begin{bmatrix} (I_2 - I_3)\nu^2 & 0 \\ 0 & (I_2 - I_1)\nu^2 \end{bmatrix} \tag{15}$$

The canonical form includes the conditions $\mathscr{M}^T = \mathscr{M} > 0$, $\mathscr{G}^T = -\mathscr{G}$, and $\mathscr{K}^T = \mathscr{K}$, which are here clearly met. To examine the stability properties of (14), several of the theorems of Appendix A are applicable. One could either (i) find the characteristic equation and use the Routh–Hurwitz criteria of Section A.3; (ii) use the general matrix-second-order stability theory of Section A.6; or (iii) solve the characteristic equation explicitly in this fourth-order case. Each of these three approaches is more specialized than its predecessors. We shall not use the first (Routh) possibility at all for this problem but approaches (ii) and (iii) are now dealt with in turn.

According to matrix-second-order stability theory, the motion $\mathbf{q} \equiv \mathbf{0}$ in (14) cannot be asymptotically stable; this would require a damping term (to be added in Chapter 5). The motion $\mathbf{q}(t)$ will be stable, however, if $\mathscr{K} > 0$, which we refer to as the condition for *static stability*. It is evident from (15) that $\mathscr{K} > 0$ if and only if $I_2 > I_3$ and $I_2 > I_1$, that is, if $\hat{\mathbf{p}}_2$ is the axis of maximum inertia. There may be other possibilities, however, since $\mathscr{K} > 0$ is not *necessary* for stability. It is possible that $\mathscr{G}$ might stabilize the motion even if $\mathscr{K} \ngtr 0$, an eventuality we call *gyric stabilization*.* To investigate this possibility, one forms the characteris-

*See footnote, p. 511

tic equation

$$\det[\mathscr{M}s^2 + \mathscr{G}s + \mathscr{K}] = 0 \tag{16}$$

This determinant can be found quite simply in the present case, and the characteristic equation has the simple form

$$b_0 s^4 + b_1 s^2 + b_2 = 0 \tag{17}$$

where

$$\begin{aligned} b_0 &\triangleq I_1 I_3 \\ b_1 &\triangleq (I_2^2 + 2I_1I_3 - I_1I_2 - I_3I_2)\nu^2 \\ b_2 &\triangleq (I_2 - I_1)(I_2 - I_3)\nu^4 \end{aligned} \tag{18}$$

According to Table A.2, the **q**-motion is stable if and only if

(i) $b_0 > 0$
(ii) $b_1 > 0$
(iii) $b_2 > 0$
(iv) $b_1^2 - 4b_0b_2 > 0$ (19)

The first of these is always satisfied, and the third requires that $\hat{\mathbf{p}}_2$ be either a major *or a minor* axis of inertia. Before imposing the remaining two conditions, an ancillary requirement is now introduced, which recognizes that not all values of I_1, I_2, and I_3 are admissible.

The Inertia Ratios k_1 and k_3

It is already clear form (14) and (15) that only the *ratios* $I_1 : I_2 : I_3$, and not their individual magnitudes, govern the characteristics of the motion. Therefore, matters can be clarified considerably by the introduction of suitable inertia ratios. We choose

$$k_1 \triangleq \frac{I_2 - I_3}{I_1}; \qquad k_3 \triangleq \frac{I_2 - I_1}{I_3} \tag{20}$$

There are two advantages to this device. First, it can be shown (Problem 4.11) that

$$|k_1| < 1; \qquad |k_3| < 1 \tag{21}$$

for real bodies, with $|k_1| = 1$ or $|k_3| = 1$ as limiting cases. These conditions therefore guarantee that only physically meaningful values of I_1, I_2, and I_3 are used. The second advantage of using k_1 and k_3 is that the number of inertial parameters needed is reduced from three to two, thereby facilitating simple stability diagrams in parameter space.

In terms of $\{k_1, k_3\}$, the characteristic equation for (17) takes the form

$$\left(\frac{s}{\nu}\right)^4 + \hat{b}_1\left(\frac{s}{\nu}\right)^2 + \hat{b}_2 = 0 \tag{22}$$

where

$$\hat{b}_1 \triangleq 1 + k_1 k_3$$

$$\hat{b}_2 \triangleq k_1 k_3 \qquad (23)$$

The form of (22) also emphasizes that the role of ν is to establish a characteristic time for the problem; it has no influence on stability conditions. Expressed in terms of $\hat{b}_1$ and $\hat{b}_2$, the stability conditions are

(i) $\hat{b}_1 = 1 + k_1 k_3 > 0$
(ii) $\hat{b}_2 = k_1 k_3 > 0$
(iii) $\hat{b}_1^2 - 4\hat{b}_2 = (1 - k_1 k_3)^2 > 0$ (24)

Stated in this manner, the conditions for stability are considerably simplified because, in view of (21), the only condition not automatically satisfied except for limiting cases is (24ii). This condition, $k_1 k_3 > 0$, specifies that the axis of the reference spin $\hat{\mathbf{p}}_2$ must be either the major or minor axis of $\mathscr{R}$. The major-axis spin is statically stable, as proved earlier, and the minor-axis spin, while *statically unstable*, is *dynamically stable*. It is gyrically stabilized by $\mathscr{G}$. At the moment, there is no reason to suspect gyric stabilization; but, as we shall see in Chapter 5, it is a peculiarity of the *rigidity* assumption for $\mathscr{R}$ and cannot actually occur for real bodies.

Use will frequently be made of the inertia ratios k_1 and k_3, so it is worthwhile pausing to understand their significance (Fig. 4.14). As noted in (21), physically real bodies are associated with points in the k_1–k_3 plane that lie within the square whose vertices are located at a unit distance from the axes. Within this square may be found points corresponding to any desired ordering of the principal inertias $\{I_1, I_2, I_3\}$. Note that in the {first, second, third, fourth}

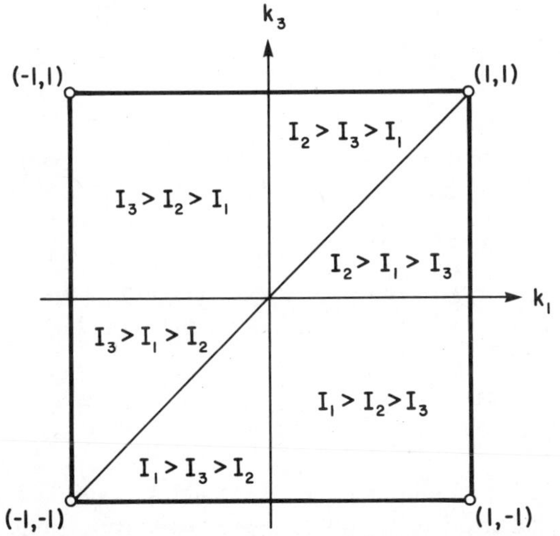

FIGURE 4.14 The $k_1 - k_3$ diagram. Note: $k_1 = (I_2 - I_3)/I_1$, $k_3 = (I_2 - I_1)/I_3$.

quadrant, $\hat{\mathbf{p}}_2$ is respectively the {major, intermediate, minor, intermediate} axis of inertia.

Of course, in the present relatively simple context—the stability properties of a single spinning rigid body—the characteristic equation (17) is so simple that it can be solved in closed form. This simplicity cannot be expected to continue however, as more complicated models are dealt with (e.g., spinning flexible bodies). The characteristic equation (17) factors neatly:

$$(s^2 + \nu^2)\left[I_1 I_3 s^2 + (I_2 - I_1)(I_2 - I_3)\nu^2\right] = 0 \tag{25}$$

showing that two zeros are $s = \pm j\nu$, independently of the inertia ratios. The other two zeros are $s = \pm j\Omega$, where

$$\Omega = \left[\frac{(I_2 - I_1)(I_2 - I_3)}{I_1 I_3}\right]^{1/2} \nu \tag{26}$$

This is an obvious extension of the definition of Ω for a symmetrical rigid body, given by (4.2, 5). The above definition is less general in one respect however: while (4.2, 5) applies to finite rotations, (26) is restricted to infinitesimal attitude excursions (with respect to a reference frame rotating uniformly with constant angular velocity $\boldsymbol{\nu}$). It is also clear that the motion is oscillatory only if Ω is real, that is, if I_2 is either the greatest or least moment of inertia, in accordance with Figs. 4.14 and 4.15. Note that $\Omega(k_1, k_3)$ is simply

$$\Omega = (k_1 k_3)^{1/2} \nu \tag{27}$$

providing the basis for curves of constant Ω/ν, which could readily be drawn (rectangular hyperbolas) in the first and third quadrants of the k_1–k_3 plane, with $\Omega/\nu \to 0$ as the axes are approached, and $\Omega/\nu \to 1$ as the $(1,1)$ and $(-1,-1)$ points are approached. In the second and third quadrants, the motion is unstable. The unstable mode grows exponentially with a time-to-double t_D, given by

$$\nu t_D = \left[\frac{I_1 I_3}{(I_1 - I_2)(I_2 - I_3)}\right]^{1/2} \ln 2 = (-k_1 k_3)^{-1/2} \ln 2 \tag{28}$$

which is valid within the linear approximation.

To more completely understand the motion, one must know the eigencolumns associated with each eigenvalue. Denoting eigencolumns by $\mathbf{q}_e$, that is,

$$\mathbf{q}(t) = \mathbf{q}_e e^{st}; \qquad \mathbf{q}_e \equiv [\alpha_{1e} \quad \alpha_{3e}]^T \tag{29}$$

the eigencolumns corresponding to $s = \pm j\nu$ have $\alpha_{1e} = \pm j\alpha_{3e}$, indicating motion on a circular cone, clockwise or counterclockwise depending on the sign chosen. For the other two eigenvalues ($s = \pm j\Omega$),

$$\frac{\alpha_{1e}}{\alpha_{3e}} = \frac{\pm\nu}{j\Omega}\left(\frac{I_2 - I_1}{I_1}\right) \equiv \frac{j\Omega}{\pm\nu}\left(\frac{I_3}{I_3 - I_2}\right) \tag{30}$$

indicating (at least for real Ω) motion on an elliptic cone. The general motion is a superposition of these constituent characteristic motions, the relative proportions being determined by the initial conditions.

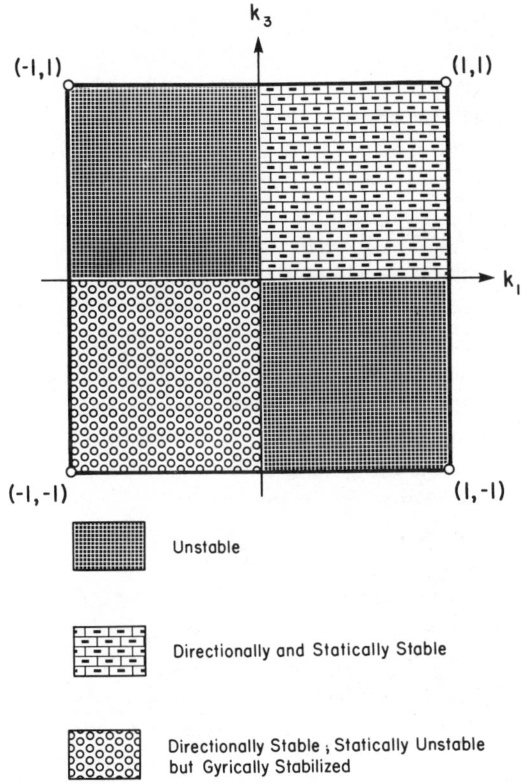

FIGURE 4.15 Stability diagram for simple spin of a rigid body about one of its principal axes.

Directional Stability

We have seen that the torque-free motion of a rigid body, spinning about any of its three principal axes, is unstable. The nature of this instability depends on the principal axis about which the spin occurs. The intermediate-axis instability is traceable simply to the stationarity of the energy and angular-momentum integrals with respect to this axis (see Figs. 4.9 and 4.11). For the other two principal axes, however, the instability is a consequence of the inability of the angular momentum vector $\underset{\rightarrow}{\mathbf{h}}$ to resist angular-rate perturbations about itself. This observation remains valid when damping, spinning wheels, or both are added to $\mathscr{R}$, as will become plain in succeeding chapters.

From this general condition of instability, a restricted but important kind of stability can be salvaged. It is a genus of stability specially adapted to attitude dynamics: *directional stability*. It is based on the fact that while the angular momentum vector $\underset{\rightarrow}{\mathbf{h}}$ cannot resist angular rate perturbations *about* itself, it can bring about stability with respect to perturbations *perpendicular to* itself. Thus, under certain conditions, a body-fixed axis aligned in the reference motion $\underset{\rightarrow}{\mathbf{h}}$ will deviate arbitrarily slightly from this direction, provided the initial disturbances are sufficiently small. In other words, the *direction* of this body-fixed axis is Liapunov stable. Couched in these terms, it has been demonstrated in the foregoing development that the major axis is directionally stable (at least infinitesimally) for a major-axis spin and that the minor axis is similarly directionally stable for a minor-axis spin (see Problem 4.12). The relevant stability condition (24ii) is portrayed in Fig. 4.15.

It will now be shown that these two cases of directional stability prevail also with noninfinitesimal initial disturbances. This can be done with the aid of the analytical solutions available in (4.3, 10 and 11), *for which the convention* $I_1 > I_2 > I_3$ *applies*; this is the approach taken in Problem 4.13. Instead, we shall demonstrate directional stability for major- and minor-axis spins directly from the energy and angular-momentum integrals (4.3, 3). To be precise, it must be admitted that the angular momentum vector subsequent to an initial disturbance is not identical to the angular momentum vector prior to that disturbance, either in magnitude or in direction, Insofar as the *magnitude* of $\underset{\rightarrow}{\mathbf{h}}$ changes, there is no impact on the argument for directional stability. As for the change in the *direction* of $\underset{\rightarrow}{\mathbf{h}}$, one must recognize that the difference in direction between the "predisturbance" $\underset{\rightarrow}{\mathbf{h}}_\nu$ and the "postdisturbance" $\underset{\rightarrow}{\mathbf{h}}$ can be made arbitrarily small by making the disturbance itself arbitrarily small (Problem 4.13*a*). Thus, it is sufficient to show directional stability with respect to the "postdisturbance" $\underset{\rightarrow}{\mathbf{h}}$.

To measure the deviation of $\underset{\rightarrow}{\mathbf{h}}$ from the spin axis $\hat{\mathbf{p}}_2$, we shall use the appropriate direction cosine, c_2. The expression of $\underset{\rightarrow}{\mathbf{h}}$ in $\mathscr{F}_p$ given by (4.3, 20), namely, $hc_i = I_i\omega_i$, is still valid, except for the caution that the *convention* $I_1 > I_2 > I_3$ *is no longer in force*. The three direction cosines of $\underset{\rightarrow}{\mathbf{h}}$ expressed in $\mathscr{F}_p$ still satisfy (4.3, 21 and 22). In particular,

$$c_1^2 + c_2^2 + c_3^2 = 1 \tag{31}$$

so that our desire to be able to make the angle between $\hat{\mathbf{p}}_2$ and $\underset{\rightarrow}{\hat{\mathbf{h}}}$ arbitrarily small translates into making $c_1^2 + c_3^2$ arbitrarily small. We choose as a Liapunov function

$$v_M \triangleq 2(T - T_h) \tag{32}$$

where, as usual,

$$2T = h^2\left(\frac{c_1^2}{I_1} + \frac{c_2^2}{I_2} + \frac{c_3^2}{I_3}\right) \tag{33}$$

and

$$2T_h \triangleq \frac{h^2}{I_2} \tag{34}$$

Substituting (31) and (33) through (34) in (32), we can express v_M in terms of c_1 and c_3, that is, in terms of the deviations of $\underset{\rightarrow}{\hat{\mathbf{p}}_2}$ from $\underset{\rightarrow}{\hat{\mathbf{h}}}$:

$$I_1 I_2 I_3 v_M = h^2\left[I_3(I_2 - I_1)c_1^2 + I_1(I_2 - I_3)c_3^2\right] \tag{35}$$

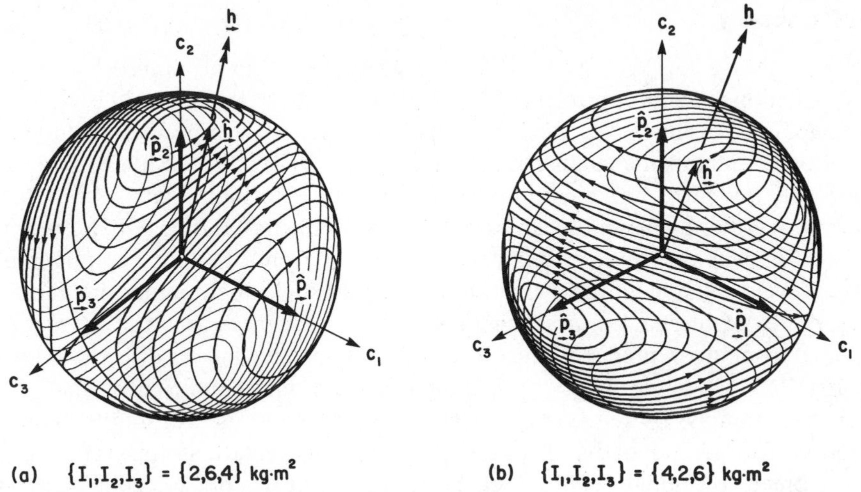

FIGURE 4.16 Motion of a rigid body with respect to the angular momentum vector $\underset{\rightarrow}{\mathbf{h}}$, indicating directional stability. (See also Fig. 4.13.)

This expression is the key to providing directional stability for a spinning rigid body $\mathscr{R}$.

Suppose first that $\hat{\mathbf{p}}_2$ is the major axis of inertia. From (32) it is evident that v_M is constant throughout the motion, and from (35) it is also seen that v_M is positive-definite in $\{c_1, c_3\}$. Therefore, since v_M can be made arbitrarily small by making the initial directional disturbances $\{c_1, c_3\}$ sufficiently small, and since $\underset{\rightarrow}{\mathbf{h}}$ can be made arbitrarily close to $\underset{\rightarrow}{\mathbf{h}}_\nu$, it can be concluded that a major-axis spin is directionally stable. This behavior is depicted in Fig. 4.16*a*, which shows the unit sphere fixed in $\mathscr{R}$, given by (31), and the curves of intersection with the energy ellipsoid (33) for $\{I_1, I_2, I_3\} = \{2, 6, 4\}$ kg $\cdot$ m^2. Thus, $\hat{\mathbf{p}}_2$ is the major axis for $\mathscr{R}$, and the curves on the surface correspond to $v_M =$ constant. The deviation of $\hat{\mathbf{p}}_2$ from the inertially fixed $\underset{\rightarrow}{\mathbf{h}}$ can be made arbitrarily small for all $t > 0$ by making this deviation small enough at $t = 0$. Thus, directional stability for the major-axis spin is geometrically evident.

Similarly, by defining the Liapunov function

$$v_m \triangleq -v_M \tag{36}$$

it can be concluded that a minor-axis spin is also directionally stable. As a geometrical illustration, Fig. 4.16*b* shows the motion of the $\mathscr{R}$-fixed unit sphere for $\{I_1, I_2, I_3\} = \{4, 2, 6\}$ kg $\cdot$ m^2. We need not be concerned again about the possibility of an intermediate-axis spin being directionally stable because it is intuitively obvious from Fig. 4.13 [and proved by the Liapunov function (4)] that an attitude deviation from the intermediate-axis spin cannot be made arbitrarily small for all subsequent time merely by making initial deviations sufficiently small.

Summary

To conclude this section, it can be stated that all "simple spins" or "permanent rotations" about a principal axis of a rigid body $\mathscr{R}$ are unstable in the sense of Liapunov, owing to the lack of attitude stability with respect to attitude rate disturbances about the angular-momentum vector $\underset{\rightarrow}{\mathbf{h}}$. This instability can be dealt with only by providing an appropriate external restoring torque (about the $\underset{\rightarrow}{\mathbf{h}}$ direction) either by exploiting an external (environmental) torque (Chapters 8 and 9) or by means of active sensing and actuation (beyond the scope of this book). The above development has shown, however, that spins about either the minor or major axes of inertia are *directionally* stable; that is, these axes can be made to deviate from their predisturbance orientation (namely, the $\underset{\rightarrow}{\mathbf{h}}_\nu$ direction) by an arbitrarily small angle simply by making the disturbances sufficiently small.

Lastly, it must be remarked that although these results are (mathematically) rigorous for a rigid body, their application to spacecraft at this stage in the development is premature. Much of the remainder of this book is concerned with specifying the caveats and provisos that must be added to these essentially classical results.

4.5 MOTION OF A RIGID BODY UNDER TORQUE

In the foregoing analysis, the external torque impressed on the rigid body $\mathscr{R}$ was zero. As a consequence, the angular momentum and the kinetic energy were constant, providing valuable integrals of the motion. These motion integrals facilitated an explicit analytical solution for the motion in terms of elliptic functions, reducing to circular functions in the case of axially symmetrical bodies. Even when external torques are present, as they are in general for spacecraft, these solutions provide a reference behavior in terms of which the actual motion can be interpreted. Especially when the external torques are small, as they are in space, it is often helpful both intuitively and analytically to treat the actual motion as a slowly evolving torque-free motion. This general viewpoint is expounded more fully in Chapter 10.

A Special Case: Self-Excited Motion

The equations for the attitude motion of a rigid body in response to external torques are given by (4.1, 7), together with some appropriately chosen kinematical equations (see the last column in Table 2.3). In view of the complexity attending the solution of these equations even for zero torque, it is not surprising that a general analytical solution does not exist when arbitrary torques are acting. In general, these differential equations of motion can be integrated only numerically. Some analytical progress can be made, however, under special assumptions, as the following brief treatment shows.

One special assumption which permits analytical progress is that the torque, expressed in body-fixed axes, is an explicit function of time, $\mathbf{g} = \mathbf{g}(t)$. Such a motion is said to be *self-excited*. A thorough review of the self-excited motions of a rigid body is the subject of Chapter 2 in [Leimanis]. We shall be content here to note that a closed-form solution is possible for *inertially axisymmetrical* bodies, as shown below, following [Wittenburg]. When the axis of symmetry is chosen to be along $\vec{\hat{\mathbf{p}}}_3$, the motion equations become, in place of (4.2, 2),

$$
\begin{aligned}
I_t\dot{\omega}_1 &= (I_t - I_a)\omega_2\omega_3 + g_1(t)\\
I_t\dot{\omega}_2 &= (I_a - I_t)\omega_3\omega_1 + g_2(t)\\
I_a\dot{\omega}_3 &= g_3(t)
\end{aligned}
\tag{1}
$$

First, assume no axial torque ($g_3 \equiv 0$). The solution for $\boldsymbol{\omega}(t)$, as studied in Problem 4.14, is

$$
\begin{aligned}
\omega_1(t) &= \omega_{1_o}\cos\Omega t + \omega_{2_o}\sin\Omega t\\
&\quad + \frac{1}{I_t}\int_0^t [g_1(\tau)\cos\Omega(t-\tau) + g_2(\tau)\sin\Omega(t-\tau)]\,d\tau\\
\omega_2(t) &= -\omega_{1_o}\sin\Omega t + \omega_{2_o}\cos\Omega t\\
&\quad + \frac{1}{I_t}\int_0^t [-g_1(\tau)\sin\Omega(t-\tau) + g_2(\tau)\cos\Omega(t-\tau)]\,d\tau\\
\omega_3(t) &\equiv \nu
\end{aligned}
\tag{2}
$$

where Ω is given in (4.2, 5).

Even when $g_3 \not\equiv 0$, however, a closed-form solution still exists. The third expression of (1) is independently integrable:

$$
\omega_3 = \omega_{3_o} + \frac{1}{I_a}\int_0^t g_3(\tau)\,d\tau \tag{3}
$$

Next, define the auxiliary variable

$$
\dot{\alpha} \triangleq \omega_3(t) \tag{4}
$$

and denote differentiation with respect to α by a prime. Then, (1) becomes

$$
\begin{aligned}
I_t\omega_1'\omega_3 &= (I_t - I_a)\omega_2\omega_3 + g_1(t)\\
I_t\omega_2'\omega_3 &= (I_a - I_t)\omega_1\omega_3 + g_2(t)
\end{aligned}
\tag{5}
$$

As a restriction on our solution, we can proceed further only if $\omega_3(t) > 0$ for all $t > 0$. This condition has important implications: $\alpha(t)$, as integrated from (4), is a strictly increasing function of time and can therefore be inverted, at least in principle, to find $t(\alpha)$. Thus, g_1 and g_2 are available as functions of α. The condition $\omega_3 > 0$ also permits us to divide (5) by ω_3:

$$
\begin{aligned}
I_t\omega_1' &= (I_t - I_a)\omega_2 + \frac{g_1(\alpha)}{\omega_3(\alpha)}\\
I_t\omega_2 &= (I_a - I_t)\omega_1 + \frac{g_2(\alpha)}{\omega_3(\alpha)}
\end{aligned}
\tag{6}
$$

In terms of the new independent variable α, the expressions in (6) are now in the same form as those in (1) after setting $\omega_3 \equiv 1$ in the latter; they can therefore be integrated to produce closed-form solutions similar to (2). For other cases where analytical progress can be made, the reader is referred to [Leimanis].

Unfortunately, the engineering significance of these developments is dubious. In the first place, expressions for torque normally depend on body attitude and angular velocity as well as time. It is also clear from the expressions for spacecraft torques developed in Chapter 8 that they are usually complicated functions which preclude literal integration of the motion equations, except in a few special cases. In the second place, mathematicians and engineers do not mean quite the same thing when they say they have "found a solution." Engineers must ultimately have in mind a numerically accessible result. To the extent that an engineer uses mathematics, he or she does so as a means to facilitate calculations. Judged by this criterion, it is not clear what role the "solutions" (2) might play. If $g_1(t)$ and $g_2(t)$ are such that (2) can only be integrated numerically, then the "solution" is just another way of stating the problem! The numerical integration of the differential equations (1) may well be as attractive as the numerical evaluation of the integrals (2).

For this reason, the two most common procedures are linearization and numerical integration. The former is particularly helpful in providing approximate design guidelines, while the latter is used in simulation and verification.

Impulsive Torques

The concept of *impulsive torque* is now examined. Like "point mass" and "rigid body," "impulsive torque" is an idealization prompted by the prospect of mathematical simplification. It can be used to advantage provided its underlying assumptions are appropriate to the circumstances. For spacecraft, the idea of impulsive torque is especially helpful in connection with the short-duration use of jet thrusters. In this respect, the situation for spacecraft attitude dynamics is analogous to spacecraft orbital dynamics, where 'guidance corrections," "mid-course maneuvers," and "perigee burns" can, for all but the most precise calculations, be treated as impulsive forces. The nature of the approximation is the same, and the analytical benefits are the same.

Mathematically, the Dirac impulse function $\delta(t)$ has the following definition [Korn and Korn]:

$$\int_{t_1}^{t_2} f(t)\,\delta(t-t_i)\,dt \triangleq \begin{cases} 0 & (t_i < t_1 \text{ or } t_i > t_2) \\ \tfrac{1}{2} f(t_i) & (t_i = t_1 \text{ or } t_i = t_2) \\ f(t_i) & (t_1 < t_i < t_2) \end{cases} \tag{7}$$

where $f(t)$ is an arbitrary function continuous for $t = t_i$; $\delta(t - t_i)$ is not a true function, but is a symbolic function, or *distribution*. It cannot be plotted, except

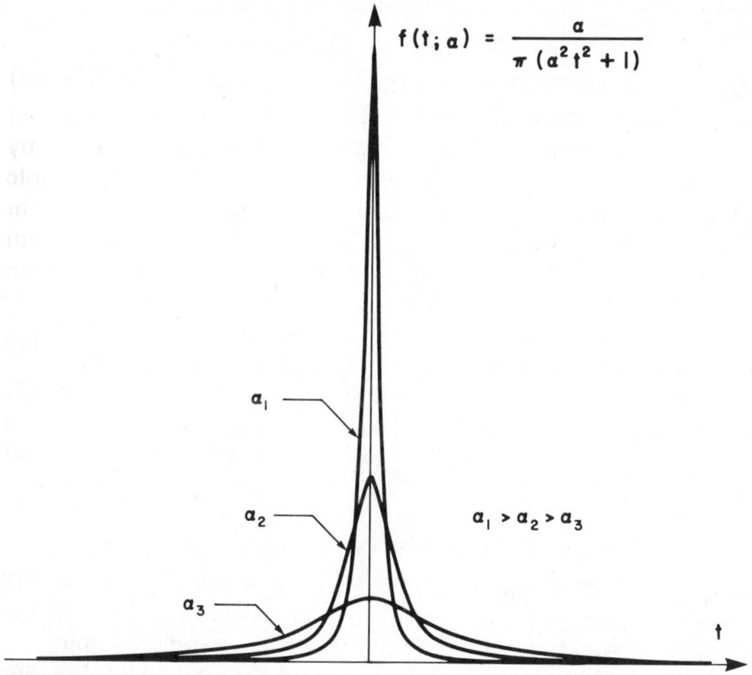

FIGURE 4.17 Example of a family of functions that asymptotes to the $\delta(t)$ function as $\alpha \to \infty$.

symbolically by an "arrow" at $t = t_i$. For example, one has

$$\lim_{\alpha \to \infty} \frac{\alpha}{\pi(\alpha^2 t^2 + 1)} = \delta(t)$$

as plotted in Fig. 4.17.

Dynamicists are interested in the δ-function, or *impulse function*, because the system response is simply a step adjustment to the system momenta at the instant the impulse acts. To discuss this matter in some generality, consider the system of motion equations (3.7, 9),

$$\dot{\boldsymbol{p}} = \boldsymbol{n}_p + \boldsymbol{f}(t) \tag{8}$$

These equations represent an arbitrarily connected set of rigid bodies and are in fact typical of an even wider class of dynamical systems. Here, $\boldsymbol{p}(t)$ and $\boldsymbol{f}(t)$ signify the momenta and external inputs.

Suppose now that $\boldsymbol{f}(t)$ is impulsive at $t = t_i$:

$$\boldsymbol{f}(t) = \boldsymbol{i}\delta(t - t_i) \tag{9}$$

and let t_i^- and t_i^+ denote instants just before and just after t_i, respectively. Now, integrate the motion equations (8) from t_i^- to t_i^+ to obtain

$$\int_{t_i^-}^{t_i^+} \dot{\boldsymbol{p}}(t)\, dt = \int_{t_i^-}^{t_i^+} \boldsymbol{n}_p(t)\, dt + \int_{t_i^-}^{t_i^+} \boldsymbol{i}\delta(t - t_i)\, dt$$

so that

$$\mathscr{p}(t_i^+) = \mathscr{p}(t_i^-) + \mathscr{i} \tag{10}$$

Since $t_i^+ - t_i^-$ is arbitrarily small, the mean-value theorem causes the first integral above to make a vanishingly small contribution. The system momenta "jump" by an amount $\mathscr{i}$ at $t = t_i$. For impulsive inputs, one merely makes at t_i the simple algebraic adjustment in the momenta indicated by the above equation. Or, in terms of velocities, since the system velocities $\mathscr{v}$ are related to the system momenta $\mathscr{p}$ via the system inertia matrix $\mathscr{M}$, $\mathscr{p} = \mathscr{M}\mathscr{v}$, the jump in system velocities is simply

$$\mathscr{v}(t_i^+) = \mathscr{v}(t_i^-) + \mathscr{M}^{-1}\mathscr{i} \tag{11}$$

The system configuration itself however does not change from $t = t_i^-$ to $t = t_i^+$. Thus, for a system of rigid bodies whose configuration is specified by the position of a reference point $\mathbf{R}_o$, plus orientation variables represented by $\mathbf{C}_n$, (3.7, 11 and 12) still hold:

$$\dot{\mathbf{R}}_o = \boldsymbol{\phi}_o(\mathbf{R}_o, \mathscr{v}); \qquad \dot{\mathbf{C}}_n = \boldsymbol{\phi}_n(\mathbf{C}_m, \mathscr{v})$$

with n and m varying from 1 to N, where N is the number of bodies. Thus, there is no discontinuity at t_i in the configuration variables represented by $\mathbf{R}_o$ and $\mathbf{C}_n$, only in their time derivatives.

This last sentence provides the criterion for when an impulsive-input approximation can be used in dynamics. If the duration of the input, which cannot of course in practice be precisely zero, is nevertheless so short that the system configuration changes negligibly during its application, then the impulsive approximation can be used to avoid integrating the differential equations of motion. To return to the general system (8), suppose that the response to an input $\mathscr{f}(t)$ is being considered over $t_i - \varepsilon \le t \le t_i + \varepsilon$, where ε is now not vanishingly small, but small enough compared to the characteristic times of the system to permit the view that the system configuration (displacements) changes by negligible amounts during the interval 2ε. Then, integrating the motion equations (8),

$$\mathscr{p}(t_i + \varepsilon) = \mathscr{p}(t_i - \varepsilon) + \mathscr{i} + 2\varepsilon\bar{\mathscr{n}}_p \tag{12}$$

where

$$\mathscr{i} \triangleq \int_{t_i - \varepsilon}^{t_i + \varepsilon} \mathscr{f}(t)\, dt \tag{13}$$

and $\bar{\mathscr{n}}_p$ is some average value of $\mathscr{n}_p$ over $t_i - \varepsilon \le t \le t_i + \varepsilon$, as suggested by the mean-value theorem. The quantity $\mathscr{i}$ is called the *system impulse*.

Note that (12) is still *exact*. The impulsive *approximation* is viable when the input $\mathscr{f}(t)$ approaches the characteristics of the ideal impulse function, that is, when it occurs over a very brief interval ($\varepsilon \to 0$) compared to the other characteristic times of the system, and when it is much larger than the other dynamical terms in the system, $\|\mathscr{f}\| \gg \|\mathscr{n}_p\|$. Under these conditions, one is justified in dropping the last term in (12), thereby producing the impulsive approximation

$$\mathscr{p}(t_i + \varepsilon) = \mathscr{p}(t_i - \varepsilon) + \mathscr{i} \tag{14}$$

where $\boldsymbol{\imath}$ is, recall, defined by (13). The above approximation is now in exactly the same form as the ideal case, (10), except for the small but finite time interval 2ε (which is frequently ignored in practice), and therefore the remarks following (10) apply here as well.

To reiterate, the impulsive approximation is available when relatively large external forces and torques of relatively brief duration are exerted on the system. The motivation for permitting this slight inaccuracy is to avoid integrating the motion equations during the tiny interval 2ε. One is instead satisfied with an approximate but simple algebraic connection between the state at $t_i + \varepsilon$ and the state at $t_i - \varepsilon$.

There are many conceivable applications of the impulsive-input approximation to spacecraft attitude dynamics. The most common example in practice arises in connection with the use of gas jet thrusters to control spacecraft attitude. We can illustrate these ideas by a spacecraft modeled as a single rigid body $\mathscr{R}$. If $\mathscr{R}$ is not nominally spinning, the equations governing small attitude excursions are simply

$$\mathbf{I}\dot{\boldsymbol{\omega}} \equiv \mathbf{I}\ddot{\boldsymbol{\theta}} = \mathbf{g}(t) \qquad (\|\boldsymbol{\theta}\| \ll 1) \tag{15}$$

to first order. The solution of this matrix equation is

$$\boldsymbol{\theta}(t) = \boldsymbol{\theta}(0) + t\dot{\boldsymbol{\theta}}(0) + \int_0^t (t - \tau)\mathbf{I}^{-1}\mathbf{g}(\tau)\, d\tau \tag{16}$$

although, as mentioned earlier, a "solution" expressed as an integral may or may not be numerically a step forward, depending on $\mathbf{g}(t)$. If we assume that an angular impulse $\mathbf{i}$ is applied at $t = t_i$, then we write, under the impulsive approximation,

$$\mathbf{g}(t) = \mathbf{i}\delta(t - t_i) \tag{17}$$

where the impulse $\mathbf{i}$ is calculated from (13). The motion proceeds according to (16):

$$\boldsymbol{\theta}(t) = \boldsymbol{\theta}(0) + t\dot{\boldsymbol{\theta}}(0) + \begin{cases} 0 & (t < t_i) \\ (t - t_i)\mathbf{I}^{-1}\mathbf{i} & (t > t_i) \end{cases} \tag{18}$$

The angular momentum $\mathbf{h} = \mathbf{I}\boldsymbol{\omega}$ undergoes a jump at t_i given by $\Delta\mathbf{h} = \mathbf{i}$, in accordance with the momentum-impulse theorem.

When the angular motion of $\mathscr{R}$ is large, precluding the approximation $\|\boldsymbol{\theta}\| \ll 1$, the motion equation $\mathbf{I}\dot{\boldsymbol{\omega}} = -\boldsymbol{\omega}^{\times}\mathbf{I}\boldsymbol{\omega} + \mathbf{g}(t)$ must be used. Assuming that the impulse represents the dominant torque, the motion both before and after $t = t_i$ is the torque-free motion of Section 4.3. At $t = t_i$ however the angular momentum $\mathbf{h}$ undergoes an augmentation $\Delta\mathbf{h} = \mathbf{i}$. The kinetic energy is also changed at $t = t_i$ (see Problem 4.18).

4.6 PROBLEMS

4.1 An inertially symmetrical rigid body has principal moments of inertia I_t, I_t, and I_a. During its motion, its rotational kinetic energy and angular-momentum magnitude are T and h, respectively.

(*a*) Show that, when $I_a < I_t$,

$$I_a \le \frac{h^2}{2T} \le I_t \tag{1a}$$

and that, when $I_a > I_t$,

$$I_a \ge \frac{h^2}{2T} \ge I_t \tag{1b}$$

(*b*) Show that the nutation angle γ, defined as the angle between the symmetry axis and $\underset{\to}{\mathbf{h}}$, is given by

$$\begin{aligned} \sin^2\gamma &= \left(\frac{I_t}{I_t - I_a}\right)\left(1 - \frac{2I_aT}{h^2}\right) \\ \cos^2\gamma &= \left(\frac{I_a}{I_a - I_t}\right)\left(1 - \frac{2I_tT}{h^2}\right) \end{aligned} \tag{2}$$

(*c*) Show that

$$h^2 \sin\gamma\cos\gamma \frac{\partial\gamma}{\partial T} = \frac{I_aI_t}{I_a - I_t} \tag{3}$$

and comment on the significance of this result for directional stability (a hint of things to come in Chapter 5).

4.2 For the torque-free motion of an inertially axisymmetrical rigid body, it is shown in (4.2, 21) that

$$\nu = \Omega + \Omega_p \cos\gamma \tag{4}$$

where Ω is the rate of rotation of the transverse component of $\underset{\to}{\boldsymbol{\omega}}$ as seen in body axes, ν is the (constant) axial component of $\underset{\to}{\boldsymbol{\omega}}$, Ω_p is the frequency of precession, and γ is the angle between $\hat{\mathbf{p}}_3$ and $\underset{\to}{\hat{\mathbf{h}}}$. Explain this equation geometrically. Note that when the nutation angle γ is infinitesimally small, $\gamma \to 0$, (4) reduces to

$$\nu = \Omega + \Omega_p \tag{5}$$

4.3 Consider the symmetrical rigid disk shown in Fig. 4.18. Neglecting the thickness of the disk, $I_t \doteq \frac{1}{2}I_a$. For $t < 0$, the disk spins about an axis lying in the plane of the disc, as shown. Assuming the disk is slightly disturbed at $t = 0$, describe its resulting attitude motion.

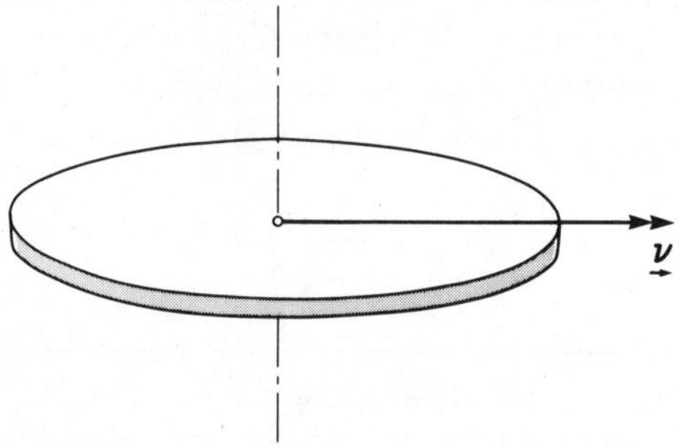

FIGURE 4.18 Rigid disk. (See Problem 4.3.)

4.4 The attitude motion of an inertially symmetrical rigid body is expressed in (4.2, 32) in terms of the Euler parameters $\boldsymbol{\varepsilon}$ and η.

(*a*) Show that this solution can be written explicitly as a function of t, as follows:

$$\begin{aligned} \varepsilon_1(t) &= \varepsilon_1(0)\cos\Delta - \varepsilon_2(0)\sin\Delta \\ \varepsilon_2(t) &= \varepsilon_1(0)\sin\Delta + \varepsilon_2(0)\cos\Delta \\ \varepsilon_3(t) &= \varepsilon_3(0)\cos\Sigma + \eta(0)\sin\Sigma \\ \eta(t) &= -\varepsilon_3(0)\sin\Sigma + \eta(0)\cos\Sigma \end{aligned} \tag{6}$$

where

$$\Delta \triangleq \left(\frac{\Omega_p - \Omega}{2}\right)t; \qquad \sigma \triangleq \left(\frac{\Omega_p + \Omega}{2}\right)t \tag{7}$$

(*b*) Show further that $\boldsymbol{\varepsilon}$ and η, given by (6), and $\boldsymbol{\omega}$, given by (4.2, 8), satisfy

$$\begin{aligned} \dot{\boldsymbol{\varepsilon}} &= \tfrac{1}{2}(\boldsymbol{\varepsilon}^{\times} + \eta\mathbf{1})\boldsymbol{\omega} \\ \dot{\eta} &= -\tfrac{1}{2}\boldsymbol{\varepsilon}^{T}\boldsymbol{\omega} \end{aligned} \tag{8}$$

as required by (2.3, 22).

4.5 For a rigid body in torque-free motion, it is known that the components of angular velocity in principal axes $\{\omega_1, \omega_2, \omega_3\}$ satisfy the energy and angular-momentum integrals (4.3, 3). Show that the maximum values of $\{\omega_1, \omega_2, \omega_3\}$ are given by (4.3, 7 through 9).

4.6 Consider the integrations of Euler's attitude motion equations for a torque-free rigid body, (4.3, 1), when $h^2 = 2I_2T$, that is, when $I = I_2$.

(*a*) Show from the energy and angular-momentum integrals that

$$\left(\frac{\omega_1}{\omega_{1m}}\right)^2 + \left(\frac{\omega_2}{\omega_{2m}}\right)^2 = 1; \qquad \left(\frac{\omega_2}{\omega_{2m}}\right)^2 + \left(\frac{\omega_3}{\omega_{3m}}\right)^2 = 1 \tag{9}$$

where $\{\omega_{1m}, \omega_{2m}, \omega_{3m}\}$ are given in (4.3, 7 through 9). Interpret these results geometrically.

(*b*) Using (9), substitute ω_1 and ω_3 into the second of Euler's equations (4.3, 1) to derive the following differential equation for $\omega_2(t)$:

$$\frac{d}{d\tau}\left(\frac{\omega_2}{\omega_{2m}}\right) = s_2\left(1 - \frac{\omega_2^2}{\omega_{2m}^2}\right) \tag{10}$$

where τ is given by (4.3, 10), $s_2 = -s_1s_3$, and s_1 and s_3 are the ± 1 factors introduced in extracting the radicals in (9).

(*c*) Integrate (10) to find the solution for $\omega_2(t)$ given in (4.3, 12): $\omega_2 = s_2\omega_{2m}\tanh\tau$. Finally, use (9) to verify that $\omega_1 = s_1\omega_{1m}\operatorname{sech}\tau$ and $\omega_3 = s_3\omega_{3m}\operatorname{sech}\tau$.

(*d*) Derive the solution (4.3, 12) also by letting $k \to 1$ in (4.3, 10) or (4.3, 11).

4.7 (*a*) Verify the following identities in connection with the torque-free motion of a tri-inertial rigid body (Section 4.3):

$$I_1\omega_{1m}^2 + I_3\omega_{3m}^2 = 2T \tag{11}$$

$$I_1^2\omega_{1m}^2 + I_3^2\omega_{3m}^2 = h^2 \tag{12}$$

$$I_2\omega_{2m}^2 = \begin{cases} k^2 I_1\omega_{1m}^2 + I_3\omega_{3m}^2 & (I_1 \geq I > I_2) \\ I_1\omega_{1m}^2 + k^2 I_3\omega_{3m}^2 & (I_2 > I \geq I_3) \end{cases} \tag{13}$$

$$I_2^2\omega_{2m}^2 = \begin{cases} k^2 I_1^2\omega_{1m}^2 + I_3^2\omega_{3m}^2 & (I_1 \geq I > I_2) \\ I_1^2\omega_{1m}^2 + k^2 I_3^2\omega_{3m}^2 & (I_2 > I \geq I_3) \end{cases} \tag{14}$$

where $\{\omega_{1m}, \omega_{2m}, \omega_{3m}\}$ and k are given by (4.3, 7 through 11).

(*b*) Using these identities and the further identities

$$sn^2\tau + cn^2\tau = 1; \qquad k^2 sn^2\tau + dn^2\tau = 1 \tag{15}$$

verify that the analytical solution in terms of elliptic functions, namely, (4.3, 10 and 11), satisfies the energy and angular-momentum integrals (4.1, 3).

(*c*) Verify that the solution corresponding to $I = I_2$, namely, (4.3, 12), also satisfies these integrals.

4.8 Euler's motion equations for a torque-free rigid body expressed in matrix form are

$$\mathbf{I}\dot{\boldsymbol{\omega}} + \boldsymbol{\omega}^{\times}\mathbf{I}\boldsymbol{\omega} = \mathbf{0} \tag{16}$$

(*a*) Show that the only possible motion for which $\boldsymbol{\omega}$ is fixed in space is one for which $\underset{\rightarrow}{\omega} \parallel \underset{\rightarrow}{h}$, that is, one for which $\boldsymbol{\omega} \equiv \boldsymbol{\nu} = \text{constant}$, where

$$\mathbf{I}\boldsymbol{\nu} = \lambda\boldsymbol{\nu} \tag{17}$$

where λ is a scalar.

(*b*) Recognizing (17) as a linear eigenvalue problem, explain why the eigenvalues are $\lambda = I_1$, I_2, and I_3 (the principal moments of inertia), with the associated eigencolumns corresponding to the so-called "simple spin" motions.

4.9 This problem studies the attitude motion of a rigid body expressed in terms of Euler parameters. Assume that I_1 is the largest moment of inertia, and align the inertial axis $\hat{\underset{\rightarrow}{i}}_1$ with $\hat{\underset{\rightarrow}{h}}$. In other words, $\mathbf{h} = \mathbf{C}\mathbf{1}_1 h$.

(*a*) Recalling from Chapter 2 the expression for $\mathbf{C}(\boldsymbol{\varepsilon}, \eta)$, namely, (2.2, 14), show that

$$\varepsilon_2^2 + \varepsilon_3^2 = \frac{h - I_1\omega_1}{2h}$$

$$\varepsilon_2\varepsilon_1 - \varepsilon_3\eta = \frac{I_2\omega_2}{2h}$$

$$\varepsilon_3\varepsilon_1 + \varepsilon_2\eta = \frac{I_3\omega_3}{2h} \tag{18}$$

where, as usual, $\boldsymbol{\omega}(t)$ is given by (4.3, 10 through 12).

(*b*) Verify that the following expressions satisfy (18), where $\lambda(t)$ is yet to be determined:

$$\varepsilon_1 = \frac{I_2\omega_2 \cos\lambda + I_3\omega_3 \sin\lambda}{[2h(h - I_1\omega_1)]^{1/2}}$$

$$\varepsilon_2 = \left[\frac{h - I_1\omega_1}{2h}\right]^{1/2} \cos\lambda$$

$$\varepsilon_3 = \left[\frac{h - I_1\omega_1}{2h}\right]^{1/2} \sin\lambda$$

$$\eta = \frac{-I_2\omega_2 \sin\lambda + I_3\omega_3 \cos\lambda}{[2h(h - I_1\omega_1)]^{1/2}} \tag{19}$$

[Slightly harder problem: *Derive* (19).]

(*c*) There are several ways to determine $\lambda(t)$. For example, from (2.3, 22), use

$$2\dot{\varepsilon}_2 = \varepsilon_3\omega_1 - \varepsilon_1\omega_3 + \eta\omega_2 \tag{20}$$

to show that

$$\lambda(t) = \frac{h(h - I\omega_1)}{2I(h - I_1\omega_1)}(t - t_o) + \lambda_o \tag{21}$$

where $\lambda_o \triangleq \lambda(t_o)$.

4.10 Consider the polhodes shown in Fig. 4.9 and projected onto the principal planes in Fig. 4.10. Each polhode is all, or part, or a conic section. Demonstrate this in the following examples.

(*a*) For $I_1 \geq I > I_2$, show that each polhode projection onto the ω_2–ω_3 plane is an ellipse:

$$\left(\frac{\omega_2}{\omega_{2m}}\right)^2 + \left(\frac{\omega_3}{\omega_{3m}}\right)^2 = 1 \tag{22}$$

[*Hint*: This may be shown from the analytical solution (4.3, 10) and the identity (15*a*).]

(*b*) For $I_2 > I \geq I_3$, show that each polhode projection onto the ω_1–ω_3 plane is part of the hyperbola

$$\left(\frac{\omega_3}{\omega_{3m}}\right)^2 - k^2\left(\frac{\omega_1}{\omega_{1m}}\right)^2 = 1 - k^2 \tag{23}$$

(*Hint*: See Problem 4.7.)

(*c*) Show that the projections of the separatrices ($I = I_2$) onto the three principal planes are

$$\left(\frac{\omega_1}{\omega_{1m}}\right)^2 + \left(\frac{\omega_2}{\omega_{2m}}\right)^2 = 1$$

$$\left(\frac{\omega_2}{\omega_{2m}}\right)^2 + \left(\frac{\omega_3}{\omega_{3m}}\right)^2 = 1$$

$$\left(\frac{\omega_3}{\omega_{3m}}\right)^2 - \left(\frac{\omega_1}{\omega_{1m}}\right)^2 = 0 \tag{24}$$

(*Hint*: This may be shown from the analytical solution (4.3, 12) and the identity $\cosh^2\tau - \sinh^2\tau = 1$.)

4.11 Consider the inertia ratio parameters k_1 and k_3 defined by (4.4, 20):

$$k_1 \triangleq \frac{I_2 - I_3}{I_1}; \qquad k_3 \triangleq \frac{I_2 - I_1}{I_3} \tag{25}$$

Noting from Problem 3.8 the physical restrictions that

$$I_1 + I_2 > I_3; \qquad I_2 + I_3 > I_1; \qquad I_3 + I_1 > I_2 \tag{26}$$

show that

$$|k_1| < 1; \qquad |k_3| < 1 \tag{27}$$

for real bodies, with equalities in (26) and (27) as limiting cases.

4.12 With reference to the infinitesimal stability of the "simple spin" motions for a rigid body $\mathscr{R}$ (see, for example, (4.4, 14) et seq.), let γ denote the angle between the "spin" axis $\vec{\hat{\mathbf{p}}}_2$ and the angular momentum vector $\vec{\mathbf{h}}$:

$$\cos\gamma \triangleq \vec{\hat{\mathbf{p}}}_2 \cdot \vec{\hat{\mathbf{h}}} \tag{28}$$

where $\vec{\hat{\mathbf{h}}} \triangleq \vec{\mathbf{h}}/h$. Furthermore, let $\vec{\hat{\mathbf{h}}}$ be aligned with $\vec{\hat{\mathbf{i}}}_2$, so that

$$\vec{\hat{\mathbf{p}}}_2 = \mathscr{F}_p^T \mathbf{1}_2; \qquad \vec{\hat{\mathbf{h}}} = \mathscr{F}_i^T \mathbf{1}_2 \tag{29}$$

(*a*) Show that

$$\cos\gamma = \mathbf{1}_2^T \mathbf{C}_{pi} \mathbf{1}_2 \tag{30}$$

indicating that $\cos\gamma$ is the "22" element in the matrix $\mathbf{C}_{pi} = \mathscr{F}_p \cdot \mathscr{F}_i^T$.

(*b*) Show from (2.2, 20) that

$$\mathbf{C}_{pi} = \mathbf{1} - \boldsymbol{\alpha}^{\times} + \tfrac{1}{2}(\boldsymbol{\alpha}\boldsymbol{\alpha}^T - \alpha^2 \mathbf{1}) \tag{31}$$

correct to second order in $\|\boldsymbol{\alpha}\|$, where $\{\alpha_1, \alpha_2, \alpha_3\}$ have the meanings used in (4.4, 11).

(*c*) Then show that

$$\gamma = \left(\alpha_1^2 + \alpha_3^2\right)^{1/2} \tag{32}$$

correct to first order in γ and $\|\boldsymbol{\alpha}\|$. Note that it follows that $\gamma = \|\mathbf{q}\|$, where $\mathbf{q}$ contains the configuration coordinates governed by (4.4, 14).

(*d*) Explain why the directional stability of the spin axis coincides with the stability of (4.4, 14); that is, explain why a major- or a minor-axis spin for $\mathscr{R}$ is directionally stable.

4.13 Consider the major-axis spin of a rigid body, with $I_1 > I_2 > I_3$. The angular momentum vector is $\vec{\mathbf{h}}_\nu$. After a disturbance, the new angular momentum is $\vec{\mathbf{h}}$.

(*a*) Explain why $\vec{\mathbf{h}}$ can be made to satisfy $(\vec{\mathbf{h}} - \vec{\mathbf{h}}_\nu) \cdot (\vec{\mathbf{h}} - \vec{\mathbf{h}}_\nu) < \varepsilon^2$ for any preassigned $\varepsilon > 0$ by making the disturbance sufficiently small.

(*b*) Align the inertial axis $\vec{\hat{\mathbf{i}}}_1$ with $\vec{\mathbf{h}}$. Show from (4.3, 10) that the angle γ between the spin axis $\vec{\hat{\mathbf{p}}}_1$ and $\vec{\hat{\mathbf{i}}}_1$ is given by

$$\cos\gamma = \frac{I_1\omega_{1m}}{h}\, dn(\tau; k) \tag{33}$$

provided the disturbance is not great enough to violate $I > I_2$.

(*c*) Noting that $dn^2(\tau; k) \geqq 1 - k^2$, prove that

$$\sin^2 \gamma \leq \frac{I_2(I_1 - I)}{I(I_1 - I_2)} \tag{34}$$

(*d*) Note that $(I_1 - I)$ is a measure of the size of the disturbance. Define δ by

$$0 \leq \delta \triangleq I_1 - I < I_1 - I_2 < I_1$$

Hence, show that

$$\sin^2 \gamma \leq \frac{I_2 \delta}{(I_1 - \delta)(I_1 - I_2)} < 1 \tag{35}$$

Verify that for any $\varepsilon > 0$, there exists a δ such that $\sin^2 \gamma < \varepsilon$ and hence that the motion is directionally stable. Note that this stability analysis is an exact (nonlinear) one.

(*e*) Repeat the steps followed in (*b*) through (*d*) above to demonstrate the directional stability of a minor-axis spin for a rigid body. [*Hint*: use (4.3, 11).]

4.14 Show that the solution for the motion equations for the self-excited motion of an axially symmetrical rigid body, (4.5, 1), is given by (4.5, 2) when there is no torque component about the symmetry axis. Compare your technique with that in Problem 4.16.

4.15 Shown in Fig. 4.19 is a rigid body $\mathscr{R}$ rotating about an axis whose direction is fixed both in space and in $\mathscr{R}$. The point O is fixed on this spin axis. A reference frame $\mathscr{F}_b$ is located at O in $\mathscr{R}$. Although body-fixed, $\mathscr{F}_b$ is not necessarily aligned with the principal axes for $\mathscr{R}$ at O. The direction cosines of the spin axis with respect to $\mathscr{F}_b$ are $\mathbf{a} \triangleq [a_1 \quad a_2 \quad a_3]^T$, and the spin rate is $\omega(t)$.

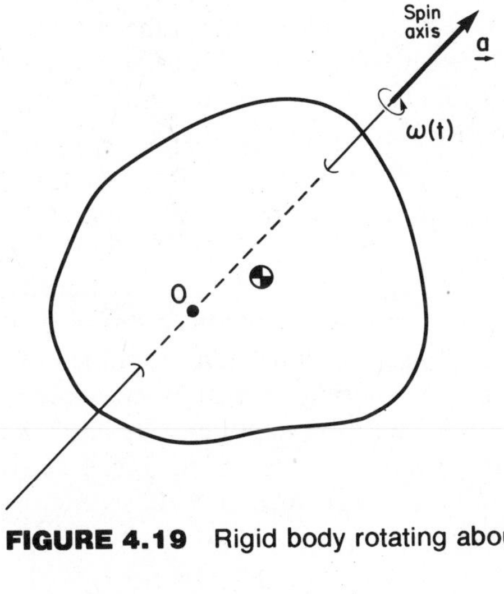

FIGURE 4.19 Rigid body rotating about a fixed axis. (See Problem 4.15.)

(a) Show that the torque on $\mathscr{R}$ (about O) to sustain this motion is

$$\mathbf{g}_o(t) = (\mathbf{Ja})\dot{\omega}(t) + (\mathbf{a}^{\times}\mathbf{Ja})\omega^2(t) \tag{36}$$

where $\mathbf{J}$ is the inertia matrix for $\mathscr{R}$ at O, expressed in $\mathscr{F}_b$.

(b) Assuming that $\mathscr{R}$ rotates about $\underrightarrow{\mathbf{a}}$ by being mounted on ideal (frictionless) bearings, what can be concluded about $\omega(t)$?

(c) If the point O is also inertially fixed, what force $\mathbf{f}(t)$ must exist on $\mathscr{R}$?

4.16 This problem deals with the self-excited motion of an inertially axisymmetrical rigid body, governed by (4.5, 1), when there is no torque along the symmetry axis, so that $\omega_3(t) \equiv \nu$.

(a) Show that the motion equations for ω_1 and ω_2 may be written

$$\dot{\omega}_1 = \Omega\omega_2 + \frac{g_1(t)}{I_t}$$

$$\dot{\omega}_2 = -\Omega\omega_1 + \frac{g_2(t)}{I_t} \tag{37}$$

(b) The solution of (37) is easily found by defining the complex variables

$$\omega^* \triangleq \omega_1 + j\omega_2$$

$$g^* \triangleq \frac{g_1 + jg_2}{I_t} \tag{38}$$

Show that in terms of these new quantities, (37) has the form

$$\dot{\omega}^* = -j\Omega\omega^* + g^*(t) \tag{39}$$

(c) By noting the standard solution of this differential equation,

$$\omega^*(t) = e^{-j\Omega t}\omega^*(0) + \int_0^t e^{-j\Omega(t-\tau)}g^*(\tau)\,d\tau \tag{40}$$

infer the solution (4.5, 2) for $\omega_1(t)$, $\omega_2(t)$ given in the text. [The device (38), although hardly an impressive utilization of the theory of complex variables, has its merits in this case. To treat (37) as a pair of scalar equations, it is hard to avoid a differentiation-and-substitution process, with the concomitant necessity of evaluating the arbitrary constant introduced by the differentiation. Or, to consider a more sophisticated approach, the pair (37) may be viewed as a single matrix equation. Finding an expression for the matrix exponential, while not difficult, is still not quite as direct as interpreting (40).]

4.17 An inertially axisymmetrical spacecraft in transfer orbit is in a simple spin about its axis of symmetry. The spin rate is 6 rad/s, and the axial and transverse moments of inertia are $I_a = 54$ kg · m^2, $I_t = 50$ kg · m^2. Located on the outer bottom rim is a small gas jet capable of exerting a thrust of 10 N in the forward direction (see Fig. 4.20) along a line 0.6 m from the mass center. The gas jet is turned on for 0.05 s during each of 10 consecutive rotations, at precisely the same azimuthal position each time. Describe the resulting motion of the vehicle after the tenth pulse, assuming a small amount of energy dissipation and that environmental torques can be neglected.

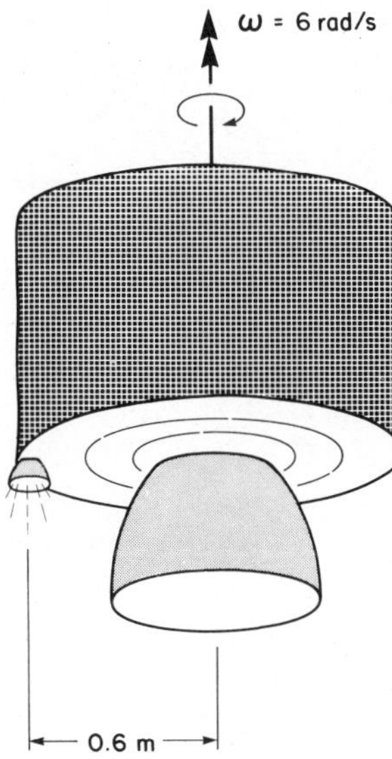

FIGURE 4.20 Spin-stabilized spacecraft. (See Problem 4.17.)

4.18 Consider the system of equations (3.7, 9 through 12), describing the motion of a quite general class of spacecraft. In addition to any other continuous inputs to the system, an impulsive input $\mathscr{f} = \mathscr{i}\delta(t - t_i)$ is applied at $t = t_i$. As shown in the text, see (4.5, 10), the system momenta jump at $t = t_i$ by an amount

$$\Delta \mathscr{p} \triangleq \mathscr{p}(t_i^+) - \mathscr{p}(t_i^-) = \mathscr{i} \tag{41}$$

Show that the system kinetic energy jumps at $t = t_i$ by an amount

$$\Delta T \triangleq T(t_i^+) - T(t_i^-) = \left(\mathscr{p}(t_i^-) + \tfrac{1}{2}\mathscr{i}\right)^T \mathscr{M}^{-1}\mathscr{i} \tag{42}$$

4.19 An inertially axisymmetrical rigid body $\mathscr{W}$ has principal moments of inertia I_a, I_t, I_t. As discussed in Section 4.2, its kinetic energy and angular momentum in torque-free motion are given by

$$\begin{aligned} 2T &= I_a \nu^2 + I_t \omega_t^2 \\ h^2 &= I_a^2 \nu^2 + I_t^2 \omega_t^2 \end{aligned} \tag{43}$$

where ν and ω_t are components of the angular velocity of $\mathscr{W}$ along, and transverse to, the symmetry axis.

(*a*) Show that, for physically real motions,

$$4T^2 > h^2 \nu^2 \tag{44}$$

and hence that α and β, given respectively by (4.2, 24) and (4.2, 26), must satisfy

$$\beta > \alpha > 0 \tag{45}$$

(*b*) Explain why this condition eliminates the "case C" interpretation of (4.2, 29).

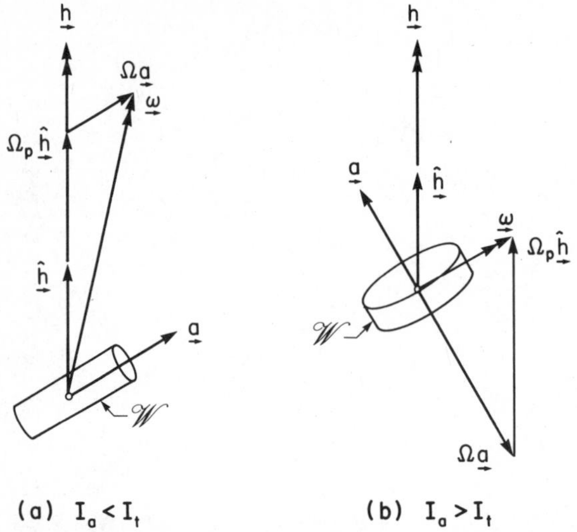

FIGURE 4.21 Symmetrical spinning rigid body. (See Problem 4.20.)

4.20 An inertially axisymmetrical rigid body $\mathscr{W}$ rotates in torque-free motion, as shown in Fig. 4.21. The angular momentum about its mass center is $\underset{\rightarrow}{\mathbf{h}}$, and $\underset{\rightarrow}{\hat{\mathbf{h}}}$ is a unit vector in the same direction. Show that the absolute angular velocity of $\mathscr{W}$ is given by

$$\underset{\rightarrow}{\boldsymbol{\omega}} = \Omega_p \underset{\rightarrow}{\hat{\mathbf{h}}} + \Omega \underset{\rightarrow}{\mathbf{a}} \tag{46}$$

where Ω_p is the precession frequency ($\Omega_p = h/I_t$), Ω is the relative spin rate defined in (4.2, 5), and $\underset{\rightarrow}{\mathbf{a}}$ is a unit vector aligned with the symmetry axis. (Compare also Problem 4.2 and Fig. 4.5.)

4.21 The motion for $t < 0$ of a certain reference frame, $\mathscr{F}_b$, consists of a simple spin in which one of the three unit vectors of $\mathscr{F}_b$, namely, $\underset{\rightarrow}{\hat{\mathbf{b}}}_1$, rotates in coincidence with the unit vector $\underset{\rightarrow}{\mathbf{a}}$. Naturally, the tips of the other two unit vectors of $\mathscr{F}_b$ trace out a circle $\mathscr{C}$ perpendicular to $\underset{\rightarrow}{\mathbf{a}}$. If, after a disturbance at $t = 0$, $\underset{\rightarrow}{\hat{\mathbf{b}}}_1$ is directionally stable with respect to $\underset{\rightarrow}{\mathbf{a}}$ for $t > 0$, show that the tips of $\underset{\rightarrow}{\hat{\mathbf{b}}}_2$ and $\underset{\rightarrow}{\hat{\mathbf{b}}}_3$ are stable orbitally to $\mathscr{C}$. (The definition of orbital stability is given in Appendix A.)

CHAPTER 5

EFFECT OF INTERNAL ENERGY DISSIPATION ON THE DIRECTIONAL STABILITY OF SPINNING BODIES

A rigid body is a mathematical abstraction; as a model for the attitude motion of real (physical) bodies, it has its strengths and limitations. Its strength lies chiefly in its simplicity and in the availability of well-understood analytical solutions for its torque-free motion. The accompanying geometrical interpretation of this motion provides further valuable insight. These topics dominated the discussion in Chapter 4.

Unfortunately, no real physical body is truly rigid. The rotational motion of $\mathscr{R}$ in general implies a time-dependent force field within $\mathscr{R}$. An element of mass dm at position $\underrightarrow{\mathbf{r}}$ within $\mathscr{R}$ experiences an acceleration relative to the mass center given by

$$\underrightarrow{\mathbf{a}}(\underrightarrow{\mathbf{r}}, t) = \overset{\circ\circ}{\underrightarrow{\mathbf{r}}} + 2\underrightarrow{\boldsymbol{\omega}} \times \overset{\circ}{\underrightarrow{\mathbf{r}}} + \overset{\circ}{\underrightarrow{\boldsymbol{\omega}}} \times \underrightarrow{\mathbf{r}} + \underrightarrow{\boldsymbol{\omega}} \times (\underrightarrow{\boldsymbol{\omega}} \times \mathbf{r})$$

as is well known from mechanics, see, for example, (B.4, 12). Within the framework of the rigidity assumption, $\overset{\circ}{\underrightarrow{\mathbf{r}}} = \overset{\circ\circ}{\underrightarrow{\mathbf{r}}} = \underrightarrow{\mathbf{0}}$; hence, expressed in $\mathscr{F}_b$,

$$\mathbf{a}(\mathbf{r}, t) = (\dot{\boldsymbol{\omega}}^{\times} + \boldsymbol{\omega}^{\times}\boldsymbol{\omega}^{\times})\mathbf{r}$$

The implied internal force field on $dm \equiv \sigma(\mathbf{r})\,dV$ is

$$d\mathbf{f} = -\sigma(\mathbf{r})(\dot{\boldsymbol{\omega}}^{\times} + \boldsymbol{\omega}^{\times}\boldsymbol{\omega}^{\times})\mathbf{r}\,dV$$

implying a state of internal stress which, in turn, produces material deformations in real bodies. The more rigid the vehicle is, the smaller these deformations are. But they are never precisely zero. Moreover, unless $\boldsymbol{\omega}(t)$ is constant, as it is only for a spin about a principal axis, the internal inertial force field varies with time. It must be recognized that the consequent time-dependent deformations are inevitably accompanied by energy dissipation although, when the spacecraft is nearly rigid, the rate of energy dissipation is nearly zero. The degrees of freedom associated with material deformation are themselves governed by differential equations of motion, and deformational and rotational motions will in general be coupled—an important subject, but beyond the scope of this text.

It is found that when the spacecraft structure is very nearly rigid, material strains can be neglected in all respects save one: energy dissipation. The *vibrational* character of the small structural motions tends to have little net effect whenever the spacecraft is rigid enough that a natural frequency of structural vibration is much higher than a characteristic frequency of the rotational motion (e.g., the spin rate). The *dissipative* character of the time-varying deformations, however, cannot always be safely neglected, regardless of how small this dissipation may be. The reason for this, in qualitative terms, is that the directional stability proved in Section 4.4 is not asymptotic. A monotonic influence such as dissipation cannot be withstood, so that the cumulative effect on the motion can be dramatic. Of course, the smaller the rate of dissipation, the longer it will take to manifest its influence.

The effects of energy dissipation are studied below in several stages. The first analysis, based on the so-called *energy sink hypothesis*, is semiquantitative in nature. The conclusions tentatively reached are then elaborated upon by adding damping terms to the motion equations for a rigid body. Finally, a more specific system is studied in which a mechanical damper is added to a rigid body. This choice augments (by one) the number of system degrees of freedom, and relatively simple stability conditions are derived.

The primary objective of this chapter is to examine the so-called "major-axis rule" for spinning spacecraft.

5.1 QUASI-RIGID BODY WITH AN ENERGY SINK, $\mathscr{Q}$

The preceding introductory remarks suggest the concept of a *quasi-rigid body*—a body $\mathscr{Q}$ that is "almost rigid" in the sense that extra degrees of freedom associated with material deformation (and the associated motion equations) need not be introduced to adequately represent the dynamics. A companion concept that complements the "quasi-rigid" idea is the *energy sink hypothesis*, which states that during the motion of any real body the kinetic energy will tend to be converted—slowly—to heat energy. The modifier "slowly" is inserted to maintain consistency with the assumption of quasi-rigidity; any mechanical device capable of "rapidly" dissipating energy would be inconsistent with the quasi-rigid assumption. The details of how the energy is dissipated are not specified, except to refer to the process as an energy "sink."

Consider now the torque-free motion of $\mathscr{Q}$. As a consequence of quasi-rigidity, the system kinetic energy is closely approximated by

$$T = \tfrac{1}{2}\left(I_1\omega_1^2 + I_2\omega_2^2 + I_3\omega_3^2\right) \tag{1}$$

At the same time, the energy sink hypothesis indicates that

$$\dot{T} \leq 0 \tag{2}$$

More precisely, $\dot{T} < 0$ unless the internal inertial force field specified by (3) is stationary. This condition is possible only when $\dot{\boldsymbol{\omega}} = \mathbf{0}$, in other words, when $\boldsymbol{\omega}^{\times}\mathbf{I}\boldsymbol{\omega} = \mathbf{0}$, corresponding to a simple spin about a principal axis (see Problem 4.8). Thus, while $\mathbf{h}$ is still constant (zero external torque), T is slowly decreasing. Moreover, it is useful to note the change in the parameter I, defined in (4.3, 4):

$$\dot{I} = -\left(\frac{I}{T}\right)\dot{T} \geq 0 \tag{3}$$

To ensure that these time derivatives are small enough to correspond to a *quasi-rigid* body, we insist that

$$\left|\frac{\dot{I}}{\omega I}\right| = \left|\frac{\dot{T}}{\omega T}\right| \ll 1 \tag{4}$$

where $\omega \triangleq \|\boldsymbol{\omega}\|$.

Within the framework of these assumptions, $I(t)$ is a slowly increasing function of time. It may be anticipated that the motion $\boldsymbol{\omega}(t)$ is still given to a good approximation by (4.3, 10 through 12), for which the convention $I_1 > I_2 > I_3$ applies, but with $I(t)$ slowly increasing. Of course, I cannot exceed its maximum permissible value, $I \leq I_1$. When I eventually reaches I_1, perhaps after an infinitely long time, T has reached its minimum value $T_{\min} = \frac{1}{2}h^2/I_1$. In this final state, $\mathscr{Q}$ spins about its axis of maximum inertia ($I = I_1$), and no further decrease in T is possible. This is consistent with the observation that for this ultimate simple spin about the major axis, the internal force field is stationary; the excitation to which the dissipation is attributed no longer exists.

This energy sink analysis implies that the only motion of $\mathscr{Q}$ which is directionally stable is a spin about the major axis of inertia. Specifically, *the minor-axis spin is unstable*. On the other hand, the energy sink hypothesis renders the major-axis spin *asymptotically* directionally stable in a special sense: the spin axis asymptotically approaches not its original direction $\underrightarrow{\mathbf{h}}_\nu$ but the direction of the new ($t > 0$) angular momentum vector $\underrightarrow{\mathbf{h}}$. This conclusion is substantiated in detail in the following paragraphs.

By making appropriate changes in the development of Chapter 4 for a rigid body $\mathscr{R}$, conclusions can be inferred concerning the consequences of permitting small material deformations, $\mathscr{R} \to \mathscr{Q}$. In the closed-form solution (4.3, 10 through 12), for example, one can reasonably allow I to be a slowly increasing function of time, so long as $\dot{I} \ll \|\boldsymbol{\omega}\|I$. The $I(t)$ history is of course not specified by the energy sink hypothesis, which is inherently vague about the mechanism providing the "sink." For qualitative demonstrations, the following is reasonable:

$$I(t) = I_1 + \left[I(0) - I_1\right]e^{-\varepsilon t} \tag{5}$$

where $\varepsilon \ll \|\boldsymbol{\omega}\|$.

Figure 5.1 shows a typical $\boldsymbol{\omega}(t)$ history for $\mathscr{Q}$, which should be compared with the corresponding $\boldsymbol{\omega}(t)$ history for $\mathscr{R}$ in Fig. 4.11. In both cases, $h = 20$ N · m · s, a constant in the absence of external torque. Thus, the tip of $\boldsymbol{\omega}(t)$ lies always on the constant-h ellipsoid $\mathscr{H}$. If T were constant also, one would obtain

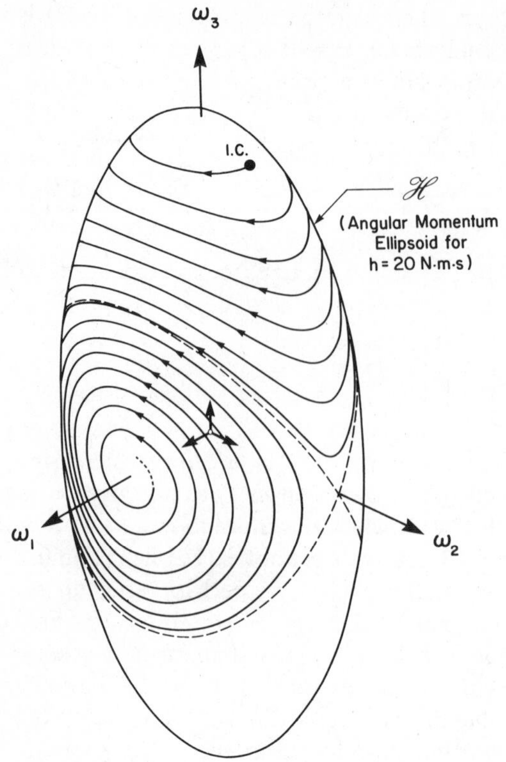

FIGURE 5.1 Motion $\boldsymbol{\omega}(t)$ for a quasi-rigid body with $\{I_1, I_2, I_3\} = \{6, 4, 2\}$ kg · m^2.

a single closed polhode, several cases of which were shown in Fig. 4.11. With T slowly decreasing, however, the motion is typified by the spiral-like polhode in Fig. 5.1. This polhode is obtained by constructing the intersection of $\mathscr{T}$ and $\mathscr{H}$ (Fig. 4.8) as the size of $\mathscr{T}$ shrinks because of dissipation.

This geometrical insight is confirmed by an examination of the Liapunov function (4.4, 8), written in the context that $I_1 > I_2 > I_3$. For a major-axis spin, one finds by differentiation

$$\dot{v} = 2(I_1 - I)\dot{T} - 2\dot{I}T + 8(T - T_\nu)\dot{T} \tag{6}$$

so that, since $I_1 > I$, $\dot{T} < 0$, $\dot{I} > 0$, and $T > T_\nu > 0$,

$$\dot{v} < 0 \tag{7}$$

unless $\dot{T} = 0$, which is possible only for the simple spin itself. Therefore it can be concluded from Liapunov's indirect method that a major-axis spin is asymptotically $\boldsymbol{\omega}$-stable with respect to perturbations $\boldsymbol{\xi} \triangleq \boldsymbol{\omega} - \boldsymbol{\nu}$.

To investigate a simple spin about the minor axis, we use the Liapunov function (4.4, 10), still with $I_1 > I_2 > I_3$. Evidently,

$$\dot{v} = 2(I - I_3)\dot{T} + 2\dot{I}T + 8(T - T_\nu)\dot{T} \tag{8}$$

and it appears that the signs of the terms are in conflict. It must be remembered however that $\dot{T}$ and $\dot{I}$ are related by (3), whereupon

$$\dot{v} = -2I_3\dot{T} + 8(T - T_\nu)\dot{T} \tag{9}$$

Recognizing that $\dot{T} < 0$ and $T < T_\nu$, one infers that $\dot{v} > 0$ unless $\dot{T} = 0$ (only possible for the reference simple spin). According to Theorem A.20 in Appendix A, the minor-axis spin is $\boldsymbol{\omega}$-unstable, and therefore unstable.

Directional Stability; The Major-Axis Rule

Thus far, our stability investigation for $\mathscr{Q}$ has been couched in terms of the *attitude rate* variables $\{\omega_1, \omega_2, \omega_3\}$. It has been learned that of the three simple spins, only the major-axis spin is $\boldsymbol{\omega}$-stable. This addresses only indirectly our real objective, however, which is to assess the possibilities for *attitude* stability. To this end, we revisit the Liapunov function v_M defined in (4.4, 35), repeated here for convenience:

$$v_M \triangleq 2(T - T_h) = h^2 \frac{I_3(I_2 - I_1)c_1^2 + I_1(I_2 - I_3)c_3^2}{I_1 I_2 I_3} \tag{10}$$

where $2T_h \triangleq h^2/I_2$. Behind this expression lies the assumption that $\hat{\mathbf{p}}_2$ is coincident with the reference simple spin, and the relative sizes of $\{I_1, I_2, I_3\}$ are arbitrary, subject to the usual limitations for a real body (Problem 3.8). The direction cosines of $\underrightarrow{\mathbf{h}}$ in $\mathscr{F}_p$, namely, c_1 and c_3, measure the difference in direction between $\underrightarrow{\mathbf{h}}$ and $\vec{\hat{\mathbf{p}}}_2$; for the simple spin itself, $c_1 = c_3 = 0$ and $\underrightarrow{\mathbf{h}} \| \vec{\hat{\mathbf{p}}}_2$. The Liapunov function above is helpful when $\vec{\hat{\mathbf{p}}}_2$ is the major axis, that is, when $I_2 > \{I_1, I_3\}$, since it is then clearly positive-definite in $\{c_1, c_3\}$. Moreover, the energy sink ensures that $\dot{v}_M \le 0$, with $\dot{v}_M = 0$ only when $T = T_h$, namely, when $c_1 = c_3 = 0$. We conclude that $\vec{\hat{\mathbf{p}}}_2$ asymptotically approaches $\underrightarrow{\mathbf{h}}$, and thus a major-axis spin for $\mathscr{Q}$ is directionally stable, asymptotically to $\underrightarrow{\mathbf{h}}$. The directional stability of a major-axis spin for a *rigid* body derived in Section 4.4 now has for a quasi-rigid body the added property that the spin axis asymptotes to $\underrightarrow{\mathbf{h}}$.

Some care should be taken, however, to avoid the statement "a major-axis spin is asymptotically directionally stable." This statement is certainly true in that the major axis asymptotically approaches $\underrightarrow{\mathbf{h}}$. It does not however asymptotically approach the *original* (i.e., predisturbance) direction, $\mathbf{h}_\nu$. Thus, in the presence of energy dissipation, a spin about the major axis of inertia is directionally stable, asymptotically to the *new* $\underrightarrow{\mathbf{h}}$. The latter will lie arbitrarily close to $\mathbf{h}_\nu$ if the initial disturbances are sufficiently small; but this is stability, not asymptotic stability. Thus, strictly speaking, even with energy dissipation, the major-axis spin is not asymptotically stable, even directionally. This seemingly picayune point is crucial to the behavior of spin-stabilized spacecraft, which, even though effectively stabilized (directionally), in the sense that $\underrightarrow{\boldsymbol{\omega}}$ is maintained essentially parallel to $\underrightarrow{\mathbf{h}}$, nevertheless experience arbitrarily large deviations in the direction of $\underrightarrow{\mathbf{h}}$ caused by the actions of small external torques over a long period of time. If a spacecraft were truly *asymptotically* directionally stable, the excursions of the spin axis, even in the presence of persistent small torques, would be limited.

Another point to note is that $\underset{\rightarrow}{\mathbf{h}}$ is not, strictly speaking, constant during the energy decay process. Although explanations of the energy sink hypothesis often claim that $\mathscr{Q}$ is in torque-free motion and that $\underset{\rightarrow}{\mathbf{h}}$ is therefore constant, this is inconsistent with the mathematics. In reviewing the energy sink development (above), only the *magnitude* of $\underset{\rightarrow}{\mathbf{h}}$ was assumed constant. At no time was its direction fixed. Indeed, the energy $2T = \boldsymbol{\omega}^T\mathbf{I}\boldsymbol{\omega}$ cannot change unless a torque is applied so that $\dot{T} = \boldsymbol{\omega}^T\mathbf{g}$; see (4.1,9). This torque is not an "external" torque in the sense that it is environmental in origin; it should be viewed as arising from the "internal" damping mechanism. But it is external to the part of $\mathscr{Q}$ whose energy and angular momentum are represented by $\frac{1}{2}\boldsymbol{\omega}^T\mathbf{I}\boldsymbol{\omega}$ and $\mathbf{I}\boldsymbol{\omega}$. To ensure that the magnitude h remains constant during energy dissipation, it is only necessary that $\mathbf{h}^T\mathbf{g} \equiv 0$; see (4.1,11). Thus, it is again clear that the spin axis $\hat{\mathbf{p}}_2$, although asymptotically approaching the direction of $\mathbf{h}$, does not asymptotically approach the original predisturbance direction $\underset{\rightarrow}{\mathbf{h}}_\nu$. To accomodate this observation, we say that a major-axis spin for $\mathscr{Q}$ is *directionally stable*, *asymptotically to* $\underset{\rightarrow}{\mathbf{h}}$.

For a minor-axis spin, it is observed that the Liapunov function

$$v_m \triangleq 2(T_h - T) = h^2\frac{I_3(I_1 - I_2)c_1^2 + I_1(I_3 - I_2)c_3^2}{I_1I_2I_3} \tag{11}$$

is positive-definite in $\{c_1, c_3\}$ and that $\dot{v}_m \geq 0$. This proves that a minor-axis spin, when disturbed, wanders away from $\underset{\rightarrow}{\mathbf{h}}$ as a consequence of dissipation. Geometrically, the situation shown in Fig. 4.16*a* is transformed by the energy sink hypothesis for $\mathscr{Q}$ into the quite different behavior exhibited in Fig. 5.2. Regardless of initial conditions, all rotational motions for $\mathscr{Q}$ eventually arrive at a simple spin about the major axis. This is the *major-axis rule*. The only remaining question, and it is sometimes an important one, is the *sense* of the final spin. It can be seen from Fig. 5.2 that under slightly different initial conditions, or a slightly different history of dissipation, the final spin is about $-\underset{\rightarrow}{\hat{\mathbf{p}}}_2$, not $+\underset{\rightarrow}{\hat{\mathbf{p}}}_2$.

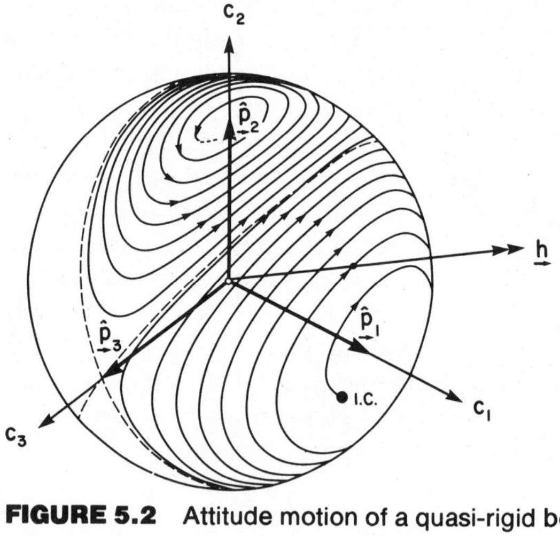

FIGURE 5.2 Attitude motion of a quasi-rigid body $\mathscr{Q}$. The unit sphere is fixed in $\mathscr{Q}$.

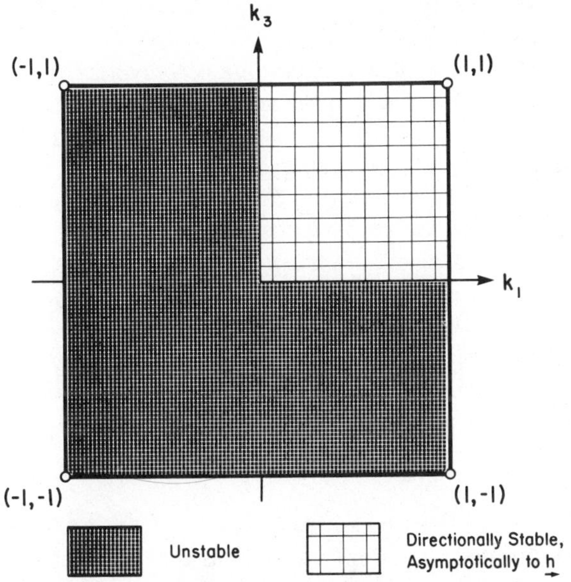

FIGURE 5.3 Stability diagram for simple spin of a quasi-rigid body about one of its principal axes. (Compare with Fig. 4.15.)

The major-axis rule is summarized in Fig. 5.3 in terms of the $\{k_1, k_3\}$ inertia parameters. (The latter were explained in Fig. 4.14.) This diagram, when compared with Fig. 4.15, shows graphically the influence of dissipation on stability. It should be observed that because of its rather heuristic nature, the major-axis rule provides only a guideline, not a precise statement. Exact stability conditions should be rigorously derived for each system of interest. The total kinetic energy of the system (including the source of dissipation) will in fact differ slightly from the rotational kinetic-energy expression $\boldsymbol{\omega}^T \mathbf{I} \boldsymbol{\omega}$. In spite of its approximate nature however, predictions for spinning spacecraft based on the major-axis rule are very different from, and more correct than, those based on the "exact" rigid-body analysis. The lesson seems to be that it is better to make mild approximations in a mathematical model such as $\mathscr{Q}$, which includes all the essential physics, than to determine an "exact" solution for a mathematical model such as $\mathscr{R}$, which omits important physical effects.

Another perspective on the major-axis rule is available by adding appropriate damping terms to the motion equations (4.4, 13), which describe, through the variables $\{\alpha_1, \alpha_3\}$, the deviation of the $\hat{\mathbf{p}}_2$ direction from its original, predisturbance direction. One might, for example, add a term $\mathscr{D}\dot{\mathbf{q}}$ to the left side of (4.4, 14) so that it becomes

$$\mathscr{M}\ddot{\mathbf{q}} + \mathscr{D}\dot{\mathbf{q}} + \mathscr{G}\dot{\mathbf{q}} + \mathscr{K}\mathbf{q} = \mathbf{0} \tag{12}$$

where the damping matrix is positive-definite, $\mathscr{D} > 0$. This is equivalent to adding damping torques

$$[g_1 \quad g_3]^T = -\mathscr{D}[\alpha_1 \quad \alpha_3]^T \tag{13}$$

to the right side of (4.4, 14). Note that, to first order, $\mathbf{h}^T\mathbf{g} = 0$, so that, to first order, $h =$ constant. According to Theorems A.32 and A.33 (Appendix A), the $\{\alpha_1, \alpha_3\}$ motion is stable if and only if $\mathscr{K} > 0$. This condition is equivalent to the major-axis rule; see (4.4, 15). In comparing this analysis to the earlier energy sink derivation of the major-axis rule, it should be observed that (i) the present analysis is a linear (infinitesimal) one, while the energy sink development permitted large deviations; (ii) the physical origin of the $\mathscr{D}\dot{\mathbf{q}}$ damping terms is as obscure as the energy "sink"; and (iii) the present analysis shows that, since $\alpha_1 \to 0$ and $\alpha_3 \to 0$, the $\hat{\mathbf{p}}_2$ axis (i.e., the major axis) *returns to its original, inertially fixed direction*. Thus, for the damping law specified by (13), one does in fact have asymptotic directional stability for the major axis. This is a stronger result than obtained earlier with the energy sink model. In Section 5.2, a specific mechanical damper will be analyzed rigorously; for that system, there will be no doubt about the nature of the "sink" or the "damping torques."

Landon's Rule

Based essentially on physical intuition, [Landon] pointed out that an inertially symmetrical spinning body containing a liquid-filled toroidal damper will tend to be directionally stabilized if the rate of precession exceeds the spin rate, and will be destabilized otherwise. In symbols, Landon's stability condition is

$$\Omega_p > \nu \tag{14}$$

where Ω_p is the precession rate, given earlier by (4.2, 21). For small directional deviations from a spin about the symmetry axis, $h \doteq I_a \nu$, and

$$\Omega_p = \frac{h}{I_t} \doteq \frac{I_a \nu}{I_t} \tag{15}$$

To the extent that this approximation holds, Landon's criterion for asymptotic directional stability is equivalent to the major-axis rule. Moreover, although Landon's argument for $\mathscr{Q}$ is limited to symmetrical bodies and small nutation angles, it can be generalized to apply to gyrostats (Chapter 7). A more detailed physical explanation of Landon's rule has been given by [Iorillo, 1].

5.2 RIGID BODY WITH A POINT MASS DAMPER, $\mathscr{R} + \mathscr{P}$

A dynamical system consisting of a rigid body $\mathscr{R}$ with an internal mechanical damper was analyzed in Section 3.4. The internal damper $\mathscr{P}$ consists of a linear damped harmonic oscillator. This system provides an opportunity to assess rigorously the effect of energy dissipation on directional stability for at least one kind of spinning spacecraft in torque-free motion. Several other important concepts are also illustrated by this system: linearization, static stability, damper tuning, and the adverse effect of nonrigidity on the major-axis rule.

The equations of motion formulated for $\mathscr{R} + \mathscr{P}$ in Section 3.4 are quite general, permitting arbitrary $\boldsymbol{\omega}(t)$ and damper mass excursions $\xi(t)$. The reference position of the damper mass, **b**, and its direction of travel, **n**, are also general. In this form, an analytical solution is highly unlikely. Therefore, we shall be content with an infinitesimal (linear) analysis of the deviations about a reference solution of interest. In this stability investigation, it is assumed that the external force and torque are zero, $\mathbf{f} \equiv \mathbf{0} \equiv \mathbf{g}$.

The inertia distribution associated with $\mathscr{R} + \mathscr{P}$ is complicated by the fact that the first and second moments of inertia depend on $\xi(t)$. Our plan is to consider reference motions in which $\xi \equiv 0$, by appropriately choosing **b** and **n**. We also select the reference point O so that the first moment of inertia is zero when $\xi = 0$. In (3.4, 1) and (3.4, 3), then,

$$\mathbf{c}_b + m_d\mathbf{b} = \mathbf{0}; \qquad \mathbf{c} = m_d\xi\mathbf{n} \tag{1}$$

Similarly, we choose the orientation of $\mathscr{F}_b$ to coincide with the principal axes of $\mathscr{R} + \mathscr{P}$ when $\xi = 0$:

$$\mathbf{I} = \mathbf{J}_b + m_d(b^2\mathbf{1} - \mathbf{b}\mathbf{b}^T) = \operatorname{diag}\{I_1, I_2, I_3\} \tag{2}$$

$$\mathbf{J}(t) = \mathbf{I} + m_d(2\mathbf{b}^T\mathbf{n}\mathbf{1} - \mathbf{b}\mathbf{n}^T - \mathbf{n}\mathbf{b}^T)\xi + m_d(\mathbf{1} - \mathbf{n}\mathbf{n}^T)\xi^2 \tag{3}$$

It will now be shown that major- and minor-axis spins with $\xi \equiv 0$ are possible and that neighboring motions excite the damper so that $\xi \not\equiv 0$. The chief objective is to determine the effect of energy dissipation and nonrigidity on the stability of the reference motion.

Consider a spin about $\underrightarrow{\hat{\mathbf{p}}}_2$ and select **b** and **n** as follows:

$$\boldsymbol{\omega} = \nu\mathbf{1}_2; \qquad \mathbf{b} = b\mathbf{1}_1; \qquad \mathbf{n} = \mathbf{1}_2 \tag{4}$$

Figure 5.4 is helpful. The relevant inertia. quantities are, from (1) through (3),

$$\mathbf{c} = m_d\xi\mathbf{1}_2$$

$$J_{11} = I_1 + m_d\xi^2; \qquad J_{22} = I_2; \qquad J_{33} = I_3 + m_d\xi^2$$

$$J_{13} = J_{31} = 0; \qquad J_{23} = J_{32} = 0; \qquad J_{21} = J_{12} = -m_d b\xi \tag{5}$$

We do not assume that $I_1 > I_2 > I_3$.

In deriving a linearized version of the motion equations, we need not make any assumption about the smallness of **v**, the velocity of the mass center of $\mathscr{R} + \mathscr{P}$. For the reference motion, $\boldsymbol{\omega} = \nu\mathbf{1}_2$ and $\xi = 0$, while for the disturbed motion, $\boldsymbol{\omega}$ is given to first order by (2.3, 33):

$$\omega_1 = \dot{\alpha}_1 + \nu\alpha_3$$

$$\omega_2 = \dot{\alpha}_2 + \nu$$

$$\omega_3 = \dot{\alpha}_3 - \nu\alpha_1 \tag{6}$$

where $\boldsymbol{\alpha}$ and $\dot{\boldsymbol{\alpha}}$ are treated as first-order infinitesimals. Calculating expressions next for the momenta, we linearize (3.4, 20 through 22). To first order in $\boldsymbol{\alpha}$ and ξ

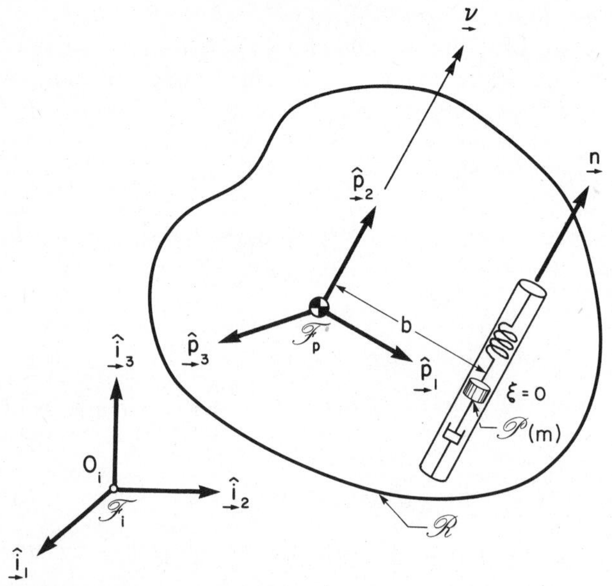

FIGURE 5.4 Rigid body ℛ with an internal point mass damper 𝒫.

and their time derivatives,

$$p_1 = mv_1$$
$$p_2 = mv_2 + m_d\dot{\xi}$$
$$p_3 = mv_3 \tag{7}$$

$$h_1 = I_1(\dot{\alpha}_1 + \nu\alpha_3) + m_d\xi(v_3 - \nu b)$$
$$h_2 = I_2(\dot{\alpha}_2 + \nu)$$
$$h_3 = I_3(\dot{\alpha}_3 - \nu\alpha_1) - m_d(\xi v_1 - b\dot{\xi}) \tag{8}$$

$$p_n = m_d\left[v_2 + b(\dot{\alpha}_3 - \nu\alpha_1) + \dot{\xi}\right] \tag{9}$$

The foundation has now been laid for writing out the motion equations (3.4, 31–33). The translational equations become

$$\dot{v}_1 + \omega_2 v_3 - \omega_3 v_2 = 0$$
$$\frac{m_d}{m}\ddot{\xi} + \dot{v}_2 + \omega_3 v_1 - \omega_1 v_3 = 0$$
$$\dot{v}_3 + \omega_1 v_2 - \omega_2 v_1 = 0 \tag{10}$$

where $\{\omega_1, \omega_2, \omega_3\}$ were given in (6). Similarly, the rotational equations (3.4, 32) become, after inserting (10),

$$I_2\ddot{\alpha}_2 = 0$$
$$I_1\ddot{\alpha}_1 + (I_1 + I_3 - I_2)\nu\dot{\alpha}_3 + (I_2 - I_3)\nu^2\alpha_1 = 0$$
$$I_3\ddot{\alpha}_3 - (I_1 + I_3 - I_2)\nu\dot{\alpha}_1 + (I_2 - I_1)\nu^2\alpha_3 + m_d b(\ddot{\xi} + \nu^2\xi) = 0 \tag{11}$$

The last motion equation, for ξ, is found by reducing (3.4, 33) to first order and imposing (10):

$$\frac{m_d m_b}{m}\ddot{\xi} + c_d\dot{\xi} + k_d\xi + m_d b\left(\ddot{\alpha}_3 + \nu^2\alpha_3\right) = 0 \tag{12}$$

The alleged reference motion ($\xi \equiv 0$, $\boldsymbol{\alpha} \equiv \mathbf{0}$) clearly satisfies these equations.

The stability characteristics of a principal-axis spin for $\mathscr{R}+\mathscr{P}$ can be assessed from (11) and (12). As usual, the motion is unstable in the sense that α_2 cannot be kept within bounds merely by choosing $\alpha_2(0)$ sufficiently small. *Directional* stability is still a possibility however. To derive the appropriate stability conditions, we note that (11) and (12) can be written, omitting the α_2 equation, in the form

$$\mathscr{M}\ddot{\mathbf{q}} + (\mathscr{D}+\mathscr{G})\dot{\mathbf{q}} + \mathscr{K}\mathbf{q} = \mathbf{0} \tag{13}$$

where

$$\mathbf{q} \triangleq \begin{bmatrix} \alpha_1 \\ \alpha_3 \\ \xi \end{bmatrix}; \qquad \mathscr{M} \triangleq \begin{bmatrix} I_1 & 0 & 0 \\ 0 & I_3 & m_d b \\ 0 & m_d b & \dfrac{m_d m_b}{m} \end{bmatrix}$$

$$\mathscr{D} \triangleq \operatorname{diag}\{0, 0, c_d\}; \qquad \mathscr{G} \triangleq (I_1 + I_3 - I_2)\nu \begin{bmatrix} 0 & 1 & 0 \\ -1 & 0 & 0 \\ 0 & 0 & 0 \end{bmatrix}$$

$$\mathscr{K} \triangleq \begin{bmatrix} (I_2 - I_3)\nu^2 & 0 & 0 \\ 0 & (I_2 - I_1)\nu^2 & m_d b\nu^2 \\ 0 & m_d b\nu^2 & k_d \end{bmatrix} \tag{14}$$

It is readily demonstrated that $\mathscr{M} > 0$. Certainly $I_1 > 0$ and $I_3 > 0$, and the last of Sylvester's criteria, $\det\mathscr{M} > 0$, requires that

$$I_3 m_b > m m_d b^2 \tag{15}$$

which, according to Problem 5.3, is always satisfied. The matrix-second-order system (13) is therefore in canonical form, since $\mathscr{M}^T = \mathscr{M} > 0$, $\mathscr{G}^T = -\mathscr{G}$, $\mathscr{D}^T = \mathscr{D}$, and $\mathscr{K}^T = \mathscr{K}$. If $\mathscr{D}$ were positive-definite, the necessary and sufficient stability condition would be $\mathscr{K} > 0$, as stated in the Kelvin–Tait–Chetayev theorem (Theorems A.32 and A.33). Unfortunately, $\mathscr{D}$ is not positive-definite, although it is positive-semidefinite. Nevertheless, $\mathscr{K} > 0$ is a sufficient condition for stability, although it may not be a necessary one. Therefore, we examine the impact of the condition $\mathscr{K} > 0$ on our system parameters. From $\mathscr{K} > 0$, it follows that

(i) $I_2 - I_3 > 0$
(ii) $I_2 - I_1 > 0$

(iii) $(I_2 - I_1)k_d > m_d^2 b^2 \nu^2$ (16)

Clearly, (ii) is subsumed by (iii), so conditions (i) and (iii) are the operative ones. We learn, in particular, that it is *not sufficient* that $\hat{\mathbf{p}}_2$ be the major axis. This lack of sufficiency of the major-axis rule is generally typical of nonrigid spinning spacecraft.

To present the stability conditions graphically with as few parameters as possible, and to facilitate comparison with the earlier results for a quasi-rigid body $\mathscr{Q}$, we define

$$\omega_d \triangleq \left(\frac{k_d}{m_d}\right)^{1/2}; \qquad \hat{\omega}_d \triangleq \frac{\omega_d}{\nu}$$

$$I_d \triangleq m_d b^2; \qquad \hat{I}_d \triangleq \frac{I_d}{I_3}; \qquad \Xi_d \triangleq \frac{\hat{I}_d}{\hat{\omega}_d^2} \tag{17}$$

and use also the inertia ratios defined earlier by (4.4, 20). Note that ω_d is the undamped natural frequency of $\mathscr{P}$ with $\mathscr{R}$ held motionless, I_d is the moment of inertia of $\mathscr{P}$ about the spin axis, and $\hat{\omega}_d$ and $\hat{I}_d$ are dimensionless versions of ω_d and I_d.

In terms of these parameters, the stability conditions (16i) and (16iii) are

$$k_{1d} \triangleq k_1 > 0; \qquad k_{3d} \triangleq k_3 - \Xi_d > 0 \tag{18}$$

It is interesting and elegant that, of the four damper parameters m_d, c_d, k_d, and b, only the combination Ξ_d is involved in defining the stability boundary. It is also noteworthy that, as $\Xi_d \to 0$, which may happen because the damper mass $m_d \to 0$, or because the spring becomes very stiff, $k_d \to \infty$, the stability conditions (18) revert to the simple major-axis rule predicted by energy sink analysis (depicted in Fig. 5.3). It is equally noteworthy however that for finite Ξ_d, the major-axis rule is an insufficient stability condition. For example, Fig. 5.5 shows the region of stability for $\Xi_d = 0.5$. Comparing with Figs. 4.15 and 5.3, it can be seen that some of the major-axis spins predicted to be directionally stable (asymptotically to $\underrightarrow{\mathbf{h}}$) by the heuristic major-axis rule do not in fact have this property. In practice, the value of Ξ_d would be kept to much less than 50% by keeping $\mathscr{P}$ small and by making the spring stiff. Nevertheless, the need to replace the major-axis rule with precise stability conditions for specific systems is amply evident from this example.

The one loose end yet to be tied in this stability analysis is whether the above conditions, which correspond to $\mathscr{K} > 0$, are necessary or just sufficient. They are in fact also necessary, as will now be demonstrated. The characteristic equation

$$\det\left[\mathscr{M}s^2 + (\mathscr{D} + \mathscr{G})s + \mathscr{K}\right] = 0 \tag{19}$$

with $\{\mathscr{M}, \mathscr{D}, \mathscr{G}, \mathscr{K}\}$ specified by (14), is

$$(s^2 + \nu^2)\sum_{n=0}^{4} a_n \left(\frac{s}{\nu}\right)^{4-n} = 0 \tag{20}$$

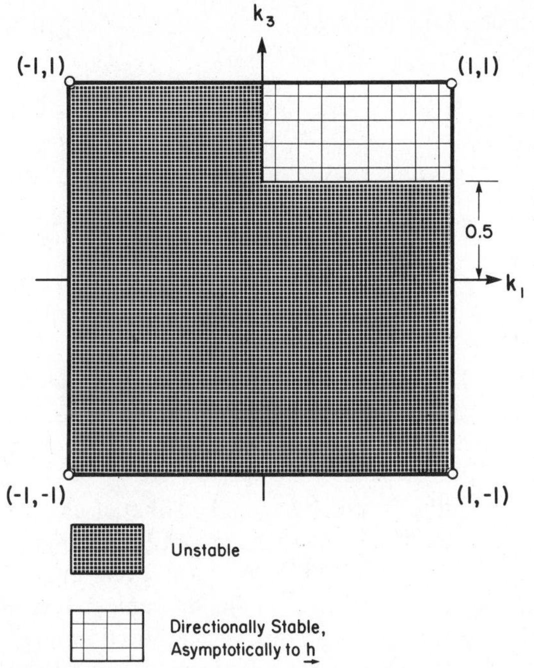

FIGURE 5.5 Stability diagram for simple spin of $\mathcal{R} + \mathcal{P}$ when $\Xi_d = 0.5$. (Compare with Figs. 4.15 and 5.2.)

where

$$\begin{aligned} a_0 &= \hat{m}_b - \hat{I}_d \\ a_1 &= \hat{c}_d \\ a_2 &= \hat{m}_b k_1 k_3 + \hat{\omega}_d^2 - (1 + k_1)\hat{I}_d \\ a_3 &= \hat{c}_d k_1 k_3 \\ a_4 &= k_1\left(k_3 \hat{\omega}_d^2 - \hat{I}_d\right) \end{aligned} \tag{21}$$

and

$$\hat{m}_b \triangleq \frac{m_b}{m}; \qquad \hat{c}_d = \frac{c_d}{\nu m_d} \tag{22}$$

The obvious roots, $s = \pm j\nu$, are a reflection of the nonpervasiveness of the damping and are associated with a (small) change in the direction of the spin axis, as can be verified from the fact that the eigencolumns corresponding to $s = \pm j\nu$ are $\{\xi_o = 0,\ \alpha_{3_o} = \pm j\alpha_{1_o}\}$. (Physically, one would not expect a neighboring simple spin to be damped.) Of greater interest are the other four roots. They will be in the left half-plane, provided the Routh–Hurwitz criteria are satisfied. The condition $a_0 > 0$ is merely a dimensionless version of (15) and is always satisfied. From $a_3 > 0$, we learn that only major- and minor-axis spins are permissible. To find necessary and sufficient conditions, we refer to the succinct summary given in

Table A.1 in Appendix A. The stability boundaries (Theorem A.6) are defined by

$$a_4 > 0$$
$$\Delta_3 \triangleq a_3(a_1a_2 - a_0a_3) - a_1^2a_4 > 0 \tag{23}$$

Expressed in terms of dimensionless parameters according to (21), these conditions are equivalent to

$$k_1(k_3 - \Xi_d) > 0; \qquad k_1(1 - k_3)(1 - k_1k_3) > 0 \tag{24}$$

In view of the fact that $1 - k_3 > 0$ and $1 - k_1k_3 > 0$ for real bodies, these conditions reduce to

$$k_1 > 0; \qquad k_3 - \Xi_d > 0 \tag{25}$$

The conditions for $\mathscr{K} > 0$, namely, (18), are apparently not only sufficient for stability but necessary as well. For example, with $\Xi_d = 0.5$, Fig. 5.5 shows the necessary and sufficient conditions that a simple spin for $\mathscr{R} + \mathscr{P}$ be directionally stable, asymptotically to $\underset{\rightarrow}{\mathbf{h}}$.

To generalize these conclusions slightly, it can be remarked that if the rest location of $\mathscr{P}$ were chosen to lie not on the 1-axis but at some more general point in the 1–3 plane, the condition $I_2 - I_3 > 0$ would also have been affected (Problem 5.5). Furthermore, the parameters $\{m_d, c_d, k_d\}$ of the damper can be judiciously chosen to maximize its effectiveness in removing the nutation angle. This is called *tuning* the damper. A study of damper tuning for essentially the $\mathscr{R} + \mathscr{P}$ system has been made by [Sarychev and Sazonov, 1].

Note that the energy-like quantity

$$\begin{aligned}\mathscr{E} \triangleq \frac{1}{2}\left(I_1\dot{\alpha}_1^2 + I_3\dot{\alpha}_3^2 + \frac{m_dm_b}{m}\dot{\xi}^2 + 2m_db\dot{\alpha}_3\dot{\xi}\right) \\ + \tfrac{1}{2}\left[(I_2 - I_3)\alpha_1^2 + (I_2 - I_1)\alpha_3^2 + 2m_db\alpha_3\xi\right]\nu^2 + \tfrac{1}{2}k_d\xi^2\end{aligned} \tag{26}$$

which may be written

$$\mathscr{E} = \tfrac{1}{2}\dot{\mathbf{q}}^T\mathscr{M}\dot{\mathbf{q}} + \tfrac{1}{2}\mathbf{q}^T\mathscr{K}\mathbf{q} \tag{27}$$

satisfies

$$\dot{\mathscr{E}} = -\tfrac{1}{2}c_d\dot{\xi}^2 \tag{28}$$

Viewed from another perspective, damper tuning has the aim of choosing $\{m_d, c_d, k_d\}$, within certain constraints, so that the damped eigenvalues corresponding to (13), that is, the left-half-plane zeros of (19), are optimal according to some suitable criterion.

5.3 PROBLEMS

5.1 Problem 4.13 dealt with the major-axis spin of a rigid body for which $I_1 \geq I > I_2 > I_3$. It was shown there that the nutation angle γ between the major axis and $\underset{\rightarrow}{\mathbf{h}}$ satisfies $\gamma < \gamma_{max}$, where

$$\sin^2\gamma_{max} = \frac{I_2(I_1 - I)}{I(I_1 - I_2)} \tag{1}$$

In the present problem, the body is assumed to be quasi-rigid, $\mathscr{R} \to \mathscr{Q}$. Thus, I is no longer constant but increases slowly with time, bounded by its maximum, $I \leq I_1$.

(*a*) By differentiating (1), show that

$$\dot{\gamma}_{\max} = \frac{-\dot{I}}{2I}\left[\frac{I_1 I_2}{(I_1 - I)(I - I_2)}\right]^{1/2} \tag{2}$$

and explain why this means that $\dot{\gamma}_{\max} < 0$ until I reaches I_1.

(*b*) By applying the result of Problem 4.13*c*, show in a parallel fashion that $\dot{\gamma}_{\min} > 0$ for a minor-axis spin.

(*c*) What do you conclude about the directional stability properties of major- and minor-axis spins? (See also Figs. 5.1 and 5.2 and [Thompson and Reiter].)

5.2 Consider the torque-free motion of a rigid body $\mathscr{R}$ in the special case where $\mathscr{R}$ possesses an axis of inertial symmetry.

(*a*) Sketch a family of polhodes on a constant-momentum ellipsoid. In other words, how is Fig. 4.11 modified by axial symmetry? Discuss both $I_a < I_t$ and $I_a > I_t$.

(*b*) If $\mathscr{R}$ is now taken to be quasi-rigid, $\mathscr{R} \to \mathscr{Q}$, sketch a typical decaying polhode. That is, modify Fig. 5.1 to account for axial symmetry. Again, discuss both $I_a < I_t$ and $I_a > I_t$.

5.3 Section 5.2 of the text contains an infinitesimal stability analysis of the $\mathscr{R} + \mathscr{P}$ system defined in Section 3.4.

(*a*) Work through each of the equations in Section 5.2, noting how terms higher than first order arise (these terms are dropped) and the care that must be taken with the *zeroth*-order term in ν.

(*b*) Show from the definitions of the symbols involved that $I_3 m_b > m m_d b^2$ always, thus guaranteeing that the system inertia matrix $\mathscr{M}$ is positive-definite.

(*c*) Show analytically that $\mathscr{D} = \text{diag}\{0, 0, c_d\}$ is pervasive for the system of equations (5.2, 13 and 14).

5.4 Consider the torque-free motion of a rigid body $\mathscr{R}$ containing a point mass damper $\mathscr{P}$ in the neighborhood of a simple spin about the 2-axis, as discussed in Section 5.2.

(*a*) Assume, as a special case of the results derived earlier, that the 2-axis is an axis of inertial symmetry, so that $I_1 = I_3 = I_t$ and $I_2 = I_a$. Show that the necessary and sufficient condition for asymptotic directional stability is

$$\frac{I_a}{I_t} > 1 + \frac{\hat{I}_d}{\hat{\omega}_d^2} \tag{3}$$

where $\hat{I}_d \triangleq m_d b^2 / I_t$ and $\hat{\omega}_d \triangleq (k_d / m_d)^{1/2} / \nu$. How does the major-axis rule relate to this condition?

(*b*) Now, assume no longer that $I_1 = I_3$, and explicitly expand the determinant

$$\det\left[\mathcal{M}s^2 + (\mathcal{D} + \mathcal{G})s + \mathcal{K}\right] = 0 \tag{4}$$

to find the sixth-order characteristic equation for the system, where $\mathcal{M}$, $\mathcal{D}$, $\mathcal{G}$, and $\mathcal{K}$ are given in (5.2, 14). Apply the Routh–Hurwitz stability criteria (Table A.1) to verify the stability diagram of Fig. 5.5 at several points. (A computer may be helpful here.)

(*c*) Compare the ease and generality of the two methods: (i) matrix-second-order theory (as in text), and (ii) Routh–Hurwitz criteria (as in this problem).

5.5 In Section 5.2, a rigid body $\mathcal{R}$ containing a point mass damper $\mathcal{P}$ was studied. As shown in Fig. 5.4, it was assumed that the rest location of $\mathcal{P}$ was on the 1-axis. In this problem the more general case is studied in which the rest location of $\mathcal{P}$ is in the $\vec{\hat{\mathbf{p}}}_1$–$\vec{\hat{\mathbf{p}}}_3$ plane. That is, it is assumed that

$$\mathbf{b} = [b_1 \quad 0 \quad b_3]^T \tag{5}$$

(*a*) Rework the derivation of Section 5.2 to show that the equations of motion for small deviations from a simple spin are still given by $\ddot{\alpha}_2 = 0$ and (5.2, 13), with the following changes in $\mathcal{M}$ and $\mathcal{K}$:

$$m_{13} = m_{31} = -m_d b_3; \qquad m_{23} = m_{32} = m_d b_1$$

$$k_{13} = k_{31} = -m_d b_3 \nu^2; \qquad k_{23} = k_{32} = m_d b_1 \nu^2 \tag{6}$$

(*b*) Show that the system inertia matrix $\mathcal{M}$ is still positive-definite.

(*c*) Show that the conditions for asymptotic directional stability are

(i) $I_2 - I_3 > 0$

(ii) $I_2 - I_1 > 0$

(iii) $k_d(I_2 - I_3)(I_2 - I_1) - (I_2 - I_3)m_d^2 b_1^2 \nu^2$

$$-(I_2 - I_1)m_d^2 b_3^2 \nu^2 > 0 \tag{7}$$

Comment on the sufficiency of the major-axis rule for this system.

(*d*) Define the following modifications to the inertial parameters k_1 and k_3:

$$k_{1d} \triangleq k_1 - \frac{I_{2d} - I_{3d}}{I_1 \hat{\omega}_d^2}$$

$$k_{3d} \triangleq k_3 - \frac{I_{2d} - I_{1d}}{I_3 \hat{\omega}_d^2}$$

$$k_{13d} \triangleq \frac{I_{13d}}{(I_1 I_3)^{1/2} \hat{\omega}_d^2} \tag{8}$$

where $I_{1d} \triangleq m_d b_3^2$, $I_{2d} \triangleq m_d(b_1^2 + b_3^2)$, $I_{3d} \triangleq m_d b_1^2$, $I_{13d} \triangleq -m_d b_1 b_3$, and $\hat{\omega}_d \triangleq \omega_d/\nu$ as before. Show that the conditions for asymptotic

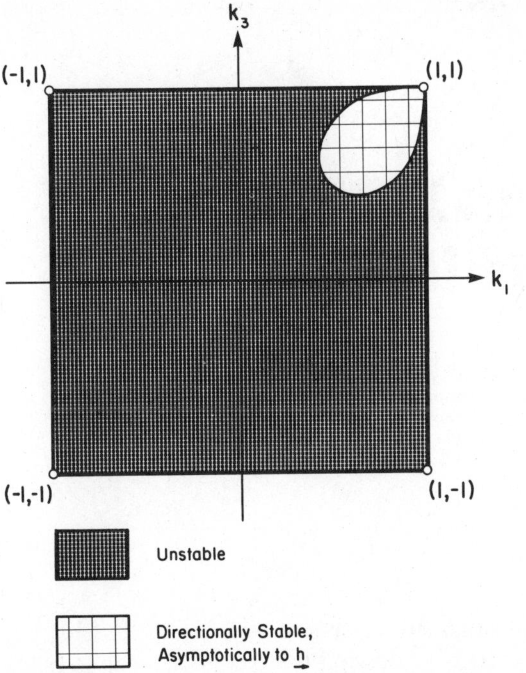

FIGURE 5.6 Stability diagram for simple spin of $\mathscr{R} + \mathscr{P}$ under conditions specified in Problem 5.5*e*. (Compare with Fig. 5.5.)

directional stability can be rewritten thus:

$$k_{1d} > 0$$
$$k_{3d} > 0$$
$$k_{1d}k_{3d} - k_{13d}^2 > 0 \tag{9}$$

(*e*) Verify the stability diagram shown in Fig. 5.6 for the special case $b_1 = b_3 = b/\sqrt{2}$, $m_d b^2 = \frac{1}{2}(I_1 I_3)^{1/2}\hat{\omega}_d^2$. Compare this diagram with Fig. 5.5 and comment.

CHAPTER 6

DIRECTIONAL STABILITY OF MULTISPIN VEHICLES

Reflecting on the stability results obtained for spinning spacecraft in Chapters 4 and 5, one might broadly summarize by observing that stability tends to be an elusive quality for spinning bodies. In the absence of external torques, three-axis attitude stability is impossible due to the inability of the body to resist changes in angular rate about the nominal spin axis. Directional stability of the spin axis however is possible under special circumstances. For a truly rigid body (an abstraction), spins about the major and minor axes are directionally stable (Chapter 4). Of these two cases, the directional stability of the minor-axis spin vanishes when energy dissipation is introduced (Chapter 5). Thus, directional stability is attainable only about the major axis of inertia (the major-axis rule), although this stability is asymptotic. Moreover, the major-axis rule does not guarantee stability for structurally flexible spacecraft, as demonstrated by the relatively simple $\mathscr{R} + \mathscr{P}$ system considered in Section 5.2. New possibilities for stabilization are made possible however by reconfiguring the spacecraft in a more general manner—by constructing the vehicle from two or more parts that spin one with respect to the other.

In this chapter we shall consider a vehicle composed of two or more *rigid* bodies in relative rotation. The all-important influence of energy dissipation is examined in the next chapter, including a rigorous analysis of the effects of one or more point mass dampers.

6.1 THE $\mathscr{R} + \mathscr{W}$ GYROSTAT

In general, one might inquire into the motion of an arbitrary system of rigid bodies, all spinning about arbitrary axes and at arbitrary rates with respect to one another. Although the analysis of such a system would be challenging and interesting (to a dynamicist at least), such an ambitious undertaking will not be attempted here because one tends to become embroiled in resolving notational problems, and the law of diminishing returns applies. It is also difficult to make fundamental inferences. Besides, actual spacecraft whose stabilization has been predicated on the interaction of several bodies in relative spin have had two, or at most three, such bodies. The principal concepts to be understood are therefore best approached via a model comprising two rigid bodies in relative spin.

We shall specialize slightly further and assume that one of the bodies is inertially axisymmetrical and spinning about its symmetry axis with respect to the second body, as in Fig. 3.8. This is the condition normally sought in practice. The symmetrical body may be called a *wheel*, a *rotor*, or a *gyro*, depending upon its physical embodiment, the application, and the writer. We shall for the present call it a wheel and denote it by $\mathscr{W}$. Thus, the system under immediate consideration is the $\mathscr{R} + \mathscr{W}$ system whose motion equations were derived in Section 3.5. As explained in that section, the $\mathscr{R} + \mathscr{W}$ system is called a *gyrostat*. Following customary terminology, we shall also refer to $\mathscr{R}$ as the *carrier*, or *platform*.

Motion Equations

Without loss of generality the system mass center can be selected as the origin of the reference frame in $\mathscr{R}$. The, $\mathbf{c} = \mathbf{0}$. According to (3.5, 25 through 27), the system momenta are

$$\mathbf{p} = m\mathbf{v}$$

$$\mathbf{h} = \mathbf{I}\boldsymbol{\omega} + I_s\omega_s\mathbf{a}$$

$$h_a = I_s\left(\mathbf{a}^T\boldsymbol{\omega} + \omega_s\right) \tag{1}$$

where, recall, m is the mass of $\mathscr{R} + \mathscr{W}$, $\mathbf{I}$ is the inertia moment matrix of $\mathscr{R} + \mathscr{W}$ about the composite mass center, and I_s is the moment of inertia of $\mathscr{W}$ about its symmetry axis, $\mathbf{a}$. The absolute angular velocity of $\mathscr{R}$, expressed in $\mathscr{F}_b$, is $\boldsymbol{\omega}$, and ω_s is the spin of $\mathscr{W}$ with respect to $\mathscr{R}$.

The motion equations, given earlier by (3.5, 36 through 38), are

$$\dot{\mathbf{p}} = -\boldsymbol{\omega}^{\times}\mathbf{p} + \mathbf{f}$$

$$\dot{\mathbf{h}} = -\boldsymbol{\omega}^{\times}\mathbf{h} + \mathbf{g}$$

$$\dot{h}_a = g_a \tag{2}$$

where $\mathbf{p}$ and $\mathbf{h}$ are respectively the absolute momentum and the absolute angular momentum (about ⊕) of $\mathscr{R} + \mathscr{W}$, $\mathbf{f}$ and $\mathbf{g}$ are the external force and the torque on $\mathscr{R}$, and g_a is the torque from $\mathscr{R}$ on $\mathscr{W}$. It can be readily shown that the form of (2) remains unchanged if there are external forces and torques on $\mathscr{W}$, in which case $\mathbf{f}$ and $\mathbf{g}$ are the external force and torque on $\mathscr{R} + \mathscr{W}$. This is an especially important point to note if $\mathscr{W}$ is physically large, or external to $\mathscr{R}$, or both. The translational and rotational motion equations are seen to be uncoupled, unless coupled by $\mathbf{f}$ or $\mathbf{g}$.

Discussion is facilitated by defining the angular momentum associated with $\mathscr{W}$ over and above its contribution to $\mathbf{I\omega}$:

$$\mathbf{h}_s \triangleq h_s\mathbf{a}; \qquad h_s \triangleq I_s\omega_s \tag{3}$$

As pointed out by [Kane and Fowler] and [Roberson, 9], there are two possibilities of practical importance which can be treated concurrently with an appropriate interpretation of the symbols. These two cases are characterized by

$$\text{(i) } h_s = \text{constant} \tag{4a}$$

$$\text{(ii) } h_a = \text{constant} \tag{4b}$$

Neither of these cases is possible in a strictly passive situation. In the former, termed *Kelvin's gyrostat* by [Roberson, 9], it is implied that $g_a \neq 0$ in a special way. In fact,

$$g_a = \dot{h}_a = I_s\mathbf{a}^T\dot{\boldsymbol{\omega}} \tag{5}$$

which in practice must be maintained by an appropriately controlled actuator (e.g., an electric motor). Substituting the second equation of (1) into the second equation of (2) and observing (4a) leads to the matrix motion equation

$$\mathbf{I}\dot{\boldsymbol{\omega}} = -\boldsymbol{\omega}^{\times}(\mathbf{I}\boldsymbol{\omega} + h_s\mathbf{a}) + \mathbf{g} \tag{6}$$

Written out in full,

$$\begin{aligned} I_1\dot{\omega}_1 &= (I_2 - I_3)\omega_2\omega_3 + h_s(a_2\omega_3 - a_3\omega_2) + g_1 \\ I_2\dot{\omega}_2 &= (I_3 - I_1)\omega_3\omega_1 + h_s(a_3\omega_1 - a_1\omega_3) + g_2 \\ I_3\dot{\omega}_3 &= (I_1 - I_2)\omega_1\omega_2 + h_s(a_1\omega_2 - a_2\omega_1) + g_3 \end{aligned} \tag{7}$$

where it has been assumed that the $\mathscr{R}$-fixed frame $\mathscr{F}_b$ is chosen to be $\mathscr{F}_p$, whose axes coincide with the principal axes for $\mathscr{R} + \mathscr{W}$. An alternative would be to align one of the three axes of $\mathscr{F}_b$ with $\vec{\mathbf{h}}_s$, so that a single constant h_s would appear in the analysis instead of four: $\{h_s, a_1, a_2, a_3\}$. Unfortunately, if one wishes to retain the generality of h_s not coinciding with any of the three principal axes, this choice also implies the introduction of extra $\{I_{12}, I_{23}, I_{31}\}$ terms in the motion equations (see Problem 3.10c).

Equations (6 and 7) apply, as mentioned, to Kelvin's gyrostat, for which $\mathbf{h}_s$ is constant. A parallel situation, which [Roberson, 9] calls the *apparent gyrostat*, occurs when $g_a = 0$. This case of interest also requires a motor, to compensate for the inevitable bearing friction. According to the third equation in (2), h_a is constant:

$$h_a \equiv I_s\mathbf{a}^T\boldsymbol{\omega} + h_s = \text{constant} \tag{8}$$

Recall from (3.5, 9) that $\mathbf{h}_w = \mathbf{I}_w\boldsymbol{\omega}_w$, where all quantities have been expressed in $\mathscr{F}_b$, and that

$$\mathbf{I}_w = I_t\mathbf{1} + (I_s - I_t)\mathbf{a}\mathbf{a}^T \tag{9}$$

$$\boldsymbol{\omega}_w = \boldsymbol{\omega} + \mathbf{a}\omega_s \tag{10}$$

from (3.5, 4) and (3.5, 11). Then,

$$\mathbf{h}_w^T\mathbf{a} = I_s(\mathbf{a}^T\boldsymbol{\omega} + \omega_s) \equiv h_a \tag{11}$$

in accordance with (3.5, 13). To establish the correspondence between the two types of gyrostats defined by (4), we rearrange the angular momentum, given by the second equation in (1), to have the following form:

$$\mathbf{h} = (\mathbf{I} - I_s\mathbf{a}\mathbf{a}^T)\boldsymbol{\omega} + \mathbf{a}h_a \tag{12}$$

Now all quantities on the right side are constant, except $\boldsymbol{\omega}(t)$. Comparing this with the second equation in (1), we see that the two idealized cases defined by (4) are equivalent, provided we observe the correspondence

$$\begin{aligned} \mathbf{I} &\leftrightarrow (\mathbf{I} - I_s\mathbf{a}\mathbf{a}^T) \\ h_s &\leftrightarrow h_a \end{aligned} \tag{13}$$

The left sides of (13) correspond to "Kelvin's gyrostat" and the right sides correspond to the "apparent gyrostat." Having examined the former, as we shall do below, the corresponding deductions for the latter follow from (13).

Extension to Several Wheels

A second extension of great practical importance is the application ([Wittenburg]) of the findings below for Kelvin's gyrostat to the system $\mathscr{R} + \Sigma\mathscr{W}_n$, that is, to a rigid body in which are located several wheels, $\mathscr{W}_1, \mathscr{W}_2, \ldots, \mathscr{W}_N$. The angular momentum of this system about the composite mass center is

$$\mathbf{h} = \mathbf{I}\boldsymbol{\omega} + \sum_{n=1}^{N} h_{sn}\mathbf{a}_n \tag{14}$$

where obviously $\mathscr{W}_n$ has an angular momentum $h_{sn}\mathbf{a}_n$, relative to $\mathscr{R}$, in the direction defined by the unit column $\mathbf{a}_n$; $\boldsymbol{\omega}$ is the absolute angular velocity of $\mathscr{R}$; and $\mathbf{I}$ is the moment-of-inertia matrix of $\mathscr{R} + \Sigma\mathscr{W}_n$. If each $\mathscr{W}_n$ obeyed the Kelvin assumption of constant $h_{sn}\mathbf{a}_n$, then it is plain that the results for a simple wheel are immediately applicable to N wheels, with the equivalent simple wheel having h_s and $\mathbf{a}$ given by

$$h_s\mathbf{a} = \sum_{n=1}^{N} h_{sn}\mathbf{a}_n \tag{15}$$

Unfortunately, in practical applications where many wheels are present, this assumption is not strictly valid. The spacecraft attitude (the attitude of $\mathscr{R}$) is controlled by absorbing the angular impulse from external torques into several wheels. By sensing the orientation of $\mathscr{R}$, the control system seeks to deal with the unwanted angular momentum contributed by external torques by transferring it to (or "storing" it in) the wheels. This transferral is accomplished by applying control torques to the wheels. In one version of this process, the wheels are fixed in direction (fixed $\mathbf{a}_n$), and h_{sn} is altered. Such wheels are called *reaction wheels* because the equal and opposite torque from the wheels on $\mathscr{R}$ tends to cancel the external torque, leaving the momentum of $\mathscr{R}$ unchanged. In another version, the magnitudes of the wheel momenta are fixed (fixed h_{sn}) but their axes $\mathbf{a}_n$ are rotated. The wheels are then called *control moment gyros*.

In either case, reaction wheels or control moment gyros, it is not possible to assume that $h_s\mathbf{a}$ is constant in (15). Indeed, the function of the wheels is precisely to absorb the angular impulse from external torques so as to maintain a desired spacecraft attitude. Nevertheless, because torques from the environment are normally quite weak, it is usually possible to assume that $h_a\mathbf{a}$ changes *quasi-statically*. In other words, one may assume that the time periods which characterize the motions studied below are quite short compared to the time it takes for a typical environmental torque to cause a significant change in $h_s\mathbf{a}$. With this assumption, the results obtained in this section become applicable to the more general gyrostat $\mathscr{R} + \Sigma\mathscr{W}_n$.

Torque-Free Gyrostat

The motion of a torque-free (Kelvin) gyrostat is governed by (6), with $\mathbf{g} \equiv \mathbf{0}$ and $\mathbf{h}_s \equiv$ constant. As scalar equations, we have (7), with $g_1 = g_2 = g_3 = 0$. It is not difficult to establish two integrals of the motion:

$$2T_o \triangleq \boldsymbol{\omega}^T\mathbf{I}\boldsymbol{\omega} \equiv I_1\omega_1^2 + I_2\omega_2^2 + I_3\omega_3^2 \tag{16}$$

$$h^2 = (\mathbf{I}\boldsymbol{\omega} + \mathbf{h}_s)^T(\mathbf{I}\boldsymbol{\omega} + \mathbf{h}_s)$$

$$\equiv (I_1\omega_1 + h_s a_1)^2 + (I_2\omega_2 + h_s a_2)^2 + (I_3\omega_3 + h_s a_3)^2 \tag{17}$$

where h is the magnitude of the total angular momentum about $\oplus$. The energy-like quantity T_o is not the total rotational kinetic energy however. From (3.5, 28), the latter is

$$T = T_o + \tfrac{1}{2}I_s\omega_s^2 + \mathbf{h}_s^T\boldsymbol{\omega} \tag{18}$$

The first two terms in this sum are constant, but the third varies. Indeed,

$$\dot{T} = \mathbf{h}_s^T\dot{\boldsymbol{\omega}} = I_s\omega_s\mathbf{a}^T\dot{\boldsymbol{\omega}} = g_a\omega_s \tag{19}$$

in agreement with (3.5, 29). In other words, power is being expended by the wheel speed control motor to maintain the gyrostat assumption $\mathbf{h}_s$ = constant. (For the "apparent gyrostat" however, T is constant.)

It is also important to find the conditions for relative equilibrium, that is, cases where $\boldsymbol{\omega}$ = constant. In such instances, called "pure spins," or "permanent rotations," $\vec{\omega}$ is fixed in $\mathscr{R}$; moreover, $\vec{\omega}$ is also fixed with respect to $\mathscr{F}_i$ since, if $\overset{\circ}{\vec{\omega}} = \vec{0}$, then $\dot{\vec{\omega}} = \vec{0}$ also. We recall that for $\mathscr{R}$ alone, the condition is that $\vec{\omega}$ be aligned with one of the principal axes of $\mathscr{R}$. We now ask what the corresponding situation is when $h_s \neq 0$. According to the matrix motion equation (6), $\boldsymbol{\omega} \equiv \boldsymbol{\nu} =$ constant if and only if

$$\boldsymbol{\nu}^\times(\mathbf{I}\boldsymbol{\nu} + h_s\mathbf{a}) = \mathbf{0} \tag{20}$$

which implies one or more of the following conditions:

$$\boldsymbol{\nu} = \mathbf{0} \qquad \text{(case A)} \tag{21}$$

$$\mathbf{h} \equiv \mathbf{I}\boldsymbol{\nu} + h_s\mathbf{a} = \mathbf{0} \qquad \text{(case B)} \tag{22}$$

$$\mathbf{h} \equiv \mathbf{I}\boldsymbol{\nu} + h_s\mathbf{a} = \lambda\boldsymbol{\nu} \qquad \text{(case C)} \tag{23}$$

The first case is somewhat degenerate and not the most interesting mathematically; but of the three, it is the most important in practice inasmuch as most spacecraft stabilized by a wheel or rotor closely approach this condition (see Chapter 11). In the second case, the total angular momentum of $\mathscr{R} + \mathscr{W}$ is zero; while in case C, the total angular momentum $\underrightarrow{\mathbf{h}}$ is parallel to $\underrightarrow{\boldsymbol{\nu}}$.

6.2 GYROSTAT WITH NONSPINNING CARRIER

We now consider the three types of relative equilibrium (6.1, 21, 22, and 23), one at a time. In case A, equation (21), $\underrightarrow{\boldsymbol{\omega}} \equiv \underrightarrow{\mathbf{0}}$ and $\mathscr{R}$ is not rotating. The wheel persists in a constant relative spin ω_s about an axis $\underrightarrow{\mathbf{a}}$, which is now fixed in space as well as in $\mathscr{R}$. The integrals of the motion are trivial: $T_o = 0$; $h = h_s$. The only remaining question of interest is to describe the motion in the neighborhood of this relative equilibrium and hence determine stability properties. This question is now addressed with the aid of an infinitesimal (linear) analysis.

If it is assumed that some initial disturbance causes $\boldsymbol{\omega}$ to differ infinitesimally from zero, then one can set

$$\boldsymbol{\omega} = \begin{bmatrix} \dot{\theta}_1 & \dot{\theta}_2 & \dot{\theta}_3 \end{bmatrix}^T \equiv \dot{\boldsymbol{\theta}} \tag{1}$$

as explained on page 27. Here, $\boldsymbol{\theta}$ represents three infinitesimal attitude parameters such that $\mathbf{C}_{bi} = \mathbf{1} - \boldsymbol{\theta}^{\times}$. Inserting (1) in (6.1, 6) and discarding second-order infinitesimals, we have

$$\mathbf{I}\ddot{\boldsymbol{\theta}} - h_s\mathbf{a}^{\times}\dot{\boldsymbol{\theta}} = \mathbf{0} \tag{2}$$

which is clearly of the form

$$\mathscr{M}\ddot{\mathbf{q}} + \mathscr{G}\dot{\mathbf{q}} = \mathbf{0} \tag{3}$$

with $\mathscr{M} = \mathbf{I}$ and $\mathscr{G} = -h_s\mathbf{a}^{\times}$. According to Theorem A.25, this system is stable if and only if $\det\mathscr{G} \neq 0$. Unfortunately, here $\det\mathscr{G} = 0$, and hence stability, is precluded. A form of *directional* stability can be obtained however, as will become apparent as (3) is solved.

The general solution to (2) is now found. The associated eigenvalues are the roots of

$$\det\left(s^2\mathbf{I} - sh_s\mathbf{a}^{\times}\right) = 0 \tag{4}$$

Removing the factor $s^3 = 0$, which corresponds to an arbitrary (but infinitesimal) reference value of $\boldsymbol{\theta}$, we have

$$\det\left(s\mathbf{I} - h_s\mathbf{a}^{\times}\right) = 0 \tag{5}$$

for the other three values of s. One of these is clearly $s = 0$, since the determinant of the skew-symmetric matrix $\mathbf{a}^{\times}$ is zero. In expanding this determinant, we shall assume that $\mathscr{F}_b$ is the principal-axis frame $\mathscr{F}_p$, so that $I = \text{diag}\{I_1, I_2, I_3\}$. (The extra effort for $\mathscr{F}_b \neq \mathscr{F}_p$ is taken in Problem 6.2.) By direct expansion of (5), the characteristic equation is

$$s\left(s^2 + \Omega_p^2\right) = 0 \tag{6}$$

where

$$\Omega_p^2 \triangleq \frac{\mathbf{a}^T\mathbf{I}\mathbf{a}}{\det \mathbf{I}} h_s^2 \tag{7}$$

and Ω_p is the *precession frequency*. Based on Problem 3.7*a*, we can rewrite this definition

$$\Omega_p \triangleq \left(\frac{I_a}{I_1 I_2 I_3}\right)^{1/2} h_s \tag{8}$$

where I_a is the moment of inertia of $\mathscr{R} + \mathscr{W}$ about an axis through the mass center and parallel to $\underset{\rightarrow}{\mathbf{a}}$.

The formula (8) is in one respect a generalization of the definition of Ω_p obtained earlier for a symmetrical spinning rigid body; see (4.2, 21). To see how the present definition reduces to the earlier one as a special case, assume that in the present $\mathscr{R} + \mathscr{W}$ system, $\mathscr{R}$ shrinks in inertia down to zero. Then, one of $\{I_1, I_2, I_3\}$ approaches I_s in the limit (where I_s is the axial moment of inertia of $\mathscr{W}$), as does I_a. The second and third members of $\{I_1, I_2, I_3\}$ approach I_t, where I_t is the transverse moment of inertia of $\mathscr{W}$. Thus, in the limit as $\mathscr{R} + \mathscr{W} \to \mathscr{W}$, the definition (8) shows that $\Omega_p \to h_s/I_t$, in agreement with the earlier definition. The earlier definition, however, was valid for arbitrary angular displacements, while (8) is restricted to small $\boldsymbol{\theta}$.

Based on (6), and recalling that $s^3 = 0$ has already been factored out, we infer that the general solution for $\dot{\boldsymbol{\theta}}(t)$ must be of the form

$$\dot{\boldsymbol{\theta}}(t) = \mathbf{c}_1 + \mathbf{c}_2 \cos \Omega_p t + \mathbf{c}_3 \sin \Omega_p t \tag{9}$$

where $\{\mathbf{c}_1, \mathbf{c}_2, \mathbf{c}_3\}$ are constant. In fact, it can be shown that $\dot{\boldsymbol{\theta}}(t)$, expressed in terms of its initial value $\dot{\boldsymbol{\theta}}(0)$, is given by

$$\dot{\boldsymbol{\theta}}(t) = \dot{\boldsymbol{\theta}}(0) \cos \Omega_p t + \frac{\mathbf{a}\mathbf{a}^T\mathbf{I}\dot{\boldsymbol{\theta}}(0)}{\mathbf{a}^T\mathbf{I}\mathbf{a}}(1 - \cos \Omega_p t) + h_s(\Omega_p \mathbf{I})^{-1}\mathbf{a}^{\times}\dot{\boldsymbol{\theta}}(0) \sin \Omega_p t \tag{10}$$

To verify this, note that it is true at $t = 0$ and that it satisfies the required differential equation (2). The latter demonstration is facilitated by the identity (Problem 6.3):

$$\mathbf{I}\mathbf{a}\mathbf{a}^T\mathbf{I} = (\mathbf{a}^T\mathbf{I}\mathbf{a})\mathbf{I} + (\det \mathbf{I})\mathbf{a}^{\times}\mathbf{I}^{-1}\mathbf{a}^{\times} \tag{11}$$

It follows that the attitude history is (Problem 6.4)

$$\boldsymbol{\theta}(t) = \boldsymbol{\theta}(0) + \Omega_p^{-1}\dot{\boldsymbol{\theta}}(0)\sin\Omega_p t + h_s\left(\Omega_p^2\mathbf{I}\right)^{-1}\mathbf{a}^{\times}\dot{\boldsymbol{\theta}}(0)(1 - \cos\Omega_p t)$$
$$+\Omega_p^{-1}\frac{\mathbf{a}\mathbf{a}^T\mathbf{I}\dot{\boldsymbol{\theta}}(0)}{\mathbf{a}^T\mathbf{I}\mathbf{a}}(\Omega_p t - \sin\Omega_p t) \tag{12}$$

This is the complete solution to the attitude motion equation (2).

Given some thought, the solutions (10) and (12) reveal a geometrical interpretation for the motion. For example, we see from the expression for $\boldsymbol{\theta}(t)$ that the source of the instability is the term that is $\sim t$. This term will be zero if and only if the initial disturbance $\dot{\boldsymbol{\theta}}(0)$ satisfies $\mathbf{a}^T\mathbf{I}\dot{\boldsymbol{\theta}}(0) = 0$. Note that to avoid this unbounded growth in attitude deviation, it is $\mathbf{I}\dot{\boldsymbol{\theta}}(0)$ and not $\dot{\boldsymbol{\theta}}(0)$ which must be perpendicular to $\mathbf{a}$. A special case of this solution is the subject of Problem 6.5.

For general (but still infinitesimal) initial disturbances $\boldsymbol{\theta}(0)$ and $\dot{\boldsymbol{\theta}}(0)$, there is only one axis fixed in $\mathscr{R}$ which remains directionally stable. Let this axis be defined by the unit vector, $\underrightarrow{a}_1$, and denote its components in $\mathscr{F}_b$ before and after initial disturbances by $\mathbf{a}_1$ and $(\mathbf{1} - \boldsymbol{\theta}^{\times})\mathbf{a}_1$, respectively, to first order. The angle γ between $\mathbf{a}_1$ and $(\mathbf{1} - \boldsymbol{\theta}^{\times})\mathbf{a}_1$ is, to first order,

$$\gamma = \|\mathbf{a}_1^{\times}\boldsymbol{\theta}\| \tag{13}$$

From (12), the only way to avoid a boundless buildup in γ is to have $\mathbf{a}^{\times}\mathbf{a}_1 = 0$. In other words, $\underrightarrow{a}_1 \equiv \pm\underrightarrow{a}$, and we have proved that the axis of $\mathscr{W}$ in $\mathscr{R}$ is

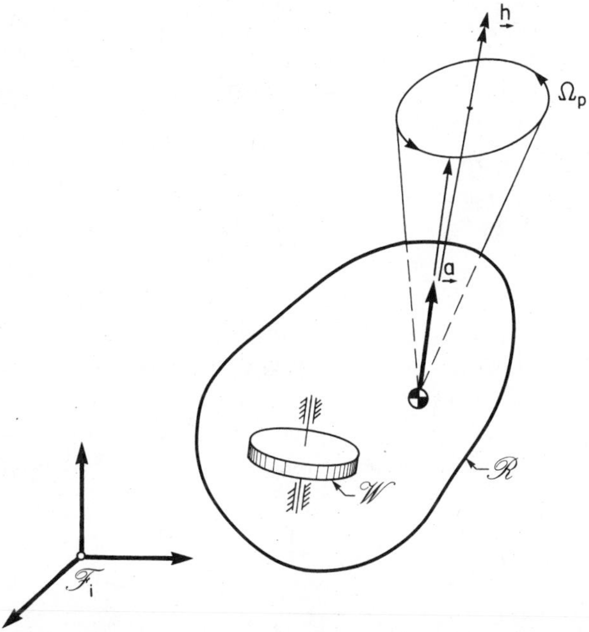

FIGURE 6.1 Coning motion of a rigid gyrostat with a nominally nonspinning carrier. (Note: The cone is not generally circular.)

directionally stable. Of course, it is not asymptotically so, owing to the $\sin \Omega_p t$ and $\cos \Omega_p t$ terms. In general, the $\underset{\rightarrow}{a}$ axis precesses about a direction fixed in $\mathscr{F}_i$, as shown in Fig. 6.1. In fact, $\underset{\rightarrow}{a}$ moves around the surface of an inertially fixed cone (although this cone is elliptic, not circular). When, as a limiting case, the inertia of $\mathscr{R}$ shrinks to zero, this elliptic cone becomes the circular cone of Section 4.2; the system is reduced to $\mathscr{W}$ alone, a single symmetrical spinning body.

6.3 THE ZERO MOMENTUM GYROSTAT

Of lesser importance but not without interest is the second class of relative equilibrium for $\mathscr{R} + \mathscr{W}$. From (6.1, 22), $\boldsymbol{\omega} \equiv \boldsymbol{\nu}$ if $\mathbf{I}\boldsymbol{\nu} + h_s\mathbf{a} = \mathbf{0}$, that is, if

$$\boldsymbol{\nu} = -h_s \mathbf{I}^{-1}\mathbf{a} \tag{1}$$

In this case, the angular momentum of $\mathscr{R}$ is equal and opposite to the angular momentum of $\mathscr{W}$, thus engendering what might be called a *zero momentum gyrostat*. Of course, if there are initial disturbances, these change the angular momentum; the delicate balance is spoiled, and a motion which we shall now study ensues.

As usual for spinning bodies, we choose as attitude variables the infinitesimal deviations $\{\alpha_1, \alpha_2, \alpha_3\}$ between the $\mathscr{R}$-fixed frame $\mathscr{F}_b$ and the nominal frame spinning at angular speed ν. Thus, $\underset{\rightarrow}{\omega}$, expressed in $\mathscr{F}_b$, is

$$\boldsymbol{\omega} = \dot{\boldsymbol{\alpha}} + (\mathbf{1} - \boldsymbol{\alpha}^{\times})\boldsymbol{\nu} \tag{2}$$

Note that we now have five "directions of interest" in $\mathscr{R}$: $\{\underset{\rightarrow}{\hat{p}}_1, \underset{\rightarrow}{\hat{p}}_2, \underset{\rightarrow}{\hat{p}}_3, \underset{\rightarrow}{a}, \underset{\rightarrow}{\hat{\nu}}\}$. In general, these are five *different* directions. Substituting (2) in the motion equation $\mathbf{I}\dot{\boldsymbol{\omega}} + \boldsymbol{\omega}^{\times}(\mathbf{I}\boldsymbol{\omega} + h_s\mathbf{a}) = \mathbf{0}$, invoking (1), and neglecting second-order infinitesimals, we find that the motion equation for $\boldsymbol{\alpha}(t)$ is

$$\mathscr{M}\ddot{\boldsymbol{\alpha}} + \mathscr{G}\dot{\boldsymbol{\alpha}} + \mathscr{K}\boldsymbol{\alpha} = \mathbf{0} \tag{3}$$

where

$$\mathscr{M} \triangleq I; \qquad \mathscr{G} \triangleq \mathbf{I}\boldsymbol{\nu}^{\times} + \boldsymbol{\nu}^{\times}\mathbf{I}; \qquad \mathscr{K} \triangleq \boldsymbol{\nu}^{\times}\mathbf{I}\boldsymbol{\nu}^{\times} \tag{4}$$

To confirm that (3) is in canonical form, note that $\mathscr{M}^T = \mathscr{M} > 0$, $\mathscr{G}^T = -\mathscr{G}$, and $\mathscr{K}^T = \mathscr{K}$.

The relative equilibrium $\boldsymbol{\nu} = -h_s\mathbf{I}^{-1}\mathbf{a}$ would be stable if $\mathscr{K} > 0$ (Theorem A.29). Unfortunately, $\mathscr{K} \leq 0$. To prove this, let $\mathbf{u}$ be any 3×1 column; then, $\mathbf{u}^T\mathscr{K}\mathbf{u} = -(\boldsymbol{\nu}^{\times}\mathbf{u})^T\mathbf{I}(\boldsymbol{\nu}^{\times}\mathbf{u}) < 0$ unless $\mathbf{u} = \mathbf{0}$ *or is parallel to* $\boldsymbol{\nu}$. Just because $\mathscr{K} \leq 0$ however, we cannot rule out stability. To pursue the matter further, we determine the system eigenvalues from the characteristic equation. Denoting an eigenvalue by s, we have

$$\det\left[s^2\mathbf{I} + s(\mathbf{I}\boldsymbol{\nu}^{\times} + \boldsymbol{\nu}^{\times}\mathbf{I}) + \boldsymbol{\nu}^{\times}\mathbf{I}\boldsymbol{\nu}^{\times}\right] = 0 \tag{5}$$

which factors smartly into

$$\det\left(s\mathbf{I}^{1/2} + \boldsymbol{\nu}^{\times}\mathbf{I}^{1/2}\right)\det\left(s\mathbf{I}^{1/2} + \mathbf{I}^{1/2}\boldsymbol{\nu}^{\times}\right) = 0 \tag{6}$$

showing that s satisfies

$$\det{}^2\left(s\mathbf{1} + \boldsymbol{\nu}^{\times}\right) = 0 \tag{7}$$

indicating double roots at $s = 0, +j\nu, -j\nu$. Apparently, Liapunov stability is precluded. In fact, three distinct constraints on the initial conditions, $\boldsymbol{\alpha}(0)$ and $\dot{\boldsymbol{\alpha}}(0)$, must be satisfied to avoid growths $\sim t$, $\sim t\cos\nu t$, and $\sim t\sin\nu t$. More important, there is no axis within $\mathscr{R}$ that is directionally stable for general initial conditions. Therefore, we conclude that the zero momentum gyrostat does not seem to provide any new possibilities for spacecraft stabilization since it has qualitatively the same behavior as a nonspinning $\mathscr{R}$ (without $\mathscr{W}$). It has an instability, and no axis is directionally stable. A special case of this solution is considered in Problem 6.6.

6.4 THE GENERAL CASE

Returning to the general gyrostat, in which the nominal motion has $\mathscr{R}$ spinning and nonzero total momentum, we recall from (6.1, 23) the condition for a relative equilibrium:

$$\mathbf{h} \equiv \mathbf{I}\boldsymbol{\nu} + h_s\mathbf{a} = \lambda\boldsymbol{\nu} \tag{1}$$

where λ is constant to be determined. This condition is of greater mathematical interest than the preceding two special cases ($\boldsymbol{\nu} = \mathbf{0}$ and $\mathbf{h} = \mathbf{0}$), but has not been applied to actual spacecraft as frequently as the nonspinning case $\boldsymbol{\nu} \doteq \mathbf{0}$. The mathematical interest lies in the increased complexity of the relative equilibria and their stability properties.

To understand the role of λ (which, by the way, has the dimensions of moment of inertia), it is helpful to recall the development for $\mathscr{R}$ alone (Section 4.3). The gyrostat reduces in effect to $\mathscr{R}$ alone when the relative angular momentum of $\mathscr{W}$ is zero, that is, when $h_s = 0$. Equation (1) then becomes an eigenvalue problem for λ, as studied in Problem 4.8. The eigenvalues λ reduce to the principal moments of inertia $\{I_1, I_2, I_3\}$, and the eigencolumns $\boldsymbol{\nu}$ lie along the principal axes and possess the property that if the equilibrium spin rate ν is changed, the direction of $\boldsymbol{\nu}$ is unaffected. The present relationship for $\boldsymbol{\nu}$, given by (1), is not an eigenvalue problem, and a change in the spin rate ν changes the direction of $\boldsymbol{\nu}$.

A comparison of the $\mathscr{R}$ and $\mathscr{R} + \mathscr{W}$ systems in $\boldsymbol{\omega}$-space is also helpful. For $\mathscr{R}$ alone, there are two integrals of the motion, T and h, producing ellipsoids $\mathscr{T}$ and $\mathscr{H}$ in $\boldsymbol{\omega}$-space, as shown in Fig. 4.8. These ellipsoids are parallel and have a common center at the origin. If one arbitrarily specifies T and h (subject of course to the restriction that $I_{\min} \le h^2/2T \le I_{\max}$), the result is not generally a relative equilibrium but a motion in which $\boldsymbol{\omega}(t)$ follows the polhode defined by the curve common to $\mathscr{T}$ and $\mathscr{H}$. Only for special combinations of T and h is the motion $\boldsymbol{\omega}(t) = \boldsymbol{\nu} =$ constant. These combinations, for $\mathscr{R}$, are specified by

$$I \triangleq \frac{h^2}{2T} = I_1, \qquad I = I_2, \qquad I = I_3 \tag{2}$$

Geometrically, the intersection of $\mathscr{T}$ and $\mathscr{H}$ consists either of a single point or a curve of bifurcation. With suitable generalizations, these ideas can be extended to apply to a gyrostat as well. The ellipsoid $\mathscr{T}_o$, representing the "energy" integral $2T_o = \boldsymbol{\omega}^T \mathbf{I} \boldsymbol{\omega} =$ constant, is still centered at the origin. However, the ellipsoid $\mathscr{H}$, representing $h^2 =$ constant, is centered at $\boldsymbol{\omega} = -h_s \mathbf{I}^{-1} \mathbf{a}$. The axes of $\mathscr{H}$ are still parallel to the axes of $\mathscr{T}_o$. Note that in addition to the two motion integrals T_o and h^2, we also have the wheel momentum parameter h_s. Only for special combinations of $\{T_o, h, h_s\}$ can relative equilibria be expected. Geometrically, these special (constant) solutions for $\boldsymbol{\nu}$ will occur when $\mathscr{T}_o$ and $\mathscr{H}$ intersect in a point or a curve of bifurcation. These conditions can be achieved by changing the size of $\mathscr{T}_o$, the size of $\mathscr{H}$, or the location of the center of $\mathscr{H}$—that is, by changing T_o, h, or h_s, respectively. It is also assumed in this section that the wheel axis $\mathbf{a}$ is fixed. Other possibilities arise if $\mathscr{W}$ is mounted on gimbals that are connected to $\mathscr{R}$ by springs; see, for example, [Beusch and Smith].

Energy and Wheel Speed Fixed

Suppose that T_o and h_s are chosen. Expressing (1) in $\mathscr{F}_p$ and solving for $\boldsymbol{\nu}$, we find

$$\nu_i = \frac{h_s a_i}{\lambda - I_i} \qquad (i = 1, 2, 3) \tag{3}$$

Now λ can be determined by imposing the conditions $\boldsymbol{\nu}^T \mathbf{I} \boldsymbol{\nu} = 2T_o$. That is, λ is found by solving the sixth-order equation

$$\phi_T(\lambda) = 2\frac{T_o}{h_s^2} \tag{4}$$

for real roots, where

$$\phi_T(\lambda) \triangleq \frac{I_1 a_1^2}{(\lambda - I_1)^2} + \frac{I_2 a_2^2}{(\lambda - I_2)^2} + \frac{I_3 a_3^2}{(\lambda - 1_3)^2} \tag{5}$$

Figure 6.2 shows a semilog plot of the function $\phi_T(\lambda)$ for $\{I_1, I_2, I_3\} = \{6, 4, 2\}$ $\text{kg} \cdot \text{m}^2$ and $a_1 = a_2 = a_3 = 1/\sqrt{3}$. The characteristics of $\phi_T(\lambda)$ can be deduced quite simply from (5): (i) $\phi_T \to \infty$ as $\lambda \to I_i$; (ii) $\phi_T'(\lambda) > 0$ for $\lambda < I_{\min}$; (iii) $\phi_T'(\lambda) < 0$ for $\lambda > I_{\max}$; (iv) ϕ_T is always concave upward because $\phi_T''(\lambda) > 0$; and (v) $\phi_T \to 0$ as $\lambda \to \pm\infty$. It is therefore evident that there may be up to six values of λ that satisfy (4), which we denote by $\{\lambda_1^-, \lambda_1^+, \lambda_2^-, \lambda_2^+, \lambda_3^-, \lambda_3^+\}$. As the wheel speed diminishes ($h_s \to 0$), $\lambda_i^\pm \to I_i$, and the situation reverts to the simpler characteristics of $\mathscr{R}$ alone. With λ thus determined, the (constant) angular velocity of $\mathscr{R}$ is given by (3) and the total angular momentum by (1). In particular,

$$h = |\lambda| \nu \tag{6}$$

where λ is one of the two (or four, or six) values found from (4). It is necessary to take the absolute value of λ in (6) because, while $h \geq 0$ and $\nu \geq 0$ by definition,

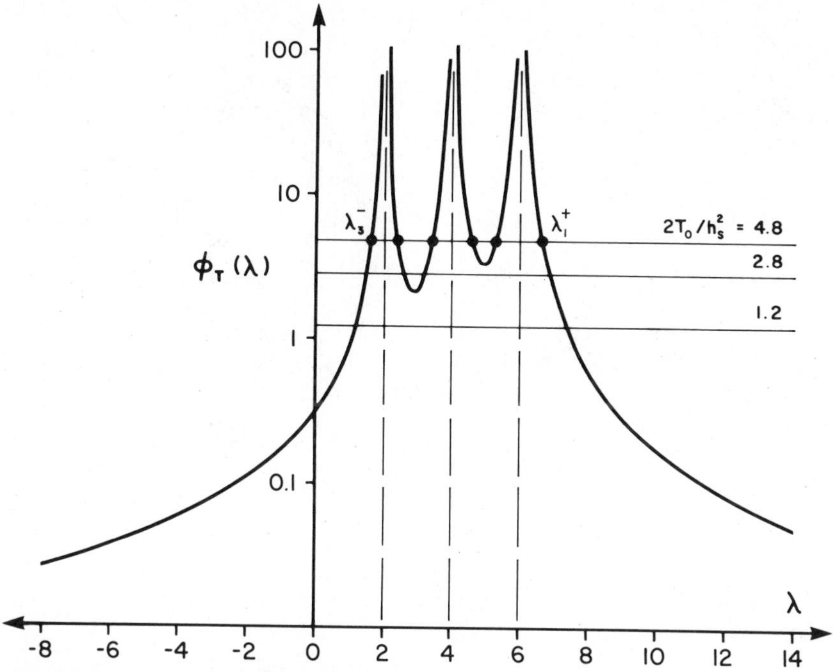

FIGURE 6.2 The function $\phi_T(\lambda)$, given by equation (6.4, 5).

λ is restricted only to real values, which may not be positive (see Fig. 6.2). The meaning of the sign of λ is the subject of Problem 6.8.

When h differs from the special values indicated in (6), the angular velocity $\boldsymbol{\omega}(t)$ is not stationary. To obtain geometrical insight into this motion, consider Fig. 6.3, drawn for $\mathscr{R} + \mathscr{W}$ with $\{I_1, I_2, I_3\} = \{6, 4, 2\}$ kg · m^2 and a wheel $\mathscr{W}$ whose spin axis is defined by the direction cosines $a_1 = a_2 = a_3 = 1/\sqrt{3}$. Shown in Fig. 6.3 are four families of polhodes, lying on the energy ellipsoid $\mathscr{T}_o$ associated with $T_o = 50$ N · m. The three families differ with respect to the spin rate assigned to $\mathscr{W}$. In Figs. 6.3b, c, and d, $h_s = 10/\sqrt{4.8}$, $10/\sqrt{2.8}$, and $10/\sqrt{1.2}$ N · m · s, respectively, corresponding to the three values of $2T_o/h_s^2$ shown in Fig. 6.2. Figure 6.3a, included for comparison, is a repeat of Fig. 4.9, which is applicable as $h_s \to 0$. On each ellipsoid can be detected the constant-$\boldsymbol{\omega}$ solutions ("simple spins," or "permanent rotations," or "equilibria"). Commencing with Fig. 6.3a, we recall the six pure spins to be expected for $\mathscr{R} + \mathscr{W}$ when $\mathscr{W}$ is frozen relative to $\mathscr{R}$. These six spins involve only three axes in $\mathscr{R} + \mathscr{W}$, the principal axes. As the wheel is given some spin, it is seen from Fig. 6.3b that these six equilibria are still identifiable, but they have migrated to new (asymmetrical) locations on $\mathscr{T}_o$. The pure spins $\boldsymbol{\nu}_2^+$ and $\boldsymbol{\nu}_2^-$ continue to be at saddle points and are unstable ([Wittenburg]). The other four pure spins, which are generalizations for $\mathscr{R} + \mathscr{W}$ of the major- and minor-axis spins for $\mathscr{R}$, continue to be $\boldsymbol{\omega}$-stable ([Wittenburg]) under the present assumption of no energy dissipation. This assumption will be removed in Chapter 7.

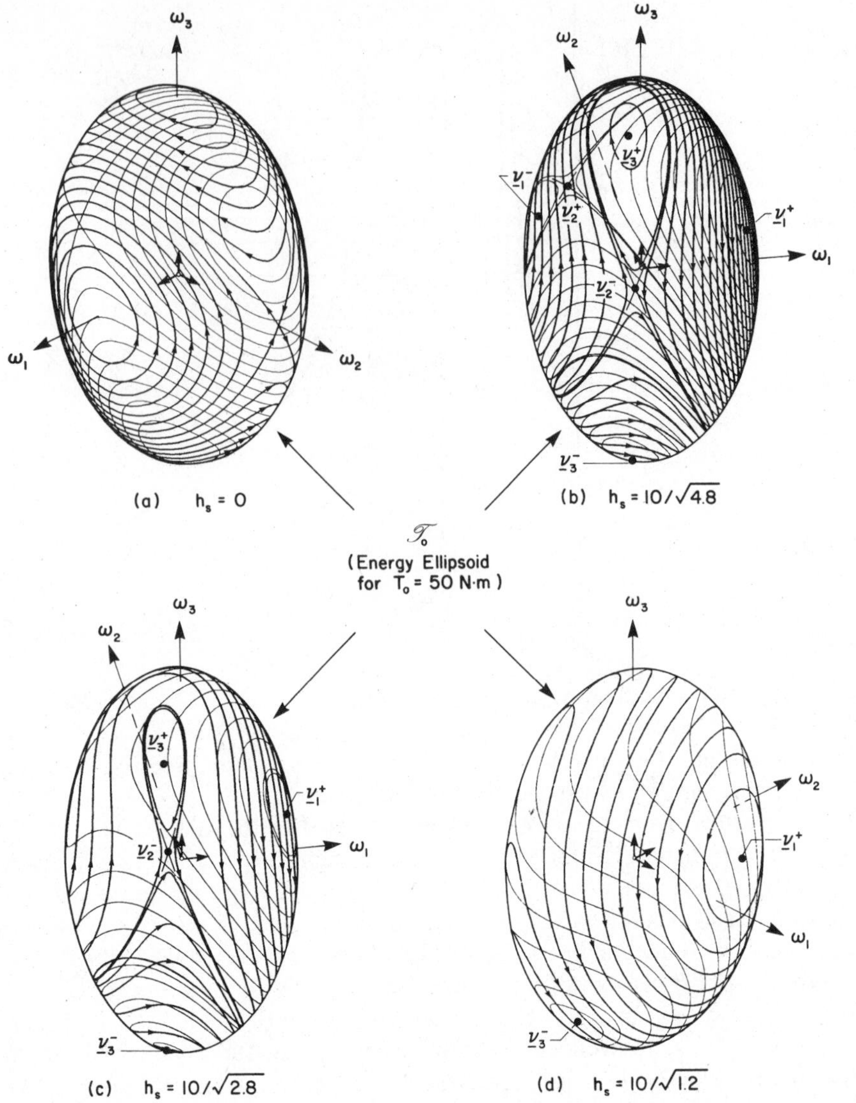

FIGURE 6.3 Families of polhodes, for varying angular momentum h, on the energy ellipsoid $\mathscr{T}_o$ corresponding to $T_o = 50$ N · m. Each family is associated with a fixed relative wheel speed, and $\{I_1, I_2, I_3\} = \{6, 4, 2\}$ kg · m^2.

As the wheel spin rate is increased still further, the simple spins for $\boldsymbol{\nu}_1^-$ and $\boldsymbol{\nu}_2^-$ coalesce on $\mathscr{T}_o$ and annihilate each other (see Fig. 6.3*c* and Fig. 6.2). Thus, the number of pure-spin possibilities is reduced from six to four. Three of these, $\boldsymbol{\nu}_3^-$, $\boldsymbol{\nu}_3^+$, and $\boldsymbol{\nu}_1^+$, are $\boldsymbol{\omega}$-stable (with the rigidity assumption), and the saddle case $\boldsymbol{\nu}_2^-$ is unstable. Increasing the wheel spin rate still further leads to a mutual cancellation of $\boldsymbol{\nu}_2^-$ and $\boldsymbol{\nu}_3^+$ (Fig. 6.3*d*), so that only two simple spins remain, namely, $\boldsymbol{\nu}_3^-$ and

$\mathbf{v}_1^+$. Both are $\boldsymbol{\omega}$-stable. If h_s were increased still further, the directions of the simple spins would approach $\pm\mathbf{a}$ (Problem 6.9).

Angular Momentum and Wheel Speed Fixed

When discussing the motion of a rigid body, treatises on classical mechanics describe polhodes on the energy ellipsoid, as shown in Figs. 4.9 and 4.10. The motivation for keeping T fixed is the elegant geometrical interpretation of Poinsot for the attitude motion of $\mathscr{R}$ [see (4.3, 19) et seq.], which depends on the observation that $\mathbf{h}^T\boldsymbol{\omega} = 2T$ throughout the motion. However, for $\mathscr{R} + \mathscr{W}$, $\mathbf{h}^T\boldsymbol{\omega} = 2T_o + h_s\mathbf{a}^T\boldsymbol{\omega}$, which is not constant during the motion. The primary motivation for emphasizing the $\mathscr{T}$ ellipsoids over the $\mathscr{H}$ ellipsoids thus disappears when one generalizes from $\mathscr{R}$ to $\mathscr{R} + \mathscr{W}$. Even more important, the treatment of energy dissipation in Chapter 7 assumes h^2 fixed and T_o slowly decreasing. Therefore, from the standpoint of spacecraft attitude dynamics, the $\mathscr{H}$ ellipsoid is more fundamental than the $\mathscr{T}_o$ ellipsoid.

Suppose that h and h_s are chosen. We first solve for the simple spin solutions and subsequently examine the general solution for $\boldsymbol{\omega}(t)$. For a permanent rotation (simple spin), $\mathbf{v}$ is still given by (3), with λ to be determined from the chosen h. By imposing the condition that $\|\mathbf{I}\mathbf{v} + h_s\mathbf{a}\| = h$, λ is found by solving the sixth-order equation

$$\phi_h(\lambda) = \frac{h^2}{h_s^2} \tag{7}$$

for real roots, where

$$\phi_h(\lambda) \triangleq \frac{\lambda^2 a_1^2}{(\lambda - I_1)^2} + \frac{\lambda^2 a_2^2}{(\lambda - I_2)^2} + \frac{\lambda^2 a_3^2}{(\lambda - I_3)^2} \tag{8}$$

Figure 6.4 shows a semilog plot of the function $\phi_h(\lambda)$ for $\{I_1, I_2, I_3\} = \{6, 4, 2\}$ kg · m^2 and $a_1 = a_2 = a_3 = 1/\sqrt{3}$. From (8), the properties of $\phi_h(\lambda)$ include (i) $\phi_h \to \infty$ as $\lambda \to I_i$; (ii) $\phi_h' < 0$ for $-\infty < \lambda < 0$; (iii) $\phi_h' > 0$ for $0 < \lambda < I_{\min}$; (iv) $\phi_h' < 0$ for $I_{\max} < \lambda < \infty$; and (v) $\phi_h \to 1$ for $\lambda \to \pm\infty$. The significance of the branch of ϕ_h for which $\lambda < 0$ is that the origin, $\boldsymbol{\omega} = \mathbf{0}$, lies outside $\mathscr{H}$; see also Problem 6.8. It is once again evident that there are six, four, or two roots of (7), depending on the portion of h due to h_s. With $h_s \ll h$, the situation is nearly identical to that for a single rigid body, as expected. With λ determined from (7), $\mathbf{v}$ is given by (3), and $2T_o = \mathbf{v}^T\mathbf{I}\mathbf{v}$.

When $2T_o$ is different from these special values, the angular velocity $\boldsymbol{\omega}(t)$ varies with time. Figure 6.5 shows four families of polhodes drawn for $\mathscr{R} + \mathscr{W}$ with $\{I_1, I_2, I_3\} = \{6, 4, 2\}$ kg · m^2, the spin axis of $\mathscr{W}$ being specified by $a_1 = a_2 = a_3 = 1/\sqrt{3}$. The size of $\mathscr{H}$ corresponds to $h = 20$ N · m · s. Different in the four cases is the wheel momentum h_s, which determines how far the centers of these four (identical) ellipsoids are from the origin. When $h_s = 0$, $\mathscr{H}$ is centered at the origin (Fig. 4.11); as h_s increases, $\mathscr{H}$ moves away from the origin.

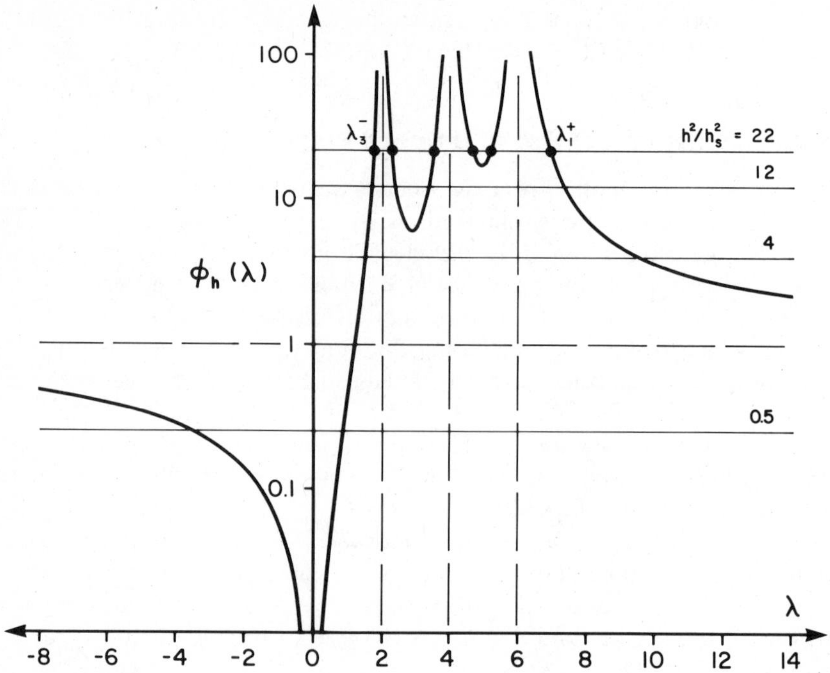

FIGURE 6.4 The function $\phi_h(\lambda)$, given by equation (6.4, 8).

Each polhode on $\mathscr{H}$ is associated with a value of $2T_o$. In Figs. 6.5*a* through *d* $h_s^2/h^2 = 1/22$, $1/12$, $1/4$, and 2, respectively; each family of polhodes contains simple spins as special cases, as indicated in Fig. 6.5. In Fig. 6.5*a*, the six simple spins of Fig. 4.11 have migrated slightly on the surface of $\mathscr{H}$. Two of these (the saddle points) correspond to $\lambda = \lambda_2^+, \lambda_2^-$ in Fig. 6.4 and are unstable. The other four are stable. As h_s increases, the simple spins $\boldsymbol{\nu}_1^-$ and $\boldsymbol{\nu}_2^+$ coalesce on $\mathscr{H}$ and cancel, until only four simple spins, namely, $\{\boldsymbol{\nu}_1^+, \boldsymbol{\nu}_2^-, \boldsymbol{\nu}_3^+, \boldsymbol{\nu}_3^-\}$ remain. As suggested by Fig. 6.5*b*, one of these, $\boldsymbol{\nu}_2^-$, is unstable; the other three are stable. When h_s increases still further, $\boldsymbol{\nu}_2^-$ and $\boldsymbol{\nu}_3^+$ advance toward each other until they are extinct. The remaining two simple spins, $\boldsymbol{\nu}_3^-$ and $\boldsymbol{\nu}_1^+$, are stable (Fig. 6.5*c*). As h_s passes through the value $h_s = h$, the origin, $\boldsymbol{\omega} = \mathbf{0}$, passes through $\mathscr{H}$ and remains outside for $h_s > h$ (Problem 6.8). However, as can be seen by comparing Figs. 6.5*c* and *d*, there is no qualitative change in the arrangement of the polhodes on $\mathscr{H}$.

With the geometrical interpretation of $\boldsymbol{\omega}(t)$ now complete (Figs. 6.3 and 6.5), the two special cases considered previously (nonspinning carrier and zero momentum gyrostat) can be visualized geometrically as special cases. When $\mathscr{R}$ is nominally nonspinning (Section 6.2), $2T_o = 0$, and $\mathscr{T}_o$ shrinks to a point at the origin. Therefore, $\mathscr{H}$ must pass through the origin. For the zero momentum gyrostat (Section 6.3), it is $\mathscr{H}$ that shrinks to a point (not the origin, however, unless the wheel isn't spinning either).

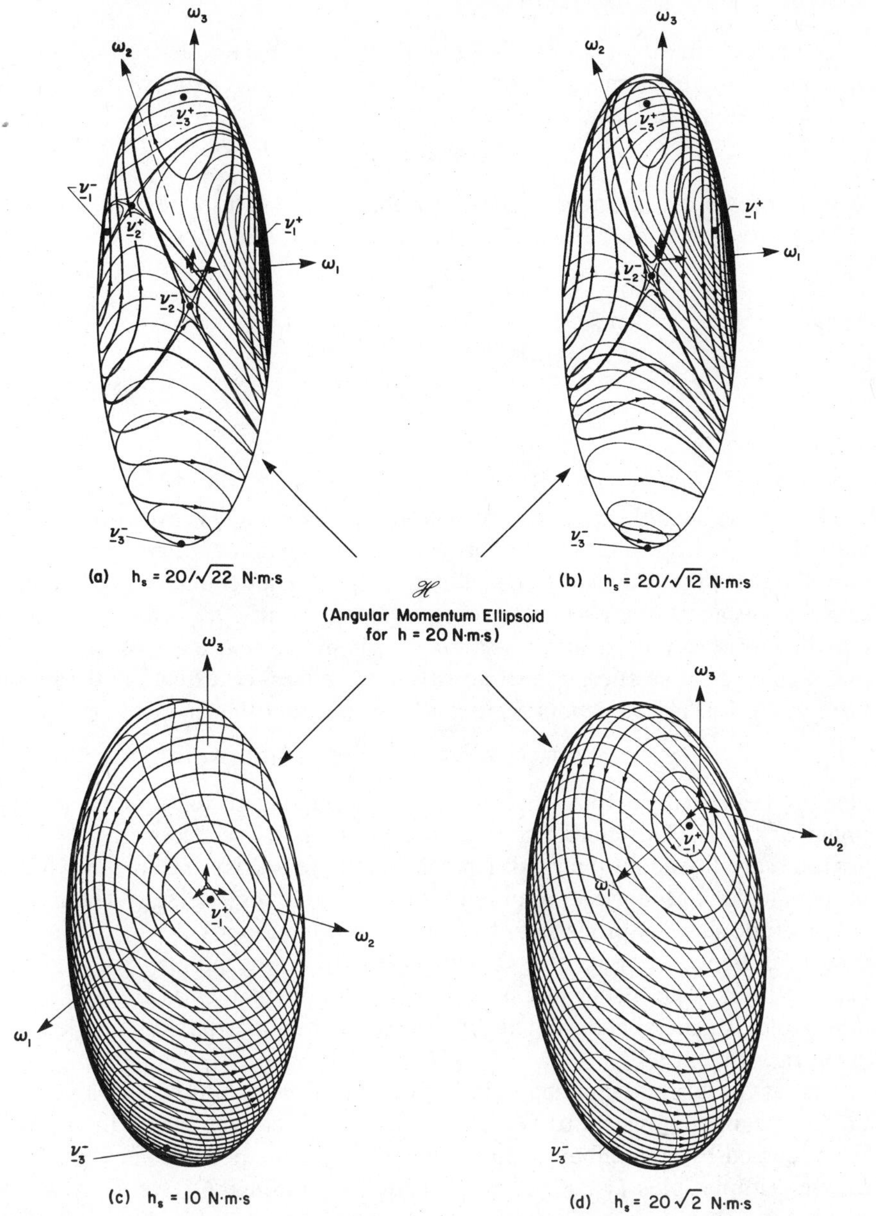

FIGURE 6.5 Families of polhodes, for varying energy T_o, on the angular momentum ellipsoid $\mathscr{H}$ corresponding to $h = 20$ N · m · s. Each family is associated with a fixed relative wheel speed, and $\{I_1, I_2, I_3\} = \{6, 4, 2\}$ kg · m^2.

Stability of Simple Spins

With the simple spins for a general gyrostat in torque-free motion now identified, the next topic is their stability. To address this issue, we begin with an infinitesimal stability analysis. We set

$$\boldsymbol{\omega} = \dot{\boldsymbol{\alpha}} + (\mathbf{1} - \boldsymbol{\alpha}^{\times})\boldsymbol{\nu} \tag{9}$$

as done previously in similar situations. Substituting (9) into the motion equation, (6.1, 6) with $\mathbf{g} \equiv \mathbf{0}$, and deleting second-order terms, one finds

$$\mathscr{M}\ddot{\boldsymbol{\alpha}} + \mathscr{G}\dot{\boldsymbol{\alpha}} + \mathscr{K}\boldsymbol{\alpha} = \mathbf{0} \tag{10}$$

where

$$\mathscr{M} \triangleq \mathbf{I} \tag{11}$$

$$\mathscr{G} \triangleq \mathbf{I}\boldsymbol{\nu}^{\times} + \boldsymbol{\nu}^{\times}\mathbf{I} - \lambda\boldsymbol{\nu}^{\times} \tag{12}$$

$$\mathscr{K} \triangleq \boldsymbol{\nu}^{\times}\mathbf{I}\boldsymbol{\nu}^{\times} - \lambda\boldsymbol{\nu}^{\times}\boldsymbol{\nu}^{\times} \tag{13}$$

Note that $\mathscr{G} = \mathscr{G}(\lambda)$, $\mathscr{K} = \mathscr{K}(\lambda)$ and that $\mathscr{M}^T = \mathscr{M} > 0$, $\mathscr{G}^T = -\mathscr{G}$, and $\mathscr{K}^T = \mathscr{K}$, as one should expect. The system (10) through (13) is evidently a generalization of (6.3, 3) and (6.3, 4) for $\lambda \neq 0$. It is known from Theorem A.29 (Appendix A) that the motion is stable if $\mathscr{K} > 0$. Sign definiteness is not possible however because $\boldsymbol{\nu}^T\mathscr{K}\boldsymbol{\nu} = 0$. However, if $\lambda > I_{max}$, it is true that $(\lambda\mathbf{1} - \mathbf{I}) > 0$; and thus $\mathbf{u}^T\mathscr{K}\mathbf{u} > 0$ for all $\mathbf{u}$ except $\mathbf{u} \sim \boldsymbol{\nu}$, implying that $\mathscr{K}$ has, for $\lambda > I_{max}$, one zero eigenvalue and two real positive eigenvalues. By extending the principle used in (A.6, 31) in Appendix A, it can be shown that the characteristic equation

$$\det(\mathscr{M}s^2 + \mathscr{G}s + \mathscr{K}) = 0 \tag{14}$$

always has two zero roots and two complex conjugate pairs of roots on the imaginary axis. The double root at the origin is associated with the lack of attitude stability about the $\boldsymbol{\nu}$-axis (that is, about the $\mathbf{h}$-axis) since $\mathbf{h} = \lambda\boldsymbol{\nu}$. Motions perpendicular to $\mathbf{h}$ however are infinitesimally stable for $\lambda > I_{max}$. The infinitesimal attitude stability properties of *all* the pure spin motions of $\mathscr{R} + \mathscr{W}$ (i.e., all λ cases) can be worked out using (10) through (13). To the author's knowledge, this has not been done. It has just been shown however that *attitude* stability is not attainable. A general and explicit expression for the characteristic equation (14) is given in Problem 6.12.

Just as for a single spinning rigid body, *directional* stability is still a possibility under certain circumstances. Directional stability is closely related to $\boldsymbol{\omega}$-stability. As discussed by [Wittenburg], and as suggested by the polhodes in Figs. 6.3 and 6.5, the simple spins $\{\boldsymbol{\nu}_1^+, \boldsymbol{\nu}_1^-, \boldsymbol{\nu}_3^+, \boldsymbol{\nu}_3^-\}$ in Figs. 6.2 through 6.5 are $\boldsymbol{\omega}$-stable, while those for $\{\boldsymbol{\nu}_2^+, \boldsymbol{\nu}_2^-\}$ are $\boldsymbol{\omega}$-unstable (and therefore unstable). This information concerning $\boldsymbol{\omega}$-stability can be utilized to derive the conditions for directional stability. In parallel with the development for a rigid body alone [equation (4.4, 31) et seq.], it will be shown that there exists a body-fixed direction which can, under certain circumstances, be made to remain arbitrarily close to its direction in the reference simple spin. The direction in question is $\underrightarrow{\nu}$, the axis of the simple spin. This is also the direction of the angular momentum vector $\underrightarrow{\mathbf{h}}_\nu$ for the simple spin, because prior to disturbance, $\underset{\rightarrow}{\mathbf{h}}_\nu = \lambda\underset{\rightarrow}{\boldsymbol{\nu}}$. The direction cosines of $\underset{\rightarrow}{\boldsymbol{\nu}}$ in $\mathscr{F}_p$ are

$$c_{i\nu} = \frac{\nu_i}{\nu} \qquad (i = 1, 2, 3) \tag{15}$$

That is, $\underset{\rightarrow}{\boldsymbol{\nu}} = \nu[c_{1\nu} \quad c_{2\nu} \quad c_{3\nu}]\mathscr{F}_p$. Note that $\underset{\rightarrow}{\boldsymbol{\nu}}$ defines a direction fixed in $\mathscr{F}_p$. The question at issue is whether, after a disturbance has altered the "initial" conditions, this direction in $\mathscr{F}_p$ can be made to lie arbitrarily close to its predisturbance direction for all $t > 0$, by making the initial disturbances sufficiently small. If so, the $\underset{\rightarrow}{\boldsymbol{\nu}}$ direction is directionally stable in the Liapunov sense.

The initial disturbance will in general change the angular momentum vector from $\underset{\rightarrow}{\mathbf{h}}_\nu$ to a new value, $\underset{\rightarrow}{\mathbf{h}}$. However, it is straightforward to show that the angle between $\underset{\rightarrow}{\mathbf{h}}$ and $\underset{\rightarrow}{\mathbf{h}}_\nu$ can be made arbitrarily small by making the initial disturbance sufficiently small. (This was done for a rigid body in Problem 4.13*a*.) It therefore remains to be shown whether the (body-fixed) $\underset{\rightarrow}{\boldsymbol{\nu}}$-direction can be made to lie arbitrarily close to the (inertially fixed) $\underset{\rightarrow}{\mathbf{h}}$ direction. Let the direction cosines of $\underset{\rightarrow}{\mathbf{h}}$ in $\mathscr{F}_p$ be $\{c_1, c_2, c_3\}$. Thus,

$$c_i = \frac{h_i}{h}; \qquad h_i = I_i\omega_i + h_s a_i \qquad (i = 1, 2, 3) \tag{16}$$

and $\underset{\rightarrow}{\mathbf{h}} = h[c_1 \quad c_2 \quad c_3]\mathscr{F}_p$. Of course, the c_i vary as the motion proceeds ($t > 0$). (For $t < 0$, $c_i = c_{i\nu}$ if $\lambda > 0$, and $c_i = -c_{i\nu}$ if $\lambda < 0$.)

Let the angle between $\underset{\rightarrow}{\boldsymbol{\nu}}$ and $\underset{\rightarrow}{\mathbf{h}}$ be γ. Noting the vectorial interpretation

$$\nu^2 h^2 \sin^2\gamma = \|\underset{\rightarrow}{\boldsymbol{\nu}} \times \underset{\rightarrow}{\mathbf{h}}\|^2 = \|\boldsymbol{\nu}^{\times}\mathbf{h}\|^2 \tag{17}$$

and using direction cosines, we have

$$\sin^2\gamma = (c_2c_{3\nu} - c_3c_{2\nu})^2 + (c_3c_{1\nu} - c_1c_{3\nu})^2 + (c_1c_{2\nu} - c_2c_{1\nu})^2 \tag{18}$$

This is a generalization of the corresponding result for a rigid body alone, for which the direction $\underset{\rightarrow}{\boldsymbol{\nu}}$ is simply a principal inertia axis (e.g., the $\underset{\rightarrow}{\hat{\mathbf{p}}}_2$ axis, as in (4.4, 31) et seq.), so that $c_{1\nu} = 0 = c_{3\nu}$ and $c_{2\nu} = 1$, and (18) reduces to $\sin^2\gamma = c_1^2 + c_3^2$. For the gyrostat, (15) and (16) are substituted in (18) to yield

$$h^2\nu^2\sin^2\gamma = (I_2\nu_3\xi_2 - I_3\nu_2\xi_3)^2 + (I_3\nu_1\xi_3 - I_1\nu_3\xi_1)^2 + (I_1\nu_2\xi_1 - I_2\nu_1\xi_2)^2 \tag{19}$$

where the perturbation to $\boldsymbol{\nu}$ has been defined as

$$\boldsymbol{\xi} \triangleq \boldsymbol{\omega} - \boldsymbol{\nu} \tag{20}$$

A basis now exists for interpreting the classical results for $\boldsymbol{\omega}$-stability in terms of the directional stability of $\underset{\rightarrow}{\boldsymbol{\nu}}$. The classical results state that the cases $\{\boldsymbol{\nu}_1^+, \boldsymbol{\nu}_1^-, \boldsymbol{\nu}_3^+, \boldsymbol{\nu}_3^-\}$ are $\boldsymbol{\omega}$-stable (when they exist—see Figs. 6.2 through 6.5). Thus, by making the initial disturbances sufficiently small, $\|\Delta\boldsymbol{\omega}\|$ can be kept smaller than ε for all $t > 0$. Since $\sin^2\gamma$ is known from (19) to be positive-semidefinite in $\Delta\boldsymbol{\omega}$, one can also maintain γ as close to 0 or π as desired. This completes the demonstration that the simple spins $\{\boldsymbol{\nu}_1^+, \boldsymbol{\nu}_1^-, \boldsymbol{\nu}_3^+, \boldsymbol{\nu}_3^-\}$ are directionally stable—

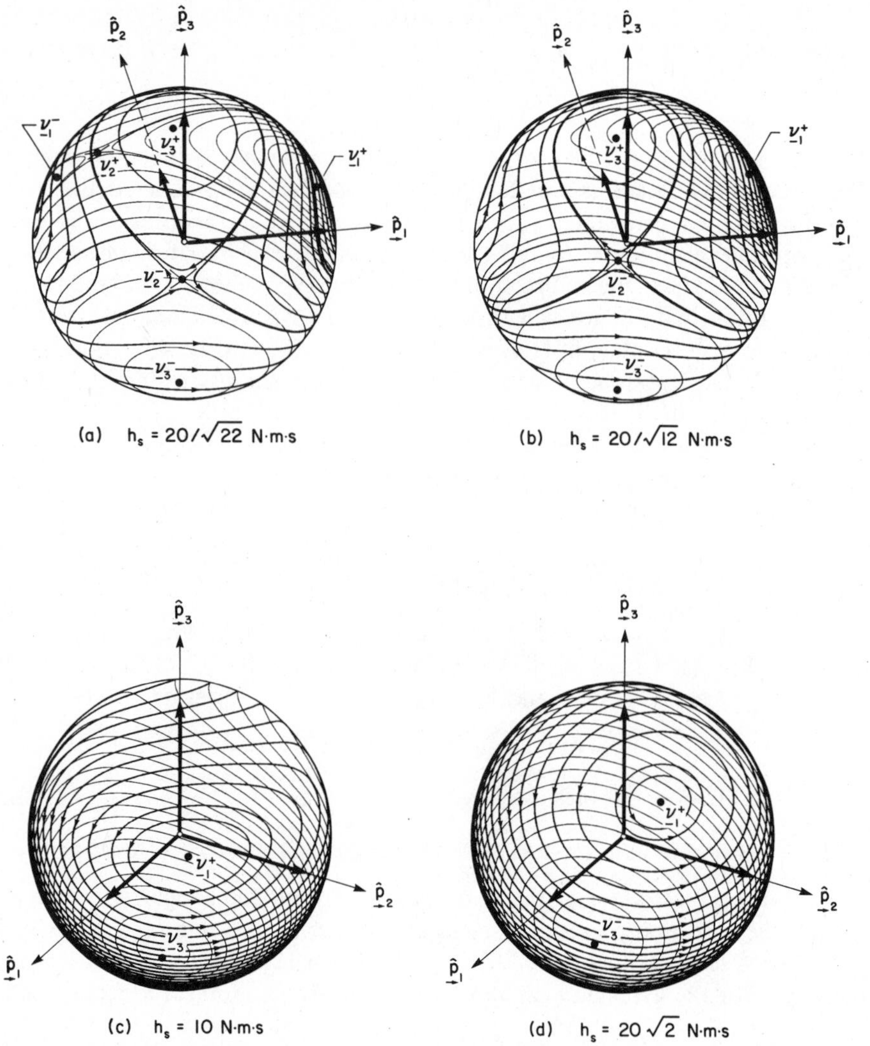

FIGURE 6.6 Attitude motion of a gyrostat with respect to the angular momentum vector **h**.

although the middle two may fail to exist for certain ranges of $\{2T_o, h_\nu, h_s\}$. The other two simple spins, $\{\boldsymbol{\nu}_2^+, \boldsymbol{\nu}_2^-\}$, may or may not exist, and when they do exist, they are directionally unstable. Further stability information is gleaned from Liapunov functions in Problems 6.10 and 6.11.

Geometrical evidence for these assertions is offered in Fig. 6.6. For each of the four diagrams, it should be borne in mind that the unit sphere is fixed in $\mathscr{R}$, so that the orientation of the axes indicates the orientation of $\mathscr{R}$. The angular momentum $\underset{\rightarrow}{\mathbf{h}}$ is fixed inertially. The attitude history of $\mathscr{R}$ is constructed by keeping $\underset{\to}{\mathbf{h}}$ fixed in direction and rotating the unit sphere (and hence $\mathscr{R}$) so that $\underset{\to}{\mathbf{h}}$ continues to pass through one of the curves shown on the surface. Each curve corresponds to specific values of $\{T_o, h, h_s\}$. Figure 6.6 is a generalization (for $\mathscr{R} + \mathscr{W}$) of Figs. 4.13 and 4.16 (for $\mathscr{R}$). They correspond, respectively, to Figs. 6.5*a* through *d*, after application of the relationships

$$c_i = \frac{I_i \omega_i + h_s a_i}{h} \qquad (i = 1, 2, 3) \tag{21}$$

which permit a projection of the ellipsoids $\mathscr{H}$ in Fig. 6.5 onto the unit sphere.

It should be recognized that two aspects of the motion cannot be determined by this construction. One is the time dependence of the motion, although increasing time is indicated by arrows. The other is the attitude motion *about* the $\underset{\to}{\mathbf{h}}$ axis; only attitude motion *perpendicular to* $\underset{\to}{\mathbf{h}}$ can be obtained. All the same, these diagrams are useful because neither of these two shortcomings affect the implications for directional stability. By nature, directional stability is not concerned with the rotation *about* the $\underset{\to}{\mathbf{h}}$ axis—only angular deviations of $\underset{\to}{\boldsymbol{\nu}}$ (fixed in $\mathscr{R}$) away from $\underset{\to}{\mathbf{h}}$. Similarly, the details of the time dependence are also unimportant for directional stability, so long as the maximum deviation for all $t > 0$ can be arbitrarily bounded by a sufficient initial bound.

Wheel Axis Parallel to a Principal Axis

Matters are considerably simplified if the axis of $\mathscr{W}$ is aligned with one of the three principal axes of $\mathscr{R} + \mathscr{W}$. To be specific, let $\underset{\to}{\mathbf{a}}$ be parallel to the $\hat{\mathbf{p}}_2$ axis. Instead of retaining the convention $h_s \geq 0$ and thereby having to consider $\underset{\to}{\mathbf{a}} = \pm \hat{\mathbf{p}}_2$, it is now simpler to take $\underset{\to}{\mathbf{a}} = +\hat{\mathbf{p}}_2$ and allow h_s to be of either sign. Expressed in $\mathscr{F}_p$, $\mathbf{a} = \mathbf{1}_2^T \mathscr{F}_p$. Moreover, we do not assume that $\hat{\mathbf{p}}_2$ is necessarily the major axis of inertia; *the convention* $I_1 > I_2 > I_3$ *will not be applied*. The simple-spin solution is now simply

$$\boldsymbol{\nu} = \nu \mathbf{1}_2 \tag{22}$$

and, with no loss in generality, we take $\nu > 0$. Note that, unlike the general gyrostat, the direction in $\mathscr{F}_p$ of the simple spin is independent of ν. Note also that the value of λ is now

$$\lambda = I_2 + \frac{h_s}{\nu} \tag{23}$$

As we shall see, λ plays the role of an "effective" I_2 in the motion; I_2 can in effect be increased or decreased by adjusting h_s/ν.

The motion equations for small excursions are still given by (10) through (13), but now they have the relatively simple form

$$I_2 \ddot{\alpha}_2 = 0 \tag{24}$$

$$\begin{aligned} I_1 \ddot{\alpha}_1 + (I_3 + I_1 - \lambda)\nu \dot{\alpha}_3 + (\lambda - I_3)\nu^2 \alpha_1 &= 0 \\ I_3 \ddot{\alpha}_3 - (I_3 + I_1 - \lambda)\nu \dot{\alpha}_1 + (\lambda - I_1)\nu^2 \alpha_3 &= 0 \end{aligned} \tag{25}$$

As usual, a secular increase in α_2 is possible. Directional stability is possible for the $\hat{\mathbf{p}}_2$-axis, as can be shown from (25). Comparing (25) with (4.4, 13), the latter being valid when $\mathscr{W}$ does not spin relative to $\mathscr{R}$, we find that the equations are identical, *provided* I_2 *in (4.4, 13) is replaced by* λ. Our earlier stability conditions for $\mathscr{R}$ are thus immediately usable with the understanding that $I_2 \to \lambda$. And, of course, **I** now refers not to $\mathscr{R}$ alone but to $\mathscr{R} + \mathscr{W}$.

The single-rigid-body results imply that $\mathscr{R} + \mathscr{W}$ is unstable if $I_1 < \lambda < I_3$ or $I_3 < \lambda < I_1$. An inappropriate value of h_s/ν *can be destabilizing*. More to the point is the potential for $\mathscr{W}$ to stabilize an otherwise unstable $\mathscr{R}$: when $\lambda > \{I_1, I_3\}$, the motion is directionally stable. Similarly, when $\lambda < \{I_1, I_3\}$, the motion is also directionally stable although experience with minor-axis spins (Section 4.4) with energy dissipation added (Section 5.1) leads one to suspect this "gyrically stabilized" spin. The characteristic frequencies and eigencolumns derived for $\mathscr{R}$ in (4.4, 26 through 30) can also be adapted to $\mathscr{R} + \mathscr{W}$ by the substitution $I_2 \to \lambda$.

$k_1 - k_3$ Stability Diagram

The influence of a spinning wheel or rotor on the directional stability of $\mathscr{R}$ can be portrayed effectively on a k_1–k_3 stability diagram. Recall that $k_1 = (I_2 - I_3)/I_1$; $k_3 = (I_2 - I_1)/I_3$. For $\mathscr{R}$ alone, the stability conditions were shown in Fig. 4.15. This diagram can now be modified for $h_s \neq 0$. To this end, let

$$k_{1h} \triangleq \frac{\lambda - I_3}{I_1} = k_1 + \hat{\Omega}_{po}\left[\frac{1 - k_1}{1 - k_3}\right]^{1/2}$$

$$k_{3h} \triangleq \frac{\lambda - I_1}{I_3} = k_3 + \hat{\Omega}_{po}\left[\frac{1 - k_3}{1 - k_1}\right]^{1/2} \tag{26}$$

$$\hat{\Omega}_{po} \triangleq \frac{\Omega_{po}}{\nu}; \qquad \Omega_{po} \triangleq \frac{h_s}{(I_1 I_3)^{1/2}} \tag{27}$$

The new parameters k_{1h} and k_{3h} have the same significance for $\mathscr{R} + \mathscr{W}$ as k_1 and k_3 do for $\mathscr{R}$ alone. For stability (directional stability for $\mathscr{R} + \mathscr{W}$), it is necessary and sufficient that $k_{1h}k_{3h} > 0$. When k_{1h} and k_{3h} are both positive, one has static stability; when they are both negative, one has static instability, but the gyric effects are stabilizing (no dissipation). The frequency parameter Ω_{po} can be interpreted as the precessional frequency for $\mathscr{R} + \mathscr{W}$ if $\boldsymbol{\nu}$ were $\mathbf{0}$ [see (6.2, 8)], and $\hat{\Omega}_{po}$ is a dimensionless version of Ω_{po}.

Figure 6.7 displays k_1–k_3 stability diagrams for four values of $\hat{\Omega}_{po}$. Since h_s can be positive or negative, $\hat{\Omega}_{po}$ can be of either sign also. When $\mathscr{R}$ and $\mathscr{W}$ have spins in the same direction, $\hat{\Omega}_{po} > 0$; when the wheel spins in the direction opposite to $\mathscr{R}$, $\hat{\Omega}_{po} < 0$. It is clear that $\mathscr{W}$ can have a substantial stabilizing effect on $\mathscr{R}$, particularly if h_s is sufficiently large and positive. Indeed, for $\hat{\Omega}_{po} > 1$, directional stability is obtained (Problem 6.14) for all values of $\{I_1, I_2, I_3\}$. In

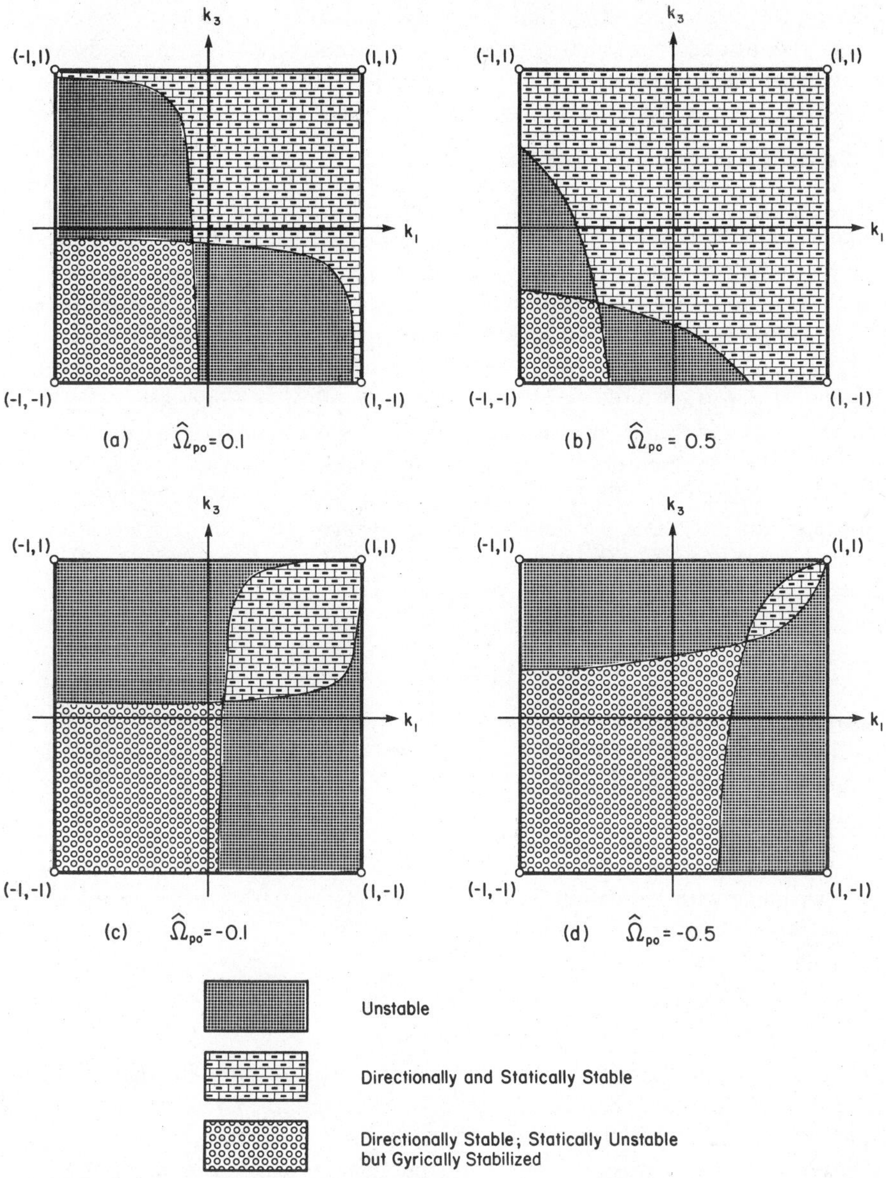

FIGURE 6.7 Stability diagrams for simple spin of a rigid gyrostat in which the rotor axis is parallel to a principal axis. (Compare with Fig. 4.15.)

view of the profound effect that damping can exert on stability, however, the reader is cautioned not to apply these results to spacecraft until energy dissipation is taken into account (Chapter 7).

Two Equivalent Gyrostats

As mentioned near the beginning of Section 6.1, there are two "gyrostats" that can be dealt with simultaneously by an appropriate interpretation of symbols. Attention in the foregoing has been paid primarily to "Kelvin's gyrostat," for which $\mathbf{I}$ is the inertia matrix for $\mathscr{R} + \mathscr{W}$, the *relative* wheel momentum $\underrightarrow{\mathbf{h}}_s$ is constant in magnitude, and the true rotational kinetic energy mentioned in (6.1, 18) varies. All of the preceding development can be reinterpreted for the "apparent gyrostat" however by means of the equivalence noted in (6.1, 13). For the apparent gyrostat, $\mathbf{I}$ must not include the polar moment of inertia of $\mathscr{W}$, the *absolute* wheel momentum $\underrightarrow{\mathbf{h}}_a$ is constant in magnitude, and the rotational kinetic energy is constant. While for Kelvin's gyrostat there is assumed to be an axial torque g_a on the wheel, for the apparent gyrostat $g_a \equiv 0$.

To be consistent with our previous notation, we continue to denote the centroidal inertia matrix for $\mathscr{R} + \mathscr{W}$ by $\mathbf{I}$ and accommodate the transition to the apparent gyrostat by the inertia matrix

$$\mathbf{I}_A \triangleq \mathbf{I} - I_s \mathbf{a}\mathbf{a}^T \tag{28}$$

Note that if $\mathbf{I}$ is diagonal, $\mathbf{I}_A$ will not be, and vice versa, unless the wheel axis $\mathbf{a}$ is parallel to a principal axis of $\mathscr{R} + \mathscr{W}$. It is not difficult to demonstrate that $\mathbf{I}_A$ is a symmetric positive-definite matrix. The angular momentum of $\mathscr{R} + \mathscr{W}$ is now written

$$\mathbf{h} = \mathbf{I}\boldsymbol{\omega} + h_s\mathbf{a} = \mathbf{I}_A\boldsymbol{\omega} + h_a\mathbf{a} \tag{29}$$

in accordance with the definition of h_a, (6.1, 11). Similarly, the rotational kinetic energy is also rewritten in suitable form; from (3.5, 28),

$$\begin{aligned} 2T &= \boldsymbol{\omega}^T\mathbf{I}\boldsymbol{\omega} + 2h_s\mathbf{a}^T\boldsymbol{\omega} + I_s\omega_s^2 \\ &= \boldsymbol{\omega}^T\mathbf{I}_A\boldsymbol{\omega} + I_s\omega_w^2 \end{aligned} \tag{30}$$

where $\omega_w \triangleq \omega_s + \mathbf{a}^T\boldsymbol{\omega}$ is the axial component of the absolute angular velocity of $\mathscr{W}$. For the apparent gyrostat, $h_a = I_s\omega_w =$ constant and $T =$ constant. With this equivalence understood, all the results of Sections 6.2 (nonspinning carrier), 6.3 (zero momentum gyrostat), and 6.4 (the general case) become immediately applicable to the apparent gyrostat. The distinction vanishes if the carrier is nominally nonspinning.

6.5 SYSTEM OF COAXIAL WHEELS

Another system of considerable practical interest consists of a series of N wheels, designated $\Sigma\mathscr{W}_n$ for brevity. The wheels are inertially axisymmetrical and are mounted on a common axis (unit vector $\underrightarrow{\mathbf{a}}$) which is coincident with their axes

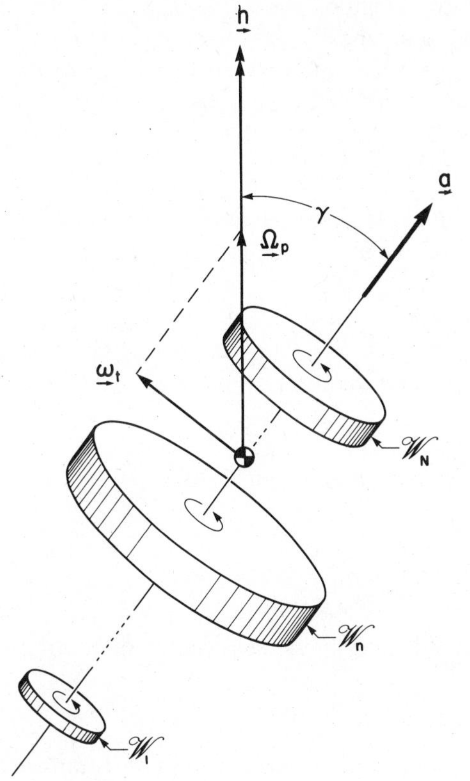

FIGURE 6.8 System of coaxial wheels.

of symmetry (Fig. 6.8). It is assumed that no torque component about $\underset{\rightarrow}{a}$ exists on any of the wheels. This stipulation corresponds to the neglect of bearing friction or, equivalently, to the assumption of axial torquing motors in each wheel which, ideally controlled, exactly compensate for bearing friction. Internal torque components perpendicular to $\underset{\rightarrow}{a}$ will occur, however, to enforce the constraint that the axis of each wheel continues to point in the direction $\underset{\rightarrow}{a}$. No external torques are present. The mass center of $\Sigma\mathscr{W}_n$, denoted $\oplus$, lies on the shared axis of symmetry and will be assumed motionless, although its translational motion cannot affect the rotational motion under study. The inertia dyadic for $\mathscr{W}_n$ with respect to $\oplus$ is

$$\underset{\rightarrow\rightarrow}{\mathbf{J}}_n = J_{tn}\underset{\rightarrow\rightarrow}{\mathbf{1}} + (J_{tn} - I_{an})\underset{\rightarrow}{\mathbf{a}}\underset{\rightarrow}{\mathbf{a}} \tag{1}$$

(see Problem 3.7b); that is, J_{tn} is the moment of inertia of $\mathscr{W}_n$ about any axis through $\oplus$ transverse to $\underset{\rightarrow}{a}$, and I_{an} is the polar moment of inertia of $\mathscr{W}_n$ about $\underset{\rightarrow}{a}$.

The absolute angular velocity of $\mathscr{W}_n$ will be denoted $\underset{\rightarrow}{\omega}_n$. We can resolve $\underset{\rightarrow}{\omega}_n$ into an axial component $\underset{\rightarrow}{\nu}_n = \nu_n\underset{\rightarrow}{a}$ and a transverse component $\underset{\rightarrow}{\omega}_t$. It is important to observe that $\underset{\rightarrow}{\omega}_t$ is the same for all the $\mathscr{W}_n$. This observation suggests the utility of a reference frame $\mathscr{F}_\Omega$, which has its 1-axis in the $\underset{\rightarrow}{a}$-direction and its 2-axis in the $\vec{\boldsymbol{\omega}}_t$ direction. This frame will have an absolute angular velocity, which we denote by $\vec{\boldsymbol{\Omega}}_p$. For reasons to become clear shortly, $\mathscr{F}_\Omega$ will be called the *precessing frame*. Furthermore, because the 1-axis of $\mathscr{F}_\Omega$ remains along $\vec{\mathbf{a}}$, the angular velocity of $\mathscr{W}_n$ with respect to $\mathscr{F}_\Omega$ must be in the $\vec{\mathbf{a}}$-direction. Thus, we write

$$\vec{\boldsymbol{\omega}}_n = \vec{\boldsymbol{\Omega}}_p + \vec{\mathbf{a}}\Omega_n \tag{2}$$

where $\vec{\boldsymbol{\Omega}}_p$ and the Ω_n are yet to be determined. It is also plain from the definition of $\mathscr{F}_\Omega$ that the component of $\vec{\boldsymbol{\Omega}}_p$ transverse to $\vec{\mathbf{a}}$ is $\vec{\boldsymbol{\omega}}_t$. To summarize the situation in symbols, we have

$$\vec{\boldsymbol{\Omega}}_p = \begin{bmatrix}\Omega_p \cos\gamma & \Omega_p \sin\gamma & 0\end{bmatrix}\mathscr{F}_\Omega \tag{3}$$

$$\vec{\boldsymbol{\omega}}_t = \begin{bmatrix}0 & \Omega_p \sin\gamma & 0\end{bmatrix}\mathscr{F}_\Omega \tag{4}$$

$$\vec{\boldsymbol{\omega}}_n = \begin{bmatrix}\Omega_p \cos\gamma + \Omega_n & \Omega_p \sin\gamma & 0\end{bmatrix}\mathscr{F}_\Omega \tag{5}$$

where γ is the angle between $\vec{\boldsymbol{\Omega}}_p$ and $\vec{\mathbf{a}}$, and the useful contractions

$$\omega_t \triangleq \Omega_p \sin\gamma \tag{6}$$

$$\nu_n \triangleq \Omega_p \cos\gamma + \Omega_n \tag{7}$$

have been made.

The angular momentum of $\mathscr{W}_n$ about the motionless mass center $\oplus$ is, from (3.3, 9),

$$\vec{\mathbf{h}}_n = \vec{\mathbf{J}}_n \cdot \vec{\boldsymbol{\omega}}_n = [I_{an}\nu_n \quad J_{tn}\omega_t \quad 0]\mathscr{F}_\Omega \tag{8}$$

upon insertion of the symbols defined above. The total angular momentum of $\Sigma\mathscr{W}_n$ about $\oplus$ is

$$\vec{\mathbf{h}} \triangleq \sum_{n=1}^{N} \vec{\mathbf{h}}_n = [I_a\nu \quad I_t\omega_t \quad 0]\mathscr{F}_\Omega \tag{9}$$

where

$$I_t \triangleq \sum_{n=1}^{N} J_{tn} \tag{10}$$

is the total transverse moment of inertia of $\Sigma\mathscr{W}_n$ about $\oplus$,

$$I_a \triangleq \sum_{n=1}^{N} I_{an} \tag{11}$$

is the total polar moment of inertia of $\Sigma\mathscr{W}_n$, and ν is the average of the ν_n, weighted by I_{an},

$$I_a\nu \triangleq \sum_{n=1}^{N} I_{an}\nu_n \tag{12}$$

The subtle benefit of axial symmetry is that it permits the expression of momentum components in a frame that is not body fixed, namely $\mathscr{F}_\Omega$, with the retained benefit of a constant inertia matrix.

A Single Wheel

The preceding development is now specialized, temporarily, to a single-wheel system. Therefore, the problem is to describe the motion of a single inertial axisymmetrical rigid body in torque-free motion—precisely the problem solved in Section 4.2. Nevertheless, it will now be solved again, from a different perspective. This alternate derivation will not only aid the understanding of this important basic system but will presently provide a stepping stone toward a treatment of the more general case, $\Sigma \mathscr{W}_n$.

The angular momentum of $\mathscr{W}$ is given by (9), where now the inertial matrix of $\mathscr{W}$ in $\mathscr{F}_\Omega$ is diag$\{I_a, I_t, I_t\}$ and ν and ω_t are the axial and transverse components of $\underset{\rightarrow}{\boldsymbol{\omega}}$, the angular velocity of $\mathscr{W}$. Denoting the time derivative of $\underset{\rightarrow}{\mathbf{h}}$ in $\mathscr{F}_\Omega$ by $\underset{\rightarrow}{\mathring{\mathbf{h}}}$, one can interpret the motion equations (3.3, 5) thus:

$$\underset{\rightarrow}{\mathring{\mathbf{h}}} + \underset{\rightarrow}{\boldsymbol{\Omega}}_p \times \underset{\rightarrow}{\mathbf{h}} = \underset{\rightarrow}{\mathbf{0}} \tag{13}$$

since there is no external torque and $\underset{\rightarrow}{\mathbf{v}}_c = \underset{\rightarrow}{\mathbf{0}}$. Expressing (13) in $\mathscr{F}_\Omega$ and integrating where needed, one finds

$$\nu = \text{constant}; \qquad \omega_t = \text{constant} \tag{14}$$

$$I_a \nu = I_t \Omega_p \cos\gamma \tag{15}$$

It follows that ν, $\Omega_p \cos\gamma$, and $\Omega_p \sin\gamma$ are constant, and hence that Ω_p and γ are constant. Moreover, it is straightforward to show that $\boldsymbol{\Omega}_p^{\times}\mathbf{h} = \mathbf{0}$, indicating that the directions of $\underset{\rightarrow}{\boldsymbol{\Omega}}_p$ and $\underset{\rightarrow}{\mathbf{h}}$ are identical.

The picture that emerges from the foregoing is fully in agreement with Section 4.2: the axis of $\mathscr{W}$ precesses at a uniform rate Ω_p about a direction $\underset{\rightarrow}{\mathbf{h}}$ fixed in inertial space; in addition, $\mathscr{W}$ has a relative spin (i.e., a spin relative to $\mathscr{F}_\Omega$) of amount Ω about the symmetry axis. A review of Figs. 4.3 through 4.6 is opportune. The values of Ω_p and Ω are also determined from the above derivation:

$$\Omega_p = \frac{\omega_t}{\sin\gamma} = \frac{h\omega_t}{h\sin\gamma} = \frac{h\omega_t}{h_t} = \frac{h\omega_t}{I_t\omega_t} = \frac{h}{I_t} \tag{16}$$

$$\Omega = \nu - \Omega_p \cos\gamma = \left(\frac{I_t - I_a}{I_t}\right)\nu \tag{17}$$

in accordance with (4.2, 21) and (4.2, 5). Incidentally, the kinetic energy can be expressed in the following three equivalent ways:

$$\begin{aligned} 2T &= I_a \nu^2 + I_t \omega_t^2 \\ &= \frac{h_a^2}{I_a} + \frac{h_t^2}{I_t} \\ &= h^2\left(\frac{\cos^2\gamma}{I_a} + \frac{\sin^2\gamma}{I_t}\right) \end{aligned} \tag{18}$$

These results are now considerably broadened to include a system of N wheels.

Extension to N Wheels

For a series of N wheels (Fig. 6.8), the total angular momentum $\mathbf{h}$ is the sum of the angular momenta of the individual wheels, as indicated in (9). And, since there is no external torque on $\Sigma\mathscr{W}_n$, $\mathbf{h}$ is a constant vector. The individual $\mathbf{h}_n$ vectors which $\mathbf{h}$ comprises however are not generally aligned with $\mathbf{h}$, nor are they constant when viewed inertially. This is due to the internal torques impressed on $\mathscr{W}_n$ to maintain the common axis $\mathbf{a}$. To ascertain the overall motion of $\Sigma\mathscr{W}_n$, we begin by observing that

$$\mathring{\mathbf{h}} + \boldsymbol{\Omega}_p \times \mathbf{h} = \mathbf{0} \tag{19}$$

where $\mathring{\mathbf{h}}$ means "time derivative measured in $\mathscr{F}_\Omega$." The right side of (19) is zero because there are no external torques; the internal torques can make no net alteration to $\mathbf{h}$. Inserting (9) and (3) in (19), we find

$$\nu = \text{constant}; \qquad \omega_t = \text{constant} \tag{20}$$

$$I_a \nu = I_t \Omega_p \cos\gamma \tag{21}$$

showing that ν, Ω_p, and γ are constants of the motion. It is easily demonstrated from (3), (9), and (21) that $\boldsymbol{\Omega}_p^{\times}\mathbf{h} = \mathbf{0}$.

Comparing these conclusions for $\Sigma\mathscr{W}_n$ with those for a single wheel, it is evident that *the gross motion is as though* $\Sigma\mathscr{W}_n$ *were a single wheel* with an angular velocity component ν along $\mathbf{a}$ such that the angular momentum about $\mathbf{a}$ is the same in both cases, as indicated in (12). The system precesses about $\mathbf{h}$ with an angular speed $\Omega_p = h/I_t$ and a constant nutation angle $\gamma = \sin^{-1}(I_t\omega_t/h)$. An "average" or "equivalent" relative spin rate can also be defined:

$$I_a\Omega \triangleq \sum_{n=1}^{N} I_{an}\Omega_n \tag{22}$$

It follows from (12) and (21) that

$$\Omega = \left(\frac{I_t - I_a}{I_t}\right)\nu \tag{23}$$

again showing the equivalence of $\Sigma\mathscr{W}_n$ to a single wheel, at least as far as overall motion is concerned.

To find out what is happening to an individual wheel in the system, we return to the expression (8) for the angular momentum of $\mathscr{W}_n$ about the mass center $\oplus$. Still denoting the time derivative associated with the precessing frame $\mathscr{F}_\Omega$ by an overcircle, we have

$$\mathring{\mathbf{h}}_n + \boldsymbol{\Omega}_p \times \mathbf{h}_n = \mathbf{g}_n \tag{24}$$

where $\mathbf{g}_n$ is the torque on $\mathscr{W}_n$, calculated about the mass center. There are no axial torques; $\mathbf{g}_n \cdot \mathbf{a} = 0$. With the recognition of (3) and (8) and recalling that ω_t is

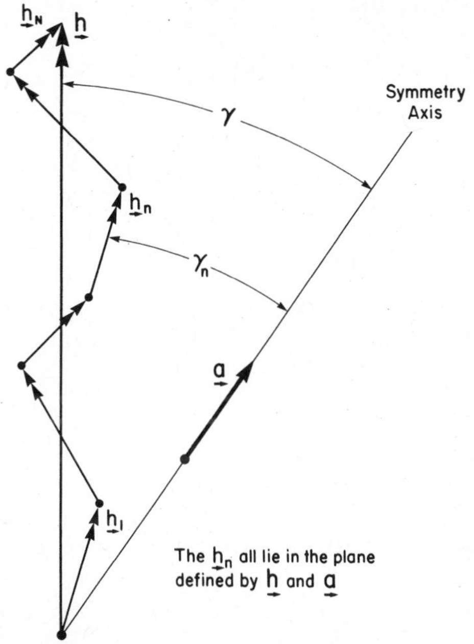

FIGURE 6.9 Angular momenta of individual wheels. (Refer to Fig. 6.8.)

constant, the motion equations indicate that

$$\nu_n = \text{constant} \tag{25}$$

$$g_{tn} = 0 \tag{26}$$

$$g_{bn} = \Omega_p(J_{tn}\omega_t \cos\gamma - I_{an}\nu_n \sin\gamma) \tag{27}$$

where g_{tn} is the torque component in the "transverse" direction (i.e., the $\underset{\rightarrow}{\omega}_t$ direction) and g_{bn} is the "binormal" torque component (i.e., the component perpendicular to both $\underset{\rightarrow}{\mathbf{a}}$ and $\underset{\rightarrow}{\omega}_t$). The latter torque, g_{bn}, which must be supplied by the bearings in $\mathscr{W}_n$, does no work because $\mathbf{g}_n \cdot \underset{\rightarrow}{\omega}_n = 0$; as shown in Fig. 6.9, the effect of g_{bn} is to precess $\underset{\rightarrow}{\mathbf{h}}_n$ about $\underset{\rightarrow}{\mathbf{h}}$ at the rate $\vec{\Omega}_p$ so that although $\underset{\rightarrow}{\mathbf{h}}_n$ is not in general aligned with $\underset{\rightarrow}{\mathbf{h}}$, the individual $\underset{\rightarrow}{\mathbf{h}}_n$ have $\underset{\rightarrow}{\mathbf{h}}$ as their sum. Moreover, because $\Omega_p \sin\gamma$ and the ν_n are constants, the $\vec{\Omega}_n$ are constants also. Furthermore, denote by γ_n the angle in the transverse plane between $\underset{\rightarrow}{\mathbf{h}}_n$ and $\underset{\rightarrow}{\mathbf{a}}$. This angle can be calculated from (8):

$$h_{an} \triangleq I_{an}\nu_n = h_n \cos\gamma_n; \qquad h_{tn} \triangleq J_{tn}\omega_t = h_n \sin\gamma_n \tag{28}$$

Note that the γ_n are constants. The introduction of γ_n also permits an alternative expression for the lateral bearing torque; from (27),

$$g_{bn} = h_n\Omega_p \sin(\gamma_n - \gamma) \tag{29}$$

It is shown in Problem 6.15*b* that the g_{bn} sum to zero, as they should.

Lastly, notice is taken of the kinetic energy. For $\mathscr{W}_n$,

$$2T_n = I_{an}\nu_n^2 + J_{tn}\omega_t^2 = h_n^2\left(\frac{\cos^2\gamma_n}{I_{an}} + \frac{\sin^2\gamma_n}{J_{tn}}\right) \tag{30}$$

and the total kinetic energy of the system is

$$2T = \sum_{n=1}^{N} 2T_n = \sum_{n=1}^{N} I_{an}\nu_n^2 + I_t\omega_t^2 \tag{31}$$

Another expression for $2T$ is mentioned in Problem 6.15d. Although the kinetic energy is something of a side issue when the $\mathscr{W}_n$ are rigid bodies, as assumed here, it takes on great importance when energy dissipation is present in some or all of the $\mathscr{W}_n$. To this subject we now turn.

6.6 PROBLEMS

6.1 For a rigid body (alone), a parameter I was defined in (4.3, 4) as follows: $I \triangleq h^2/2T$. Suppose the definition of I is extended to apply to a gyrostat as follows:

$$I \triangleq \frac{h^2}{2T_o + \mathbf{h}_s^T\boldsymbol{\omega}} \tag{1}$$

Is I a constant of the motion? Show that $I = \lambda$ when the gyrostat rotates in a simple spin.

6.2 In connection with the rigid gyrostat with a nominally nonspinning carrier, the characteristic equation was shown in Section 6.2 to be

$$s^3\det\left(s\mathbf{I} - h_s\mathbf{a}^\times\right) = 0 \tag{2}$$

Let us suppose that an $\mathscr{R}$-fixed reference frame $\mathscr{F}_b$ is being used for $\mathbf{I}$ and $\mathbf{a}$ such that $\mathscr{F}_b$ does not coincide with the principal-axis frame $\mathscr{F}_p$, and let $\mathbf{C}_{pb} = \mathscr{F}_p \cdot \mathscr{F}_b^T$.

(a) Show that the inertia matrix for $\mathscr{F}_p$ is

$$\mathbf{I}_p = \mathrm{diag}\{I_1, I_2, I_3\} = \mathbf{C}_{pb}\mathbf{I}\mathbf{C}_{bp} \tag{3}$$

See also Problem 3.9b.

(b) Hence, show that the characteristic equation is

$$s^3\det\left(s\mathbf{I}_p - h_s\mathbf{a}_p^\times\right) = 0 \tag{4}$$

where $\mathbf{a}_p \triangleq \mathbf{C}_{pb}\mathbf{a}$. That is, $\mathbf{a}_p$ contains the components of $\underrightarrow{\mathbf{a}}$ in $\mathscr{F}_p$.

(c) By directly expanding the 3×3 determinant in (4), show that the characteristic equation is

$$s^4\left(s^2 + \Omega_p^2\right) = 0 \tag{5}$$

where

$$\Omega_p \triangleq \left(\frac{\mathbf{a}^T\mathbf{I}\mathbf{a}}{\det\mathbf{I}}\right)^{1/2} h_s \tag{6}$$

is the precession frequency, as indicated in (6.2, 8).

6.3 Prove the identity (6.2, 11).

6.4 By substitution, verify that (6.2, 12) satisfies the motion equations (6.2, 2) for a gyrostat with a nonspinning carrier. *Harder problem*: *Derive* (6.2, 12).

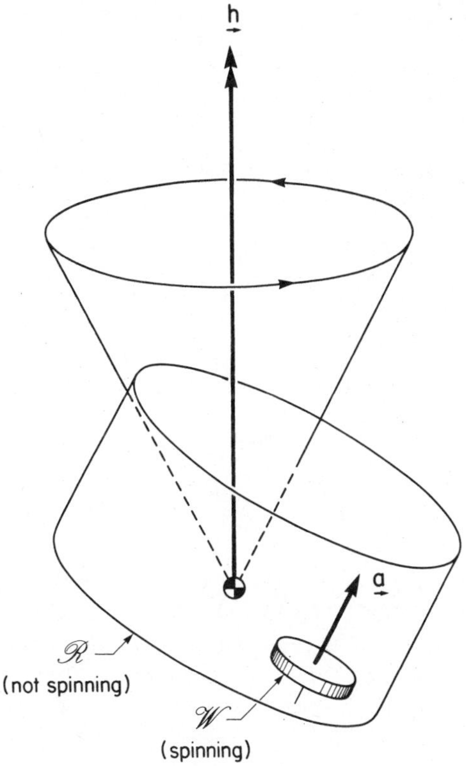

FIGURE 6.10 Symmetrical gyrostat with nonspinning carrier. (See Problem 6.5.)

6.5 A rigid gyrostat for which the carrier (platform) $\mathscr{R}$ is nominally rotationally motionless was studied in Section 6.2. The solution for small deviations $\boldsymbol{\theta}(t)$ was given in (6.2, 12), assuming $\mathscr{R} + \mathscr{W}$ to have three principal moments of inertia $\{I_1, I_2, I_3\}$, not necessarily equal. In this problem, two additional assumptions are made, as shown in Fig. 6.10: (i) $\mathscr{R} + \mathscr{W}$ is inertially axisymmetrical, with $\mathbf{I} = \text{diag}\{I_t, I_a, I_t\}$, and (ii) the axis of symmetry of $\mathscr{R} + \mathscr{W}$ is parallel to $\underset{\rightarrow}{\mathbf{a}}$, the axis of $\mathscr{W}$. (Note that it is not necessary to assume that $\mathscr{R}$ and $\mathscr{W}$ have a common symmetry axis.)

(*a*) Show that the motion of $\mathscr{R}$ with respect to the inertial frame $\mathscr{F}_i$ is given by

$$
\begin{aligned}
\theta_1(t) &= \theta_1(0) + \frac{\dot{\theta}_1(0)\sin\Omega_p t + \dot{\theta}_3(0)(1 - \cos\Omega_p t)}{\Omega_p} \\
\theta_2(t) &= \theta_2(0) + \dot{\theta}_2(0)t \\
\theta_3(t) &= \theta_3(0) + \frac{-\dot{\theta}_1(0)(1 - \cos\Omega_p t) + \dot{\theta}_3(0)\sin\Omega_p t}{\Omega_p}
\end{aligned} \tag{7}
$$

where $\Omega_p \triangleq h_s/I_t$, by solving the motion equation (6.2, 2).

(*b*) Does the more general solution given in (6.2, 12) reduce to the above motion when $\mathbf{I} = \mathrm{diag}\{I_t, I_a, I_t\}$?

(*c*) Interpret the solution (7) geometrically.

6.6 As in the last problem, assume that a rigid gyrostat $\mathscr{R} + \mathscr{W}$ is inertially axisymmetrical, with the axis of symmetry parallel to $\underset{\rightarrow}{\mathbf{a}}$. Figure 6.10 still applies, except that instead of $\nu = 0$, we now have $\nu = -h_s/I_a$. In other words, $\mathscr{R} + \mathscr{W}$ is a symmetrical zero momentum gyrostat, a special case of the development in Section 6.3. Denoting by $\{\alpha_1, \alpha_2, \alpha_3\}$ the small deviations of $\mathscr{R}$ from its simple-spin reference motion, show that the equations of motion are

$$\ddot{\alpha}_2 = 0$$
$$\ddot{\alpha}_1 + 2\nu\dot{\alpha}_3 - \nu^2\alpha_1 = 0$$
$$\ddot{\alpha}_3 - 2\nu\dot{\alpha}_1 - \nu^2\alpha_3 = 0 \qquad (8)$$

and discuss the perturbed motion.

6.7 For the general rigid gyrostat of Section 6.4, the following two motion integrals are known:

$$\mathscr{T}_o: \quad \boldsymbol{\omega}^T\mathbf{I}\boldsymbol{\omega} = 2T_o \qquad (9)$$
$$\mathscr{H}: \quad (\mathbf{I}\boldsymbol{\omega} + h_s\mathbf{a})^T(\mathbf{I}\boldsymbol{\omega} + h_s\mathbf{a}) = h^2 \qquad (10)$$

Geometrically, these integrals correspond to ellipsoids $\mathscr{T}_o$ and $\mathscr{H}$ in $\boldsymbol{\omega}$-space (see Figs. 6.3 and 6.5, for example). It is known from analytical geometry that the tangent plane to $\mathscr{T}_o$ at the point $\boldsymbol{\omega} = \boldsymbol{\nu}$ is given by

$$\boldsymbol{\nu}^T\mathbf{I}\mathbf{x} = 2T_o$$

and the tangent plane to $\mathscr{H}$ at $\boldsymbol{\omega} = \boldsymbol{\nu}$ is given by

$$(\mathbf{I}\boldsymbol{\nu} + h_s\mathbf{a})^T(\mathbf{I}\mathbf{x} + h_s\mathbf{a}) = h^2$$

where $\mathbf{x}$ is a dummy variable associated with an arbitrary point on a tangent plane. Let $\boldsymbol{\omega} = \boldsymbol{\nu}$ be a point common to $\mathscr{T}_o$ and $\mathscr{H}$. Show that the tangent planes to $\mathscr{T}_o$ and $\mathscr{H}$ at $\boldsymbol{\omega} = \boldsymbol{\nu}$ are coincident if and only if

$$\mathbf{I}\boldsymbol{\nu} + h_s\mathbf{a} = \lambda\boldsymbol{\nu} \qquad (11)$$

where λ is a real constant. Compare this condition to the condition derived for a simple spin, namely, (6.1, 23), and comment. See also [Roberson, 8].

6.8 For the general rigid gyrostat $\mathscr{R} + \mathscr{W}$, the momentum ellipsoid in $\boldsymbol{\omega}$-space is given by

$$\mathscr{H}: \quad (\mathbf{I}\boldsymbol{\omega} + h_s\mathbf{a})^T(\mathbf{I}\boldsymbol{\omega} + h_s\mathbf{a}) = h^2$$

(*a*) Show that the origin, $\boldsymbol{\omega} = \mathbf{0}$, lies inside (outside) $\mathscr{H}$ accordingly as

$$h_s < h \ (h_s > h).$$

(*b*) Show that for a simple spin of $\mathscr{R} + \mathscr{W}$, the angle between $\underset{\rightarrow}{\mathbf{h}}$ and $\underset{\rightarrow}{\boldsymbol{\nu}}$ is acute (obtuse) accordingly as $\lambda > 0$ ($\lambda < 0$).

(*c*) Using Fig. 6.4, explain why λ can be negative only if $h < h_s$. Hence, show that λ can be negative only if the origin lies outside $\mathscr{H}$.

6.9 The existence and stability of equilibria for the general gyrostat $\mathscr{R} + \mathscr{W}$ were treated in Section 6.4. Suppose, as a limiting case, that the spin rate of

$\mathscr{W}$ is allowed to become arbitrarily large, with the spin rate of $\mathscr{R}$ limited. In symbols, $h_s \to \infty$, while ν remains finite.

(a) With reference to Figs. 6.2 and 6.4, show that $\phi_T \to 0$ and $\phi_h \to 1$. Show that $|\lambda| \to \infty$.

(b) Show that the direction of the pure spin of $\mathscr{R}$ tends to become aligned with $\mathbf{a}$; that is, show that $\hat{\boldsymbol{\nu}} \to \mathbf{a}$ as $\lambda \to \infty$, and $\hat{\boldsymbol{\nu}} \to -\mathbf{a}$ when $\lambda \to -\infty$, where $\hat{\boldsymbol{\nu}} = \boldsymbol{\nu}/\nu$.

(c) Based on (6.4, 12 and 13), explain why the characteristic equation asymptotes to

$$\det\left[s^2\mathbf{I} - \lambda\boldsymbol{\nu}^{\times}(s\mathbf{1} + \boldsymbol{\nu}^{\times})\right] = 0 \tag{12}$$

(d) By direct expansion or otherwise, show that the asymptotic characteristic equation (12) is

$$s^2(s^2 + \nu^2)\left(s^2 + \Omega_p^2\right) = 0 \tag{13}$$

where Ω_p is given by (6.2, 8), and interpret this finding.

6.10 For the general rigid gyrostat of Section 6.4, let the direction cosines of the angular velocity of pure spin, $\underset{\rightarrow}{\nu}$, with respect to $\mathscr{F}_p$ be denoted $c_{i\nu}$ $(i = 1, 2, 3)$. That is,

$$\boldsymbol{\nu} = \nu[c_{1\nu} \quad c_{2\nu} \quad c_{3\nu}]\mathscr{F}_p \tag{14}$$

Note that, since $\underset{\rightarrow}{\nu}$ is fixed in $\mathscr{R}$, the $c_{i\nu}$ are constants.

(a) Show that

$$h_s a_i = \frac{(\lambda - I_i)hc_{i\nu}}{|\lambda|} \qquad (i = 1, 2, 3) \tag{15}$$

for the reference simple-spin motion.

(b) For perturbed motions in the neighborhood of the simple spin, $\boldsymbol{\omega}(t) = \boldsymbol{\nu} + \boldsymbol{\xi}(t)$, let the direction cosines of the angular momentum vector, $\underset{\rightarrow}{\mathbf{h}}$, with respect to $\mathscr{F}_p$ be denoted c_i $(i = 1, 2, 3)$. That is,

$$\underset{\rightarrow}{\mathbf{h}} = h[c_1 \quad c_2 \quad c_3]\mathscr{F}_p \tag{16}$$

Note that, since $\underset{\rightarrow}{\mathbf{h}}$ is fixed in $\mathscr{F}_i$ (an inertial frame), the c_i will vary during the motion. Explain why $c_i = c_{i\nu}$ for the reference simple spin.

(c) For the perturbed motion, consider the candidate Liapunov function

$$v \triangleq 2(T_o - T_h) \tag{17}$$

where $T_h = T_o(c_i = c_{i\nu}, i = 1, 2, 3)$. Show that

$$v = \sum_{i=1}^{3} \frac{(hc_i - h_s a_i)^2}{I_i} - \sum_{i=1}^{3} \frac{\left[hc_{i\nu}\,\mathrm{sign}(\lambda) - h_s a_i\right]^2}{I_i} \tag{18}$$

Note that $v = 0$ for the reference motion.

(d) Using the result of part (a) above, show that

$$v = h^2 \sum_{i=1}^{3} \frac{[c_i - c_{i\nu}\,\mathrm{sign}(\lambda)]^2}{I_i}$$
$$-2h^2\frac{\mathrm{sign}(\lambda) - \cos\gamma}{|\lambda|} \tag{19}$$

where $\gamma(t)$ is the angle between $\underrightarrow{\mathbf{h}}$ and $\underrightarrow{\boldsymbol{\nu}}$:

$$\cos\gamma = c_1 c_{1\nu} + c_2 c_{2\nu} + c_3 c_{3\nu} \tag{20}$$

(*e*) Show that, as should be expected, v reduces to the function v_M used for a single rigid body [see (4.4, 35)] when $c_{1\nu} = c_{3\nu} = 0$, $c_{2\nu} = 1$, $\lambda = I_2$.

(*f*) Bearing in mind that $\Sigma c_{i\nu}^2 = 1$ and $\Sigma c_i^2 = 1$, what are the conditions that v, given by (19), is positive-definite (thus indicating directional stability)?

(*g*) Explain why $|\lambda| \to \infty$ as the reference spin rate of the carrier, ν, is taken to be ever smaller. What do you conclude about the directional stability of a gyrostat with a nonspinning carrier? Compare this result with that of Problem 6.11*c*.

6.11 Consider the motion of the rigid gyrostat $\mathscr{R} + \mathscr{W}$, discussed in Section 6.4, and let a change in $\boldsymbol{\omega}$ from the single-spin value $\boldsymbol{\nu}$ be denoted $\boldsymbol{\xi}$:

$$\boldsymbol{\omega} = \boldsymbol{\nu} + \boldsymbol{\xi} \tag{21}$$

Note the following constants of the motion:

$$v_1 \triangleq 2(T_o - T_\nu) = \boldsymbol{\xi}^T\mathbf{I}\boldsymbol{\xi} + 2\boldsymbol{\nu}^T\mathbf{I}\boldsymbol{\xi}$$
$$v_2 \triangleq h^2 - h_\nu^2 = \boldsymbol{\xi}^T\mathbf{I}^2\boldsymbol{\xi} + 2\boldsymbol{\nu}^T\mathbf{I}^2\boldsymbol{\xi} + 2h_s\mathbf{a}^T\mathbf{I}\boldsymbol{\xi} \tag{22}$$

where $2T_\nu \triangleq \boldsymbol{\nu}^T\mathbf{I}\boldsymbol{\nu}$ and $h_\nu \triangleq \|\mathbf{I}\boldsymbol{\nu} + h_s\mathbf{a}\|$. Following [Rumyentsev], define the Liapunov function

$$v \triangleq \lambda v_1 - v_2 \tag{23}$$

where, recall, $\lambda = I_i + h_s a_i/\nu_i$ $(i = 1, 2, 3)$.

(*a*) Show that

$$v = h_s \sum_{i=1}^{3} \frac{I_i a_i \xi_i^2}{\nu_i} \tag{24}$$

(*b*) Explain why v is a constant of the motion and why $v = 0$ for the reference motion. Demonstrate that the reference simple spin is $\boldsymbol{\omega}$-stable and hence directionally stable, provided a_i and ν_i have the same sign for $i = 1, 2, 3$.

(*c*) Explain why a_i/ν_i become asymptotically equal and positive when the reference spin rate becomes arbitrarily small ($\nu \to 0$). Show that when the platform $\mathscr{R}$ has a sufficiently small spin, the reference motion of $\mathscr{R} + \mathscr{W}$ is stable.

(*d*) Compare the result in part (*c*) with that in Problem 6.10*g*.

6.12 For the general $\mathscr{R} + \mathscr{W}$ gyrostat, the characteristic equation for infinitesimal derivations from a simple spin, $\boldsymbol{\omega}(t) \equiv \boldsymbol{\nu} =$ constant, is given by (6.4, 14), with $\{\mathscr{M}, \mathscr{G}, \mathscr{K}\}$ defined in (6.4, 11 through 13).

(*a*) By direct expansion or otherwise, show that the explicit form of this characteristic equation is

$$s^2\big[(\det\mathbf{I})s^4 + \big(\boldsymbol{\nu}^T\mathbf{I}\boldsymbol{\nu}\lambda^2 - \Gamma_1\lambda + 2\nu^2\det\mathbf{I}\big)s^2 + \big(\boldsymbol{\nu}^T\mathbf{I}\boldsymbol{\nu}\nu^2\lambda^2 - \Gamma_2\lambda + \nu^4\det\mathbf{I}\big)\big] = 0 \tag{25}$$

where

$$\Gamma_1 \triangleq I_1 I_2(\nu_1^2 + \nu_2^2) + I_2 I_3(\nu_2^2 + \nu_3^2) + I_3 I_1(\nu_3^2 + \nu_1^2)$$

$$\Gamma_2 \triangleq \left[I_1 I_2(\nu_1^2 + \nu_2^2)^2 + I_3(I_1 + I_2)\nu_1^2\nu_2^2\right] + (2 \text{ cyclic terms}) \quad (26)$$

and λ is determined by (6.4, 4) or (6.4, 7), depending on whether the energy or the momentum is specified.

(*b*) Interpret the $s^2 = 0$ roots, and state the stability conditions for the remaining four roots. Using numerical examples, demonstrate that these stability conditions are in accordance with the stability properties ascribed to the solutions $\{\boldsymbol{\nu}_1^{\pm}, \boldsymbol{\nu}_2^{\pm}, \boldsymbol{\nu}_3^{\pm}\}$ when they exist (see Figs. 6.2 through 6.5 and accompanying discussion).

(*c*) As a limiting case, consider the zero momentum gyrostat, $\mathbf{h} = \mathbf{0}$. Explain why $\lambda = 0$. Show that the characteristic equation for $\lambda = 0$ is

$$s^2(s^2 + \nu^2)^2 = 0 \quad (27)$$

and comment on the relationship of this result to those of Problem 6.6.

(*d*) As another limiting case, suppose $h_s \gg \|\mathbf{I}\boldsymbol{\nu}\|$. It was shown in Problem 6.9 that, under this circumstance, $|\lambda| \to \infty$. Show that, as $|\lambda| \to \infty$, the characteristic equation (25) has the asymptotic form

$$s^2\left[(\det \mathbf{I})\left(\frac{s}{\nu}\right)^4 + \mathbf{a}^T\mathbf{I}\mathbf{a}\lambda^2\left(\frac{s}{\nu}\right)^2 + \mathbf{a}^T\mathbf{I}\mathbf{a}\lambda^2\right] \equiv 0 \quad (28)$$

Show that this agrees with (13) of Problem 6.9.

(*e*) As a final check case, assume that the wheel axis is parallel to the principal 2-axis of $\mathscr{R} + \mathscr{W}$, so that $a_1 = 0 = a_3$ and $a_2 = 1$. Explain why $\lambda = I_2 + (h_s/\nu)$. Show that the characteristic equation under these circumstances is identical to that for the system (6.4, 24 and 25).

6.13 Consider an inertially axisymmetrical rigid body $\mathscr{R}$ which contains a spinning wheel $\mathscr{W}$ whose axis is parallel to the symmetry axis. No external torque is acting. The principal moments of inertia of the composite system are $\{I_t, I_a, I_t\}$, and h_s is the relative angular momentum of the wheel.

(*a*) With these assumptions, show that the motion equations (6.1, 7) reduce to

$$\dot{\omega}_1 = -(\Omega - \Omega_{po})\omega_3$$

$$\dot{\omega}_3 = (\Omega - \Omega_{po})\omega_1 \quad (29)$$

where $\omega_2 \equiv \nu = $ constant, $\Omega \triangleq [1 - (I_a/I_t)]\nu$, and $\Omega_{po} \triangleq h_s/I_t$. Note that Ω_{po} would be the precession rate if ν was zero (see Problem 6.5).

(*b*) Let $\{\alpha_1, \alpha_2, \alpha_3\}$ denote small angular deviations from a simple spin about the symmetry axis. Show that the motion equations (6.4, 10 through 13) reduce to

$$I_t\ddot{\alpha}_1 + (2I_t - \lambda)\nu\dot{\alpha}_3 + (\lambda - I_t)\nu^2\alpha_1 = 0$$

$$I_t\ddot{\alpha}_3 - (2I_t - \lambda)\nu\dot{\alpha}_1 + (\lambda - I_t)\nu^2\alpha_3 = 0 \quad (30)$$

and $\ddot{\alpha}_2 = 0$, where $\lambda = I_a + (h_s/\nu)$. Show that (6.4, 25) also reduce to (30) when $I_1 = I_3 = I_t$ and $I_2 = I_a$.

(*c*) Verify that the characteristic equation is

$$(s^2 + \nu^2)\left[s^2 + (\Omega - \Omega_{po})^2\right] = 0 \tag{31}$$

and explain why directional stability is indicated.

(*d*) Show that the symmetry axis precesses about the angular momentum vector at a rate

$$\Omega_p = \Omega_{po} + \nu - \Omega \tag{32}$$

(*e*) Consider the limiting case $\nu \to 0$. Explain why $\boldsymbol{\alpha}(t) \to \boldsymbol{\theta}(t)$, where $\boldsymbol{\theta}(t)$ measures the deviation of $\mathscr{R}$ from an inertial reference frame. Explain further why the solution $\alpha(t)$ becomes identical with $\boldsymbol{\theta}(t)$ given by (7), where now $\Omega_{po} \equiv \Omega_p$.

6.14 At the end of Section 6.4, directional stability conditions were derived for a gyrostat in which the wheel axis was aligned with a principal axis (the 2-axis). The carrier has nominal spin ν and transverse inertias I_1, I_3, and the relative momentum of the wheel is h_s. Show that this simple spin is directionally and statically stable if

$$h_s > (I_1 I_3)^{1/2}\nu \tag{33}$$

that is, if $\Omega_{po} > 1$. See also Fig. 6.7.

6.15 Consider the system of coaxial wheels $\Sigma\mathscr{W}_n$ in torque-free motion, discussed in Section 6.5 and pictured in Fig. 6.8. It is assumed that the net axial torque on each wheel is zero.

(*a*) It was shown in (6.5, 24 through 27) that the only torque component acting on $\mathscr{W}_n$ is perpendicular to both $\underset{\rightarrow}{\mathbf{a}}$ and $\underset{\rightarrow}{\mathbf{h}}$. Explain why this is consistent with the fact that the angular momentum for $\mathscr{W}_n$, namely, $\underset{\rightarrow}{\mathbf{h}}_n$, is fixed in magnitude but precesses about the total angular momentum vector $\underset{\rightarrow}{\mathbf{h}}$.

(*b*) Still with reference to (6.5, 27), show that the total external torque on $\Sigma\mathscr{W}_n$ is zero, as required.

(*c*) Show that

$$I_a\Omega = (I_t - I_a)\Omega_p \cos\gamma \tag{34}$$

$$\nu = \Omega_p \cos\gamma + \Omega \tag{35}$$

where ν is the mean spin rate, given by (6.5, 12); Ω is the mean relative spin rate, given by (6.5, 22); γ is the nutation angle, see Fig. 6.8; Ω_p is the precession rate, given by h/I_t; and the inertias I_t and I_a are defined in (6.5, 10 and 11).

(*d*) An expression for the rotational kinetic energy of $\Sigma\mathscr{W}_n$ was given in (6.5, 31). Show that this can also be written

$$2T = I_a\nu^2 + I_t\omega_t^2 + \sum_{n=1}^{N} I_{an}(\Omega_n - \Omega)^2 \tag{36}$$

and comment.

6.16 In Section 6.5, a system of N symmetrical coaxial wheels was studied under the condition of no external torque acting. In this problem, that development is generalized to deal with the motion in response to external torques. The appropriate vectorial motion equation is (6.5, 19), but with $\underset{\rightarrow}{\mathbf{0}}$ on the right side replaced by $\underset{\rightarrow}{\mathbf{g}}$, the external torque. Once again using the precessing frame $\mathscr{F}_\Omega$, we express the components of $\underset{\rightarrow}{\mathbf{g}}$ thus:

$$\underset{\rightarrow}{\mathbf{g}} = [g_a \quad g_t \quad g_b]\mathscr{F}_\Omega \tag{37}$$

The vectors $\underset{\rightarrow}{\boldsymbol{\Omega}}_p$ and $\underset{\rightarrow}{\mathbf{h}}$ were similarly expressed by (6.5, 3 and 9). Note that by construction we still have $\underset{\rightarrow}{\mathbf{h}}$, $\underset{\rightarrow}{\boldsymbol{\Omega}}_p$, and $\underset{\rightarrow}{\mathbf{a}}$ coplanar, although $\underset{\rightarrow}{\mathbf{h}}$ and $\underset{\rightarrow}{\boldsymbol{\Omega}}_p$ are no longer constant coaxial vectors. The angle between $\underset{\rightarrow}{\boldsymbol{\Omega}}_p$ and $\underset{\rightarrow}{\mathbf{a}}$ is still denoted by γ, and γ_h signifies the angle between $\underset{\rightarrow}{\mathbf{h}}$ and $\underset{\rightarrow}{\mathbf{a}}$.

(*a*) Show that the mean angular velocity of $\Sigma\mathscr{W}_n$ is found from

$$\nu(t) = \nu(0) + \frac{1}{I_a}\int_0^t g_a(\tau)\,d\tau$$

$$\omega_t(t) = \omega_t(0) + \frac{1}{I_t}\int_0^t g_t(\tau)\,d\tau \tag{38}$$

and that

$$\gamma(t) = \tan^{-1}(I_t\omega_t, I_a\nu) \tag{39}$$

(*b*) Show that the angular velocity of precession, $\underset{\rightarrow}{\boldsymbol{\Omega}}_p$, is inertially fixed if and only if

$$g_t \equiv 0; \qquad \dot{g}_b + \omega_t g_a \equiv 0 \tag{40}$$

and comment on the sufficient conditions

$$g_t \equiv 0; \qquad g_a \equiv 0; \qquad g_b = \text{constant} \tag{41}$$

(*c*) Show that $\underset{\rightarrow}{\boldsymbol{\Omega}}_p$ and $\underset{\rightarrow}{\mathbf{h}}$ are coaxial if and only if $g_b \equiv 0$, and that in fact

$$\Omega_p^2 = \omega_t^2 + \frac{[(g_b/\omega_t) + I_a\nu]^2}{I_t^2}$$

$$\sin(\gamma_h - \gamma) = \frac{g_b}{\Omega_p h} \tag{42}$$

thus giving a solution for $\gamma_h(t)$ and $\Omega_p(t)$.

(*d*) Particularize these results so that they apply to a single wheel (i.e., to a single, inertially axisymmetrical rigid body), and compare with the development in Section 4.5.

CHAPTER 7

EFFECT OF INTERNAL ENERGY DISSIPATION ON THE DIRECTIONAL STABILITY OF GYROSTATS

For reasons explained at the beginning of Chapter 5, the concept of a rigid body $\mathscr{R}$ has a significant defect when used to model physical bodies in circumstances where energy dissipation is important. It was found there that for the torque-free motion of a single body, the idea of a quasi-rigid body $\mathscr{Q}$, in which there is in general a slow loss of energy, leads to markedly different (and more accurate) predictions concerning directional stability. In particular, it was found that minor-axis spins, which for $\mathscr{R}$ are statically unstable but directionally stabilized by gyric effects, become unstable when the spacecraft is modeled by $\mathscr{Q}$ instead of $\mathscr{R}$. The heuristic analysis was made precise in Section 5.2 for a special class of models, namely, $\mathscr{R} + \mathscr{P}$, where $\mathscr{P}$ is a point mass damper. It was found that the major-axis rule—the heuristic guideline furnished by a spinning-$\mathscr{Q}$ model—is necessary but not sufficient for the directional stability of spin about the major axis of $\mathscr{R} + \mathscr{P}$.

These same issues are now addressed for a gyrostat. Many cases can be studied, each case depending on the limiting assumptions made: sometimes only one of the constituent bodies is spinning; sometimes they both are; one or both bodies may be assumed inertially axisymmetrical; the spin axis of the wheel may or may not be parallel to a principal axis; dissipation can be handled either heuristically (using quasi-rigid bodies with energy sinks) or more exactly (by explicitly expanding the order of the model to accommodate the degrees of freedom associated with damping devices). An assumption of intermediate accuracy is to neglect damper mass in the overall rotational motion equations, with motion equations for the damping devices added later to complete the system description. Further (nonlinear) phenomena have been demonstrated in the literature for certain types of nonlinear damping laws. Clearly, the number of combinations of assumptions quickly mounts up, although space allows only a subset of these cases to be considered here.

The first half of this chapter examines the implications for directional stability of gyrostats in which one or both bodies are quasi-rigid. The energy sink hypothesis is therefore applied. The several cases treated are portrayed in Figs. 7.1 through 7.11. In the second half of this chapter, discrete dampers are included in the model; the results obtained are in general agreement with, but more precise than, the energy sink results.

7.1 ENERGY SINK ANALYSES

To begin our study of the implications of energy dissipation on the directional stability of gyrostats, we replace the rigid carrier or platform $\mathscr{R}$ by a quasi-rigid body $\mathscr{Q}$. The model is still as shown in Fig. 3.8, except that now $\mathscr{R} \to \mathscr{Q}$. The reader is referred to Section 5.1 for a description of the quasi-rigid body $\mathscr{Q}$. The mechanism for energy dissipation is not explicitly taken into account in the expressions for T_o and h, but it is assumed that T_o slowly decreases until it reaches some lower bound where dissipation ceases. For convenience, the expressions for T_o and h^2 are repeated here:

$$2T_o = \boldsymbol{\omega}^T\mathbf{I}\boldsymbol{\omega} \equiv I_1\omega_1^2 + I_2\omega_2^2 + I_3\omega_3^2 \tag{1}$$

$$h^2 = (\mathbf{I}\boldsymbol{\omega} + h_s\mathbf{a})^T(\mathbf{I}\boldsymbol{\omega} + h_s\mathbf{a}) \tag{2}$$

We assume that h is constant and that the energy sink in $\mathscr{Q}$ slowly reduces T_o.

The discussion naturally focuses on motions for which T_o is a (local) minimum, given h. These motions are (locally) asymptotically $\boldsymbol{\omega}$-stable since any neighboring motion with the same h has a higher value of T_o and the excess T_o would be removed by the energy sink. Mathematically, the question is posed thus: find $\boldsymbol{\omega}$ such that T_o is minimized for a given h. In other words, find

$$\min_{\boldsymbol{\omega}} \left\{ \boldsymbol{\omega}^T\mathbf{I}\boldsymbol{\omega} - \frac{(\mathbf{I}\boldsymbol{\omega} + h_s\mathbf{a})^T(\mathbf{I}\boldsymbol{\omega} + h_s\mathbf{a})}{\lambda} \right\} \tag{3}$$

where λ is a Lagrange multiplier. This λ is not (yet) identified as being identical with the λ used in the determination of pure-spin solutions for rigid gyrostats. Substituting (1) and (2) in (3) and setting to zero the partial derivatives with respect to $\{\omega_1, \omega_2, \omega_3\}$, we find that stationary points must satisfy the following condition, expressed in matrix notation:

$$\frac{\partial}{\partial \boldsymbol{\omega}}\left(2T_o - h^2/\lambda\right) = \frac{\partial}{\partial \boldsymbol{\omega}}\left[\boldsymbol{\omega}^T\mathbf{I}\boldsymbol{\omega} - \frac{(\mathbf{I}\boldsymbol{\omega} + h_s\mathbf{a})^T(\mathbf{I}\boldsymbol{\omega} + h_s\mathbf{a})}{\lambda}\right]$$

$$= 2\mathbf{I}\left[\boldsymbol{\omega} - \frac{(\mathbf{I}\boldsymbol{\omega} + h_s\mathbf{a})}{\lambda}\right] = \mathbf{0} \tag{4}$$

Thus, subject to the constraint h = constant, stationary values of T_o occur for

$\boldsymbol{\omega} = \boldsymbol{\nu}$, where

$$\mathbf{h} = \mathbf{I}\boldsymbol{\nu} + h_s\mathbf{a} = \lambda\boldsymbol{\nu} \tag{5}$$

Comparing with (4.3, 19), we see that stationary values of T_o correspond to simple spins for $\mathscr{Q} + \mathscr{W}$ and that the Lagrangian multiplier λ is precisely the λ used in Chapter 6. It is also observed that the energy sink ceases to extract energy when the motion of $\mathscr{Q}$ is a simple spin, corresponding to a stationary (centrifugal) inertial force field in $\mathscr{Q}$.

Asymptotic ω-Stability

To ascertain which of these stationary points are local minima, one must examine the matrix of second derivatives (the Hessian matrix). In view of the momentum constraint however, the appropriate matrix of second derivatives is a *constrained* Hessian, derived as follows. Let the change in $\boldsymbol{\omega}$, referred to the stationary value $\boldsymbol{\omega} = \boldsymbol{\nu}$, be denoted $\boldsymbol{\xi}$:

$$\boldsymbol{\omega} = \boldsymbol{\nu} + \boldsymbol{\xi} \tag{6}$$

Then,

$$\begin{aligned} \Delta T_o &= \tfrac{1}{2}\boldsymbol{\omega}^T\mathbf{I}\boldsymbol{\omega} - \tfrac{1}{2}\boldsymbol{\nu}^T\mathbf{I}\boldsymbol{\nu} \\ &= \boldsymbol{\nu}^T\mathbf{I}\boldsymbol{\xi} + \tfrac{1}{2}\boldsymbol{\xi}^T\mathbf{I}\boldsymbol{\xi} \end{aligned} \tag{7}$$

However, the constraint that $\|\mathbf{I}\boldsymbol{\omega} + h_s\mathbf{a}\|^2 = h^2$ reveals that

$$2\lambda\boldsymbol{\nu}^T\mathbf{I}\boldsymbol{\xi} + \boldsymbol{\xi}^T\mathbf{I}^2\boldsymbol{\xi} = 0 \tag{8}$$

after the insertion of (5). To first order in $\|\boldsymbol{\xi}\|$, (8) expresses the fact that changes $\boldsymbol{\xi}$ in $\boldsymbol{\omega}$ must lie in the plane tangent to $\mathscr{H}$ at $\boldsymbol{\omega} = \boldsymbol{\nu}$. (The normal to this plane is in the direction $\mathbf{I}\boldsymbol{\nu}$.) In view of (8), the new form for (7) is

$$2\lambda\,\Delta T_o = \boldsymbol{\xi}^T\mathbf{I}(\lambda\mathbf{1} - \mathbf{I})\boldsymbol{\xi} \tag{9}$$

Thus far, all expressions have been exact. We now examine the positivity properties of ΔT_o by expressing ΔT_o exclusively in terms of *independent* variables, keeping only terms up to second order in $\|\boldsymbol{\xi}\|$. This can be done by choosing one of the ξ_i, say, ξ_3, and solving (8) for ξ_3 in terms of the other two. This latter solution need only be found to first order in $\|\boldsymbol{\xi}\|$ because (8) has no zeroth-order terms, and consequently a first-order expression, when inserted in (9), contributes all the required second-order terms. For example, if ξ_3 is selected, it is sufficient to insert

$$\xi_3 = -\frac{I_1\nu_1\xi_1 + I_2\nu_2\xi_2}{I_3\nu_3} \tag{10}$$

After this substitution, T_o is a quadratic form accurate to second order in the remaining two elements of $\boldsymbol{\xi}$. When the symmetric matrix associated with this quadratic form is positive-definite, one concludes that the simple spin $\boldsymbol{\omega} = \boldsymbol{\nu}$ indeed corresponds to a local minimum in the energy T_o. (The details are similar to those in Problem 6.10.) This analysis reveals the following conditions for a relative minimum at $\boldsymbol{\omega} = \boldsymbol{\nu}$:

$$\lambda\left[I_1(\lambda - I_2)\nu_1^2 + I_2(\lambda - I_1)\nu_2^2\right] > 0$$
$$\lambda\left[I_2(\lambda - I_3)\nu_2^2 + I_3(\lambda - I_2)\nu_3^2\right] > 0$$
$$\lambda\left[I_3(\lambda - I_1)\nu_3^2 + I_1(\lambda - I_3)\nu_1^2\right] > 0 \tag{11}$$

$$I_1(\lambda - I_2)(\lambda - I_3)\nu_1^2 + I_2(\lambda - I_3)(\lambda - I_1)\nu_2^2 + I_3(\lambda - I_1)(\lambda - I_2)\nu_3^2 > 0 \tag{12}$$

Any one of the three conditions (11), together with (12), implies the other two (a good exercise). Conditions (11) and (12) represent the *Major-Axis Rule for gyrostats with energy dissipation in the carrier*. Note that any simple spin for which $\lambda < 0$ is asymptotically $\boldsymbol{\omega}$-stable.

These conditions are illustrated in Fig. 7.1 for the parameters $\{I_1, I_2, I_3\} = \{6, 4, 2\}$ kg · m^2 and $a_1 = a_2 = a_3 = 1/\sqrt{3}$. The simple spins $\boldsymbol{\nu}_1^{\pm}$, corresponding to $\lambda = \lambda_1^{\pm}$, respectively, satisfy the conditions (11) and (12) and are thus asymptotically $\boldsymbol{\omega}$-stable (although the solution $\boldsymbol{\nu}_1^{-}$ does not exist when h_s is sufficiently large). For the simple spins $\boldsymbol{\nu}_3^{\pm}$, on the other hand, condition (12) is satisfied, but the inequalities (11) are reversed, indicating a local maximum for the energy T_o. These spins, which are stable without dissipation, are unstable with dissipation. This disappearance of stability parallels the instability of minor-axis spins for a single quasi-rigid body (Chapter 5).

An alternative form for the major-axis rule for gyrostats is useful. By noting that

$$\lambda - I_i = \frac{h_s a_i}{\nu_i} \qquad (i = 1, 2, 3) \tag{13}$$

one finds the necessary and sufficient conditions

$$\lambda\left[\left(\frac{a_2}{\nu_2}\right)I_1\nu_1^2 + \left(\frac{a_1}{\nu_1}\right)I_2\nu_2^2\right] > 0$$
$$\lambda\left[\left(\frac{a_3}{\nu_3}\right)I_2\nu_2^2 + \left(\frac{a_2}{\nu_2}\right)I_3\nu_3^2\right] > 0$$
$$\lambda\left[\left(\frac{a_1}{\nu_1}\right)I_3\nu_3^2 + \left(\frac{a_3}{\nu_3}\right)I_1\nu_1^2\right] > 0 \tag{14}$$

$$\left(\frac{a_2 a_3}{\nu_2 \nu_3}\right)I_1\nu_1^2 + \left(\frac{a_3 a_1}{\nu_3 \nu_1}\right)I_2\nu_2^2 + \left(\frac{a_1 a_2}{\nu_1 \nu_2}\right)I_3\nu_3^2 > 0 \tag{15}$$

For asymptotic $\boldsymbol{\omega}$-stability, it is sufficient but not necessary to have

$$\frac{\lambda a_i}{\nu_i} > 0 \qquad (i = 1, 2, 3) \tag{16}$$

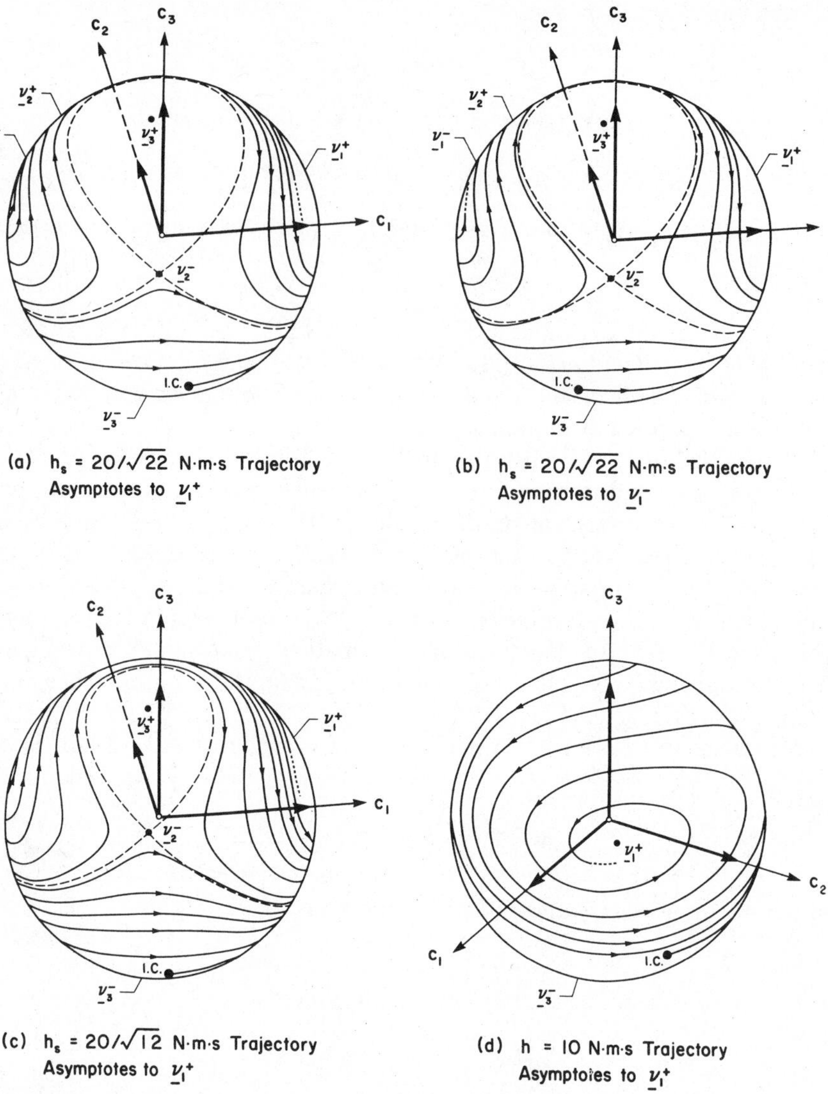

(a) $h_s = 20/\sqrt{22}$ N·m·s Trajectory Asymptotes to $\underline{\nu}_1^+$

(b) $h_s = 20/\sqrt{22}$ N·m·s Trajectory Asymptotes to $\underline{\nu}_1^-$

(c) $h_s = 20/\sqrt{12}$ N·m·s Trajectory Asymptotes to $\underline{\nu}_1^+$

(d) $h = 10$ N·m·s Trajectory Asymptotes to $\underline{\nu}_1^+$

FIGURE 7.1 Attitude motion of $\mathscr{Q} + \mathscr{W}$ system with respect to $\underset{\rightarrow}{\mathbf{h}}$, as implied by energy sink analysis. (Compare with Fig. 6.6.)

This is in agreement with the Liapunov function derived in Problem 6.11, namely,

$$v \triangleq 2\lambda(T_o - T_\nu) - \left(h^2 - h_\nu^2\right) = h_s \sum_{i=1}^{3} \frac{I_i a_i \xi_i^2}{\nu_i} \tag{17}$$

where $\xi_i = \omega_i - \nu_i$. When $a_i/\nu_i > 0$, v is positive-definite; because $\dot{T}_o \leq 0$ (the equality holding only at $\boldsymbol{\xi} = \mathbf{0}$), $\dot{v} \leq 0$ if $\lambda > 0$, indicating asymptotic ω-stability.

Similarly, if $\lambda < 0$, choose $-\upsilon$ as the Liapunov function. Then, when $a_i/\nu_i < 0$, $-\upsilon$ is positive-definite and $-\dot{\upsilon} \le 0$, once again implying asymptotic $\boldsymbol{\omega}$-stability.

Asymptotic Directional Stability

To investigate directional stability, one must convert from the variables $\{\omega_1, \omega_2, \omega_3\}$ to attitude variables. This is readily accomplished ([Pringle, 2]) by utilizing the direction cosines of $\underrightarrow{\mathbf{h}}$ in $\mathscr{F}_p$:

$$\underrightarrow{\mathbf{h}} = h[c_1 \quad c_2 \quad c_3]\mathscr{F}_p \tag{18}$$

The other direction of interest is the direction about which the reference spin occurs:

$$\underrightarrow{\boldsymbol{\nu}} = \nu[c_{1\nu} \quad c_{2\nu} \quad c_{3\nu}]\mathscr{F}_p \tag{19}$$

For the reference spin itself, $\underrightarrow{\mathbf{h}} = \underrightarrow{\mathbf{h}}_\nu = \lambda\underrightarrow{\boldsymbol{\nu}}$, and $c_i = c_{i\nu}\operatorname{sign}(\lambda)$. After a disturbance at $t = 0$ however, $c_i \ne c_{i\nu}\operatorname{sign}(\lambda)$, and there is an angle $\gamma \ne 0$ or π between $\underrightarrow{\mathbf{h}}$ and $\underrightarrow{\boldsymbol{\nu}}$. As indicated by (6.4, 19), the angle $\gamma(t)$ is found from

$$h^2\nu^2\sin^2\gamma = \|\mathbf{h}^\times\boldsymbol{\nu}\|^2 = \|\boldsymbol{\nu}^\times(\mathbf{I}\boldsymbol{\omega} + h_s\mathbf{a})\|^2 \tag{20}$$

as the motion progresses. The significance of asymptotic $\boldsymbol{\omega}$-stability is that $\boldsymbol{\omega} \to \boldsymbol{\nu}$ as $t \to \infty$. Hence,

$$\lim_{t\to\infty} h^2\nu^2\sin^2\gamma = \|\boldsymbol{\nu}^\times(\mathbf{I}\boldsymbol{\nu} + h_s\mathbf{a})\|^2 = \|\lambda\boldsymbol{\nu}^\times\boldsymbol{\nu}\|^2 = 0 \tag{21}$$

demonstrating that $\sin\gamma \to 0$. We conclude that energy sink analysis shows that, provided the conditions (11) and (12) are satisfied, or equivalently the conditions (14) and (15), the direction $\underrightarrow{\boldsymbol{\nu}}$ in $\mathscr{F}_p$ asymptotically approaches the direction $\underrightarrow{\mathbf{h}}$. It does not follow that the direction $\underrightarrow{\boldsymbol{\nu}}$ is *asymptotically stable*, unfortunately, because $\underrightarrow{\mathbf{h}}$ is not generally the same as the original direction $\underrightarrow{\mathbf{h}}_\nu$. This is partly because the initial disturbance may alter the direction of th angular momentum vector and partly because the direction of $\underrightarrow{\mathbf{h}}$ (as defined by $\mathbf{h} = \mathbf{I}\boldsymbol{\omega} + h_s\mathbf{a}$) changes slightly during the decay process due to the internally generated energy sink torques. Thus, a precise statement for the energy sink stability result for a general gyrostat would be: "If the conditions (11) and (12) are satisfied for the reference spin $\underrightarrow{\boldsymbol{\nu}}$, then this reference spin direction is directionally stable, asymptotically to $\underrightarrow{\mathbf{h}}$."

Another viewpoint on directional stability is offered by the Liapunov function defined in Problem 6.10:

$$\upsilon \triangleq 2(T_o - T_h) = -2h^2\frac{\operatorname{sign}(\lambda) - \cos\gamma}{|\lambda|} + h^2\sum_{i=1}^{3}\frac{[c_i - c_{i\nu}\operatorname{sign}(\lambda)]^2}{I_i} \tag{22}$$

where

$$\cos\gamma = c_1c_{1\nu} + c_2c_{2\nu} + c_3c_{3\nu} \tag{23}$$

Alternatively, $\sin\gamma$ is expressed in (6.4, 18). To measure the deviation of c_i from $c_{i\nu}\operatorname{sign}(\lambda)$, we define

$$c_i = c_{i\nu}\operatorname{sign}(\lambda) + \delta_i \qquad (i = 1, 2, 3) \tag{24}$$

It follows that

$$\cos\gamma = \operatorname{sign}(\lambda) + c_{1\nu}\delta_1 + c_{2\nu}\delta_2 + c_{3\nu}\delta_3 \tag{25}$$

and that

$$v = h^2\left[\frac{\delta_1^2}{I_1} + \frac{\delta_2^2}{I_2} + \frac{\delta_3^2}{I_3} + \frac{2}{|\lambda|}(c_{1\nu}\delta_1 + c_{2\nu}\delta_2 + c_{3\nu}\delta_3)\right] \tag{26}$$

However, the δ_i are not independent. Setting $\Sigma c_i^2 = 1$ reveals that

$$2(c_{1\nu}\delta_1 + c_{2\nu}\delta_2 + c_{3\nu}\delta_3)\operatorname{sign}(\lambda) = -(\delta_1^2 + \delta_2^2 + \delta_3^2) \tag{27}$$

and hence that

$$v = h^2\left[\left(\frac{1}{I_1} - \frac{1}{\lambda}\right)\delta_1^2 + \left(\frac{1}{I_2} - \frac{1}{\lambda}\right)\delta_2^2 + \left(\frac{1}{I_3} - \frac{1}{\lambda}\right)\delta_3^2\right] \tag{28}$$

For v to be positive-definite in $\boldsymbol{\delta}$, it is clearly sufficient that $\lambda > \{I_1, I_2, I_3\}$ or that $\lambda < 0$. To find sufficient conditions that are also necessary, the constraint on $\{\delta_1, \delta_2, \delta_3\}$ exhibited in (27) must be used. The resulting expression, accurate to second order, shows that $v > 0$ in some neighborhood of $\sin\gamma = 0$, provided

$$\begin{gathered}\lambda\left[I_1(\lambda - I_2)c_{1\nu}^2 + I_2(\lambda - I_1)c_{2\nu}^2\right] > 0\\ \lambda\left[I_2(\lambda - I_3)c_{2\nu}^2 + I_3(\lambda - I_2)c_{3\nu}^2\right] > 0\\ \lambda\left[I_3(\lambda - I_1)c_{3\nu}^2 + I_1(\lambda - I_3)c_{1\nu}^2\right] > 0\end{gathered} \tag{29}$$

$$(\lambda - I_2)(\lambda - I_3)c_{1\nu}^2 + (\lambda - I_3)(\lambda - I_1)c_{2\nu}^2 + (\lambda - I_1)(\lambda - I_2)c_{3\nu}^2 > 0 \tag{30}$$

Any one of (29), with (30), implies the other two. Of course, (29) is just another form for (11), the "major-axis rule for gyrostats," because $c_{i\nu} = \nu_i/\nu$. We infer that v, defined by (22), is positive-definite in the neighborhood of $\sin\gamma = 0$ if and only if conditions (29) and (30) are satisfied. It is equally clear from the definition of v that $\dot{v} \le 0$ with an energy sink, the equality applying only when $\sin\gamma = 0$. This concludes another version of the proof that if the major-axis rule for gyrostats is obeyed, the reference spin direction $\underset{\rightarrow}{\boldsymbol{\nu}}$ is stable, asymptotically to $\underset{\rightarrow}{\mathbf{h}}$.

Nominally Nonspinning Carrier

Energy sink analysis is particularly straightforward when the carrier is nominally nonspinning. Our model thus consists of a quasi-rigid body $\mathscr{Q}$ in which a spinning rotor or wheel $\mathscr{W}$ is placed so that the axis of $\mathscr{W}$ is fixed in $\mathscr{Q}$—and, for the reference motion, $\mathscr{Q}$ is not rotating. This model would serve reasonably well for an essentially rigid interplanetary spacecraft with an internal wheel (or wheels; see Section 6.1) or for a geostationary "bias momentum" Earth-pointing satellite whose once-per-day period of rotation is negligibly small, whose carrier contains significant damping, and whose momentum wheel (Fig. 7.2) contains

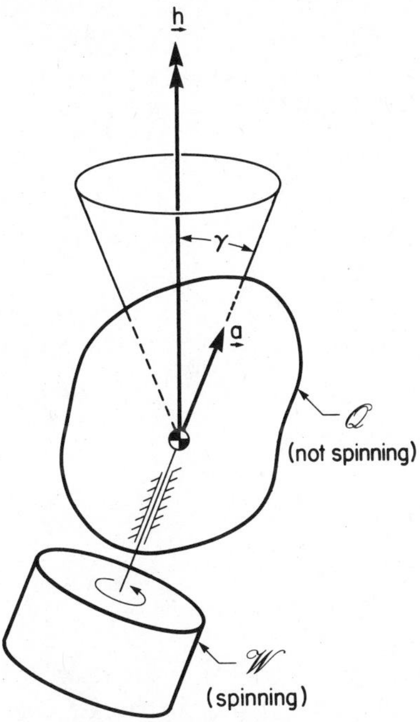

FIGURE 7.2 Attitude motion of $\mathscr{Q} + \mathscr{W}$ when $\mathscr{Q}$ is nominally nonspinning. (The cone is not generally circular.)

negligible dissipation. The motion with no dissipation was the subject of Section 6.2. It was found there that the disturbed motion consists of a precession of the wheel axis about the angular momentum vector at a rate Ω_p which depends on the wheel momentum and the system inertias (see Fig. 6.1). The special case of an inertially axisymmetrical carrier, with the symmetry axis and the wheel axis parallel, is studied further in Problem 6.5. The present objective is to ascertain the influence of energy dissipation in the carrier. Does the cone of precession shown in Fig. 6.1 expand or contract?

The major-axis rule for gyrostats, represented by (11) and (12), is not directly applicable because the reference spin $\boldsymbol{\nu} = \mathbf{0}$ and $\lambda = \infty$. We return instead to first principles. The reference value of $\boldsymbol{\omega}$ is $\boldsymbol{\omega} = \boldsymbol{\nu} = \mathbf{0}$; and hence, for disturbed motion,

$$\Delta T_o = T_o = \tfrac{1}{2}\boldsymbol{\xi}^T \mathbf{I} \boldsymbol{\xi} \tag{31}$$

It is immediately clear that ΔT_o is positive-definite in $\boldsymbol{\xi}$, even before the constant-h^2 constraint is imposed; any motion by $\mathscr{Q}$ contributes an increase in energy. As for attitude variables, the direction cosines of $\underrightarrow{\mathbf{h}}$ in $\mathscr{F}_p$ are denoted, as usual, by $\{c_1, c_2, c_3\}$. Thus, the angular momentum components in $\mathscr{F}_p$ are

$$I_i \xi_i + h_s a_i = h_s c_i \qquad (i = 1, 2, 3) \tag{32}$$

(It has been noted that the magnitude of $\underrightarrow{\mathbf{h}}$ is unchanged: $h = h_s$.) Combining (31) and (32), one may express the energy as

$$\Delta T_o = h_s^2 \left[\frac{(c_1 - a_1)^2}{I_1} + \frac{(c_2 - a_2)^2}{I_2} + \frac{(c_3 - a_3)^2}{I_3} \right] \tag{33}$$

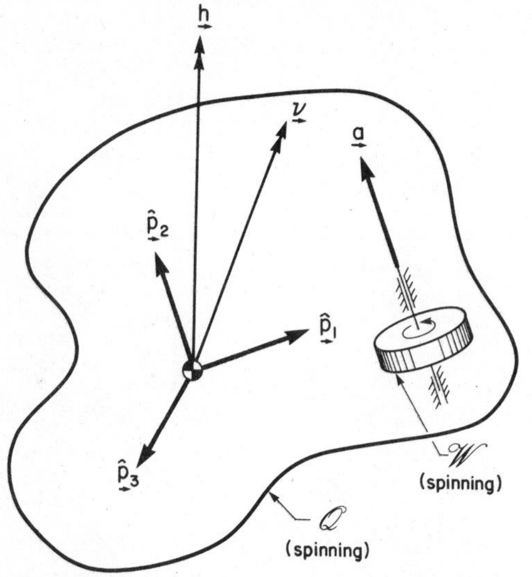

FIGURE 7.3 $\mathscr{Q} + \mathscr{W}$ system with axis of $\mathscr{W}$ parallel to a principal axis.

So again it is seen that ΔT_o is positive for all $\{c_1, c_2, c_3\}$, except at the reference condition, $c_i = a_i$, where $\Delta T_o = 0$. Therefore, since $(\Delta T_o)^{\cdot} < 0$, it can be concluded that the energy sink causes $c_i \to a_i$; in other words, the wheel axis direction $\underset{\to}{\mathbf{a}}$ asymptotically approaches the angular-momentum direction $\underset{\to}{\mathbf{h}}$. With energy dissipation, then, the cones shown in Figs. 6.1 and 7.2 shrink to zero, independently of the relative sizes of $\{I_1, I_2, I_3\}$ and independently of the orientation of $\mathscr{W}$ in $\mathscr{Q}$.

Wheel Axis Parallel to a Principal Axis

In another specialized but common case deserving of special mention, the axis of $\mathscr{W}$ is parallel to a principal axis (Fig. 7.3). To be specific, we choose $\underset{\to}{\mathbf{a}} \parallel \underset{\to}{\hat{\mathbf{p}}}_2$. Simple spins occur when

$$I_1\nu_1 = \lambda\nu_1$$

$$I_2\nu_2 + h_s = \lambda\nu_2$$

$$I_3\nu_3 = \lambda\nu_3 \tag{34}$$

and we again adopt the convention that h_s may be of either sign rather than considering the two cases $a_2 = \pm 1$. Evidently, $\nu_1 = 0 = \nu_3$, and we may with no loss in generality pick $\nu_2 = +\nu > 0$. As in (6.4, 23), $\lambda = I_2 + h_s/\nu$. An energy sink analysis proceeds along the same lines as in (6) through (12), except that now the constant-h constraint reduces simply to $\xi_2 = 0$ (to first order). The expression for ΔT_o, given in general by (9), is now

$$2\lambda\,\Delta T_o = I_1(\lambda - I_1)\xi_1^2 + I_3(\lambda - I_3)\xi_3^2 \tag{35}$$

correct to second order. The condition for asymptotic $\boldsymbol{\omega}$-stability is therefore

$$\lambda(\lambda - I_1) > 0; \qquad \lambda(\lambda - I_3) > 0 \tag{36}$$

which guarantees that the constrained ΔT_o is positive-definite. These two conditions can be restated as follows: either $\lambda > \max\{I_1, I_3\}$ or $\lambda < 0$. It has thus been demonstrated that a sufficiently large and positive wheel momentum can stabilize a minor-axis or intermediate-axis spin. This is in accord with the results for a rigid gyrostat whose motion equations, (6.4, 25), were analyzed near the end of Section 6.4. Furthermore, this stability has now been found to be asymptotic, in the sense that $\boldsymbol{\omega} \to \nu \mathbf{1}_2$, within the limitations of energy sink analysis. For λ in the range $0 < \lambda < \min\{I_1, I_3\}$—shown earlier to produce stability in the absence of dissipation—we now have instability. Finally, for $\lambda < 0$, the $\boldsymbol{\omega}$-stability proved earlier for a rigid gyrostat becomes asymptotic.

As in several previous instances, the property of $\boldsymbol{\omega}$-stability can be reinterpreted as directional stability by introducing appropriate direction cosines. To restate the above conclusions in this more important manner, the direction $\underset{\rightarrow}{\mathbf{a}}$ shown in Fig. 7.3 is directionally stable, asymptotically to $\underset{\rightarrow}{\mathbf{h}}$ if $\lambda > \max\{I_1, I_3\}$ or if $\lambda < 0$. For λ in the range $0 < \lambda < \max\{I_1, I_3\}$ the motion is unstable. To facilitate comparison with earlier results, these stability conditions can be rewritten (Problem 6.10)

$$k_1(1 - k_3)^{1/2} + \hat{\Omega}_{po}(1 - k_1)^{1/2} > 0$$

$$k_3(1 - k_1)^{1/2} + \hat{\Omega}_{po}(1 - k_3)^{1/2} > 0 \tag{37}$$

$$(1 - k_1 k_3) + \hat{\Omega}_{po}[(1 - k_1)(1 - k_3)]^{1/2} < 0 \tag{38}$$

It is to be understood that *either* the condition (38) or *both* the conditions (37) are necessary and sufficient for directional stability. These conditions are exhibited in Fig. 7.4 for four values of $\hat{\Omega}_{po}$. Comparing the dissipation-free results in Fig. 6.7, one observes that the region designated "directionally and statically stable" has now become "directionally stable, asymptotically to $\underset{\rightarrow}{\mathbf{h}}$"; the unstable regions remain unstable; and some of the gyrically stabilized region is now unstable. Also revealing is a comparison with Fig. 5.3, for which $h_s = 0$. In general, energy sink analysis indicates that a simple spin of a quasi-rigid body about any principal axis can be made directionally stable, asymptotically to $\underset{\rightarrow}{\mathbf{h}}$, by including a rotor or wheel with a sufficiently great momentum, positive or negative, parallel to that axis.

It is also instructive to note that although these stability conditions have been derived for "Kelvin's gyrostat," their reinterpretation for the "apparent gyrostat" is a simple matter, because the wheel axis is parallel to a principal axis. According to (6.4, 28), I_1 and I_3 are still for $\mathscr{Q} + \mathscr{W}$ combined, while I_2 should now, for the apparent gyrostat, be the moment of inertia of $\mathscr{Q}$ alone. On the other hand, h_s, the component of *relative* angular momentum along the wheel axis, should likewise be replaced by h_a, the component of *absolute* angular momentum. Thus, λ, which was given by $\lambda = I_2 + (h_s/\nu)$ for Kelvin's gyrostat, must now be

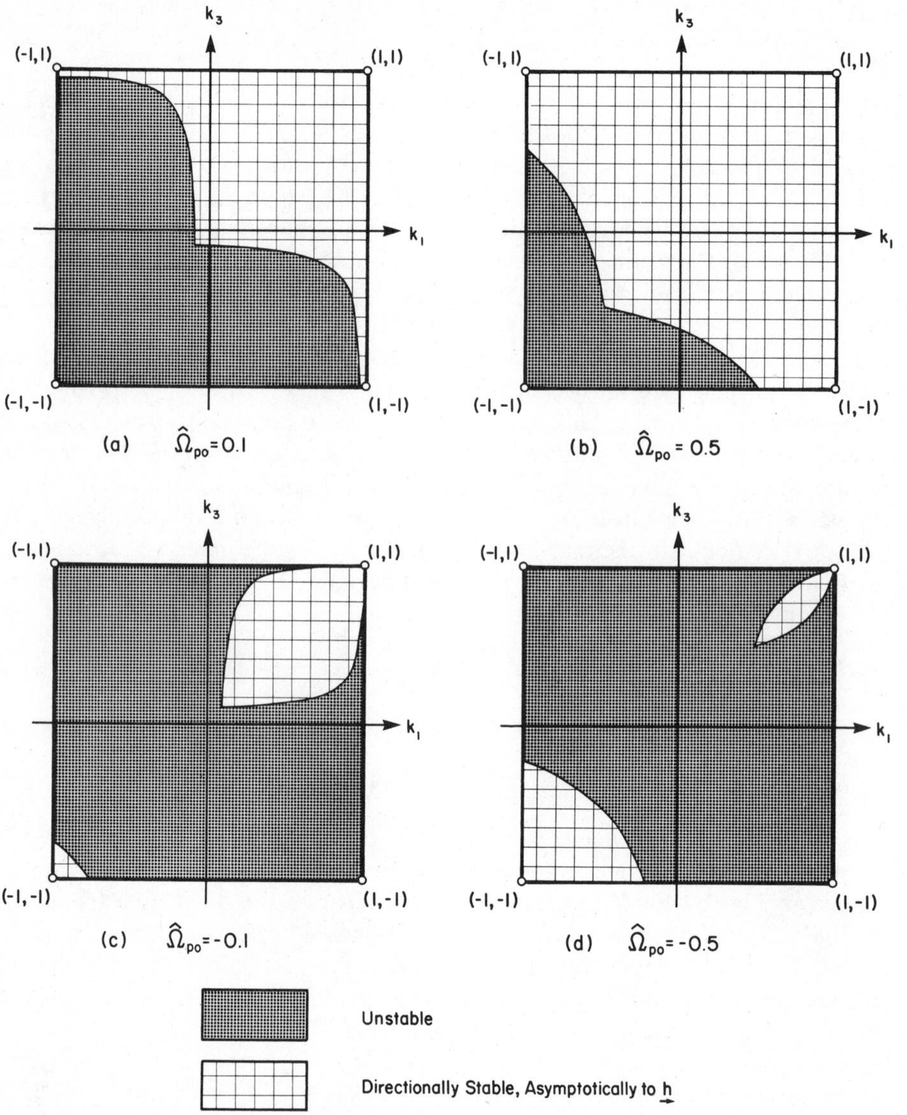

FIGURE 7.4 Stability diagrams for system of Fig. 7.3, obtained from energy sink analysis. (Compare with Figs. 5.3 and 6.7.)

defined as $\lambda = I_{2b} + (h_a/\nu)$, where $I_{2b} = I_2 - I_s$. Also,

$$h_a = I_s(\nu + \omega_s) \tag{39}$$

and so

$$\lambda = I_{2b} + I_s \frac{\nu + \omega_s}{\nu} = I_2 + \frac{h_s}{\nu} \tag{40}$$

That is, λ has the same value for both types of gyrostat. Furthermore, I_1 and I_3

are identical for both types, referring to the transverse inertias of $\mathcal{Q} + \mathcal{W}$. Therefore, the energy sink stability conditions $\lambda < 0$ and $\lambda > \max\{I_1, I_3\}$ are identical in both cases. The reason for this is that the change in ω_2 is zero ($\xi_2 = 0$) under the constant-h assumption. The distinction between the two types of gyrostat disappears.

The Symmetrical Gyrostat

An additional restriction is now imposed, namely, that $\mathcal{Q}$ is inertially axisymmetrical about $\hat{\mathbf{p}}_2$. In other words, $\mathcal{Q}$ is a quasi-rigid wheel $\mathcal{Q}_w$ (with an energy sink), containing a rigid wheel $\mathcal{W}$ (Fig. 7.5). The axes of symmetry of $\mathcal{Q}_w$ and $\mathcal{W}$ are parallel, and the inertias $\{I_1, I_2, I_3\}$ are now denoted respectively by $\{I_t, I_a, I_t\}$. According to (36), an energy sink in $\mathcal{Q}_w$ produces stability of $\underset{\rightarrow}{\mathbf{a}}$, asymptotically to $\underset{\rightarrow}{\mathbf{h}}$, provided

$$\left(I_a + \frac{h_s}{\nu}\right)\left(I_a + \frac{h_s}{\nu} - I_t\right) > 0 \tag{41}$$

where h_s may be of either sign.

Consider first $h_s > 0$; the spin of $\mathcal{W}$ and the spin of $\mathcal{Q}_w$ have the same sense. Condition (41) reduces to

$$I_a + \frac{h_s}{\nu} > I_t \tag{42}$$

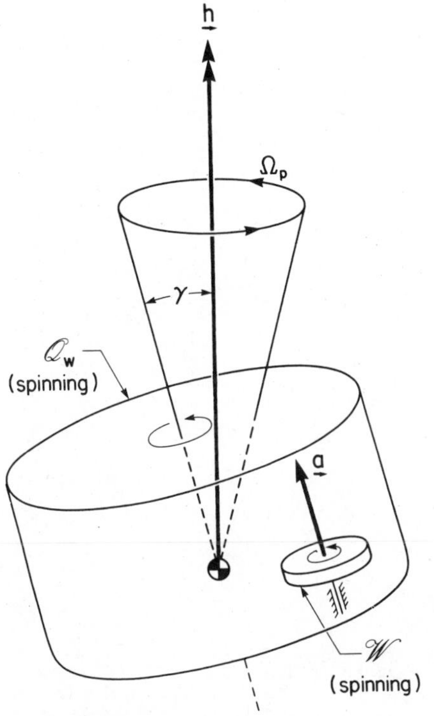

FIGURE 7.5 Symmetrical gyrostat with a quasi-rigid carrier.

It is apparent that $\mathscr{Q}_w + \mathscr{W}$ does not have to be in a major-axis spin to achieve the asymptotic stability of $\underset{\rightarrow}{\mathbf{a}}$ to $\underset{\rightarrow}{\mathbf{h}}$. Energy sink analysis predicts that if the axial momentum of $\mathscr{Q}_w$ is sufficiently augmented by the axial momentum of a wheel spinning in the same direction, a minor-axis spin can be stabilized. This may also be interpreted in terms of Landon's rule for a symmetrical quasi-rigid body extended to a symmetrical gyrostat. For a quasi-rigid body above, Landon's rule is given in (5.1, 14): $\Omega_p > \nu$, where Ω_p is the precession frequency. Since the rate of precession for $\mathscr{Q}_w + \mathscr{W}$ is $\Omega_p = h/I_t$, and $h = I_a\nu + h_s$, condition (42) is in accordance with Landon's rule extended to symmetrical gyrostats.

Next, let h_s lie in the range $-I_a\nu < h_s < 0$. Now $\mathscr{W}$ spins (relative to $\mathscr{Q}_w$) in the opposite direction to the spin of $\mathscr{Q}_w$, but the total angular momentum is still in the $+\underset{\rightarrow}{\hat{\mathbf{p}}}_2$ direction. The stability condition is still (42), but now, due to the adverse sign of the wheel momentum, even $I_a > I_t$ does not ensure stability. The wheel makes a negative contribution to stability.

Finally, let $h_s < -I_a\nu$. Now the total momentum $\underset{\rightarrow}{\mathbf{h}}$ is in the $-\hat{\mathbf{p}}_2$ direction. According to (41), this condition is asymptotically stable. However, the angle between $\underset{\rightarrow}{\hat{\mathbf{p}}}_2$ and $\underset{\rightarrow}{\mathbf{h}}$ is not 0 but π.

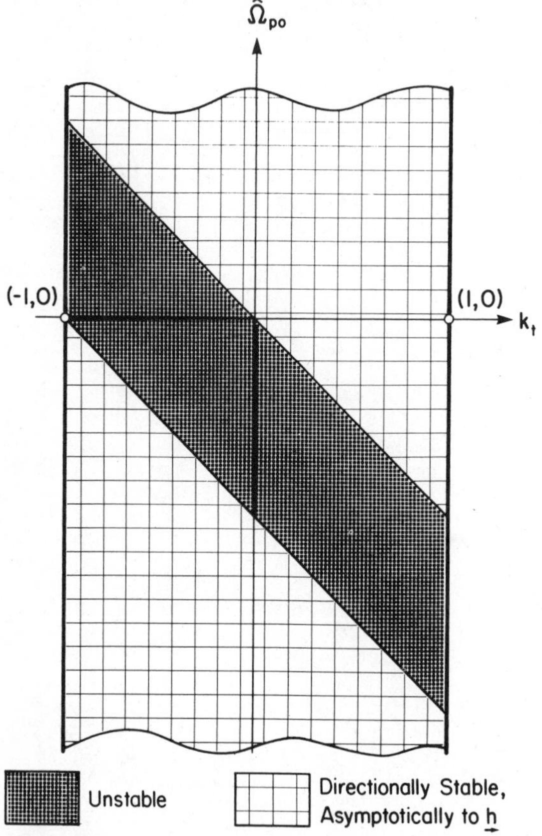

FIGURE 7.6 Stability diagram for system shown in Fig. 7.5.

The stability conditions for $\mathscr{Q}_w + \mathscr{W}$ are summarized in Fig. 7.6. Here,

$$k_1 = k_3 = k_t \triangleq \frac{I_a - I_t}{I_t}; \qquad \hat{\Omega}_{po} = \frac{h_s}{(I_1 I_3)^{1/2}\nu} = \frac{h_s}{I_t \nu} \tag{43}$$

Thus, condition (41) may be succinctly expressed in terms of the two parameters k_t and $\hat{\Omega}_{po}$ as follows:

$$(k_t + 1 + \hat{\Omega}_{po})(k_t + \hat{\Omega}_{po}) > 0 \tag{44}$$

It can be deduced that one of the following two conditions must hold:

$$k_t + \hat{\Omega}_{po} > 0; \qquad k_t + 1 + \hat{\Omega}_{po} < 0 \tag{45}$$

This is shown in Fig. 4.37.

Single Quasi-Rigid Wheel $\mathscr{Q}_w$

In Section 6.5, the system of inertially axisymmetrical wheels shown in Fig. 6.8 was studied. The wheels spin at individual rates about their common axis of symmetry. It was shown that this system, designated $\Sigma\mathscr{W}_n$, behaves in its overall motion as though it were a single wheel with axial and transverse inertias given by $I_a = \Sigma I_{an}$ and $I_t = \Sigma J_{tn}$, where I_{an} and J_{tn} are the axial and transverse moments of inertia of $\mathscr{W}_n$ about the system mass center. In other words, the relative motion of the wheels is immaterial in the calculation of I_a and I_t. The equivalent spin rate ν, on the other hand, depends on the individual spin rates ν_n (which may be of either sign), as indicated by (6.5, 12):

$$I_a \nu = \sum_{n=1}^{N} I_{an} \nu_n \tag{46}$$

We choose the sense of $\underline{\mathbf{a}}$ so that $\nu > 0$. The equivalent relative spin rate Ω is defined in the same fashion in terms of the individual relative spin rates Ω_n. The axis of symmetry is always directionally stable, just as it is for a single isolated wheel, in the sense that a disturbance to a simple spin ($\underline{\mathbf{h}}$ aligned with $\underline{\boldsymbol{\nu}}$) causes a precession, at rate $\Omega_p = h/I_t$, about $\underline{\mathbf{h}}$. After the disturbance, the inclination of the symmetry axis $\underline{\mathbf{a}}$ to $\underline{\mathbf{h}}$, termed the nutation angle γ, is constant and can be made arbitrarily small by making the initial disturbance sufficiently small. The objective of the following discussion is to ascertain the consequences when one or more of the wheels are quasi-rigid (i.e., contain an energy sink).

To make the path perfectly clear, the analysis will be constructed in a step-by-step fashion, beginning once again with a brief review of the torque-free motion of a single rigid wheel $\mathscr{W}$. The results of this paragraph should be compared with Section 4.2 and equations (6.5, 13 through 18). We begin with the angular momentum $\underline{\mathbf{h}}$:

$$\underline{\mathbf{h}} = \underline{\mathbf{I}} \cdot \underline{\boldsymbol{\omega}} \tag{47}$$

where the inertia dyadic for $\mathscr{W}$ is given by (see Problem 3.7*b*)

$$\underline{\mathbf{I}} = I_t \underline{\mathbf{1}} + (I_a - I_t)\underline{\mathbf{a}}\underline{\mathbf{a}} \tag{48}$$

The vector $\underset{\rightarrow}{\mathbf{h}}$ can be broken into its axial and transverse components. Setting

$$\underset{\rightarrow}{\omega} = \nu \underset{\rightarrow}{\mathbf{a}} + \underset{\rightarrow}{\omega}_t \tag{49}$$

indicates that

$$\underset{\rightarrow}{\mathbf{h}} = I_a \nu \underset{\rightarrow}{\mathbf{a}} + I_t \underset{\rightarrow}{\omega}_t \tag{50}$$

In the absence of external torque, $\underset{\rightarrow}{\mathbf{h}}$ is constant in an inertial frame $\mathscr{F}_i$. Now define

$$\underset{\rightarrow}{\Omega}_p \triangleq \frac{\underset{\rightarrow}{\mathbf{h}}}{I_t} \tag{51}$$

which, from its dimensions alone, can be interpreted as an angular velocity; $\underset{\rightarrow}{\Omega}_p$ is also constant in $\mathscr{F}_i$. Next, define the relative angular velocity $\underset{\rightarrow}{\Omega}$ by

$$\underset{\rightarrow}{\omega} = \underset{\rightarrow}{\Omega}_p + \underset{\rightarrow}{\Omega} \tag{52}$$

and substitute in (47) and (48) to obtain

$$I_t \underset{\rightarrow}{\Omega} + (I_a - I_t)(\underset{\rightarrow}{\mathbf{a}} \cdot \underset{\rightarrow}{\Omega} + \Omega_p \cos\gamma)\underset{\rightarrow}{\mathbf{a}} = \underset{\rightarrow}{\mathbf{0}} \tag{53}$$

where γ is the angle between $\underset{\rightarrow}{\mathbf{a}}$ and $\underset{\rightarrow}{\mathbf{h}}$ (i.e., the angle between $\underset{\rightarrow}{\mathbf{a}}$ and $\underset{\rightarrow}{\Omega}_p$). Clearly, $\underset{\rightarrow}{\Omega}$ is in the $\underset{\rightarrow}{\mathbf{a}}$-direction, so let $\underset{\rightarrow}{\Omega} = \Omega \underset{\rightarrow}{\mathbf{a}}$, whence (53) indicates that

$$I_t \Omega + (I_a - I_t)(\Omega + \Omega_p \cos\gamma) = 0 \tag{54}$$

However, from (49) and (52),

$$\nu = \Omega + \Omega_p \cos\gamma \tag{55}$$

and so the now familiar relationship between Ω and ν is obtained. Lastly, the kinetic energy is $T = \frac{1}{2}\underset{\rightarrow}{\omega} \cdot \underset{\rightarrow}{\mathbf{I}} \cdot \underset{\rightarrow}{\omega} = \frac{1}{2}\underset{\rightarrow}{\omega} \cdot \underset{\rightarrow}{\mathbf{h}}$, which can be expressed in many ways, including (6.5, 18). Thus, in a few lines, and without explicitly integrating motion equations, most of the essential earlier results have been reproduced from a different viewpoint.

If $\mathscr{W}$ is quasi-rigid ($\mathscr{W} \to \mathscr{Q}_w$), $\dot{T} < 0$. In (6.5, 18), T was expressed as follows:

$$2T = h^2\left(\frac{\cos^2\gamma}{I_a} + \frac{\sin^2\gamma}{I_t}\right) \tag{56}$$

According to the energy sink hypothesis, h is constant. Therefore (see Fig. 7.7 and Problem 4.1),

$$\dot{T} = h^2 \sin\gamma\cos\gamma\left(\frac{I_a - I_t}{I_a I_t}\right)\dot{\gamma} \tag{57}$$

indicating that for $\underset{\rightarrow}{\mathbf{a}}$ to be asymptotic to $\underset{\rightarrow}{\mathbf{h}}$ ($\dot{\gamma} < 0$), one must have $I_a > I_t$ (the major-axis rule). Another succinct statement of (57), since $h\cos\gamma = I_a\nu$, $h\sin\gamma = I_t\omega_t$, is

$$\dot{\gamma} = -\frac{\dot{T}}{h_t \Omega} \tag{58}$$

implying the stability condition

$$\Omega < 0 \tag{59}$$

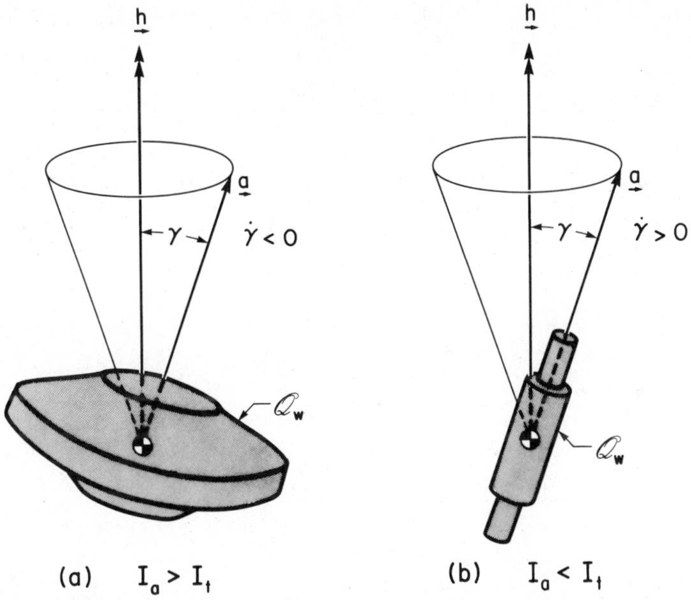

FIGURE 7.7 Directional stability of a quasi-rigid wheel when the wheel is (*a*) oblate or (*b*) prolate.

Thus, for dissipation to produce a stabilizing effect, the relative spin must be negative. Noting (55), we see that this is equivalent to

$$\Omega_p \cos\gamma > \nu \tag{60}$$

which is a more precise (nonlinear) statement of Landon's rule than the linearized version, $\Omega_p > \nu$. For a single wheel, Ω_p, $\cos\gamma$, and ν are all positive, and the statement (60) is easily interpreted. When two or more wheels are involved, as they will be momentarily, even (60) is not sufficiently general. *The most general form of Landon's idea is to require that*

$$\underset{\rightarrow}{\boldsymbol{\Omega}}_p \cdot \underset{\rightarrow}{\boldsymbol{\Omega}} < 0 \tag{61}$$

for energy dissipation in a symmetrical quasi-rigid body to be stabilizing.

In preparation for a treatment of the N-wheel system, it is helpful to calculate the torques on $\mathcal{Q}_w$ which produce the energy changes. As in Section 6.5, the most convenient reference frame is the precessing frame $\mathcal{F}_\Omega$, in which the key vectors are expressed as follows:

$$\underset{\rightarrow}{\boldsymbol{\omega}} = [\nu \quad \omega_t \quad 0]\mathcal{F}_\Omega \tag{62}$$

$$\underset{\rightarrow}{\mathbf{h}} = [I_a\nu \quad I_t\omega_t \quad 0]\mathcal{F}_\Omega \tag{63}$$

$$\underset{\rightarrow}{\mathbf{g}} = [g_a \quad g_t \quad g_b]\mathcal{F}_\Omega \tag{64}$$

The motion equation for $\mathcal{Q}_w$ is

$$\overset{\circ}{\underset{\rightarrow}{\mathbf{h}}} + \underset{\rightarrow}{\boldsymbol{\Omega}}_p \times \underset{\rightarrow}{\mathbf{h}} = \underset{\rightarrow}{\mathbf{g}} \tag{65}$$

where the overcircle means differentiation measured in $\mathscr{F}_\Omega$. Thus, since $\underset{\rightarrow}{\mathbf{\Omega}}_p = \underset{\rightarrow}{\mathbf{h}}/I_t$,

$$I_a\dot{\nu} = g_a; \qquad I_t\dot{\omega}_t = g_t; \qquad 0 = g_b \tag{66}$$

The energy sink hypothesis assumes that h = constant, which implies that

$$h\dot{h} = I_a^2\nu\dot{\nu} + I_t^2\omega_t\dot{\omega}_t = 0 \tag{67}$$

and that the rate of energy change is

$$\dot{T} = I_a\nu\dot{\nu} + I_t\omega_t\dot{\omega}_t < 0 \tag{68}$$

Thus,

$$\dot{T} = I_a\nu\dot{\nu}\frac{I_t - I_a}{I_t} = \Omega g_a \tag{69}$$

showing that the axial torque is

$$g_a = \frac{\dot{T}}{\Omega} \tag{70}$$

Similarly, the transverse torque is

$$g_t = -\left(\frac{\dot{T}}{\Omega}\right)\cot\gamma \tag{71}$$

Since we are already aware that $\Omega < 0$ for stability, it is evident that $g_a > 0$ and $g_t < 0$.

Quasi-Rigid Wheel with Rigid Wheel, $\mathscr{Q}_{w1} + \mathscr{W}_2$

To proceed methodically, we consider next a system of two wheels, $\mathscr{W}_1 + \mathscr{W}_2$, with $\mathscr{W}_1$ quasi-rigid, $\mathscr{W}_1 \to \mathscr{Q}_{w1}$ (Fig. 7.8). Stability conditions for this system, which is similar to the symmetrical gyrostat with dissipation in the carrier ($\mathscr{Q}_w + \mathscr{W}$, Fig. 7.5) studied earlier in this section, were first obtained by [Landon and Stewart]. Drawing on the development for $\Sigma\mathscr{W}_n$ in Section 6.5, the nutation angle γ is found from

$$I_a\nu \equiv I_{a1}\nu_1 + I_{a2}\nu_2 = h\cos\gamma \tag{72}$$

$$I_t\omega_t = h\sin\gamma \tag{73}$$

where ν_1 and ν_2 are the individual spin rates of $\mathscr{Q}_{w1}$ and $\mathscr{W}_2$ and ω_t is the shared transverse component of angular velocity. By convention, $\nu > 0$ and $h > 0$; thus, $\cos\gamma > 0$. The total energy is

$$T = T_1 + T_2 = \tfrac{1}{2}\left(I_{a1}\nu_1^2 + I_{a2}\nu_2^2 + I_t\omega_t^2\right) \tag{74}$$

and the magnitude of the angular momentum is

$$h^2 = \left(I_{a1}\nu_1 + I_{a2}\nu_2\right)^2 + I_t^2\omega_t^2 \tag{75}$$

As usual in energy sink analysis, we take $\dot{h} = 0$ and $\dot{T} < 0$. Consequently,

$$h\dot{h} = \left(I_{a1}\nu_1 + I_{a2}\nu_2\right)I_{a1}\dot{\nu}_1 + I_t^2\omega_t\dot{\omega}_t = 0 \tag{76}$$

$$\dot{T} = I_{a1}\nu_1\dot{\nu}_1 + I_t\omega_t\dot{\omega}_t < 0 \tag{77}$$

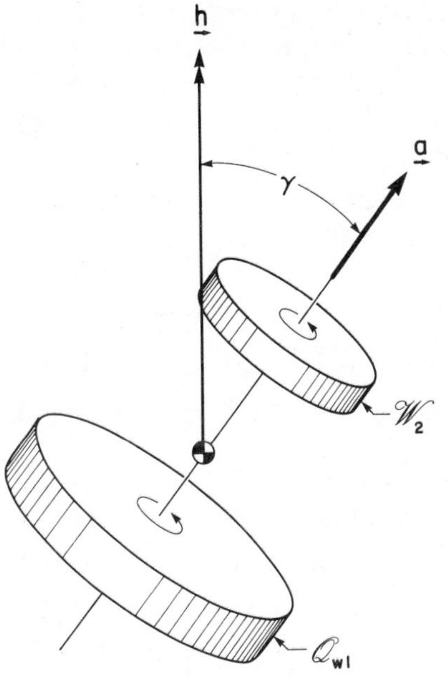

FIGURE 7.8 Two coaxial wheels, one quasi-rigid.

It has been assumed in these relations that ν_2 = constant. This is consistent with the absence of both bearing friction and internal energy dissipation in $\mathscr{W}_2$; see (6.5, 25) and (70). It follows from (72), (73), and (76) that

$$I_t\dot{\omega}_t = (h\cos\gamma)\dot{\gamma} = h_a\dot{\gamma}; \qquad I_{a1}\dot{\nu}_1 = -(h\sin\gamma)\dot{\gamma} = -h_t\dot{\gamma} \tag{78}$$

These results allow one to relate the change in kinetic energy to the change in nutation angle:

$$\dot{\gamma} = -\frac{\dot{T}}{h_t\Omega_1} \tag{79}$$

where $h_t = I_t\omega_t$ and $\Omega_1 = \nu_1 - \Omega_p\cos\gamma$. This result should be compared to (58), derived for $\mathscr{Q}$ alone. For asymptotic directional stability (in the limited sense that $\underset{\rightarrow}{\mathbf{a}} \to \underset{\rightarrow}{\mathbf{h}}/h$) the appropriate condition can clearly be expressed in either of the following two equivalent ways:

$$\Omega_1 < 0; \qquad \Omega_p\cos\gamma > \nu_1 \tag{80}$$

This is basically Landon's rule as applied to $\mathscr{Q}_{w1} + \mathscr{W}_2$.

There is an equivalence, although perhaps not an immediately obvious one, between the stability condition (80) and the condition (41) derived earlier for the symmetrical gyrostat with dissipation in the carrier. To establish this equivalence requires a clear understanding of the meaning of the symbols. Thus, in (41), one must realize that

$$I_a = I_{a1} + I_{a2}; \qquad h_s = I_{a2}(\nu_2 - \nu_1) \tag{81}$$

and that ν in (41) is not the mean value of spin for $\mathscr{Q}_{w1} + \mathscr{W}_2$ as expressed in

(72), but ν_1 instead. With these notational identifications made, one finds that (41), written for the $\mathscr{Q}_w + \mathscr{W}$ system of Fig. 7.5, can be rewritten, for the $\mathscr{Q}_{w1} + \mathscr{W}_2$ system of Fig. 7.8, as follows:

$$(I_{a1}\nu_1 + I_{a2}\nu_2)(I_{a1}\nu_1 + I_{a2}\nu_2 - I_t\nu_1) > 0 \tag{82}$$

However, from (72), $I_{a1}\nu_1 + I_{a2}\nu_2 = h\cos\gamma$, which can be taken positive by convention. Finally, then, this condition is identical to (80), as it should be.

These energy sink stability conditions for $\mathscr{Q}_{w1} + \mathscr{W}_2$ are depicted in Fig. 7.9. A simple diagram suffices if Ω_p and ν_1 are used as variables (Fig. 7.9*a*). Perhaps more useful for design purposes is the type of diagram shown in Figs. 7.9*b* through *d*. For example, if $\{I_{a1}, I_{a2}, I_t, \nu_1\}$ are specified and one wishes to know the range of appropriate spin rates for $\mathscr{W}_2$, this second kind of stability diagram can be used directly.

As a final note on the condition for stability, we remark that it is a matter of indifference whether the quasi-rigid and rigid wheels are respectively labeled "1" and "2," or vice versa. Thus, for a system designated $\mathscr{W}_1 + \mathscr{Q}_{w2}$, the condition for stability is

$$\Omega_2 < 0; \qquad \Omega_p\cos\gamma > \nu_2 \tag{83}$$

To conclude our study of the $\mathscr{Q}_{w1} + \mathscr{W}_2$ system, an accounting is taken of the torque on each body and of the kinetic energy of each body. This information will be especially helpful when we extend the development to encompass an arbitrary number of symmetrical rigid and quasi-rigid bodies. By applying the motion equation (6.5, 24) to $\mathscr{Q}_{w1}$ and then to $\mathscr{W}_2$, we find

$$I_{a1}\dot{\nu}_1 = g_{a1}; \qquad J_{t1}\dot{\omega}_t = g_{t1} \tag{84}$$

$$I_{a2}\dot{\nu}_2 = g_{a2}; \qquad J_{t2}\dot{\omega}_t = g_{t2} \tag{85}$$

The binormal components, g_{b1} and g_{b2}, are still given by (6.5, 27) or (6.5, 29), although now $\{\omega_t, \gamma, \nu_1, h_1, h_2, \gamma_1, \gamma_2\}$ vary as the motion progresses. Now, by hypothesis, $g_{a2} = 0$. The remaining torque components are, from (78),

$$[g_{a1} \quad g_{t1} \quad g_{t2}] = \left[-h_t \quad \left(\frac{J_{t1}}{I_t}\right)h_a \quad \left(\frac{J_{t2}}{I_t}\right)h_a\right]\dot{\gamma} \tag{86}$$

and $\dot{\gamma}$ has already been related to $\dot{T}$ by (79). The total torque on the vehicle is

$$[g_a \quad g_t \quad g_b] = [-h_t \quad h_a \quad 0]\dot{\gamma} \tag{87}$$

As for the kinetic energy residing in $\mathscr{Q}_{w1}$ and $\mathscr{W}_2$, we may use (6.5, 30) to show that

$$\dot{T}_1 = \left[\left(\frac{J_{t1}}{I_t}\right)\Omega_p\cos\gamma - \nu_1\right]h_t\dot{\gamma}; \qquad \dot{T}_2 = \left[\left(\frac{J_{t2}}{I_t}\right)\Omega_p\cos\gamma\right]h_t\dot{\gamma} \tag{88}$$

There are many other equivalent ways of writing $\dot{T}_1$ and $\dot{T}_2$. *It is interesting to*

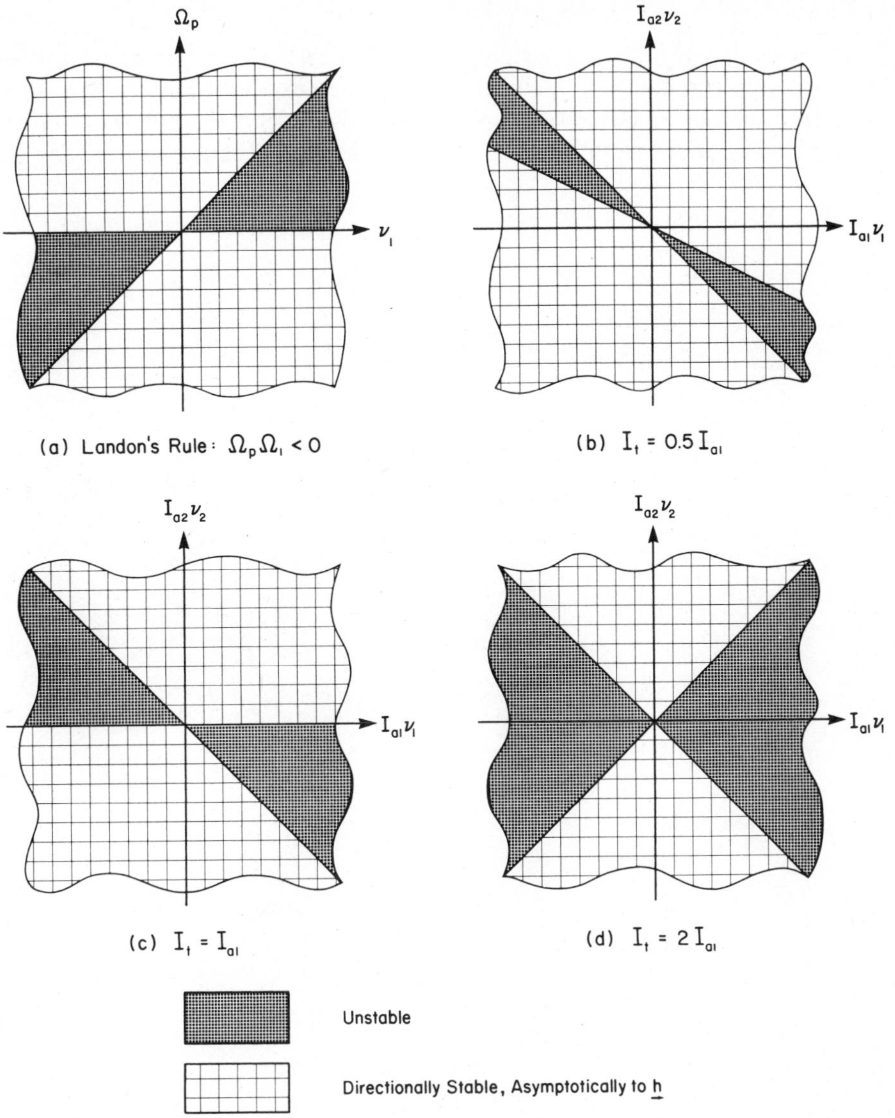

FIGURE 7.9 Stability diagrams for system shown in Fig. 7.8.

learn that $\dot{T}_2 < 0$ *although the energy sink is in* $\mathscr{Q}_{w1}$, *not in* $\mathscr{W}_2$. The agency for this reduction in energy of $\mathscr{W}_2$ is the transverse torque g_{t2}, and in fact $\dot{T}_2 = g_{t2}\omega_t$, as one should expect. As a check, note that

$$\dot{T}_1 + \dot{T}_2 = \left(\Omega_p \cos\gamma - \nu_1\right) h_t \dot{\gamma} = -\Omega_1 h_t \dot{\gamma} = \dot{T} \tag{89}$$

which shows that the energy analysis is consistent.

System of Coaxial Quasi-Rigid Wheels

We turn now to the quite general case of N inertially axisymmetrical wheels mounted on a common axis $\underset{\rightarrow}{a}$, coincident with their axes of symmetry. All wheels are permitted to be quasi-rigid and to possess an energy sink. This system, denoted $\Sigma\mathscr{Q}_{wn}$, is shown in Fig. 7.10. An extension to the earlier analysis for $\Sigma\mathscr{W}_n$ in Section 6.5 is required. The kinetic energy and the angular momentum (magnitude) are still given by

$$h^2 = h_a^2 + h_t^2; \qquad h_a = \sum_{n=1}^{N} I_{an}\nu_n; \qquad h_t = I_t\omega_t \tag{90}$$

$$2T = \sum_{n=1}^{N} 2T_n = \sum_{n=1}^{N} I_{an}\nu_n^2 + I_t\omega_t^2 \tag{91}$$

It follows that

$$h\dot{h} = h_a \sum_{n=1}^{N} I_{an}\dot{\nu}_n + h_t I_t \dot{\omega}_t = 0 \tag{92}$$

$$\dot{T} = \sum_{n=1}^{N} \dot{T}_n < 0 \tag{93}$$

$$\dot{T}_n = I_{an}\nu_n\dot{\nu}_n + J_{tn}\omega_t\dot{\omega}_t \tag{94}$$

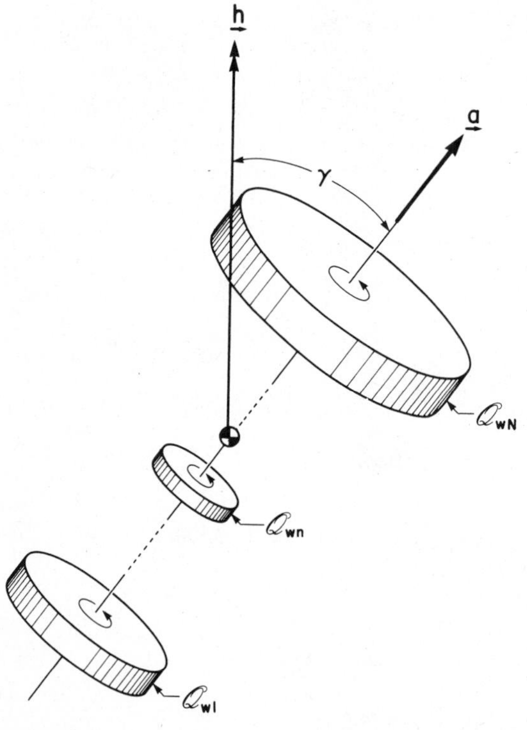

FIGURE 7.10 System of quasi-rigid coaxial wheels.

The nutation angle γ can be related directly to the (common) transverse velocity component ω_t:

$$h \sin \gamma = I_t \omega_t \tag{95}$$

Since h is assumed constant, (95) when differentiated indicates that

$$\dot{\omega}_t = \Omega_p \cos \gamma \; \dot{\gamma} \tag{96}$$

since $\Omega_p = h/I_t$. However, from (92),

$$I_t \omega_t \dot{\omega}_t = -\Omega_p \cos \gamma \sum_{n=1}^{N} I_{an} \dot{\nu}_n \tag{97}$$

So the change in nutation angle is related to the changes in spin rate in the following manner:

$$h_t \dot{\gamma} = -\sum_{n=1}^{N} I_{an} \dot{\nu}_n \tag{98}$$

The challenge is to express the right side of (98) in terms of energy changes.

An analysis by [Iorillo, 2], repeated many times in the literature, proceeds thus: from (92 through 94),

$$\begin{aligned} \dot{T} &= \sum_{n=1}^{N} I_{an} \nu_n \dot{\nu}_n + I_t \omega_t \dot{\omega}_t \\ &= \sum_{n=1}^{N} I_{an} \dot{\nu}_n \left(\nu_n - \frac{h_a}{I_t} \right) \\ &= \sum_{n=1}^{N} I_{an} \dot{\nu}_n (\nu_n - \Omega_p \cos \gamma) = \sum_{n=1}^{N} I_{an} \Omega_n \dot{\nu}_n \end{aligned} \tag{99}$$

One then compares (93) and (99) and concludes that

$$\dot{T}_n = I_{an} \Omega_n \dot{\nu}_n \tag{100}$$

on the grounds that the "bodies are assumed uncoupled" and that the "reaction torques" are $I_{an} \dot{\nu}_n$. This argument, with (98), produces nutation angle decay given by

$$h_t \dot{\gamma} = -\sum_{n=1}^{N} \frac{\dot{T}_n}{\Omega_n} \tag{101}$$

from which it appears that $\Omega_n < 0$ is stabilizing. This is also in accordance with Landon's rule.

This argument can be made more precise. If T_n is the kinetic energy of $\mathscr{Q}_{wn}$, then $\dot{T}_n$ is in fact not given by (100), but by (94). Nor are the bodies uncoupled; they are coupled by the constraint of a common axis. And the expression $I_{an} \dot{\nu}_n$, while correctly giving the *axial* torque g_{an}, ignores the *transverse* torque g_{tn}, which also contributes to $\dot{T}_n$. To fashion a more precise energy sink analysis for $\Sigma \mathscr{Q}_{wn}$, we denote by $\dot{T}_{sn}$ the consumption of energy in $\mathscr{Q}_{wn}$. Note that $\dot{T}_{sn} \neq \dot{T}_n$. Indeed, an energy sink in just one of the $\mathscr{Q}_{wn}$ (the other wheels being rigid) will generally change all the T_n (see, for example, (88), valid for the two-wheel system $\mathscr{Q}_{w1} + \mathscr{W}_2$). In general, one has

$$\dot{T}_n = \sum_{m=1}^{N} \beta_{nm}\dot{T}_{sm} \tag{102}$$

where the $\beta_{nm}(t)$ are now calculated. First, for a consistent model,

$$\dot{T} = \sum_{n=1}^{N} \dot{T}_n = \sum_{n=1}^{N}\sum_{m=1}^{N} \beta_{nm}\dot{T}_{sm} = \sum_{m=1}^{N} \dot{T}_{sm} \tag{103}$$

This implies that

$$\beta_{nn} = 1 - {\sum_{m=1}^{N}}' \beta_{mn} \qquad (n = 1, \ldots, N) \tag{104}$$

where the prime on the summation operation requires that the $m = n$ term be omitted. Next, to establish expressions for the β_{mn} $(m \neq n)$, we observe that when $\{N = 2,\ \dot{T}_{s1} = \dot{T} < 0,\ \dot{T}_{s2} = 0\}$, expressions for $\dot{T}_1$ and $\dot{T}_2$ were given earlier by (88). For present purposes, we write these expressions, noting (79) also, in this way:

$$\dot{T}_1 = (1 - \beta_{21})\dot{T}_{s1}; \qquad \dot{T}_2 = \beta_{21}\dot{T}_{s1} \tag{105}$$

where

$$\beta_{21} = -\frac{J_{t2}\Omega_p \cos\gamma}{I_t\Omega_1} \tag{106}$$

Therefore, for our present system, $\Sigma\mathscr{Q}_{wn}$, we use

$$\beta_{nm} = -\frac{J_{tn}\Omega_p \cos\gamma}{I_t\Omega_m} \qquad (n, m = 1, \ldots, N) \qquad (n \neq m) \tag{107}$$

from which, with (104),

$$\beta_{nn} = 1 + \frac{\Omega_p \cos\gamma}{I_t\Omega_n} {\sum_{m=1}^{N}}' J_{tm} = \frac{I_t\nu_n - J_{tn}\Omega_p\cos\gamma}{I_t\Omega_n} \tag{108}$$

In other words, the relationship between the energy decreases $\dot{T}_n$ and the energy sinks $\dot{T}_{sn}$, namely, (102), can now be written explicitly as

$$\dot{T}_n = \left(\frac{\nu_n}{\Omega_n}\right)\dot{T}_{sn} - \left(\frac{J_{tn}}{I_t}\right)\Omega_p\sigma\cos\gamma \tag{109}$$

where

$$\sigma \triangleq \sum_{n=1}^{N} \frac{\dot{T}_{sn}}{\Omega_n} \tag{110}$$

Equations (109) and (94) are now combined to provide an expression for $I_{an}\dot{\nu}_n$, which is then substituted into (97) to yield

$$I_t\omega_t\dot{\omega}_t = -\Omega_p\sigma\cos\gamma \tag{111}$$

after the common factor

$$I_t - \Omega_p \cos\gamma \sum_{n=1}^{N} \frac{J_{tn}}{\nu_n} \tag{112}$$

is divided out. It is clear from (97), (98) and (111) that the influence of the energy sinks $\dot{T}_{sn}$ on the nutation angle is

$$h_t \dot{\gamma} = -\sum_{n=1}^{N} \frac{\dot{T}_{sn}}{\Omega_n} \qquad (\dot{\gamma} < 0 \text{ for asymptotic stability}) \tag{113}$$

This is the principal result of this section. In spite of its superficial resemblance to (101), it is really quite different because, as amply demonstrated above, $\dot{T}_n$ depends on all the energy sinks in the system and may even be positive. Of course, $\dot{T}_{sn} < 0$ always, unless $\mathscr{Q}_{wn}$ is assumed rigid ($\mathscr{Q}_{wn} \to \mathscr{W}_n$), or unless the nutation angle has been reduced to zero. The condition (113) for directional stability of the $\underset{\rightarrow}{\mathbf{a}}$ direction, asymptotically to $\underset{\rightarrow}{\mathbf{h}}$, indicates that, in agreement with Landon's rule, $\Omega_n < 0$ is always stabilizing, that $\Omega_n > 0$ is always destabilizing, and that the balance between these conflicting tendencies will be favorable, provided $\Sigma \dot{T}_{sn}/\Omega_n > 0$. This can always be achieved by placing sufficiently strong energy sinks in bodies that have $\Omega_n < 0$.

Extension to Systems of Tri-Inertial Quasi-Rigid Bodies

It is difficult to extend energy sink analysis to systems of nonsymmetrical quasi-rigid bodies. The precessing frame, so useful an artifice for systems of inertially axisymmetrical bodies, is no longer applicable for bodies not possessing an axis of inertial symmetry (tri-inertial bodies). No longer is there a uniform precession frequency Ω_p, and even if one chooses a reference frame defined by the plane of $\underset{\rightarrow}{\mathbf{h}}$ and the nominal spin axis $\underset{\rightarrow}{\boldsymbol{\nu}}$, the inertias in this frame are in general time varying. Since the energy sink method is already heuristic in nature, a reluctance to add further approximations is probably justified. Under special circumstances however, energy sink ideas can still be applied even to tri-inertial bodies $\mathscr{Q}$. Indeed, several important cases have already been treated. A single quasi-rigid body $\mathscr{Q}$ was examined in Section 5.1, and this first led to the major-axis rule. The quasi-rigid gyrostat $\mathscr{Q} + \mathscr{W}$, studied earlier in this section, again permitted a tri-inertial $\mathscr{Q}$, but the wheel was assumed rigid and free of dissipation. Difficulty arises however when dissipation is assumed to occur on more than one body in the system and when, at the same time, all bodies are not axisymmetrical.

One of the few gaps in this barricade is the system $\mathscr{Q} + \mathscr{Q}_w$, with $\mathscr{Q}$ nominally nonspinning—a very important case in practice. This system consists (Fig. 7.11) of a general (i.e., tri-inertial) quasi-rigid body $\mathscr{Q}$ and of a rotor or wheel $\mathscr{Q}_w$; $\mathscr{Q}_w$ spins about its symmetry axis, which is fixed in $\mathscr{Q}$, but $\mathscr{Q}$ itself has no spin, $\boldsymbol{\nu} = \mathbf{0}$. It was shown in Section 6.2 that, in the absence of energy dissipation, the wheel spin axis precesses about $\underset{\rightarrow}{\mathbf{h}}$ at the rate Ω_p given by (6.2, 8). The cone traced out by

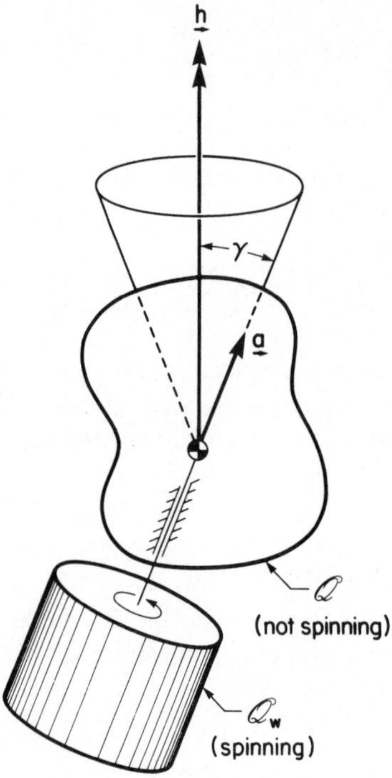

FIGURE 7.11 Gyrostat with both carrier and rotor quasi-rigid. The carrier is nominally nonspinning.

the symmetry axis $\underset{\rightarrow}{\mathbf{a}}$ is an elliptic one, implying that the angle γ between $\underset{\rightarrow}{\mathbf{h}}$ and $\underset{\rightarrow}{\mathbf{a}}$ is not constant, but varies periodically between a maximum and a minimum. It should also be mentioned that the analysis of Section 6.2 is a linearized one, accurate to first order in γ. In spite of the fact that γ itself it not constant, it is nevertheless plausible to apply Landon's rule in the form suggested by (113):

$$\frac{\dot{T}_{sc}}{\Omega_c} + \frac{\dot{T}_{sw}}{\Omega_w} > 0 \tag{114}$$

where it is now understood that this condition can be expected to produce a decrease in the *average* value of the nutation angle γ. The subscripts "c" and "w" refer to "carrier" and "wheel," respectively. The reason (114) is still applicable is that the relative spin rates Ω_c and Ω_w are constant for $\mathcal{Q} + \mathcal{Q}_w$.

Let us further assume that the spin axis of $\mathcal{Q}_w$ is aligned with $\hat{\mathbf{p}}_2$, a principal axis of the vehicle. According to (6.2, 8), and making the transition from Kelvin's gyrostat to the apparent gyrostat, one finds the precession rate

$$\Omega_p = \frac{I_{aw}\nu_w}{(I_1 I_3)^{1/2}} \tag{115}$$

where I_{aw} is the axial inertia of $\mathcal{Q}_w$, I_1 and I_3 are the transverse inertias of the entire vehicle, and ν_w is the spin rate of $\mathcal{Q}_w$. Under the current assumption that the spin rate of the carrier is zero,

$$\Omega_c = -\Omega_p \tag{116}$$

and the relative spin rate for the wheel is

$$\Omega_w = \nu_w - \Omega_p = \left[1 - \frac{I_{aw}}{(I_1 I_3)^{1/2}}\right]\nu_w \tag{117}$$

The condition for asymptotic stability (to $\underset{\rightarrow}{\mathbf{h}}$) is therefore

$$\frac{\dot{T}_{sc}}{I_{aw}} + \frac{\dot{T}_{sw}}{I_{aw} - (I_1 I_3)^{1/2}} < 0 \tag{118}$$

When there is no dissipation in the wheel ($\dot{T}_{sw} = 0$), this condition is always satisfied, in agreement with the observation made earlier in this section for an arbitrary wheel axis and noninfinitesimal nutation angles: dissipation in the carrier is always stabilizing. Dissipation in the wheel, on the other hand, can be stabilizing or destabilizing, depending on the inertias.

The stability results obtained from (118) may therefore be summarized by saying that (i) a vehicle for which

$$I_{aw} > (I_1 I_3)^{1/2} \tag{119}$$

is always stable, and (ii) a vehicle for which $I_{aw} < (I_1 I_3)^{1/2}$ is stable provided

$$|\dot{T}_{sc}| > \frac{I_{aw}}{(I_1 I_3)^{1/2} - I_{aw}} |\dot{T}_{sw}| \tag{120}$$

By "stable" we mean, as usual, "directionally stable, asymptotically to $\underset{\rightarrow}{\mathbf{h}}$." The condition (119) can be viewed as a *major-axis rule for nominally nonspinning gyrostats with energy dissipation in both carrier and wheel*. It was shown earlier that the major-axis rule for a single body can be violated if a rigid wheel of sufficiently high angular momentum is added to the system. We now understand that the stabilization provided by a wheel may not in fact be stabilizing if there is dissipation in the wheel also and (119) is not satisfied. This problem can be overcome, in turn, by providing sufficient dissipation in the carrier, as required by (120). The significance of the geometrical mean of the transverse inertias was discovered by [Spencer, 1, 3] and [Cherchas and Hughes].

7.2 GYROSTATS WITH DISCRETE DAMPERS

It should be remembered that in spite of its importance in providing guidance on how directional stability may generally be achieved, the energy sink hypothesis does not produce an exact analysis and therefore cannot be relied upon for precise stability conditions. A ready example is the single spinning body in torque-free motion (Chapter 5). Energy sink analysis produces the major-axis rule ($k_1 > 0$, $k_3 > 0$), while for the specific damping device of Section 5.2 the precise conditions are $k_1 > 0$, $k_3 > \Xi_d > 0$. The need to sharpen stability conditions derived from energy sink considerations holds equally true for gyrostats.

Many types of discrete damping device have been suggested in the literature. Combined with the already large number of gyrostat assumption sets, a great many dynamical models are possible. Limited space permits only two of these possibilities to be explored here. In the first, labeled $\mathscr{R}+\mathscr{P}+\mathscr{W}$, a rigid wheel is added to a general rigid body containing a mass spring dashpot damper. The second consists of a system of two rigid wheels in relative rotation, $\mathscr{W}_1+\mathscr{W}_2$, and in each of these is placed a mass spring dashpot damper. These two systems are sufficient to allow an appreciation for the character of typical stability conditions for gyrostats with discrete dampers.

Gyrostat with One Discrete Damper

We now investigate the $\mathscr{R}+\mathscr{P}+\mathscr{W}$ system (Fig. 7.12), in which the rigid body $\mathscr{R}$ contains a wheel $\mathscr{W}$ and a mass spring dashpot damper $\mathscr{P}$. It is helpful to think of this system as a generalization of the $\mathscr{R}+\mathscr{W}$ system of Section 6.1 (Fig. 3.8) to include damping, or of the $\mathscr{R}+\mathscr{P}$ system of Section 5.2 (Fig. 5.4) to include a wheel.

Employing the methods explained in Chapter 3, one can derive the following motion equations for $\mathscr{R}+\mathscr{P}+\mathscr{W}$:

$$\begin{aligned}
\dot{\mathbf{p}} &= -\boldsymbol{\omega}^\times\mathbf{p}+\mathbf{f}\\
\dot{\mathbf{h}} &= -\boldsymbol{\omega}^\times\mathbf{h}-\mathbf{v}^\times\mathbf{p}+\mathbf{g}\\
\dot{p}_n &= m_d\boldsymbol{\omega}^T\mathbf{n}^\times(\mathbf{v}-\mathbf{r}_d^\times\boldsymbol{\omega})-c_d\dot{\xi}-k_d\xi\\
\dot{h}_a &= g_a
\end{aligned} \tag{1}$$

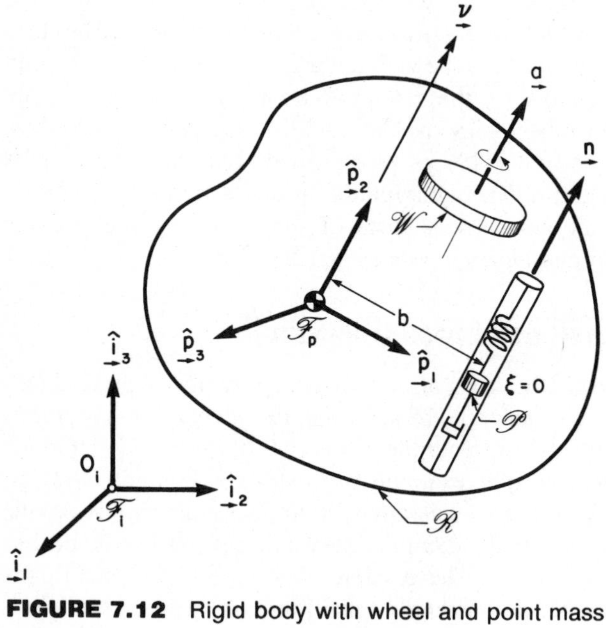

FIGURE 7.12 Rigid body with wheel and point mass damper.

The system momenta are

$$\begin{aligned}
\mathbf{p} &= m\mathbf{v} - \mathbf{c}^{\times}\boldsymbol{\omega} + m_d\dot{\xi}\mathbf{n} \\
\mathbf{h} &= \mathbf{c}^{\times}\mathbf{v} + \mathbf{J}\boldsymbol{\omega} + m_d\dot{\xi}\mathbf{b}^{\times}\mathbf{n} + \mathbf{h}_s \\
p_n &= m_d\left(\mathbf{n}^T\mathbf{v} - \mathbf{n}^T\mathbf{b}^{\times}\boldsymbol{\omega} + \dot{\xi}\right) \\
h_a &= I_a\mathbf{a}^T\boldsymbol{\omega} + h_s
\end{aligned} \tag{2}$$

The meaning of the symbols is as follows: $\{m_b, m_p, m_w\}$ are respectively the masses of $\{\mathscr{R}, \mathscr{P}, \mathscr{W}\}$ and $m \triangleq m_b + m_p + m_w$. All matrices are expressed in an $\mathscr{R}$-fixed principal-axis frame $\mathscr{F}_p$ whose absolute angular velocity $\underset{\rightarrow}{\omega} = \mathscr{F}_p^T\boldsymbol{\omega}$. When the spring is relaxed, $\mathscr{P}$ is at $\mathbf{b}$ with respect to O_p, and $\xi = 0$. Damper $\mathscr{P}$ is free to move a distance ξ in the $\mathbf{n}$-direction, so that in general its position is $\mathbf{r}_d = \mathbf{b} + \xi\mathbf{n}$. The first moments of inertia of $\{\mathscr{R}, \mathscr{P}, \mathscr{W}\}$ with respect to O_p are $\{\mathbf{c}_b, \mathbf{c}_w, m_d\mathbf{r}_d\}$, and the total first inertial moment for $\mathscr{R} + \mathscr{P} + \mathscr{W}$ is

$$\mathbf{c} = \mathbf{c}_b + \mathbf{c}_w + m_d\mathbf{r}_d \tag{3}$$

Similarly, the total (second) moment of inertia about O_p is

$$\mathbf{J} = \mathbf{J}_b + \mathbf{J}_w + m_d\left(r_d^2\mathbf{1} - \mathbf{r}_d\mathbf{r}_d^T\right) \tag{4}$$

The angular momentum of $\mathscr{W}$ relative to $\mathscr{R}$ is $\mathbf{h}_s = \mathbf{a}I_s\omega_s \triangleq \mathbf{a}h_s$, where $\mathscr{W}$ has an angular speed ω_s relative to $\mathscr{R}$ and a moment of inertia I_s about its symmetry (spin) axis, which is in the $\mathbf{a}$ direction.

Assuming that the system momenta and velocities are arranged as follows,

$$\boldsymbol{\mathscr{p}}^T = \begin{bmatrix}\mathbf{p}^T & \mathbf{h}^T & p_n & h_a\end{bmatrix}; \qquad \boldsymbol{\mathscr{v}}^T \triangleq \begin{bmatrix}\mathbf{v}^T & \boldsymbol{\omega}^T & \dot{\xi} & \omega_s\end{bmatrix} \tag{5}$$

we have

$$\boldsymbol{\mathscr{p}} = \mathscr{M}\boldsymbol{\mathscr{v}}; \qquad T = \tfrac{1}{2}\boldsymbol{\mathscr{v}}^T\mathscr{M}\boldsymbol{\mathscr{v}} \tag{6}$$

where T is the total kinetic energy and $\mathscr{M}$ is the system inertia matrix defined by

$$\mathscr{M} \triangleq \begin{bmatrix}
m\mathbf{1} & -\mathbf{c}^{\times} & m_d\mathbf{n} & \mathbf{0} \\
\mathbf{c}^{\times} & \mathbf{J} & m_d\mathbf{b}^{\times}\mathbf{n} & I_s\mathbf{a} \\
m_d\mathbf{n}^T & -m_d\mathbf{n}^T\mathbf{b}^{\times} & m_d & 0 \\
\mathbf{0}^T & I_s\mathbf{a}^T & 0 & I_s
\end{bmatrix} \tag{7}$$

As usual, the 8×8 matrix $\mathscr{M}$ is symmetric and positive-definite. The change in total mechanical energy $E = T + V$ is

$$\dot{E} = \mathbf{f}^T\mathbf{v} + \mathbf{g}^T\boldsymbol{\omega} - c_d\dot{\xi}^2 + g_a\omega_s \tag{8}$$

where V is the strain energy in the spring, $\frac{1}{2}k_d\xi^2$.

To obtain results that are literal and readily susceptible to interpretation, we again resort to a linearized (infinitesimal) analysis, consisting of two steps: (i) locating the (relative) equilibria, and (ii) examining the stability of these equilibria. Although some progress can be made with more general cases, we shall restrict attention here to the simple (and most practical) case where the spin axis of $\mathscr{W}$ and the travel of $\mathscr{P}$ are parallel to each other and to a principal axis of $\mathscr{R}$ (Fig. 7.12). In symbols,

$$\mathbf{b} = b\mathbf{1}_1; \qquad \mathbf{a} = \mathbf{n} = \mathbf{1}_2 \tag{9}$$

As in Section 5.2, we choose the origin of $\mathscr{F}_p$ to coincide with the mass center of $\mathscr{R} + \mathscr{P} + \mathscr{W}$ when $\xi = 0$:

$$\mathbf{c}_b + \mathbf{c}_w + m_d\mathbf{b} = 0; \qquad \mathbf{c} = m_d\xi\mathbf{1}_2 \tag{10}$$

Note that $\mathbf{c}$ is now a first-order quantity. The elements of $\mathbf{J}$ are still as given earlier in (5.2, 3), except that now I_{ij} includes $\mathscr{W}$ as well as $\mathscr{R}$ and $\mathscr{P}$, the latter with $\xi = 0$. Note also that a solution of the motion equations (1) (with $\mathbf{f} = \mathbf{g} = \mathbf{0}$) is

$$\dot{v}_1 = -\nu v_3; \qquad \dot{v}_2 = 0; \qquad \dot{v}_3 = \nu v_1; \qquad \omega_1 \equiv 0;$$
$$\omega_2 \equiv \nu; \qquad \omega_3 \equiv 0; \qquad h_s = \text{constant} \tag{11}$$

The equations relating to $\mathbf{v}$ state that $\dot{\underset{\rightarrow}{\mathbf{v}}} = \underset{\rightarrow}{\mathbf{0}}$. In this reference solution, $g_a = 0$ also.

To ascertain the stability of this equilibrium using infinitesimal analysis, one sets

$$\boldsymbol{\omega} = \dot{\boldsymbol{\alpha}} + (\mathbf{1} - \boldsymbol{\alpha}^{\times})\boldsymbol{\nu} \tag{12}$$

and regards $\boldsymbol{\alpha}$ and ξ as first-order quantities. In accordance with the assumptions for a Kelvin gyrostat, we assume h_s remains constant, which implies [see (6.1, 5)] that the motor controlling $\mathscr{W}$ and bearing friction contribute a net torque given by

$$g_a(t) = I_s\ddot{\alpha}_2 \tag{13}$$

Substituting these perturbation relations into the momentum expressions and (2) and retaining only first-order terms in $\boldsymbol{\alpha}$ and ξ, we have

$$p_1 = mv_1; \qquad p_2 = mv_2 + m_d\dot{\xi}; \qquad p_3 = mv_3 \tag{14}$$

$$\begin{aligned} h_1 &= I_1(\dot{\alpha}_1 + \nu\alpha_3) + m_d\xi(v_3 - \nu b) \\ h_2 &= I_2\dot{\alpha}_2 + \lambda\nu \\ h_3 &= I_3(\dot{\alpha}_3 - \nu\alpha_1) - m_d(\xi v_1 - b\dot{\xi}) \end{aligned} \tag{15}$$

$$p_n = m_d\left[v_2 + b(\dot{\alpha}_3 - \nu\alpha_1) + \dot{\xi}\right] \tag{16}$$

where $\lambda = I_2 + (h_s/\nu)$. The translational equations are still (5.2, 10), the ξ equation is still (5.2, 12), and the rotational equations are still (5.2, 11), *except that* λ *replaces* I_2. The coordinates germane to directional stability are

$$\mathbf{q}^T \triangleq [\alpha_1 \quad \alpha_3 \quad \xi] \tag{17}$$

and the relevant motion equations are of the standard form

$$\mathscr{M}\ddot{\mathbf{q}} + (\mathscr{D} + \mathscr{G})\dot{\mathbf{q}} + \mathscr{K}\mathbf{q} = \mathbf{0} \tag{18}$$

The system matrices $\{\mathcal{M}, \mathcal{D}, \mathcal{G}, \mathcal{K}\}$ are displayed in (5.2, 14), with the understanding that λ is inserted wherever I_2 occurs.

Conditions for Stability

The stability of the reference solution $\mathbf{q} \equiv \mathbf{0}$ is guaranteed if $\mathcal{K} > 0$. Referring to the earlier definitions, (5.2, 17 and 18) and (5.4, 26 and 27), $\mathcal{K} > 0$ if and only if

$$k_{1hd} \triangleq \frac{\lambda - I_3}{I_1} = k_1 + \hat{\Omega}_{po}\left(\frac{1 - k_1}{1 - k_3}\right)^{1/2} > 0 \tag{19}$$

$$k_{3hd} \triangleq \frac{\lambda - I_1}{I_3} - \Xi_d = k_3 + \hat{\Omega}_{po}\left(\frac{1 - k_3}{1 - k_1}\right)^{1/2} - \Xi_d > 0 \tag{20}$$

where $k_1 = (I_2 - I_3)/I_1$, $k_3 = (I_2 - I_1)/I_3$, and $\hat{\Omega}_{po}$ and Ξ_d are available from (6.4, 27) and (5.2, 17). These conditions are similar to the condition $\lambda > \max\{I_1, I_3\}$ obtained from energy sink analysis in (7.1, 36) *et seq.*; the only modification is the appearance of the damper-related term Ξ_d in (20). Indeed, as $\Xi_d \to 0$, that is, as the damper mass becomes vanishingly small, or as the damper spring stiffness becomes very large, where one might anticipate that the assumptions of energy sink analysis are especially appropriate, the above conditions for $\mathcal{K} > 0$ become identical to the energy sink conditions (37). Pictorially, the region of stability $\lambda > \max\{I_1, I_3\}$ is represented by the "upper right" regions shown in Fig. 7.4 for four values of $\hat{\Omega}_{po}$. The effect of a nonzero Ξ_d is apparent in Fig. 7.13. It is clear that the upper right regions have shrunk in extent owing to the flexibility term Ξ_d. Thus, the major-axis rule for gyrostats is not sufficient when significant positions of the spacecraft are flexible. (The reader may also wish to review the analogous conclusion reached earlier, before the wheel was added: the energy sink stability diagram for $\mathcal{Q}$, Fig. 5.3, becomes Fig. 5.5 for $\mathcal{R} + \mathcal{P}$, although the latter does approach the former as $\Xi_d \to 0$.)

The next question is: Is $\mathcal{K} > 0$ *necessary* for stability? The answer lies in the characteristic equation, which must be of the same form as (5.2, 20 through 22), with λ replacing I_2. That is, the characteristic equation has two factors: one is $s^2 + \nu^2$ and the other is a quartic with coefficients as follows:

$$\begin{aligned}
a_0 &= \hat{m}_b - \hat{I}_d \\
a_1 &= \hat{c}_d \\
a_2 &= \hat{m}_b k_{1h} k_{3h} + \hat{\omega}_d^2 - (1 + k_{1h})\hat{I}_d \\
a_3 &= \hat{c}_d k_{1h} k_{3h} \\
a_4 &= k_{1h}\left(k_{3h}\hat{\omega}_d^2 - \hat{I}_d\right)
\end{aligned} \tag{21}$$

and $\{k_{1h}, k_{3h}\}$ are defined in (6.4, 26). The relations governing the stability boundaries are written by making a similar notational adjustment to (5.2, 24):

$$k_{1h}(k_{3h} - \Xi_d) > 0; \qquad k_{1h}(1 - k_{3h})(1 - k_{1h}k_{3h}) > 0 \tag{22}$$

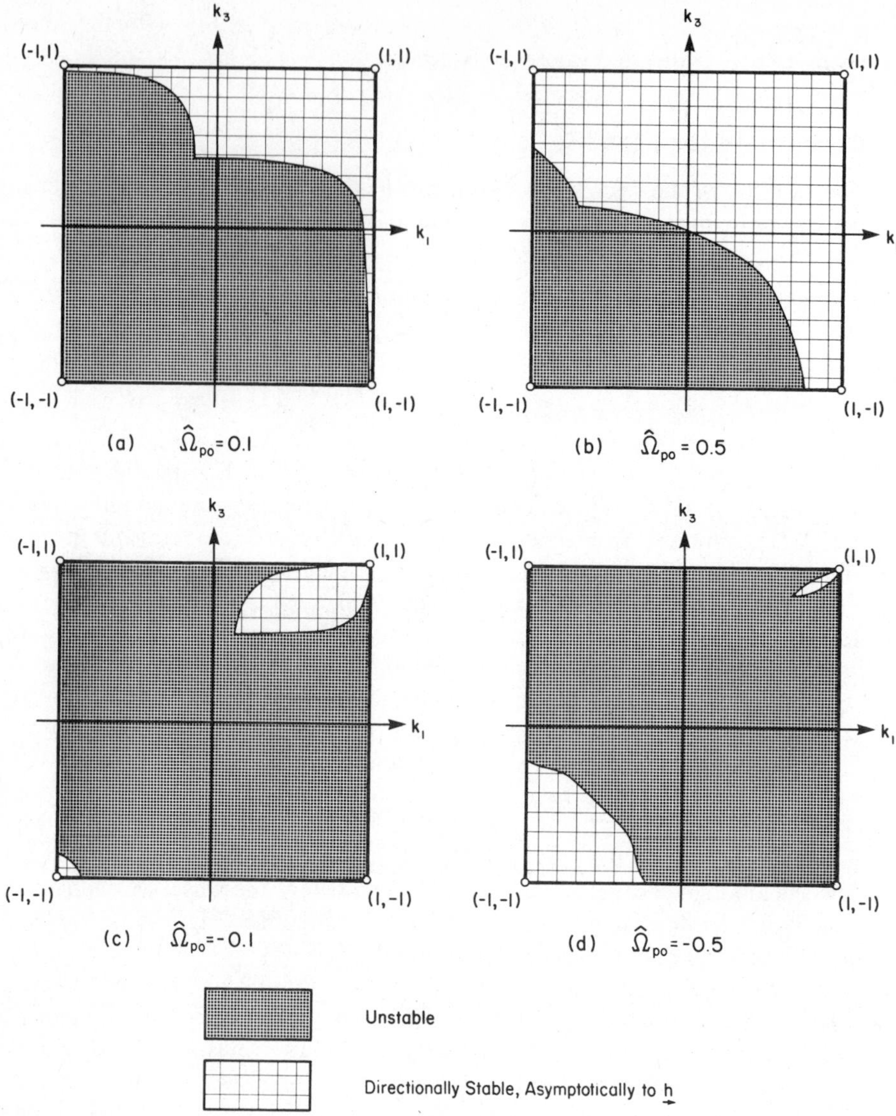

FIGURE 7.13 Stability diagrams for system of Fig. 7.12, with $\Xi_d = 0.5$, $\hat{c}_d = 0.1$, $\hat{\omega}_d = 0.5$, and $\hat{m}_b = 0.9$. Compare with Fig. 7.4.

In addition to the upper right stability regions in Fig. 7.13, which correspond as we have seen to $\{k_{1h} > 0;\ k_{3h} > \Xi_d\}$, there is another possibility when $\mathscr{W}$ spins sufficiently rapidly in the opposite direction to $\mathscr{R}$. This possibility is associated with the lower left region in Fig. 7.13 for $\hat{\Omega}_{po} < 0$. Comparing these regions with those predicted by energy sink analysis in Fig. 7.4, we see that they are essentially identical. Evidently, the damper-related term Ξ_d has no effect on the extent of this region. For a vanishingly small damper mass, $\Xi_d \to 0$, the stability conditions (22) agree with those obtained by [Likins, 2]. The special case of $\mathscr{R}$ being symmetrical is studied in more detail in Problem 7.1.

Symmetric Gyrostat with Two Discrete Dampers

To address the important issue of the relative degrees of damping in a gyrostat's two constituent bodies, we now consider the gyrostat shown in Fig. 7.14. This model, first analyzed by [Mingori], consists of two wheels in each of which is placed a mass-spring-dashpot damper. The symbols used below will be obvious extensions of those used above for the $\mathscr{R} + \mathscr{P} + \mathscr{W}$ system. Subscripts 1 and 2 refer respectively to $\mathscr{W}_1 + \mathscr{P}_1$ and $\mathscr{W}_2 + \mathscr{P}_2$. [Mingori] assumes *a priori* that the spin rate of $\mathscr{W}_2$ relative to $\mathscr{W}_1$ is constant, ω_s = constant. This assumption, which is similar to Kelvin's for the (undamped) Kelvin gyrostat, removes one degree of freedom from explicit consideration (the angle of $\mathscr{W}_2$ relative to $\mathscr{W}_1$), but produces periodic coefficients in the motion equations associated with the remaining five degrees of freedom (three angles of $\mathscr{W}_1$ and the damper motions ξ_1 and ξ_2). Mingori's equations of motion, in present

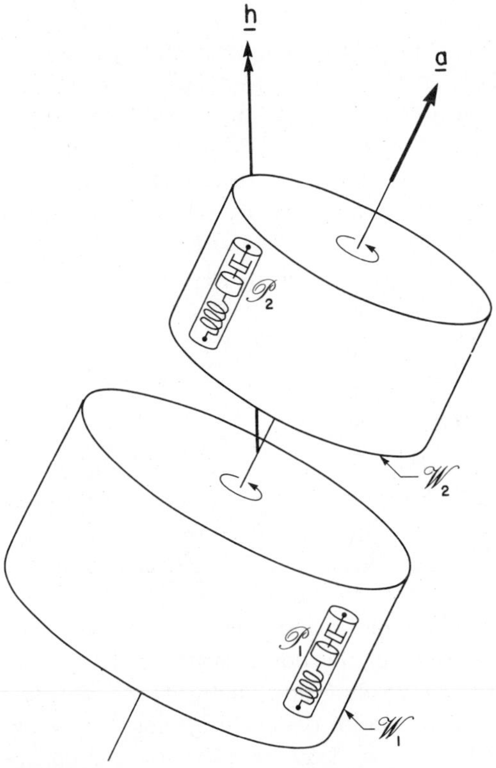

FIGURE 7.14 Mingori's gyrostat.

notation, are as follows, after linearization:

$$
\begin{aligned}
I_t\dot{\omega}_1 - (\lambda - I_t)\nu\omega_3 + f_{d1} + f_{d2}\cos\omega_s t &= 0\\
I_t\dot{\omega}_3 + (\lambda - I_t)\nu\omega_1 - f_{d2}\sin\omega_s t &= 0\\
m_{11}\ddot{\xi}_1 + m_{12}\ddot{\xi}_2 + c_{d1}\dot{\xi}_1 + k_{d1}\xi_1 + m_d b_1(\dot{\omega}_1 - \nu\omega_3) &= 0\\
m_{21}\ddot{\xi}_1 + m_{22}\ddot{\xi}_2 + c_{d2}\dot{\xi}_2 + k_{d2}\xi_2 - m_{d2}b_2\sin\omega_s t\left[\dot{\omega}_3 + (\nu + 2\omega_s)\omega_1\right]&\\
+ m_{d2}b_2\cos\omega_s t\left[\dot{\omega}_1 - (\nu + 2\omega_2)\omega_3\right] &= 0
\end{aligned}
\tag{23}
$$

where

$$
\begin{aligned}
f_{d1} &= m_{d1}b_1\left(\ddot{\xi}_1 + \nu^2\xi_1\right)\\
f_{d2} &= m_{d2}b_2\left[\ddot{\xi}_2 + (\nu + \omega_s)^2\xi_2\right]\\
\lambda &= I_a + \frac{h_s}{\nu}; \qquad h_s = I_s\omega_s\\
m_{11} &= m_{d1}\frac{m - m_{d1}}{m}; \qquad m_{22} = m_{d2}\frac{m - m_{d2}}{m}\\
m_{12} &= m_{21} = -\frac{m_{d1}m_{d2}}{m}
\end{aligned}
\tag{24}
$$

The time-varying coefficients preclude closed-form solution. Because they are periodic however, Floquet theory (Theorems A.9, A.10 and A.11 in Appendix A) can be used to generate numerical results. With such a large number of system parameters, the detailed stability results available in [Mingori] will not be repeated here, but they provide evidence that precise dynamical models are needed if precise stability boundaries are desired. However, they also show that energy sink analysis does give helpful guidelines and becomes increasingly reliable as the size of the damper(s) becomes vanishingly small. In any case, there is no doubt that suitably located dampers of appropriate dissipative capacity can indeed stabilize a gyrostat whose spin axis is the minor axis of inertia.

Related Dual-Spin Investigations

Mingori's configuration (Fig. 7.14) has been studied further by [Flatley, 1], who avoids periodic coefficients by taking the damper masses to be vanishingly small. This approach appears to be intermediate in accuracy between energy sink analysis and the "exact" equations of Mingori. It is an improvement on energy sink analysis in that an explicit nutation history $\gamma(t)$ can be calculated, although instabilities dependent on the size of the damper mass cannot be detected. The chief objective of Flatley's work was to retain the nonlinearity of the motion equations so as to discover the history of the nutation angle $\gamma(t)$ in unstable cases. Two general types of instability were found: complete inversions, $\gamma \to \pi$; and a steady-state nutation angle neither 0 nor π. This same style of analysis has also been shown by [Likins, Tseng, and Mingori] and [Mingori, Tseng, and

Likins] to reveal the possibility of limit cycle behavior when either the damper's spring stiffness or the damping law is nonlinear. Damper tuning for Mingori's configuration has been investigated by [Sarychev and Sazonov].

As pointed out above, the apparent obstacle to closed-form stability criteria presented by periodic coefficients in the motion equations for two or more coaxially spinning bodies containing dampers can be overcome if the damper masses are very small. The rotational and damper motion equations can be approximately decoupled. [Vigneron, 1, 2] has pointed out that the *method of averaging*, as expounded by [Volosov], is ideally suited to deal with this problem. The method of averaging, which can in principle be applied to any degree of accuracy desired by forming higher-order approximations, is recommended whenever the system motion can be viewed as the superposition of a "fast" motion and a "slow" motion. Differential equations are obtained for the slow motion in which the fast motion has been averaged. For dual-spin spacecraft having dampers of relatively small mass, the fast motions include the basic dissipation-free precessional motion of the vehicle and the oscillation of the dampers, which may be presumed tuned to the relative spin rate. The slow motion is associated with the evolution of this precessional motion, wherein the nutation angle slowly changes as the dampers gradually make their influence felt. Indeed, [Vigneron, 2] suggests that the energy sink philosophy can also be regarded as first-order averaging and shows that with vanishingly small damper mass the energy sink conclusions (which are also the Routh–Hurwitz conclusions) for an axisymmetrical gyrostat with damping in one of the two wheels agree with the criterion derived from first-order averaging. The method of averaging also provides an analytical alternative when the simultaneous fast and slow motions cause numerical difficulties in the application of Floquet theory.

The one-way decoupling for small damper masses has also been used by [Puri and Gido] to find literal stability criteria for axisymmetrical dual spinners with pendulum dampers. Their methodology is similar in style to that of [Flatley, 1], described above. [Alfriend and Hubert] have used the same technique to assess the adverse influence on stability of a rotor-mounted partially filled viscous ring damper. If such damping were the unavoidable consequence of using a partially filled ring as a heat pipe, the objective would be to ascertain the level of damping needed in the platform to offset the destabilizing influence of the heat pipe. Alfriend and Hubert compare their analytical results with numerical integration of the exact equations of motion, and the "small-damper decoupling" technique is clearly justified. They also note the possibility of limit cycles, which is a point of comparison with the work of [Flatley, 1] and [Likins, Tseng, and Mingori].

7.3 PROBLEMS

7.1 Consider the rigid body $\mathscr{R}$ shown in Fig. 7.12 containing a wheel $\mathscr{W}$ and point mass damper $\mathscr{P}$. As shown, the axis of the wheel and the line of axis of the damper are aligned with the principal axis $\vec{\hat{\mathbf{p}}}_2$ of the system. The stability conditions for this system were given by equations (7.2, 22) in the text. This problem examines the special case where $\mathscr{R} + \mathscr{W} + \mathscr{P}$ has $\hat{\mathbf{p}}_2$ as an axis of inertial symmetry.

(*a*) Show from first principles, that is, by using the motion equations (7.2, 18) but not the stability conditions (7.2, 22), that the conditions for asymptotic stability are

$$\text{(i)} \quad k_{th}(k_{th} - \Xi_d) > 0$$
$$\text{(ii)} \quad k_{th}(1 - k_{th})(1 - k_{th}^2) > 0 \qquad (1)$$

where

$$k_{th} = k_t + \hat{\Omega}_{po}; \qquad k_t = k_1 = k_3 \qquad (2)$$

(*b*) Deduce that the stability diagram in the k_{th}–Ξ_d plane is as shown in Fig. 7.15*a*. Note that although k_t is limited physically to $|k_t| \le 1$, no such limit applies to k_{th}.

(*c*) Show that the asymptotic stability of $\mathscr{R} + \mathscr{P} + \mathscr{W}$ in the k_t–$\hat{\Omega}_{po}$ plane is as depicted in Figs. 7.15*b* through *e* for $\hat{\Omega}_{po} = 0.1, 0.5, -0.1, -0.5$.

(*d*) Compare these results with those in Fig. 7.13 for $\Xi_d = 0.5$.

7.2 Consider a quasi-rigid body $\mathscr{Q}$ within which is a rotor $\mathscr{W}$. The $\mathscr{Q} + \mathscr{W}$ system has principal inertias $\{I_1, I_2, I_3\} = \{6, 4, 2\}$ kg · m^2.

(*a*) Assume that the system angular momentum is $h = 20$ N · m · s and that the relative angular momentum of $\mathscr{W}$ is $h_s = 20/\sqrt{22}$ N · m · s. The critical values of λ, denoted $\{\lambda_1^-, \lambda_1^+, \lambda_2^-, \lambda_2^+, \lambda_3^-, \lambda_3^+\}$, are found by solving $\phi_h(\lambda) = 0$, where $\phi_h(\lambda)$ is given by (6.4, 8). Assume $a_1 = a_2 = a_3 = 1/\sqrt{3}$. The critical spin rates for $\mathscr{Q}$ are then found from (6.4, 1). For example,

$$\boldsymbol{\nu}_1^- = \frac{20}{\sqrt{66}} \begin{bmatrix} (\lambda_1^- - 6)^{-1} \\ (\lambda_1^- - 4)^{-1} \\ (\lambda_1^- - 2)^{-1} \end{bmatrix}$$

and similarly for the other five critical (equilibrium) spin rates. (These equilibria are shown in Figs. 6.5*a* and 6.6*a*.) Calculate the energy $T_o = \frac{1}{2}\boldsymbol{\nu}^T \boldsymbol{I} \boldsymbol{\nu}$ for each of these six spin equilibria and compare their magnitudes. Plot these values on a graph similar to Fig. 6.2. What do you conclude about the stability of these equilibria? Compare with Figs. 7.1*a* and *b*.

(*b*) Repeat the steps in part (*a*) for $h_s = 10/\sqrt{3}$, 10, and $20\sqrt{2}$ N · m · s.

7.3 Consider the system $\mathscr{Q}_{w1} + \mathscr{W}_2$ discussed in Section 7.1, that is, the system of two axisymmetrical rigid bodies shown in Fig. 7.8. An energy sink is present in the quasi-rigid body $\mathscr{Q}_{w1}$. The transverse moment of inertia of the system about the system mass center is I_t, and the individual spin rates are ν_1 and ν_2. The magnitude of the angular momentum of the system is h.

(*a*) Show that the condition for directional stability of the spin axis, asymptotic to $\underset{\rightarrow}{\mathbf{h}}$, is

$$h \cos\gamma > I_t \nu_1 \qquad (3)$$

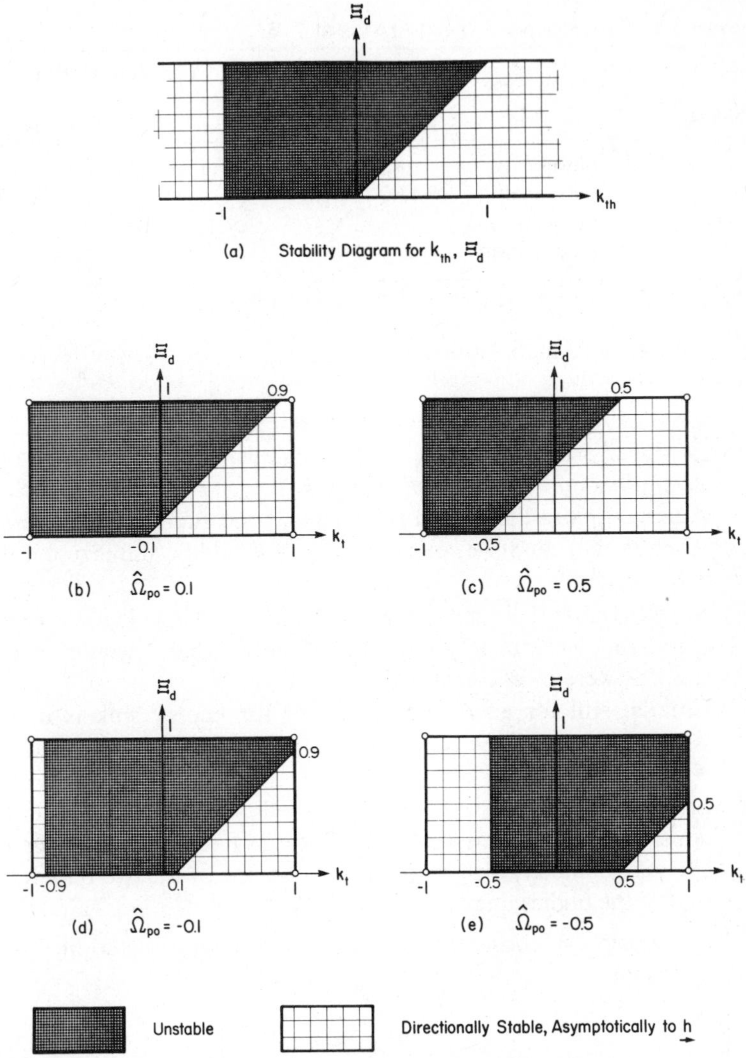

FIGURE 7.15 Stability diagram for a symmetrical spinning rigid body containing a spinning wheel and a point mass damper. (See Problem 7.1.)

Note that this condition does not assume the nutation angle γ to be small.

(*b*) If the energy sink were in $\mathscr{W}_2$ instead of $\mathscr{Q}_{w1}$ (i.e., if $\mathscr{Q}_{w1} + \mathscr{W}_2 \to \mathscr{W}_1 + \mathscr{Q}_{w2}$), explain why the appropriate stability condition would be

$$h \cos \gamma > I_t \nu_1 \tag{4}$$

(*c*) Suppose the system is *prolate* ($I_a < I_t$). Show that if an energy sink is stabilizing when in $\mathscr{W}_1$ ($\mathscr{W}_1 \to \mathscr{Q}_{w1}$), then (for the same system and

TABLE 7.1
Summary of Conclusions from Problem 7.3

If, for the System	The Energy Sink Is	Then, for the System*	The Energy Sink is $I_a < I_t$	The Energy Sink is $I_a > I_t$
$\mathcal{Q}_{w1} + \mathcal{W}_2$	stabilizing	$\mathcal{W}_1 + \mathcal{Q}_{w2}$	destabilizing	NMI†
$\mathcal{Q}_{w1} + \mathcal{W}_2$	destabilizing	$\mathcal{W}_1 + \mathcal{Q}_{w2}$	NMI†	stabilizing
$\mathcal{W}_1 + \mathcal{Q}_{w2}$	stabilizing	$\mathcal{Q}_{w1} + \mathcal{W}_2$	destabilizing	NMI†
$\mathcal{W}_1 + \mathcal{Q}_{w2}$	destabilizing	$\mathcal{Q}_{w1} + \mathcal{W}_2$	NMI†	stabilizing

*Same wheel configuration, same values of ν_1 and ν_2; only the location of the energy sink changes.
†NMI = "Need More Information" to form conclusion.

identical spin rates) it would be destabilizing if it were in $\mathcal{W}_2$ instead ($\mathcal{W}_2 \to \mathcal{Q}_{w2}$). If the energy sink is stabilizing when in $\mathcal{W}_2$, would it be destabilizing if it were in $\mathcal{W}_1$ instead?

(*d*) Again suppose the system is prolate. If the energy sink is destabilizing when in $\mathcal{W}_1$, can it be concluded that it would be stabilizing if it were in $\mathcal{W}_2$ instead?

(*e*) Suppose now that the system is *oblate* ($I_a > I_t$). If an energy sink is stabilizing when in $\mathcal{W}_1$, can it be concluded that it would be destabilizing if it were in $\mathcal{W}_2$ instead?

(*f*) Finally, still for an oblate system, if the energy sink is destabilizing when in $\mathcal{W}_1$, can it be concluded that it would be stabilizing if it were in $\mathcal{W}_2$ instead?

(*g*) Show from the above results that the following table applies to gyrostats consisting of two axisymmetric bodies spinning about their common axis of symmetry, with an energy sink in one of the two bodies (i.e., with one of the bodies quasi-rigid).

7.4 As in Problem 7.3, consider the $\mathcal{Q}_{w1} + \mathcal{W}_2$ system depicted in Fig. 4.39.

(*a*) Show that

$$g_a \triangleq g_{a1} + g_{a2} = -h_t\dot{\gamma} \tag{5}$$
$$g_t \triangleq g_{t1} + g_{t2} = h_a\dot{\gamma} \tag{6}$$
$$g_b \triangleq g_{b1} + g_{b2} = 0 \tag{7}$$

thus giving the components in the precessing frame $\mathcal{F}_\Omega$ of the "external" torque equivalent to the energy sink in $\mathcal{Q}_{w1}$.

(*b*) Show that the kinetic energy of $\mathcal{Q}_{w1}$ changes at the rate

$$\dot{T}_1 = g_{a1}\nu_1 + g_{t1}\omega_t \tag{8}$$

and interpret this equation as a power balance.

(*c*) In the limiting case as $\mathcal{W}_2$ becomes vanishingly small (the system $\mathcal{Q}_{w1} + \mathcal{W}_2$ then reverts to a single quasi-rigid body $\mathcal{Q}$), show that the components in the precessing frame $\mathcal{F}_\Omega$ of the "external" torque equivalent to the action of the energy sink reduce to those given by (7.1, 70) and (7.1, 71) for $\mathcal{Q}$.

7.5 Consider the system of N inertially axisymmetric quasi-rigid wheels mounted on a common axis, $\Sigma\mathscr{Q}_{wn}$. This system is shown in Fig. 7.10 and analyzed in (7.1, 90) *et seq.*

(a) Given the following motion equations in the precessing frame $\mathscr{F}_{\Omega}$,

$$I_{an}\dot{\nu}_n = g_{an}; \qquad J_{tn}\dot{\omega}_t = g_{tn} \tag{9}$$

Show that the equivalent "external" torque acting on $\mathscr{Q}_{wn}$ has the following components in $\mathscr{F}_{\Omega}$:

$$g_{an} = \frac{\dot{T}_{sn}}{\Omega_n} \tag{10}$$

$$g_{tn} = \frac{\Omega_p J_{tn}\sigma\cos\gamma}{I_t\omega_t} \tag{11}$$

$$g_{bn} = \Omega_p(J_{tn}\omega_t\cos\gamma - I_{an}\nu_n\sin\gamma) \tag{12}$$

where σ is given by (7.1, 110), and $\dot{T}_{sn}$ is the rate of energy extraction by the energy sink in $\mathscr{Q}_{wn}$.

(b) How much of g_{tn}, as given by (11), is transmitted by the bearings in $\mathscr{Q}_{wn}$?

(c) Show that the total "external" torque on $\Sigma\mathscr{Q}_{wn}$ equivalent to the action of all the energy sinks is, in $\mathscr{F}_{\Omega}$,

$$g_a \triangleq \sum_{n=1}^{N} g_{an} = \sigma \tag{13}$$

$$g_t = \sum_{n=1}^{N} g_{tn} = -\sigma\cot\gamma \tag{14}$$

$$g_b = \sum_{n=1}^{N} g_{bn} = 0 \tag{15}$$

(c) Explain why all the following are equivalent conditions for the directional stability of $\Sigma\mathscr{Q}_{wn}$, asymptotic to $\underrightarrow{\mathbf{h}}$: (i) $\sigma > 0$; (ii) $g_a > 0$; (iii) $g_t < 0$.

7.6 This problem is concerned with the torque-free motion of a nominally nonspinning rigid body $\mathscr{R}$ which contains a spinning wheel $\mathscr{W}$ and a mass spring dashpot damper $\mathscr{P}$. The axis of the wheel is aligned with $\underrightarrow{\hat{\mathbf{p}}}_2$, one of the system's principal axes of inertia. The axis of the damper is also aligned with $\underrightarrow{\hat{\mathbf{p}}}_2$, and the damper moves in the $\underrightarrow{\hat{\mathbf{p}}}_1$–$\underrightarrow{\hat{\mathbf{p}}}_2$ plane at a distance b from the $\underrightarrow{\hat{\mathbf{p}}}_2$-axis. In short, the system is as shown in Fig. 7.12 and as analyzed in equation (7.2, 1) *et seq.*, but with $\underrightarrow{\mathbf{v}} = \underrightarrow{\mathbf{0}}$.

(a) Let $\boldsymbol{\theta}$ represent the three infinitesimal rotational displacements of $\mathscr{R}$ with respect to an inertial frame $\mathscr{F}_i$. Show from first principles that the equations of rotational motion for this system are

$$I_1\ddot{\theta}_1 - h_s\dot{\theta}_3 = 0$$

$$I_2\ddot{\theta}_2 = 0$$

$$I_3\ddot{\theta}_3 + h_s\dot{\theta}_1 + m_d b\ddot{\xi} = 0$$

$$m_d b\ddot{\theta}_3 + \left(\frac{m_b m_d}{m}\right)\ddot{\xi} + c_d\dot{\xi} + k_d\xi = 0 \tag{16}$$

(The symbols are as defined in the text.)

(*b*) Explain why the relative angles $\boldsymbol{\alpha}$ in the text are now identical with the absolute angles $\boldsymbol{\theta}$, and show that the motion equations for $\nu \neq 0$ reduce to the above equations when $\nu = 0$.

(*c*) Note that three-axis stability is not possible owing to the evident instability with respect to perturbations in $\dot{\theta}_2$. Thus, we examine directional stability using the variables $\mathbf{q} = [\theta_1 \quad \theta_3 \quad \xi]^T$. Show that the characteristic equation associated with these variables is

$$s^2\left[\left(\hat{m}_b s^2 + 2\zeta_d\omega_d s + \omega_d^2\right)\left(s^2 + \Omega_{po}^2\right) - \hat{I}_d s^4\right] = 0 \qquad (17)$$

where $\hat{m}_b \triangleq m_b/m$, $\omega_d^2 = k_d/m_d$, $2\zeta_d\omega_d \triangleq c_d/m_d$, and $\Omega_{po}^2 \triangleq h_s^2/I_1 I_3$.

(*d*) Show that the two eigencolumns associated with the roots $s^2 = 0$ in (17) are

$$\mathbf{q} = [1 \quad 0 \quad 0]^T; \qquad [0 \quad 1 \quad 0]^T$$

and give the physical interpretation of this fact.

(*e*) Show that the remaining four roots of (17) are always in the left half-plane, implying that the $\mathscr{R} + \mathscr{P} + \mathscr{W}$ system, with $\mathscr{R}$ not spinning, is always directionally stable, asymptotically to $\underset{\rightarrow}{\mathbf{h}}$ ($\equiv \underset{\rightarrow}{\mathbf{h}}_s$).

(*f*) Show that the comparable stability conditions when $\mathscr{R}$ is spinning, namely, (7.2, 22), imply the same conclusion.

(*g*) Show that Landon's rule, as expressed by (7.1, 61), is consistent with the above conclusions.

7.7 Consider the following two torque-free spinning systems:

(i) $\mathscr{Q} + \mathscr{W}$, a spinning quasi-rigid body $\mathscr{Q}$ containing a wheel $\mathscr{W}$; see Fig. 7.3.

(ii) $\mathscr{R} + \mathscr{P} + \mathscr{W}$, a spinning rigid body $\mathscr{R}$ containing a point mass damper $\mathscr{P}$ and a wheel $\mathscr{W}$; see Fig. 7.12.

The conditions for asymptotic directional stability of the spin axis of the system $\mathscr{Q} + \mathscr{W}$ were derived in the text to be (7.1, 37 and 38), using energy sink analysis. These conditions are repeated here for clarity: either

$$k_{1h} > 0 \text{ and } k_{3h} > 0 \qquad (18)$$

or

$$(1 - k_1 k_3) + \hat{\Omega}_{po}[(1 - k_1)(1 - k_3)]^{1/2} < 0 \qquad (19)$$

where

$$k_{1h} \triangleq k_1 + \hat{\Omega}_{po}\left(\frac{1 - k_1}{1 - k_3}\right)^{1/2}$$

$$k_{3h} \triangleq k_3 + \hat{\Omega}_{po}\left(\frac{1 - k_3}{1 - k_1}\right)^{1/2}$$

$$\hat{\Omega}_{po} \triangleq \frac{\Omega_{po}}{\nu}; \qquad \Omega_{po}^2 \triangleq \frac{h_s^2}{I_1 I_3}$$

and k_1, k_3, and ν have their usual meanings. On the other hand, the conditions for asymptotic directional stability of the spin axis of the system $\mathscr{R} + \mathscr{P} + \mathscr{W}$ were derived in the text to be (7.2, 22):

$$k_{1h}(k_{3h} - \Xi_d) > 0 \qquad \text{and} \qquad k_{1h}(1 - k_{3h})(1 - k_{1h}k_{3h}) > 0 \quad (20)$$

where Ξ_d is a dimensionless parameter signifying the dynamical influence of the damper $\mathscr{P}$ on $\mathscr{R} + \mathscr{W}$. Show that as the damper becomes less and less dynamically significant ($\Xi_d \to 0$), the exact stability conditions, (20), approach the energy sink stability conditions (18) or (19).

CHAPTER 8

SPACECRAFT TORQUES

Chapter 2 dealt with a mathematical description of the rotational motion of one reference frame with respect to another. By imbedding the first reference frame in a rigid body $\mathscr{R}$, the formulation is immediately applicable to describing the orientation of $\mathscr{R}$. In Chapter 3, the cause-and-effect relationship between torque and motion was examined for a series of assumed models that were both pedagogically instructive and directly applicable to certain types of spacecraft. It then was established in Chapters 4 through 7 that, even with external influences absent, the motion of each model is nontrivial, and that these torque-free characteristics—especially stability— are essential to a proper understanding of the dynamics in each case. A firm basis now exists for a discussion of the attitude response of spacecraft to the torques they typically encounter, provided we can model these torques also. It is the purpose of this chapter to survey spacecraft torques. Among the particulars that make spacecraft attitude dynamics distinctive in the field of applied mechanics are the torque origins and characteristics.

The most striking characteristic of spacecraft torques is their minuteness. In terms of familiar terrestrial experience, they are intuitively negligible. On closer examination however, there are no "large" torques in space, and hence minor influences play major roles in governing the attitude dynamics of spacecraft. Consider, for example, the pressure exerted on a material surface by sunlight. Depending on the surface properties, this pressure in the vicinity of Earth is about 10^{-5} N/m^2 and produces on a typical solid body a force at least nine orders of magnitude smaller than the gravitational force. Our terrestrial intuition would lead us to neglect an effect of this size. Yet, as we shall see, solar pressure is frequently the dominant external influence on spacecraft attitude.

It is helpful to classify spacecraft torques as being either *external* or *internal*. External torques arise through the interaction of a vehicle with its environment;

this category is based on the notion of a spacecraft and its contents as an identifiable *system* influenced by external agencies, such as a planetary atmosphere or magnetic field. Calculation of external torques requires a specification both of vehicle properties and of the space environment within which the vehicle is situated. Internal torques, on the other hand, are in a sense self-generated. They would persist even if the spacecraft could (conceptually) be moved to a point in space entirely removed from all external influences. Fuel sloshing, control jets, and crew motions in manned vehicles are typical sources of internal torques. As with any classification scheme, whenever the external/internal distinction leads more to obfuscation than to clarity, it should be abandoned. In Sections 8.1 through 8.4, the principal external torques are examined; internal torques are discussed in Section 8.5.

8.1 GRAVITATIONAL TORQUE

Gravitational torques are fundamental to the attitude dynamics of spacecraft. If the gravitational field were *uniform* over a material body, then the center of *mass* would become the center of *gravity*, and the gravitational torque about the mass center would be zero. In space, the gravitational field is not uniform, and the consequent variations in the specific gravitational force (in both magnitude and direction) over a material body leads, in general, to a gravitational torque about the body mass center. According to [NASA, 9], this effect was first considered in a celestial mechanics context by d'Alembert and Euler (1749), and Lagrange used it in 1780 to explain why the moon always has the same face toward Earth. At the beginning of the space age, [Roberson, 1] pointed out the importance of gravitational torques on orbiting man-made satellites, and a seminal analysis was made by [Roberson and Tatistcheff].

General Considerations

Let us begin with a body $\mathscr{B}$, not necessarily rigid, immersed in a gravitational field, as shown in Fig. 8.1, where $\mathscr{B}$ represents the spacecraft. The field is due to other bodies (only the nth one is shown), and $\underset{\rightarrow}{\boldsymbol{\rho}}_n$ is the position of dm (at $\underset{\rightarrow}{\mathbf{r}}$ in $\mathscr{B}$) with respect to dm_n in $\mathscr{B}_n$. The bodies $\mathscr{B}_n$ represent the celestial primaries. The gravitational force on dm is given by

$$d\underset{\rightarrow}{\mathbf{f}} = -G\,dm \sum_{n=1}^{N} \int_{\mathscr{B}_n} \frac{\underset{\rightarrow}{\boldsymbol{\rho}}_n\, dm_n}{\rho_n^3} \tag{1}$$

where ρ_n is the magnitude of $\underset{\rightarrow}{\boldsymbol{\rho}}_n$ and G is the universal gravitational constant:

$$G \doteq 6.67 \times 10^{-11}\ \mathrm{N \cdot m^2/kg^2} \tag{2}$$

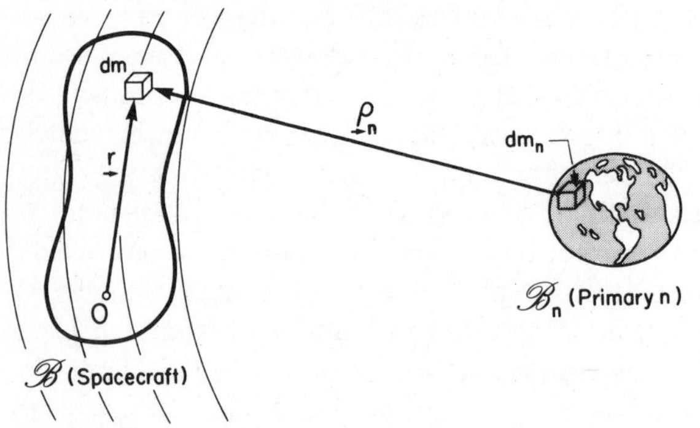

FIGURE 8.1 Spacecraft in the gravitational field of several primaries.

Therefore, the total gravitational force and torque (about O) on $\mathscr{B}$ are

$$\underset{\rightarrow}{\mathbf{f}} = \int_{\mathscr{B}} d\underset{\rightarrow}{\mathbf{f}} = -G \sum_{n=1}^{N} \int_{\mathscr{B}_n} \int_{\mathscr{B}} \frac{\underset{\rightarrow}{\boldsymbol{\rho}}_n \, dm_n \, dm}{\rho_n^3} \tag{3}$$

$$\underset{\rightarrow}{\mathbf{g}}_o = \int_{\mathscr{B}} \underset{\rightarrow}{\mathbf{r}} \times d\underset{\rightarrow}{\mathbf{f}} = -G \sum_{n=1}^{N} \int_{\mathscr{B}_n} \int_{\mathscr{B}} \frac{\underset{\rightarrow}{\mathbf{r}} \times \underset{\rightarrow}{\boldsymbol{\rho}}_n \, dm_n \, dm}{\rho_n^3} \tag{4}$$

(Self-gravity is neglected.) Alternatively, the gravitational potential energy of $\mathscr{B}$ may be employed,

$$V = \int_{\mathscr{B}} dV = -G \sum_{n=1}^{N} \int_{\mathscr{B}_n} \int_{\mathscr{B}} \frac{dm_n \, dm}{\rho_n} \tag{5}$$

and the forces and torques derived therefrom. While (3), (4), and (5) have the virtue of generality, their multiple integrations make analytical progress virtually impossible without further assumptions. Fortunately, the most important analytical result for spacecraft attitude dynamics is also the simplest, although it rests on four additional assumptions (described below). We shall now derive this result and then subsequently indicate the generalizations that result from the removal of each of these assumptions.

The Fundamental Result

The following four assumptions greatly simplify the gravitational-torque expressions; they are also excellent assumptions for most spacecraft situations. Two refer to the gravitational source, and two to the spacecraft itself:

(a) Only one celestial primary need be considered.

(b) This primary possesses a spherically symmetrical mass distribution.

(c) The spacecraft is small compared to its distance from the mass center of the primary.

(d) The spacecraft consists of a single body.

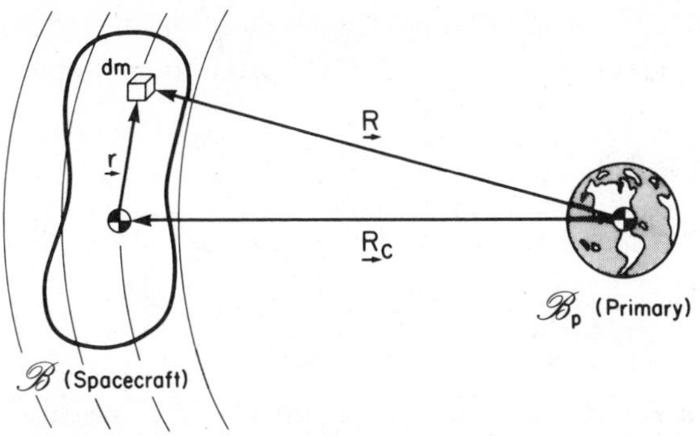

FIGURE 8.2 Spacecraft in the gravitational field of one inertially spherical primary.

These assumptions permit simple torque expressions to be derived. In the sequel, the effect of removing each assumption is discussed, thereby facilitating a quantitative assessment of its validity.

Assumption (**a**) implies that the sums in (3), (4), and (5) can each be replaced by a single term. Assumption (**b**) enables us to replace the integration over $\mathscr{B}_n$ by an equivalent point mass (as first noted by Newton) at the mass center of $\mathscr{B}_n$. Assumption (**c**) can be translated as $r/R_c \ll 1$, where r is a typical spacecraft dimension and R_c is the distance between the spacecraft and primary mass centers. Finally, assumption (**d**) allows us to choose, with no loss in generality, the spacecraft mass center as our reference point in $\mathscr{B}$ (Fig. 8.2). According to assumptions (**a**), (**b**), and (**d**), the earlier expressions, (3), (4), and (5), become

$$\underset{\rightarrow}{\mathbf{f}} = -\mu \int_{\mathscr{B}} \frac{\underset{\rightarrow}{\mathbf{R}}\, dm}{R^3} \tag{6}$$

$$\underset{\rightarrow}{\mathbf{g}}_c = -\mu \int_{\mathscr{B}} \frac{\underset{\rightarrow}{\mathbf{r}} \times \underset{\rightarrow}{\mathbf{R}}}{R^3}\, dm \tag{7}$$

$$V = -\mu \int_{\mathscr{B}} \frac{dm}{R} \tag{8}$$

and

$$\underset{\rightarrow}{\mathbf{R}} = \underset{\rightarrow}{\mathbf{R}}_c + \underset{\rightarrow}{\mathbf{r}} \tag{9}$$

where (refer to Fig. 8.2) $\underset{\rightarrow}{\mathbf{R}}$ is the location of the mass center of $\mathscr{B}$ (spacecraft) with respect to the mass center of $\mathscr{B}_p$ (primary), $\underset{\rightarrow}{\mathbf{r}}$ locates the mass element dm with respect to the mass center ⊕, and $\mu = Gm_p$, where m_p is the mass of the celestial primary. For Earth,

$$m_p \doteq 5.97 \times 10^{24}\ \text{kg}; \qquad \mu \doteq 3.986 \times 10^{14}\ \text{N} \cdot \text{m}^2/\text{kg} \tag{10}$$

It is immediately evident from (7) and (9) that

$$\underset{\rightarrow}{\mathbf{g}}_c \cdot \underset{\rightarrow}{\mathbf{R}}_c = 0 \tag{11}$$

that is, there is zero gravitational torque on $\mathscr{B}$ about the local vertical.

Upon substitution of (9) into (6), (7), and (8), we are able to implement the last assumption, $r/R_c \ll 1$. Using a binomial expansion (which certainly converges, since $r/R_c \ll 1$), we have

$$R^{-3} = R_c^{-3}\left[1 - \frac{3(\mathbf{r}\cdot\mathbf{R}_c)}{R_c^2} + \mathrm{O}\left(\frac{r^2}{R_c^2}\right)\right]$$

$$R^{-1} = R_c^{-1}\left[1 - \frac{\mathbf{r}\cdot\mathbf{R}_c}{R_c^2} - \frac{1}{2}\left(\frac{r^2}{R_c^2}\right) + \frac{3}{2}\frac{(\mathbf{r}\cdot\mathbf{R}_c)^2}{R_c^4} + \mathrm{O}\left(\frac{r^3}{R_c^3}\right)\right]$$

We have taken the expansion for R^{-1} to one order higher than for R^{-3} because, after differentiation, terms of $\mathrm{O}(r^n)$ in V become terms of order $\mathrm{O}(r^{n-1})$ in $\mathbf{f}$ and $\mathbf{g}_c$. To this degree of accuracy, the expressions for gravitational force, torque, and potential become

$$\mathbf{f} = -\left(\frac{\mu m}{R_c^3}\right)\mathbf{R}_c \tag{12}$$

$$\mathbf{g}_c = -\left(\frac{3\mu}{R_c^5}\right)\mathbf{R}_c \times \int_{\mathscr{B}} \mathbf{r}\mathbf{r}\,dm \cdot \mathbf{R}_c \tag{13}$$

$$V = -\frac{\mu m}{R_c} + \frac{1}{2}\left(\frac{\mu}{R_c^3}\right)\int_{\mathscr{B}} r^2\,dm - \frac{3}{2}\left(\frac{\mu}{R_c^5}\right)\int_{\mathscr{B}}(\mathbf{r}\cdot\mathbf{R}_c)^2\,dm \tag{14}$$

We have set $\int \mathbf{r}\,dm = \mathbf{0}$ wherever it occurred.

By keeping terms only up to $\mathrm{O}(r/R)$ in (12) and (13), we have in effect replaced the gravitational field over $\mathscr{B}$ by a Taylor expansion consisting of two terms: the value at the mass center $\oplus$ and linear terms proportional to the gradient evaluated at $\oplus$. This second term is therefore called the *gravity gradient* force field.

From (12) we see that there is no net force on $\mathscr{B}$ due to the *gravity gradient* force field; only the familiar inverse-square force remains. To this order of accuracy, then, attitude motion does not affect the orbital motion *gravitationally*. (Of course, the magnitude and direction of other forces on the spacecraft may be greatly affected by attitude.) Equation (13), on the other hand, makes it clear that the gravity gradient approximation does yield a torque. We can make this result more meaningful by defining a unit vector $\hat{\mathbf{o}}_3 \triangleq -\mathbf{R}_c/R_c$. Loosely speaking, this is a unit vector in the "down" direction; this choice of notation is in anticipation of a set of "orbiting" axes to be introduced in the next chapter, one of which will be $\hat{\mathbf{o}}_3$. As for the integral $\int \mathbf{r}\mathbf{r}\,dm$, we recall from (3.3, 2) that the (second) moment of inertia about $\oplus$ is

$$\mathbf{I} \triangleq \int_{\mathscr{B}} \left(r^2\mathbf{1} - \mathbf{r}\mathbf{r}\right) dm \tag{15}$$

Therefore, the gravity gradient torque is written

$$\mathbf{g}_c = 3\left(\frac{\mu}{R_c^3}\right)\hat{\mathbf{o}}_3 \times \mathbf{I}\cdot\hat{\mathbf{o}}_3 \tag{16}$$

At this point, the notation

$$\omega_c^2 \triangleq \frac{\mu}{R_c^3} \tag{17}$$

can be introduced, since μ/R_c^3 has the dimension $(\text{time})^{-2}$. Indeed, ω_c has the interesting physical interpretation that $\omega_c R_c$ is the speed of the spacecraft in a circular orbit of radius R_c. To avoid confusion however, it is emphasized that the instantaneous gravity gradient torque, when written as

$$\underset{\rightarrow}{\mathbf{g}}_c = 3\omega_c^2 \underset{\rightarrow}{\hat{\mathbf{o}}}_3 \times \underset{\rightarrow}{\mathbf{I}} \cdot \underset{\rightarrow}{\hat{\mathbf{o}}}_3 \tag{18}$$

is valid for any motion of ⊕ and any attitude motion of $\mathscr{B}$. Also, note again that $\underset{\rightarrow}{\mathbf{g}}_c \cdot \underset{\rightarrow}{\hat{\mathbf{o}}}_3 = 0$.

The potential function, given by (14), can be interpreted in a similar manner:

$$V = -\frac{\mu m}{R_c} - \frac{1}{2}\left(\frac{\mu}{R_c^3}\right)\text{trace }\underset{\rightarrow}{\mathbf{I}} + \frac{3}{2}\left(\frac{\mu}{R_c^3}\right)\underset{\rightarrow}{\hat{\mathbf{o}}}_3 \cdot \underset{\rightarrow}{\mathbf{I}} \cdot \underset{\rightarrow}{\hat{\mathbf{o}}}_3 \tag{19}$$

where it has been realized that $\int r^2 dm = \frac{1}{2}\text{trace }\underset{\rightarrow}{\mathbf{I}}$ and that the trace is invariant with the coordinate system used for $\underset{\rightarrow}{\mathbf{I}}$. In (19), we recognize the first term as the Keplerian potential, and the last two terms arise from the gravity gradient field.

When $\mathscr{B}$ rotates about an axis other than $\underset{\rightarrow}{\hat{\mathbf{o}}}_3$, the gravity gradient terms in (18) and (19) vary. To exhibit this variation in a useful form, we choose a set of axes $\mathscr{F}_b$, represented by the vectrix $\mathscr{F}_b$, fixed in $\mathscr{B}$. Then, $\underset{\rightarrow}{\mathbf{I}} = \mathscr{F}_b^T \mathbf{I} \mathscr{F}_b$, where

$$\mathbf{I} \triangleq \mathscr{F}_b \cdot \underset{\rightarrow}{\mathbf{I}} \cdot \mathscr{F}_b^T \tag{20}$$

is the corresponding (symmetrical, positive-definite) inertia matrix. To express $\underset{\rightarrow}{\mathbf{g}}_c$ in $\mathscr{F}_b$, denote the direction cosines $\underset{\rightarrow}{\hat{\mathbf{b}}}_i \cdot \underset{\rightarrow}{\hat{\mathbf{o}}}_3$ by c_{i3} $(i = 1,2,3)$ and let $\mathbf{c}_3 = [c_{13} \quad c_{23} \quad c_{33}]^T$. Then, $\underset{\rightarrow}{\hat{\mathbf{o}}}_3 = \mathscr{F}_b^T \mathbf{c}_3 = \mathbf{c}_3^T \mathscr{F}_b$. (Refer to Appendix B for the handling of vectrices.) From (18),

$$\mathbf{g}_c = \mathscr{F}_b \cdot \underset{\rightarrow}{\mathbf{g}}_c = 3\left(\frac{\mu}{R_c^3}\right)\mathscr{F}_b \cdot \underset{\rightarrow}{\hat{\mathbf{o}}}_3 \times \mathscr{F}_b^T \mathbf{I} \mathscr{F}_b \cdot \underset{\rightarrow}{\hat{\mathbf{o}}}_3$$

$$\mathbf{g}_c = 3\left(\frac{\mu}{R_c^3}\right)\mathscr{F}_b \cdot \mathbf{c}_3^T \mathscr{F}_b \times \mathscr{F}_b^T \mathbf{I} \mathscr{F}_b \cdot \mathscr{F}_b^T \mathbf{c}_3$$

$$\mathbf{g}_c = 3\left(\frac{\mu}{R_c^3}\right)\mathscr{F}_b \cdot \mathbf{c}_3^T \mathscr{F}_b \times \mathscr{F}_b^T \mathbf{I} \mathbf{c}_3 \qquad (\mathscr{F}_b \cdot \mathscr{F}_b^T = \mathbf{1})$$

$$\mathbf{g}_c = 3\left(\frac{\mu}{R_c^3}\right)\mathscr{F}_b \cdot \mathscr{F}_b^T \mathbf{c}_3^\times \mathbf{I} \mathbf{c}_3 \qquad [\text{see (B.2, 11)}]$$

$$\mathbf{g}_c = 3\left(\frac{\mu}{R_c^3}\right)\mathbf{c}_3^\times \mathbf{I} \mathbf{c}_3 \tag{21}$$

Written out in scalar detail,

$$g_1 = 3\left(\frac{\mu}{R_c^3}\right)\left[(I_{33} - I_{22})c_{23}c_{33} + I_{23}\left(c_{23}^2 - c_{33}^2\right) + I_{31}c_{13}c_{23} - I_{12}c_{33}c_{13}\right]$$

$$g_2 = 3\left(\frac{\mu}{R_c^3}\right)\left[(I_{11} - I_{33})c_{33}c_{13} + I_{31}\left(c_{33}^2 - c_{13}^2\right) + I_{12}c_{23}c_{33} - I_{23}c_{13}c_{23}\right]$$

$$g_3 = 3\left(\frac{\mu}{R_c^3}\right)\left[(I_{22} - I_{11})c_{13}c_{23} + I_{12}\left(c_{13}^2 - c_{23}^2\right) + I_{23}c_{33}c_{13} - I_{31}c_{23}c_{33}\right] \quad (22)$$

(The subscript c on g is omitted if it is clear from the context that the torque is about the mass center.) This agrees with equation (8) of [NASA, 1]. As usual, principal axes should be chosen as the reference axes, $\mathcal{F}_b \to \mathcal{F}_p$, unless there is some countervailing consideration. With this choice, (22) simplifies to

$$\mathcal{F}_p \cdot \underset{\rightarrow}{\mathbf{g}}_c = 3\left(\frac{\mu}{R_c^3}\right)\begin{bmatrix}(I_3 - I_2)c_{23}c_{33} \\ (I_1 - I_3)c_{33}c_{13} \\ (I_2 - I_1)c_{13}c_{23}\end{bmatrix} \quad (23)$$

in concurrence with equation (1.2.2) of [Beletskii]. It can be seen that if two principal moments of inertia are equal, the gravity gradient torque about the third axis is always zero. For a tri-inertial body (I_1, I_2, I_3 unequal), it is evident from (23) that the gravity gradient torque vanishes if and only if two of the three direction cosines are zero (in which case the third must equal ± 1). Geometrically, one of the principal axes must be aligned with the local vertical. The infinitesimal static stability of these equilibria is considered in Problem 8.3.

The expression for the gravitational potential, (19), can be similarly interpreted:

$$V = -\frac{\mu m}{R_c} - \frac{1}{2}\left(\frac{\mu}{R_c^3}\right)\operatorname{trace}\mathbf{I} + \left(\frac{3\mu}{2R_c^3}\right)\mathbf{c}_3^T\mathbf{I}\mathbf{c}_3 \quad (24)$$

Written out in scalar detail,

$$V = -\frac{\mu m}{R_c} - \frac{1}{2}\left(\frac{\mu}{R_c^3}\right)(I_{11} + I_{22} + I_{33})$$
$$+ \frac{3\mu}{2R_c^3}\left[I_{11}c_{13}^2 + I_{22}c_{23}^2 + I_{33}c_{33}^2 + 2I_{12}c_{13}c_{23} + 2I_{23}c_{23}c_{33} + 2I_{31}c_{33}c_{13}\right] \quad (25)$$

and for principal axes, $\mathcal{F}_b \to \mathcal{F}_p$, this further simplifies to

$$V = -\frac{\mu m}{R_c} - \frac{1}{2}\left(\frac{\mu}{R_c^3}\right)(I_1 + I_2 + I_3) + \frac{3\mu}{2R_c^3}\left(I_1c_{13}^2 + I_2c_{23}^2 + I_3c_{33}^2\right) \quad (26)$$

agreeing, under common assumptions, with equation (12) in [Doolin]. Note that $\underset{\rightarrow}{\hat{\mathbf{o}}}_3 = [c_{13} \quad c_{23} \quad c_{33}]\mathcal{F}_p$. Equations (23) and (26) are the principal results of this section. Several special cases are introduced in the problems at the end of this chapter.

Several Celestial Primaries

We pause now to reconsider the four assumptions made in arriving at the basic results (23) and (26). Each assumption is examined in turn, thereby extending the analysis to accommodate more general circumstances. This broadens the range of application of the results and provides quantitative criteria for the validity of each assumption.

Let us examine first the assumption that only one primary need be considered. Examining (23), it is apparent that if two primaries, $\mathscr{B}_1$ and $\mathscr{B}_2$, are candidate sources of gravity gradient torque (Fig. 8.3), then the ratio of the maximal torque from $\mathscr{B}_1$, $g_{\max}^{(1)}$, to the maximal torque from $\mathscr{B}_2$, $g_{\max}^{(2)}$, is

$$\frac{g_{\max}^{(1)}}{g_{\max}^{(2)}} = \frac{\mu_1 R_{c2}^3}{\mu_2 R_{c1}^3} \tag{27}$$

An important instance of this ratio concerns satellites orbiting Earth. The ratio of gravity gradient torque arising from the moon to that arising from Earth is bounded by

$$6.1 \times 10^{-8} \leq g_{\max}^{☾}/g_{\max}^{\oplus} \leq 2.3 \times 10^{-5} \tag{28}$$

The left limit is for near-Earth orbits; the right limit is for geostationary orbits. It is clearly advisable to neglect the lunar gravity gradient torque on Earth-orbiting spacecraft. A similar calculation for the sun's effect gives

$$4.0 \times 10^{-8} \leq g_{\max}^{\odot}/g_{\max}^{\oplus} \leq 1.1 \times 10^{-5} \tag{29}$$

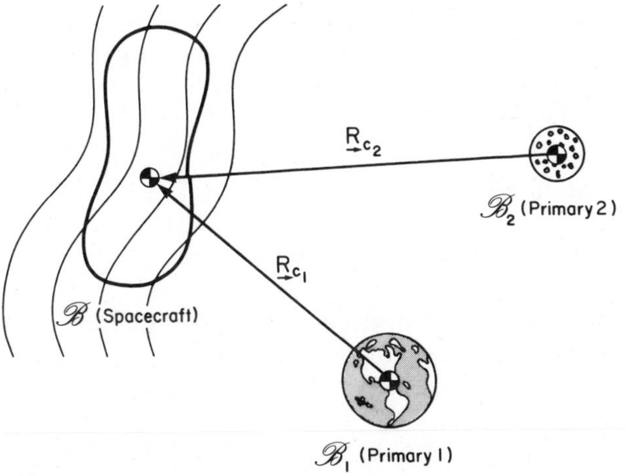

FIGURE 8.3 Spacecraft in the gravitational field of two inertially spherical primaries.

The limits in (28) and (29) are strikingly similar. Evidently the solar gravity gradient torque on an Earth-orbiting satellite is of the same magnitude as the lunar torque; while the sun is much farther away, it has a much larger mass. Both should normally be neglected for Earth-orbiting satellites.

For spacecraft farther from Earth, the situation changes somewhat. An interesting and important example is the torque on a body at one of the five points of (relative) equilibrium in the restricted problem of three bodies. A spacecraft at one of these points in the Earth–moon system is of particular interest. One way to proceed is to generalize (16) as follows:

$$\underset{\rightarrow}{\mathbf{g}}_c = 3\left(\frac{\mu_1}{R_{c1}^3}\right)\underset{\rightarrow}{\hat{\mathbf{o}}}_3^1 \times \underset{\rightarrow}{\mathbf{I}} \cdot \hat{\mathbf{o}}_3^1 + 3\left(\frac{\mu_2}{R_{c2}^3}\right)\underset{\rightarrow}{\hat{\mathbf{o}}}_3^2 \times \underset{\rightarrow}{\mathbf{I}} \cdot \underset{\rightarrow}{\hat{\mathbf{o}}}_3^2 \tag{30}$$

where 1 and 2 refer to the two primaries. Following the same development that led to (23), the gravity gradient torque about principal body axes is given by

$$g_1 = 3(I_3 - I_2)\left[\left(\frac{\mu_1}{R_{c1}^3}\right)c_{23,1}c_{33,1} + \left(\frac{\mu_2}{R_{c2}^3}\right)c_{23,2}c_{33,2}\right]$$

$$g_2 = 3(I_1 - I_3)\left[\left(\frac{\mu_1}{R_{c1}^3}\right)c_{33,1}c_{13,1} + \left(\frac{\mu_2}{R_{c2}^3}\right)c_{33,2}c_{13,2}\right]$$

$$g_3 = 3(I_2 - I_1)\left[\left(\frac{\mu_1}{R_{c1}^3}\right)c_{13,1}c_{23,1} + \left(\frac{\mu_2}{R_{c2}^3}\right)c_{13,2}c_{23,2}\right] \tag{31}$$

where $[c_{13,1} \quad c_{23,1} \quad c_{33,1}] = \mathscr{F}_{p}^{T}\underset{\rightarrow}{\hat{\mathbf{o}}}_3^1$ and $[c_{13,2} \quad c_{23,2} \quad c_{33,2}] = \mathscr{F}_{p}^{T}\underset{\rightarrow}{\hat{\mathbf{o}}}_3^2$. The attitude stability of a symmetrical rigid spacecraft at the restricted three-body "libration" points has been treated by [Kane and Marsh], and that for a tri-inertial rigid spacecraft by [W. R. Robinson].

As a useful rule of thumb, whenever the *trajectory* of a spacecraft is affected largely by a single celestial primary, the gravity gradient torque on that spacecraft arises largely from the same primary. This is seen from the ratio of gravity forces,

$$\frac{f_{\max}^{(1)}}{f_{\max}^{(2)}} = \frac{\mu_1 R_{c2}^2}{\mu_2 R_{c1}^2} \tag{32}$$

which is less selective of gravity sources than (27).

Nonspherical Primary

The familiar celestial bodies possess mass distributions that are almost spherically symmetrical. Earth, for example, whose gravity field is best known, produces a gravitational potential for an infinitesimal mass dm located at $\underset{\rightarrow}{\mathbf{R}}$ from Earth's center that can be expressed as

$$dV = -\frac{\mu\, dm}{R}\left\{1 + \sum_{i=2}^{\infty}\left(\frac{R_e}{R}\right)^i \sum_{j=0}^{i} P_i^j(\sin\lambda)(C_{ij}\cos j\phi + S_{ij}\sin j\phi)\right\} \tag{33}$$

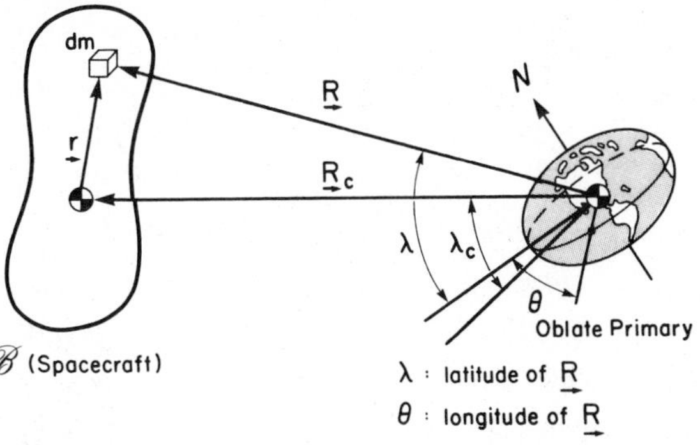

FIGURE 8.4 Spacecraft in the gravitational field of one inertially nonspherical primary.

where R_e is the mean equatorial radius and where λ and ϕ respectively are the latitude and longitude of $\underset{\rightarrow}{\mathbf{R}}$ (Fig. 8.4). The value of R_e is

$$R_e \doteq 6.378 \times 10^6 \text{ m} \tag{34}$$

The motivation for the expansion in terms of Legendre functions P_i^j is that this expansion is known to satisfy Laplace's potential equation, $\nabla^2 V = 0$, for arbitrary constants C_{ij} and S_{ij}. It is the aim of geodesy (refer, for example, to [Caputo] or [Kaula]) to determine values for these constants. For a thorough discussion of (33) in the context of gravitational forces and torques on spacecraft, the reader is referred to [Kane and Likins]. [Roberson, 2] was among the first to consider this problem.

It is clear that (33) represents a major increase in complexity over (8). Fortunately, there is a dominant nonspherical term in (33). Its physical origin lies in the observation that even though Earth is not spherical, it can, to a high degree of accuracy, be regarded as a solid of revolution about the polar axis. This assumption removes all ϕ-dependence in (33):

$$\begin{aligned} C_{ij} &= 0 \qquad (i = 2, \ldots, \infty;\ j = 1, \ldots, \infty) \\ S_{ij} &= 0 \qquad (i = 2, \ldots, \infty;\ j = 0, \ldots, \infty) \end{aligned} \tag{35}$$

Then, (33) reduces to

$$dV = -\frac{\mu\, dm}{R}\left[1 - \sum_{i=2}^{\infty} J_i \left(\frac{R_e}{R}\right)^i P_i(\sin\lambda)\right] \tag{36}$$

where $P_i \equiv P_i^0$ are Legendre polynomials and C_{i0} has been denoted by $-J_i$ to comply with convention. The first few Legendre polynomials are $P_0(y) = 1$, $P_1(y) = y$, $P_2(y) = \frac{1}{2}(3y^2 - 1)$, $P_3(y) = \frac{1}{2}(5y^2 - 3y)$. The first two do not occur in (36) because we have chosen the origin at Earth's center. Moreover, the dominant term in the sum in (36) is the first term. This has been confirmed by geodesic measurements and furnishes our final form for the potential of dm at $\underset{\rightarrow}{\mathbf{R}}$

from Earth's center:

$$dV = -\frac{\mu\, dm}{r}\left[1 + \tfrac{1}{2}J_2\left(\frac{R_e}{R}\right)^2(1 - 3\sin^2\lambda)\right] \tag{37}$$

[Kane and Likins] show that J_2 may be interpreted physically as follows:

$$J_2 \doteq \frac{I_p^e - I_e^e}{m_e R_e^2} \tag{38}$$

where m_e is Earth's mass and I_p^e, I_e^e are respectively polar and equatorial moments of inertia. The numerical value of J_2 is

$$J_2 \doteq 1.083 \times 10^{-3} \tag{39}$$

In integrating (37) over the satellite $\mathscr{B}$, it must be remembered that not only R varies ($\underset{\rightarrow}{\mathbf{R}} = \underset{\rightarrow}{\mathbf{R}}_c + \underset{\rightarrow}{\mathbf{r}}$), but also λ ($\lambda = \lambda_c + \delta\lambda$), where λ_c is the latitude of the spacecraft mass center. The result of this integration has been given by [Sarychev] and [Beletskii]. The following form is equivalent to theirs:

$$\begin{aligned} V = &-\frac{\mu m}{R_c}\left[1 + \frac{1}{2}J_2\left(\frac{R_e}{R_c}\right)^2\left(1 - 3\sin^2\lambda_c\right)\right] \\ &- \frac{\mu}{2R_c^3}\left\{(\Sigma_1 - 3\Sigma_2) + \frac{3}{2}J_2\left(\frac{R_e}{R_c}\right)^2\right. \\ &\left.\times\left[\left(1 - 5\sin^2\lambda_c\right)\Sigma_1 - 5\left(1 - 7\sin^2\lambda_c\right)\Sigma_2 + 2\Sigma_3 + 20\sin\lambda_c\Sigma_4\right]\right\} \end{aligned} \tag{40a}$$

where

$$\begin{aligned} \Sigma_1 &\triangleq I_1 + I_2 + I_3 \\ \Sigma_2 &\triangleq I_1 c_{13}^2 + I_2 c_{23}^2 + I_3 c_{33}^2 \\ \Sigma_3 &\triangleq I_1 c_{e13}^2 + I_2 c_{e23}^2 + I_3 c_{e33}^2 \\ \Sigma_4 &\triangleq I_1 c_{13} c_{e13} + I_2 c_{23} c_{e23} + I_3 c_{33} c_{e33} \end{aligned} \tag{40b}$$

Here, I_i are the principal moments of inertia of the satellite, c_{i3} are the direction cosines between the principal axes and the "down" direction $\underset{\rightarrow}{\hat{\mathbf{o}}}_3$, and c_{ei3} are the direction cosines between the principal axes and Earth's polar axis $\underset{\rightarrow}{\hat{\mathbf{e}}}_3$. In symbols,

$$\begin{aligned} \underset{\rightarrow}{\hat{\mathbf{o}}}_3 &= \begin{bmatrix} c_{13} & c_{23} & c_{33} \end{bmatrix}\mathscr{F}_p \\ \underset{\rightarrow}{\hat{\mathbf{e}}}_3 &= \begin{bmatrix} c_{e13} & c_{e23} & c_{e33} \end{bmatrix}\mathscr{F}_p \end{aligned} \tag{41}$$

To establish the equivalence between (39) and the expression given by [Sarychev] and [Beletskii], it is necessary to observe that

$$\begin{aligned} c_{13}^2 + c_{23}^2 + c_{33}^2 &= 1 \\ c_{e13}^2 + c_{e23}^2 + c_{e33}^2 &= 1 \\ c_{13}c_{e13} + c_{23}c_{e23} + c_{33}c_{e33} &= -\sin\lambda_c \end{aligned} \tag{42}$$

Note that when $J_2 = 0$, (40a) reduces to (26). In another limiting case, the body size vanishes, implying that $\Sigma_1 = \Sigma_2 = \Sigma_3 = \Sigma_4 = 0$, whence (40a) reduces to (37). The apparent simplicity of the direction cosines c_{ei3} should not belie their complexity when expressed—as they would be for many calculations—in terms of orbital inclination, right ascension, argument of perigee, true anomaly, and the c_{i3}.

Note that (40a) has arisen from only one nonspherical term in the expansion of Earth's gravitational potential. Gravity gradient torques from further terms in this expansion have not been calculated (insofar as the author is aware) since the ensuing complications scarcely seem warranted. If, for a celestial primary other than Earth, the spherical-harmonic expansion (33) is not dominated by a single term, it would appear judicious to question whether, at least from a gravitational torque standpoint, there might not be more efficacious expansions.

Similarly, the gravity gradient torques may be derived. Following [Sarychev] and [Beletskii] again, but using present notation, one may calculate the additional gravity gradient torque due to oblateness:

$$\Delta g_1 = \frac{3\mu J_2 R_e^2}{2R_c^5}(I_3 - I_2)\big[5(1 - 7\sin^2\lambda_c)c_{23}c_{33} - 10\sin\lambda_c(c_{23}c_{e33} + c_{33}c_{e23}) - 2c_{e23}c_{e33}\big]$$

$$\Delta g_2 = \frac{3\mu J_2 R_e^2}{2R_c^5}(I_1 - I_3)\big[5(1 - 7\sin^2\lambda_c)c_{33}c_{13} - 10\sin\lambda_c(c_{33}c_{e13} + c_{13}c_{e33}) - 2c_{e33}c_{e13}\big] \tag{43}$$

$$\Delta g_3 = \frac{3\mu J_2 R_e^2}{2R_c^5}(I_2 - I_1)\big[5(1 - 7\sin^2\lambda_c)c_{13}c_{23} - 10\sin\lambda_c(c_{13}c_{e23} + c_{23}c_{e13}) - 2c_{e13}c_{e23}\big]$$

These components are expressed in $\mathcal{F}_p$ and must be added to these in (23).

It can be seen from (23) and (43) that if two principal moments of inertia are equal, the gravity gradient torque about the third axis is still zero, even with oblateness accounted for. For a tri-inertial satellite (I_1, I_2, and I_3 all different), it is no longer sufficient that a principal axis point "down" (i.e., that $|c_{13}| = 1$, or $|c_{23}| = 1$, or $|c_{33}| = 1$). The special cases where the spacecraft is over the equator or over the poles are examined in Problem 8.5.

For most purposes, the correction for oblateness can be neglected. It diminishes with distance from Earth's center ($\sim R_c^{-5}$) much faster than the "spherical" torque ($\sim R_c^{-3}$). Even for near-Earth orbits, it is negligible except for high-precision calculations, being typically less than 1% of the spherical term. Several illustrative numerical examples have been worked out by [Schlengel].

General Gravity Gradient Fields

Before moving on to the last two assumptions underlying the basic result concerning gravitational torques, (23), we examine a generalized gravitational potential of the form

$$dV(\underset{\rightarrow}{\mathbf{r}}) = \left[V_c + (\underset{\rightarrow}{\nabla} V)_c \cdot \underset{\rightarrow}{\mathbf{r}} + \tfrac{1}{2}\underset{\rightarrow}{\mathbf{r}} \cdot (\underset{\rightarrow}{\nabla}\underset{\rightarrow}{\nabla} V)_c \cdot \underset{\rightarrow}{\mathbf{r}}\right] dm \tag{44}$$

This potential is valid only over the spacecraft $\mathscr{B}$; V_c is the value of V at $\mathscr{B}$'s mass center ⊕; and $\underset{\rightarrow}{\mathbf{r}}$ is the position of dm with respect to ⊕. In other words, $(\underset{\rightarrow}{\nabla} V)_c$ and $(\underset{\rightarrow}{\nabla}\underset{\rightarrow}{\nabla} V)_c$ are respectively the gradient and the Hessian of dV, evaluated at ⊕.

We have had two examples of (44) already, although the notational details may have concealed this. When N spherically symmetrical celestial primaries were present,

$$(\underset{\rightarrow}{\nabla} V)_c = -\sum_{n=1}^{N} \frac{\mu_n}{R_{cn}^2} \underset{\rightarrow}{\hat{\mathbf{o}}}{}_3^n$$

$$\underset{\rightarrow}{\mathbf{H}} \triangleq (\underset{\rightarrow}{\nabla}\underset{\rightarrow}{\nabla} V)_c = \sum_{n=1}^{N} \frac{\mu_n}{R_{cn}^3} \left[\underset{\rightarrow}{\mathbf{1}} - 3\hat{\mathbf{o}}_3^n \hat{\mathbf{o}}_3^n\right] \tag{45}$$

where μ_n is Gm_n, $\underset{\rightarrow}{\mathbf{R}}_{cn}$ is the vector from the mass center of the nth primary to ⊕, and $\underset{\rightarrow}{\hat{\mathbf{o}}}{}_3^n$ is $-\underset{\rightarrow}{\mathbf{R}}_{cn}/R_{cn}$. The second example, a single nonspherical primary, is implied by (33), although the calculation of $(\underset{\rightarrow}{\nabla} V)_c$ and $\underset{\rightarrow}{\mathbf{H}}$ is a formidable task.

In general, let us consider (44), which assumes only that $O(l^3/R_c^3)$ is negligible in dV, where l is a typical length for $\mathscr{B}$. The gravitational force $d\underset{\rightarrow}{\mathbf{f}}$ is found from

$$d\underset{\rightarrow}{\mathbf{f}} = -\underset{\rightarrow}{\nabla}(dV) = -\left[(\underset{\rightarrow}{\nabla} V)_c + \underset{\rightarrow}{\mathbf{H}} \cdot \underset{\rightarrow}{\mathbf{r}}\right] dm \tag{46}$$

The Hessian dyadic $\underset{\rightarrow}{\mathbf{H}}$ is symmetrical; this is clear from (44) and is consistent with $d\underset{\rightarrow}{\mathbf{f}}$ being a conservative force field, $\underset{\rightarrow}{\nabla} \times d\underset{\rightarrow}{\mathbf{f}} = \underset{\rightarrow}{\mathbf{0}}$. The gravitational torque on $\mathscr{B}$ due to (44) is

$$\underset{\rightarrow}{\mathbf{g}}_c = \int_{\mathscr{B}} \underset{\rightarrow}{\mathbf{r}} \times d\underset{\rightarrow}{\mathbf{f}} = -\int_{\mathscr{B}} \underset{\rightarrow}{\mathbf{r}} \times \underset{\rightarrow}{\mathbf{H}} \cdot \underset{\rightarrow}{\mathbf{r}}\, dm \tag{47}$$

since, by definition, $\int \underset{\rightarrow}{\mathbf{r}}\, dm = \underset{\rightarrow}{\mathbf{0}}$. We can learn something valuable by expressing these vectors in a principal-axis reference frame for $\mathscr{B}$, $\mathscr{F}_p$:

$$\underset{\rightarrow}{\mathbf{H}} = \mathscr{F}_p^T \mathbf{H} \mathscr{F}_p; \qquad \underset{\rightarrow}{\mathbf{r}} = \mathscr{F}_p^T \mathbf{r} \tag{48}$$

Then,

$$\mathscr{F}_p \cdot \underset{\rightarrow}{\mathbf{g}}_c = -\int_{\mathscr{B}} \mathbf{r}^{\times} \mathbf{H} \mathbf{r}\, dm = \begin{bmatrix} (I_2 - I_3)\dfrac{\partial^2 V}{\partial r_2\, \partial r_3} \\[2ex] (I_3 - I_1)\dfrac{\partial^2 V}{\partial r_3\, \partial r_1} \\[2ex] (I_1 - I_2)\dfrac{\partial^2 V}{\partial r_1\, \partial r_2} \end{bmatrix}_{\mathbf{r}=\mathbf{0}} \tag{49}$$

For a tri-inertial spacecraft, it follows that, for equilibrium, the cross derivatives $\partial^2 V/\partial r_i\, \partial r_j$, evaluated at the origin, must be zero; in other words, $\underset{\rightarrow}{\mathbf{H}}$ must be diagonal when expressed in $\mathscr{F}_p$. This almost agrees with Theorem VI in [Roberson, 5]: "If the three principal moments of inertia are unequal, the only equilibrium configurations are those for which the body principal axes coincide with *the* principal axes of the dyadic [$\underset{\rightarrow}{\mathbf{H}}$]" (emphasis added). An improved wording would replace "the" by "a set of," since there may be more than one set of principal axes for $\underset{\rightarrow}{\mathbf{H}}$. This seeming quibble is important because in the basic case (a single spherically symmetrical primary), $\underset{\rightarrow}{\mathbf{H}}$ has infinitely many principal axes (see Problem 8.7).

From the definition of $\underset{\rightarrow}{\mathbf{H}}$ in (45), we note that $\underset{\rightarrow}{\mathbf{H}} = \underset{\rightarrow}{\mathbf{H}}(\underset{\rightarrow}{\mathbf{R}}_c)$ if there is a single primary, or $\underset{\rightarrow}{\mathbf{H}} = \underset{\rightarrow}{\mathbf{H}}(\underset{\rightarrow}{\mathbf{R}}_{c1}, \ldots, \underset{\rightarrow}{\mathbf{R}}_{cN})$ if there are N primaries. It is usually more convenient to express $\underset{\rightarrow}{\mathbf{H}}$ in $\mathscr{F}_o$, the orbital frame, than in $\mathscr{F}_p$:

$$\mathbf{H}^o \triangleq \mathscr{F}_o \cdot \underset{\rightarrow}{\mathbf{H}} \cdot \mathscr{F}_o^T \tag{50}$$

However, it is straightforward to show that

$$\mathbf{H} = \mathbf{C}_{po}\mathbf{H}^o\mathbf{C}_{op} \tag{51}$$

where $\mathscr{F}_p = \mathbf{C}_{po}\mathscr{F}_o$. Typically, $\mathbf{C}_{po}$ depends on the orientation of the spacecraft. Therefore, a more practical version of our earlier result is that equilibrium for a tri-inertial vehicle is attained whenever $\mathbf{C}_{po}\mathbf{H}^o\mathbf{C}_{op}$ is diagonal. A related theme is developed in Problem 8.7. See also [Pengellay].

Very Large Spacecraft

It has been assumed in the foregoing that the spacecraft is small relative to its distance from the centers of gravitational primaries. In symbols, let R_c be (as usual) the distance between the spacecraft and the primary mass centers, and let l be the largest spacecraft dimension. Until now, we have neglected $(l/R_c)^2$ compared to l/R_c in force expressions and $(l/R_c)^3$ compared to $(l/R_c)^2$ in the potential.

Let us estimate the error thereby incurred. Earth-centered orbits of practical interest range from "near-Earth" to geostationary. In other words, $6.4 \times 10^6 \leq R_c \leq 4.2 \times 10^7$ m. Until recently, spacecraft were limited by what would fit in a nose cone or what could reasonably be deployed in space; typically, $l < 10$ m. Thus, $(l/R_c) < 1.5 \times 10^{-6}$, and the view that gravitational torques are essentially indistinguishable from gravity gradient torques was a sensible one. At the present time however, spacecraft sizes as large as $l = 20$ km are being forecast, to be constructed in space from shuttle-borne materials. For this generation of vehicle, $4.8 \times 10^{-4} < l/R_c < 3.1 \times 10^{-3}$. Though still small, these larger ratios may stimulate renewed interest in the higher-order gravitational torques.

As [Meirovitch, 1] has pointed out, the higher-order terms must become important relative to the gravity gradient terms when the principal moments of inertia of the spacecraft are equal (the *isoinertial* case), or nearly equal. We have seen this to be true not only for an inverse-square field in (23) and an oblate primary in (43), but, indeed, for *any* gravity gradient field in (49). There is however a law of diminishing returns at work here: while the higher-order terms contribute ever less to the torque on the spacecraft (relative to other sources of torque), their calculation rapidly escalates in complexity. For example, if we include terms of $O(l^2/R_c^2)$ in (46), then, expressed in $\mathscr{F}_p$, we may write

$$df_i = \left[\left(\frac{\partial V}{\partial r_i}\right)_c + H_{ij}r_j + c_{ijk}r_jr_k\right] dm \tag{52}$$

(with summation convention). Since $d\underset{\rightarrow}{\mathbf{f}}$ is conservative, only 10 of the 27 c_{ijk} are independent. The $\underset{\rightarrow}{\hat{\mathbf{p}}}_1$ component of gravitational torque is then

$$\begin{aligned} g_1 = {} & (I_2 - I_3)H_{23} + I_{113}c_{112} - I_{112}c_{113} + I_{223}c_{222} - I_{233}c_{333} + (2I_{233} - I_{222})c_{223} \\ & + (I_{333} - 2I_{223})c_{233} + 2I_{123}(c_{122} - c_{133}) + 2(I_{133} - I_{122})c_{123} \end{aligned} \tag{53}$$

where the third moments of inertia are

$$I_{ijk} \triangleq \int_{\mathscr{B}} r_ir_jr_k\, dm \tag{54}$$

Only 10 of these are independent also; and with symmetry, many of the I_{ijk} vanish. Expressions for g_2 and g_3 are similar to (54). In view of the large complexity and small magnitude of these higher-order terms, their inclusion would be exceptional.

Multibody Spacecraft

Many spacecraft comprise several constituent bodies. These bodies may rotate (or even translate) with respect to one another during maneuvers. To construct a dynamical model for the vehicle, it is necessary to know the gravitational force and torque on each body individually. An early example is the work of [Fletcher, Rongved, and Yu], who analyzed a two-body gravitationally oriented satellite. This was later generalized to N bodies by [Hooker and Margulies].

Let us assume that the spacecraft can be idealized as a connected set of N bodies, as shown in Fig. 8.5, and that the gravitational potential and force distribution are given by (44) and (46), where we have expanded about the *mass center of the whole vehicle*, ⊕. For simplicity, we treat only the case of a single inverse-square primary, for which, from (45),

$$(\nabla V)_c = -\frac{\mu}{R_c^2}\underset{\rightarrow}{\hat{\mathbf{o}}}_3; \qquad \underset{\rightarrow}{\mathbf{H}} = \frac{\mu}{R_c^3}(\underset{\rightarrow}{\mathbf{1}} - 3\underset{\rightarrow}{\hat{\mathbf{o}}}_3\underset{\rightarrow}{\hat{\mathbf{o}}}_3) \tag{55}$$

where $\underset{\rightarrow}{\mathbf{R}}_c$ is the vector from the center of the primary to ⊕, and $\underset{\rightarrow}{\hat{\mathbf{o}}}_3 = -\underset{\rightarrow}{\mathbf{R}}_c/R_c$. Now calculate the force on $\mathscr{B}_n$ (where $\mathscr{B}_n$ is, in contrast to (1), the nth component body in the spacecraft):

$$\begin{aligned} \underset{\rightarrow}{\mathbf{f}}^n \triangleq \int_{\mathscr{B}_n} d\underset{\rightarrow}{\mathbf{f}} &= \frac{\mu}{R_c^2}\underset{\rightarrow}{\hat{\mathbf{o}}}_3\int_{\mathscr{B}_n} dm - \frac{\mu}{R_c^3}(\underset{\rightarrow}{\mathbf{1}} - 3\underset{\rightarrow}{\hat{\mathbf{o}}}_3\underset{\rightarrow}{\hat{\mathbf{o}}}_3)\cdot\int_{\mathscr{B}_n}\underset{\rightarrow}{\mathbf{r}}\, dm \\ &= \frac{\mu m_n}{R_c^2}\underset{\rightarrow}{\hat{\mathbf{o}}}_3 - \frac{\mu}{R_c^3}(\underset{\rightarrow}{\mathbf{c}}^n - 3\underset{\rightarrow}{\hat{\mathbf{o}}}_3\underset{\rightarrow}{\hat{\mathbf{o}}}_3\cdot\underset{\rightarrow}{\mathbf{c}}^n) \end{aligned} \tag{56}$$

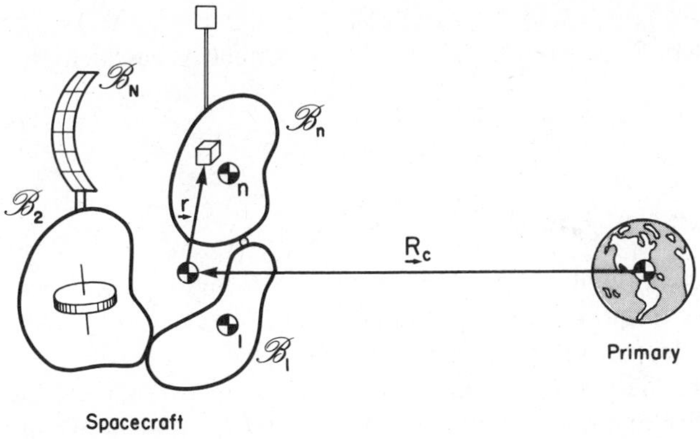

FIGURE 8.5 Gravitational field over an N-body spacecraft.

where m_n is the mass of $\mathscr{B}_n$ and $\vec{\mathbf{c}}^{\,n}$ is the first moment of inertia of $\mathscr{B}_n$ about the mass center $\oplus$. Note that the total force on the vehicle is

$$\vec{\mathbf{f}} \triangleq \sum_{n=1}^{N} \vec{\mathbf{f}}^{\,n} = \frac{\mu m}{R_c^2} \vec{\hat{\mathbf{o}}}_3 \tag{57}$$

as before, since by definition

$$m \triangleq \sum_{n=1}^{N} m_n; \qquad \vec{\mathbf{c}} \triangleq \sum_{n=1}^{N} \vec{\mathbf{c}}^{\,n} = \vec{\mathbf{0}} \tag{58}$$

The force $\vec{\mathbf{f}}^{\,n}$ can be considered to act at $\oplus_n$, the mass center of $\mathscr{B}_n$.

Similarly, we can calculate the torque about $\oplus$ on $\mathscr{B}_n$, denoted $\mathbf{g}_c^n$. By definition, from (46) and (55),

$$\begin{aligned} \vec{\mathbf{g}}_c^{\,n} &\triangleq \int_{\mathscr{B}_n} \vec{\mathbf{r}} \times d\vec{\mathbf{f}} = \frac{\mu}{R_c^2} \int_{\mathscr{B}_n} \vec{\mathbf{r}}\, dm \times \vec{\hat{\mathbf{o}}}_3 - \frac{3\mu}{R_c^3} \vec{\hat{\mathbf{o}}}_3 \times \int_{\mathscr{B}_n} \vec{\hat{\mathbf{r}}}\vec{\hat{\mathbf{r}}}\, dm \cdot \vec{\hat{\mathbf{o}}}_3 \\ &= \frac{\mu}{R_c^2} \vec{\mathbf{c}}^{\,n} \times \vec{\hat{\mathbf{o}}}_3 + \frac{3\mu}{R_c^3} \vec{\hat{\mathbf{o}}}_3 \times \vec{\mathbf{J}}^{\,n} \cdot \vec{\hat{\mathbf{o}}}_3 \end{aligned} \tag{59}$$

where $\vec{\mathbf{J}}^{\,n}$ is the moment-of-inertia dyadic of $\mathscr{B}_n$ about $\oplus$. The parallel-axis theorem asserts that

$$\vec{\mathbf{J}}^{\,n} = \vec{\mathbf{I}}^{\,n} - \frac{\vec{\mathbf{c}}^{\,n}\vec{\mathbf{c}}^{\,n}}{m_n} \tag{60}$$

where $\vec{\mathbf{I}}^{\,n}$ is the moment-of-inertia dyadic of $\mathscr{B}_n$ about its own mass center $\oplus_n$.

Expressions (56) and (59) state the gravity gradient force and torque on $\mathscr{B}_n$. Note that the latter is calculated about $\oplus$. The gravity gradient torque on $\mathscr{B}_n$, calculated about $\oplus_n$, is

$$\vec{\mathbf{g}}_{cn}^{\,n} = \frac{3\mu}{R_c^3} \vec{\hat{\mathbf{o}}}_3 \times \vec{\mathbf{I}}^{\,n} \cdot \vec{\hat{\mathbf{o}}}_3 \tag{61}$$

agreeing with [Fletcher, Rongved, and Yu] and [Hooker and Margulies]. (The derivation is left to Problem 8.9.) The torque on $\mathscr{B}_n$ can similarly be calculated about any other point of interest.

8.2 AERODYNAMIC TORQUE

Before the advent of space flight, the attitude dynamics of flight vehicles was dominated by aerodynamic torques. Aerodynamic theory had evolved to the state where in spite of the complexities of compressibility and viscosity it had become an engineering tool in the design of aircraft. The space vehicle, in contrast, operates almost beyond the realm of aerodynamics. Apart from the important phases of launch and atmospheric entry, space vehicles move either on the fringe of the sensible atmosphere or beyond it. Their energy would otherwise soon be dissipated by aerodynamic drag, and they would begin a rapid descent to Earth's surface ([King-Hele]). There is therefore an intrinsic upper bound on the magnitude of aerodynamic forces (and torques) on space vehicles.

There is a range of altitudes however at which resident satellites can have a useful lifetime, but at which aerodynamic torques are not negligible; they may, in fact, even be dominant. Although this altitude band is relatively narrow owing to the rapid decrease of density with height, the cost of placing payloads in higher orbits and the mission requirements of many Earth-observing spacecraft have led to the insertion of important classes of spacecraft at relatively low altitudes. Even for spacecraft, then, aerodynamically generated torques remain an important subject for study.

General Considerations

At altitudes where the orbit does not immediately decay from aerodynamic drag, the *molecular mean free path* (denoted λ) is large compared to the dimensions of typical satellites. As shown in Fig. 8.6, the mean free path is the

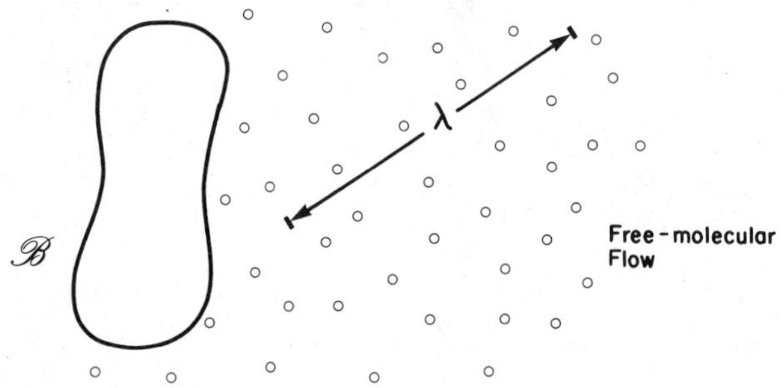

FIGURE 8.6 Spacecraft in free-molecular flow; mean free path.

average distance traveled by a molecule before collision with another molecule; it is not to be confused with the *intermolecular distance*, which is orders of magnitude smaller. Even for the lowest possible orbits, the mean free path is almost 1 km. Molecules approaching the vehicle's surface are therefore essentially unaware of other molecules coming from that surface. This leads to a treatment in which the molecules incoming to the surface and the molecules outgoing from the surface are dealt with separately—a treatment known as the *free-molecular flow* model. This model relies on the kinetic theory of gases; it is at the far end of the rarefaction spectrum from the more conventional *continum flow* model. Even for the much larger low-Earth-orbit spacecraft currently being planned, the requirement to achieve orbit will automatically place them in an atmospheric environment whose density justifies free-molecular assumptions.

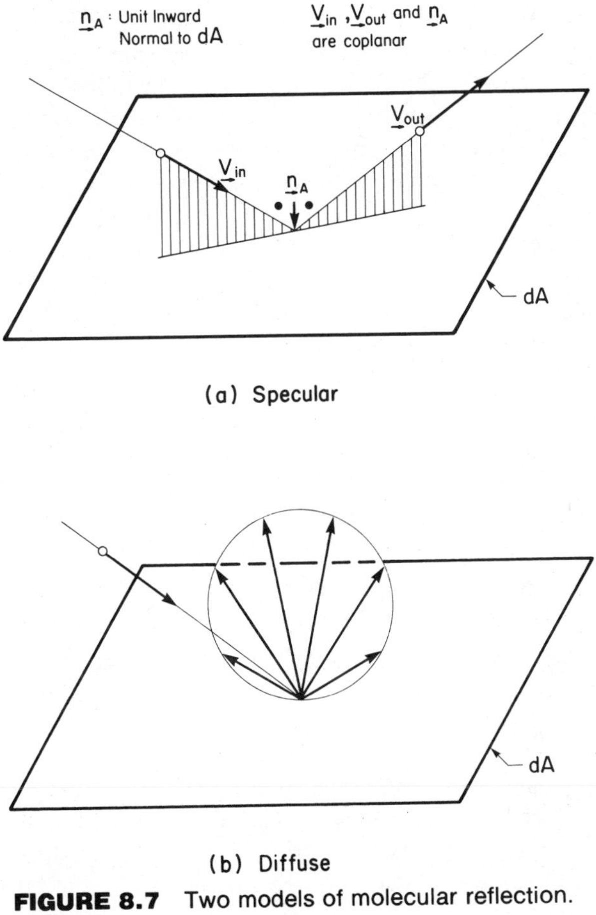

FIGURE 8.7 Two models of molecular reflection.

For aerodynamic torque calculations, one is interested in the transfer of *momentum* from atmospheric molecules to the vehicle surface. Momentum transfer occurs when molecules arrive at the surface and when molecules leave the surface. The free-molecular flow assumption allows us to calculate these two cases additively; it also permits us to regard the incoming molecules as though they were "unaware" of the impending surface. Furthermore, free-molecular flow theory makes it possible to treat a geometrically complicated surface in terms of simpler constituent surfaces—with two exceptions: "shadowing" and multiple collisions.

Two canonical limiting cases bound the molecular-momentum transfer at a physical surface: *specular* reflection, and *diffuse* reflection. Specular reflection (Fig. 8.7) is essentially a deterministic concept: each molecule bounces off the surface with no change in energy. The angle of reflection equals the angle of incidence, and the incoming velocity, the outgoing velocity, and the surface normal are coplanar. The momentum transfer is therefore normal to the surface and equals twice the normal component of the incoming momentum. As it happens, very few molecules experience specular reflection. More often, the incoming molecule becomes at least partially *accommodated* to the surface. This suggests the other limiting case: in the diffuse reflection model, the incoming molecule is completely accommodated to the surface. It loses all "memory" of its incoming direction and energy; it mingles with other molecules in the layer of surface contamination and eventually leaves with a probabilistic *kinetic energy* characteristic of the surface temperature and a probabilistic *direction* governed by a "cosine" distribution.

The discussion now proceeds by making four assumptions that greatly simplify the calculation of aerodynamic torque. After deriving this main result, the consequences of removing the assumptions are examined.

The Fundamental Result

We now assume the following:

(i) The momentum of molecules arriving at the surface is totally lost to the surface.

(ii) The mean thermal motion of the atmosphere is much smaller than the speed of the spacecraft through the atmosphere.

(iii) Momentum transfer from molecules leaving the surface is negligible.

(iv) For spinning vehicles, the relative motion between surface elements is much smaller than the speed of the mass center.

Let $\underset{\rightarrow}{\mathbf{V}}_R$ be the velocity of the local atmosphere relative to a surface element dA. Assumption (ii) permits us to view the incoming atmosphere as a collimated molecular beam of density ρ_a and velocity $\underset{\rightarrow}{\mathbf{V}}_R$. As shown in Fig. 8.8, let $\underset{\rightarrow}{\hat{\mathbf{V}}}_R \triangleq \underset{\rightarrow}{\mathbf{V}}_R / V_R$, and let $\underset{\rightarrow}{\mathbf{n}}_A$ be a unit inward normal to the surface at dA. We also

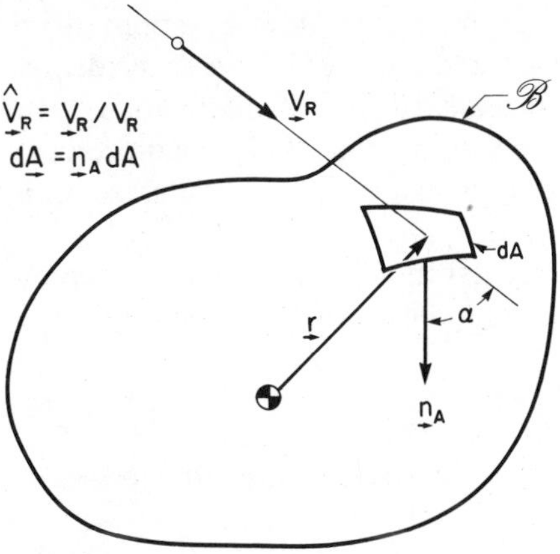

FIGURE 8.8 Molecules incident on an element of spacecraft surface.

denote $d\underset{\to}{\mathbf{A}} \triangleq \underset{\to}{\mathbf{n}}_A\, dA$. The area projected normal to $\underset{\to}{\hat{\mathbf{V}}}_R$ is $dA \cos\alpha$, where

$$\cos\alpha \triangleq \underset{\to}{\hat{\mathbf{V}}}_R \cdot \underset{\to}{\mathbf{n}}_A \tag{1}$$

and α is the *angle of attack*. The molecular momentum flux through $dA \cos\alpha$, which is the force imparted to dA when this momentum is lost, is

$$d\underset{\to}{\mathbf{f}} = \rho_a V_R^2 \cos\alpha \underset{\to}{\hat{\mathbf{V}}}_R\, dA \tag{2}$$

Care must be taken to apply this formula only where $\cos\alpha \geq 0$; when $\cos\alpha < 0$, dA cannot "see" the flow. On portions of the surface where $\cos\alpha < 0$, the aerodynamic pressure is zero.

With the force on dA known, we integrate over the vehicle surface to find the total force and torque:

$$\underset{\to}{\mathbf{f}} = \oint\!\!\oint H(\cos\alpha)\rho_a V_R^2 \cos\alpha\, dA\, \underset{\to}{\hat{\mathbf{V}}}_R \tag{3}$$

$$\underset{\to}{\mathbf{g}}_c = \oint\!\!\oint H(\cos\alpha)\rho_a V_R^2 \cos\alpha\, \underset{\to}{\mathbf{r}}\, dA \times \underset{\to}{\hat{\mathbf{V}}}_R \tag{4}$$

where $H(x)$ is the Heaviside function ($H = 1$ for $x \geq 0$, and $H = 0$ otherwise) and $\underset{\to}{\mathbf{r}}$ locates dA with respect to the mass center ⊕. We are normally justified in taking ρ_a as independent of position and neglecting "buoyancy"; and $\underset{\to}{\mathbf{V}}_R$ is also constant unless the body is rotating with respect to the atmosphere. We consider the nonspinning case first and generalize later to a spinning body.

For $\mathscr{B}$ not spinning, then,

$$\underset{\to}{\mathbf{f}} = \left(\rho_a V_R^2 A_p\right)\underset{\to}{\hat{\mathbf{V}}}_R \tag{5}$$

$$\underset{\to}{\mathbf{g}}_c = \underset{\to}{\mathbf{c}}_p \times \underset{\to}{\mathbf{f}} \tag{6}$$

where

$$A_p \triangleq \oint\!\!\oint H(\cos\alpha)\cos\alpha\, dA \tag{7}$$

$$A_p \underset{\to}{\mathbf{c}}_p \triangleq \oint\!\!\oint H(\cos\alpha)\cos\alpha\, \underset{\to}{\mathbf{r}}\, dA \tag{8}$$

These two integrals depend only on surface geometry and flow direction; A_p is the total projected area the flow "sees" and $\underset{\rightarrow}{\mathbf{c}}_p$ is the *center of pressure*. Incidently, in terms of conventional aerodynamic coefficients, (5) is equivalent to $C_L = 0$, $C_D = 2$. From (6), we discern that the torque can in principle be made zero by designing the surface and mass distribution such that $\underset{\rightarrow}{\mathbf{c}}_p = \underset{\rightarrow}{\mathbf{0}}$. In practice, only approximate coincidence is possible.

Anticipating their use in conjunction with scalar motion equations, we express the preceding vectors in a set of body axes, $\mathscr{F}_b$, whose corresponding vectrix is $\mathscr{F}_b$. Then,

$$\mathbf{g}_c = \rho_a V_R^2 A_p \mathbf{c}_p^{\times} \hat{\mathbf{V}}_R \tag{9}$$

where $\hat{\mathbf{V}}_R = [c_1^a \quad c_2^a \quad c_3^a]\mathscr{F}_b$; in other words, c_i^a are the direction cosines of $\underset{\rightarrow}{\hat{\mathbf{V}}}_R$ with respect to $\mathscr{F}_b$.

When $\mathscr{B}$ is rotating, the relative velocity varies from point to point on the surface. In fact,

$$\underset{\rightarrow}{\mathbf{V}}_R = \underset{\rightarrow}{\mathbf{V}}_{cR} - \underset{\rightarrow}{\boldsymbol{\omega}} \times \underset{\rightarrow}{\mathbf{r}} \tag{10}$$

where $\underset{\rightarrow}{\mathbf{V}}_{cR}$ is the velocity of the atmosphere with respect to the mass center and $\underset{\rightarrow}{\boldsymbol{\omega}}$ is the angular velocity of the spacecraft relative to the atmosphere. With negligible loss in accuracy, the inertial angular velocity of $\mathscr{B}$ can be used for $\underset{\rightarrow}{\boldsymbol{\omega}}$; thus, atmospheric rotation is included in $\underset{\rightarrow}{\mathbf{V}}_R$ but not in $\underset{\rightarrow}{\boldsymbol{\omega}}$. Moreover, as mentioned in assumption (iv) above, we note that $\|\underset{\rightarrow}{\boldsymbol{\omega}} \times \underset{\rightarrow}{\mathbf{r}}\| \ll \|\underset{\rightarrow}{\mathbf{V}}_{cR}\|$; their ratio is usually 10^{-3} at most. This circumstance also allows us to replace the precise condition $\underset{\rightarrow}{\mathbf{V}}_R \cdot d\underset{\rightarrow}{\mathbf{A}} \geq 0$ with the much simpler condition $\underset{\rightarrow}{\mathbf{V}}_{cR} \cdot d\underset{\rightarrow}{\mathbf{A}} \geq 0$. The force and torque expressions (3) and (4) are still valid; expressed in $\mathscr{F}_b$, they become

$$\mathbf{f} = \rho_a V_{cR} A_p \left[\left(V_{cR} - \mathbf{c}_A^T \boldsymbol{\omega} \right) \hat{\mathbf{V}}_R + \mathbf{c}_p^{\times} \boldsymbol{\omega} \right] \tag{11}$$

$$\mathbf{g}_c = \rho_a V_{cR} \left[V_{cR} A_p \mathbf{c}_p^{\times} \hat{\mathbf{V}}_R - \left(\mathbf{I}_A + \hat{\mathbf{V}}_R^{\times} \mathbf{J}_A \right) \boldsymbol{\omega} \right] \tag{12}$$

The new geometrical integrations are

$$A_p \mathbf{c}_A \triangleq \oiint H(\cos\alpha) \mathbf{r}^{\times} d\mathbf{A} \tag{13}$$

$$\mathbf{I}_A \triangleq \oiint H(\cos\alpha)(r^2 \mathbf{1} - \mathbf{r}\mathbf{r}^T) \cos\alpha \, dA \tag{14}$$

$$\mathbf{J}_A \triangleq \oiint H(\cos\alpha) \mathbf{r} \, d\mathbf{A}^T \mathbf{r}^{\times} \tag{15}$$

and we interpret $\hat{\mathbf{V}}_R$ to be $\mathbf{V}_{cR}/V_{cR}$ (not $\mathbf{V}_R/V_R$). Note especially the small torque in (12) proportional to $\boldsymbol{\omega}$. This causes the spin rate to decay.

Partial Accommodation at Surface

The first assumption made above was that all incoming molecular momentum is lost to the surface. In this view, molecules colliding with the surface stay there, at least for a short time, and become *accommodated* to the surface. They later

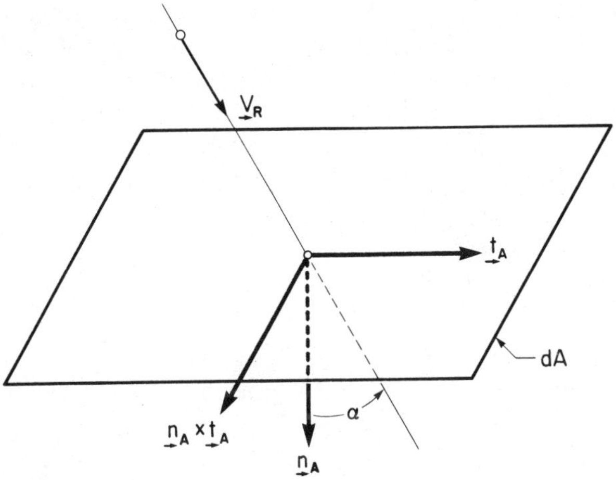

FIGURE 8.9 Orthogonal triad of surface-oriented unit vectors.

leave with a random distribution of velocities, with equal numbers in each element of solid angle subtended by the surface (the *cosine* distribution), and with a mean speed characteristic of the surface temperature T_b. We have also assumed thus far that this latter characteristic speed is small compared to the incoming speed V_R and have therefore neglected the momentum imparted to the surface by molecules leaving it. These assumptions are now removed, and expressions that include the momentum of molecules leaving the surface are derived based on *partial* surface accommodation.

Figure 8.9 shows the geometry at the surface element dA on spacecraft $\mathscr{B}$. Unit vectors normal to the surface and in the $\mathbf{V}_R$ directions are denoted $\underset{\rightarrow}{\mathbf{n}}_A$ and $\underset{\rightarrow}{\hat{\mathbf{V}}}_R$, as before. A *tangential* direction is also introduced, defined by the unit vector $\underset{\rightarrow}{\mathbf{t}}_A$, where

$$\underset{\rightarrow}{\mathbf{t}}_A = \frac{\underset{\rightarrow}{\mathbf{n}}_A \times (\underset{\rightarrow}{\hat{\mathbf{V}}}_R \times \underset{\rightarrow}{\mathbf{n}}_A)}{\|\underset{\rightarrow}{\mathbf{n}}_A \times (\underset{\rightarrow}{\hat{\mathbf{V}}}_R \times \underset{\rightarrow}{\mathbf{n}}_A)\|} \tag{16}$$

If $\underset{\rightarrow}{\hat{\mathbf{V}}}_R \times \underset{\rightarrow}{\mathbf{n}}_A = \underset{\rightarrow}{\mathbf{0}}$, $\underset{\rightarrow}{\mathbf{t}}_A$ is not defined, as should be expected. We introduce $\underset{\rightarrow}{\mathbf{t}}_A$ because it is customary to assign separate accommodation coefficients for the normal and tangential momentum components. The accommodation model we propose to use is a simple phenomenological one, and *two* coefficients permit a better fit with theoretical calculations or experimental data.

The force on dA is written

$$d\underset{\rightarrow}{\mathbf{f}} = d\underset{\rightarrow}{\mathbf{f}}_n + d\underset{\rightarrow}{\mathbf{f}}_t \tag{17}$$

where $d\underset{\rightarrow}{\mathbf{f}}_n$ and $d\underset{\rightarrow}{\mathbf{f}}_t$ are the components of $d\underset{\rightarrow}{\mathbf{f}}$ in the $\underset{\rightarrow}{\mathbf{n}}_A$ and $\underset{\rightarrow}{\mathbf{t}}_A$ directions, respectively. If, as one limiting case, we assume the molecules all accommodated,

$$\begin{aligned} d\underset{\rightarrow}{\mathbf{f}}_n^{(D)} &= H(\cos\alpha)\rho_a V_R \cos\alpha[V_R\cos\alpha + V_b]\underset{\rightarrow}{\mathbf{n}}_A\,dA \\ d\underset{\rightarrow}{\mathbf{f}}_t^{(D)} &= H(\cos\alpha)\rho_a V_R^2 \sin\alpha\cos\alpha\,\underset{\rightarrow}{\mathbf{t}}_A\,dA \end{aligned} \tag{18}$$

The factor $H(\cos\alpha)$ ensures that the forces are zero if $\cos\alpha < 0$. The incoming mass flux on dA is $\rho_a V_R \cos\alpha\, dA \triangleq dQ$. The incoming momentum flux in the $\underset{\rightarrow}{\mathbf{n}}_A$ and $\underset{\rightarrow}{\hat{\mathbf{V}}}_R$ directions is therefore $(V_R \cos\alpha)\underset{\rightarrow}{\mathbf{n}}_A\, dQ$ and $(V_R \sin\alpha)\underset{\rightarrow}{\hat{\mathbf{V}}}_R\, dQ$, respectively. Molecules leaving the wall have no mean motion in the $\underset{\rightarrow}{\mathbf{t}}_A$ direction, but they do have a mean velocity in the $-\underset{\rightarrow}{\mathbf{n}}_A$ direction, whose magnitude we have denoted by V_b. According to the kinetic theory of gases, V_b is related to the surface temperature T_b as follows:

$$V_b = \left(\frac{\pi R T_b}{2m}\right)^{1/2} \tag{19}$$

where R is the universal gas constant ($R = 8.314 \times 10^3$ J/kg · mole · °C) and m is the molecular weight of the gas. At the altitudes of interest, V_b is typically about 5% of V_R.

At the other extreme, the reflection is modeled as *specular*. Molecules are then said to leave the surface with no change in speed, at an angle of reflection equal to their angle of incidence. For specular reflection, then,

$$\begin{aligned} d\underset{\rightarrow}{\mathbf{f}}_n^{(S)} &= 2H(\cos\alpha)\rho_a V_R^2 \cos^2\alpha\, \underset{\rightarrow}{\mathbf{n}}_A\, dA \\ d\underset{\rightarrow}{\mathbf{f}}_t^{(S)} &= \underset{\rightarrow}{\mathbf{0}} \end{aligned} \tag{20}$$

No tangential momentum is lost to the surface.

Real surface interactions will tend to lie between these two extremes. Accordingly, we introduce two factors, σ_n and σ_t, called *accommodation coefficients for normal and tangential momentum exchange* (or, for brevity, *normal and tangential accommodation coefficients*), and we write

$$\begin{aligned} d\underset{\rightarrow}{\mathbf{f}}_n &= \sigma_n\, d\underset{\rightarrow}{\mathbf{f}}_n^{(D)} + (1-\sigma_n)\, d\underset{\rightarrow}{\mathbf{f}}_n^{(S)} \\ d\underset{\rightarrow}{\mathbf{f}}_t &= \sigma_t\, d\underset{\rightarrow}{\mathbf{f}}_t^{(D)} + (1-\sigma_t)\, d\underset{\rightarrow}{\mathbf{f}}_t^{(S)} \end{aligned} \tag{21}$$

Upon substitution of (18), (20), and (21) into (17), and recognizing that

$$\underset{\rightarrow}{\mathbf{t}}_A \sin\alpha = \underset{\rightarrow}{\hat{\mathbf{V}}}_R - \underset{\rightarrow}{\mathbf{n}}_A \cos\alpha \tag{22}$$

we arrive at the result

$$d\underset{\rightarrow}{\mathbf{f}} = H(\cos\alpha)\rho_a V_R^2 \cos\alpha \left\{ \left[(2-\sigma_n-\sigma_t)\cos\alpha + \sigma_n\left(\frac{V_b}{V_R}\right) \right] \underset{\rightarrow}{\mathbf{n}}_A + \sigma_t \underset{\rightarrow}{\hat{\mathbf{V}}}_R \right\} dA \tag{23}$$

This is a generalization of (2). Normally, we should expect that $\sigma_n = \sigma_n(V_R, \alpha, T_b)$ and $\sigma_t = \sigma_t(V_R, \alpha, T_b)$, although the usefulness of these coefficients is enhanced if constant (average) values can be used, especially averages with respect to α and T_b.

The force and torque on $\mathscr{B}$ are now found. Neglecting the small effects of satellite rotation, if any, we integrate (23) over the surface of $\mathscr{B}$ to get the force in body axes:

$$\mathbf{f} = \rho_a V_R^2 \left[\sigma_t A_p \hat{\mathbf{V}}_R + \sigma_n \left(\frac{V_b}{V_R} \right) \mathbf{A}_p + (2 - \sigma_n - \sigma_t) \mathbf{A}_{pp} \right] \tag{24}$$

where A_p was defined in (7), and

$$\mathbf{A}_p \triangleq \oiint H(\cos\alpha) \cos\alpha \, d\mathbf{A}$$

$$\mathbf{A}_{pp} \triangleq \oiint H(\cos^2\alpha) \cos\alpha \, d\mathbf{A} \tag{25}$$

As an exception to convention, A_p is not necessarily the magnitude of $\mathbf{A}_p$. Average values of σ_n and σ_t have been used; these coefficients usually have values in the range $0.8 < \sigma < 0.9$. Of greater interest is the aerodynamic torque:

$$\mathbf{g}_c = \iint \mathbf{r}^{\times} d\mathbf{f} = \rho_a V_R^2 \left[\sigma_t A_p \mathbf{c}_p^{\times} \hat{\mathbf{V}}_R + \sigma_n \left(\frac{V_b}{V_R} \right) \mathbf{G}_p + (2 - \sigma_n - \sigma_t) \mathbf{G}_{pp} \right] \tag{26}$$

where $\mathbf{c}_p$ was defined in (8), and

$$\mathbf{G}_p = \oiint H(\cos\alpha) \cos\alpha \, \mathbf{r}^{\times} \, dA$$

$$\mathbf{G}_{pp} = \oiint H(\cos\alpha) \cos^2\alpha \, \mathbf{r}^{\times} \, d\mathbf{A} \tag{27}$$

The torque expression (26) is a generalization of (9) to account for partial surface accommodation and the momentum of molecules leaving the surface diffusely.

Finite Speed Ratio

The atmospheric molecules through which the spacecraft passes have a random thermal motion whose intensity is directly related to the gas temperature T_a. The consequences of this motion, neglected above, are now examined. From the kinetic theory of gases we know that the gas under equilibrium conditions will have a distribution of molecular speeds given by the classical *Maxwell distribution function* ([Patterson]). Denoting random velocity by $\mathbf{u}$, the number of molecules (per unit volume) that have velocities in the range $(u_1, u_1 + du_1;\ u_2, u_2 + du_2;\ u_3, u_3 + du_3)$ is $f\, du_1\, du_2\, du_3$, where

$$f(u_1, u_2, u_3) = \frac{\rho_a}{m V_a^3 \pi^{3/2}} \exp\left(\frac{-u^2}{V_a^2} \right) \tag{28}$$

where m is the molecular weight, $u^2 = \underset{\rightarrow}{\mathbf{u}} \cdot \underset{\rightarrow}{\mathbf{u}}$, and V_a is the mean random speed

$$V_a = \left(\frac{2RT_a}{m} \right)^{1/2} \tag{29}$$

Typically, $T_a = 940°\text{K}$. The numerical factor in (28) is chosen so that

$$\int_{-\infty}^{\infty} \int_{-\infty}^{\infty} \int_{-\infty}^{\infty} f\, du_1\, du_2\, du_3 = \frac{\rho_a}{m} \tag{30}$$

that is, the total number of molecules per unit volume.

The distribution function (28) is seen by an observer at rest with respect to the mean motion of the gas. The spacecraft surface, on the other hand, sees a superimposed mean velocity $\underrightarrow{\mathbf{V}}_R$. Denoting the velocity of an individual molecule with respect to the surface by $\underrightarrow{\mathbf{v}}$, we have $\underrightarrow{\mathbf{v}} = \underrightarrow{\mathbf{V}}_R + \underrightarrow{\mathbf{u}}$. Furthermore, $\underrightarrow{\mathbf{v}}$ is expressed in a local surface reference frame $\mathscr{F}_A$, whose vectrix is $\underrightarrow{\mathscr{F}}_A \triangleq [\underrightarrow{\mathbf{n}}_A \quad \underrightarrow{\mathbf{t}}_A \quad \underrightarrow{\mathbf{n}}_A \times \underrightarrow{\mathbf{t}}_A]^T$; $\underrightarrow{\mathbf{n}}_A$ and $\underrightarrow{\mathbf{t}}_A$ were shown in Fig. 8.9. We also need

$$\underrightarrow{\mathbf{v}} \cdot \underrightarrow{\mathbf{V}}_R = \mathbf{v}^T \underrightarrow{\mathscr{F}}_A \cdot \underrightarrow{\mathbf{V}}_R = (v_1 \cos\alpha + v_2 \sin\alpha) V_R \tag{31}$$

Therefore, the distribution function seen by the surface element dA is

$$f(v_1, v_2, v_3) = \frac{\rho_a}{m V_a^3 \pi^{3/2}} \exp(-S^2) \exp\left(\frac{-v^2}{V_a^2}\right) \exp\left[2S \frac{v_1 \cos\alpha + v_2 \sin\alpha}{V_a}\right] \tag{32}$$

where S is the *molecular speed ratio*

$$S \triangleq \frac{V_R}{V_a} \tag{33}$$

Our earlier aerodynamic analysis assumed that $S \to \infty$, the so-called *hyperthermal flow* assumption, not in the sense $V_R \to \infty$, but in the sense $V_a \to 0$. In reality (for satellites in the atmospheric fringe), S is just over 5. Using f as our basic tool, we can find quantities of interest by quadrature. For example, the total mass flux to dA is given by (see Problem 8.13)

$$dQ = \left(\int_{-\infty}^{\infty} dv_3 \int_{-\infty}^{\infty} dv_2 \int_0^{\infty} v_1 f \, dv_1\right) m \, dA = \rho_a V_a \Gamma_1(S \cos\alpha) \, dA \tag{34}$$

where

$$\Gamma_1(x) \triangleq \frac{\exp(-x^2) + \sqrt{\pi}\, x(1 + \operatorname{erf} x)}{2\sqrt{\pi}} \tag{35}$$

and the error function is defined by

$$\operatorname{erf} x \triangleq \frac{2}{\sqrt{\pi}} \int_0^x \exp(-t^2) \, dt \tag{36}$$

The numerical factor in $\operatorname{erf} x$ is chosen so that $\operatorname{erf} \infty = 1$. In connection with (34), it is important to note that there is no longer any restriction that $\cos\alpha \geq 0$. Mathematically, $\Gamma_1(x)$ is defined for all arguments, positive and negative (see Fig. 8.10); physically, some molecules arrive at *all* points on the surface of $\mathscr{B}$ (because of the random molecular velocities) although, as might be expected, most arrive where $\alpha = 0$ and fewest where $\alpha = \pi$ (see Fig. 8.11). The expression (34) is immediately useful because the momentum imparted by the diffusely reflected molecules as they leave is $V_b \underrightarrow{\mathbf{n}}_A \, dQ$ (recall from (19) that V_b was the mean normal motion of the molecules leaving the surface diffusely).

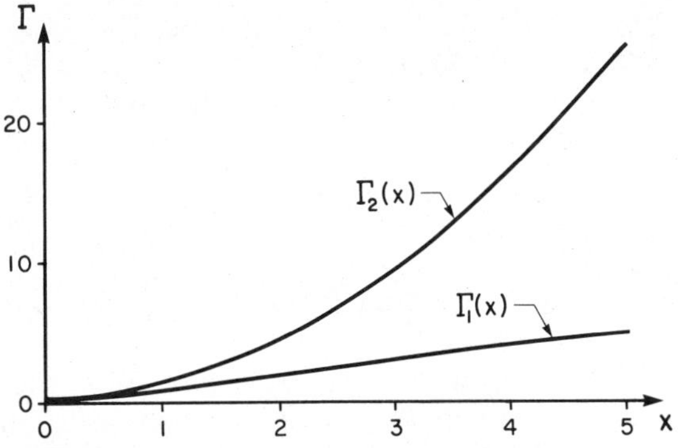

FIGURE 8.10 The functions Γ_1 and Γ_2.

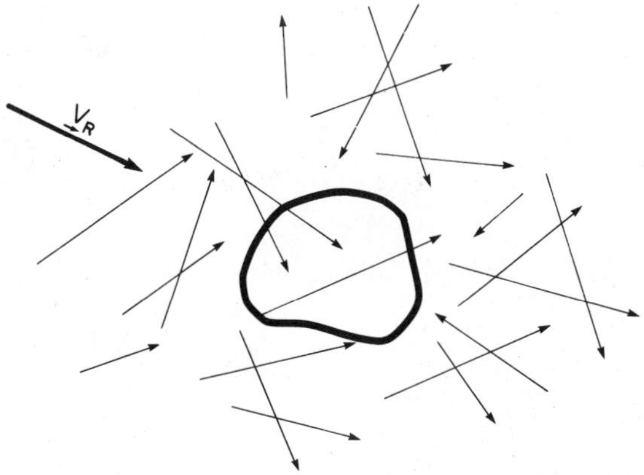

FIGURE 8.11 Body immersed in a drifting Maxwellian, free-molecular flow.

The momentum imparted by the diffusely reflected molecules as they *arrive* is

$$\left[\int_{-\infty}^{\infty} dv_3 \int_{-\infty}^{\infty} dv_2 \int_0^{\infty} v_1(v_1 \underset{\rightarrow}{\mathbf{n}}_A + v_2 \underset{\rightarrow}{\mathbf{t}}_A)\, dv_1\right] m\, dA$$

$$= \rho_a V_a^2 \left[S \sin\alpha \Gamma_1(S\cos\alpha) \underset{\rightarrow}{\mathbf{t}}_A + \Gamma_2(S\cos\alpha) \underset{\rightarrow}{\mathbf{n}}_A \right] dA \tag{37}$$

where $\Gamma_1(x)$ was given above, in (35), and $\Gamma_2(x)$ is defined as follows:

$$\Gamma_2(x) \triangleq \frac{x\exp(-x^2) + (\sqrt{\pi}/2)(1+2x^2)(1+\operatorname{erf} x)}{2\sqrt{\pi}} \tag{38}$$

Therefore, the normal and tangential force on dA if all molecules were diffusely reflected would be

$$d\underset{\rightarrow}{\mathbf{f}}_n^{(D)} = \rho_a V_a \left[V_a \Gamma_2(S\cos\alpha) + V_b \Gamma_1(S\cos\alpha) \right] \underset{\rightarrow}{\mathbf{n}}_A\, dA$$

$$d\underset{\rightarrow}{\mathbf{f}}_t^{(D)} = \rho_a V_a^2 S \sin\alpha \Gamma_1(S\cos\alpha) \underset{\rightarrow}{\mathbf{t}}_A\, dA \tag{39}$$

These formulas generalize (18) to reflect the reality of finite speed ratio $S < \infty$.

At the other extreme, we can calculate the momentum exchange as though all the molecules were reflected specularly. No momentum change occurs in the $\underset{\rightarrow}{\mathbf{t}}_A$ direction, and the change in the $\underset{\rightarrow}{\mathbf{n}}_A$ direction is twice the incoming momentum. Guided by (37), the specular force expressions are

$$d\underset{\rightarrow}{\mathbf{f}}_n^{(S)} = 2\rho_a V_a^2 \Gamma_2(S\cos\alpha)\underset{\rightarrow}{\mathbf{n}}_A\, dA$$

$$d\underset{\rightarrow}{\mathbf{f}}_t^{(S)} = \underset{\rightarrow}{\mathbf{0}} \tag{40}$$

These equations are generalizations of, and should be compared with, equation (20).

Our ultimate aim, which we are now in a position to achieve, is to find expressions for the force and torque on body $\mathscr{B}$. The notion of accommodation coefficients, introduced in (21), is now applied, although we are now more clearly committed to the idea of *average* coefficients, σ_n and σ_t. If variations in σ_n and σ_t with incoming molecular velocity (magnitude, direction, or both) are contemplated, this refinement must be introduced prior to (39) and (40)—in which case σ_t and σ_n below should be interpreted as mean values. In either case, the force on the area element dA is

$$d\underset{\rightarrow}{\mathbf{f}} = \rho_a V_a^2 \left\{ \left[(2-\sigma_n)\Gamma_2(S\cos\alpha) - \sigma_t S\cos\alpha\,\Gamma_1(S\cos\alpha) + \sigma_n\left(\frac{V_b}{V_a}\right)\Gamma_1(S\cos\alpha)\right]\underset{\rightarrow}{\mathbf{n}}_A + \sigma_t S\Gamma_1(\mathbf{S}\cos\alpha)\underset{\rightarrow}{\hat{\mathbf{V}}}_R \right\} \tag{41}$$

corresponding, for $S < \infty$, to (23).

Integrating (41) to find the total force in body axes, we find

$$\mathbf{f} = \rho_a V_R^2 \left[\left(\frac{\sigma_t}{S}\right) \oiint \Gamma_1(S\cos\alpha)\, dA\,\hat{\mathbf{V}}_R + \left(\frac{2-\sigma_n}{S^2}\right) \oiint \Gamma_2(S\cos\alpha)\, d\mathbf{A} - \left(\frac{\sigma_t}{S}\right) \oiint \Gamma_1(S\cos\alpha)\cos\alpha\, d\mathbf{A} + \left(\frac{\sigma_n}{S}\right)\left(\frac{V_b}{V_R}\right) \oiint \Gamma_1(S\cos\alpha)\, d\mathbf{A} \right] \tag{42}$$

Similarly, the torque is given by

$$\mathbf{g}_c = \rho_a V_R^2 \left[\left(\frac{\sigma_t}{S}\right) \oiint \Gamma_1(S\cos\alpha)\mathbf{r}^{\times}\, dA\,\hat{\mathbf{V}}_R + \left(\frac{2-\sigma_n}{S^2}\right) \oiint \Gamma_2(S\cos\alpha)\mathbf{r}^{\times}\, d\mathbf{A} - \left(\frac{\sigma_t}{S}\right) \oiint \Gamma_1(S\cos\alpha)\cos\alpha\,\mathbf{r}^{\times}\, d\mathbf{A} + \left(\frac{\sigma_n}{S}\right)\left(\frac{V_b}{V_R}\right) \oiint \Gamma_1(S\cos\alpha)\mathbf{r}^{\times}\, d\mathbf{A} \right] \tag{43}$$

Note that V_b has also been taken outside the integral sign. If there are great differences in T_b over the body, V_b should be left inside. To reiterate why the latter is more complicated than (9), it takes into account (i) partial accommodation at the surface ($\sigma_n < 1$), (ii) a significant wall temperature ($T_b > 0$, $V_a > 0$; $S < \infty$). If the geometry of $\mathscr{B}$ is relatively simple, the indicated integrations can often be performed analytically (see, for example, Problems 8.12 through 8.15); otherwise, one must employ numerical analysis and a computer. This is particularly true where surface concavity problems (shadowing or multiple collisions) are concerned. See, for example, [Evans].

Atmospheric Model

At the beginning of this section, four assumptions were described that led directly to the fundamental result (9). In succeeding developments, each of these assumptions was revoked until, when none of them is made, we arrive at (43) for nonspinning spacecraft. (This equation also models spinning spacecraft to essentially the same accuracy.) The increased sophistication of the quadratures called for by the last result, (43), should not occlude the profound importance of the prefix $\rho_a V_R^2$. The subscript R on V_R is a reminder that it is not enough to know the inertial motion of the spacecraft—the motion of the atmosphere must be available to comparable accuracy. In fact, a relative error ε in V_R causes a relative error 2ε in the torque. The atmospheric density ρ_a is even more difficult to model accurately. The rarefied fringe of the upper atmosphere is an exceedingly complex physical system. Many gasdynamic species interact continuously, influenced by outside energy sources including Earth's rotation, Earth's magnetic field, sunlight, and the sun's unsteady electrically charged effluent.

A simple approximate model for the speed of the atmosphere is that it rotates with Earth. This makes calculation of the local speed of the spacecraft, relative to the atmosphere, relatively simple. Note in particular that the atmospheric speed is greater at lower latitudes. Moreover, for inclined orbits, the cross-orbit component of relative velocity is larger in the neighborhood of the equatorial nodes. This leads in turn to periodic attitude excitation; indeed, for aerodynamic stabilization schemes, one must guard against the destabilizing possibilities that accompany *parametric* excitation.

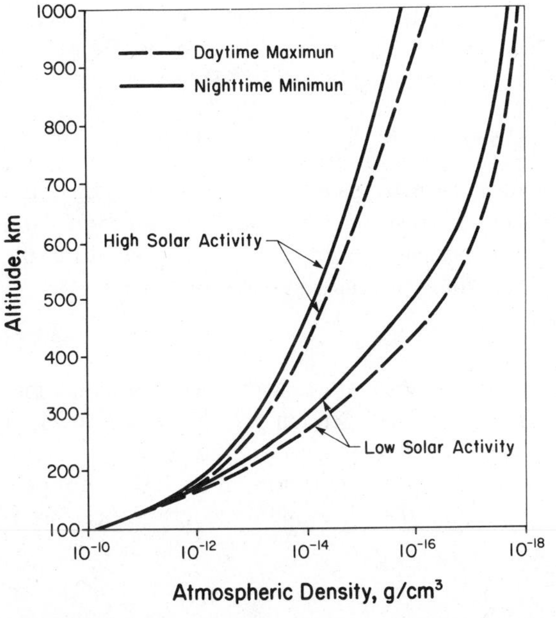

FIGURE 8.12 Daytime maximum and nighttime minimum atmospheric density profiles for high and low solar-activity levels. From [NASA, 12].

Even more important, one must be aware of large fluctuations in density ρ_a—even at fixed altitude. The two dominant causes of these fluctuations are Earth's *diurnal* (day–night) cycle and *solar activity* fluctuations. Figure 8.12, taken from [NASA, 12], summarizes these two effects. It is clear that there is scarcely any benefit from elaborate gasdynamic calculations if the atmospheric density is not known to within two orders of magnitude. For precise calculations, the chemical composition and species temperatures must also be available. A good reference for Earth's atmosphere is [NASA, 8]. Similar data for Mars and Venus are available in [NASA, 2, 4].

8.3 RADIATION TORQUES

Electromagnetic radiation is usually thought of as a mechanism for *energy* transfer. In this section however, we must also recall that this radiation will produce a pressure on any material surface that intercepts it. This pressure is normally very "small" but, as observed earlier, spacecraft torques are all "small"; their cumulative effect over sufficiently long periods however may be significant.

Radiation pressure is most easily explained in terms of the corpuscular nature of radiation. A beam of light quanta, or *photons*, possesses a momentum flux as well as an energy flux. If this momentum flux is arrested by a material surface, a corresponding pressure is exerted. The analogy with free-molecular aerodynamic pressure (Section 8.2) is well made and can frequently be useful. Radiation pressure can also be explained in terms of the wave theory of radiation. By defining the *Poynting energy flux vector* $\underset{\rightarrow}{\mathbf{S}} \triangleq \underset{\rightarrow}{\mathbf{E}} \times \underset{\rightarrow}{\mathbf{H}}$ ($\underset{\rightarrow}{\mathbf{E}}$ being the electric field strength and $\underset{\rightarrow}{\mathbf{H}}$ the magnetic excitation), it can be shown from Maxwell's field equations that the electromagnetic momentum per unit volume is S/c^2, where c is the speed of light in vacuum ($\doteq 3 \times 10^8$ m/s). Derivations from both viewpoints have been prepared and compared by [Wiggins, 1].

Force and Torque Expressions

It suffices for our work to realize that electromagnetic radiation contains, in a volume dv, a momentum $(\underset{\rightarrow}{\mathbf{S}}/c^2)\,dv$. If a surface element dA intercepts and absorbs all this momentum, it will experience a force $(\underset{\rightarrow}{\mathbf{S}}/c)\,dA$. Therefore, the pressure of radiation on a totally absorbing normal surface p has magnitude

$$p = \frac{S}{c} \tag{1}$$

and is in the direction of propagation. The analogy with hyperthermal atmospheric flow to a fully accommodating surface (Section 8.2) is immediate, with S/c and $\rho_a V_R^2$ playing parallel roles. For this reason, we shall denote by $\underset{\rightarrow}{\hat{\mathbf{S}}}$ a unit vector in the $\underset{\rightarrow}{\mathbf{S}}$ direction and by $\underset{\rightarrow}{\mathbf{n}}_A$ an *inward* surface normal of unit magnitude. If the surface dA absorbs all the radiation, the force due to radiation pressure is

$$d\underset{\rightarrow}{\mathbf{f}} = p \cos\alpha \, \underset{\rightarrow}{\hat{\mathbf{S}}} \, dA \tag{2}$$

where p is given by (1) and

$$\cos\alpha \triangleq \underset{\rightarrow}{\hat{\mathbf{S}}} \cdot \underset{\rightarrow}{\mathbf{n}}_A \tag{3}$$

Equation (2) should be compared to (8.2, 2). Note that (2) applies only if $\cos\alpha \geq 0$. If $\cos\alpha < 0$, the area element dA is not exposed to the radiation, and $d\underset{\rightarrow}{\mathbf{f}} = \underset{\rightarrow}{\mathbf{0}}$.

The force and torque on a body $\mathscr{B}$ due to radiation pressure on its surface therefore are (for a totally absorbing surface)

$$\underset{\rightarrow}{\mathbf{f}} = pA_p\underset{\rightarrow}{\hat{\mathbf{S}}} \tag{4}$$

$$\underset{\rightarrow}{\mathbf{g}}_c = \underset{\rightarrow}{\mathbf{c}}_p \times \underset{\rightarrow}{\mathbf{f}} \tag{5}$$

where

$$A_p \triangleq \oiint H(\cos\alpha)\cos\alpha\, dA \tag{6}$$

$$A_p\underset{\rightarrow}{\mathbf{c}}_p = \oiint H(\cos\alpha)\cos\alpha\, \underset{\rightarrow}{\mathbf{r}}\, dA \tag{7}$$

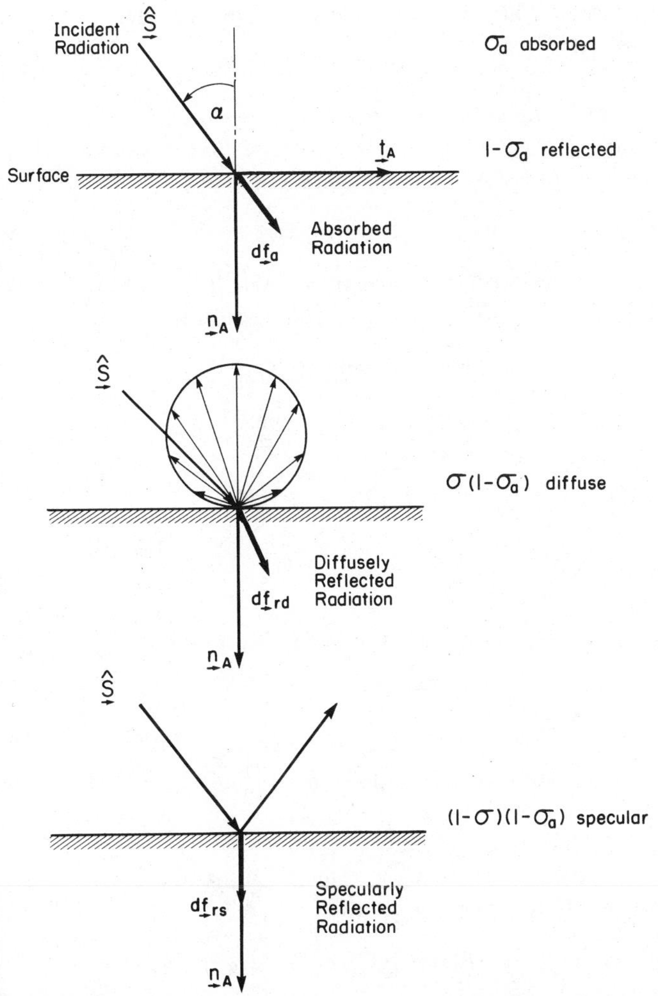

FIGURE 8.13 Three components of force from radiation pressure.

and $H(x) = 1$ for $x \geq 0$; $H(x) = 0$ for $x < 0$.

There is no point in considering spinning spacecraft separately (as is done for aerodynamic torques) because relative surface velocities are dwarfed by the speed of light. Similarly, there is no analogy to the finite-speed-ratio molecular flow of Section 8.2. There is an analogy however in the question of surface accommodation. While (4) and (5) hold for total absorption, there will in general be some reflection, and this reflection can be either specular, diffuse, or of some intermediate character. There is even the possibility of transmission of radiation through the surface. This would happen if the material were transparent or semitransparent. Spacecraft surface materials are normally opaque and transmission is therefore not included in the development below. If a nonzero transmissibility coefficient is needed for a particular application, it is not difficult to incorporate one into the analysis.

Figure 8.13 portrays the three possible destinies of incoming radiation. Let us assume a fraction σ_a is absorbed. The force on dA for total absorption is

$$d\underset{\rightarrow}{\mathbf{f}}_a = pH(\cos\alpha)\cos\alpha\underset{\rightarrow}{\hat{\mathbf{S}}}\,dA \tag{8}$$

This is consistent with the meaning of p; $df = p\,dA$ when $\alpha = 0$. Of the fraction of incoming radiation that is reflected, $1 - \sigma_a$, we assume that a fraction σ is reflected diffusely. The force expression for this case is

$$d\underset{\rightarrow}{\mathbf{f}}_{rd} = pH(\cos\alpha)\cos\alpha\left(\underset{\rightarrow}{\hat{\mathbf{S}}} + \tfrac{2}{3}\underset{\rightarrow}{\mathbf{n}}_A\right)dA \tag{9}$$

where $\underset{\rightarrow}{\mathbf{n}}_A$ is a unit inward normal to the surface at dA. Finally, of the reflected radiation, the fraction $1 - \sigma$ is reflected specularly, leading to the force

$$d\underset{\rightarrow}{\mathbf{f}}_{rs} = 2pH(\cos\alpha)\cos^2\alpha\underset{\rightarrow}{\mathbf{n}}_A\,dA \tag{10}$$

If the optical properties of the surface are not isotropic, [Wiggins, 1] can be consulted. The total radiation force on dA is

$$\begin{aligned} d\underset{\rightarrow}{\mathbf{f}} &= \sigma_a\,d\underset{\rightarrow}{\mathbf{f}}_a + \sigma(1-\sigma_a)\,d\underset{\rightarrow}{\mathbf{f}}_{rd} + (1-\sigma)(1-\sigma_a)\,d\underset{\rightarrow}{\mathbf{f}}_{rs} \\ &= pH(\cos\alpha)\cos\alpha\left[(\sigma_a + \sigma_{rd})\underset{\rightarrow}{\hat{\mathbf{S}}} + \left(\tfrac{2}{3}\sigma_{rd} + 2\sigma_{rs}\cos\alpha\right)\underset{\rightarrow}{\mathbf{n}}_A\right]dA \end{aligned} \tag{11}$$

where

$$\sigma_{rd} \triangleq \sigma(1-\sigma_a); \qquad \sigma_{rs} \triangleq (1-\sigma)(1-\sigma_a) \tag{12}$$

As a check, note that

$$\sigma_a + \sigma_{rd} + \sigma_{rs} = 1 \tag{13}$$

If transmission through the surface were included, the formula (11) would still apply; but in place of (13), one would have $\sigma_a + \sigma_{rd} + \sigma_{rs} = 1 - \sigma_t$, where σ_t is the fraction of incoming radiation transmitted.

The result for $d\underset{\rightarrow}{\mathbf{f}}$ can also be expressed in terms of $\underset{\rightarrow}{\mathbf{n}}_A$ and $\underset{\rightarrow}{\mathbf{t}}_A$, the local tangential vector, by setting $\underset{\rightarrow}{\hat{\mathbf{S}}} = \underset{\rightarrow}{\mathbf{n}}_A\cos\alpha + \underset{\rightarrow}{\mathbf{t}}_A\sin\alpha$:

$$d\underset{\rightarrow}{\mathbf{f}} = pH(\cos\alpha)\cos\alpha\left\{(\sigma_a + \sigma_{rd})\sin\alpha\underset{\rightarrow}{\mathbf{t}}_A + \left[(1+\sigma_{rs})\cos\alpha + \tfrac{2}{3}\sigma_{rd}\right]\underset{\rightarrow}{\mathbf{n}}_A\right\}dA \tag{14}$$

When integrating over the body surface, (11) may be preferable because $\underset{\rightarrow}{\hat{\mathbf{S}}}$ is a fixed direction and can be taken outside the integral sign.

Integrating (14) over the body, we find the force and torque from radiation. Expressed in body axes,

$$\mathbf{f} = p\left[(\sigma_a + \sigma_{rd})A_p\hat{\mathbf{S}} + \tfrac{2}{3}\sigma_{rd}\mathbf{A}_p + 2\sigma_{rs}\mathbf{A}_{pp}\right] \tag{15}$$

$$\mathbf{g} = p\left[(\sigma_a + \sigma_{rd})\mathbf{c}_p^{\times}\hat{\mathbf{S}} + \tfrac{2}{3}\sigma_{rd}\mathbf{G}_p + 2\sigma_{rs}\mathbf{G}_{pp}\right] \tag{16}$$

where $\mathbf{A}_p$ and $\mathbf{c}_p$ are given in (6) and (7) and $\mathbf{A}_p$, $\mathbf{A}_{pp}$, $\mathbf{G}_p$, and $\mathbf{G}_{pp}$ were shown in (25) and (27). All these quantities are, in general, time dependent. Concavity (shadowing) effects are not accounted for in these formulas.

Radiation Environment

The chief source of radiation is directly from the sun. In the vicinity of Earth, the value of p is essentially constant:

$$p \doteq 4.5 \times 10^{-6} \text{ N/m}^2 \tag{17}$$

For near-Earth orbits, p varies by less than 0.1% in magnitude and 10 arc · s in direction. Bearing in mind the other uncertainties in solar torque calculation, one should take p as constant over each orbit. This remark applies equally to geostationary orbits where the variation in magnitude and direction is about six times larger. There is however a seasonal variation in p (about 6%) owing to the eccentricity of Earth's orbit around the sun; this may be worth including in precise calculations. Elsewhere in the solar system, the solar p ranges from $\doteq 3 \times 10^{-5}$ near Mercury to $\doteq 3 \times 10^{-9}$ out at Pluto.

The possibility of Earth's shadowing should be recognized, particularly for near-Earth orbits. In the *umbra*, solar pressure vanishes, while in the *penumbra* there is a solar pressure *gradient* which may produce a torque on an otherwise balanced geometry (see Problem 8.20). For a detailed exposition of solar radiation, the reader is referred to [NASA, 15].

Earth itself is a source of radiation, and this is often significant for near-Earth satellites. Part of this radiation is reflected solar radiation. Earth's surface and atmosphere tend to reflect, on the average, about 34% of incident sunlight. The details of this reflectance are complex because the atmosphere and surface are unsteady and nonhomogeneous and the reflected radiation is therefore unsteady, nonisotropic, and wavelength dependent. Even if relatively simple models are used, the strong dependence on altitude, latitude, and longitude (with respect of local noon) complicates the geometry. The maximum reflection for a specified altitude is at the subsolar point, which is directly over the equator at local noon. The subsolar value of p from Earth's reflection ranges from $\doteq 2 \times 10^{-6}$ N/m^2 near Earth to $\doteq 3 \times 10^{-8}$ N/m^2 at geostationary altitude.

Earth not only *reflects* radiation, it also *emits* radiation. A satellite directly over the equator at local midnight would on this account receive Earth radiation. Its magnitude is typically one third of the reflected radiation discussed in the last paragraph, but it is essentially independent of latitude and longitude. It should be taken into account for precise calculations on near-Earth satellites. A comprehensive summary of Earth's reflected and emitted radiation and many further references are available in NASA [16]. The same phenomena are present near other planets, although the numerical details are less well known. For example, one might speculate that emitted radiation would be more important near Jupiter than reflected solar radiation. Many further references on spacecraft radiation torques are available in [NASA, 10].

8.4 OTHER ENVIRONMENTAL TORQUES

Three sources of torque have now been described—gravity, aerodynamics, and radiation pressure. In each instance, the environment of the spacecraft has had a potentially significant influence on its orientation. This section continues this theme by presenting, rather more briefly, three other sources of environmental torque.

Magnetic Torque

When the spacecraft is immersed in a magnetic field, several torque mechanisms can be identified. The principal mechanism occurs when the vehicle has a magnetic moment $\underset{\rightarrow}{\mathbf{m}}_m$. Denoting the external magnetic flux density by $\underset{\rightarrow}{\mathbf{B}}$, the consequent torque is

$$\underset{\rightarrow}{\mathbf{g}}_m = \underset{\rightarrow}{\mathbf{m}}_m \times \underset{\rightarrow}{\mathbf{B}} \tag{1}$$

This is the familiar effect that causes a magnetic compass needle to align itself with the local magnetic field.

The formula (1) appears quite simple, although the detailed calculation of its two constituent vectors often requires care. The magnetic moment $\underset{\rightarrow}{\mathbf{m}}_m$ refers to the *net* magnetic moment on the spacecraft. Permanent magnets contribute vectorially to $\underset{\rightarrow}{\mathbf{m}}_m$ and so do current loops; careful bookkeeping is mandatory for modern vehicles with their ubiquitous electronics and (possibly) scientific instruments. In principle, $\underset{\rightarrow}{\mathbf{m}}_m$ can be made zero by judicious relative orientation of each contributing on-board item. At the end of this process, one can also add a special permanent magnet to trim the net moment to zero. It is furthermore fortunate that when $\underset{\rightarrow}{\mathbf{m}}_m$ is zero, it is zero independently of attitude (unlike, for example, aerodynamic or radiation pressure torques, which can be balanced in one orientation but cause problems in another).

Magnetic torques are possible for Earth satellites because $\underset{\rightarrow}{\mathbf{B}}$ is nonzero in the neighborhood of Earth. The basic model for the terrestrial magnetic field is a dipole at Earth's center along an axis that pierces Earth's surface at 78.5° N, 69.0° W (in northwestern Greenland). The magnetic potential satisfies Laplace's equation and can be expanded in a series of spherical harmonics in the same manner as the gravitational potential (see 8.1, 33). The benefits of doing so are questionable because, unlike the gravitational field, the magnetic field tends to be rather unsteady at satellite altitudes. Buffeted by the outrush of charged particles from the sun's atmosphere, the magnitude field is constantly undulating, particu-

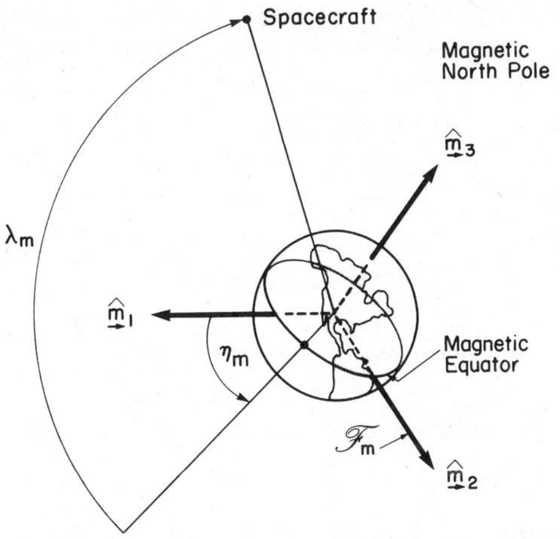

FIGURE 8.14 The geomagnetic reference frame $\mathscr{F}_m$.

larly at higher altitudes and particularly following periods of high solar activity (magnetic storms). At geostationary altitude, the field can even reverse its direction!

The basic dipole model for Earth's magnetic field is expressible using the magnetic potential

$$\phi_m = -\frac{\mu_m}{R^2}\sin\lambda_m \tag{2}$$

where R represents distance from the geocenter, λ_m is the latitude with respect to the geomagnetic equatorial plane, and μ_m is Earth's dipole strength:

$$\mu_m \doteq 1 \times 10^{17}\ \text{Wb}\cdot\text{m} \tag{3}$$

The magnetic flux density is then calculated from $\underset{\to}{\mathbf{B}} = -\underset{\to}{\nabla}\phi_m$. Expressed in a geomagnetic reference frame (see Fig. 8.14),

$$\mathbf{B} \triangleq \mathscr{F}_m \cdot \underset{\to}{\mathbf{B}} = -\frac{\mu_m}{R^3}\begin{bmatrix} 3\sin\lambda_m\cos\lambda_m\cos\eta_m \\ 3\sin\lambda_m\cos\lambda_m\sin\eta_m \\ 3\sin^2\lambda_m - 1 \end{bmatrix} \tag{4}$$

where η_m is the longitude, with respect to $\hat{\underset{\to}{\mathbf{m}}}_1$, along the magnetic equator. It remains to express $\underset{\to}{\mathbf{B}}$ in spacecraft axes. Vectrices (Appendix B) can be used to assemble the transformation $\mathbf{C}_{bm} \triangleq \mathscr{F}_b \cdot \mathscr{F}_m^T$. Assuming the magnetic moment is expressed in $\mathscr{F}_b$, $\mathbf{m}_m = \mathscr{F}_b \cdot \underset{\to}{\mathbf{m}}_m$, the torque (in $\mathscr{F}_b$) is

$$\mathbf{g}_m \triangleq \mathscr{F}_b \cdot \underset{\to}{\mathbf{g}}_m = \mathbf{m}_m^{\times}\mathbf{C}_{bm}\mathbf{B} \tag{5}$$

In principle, one could also compute a "magnetic-gradient" torque, taking into account the variation of $\mathbf{m}_m$ and $\mathbf{B}$ throughout the spacecraft, but this is a small effect and has not been analyzed in the literature.

For a more detailed examination of spacecraft magnetic torques, the reader is referred to [NASA, 6, 7]. The latter reference also discusses the magnetic fields of other celestial bodies in the solar system. It is known that the fields of the moon,

Venus, and Mars are very small and that Jupiter's field is quite large. Information on the outer planets is now becoming available from the "Grand Tour" missions.

Magnetic-field interaction can also produce dissipative torques. *Eddy current damping* is an important example. Assume the spacecraft to be rotating in a magnetic field; unless the axis of rotation is parallel to $\underset{\rightarrow}{\mathbf{B}}$, surface elements experience an unsteady magnetic field. When the surface is electrically conducting, the free electrons will be incited to form eddy currents. Since the conductivity is not infinite, these moving electrons are resisted by—and therefore exert an equal and opposite force on—the surface material. Although this dissipative torque is quite small, it is important for spinning conducting spacecraft, where it can act steadily to oppose the spin. The instantaneous value of the eddy current damping torque on a body whose angular velocity vector is $\underset{\rightarrow}{\boldsymbol{\omega}}$ in a field $\underset{\rightarrow}{\mathbf{B}}$ is

$$\underset{\rightarrow}{\mathbf{g}} = -k\underset{\rightarrow}{\mathbf{B}} \times (\underset{\rightarrow}{\boldsymbol{\omega}} \times \underset{\rightarrow}{\mathbf{B}}) \tag{6}$$

where k is a constant that depends on the geometry and electromagnetic properties of the surface. For a thin spherical shell of radius r and thickness t, [Wilson] has calculated that

$$k = \tfrac{2}{3}\sigma\pi\mu^2 r^3 t \tag{7}$$

where σ is the conductivity and μ is the effective magnetic permeability of the shell.

A second type of magnetically induced dissipative torque is *magnetic-hysteresis torque*. As discussed by [Whisnant et al.], one may append magnetically permeable rods to the spacecraft; as the rods experience a varying **B**, periodic magnetization and demagnetization results in hysteretic energy losses. This can be used to remove an initial postinjection spin at the beginning of the mission or to continually provide stabilizing damping for an Earth-pointing attitude stabilization system.

Meteoroidal Impact

Throughout its lifetime, a spacecraft is continually colliding with the tiny residents of the solar system known as *meteoroids*. These solid particles are also referred to in the literature as "meteorites" although, strictly speaking, this latter term should be reserved for meteoroids that have survived atmospheric entry and are found on Earth's surface. Spacecraft in geocentric orbits encounter meteoroids whose origin can be traced primarily to cometary breakup. As described in [NASA, 15], these meteoroids are classified either as *sporadic*, when their orbits are random, or as *showers*, when a large number of them have nearly identical orbits. The occurrence of showers is largely predictable, and calculations of greater accuracy can be made once the launch date, orbit, and mission duration of the satellite are known. For interplanetary missions out toward Jupiter, asteroidal sources also become important. The interaction must be treated as a stochastic process, because it is not possible to predict encounters with individual meteoroids. Until the space age, the statistics of meteoroids were not well known. However, with the aid of photographic and radar observations, and direct measurements made by sounding rockets and spacecraft (e.g., Pegasus), the meteoroid environment has become known well enough to permit quantitative predictions of acceptable accuracy.

It is to be recognized that an external torque is not normally the most significant consequence of spacecraft–particle encounter. The worst scenario is collision with a meteoroid large enough to destroy the usefulness of the spacecraft. Less dramatically, the cumulative effect of many collisions with smaller particles can degrade surface properties; deterioration of thermal coatings and solar cells furnish two examples.

The impulse imparted by a particle of mass m and velocity $\underset{\rightarrow}{\mathbf{v}}$ is, in principle, straightforward. The two areas of difficulty concern what precisely happens at the surface and how many incident meteoroids have mass m and velocity $\underset{\rightarrow}{\mathbf{v}}$. [Eichelberger and Gehring] have demonstrated experimentally that a small high-speed particle can cause cratering and that the momentum of the ejected surface material can be even greater than the incident momentum. As for the distribution of incident particles, a dynamicist is interested in the number of incident particles whose momentum lies between $\mathbf{p}$ and $\mathbf{p} + d\mathbf{p}$. For reasons related to the types of measurements made and to the uses to which the data are often put, reported data tend to give separate distributions of mass m and speed v, [NASA, 5], for example. Sporadic meteoroids have geocentric speeds typically about 20 km/s, although particles in retrograde heliocentric orbits cause the v-distribution to spread up to 70 km/s. Shower meteoroids tend to have a speed greater than 20 km/s. It should be noted that the speed of an Earth-orbiting satellite (3 to 7 km/s) is not really negligible compared to these values, and so an accurate calculation would presumably use the *relative* velocity between meteoroid and satellite, although this is not done in the literature. As for the mass distribution, the following empirical fit converted to SI units is derived in [NASA, 5]:

$$\log N = \begin{cases} -18.01 - 1.213\log m, & -9 \le \log m \le -3 \\ -19.66 - 1.962\log m - 0.063\log^2 m & -15 \le \log m \le -9 \end{cases} \tag{8}$$

where $N(m)$ is the cumulative distribution function for the number of particles in the immediate vicinity of Earth, per square meter per second, whose mass (in kg) is $\ge m$. Particles whose mass is greater than 1 gram are fortunately rare.

Also give in [NASA, 5] are guidelines for taking into account the gravitational and shielding influences of Earth and the moon on the meteoroid flux encountered in Earth's vicinity. The gravitational effect is termed "defocusing," and at higher attitudes the flux given by (8) is reduced by a factor G, where

$$G \doteq 0.57 + \frac{0.43}{R/R_e} \tag{9}$$

where R is the distance of the satellite from Earth's center and R_e is Earth's radius. (The factor G in [NASA, 5] is presented graphically, but (9) appears to fit the curve to within 2%.)

Unfortunately, the above description does not place us in a position to calculate accurately the torque on a spacecraft. Ideally, one would wish for a meteoroid momentum distribution function, a function to play the same role for meteoroid impact that the Maxwellian distribution function (8.2, 32) plays in free-molecular aerodynamics. Denoted by $f(\mathbf{p}; \vec{r}, t)$, the number of particles possessing momentum between $\mathbf{p}$ and $\mathbf{p} + d\mathbf{p}$ would be $f\,d\mathbf{p}$. The dependence on $\vec{r}$ would include the defocusing mentioned above and whether the vehicle was in Earth's proximity or in an interplanetary position, and the time dependence would reflect other phenomena such as meteoroid showers. The theory would then proceed in a fashion closely resembling free-molecular aerodynamics. Unfortunately, meteoroid flux theory appears to be in a more primitive state; with current predictions of large-area spacecraft in mind, a refinement in the approach to meteoroid torque prediction is warranted. The lack of fidelity in modeling meteoroid torque has not been a serious matter in the past because it is normally a minor source of torque. With the definition of $N(m)$ given above, the number of particles/m$^2 \cdot$ s having masses between m and $m + dm$ is $-(dN/dm)\,dm$; their momentum is $-mv(dN/dm)\,dm$. The total momentum/m$^2 \cdot$ s (i.e., the incident normal pressure) of meteoroids with masses between 10^{-15} and 10^{-3} kg is therefore

$$p_m = -\int_{10^{-15}}^{10^{-3}} mv\left(\frac{dN}{dm}\right)dm \tag{10}$$

A crude integration with (8) inserted in (10) indicates that $p_m < 10^{-10}$ N/m^2. This should be compared with the solar radiation pressure mentioned in (8.3, 17), which has a value $\doteq 4.5 \times 10^{-6}$ N/m^2. Meteoroidal impact torques are therefore not generally a major dynamical influence. However, for special spacecraft orientations, or farther out in the solar system, or during exceptional meteoroid showers, this environmental torque may be worthy of close scrutiny.

Thermoelastic Deformation

The temperature distribution over a spacecraft is not uniform, and the calculation and control of this distribution represents a major area of space technology. For present purposes, it suffices to point out that there are heat sources both within the vehicle (dissipation from electrical equipment, for example) and from outside (sunlight being the principal source) and that the resulting temperature distribution depends on the radiative and thermal-conductivity properties of each constituent element of the configuration. The consequent thermal strains produce structural deformations whose torque implications concern us here.

In view of the environmental torques previously considered, thermoelastic deformation constitutes not so much a new source of torque as a factor that must be considered in the calculation of other torques. For example, to the extent that the inertia distribution is altered, gravitational torques will be influenced. Similarly, torques that depend on the relationship between the distribution of area with respect to the mass center (aerodynamic and solar-pressure torques are examples) clearly will be affected. These changes can be especially dramatic when the zero deformation torque is nominally zero and the thermal deformations are quite large (long slender antennas, for example). Furthermore, these effects are time varying when the heat input is changing. A vehicle entering or leaving

Earth's shadow, for example, experiences a quite rapid change in solar input. Another common instance of unsteady heat input is a spinning satellite, whose surface sees in effect a rotating sun. Even Earth-pointing satellites rotate slowly (once per orbit) with respect to the sun. However, if we can assume that a thermal equilibrium exists and that the characteristic time to reach this equilibrium is much shorter than the orbital period, thermal distortions—and hence their effect on torques—can be regarded as quasi-steady. Even when the heat input is nominally constant, a steady thermal equilibrium is not reached for some structures. Instead, a limit-cycle-like behavior may ensue, aptly named *thermal flutter*. For example, [Frisch, 1, 2] has analyzed thermal flutter for long thin-walled open-section tubes. Whatever the cause of the thermally induced motion, the ensuing dynamic changes in vehicle geometry can have serious adverse effects on the attitude dynamics quite apart from the changes in environmental torques under discussion here.

8.5 NONENVIRONMENTAL TORQUES

The foregoing discussion has described the important interactions between a spacecraft and its environment. We turn now to torques that are *self*-generated, torques that would persist even if the vehicle were (hypothetically) placed in truly "empty" space devoid of gravitational fields, magnetic fields, radiation, tenuous atmosphere, meteoroids, and so on. Since we preclude by hypothesis the influence of an external agency, these torques may also be called *internal* torques, provided that by "spacecraft" we mean all the mass initially associated with the vehicle. In the absence of external torques, the principle of conservation of angular momentum, (3.2, 22), stipulates that the angular momentum about the mass center is constant. It is usually more convenient however to treat the mass expelled from a thruster, for example, as having become external, whereupon the torque generated by the thruster is viewed as an "external" torque even though it is not an "environmental" one. The angular momentum of the (remaining) spacecraft will change accordingly. Whenever mass is ejected from the vehicle, the reaction torque of this mass on the (remaining) vehicle produces an angular impulse. This process may be either intentional, as with attitude control jets, or unintentional, as with leaks, venting, or exhaust plumes. A complete discussion of mass expulsion torques is available in [NASA, 11].

With high-power communications satellites in mind, the torque occasioned by the radiation force on transmitting antennas should be mentioned. The physical process is identical to the environmental radiation effects described in Section 8.3. [Roberson, 4] has noted the reaction force of 0.33×10^{-5} N per kilowatt of radiated energy.

There are also many sources of strictly internal torques, wherein one part of the spacecraft or its contents exerts a force or torque on another part. These torques cannot change the total angular momentum of the system but can cause significant changes in the orientation of critical parts of the vehicle—antennas, cameras, telescopes, or attitude sensors. Such torques are to be expected whenever the vehicle is nonrigid and there is relative motion between constituent elements of the configuration. An early formulation of this problem was given by [Roberson, 3], who considered the translational and rotational motion of rigid internal bodies relative to the main vehicle. This general problem was also discussed by [Grubin, 1]. When placed on the "right-hand side" of the motion equations, the terms that result from moving parts can appear similar to external torques. This viewpoint is especially appropriate when these terms are a consequence of active control effort—control torques from reaction wheels or control moment gyros, for example. When the internal torques are of an autonomous (or passive) nature (e.g., the basic gyroscopic torques from the same reaction wheels or gyros, or torques from passive dampers), we prefer to regard them as dynamical terms in the system motion equations, as in Chapter 6.

Perhaps the most general class of internally generated torque on modern space vehicles arises from *vehicle structural flexibility*. Indeed, specific cases of nonrigidity have already been examined in detail in this book. As dealt with in Chapter 3, a basically rigid spacecraft may contain a point mass damper (Sections 3.4 and 5.2), a wheel or rotor (Section 3.5 and Chapter 6), or both a point mass damper and a wheel (Section 7.4). Moreover, it is amply plain from the discussion in Chapters 4 through 7 that even small relative motions of even small parts of the spacecraft can be significantly—even critically—important. This general observation is true also of distributed structure flexibility within the vehicle, although a representative analysis of such interactions is a major undertaking quite beyond the scope of this book. Suffice it to say that the dynamical influence of flexibility noted when there is a simple flexible degree of freedom introduced via a point mass damper $\mathscr{P}$ (Sections 3.4, 5.2, and 7.2) is indicative of the kinds of inertial and dissipative effects that can be expected. By proper design, the internal torques caused by internal dampers and by internal or external rotors can in fact often be very beneficial to attitude stabilization (Chapters 9 through 11). Distributed structural flexibility, on the other hand, is virtually always deleterious to attitude stabilization objectives.

Two other examples of internal-torque mechanisms that require special handling are crew motion and fuel sloshing. The motions of crew members are not predictable, except perhaps statistically, and yet they can have a measurable impact on the attitude control system. They are best treated stochastically; see [Davidson and Armstrong], for example. As for fuel sloshing, this poses a very difficult problem in applied mechanics. The dynamic loads from the fuel on its container have both inertial and dissipative components, and both can have a profound influence on attitude stability and control, particularly for spinning vehicles. The basic equations governing the fluid motion are available (the Navier–Stokes equations), but the closed-form solution of these nonlinear partial differential equations with boundary conditions at free surfaces appropriate to "free fall" and other boundary conditions at "walls" consisting of geometrically complex baffles is not in prospect. Recent advances in the application of the finite-element methodology to fluid mechanics suggest a promising approach. A review of slosh loads and many further references are available in [NASA, 3].

8.6 CLOSING REMARKS

Clearly, the origins of torque on spacecraft are many in number and diverse in character. Although the most important torques have been discussed, the attitude dynamics and control specialist must be alert to the special possibilities presented by each spacecraft and each mission. Past surprises and oversights provide lessons worthy of thoughtful study.

It is difficult to provide general guidelines for selecting the most important torques for particular combinations of spacecraft and mission, except that the environmental torques tend to have a simple dependence on vehicle position. Both gravity gradient and magnetic torques vary as R^{-3}, where R is the distance to the gravitational or magnetic center; aerodynamic torques, according to the simplest model atmospheres, decrease exponentially with altitude. Solar-pressure torques are essentially constant for Earth orbits, varying inversely as the square of

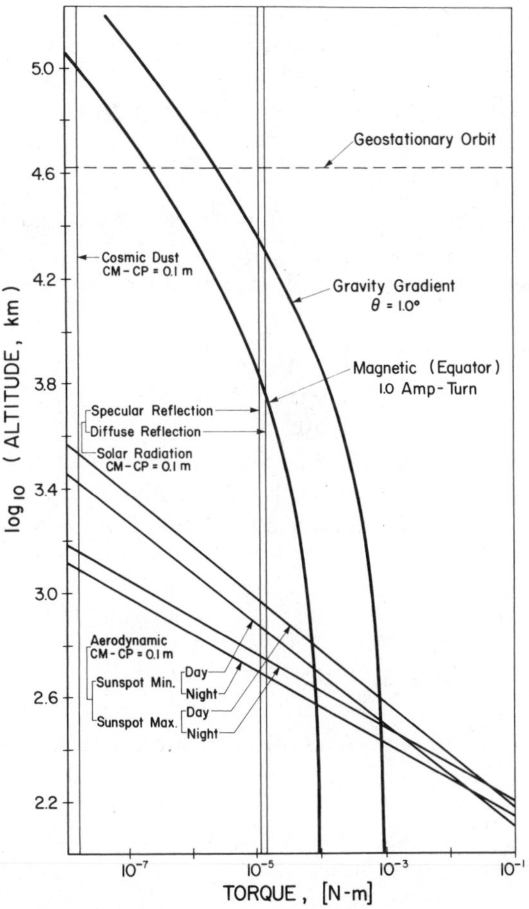

FIGURE 8.15 Comparison of several common torques on a typical spacecraft, vs. altitude.

distance from the sun. The gravitational defocusing effect on meteoroid flux has a simple dependence on R, given by (8.4, 9). All nonenvironmental torques have, by definition, no dependence on position. Figure 8.15 shows a typical comparison for Earth orbits, based on [Wiggins, 2], extended to geostationary altitudes and converted to SI units.

The attitude dynamics of any spacecraft is inevitably and adversely affected to some degree by unwanted torques. These can be compensated for only by the judicious exploitation of other torquing possibilities. A basic element in attitude control strategy is the selection of which torques to regard as perturbing torques (to be minimized) and which to choose as control torques (to be maximized). Many clever answers to this question have evolved, some of which are described in the next chapter.

8.7 PROBLEMS

8.1 Consider a body $\mathscr{B}$ under the gravitational influence of an inverse-square primary and express the centroidal inertia dyadic of $\mathscr{B}$, namely, $\underset{\rightarrow}{\mathbf{I}}$, in an "orbiting" frame $\mathscr{F}_o$. (For this problem, all that matters is that $\hat{\mathbf{o}}_3$ is a unit vector from ⊕, the mass center of $\mathscr{B}$, to the primary.) Then,

$$\mathbf{I}^o \triangleq \mathscr{F}_o \cdot \underset{\rightarrow}{\mathbf{I}} \cdot \mathscr{F}_o^T \tag{1}$$

where $\mathbf{I}^o$ is the inertia matrix in $\mathscr{F}_o$ for $\mathscr{B}$ and $\mathscr{F}_o$ is the vectrix associated with $\mathscr{F}_o$.

(*a*) Show that the gravity gradient torque on $\mathscr{B}$, expressed in $\mathscr{F}_o$, is given by

$$\mathscr{F}_o \cdot \underset{\rightarrow}{\mathbf{g}}_c = \frac{3\mu}{R_c^3}\begin{bmatrix} -I_{23}^o \\ I_{13}^o \\ 0 \end{bmatrix} \tag{2}$$

This result was obtained by [Nidey]. It appears to be simpler than when expressed in principal axes $\mathscr{F}_p$, as in (8.1, 23), until it is realized that, while I_1, I_2, and I_3 are constants (for a rigid body), I_{23}^o and I_{13}^o depend on vehicle attitude.

(*b*) To eliminate the attitude dependence in (2), show that

$$\mathscr{F}_o \cdot \underset{\rightarrow}{\mathbf{g}}_c = \frac{3\mu}{R_c^3}\begin{bmatrix} -(I_1c_{13}c_{12} + I_2c_{23}c_{22} + I_3c_{33}c_{32}) \\ I_1c_{13}c_{11} + I_2c_{23}c_{21} + I_3c_{33}c_{31} \\ 0 \end{bmatrix} \tag{3}$$

where $\{c_{ij}\} = \mathbf{C}_{po} \triangleq \mathscr{F}_p \cdot \mathscr{F}_o^T$ and I_1, I_2, and I_3 are principal moments of inertia for $\mathscr{B}$. The equivalent of this result was also obtained by [Nidey].

(*c*) In the same vein, show [presumably from (8.1, 19)] that the attitude-dependent potential energy is given by

$$V_{\text{rot}} = \frac{3\mu}{2R_c^3} I_{33}^o \tag{4}$$

8.2 One way to go from orbital axes $\mathcal{F}_o$ to principal axes for $\mathcal{B}$, $\mathcal{F}_p$, is via the yaw–pitch–roll (ψ, θ, ϕ) sequence of Euler angles familiar to airplane dynamicists ([Etkin, 2]). In the notation of Chapter 2, $\mathbf{C}_{po} = \mathbf{C}_1(\phi)\mathbf{C}_2(\theta)\mathbf{C}_3(\psi)$. Show that the gravity gradient torque on a body in an inverse square field is given [see (8.1, 23)] by

$$\mathcal{F}_p \cdot \underset{\rightarrow}{\mathbf{g}}_c = \frac{3\mu}{R_c^3}\begin{bmatrix} (I_3 - I_2)\sin\phi\cos\phi\cos^2\theta \\ (I_3 - I_1)\cos\phi\sin\theta\cos\theta \\ (I_1 - I_2)\sin\phi\sin\theta\cos\theta \end{bmatrix} \tag{5}$$

This agrees with [Thomson, 3].

8.3 Consider the expressions for gravity gradient torque given by (8.1, 23). For zero torque, one of the principal axes must be vertical (aligned with $\underset{\rightarrow}{\hat{\mathbf{o}}}_3$).

(*a*) Does it matter in what direction the other two principal axes are pointed?

(*b*) At what attitude (at what values of c_{13}, c_{23}, and c_{33}) is the gravity gradient torque a maximum?

(*c*) Referring to (5) above, a spacecraft has its principal axis $\underset{\rightarrow}{\hat{\mathbf{p}}}_3$ off the vertical $\underset{\rightarrow}{\hat{\mathbf{o}}}_3$ by an infinitesimal amount. Explain why θ and ϕ are therefore infinitesimal. Show further that the gravity gradient torque is infinitesimally statically stabilizing for $\underset{\rightarrow}{\hat{\mathbf{p}}}_3$ if I_3 is the minimum moment of inertia. Can $\underset{\rightarrow}{\hat{\mathbf{p}}}_1$ or $\underset{\rightarrow}{\hat{\mathbf{p}}}_2$ be stabilized in this way?

8.4 Consider the term $-\frac{1}{2}(\mu/R_c^3)(I_1 + I_2 + I_3)$ in the gravity gradient potential, given by (8.1, 26). Does this mean that the orbit of a spacecraft of finite size is not Keplerian? (Refer to [Beletskii].)

8.5 The gravity gradient torque on a spacecraft in the neighborhood of an oblate primary is given by (8.1, 23) and (8.1, 43).

(*a*) Verify (8.1, 43).

(*b*) Show that the gravity gradient torque vanishes if the three principal moments of inertia of the spacecraft are equal. Does it follow that the spacecraft is a sphere?

(*c*) A spacecraft is over the equator. For the gravity gradient torque to vanish, show that it is sufficient that one principal axis be vertical and a second principal axis point north/south. Is this necessary?

(*d*) A spacecraft is over one of the poles. For the gravity gradient torque to vanish, show that it is necessary and sufficient that one of the principal axes be vertical (parallel to the polar axis).

(*e*) Verify (8.1, 39).

8.6 Find expressions for the oblate-primary gravity gradient torques expressed in "orbiting" axes $\mathcal{F}_o$. In other words, generalize (3) for an oblate primary. Is it true that $\underset{\rightarrow}{\hat{\mathbf{o}}}_3 \cdot \underset{\rightarrow}{\mathbf{g}}_c = 0$?

8.7 A generalized gravity gradient field is given by

$$dV = \left(V_c + \underset{\rightarrow}{\mathbf{a}} \cdot \underset{\rightarrow}{\mathbf{r}} + \tfrac{1}{2}\underset{\rightarrow}{\mathbf{r}} \cdot \underset{\rightarrow}{\mathbf{B}} \cdot \underset{\rightarrow}{\mathbf{r}}\right) dm \tag{6}$$

where $\underset{\rightarrow}{\mathbf{B}}$ is a symmetrical dyadic and $\underset{\rightarrow}{\mathbf{r}}$ is measured from the mass center of a body $\mathscr{B}$. Let $\mathscr{F}_d$ be a frame in which $\underset{\rightarrow}{\mathbf{B}}$ is diagonal, and let $\mathscr{F}_d$ be the corresponding vectrix. Then,

$$\mathbf{B}^d \triangleq \mathscr{F}_d \cdot \underset{\rightarrow}{\mathbf{B}} \cdot \mathscr{F}_d^T \triangleq \operatorname{diag}\{b_1, b_2, b_3\} \tag{7}$$

(*a*) Show that the gravity gradient torque, expressed in $\mathscr{F}_d$, is

$$\mathscr{F}_d \cdot \underset{\rightarrow}{\mathbf{g}}_c = \begin{bmatrix} (b_3 - b_2) I_{23}^d \\ (b_1 - b_3) I_{31}^d \\ (b_2 - b_1) I_{12}^d \end{bmatrix} \tag{8}$$

where $\mathbf{I}^d \triangleq \mathscr{F}_d \cdot \underset{\rightarrow}{\mathbf{I}} \cdot \mathscr{F}_d^T$. If b_1, b_2, and b_3 are unequal, what can one conclude? If two of them are equal, what can one conclude?

(*b*) For a single inverse-square primary, explain why one may choose $\mathscr{F}_o = \mathscr{F}_d$, and why

$$b_1 = b_2 = \frac{\mu}{2R^3}; \qquad b_3 = \frac{-\mu}{R^3} \tag{9}$$

in this case.

(*c*) Returning to the general gravity gradient field represented by (6), let $\underset{\rightarrow}{\mathbf{B}}$ be expressed in an orbiting frame $\mathscr{F}_o$; thus,

$$\mathbf{B}^o \triangleq \mathscr{F}_o \cdot \underset{\rightarrow}{\mathbf{B}} \cdot \mathscr{F}_o^T \tag{10}$$

and suppose further that the principal-axis frame for $\mathscr{B}$, namely $\mathscr{F}_p$, is infinitesimally different from $\mathscr{F}_o$:

$$\mathbf{C}_{po} = \mathbf{1} - \boldsymbol{\alpha}^{\times} \tag{11}$$

Assuming $\mathscr{B}$ is tri-inertial (I_1, I_2, and I_3 all different), show that the equilibrium for $\mathscr{B}$ is given by the solution to

$$\begin{aligned} (b_{22}^o - b_{33}^o)\alpha_1 - b_{12}^o\alpha_2 + b_{31}^o\alpha_3 &= b_{23}^o \\ b_{12}^o\alpha_1 + (b_{33}^o - b_{11}^o)\alpha_2 - b_{23}^o\alpha_3 &= b_{31}^o \\ -b_{31}^o\alpha_1 + b_{23}^o\alpha_2 + (b_{11}^o - b_{22}^o)\alpha_3 &= b_{12}^o \end{aligned} \tag{12}$$

assuming the solution exists. What do you conclude if the solution to (12) does not exist? What do you conclude if the solution exists but does not satisfy $\|\boldsymbol{\alpha}\| \ll 1$?

(*d*) Now express $\mathbf{B}$ in a principal-axis frame $\mathscr{F}_p$:

$$\mathbf{B}^p \triangleq \mathscr{F}_p \cdot \underset{\rightarrow}{\mathbf{B}} \cdot \mathscr{F}_p^T \tag{13}$$

It was shown in (8.1, 49) that for equilibrium (no torque) we need $b_{23}^p = b_{31}^p = b_{12}^p = 0$. Show that this equilibrium is infinitesimally statically stable provided

$$\begin{aligned} (I_3 - I_2)(b_3^p - b_2^p) &> 0 \\ (I_1 - I_3)(b_1^p - b_3^p) &> 0 \\ (I_2 - I_1)(b_2^p - b_1^p) &> 0 \end{aligned} \tag{14}$$

(*e*) Explain why the conclusion in Problem 8.3*c* is a specific instance of (14).

8.8 Verify (8.1, 45).

8.9 From (8.1, 99), which applies to a body $\mathscr{B}_n$ in a system of N bodies, show that the total gravity gradient torque about ⊕, the mass center of the

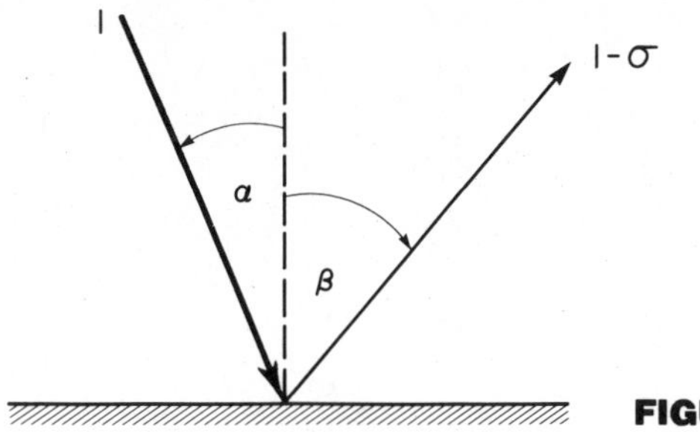

FIGURE 8.16 Geometry for Problem 8.11.

system, is given by

$$\underrightarrow{\mathbf{g}}_c = \sum_{n=1}^{N} \underrightarrow{\mathbf{g}}_c^n = \frac{3\mu}{R_c^3} \underrightarrow{\hat{\mathbf{o}}}_3 \times \underrightarrow{\mathbf{I}} \cdot \underrightarrow{\hat{\mathbf{o}}}_3 \tag{15}$$

where

$$\underrightarrow{\mathbf{I}} = \sum_{n=1}^{N} \underrightarrow{\mathbf{J}}^n \tag{16}$$

Compare this to (8.1, 16) and (8.1, 57).

8.10 The aerodynamic torque is to be calculated on a rapidly rotating spacecraft under the assumptions of (i) total momentum accommodation at the surface, (ii) negligible surface temperature, and (iii) hyperthermal incident flow. Verify that, in body axes,

$$\mathbf{g}_c \doteq \rho_a V_{cR}\left[V_{cR} A_p \mathbf{c}_p^{\times} \hat{\mathbf{V}}_R - (\mathbf{I}_A + \hat{\mathbf{V}}_R^{\times} \mathbf{J}_A)\boldsymbol{\omega}\right] \tag{17}$$

The notation is defined in Section 8.2.

8.11 Consider a surface with the following properties (Fig. 8.16). When molecules are incident at angle α, σ of them are absorbed and reemitted with negligible speed, and $1 - \sigma$ of them are reflected with undiminished speed at the reflection angle β. (This preferred direction might be due, for example, to a preferred direction of the surface molecular lattice.) Show that the tangential and normal momentum accommodation coefficients for this surface are

$$\sigma_t = 1 - (1 - \sigma)\frac{\sin\beta}{\sin\alpha}$$

$$\sigma_n = 1 - (1 - \sigma)\frac{\cos\beta}{\cos\alpha} \tag{18}$$

and note the check at $\beta = \alpha$.

8.12 Consider a flat plate of area A immersed in a free-molecular flow of density ρ_a and (mean) speed V_R (Fig. 8.17). Its *angle of attack* α_a is $(\pi/2) - \alpha$, its momentum accommodation coefficients are σ_t and σ_n, and its surface temperature is not negligible. The *lift* L and *drag* D are respectively perpendicular and parallel to $\underrightarrow{\mathbf{V}}_R$. The lift and drag coefficients are defined as follows: $C_L \triangleq L/(\frac{1}{2}\rho_a V_R^2 A)$, $C_D \triangleq D/(\frac{1}{2}\rho_a V_R^2 A)$

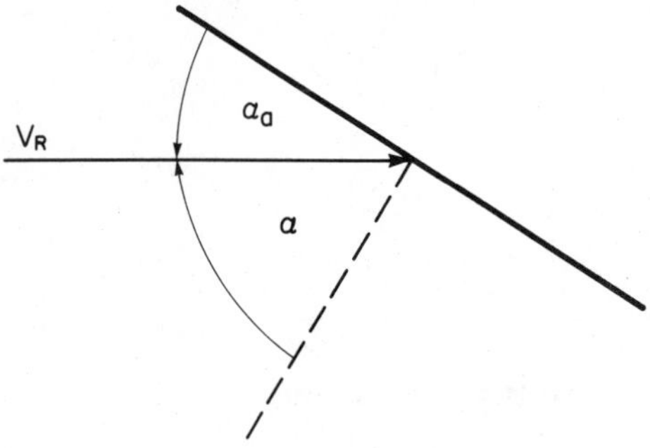

FIGURE 8.17 Flat plate in a free-molecular flow. (See Problem 8.12.)

(*a*) Using the assumption of hyperthermal flow ($S \to \infty$), show from (8.2, 24) that

$$C_L = 2\left[(2 - \sigma_n - \sigma_t)\sin\alpha_a + \sigma_n\left(\frac{V_b}{V_R}\right)\right]\sin\alpha_a\cos\alpha_a \tag{19}$$

$$C_D = 2\left[(2 - \sigma_n - \sigma_t)\sin^2\alpha_a + \sigma_n\left(\frac{V_b}{V_R}\right)\sin\alpha_a + \sigma_t\right]\sin\alpha_a \tag{20}$$

Note that the condition $\cos\alpha \geq 0$ becomes $\sin\alpha_a \geq 0$.

(*b*) For specular reflection, show that (with $\sin\alpha_a \geq 0$)

$$C_L = 4\sin^2\alpha_a\cos\alpha_a; \qquad C_D = 4\sin^3\alpha_a$$

(*c*) For diffuse reflection, show that (with $\sin\alpha_a \geq 0$)

$$C_L = 2\left(\frac{V_b}{V_R}\right)\sin\alpha_a\cos\alpha_a; \qquad C_D = 2\left(\frac{V_b}{V_R}\right)\sin^2\alpha_a + 2\sin\alpha_a$$

(*d*) Show that C_L has a maximum value at

(i) $\alpha_a = \pi/4$, for diffuse reflection.

(ii) $\alpha_a = (\pi/4) + (1/\sqrt{8})(V_R/V_b)(2 - \sigma_n - \sigma_t)$, for nearly diffuse reflection.

(iii) $\alpha_a = \sin^{-1}\sqrt{2/3}$, for specular reflection.

(*e*) Now consider the same flat plate but remove the hyperthermal assumption. Show from the force expression (8.2, 42) that

$$C_L = \left[\frac{2(2 - \sigma_n - \sigma_t)}{\sqrt{\pi}\,S}\exp\left(-S^2\sin^2\alpha_a\right) + 2\sigma_n\left(\frac{V_b}{V_R}\right)\right]\sin\alpha_a\cos\alpha_a + \left[\frac{2 - \sigma_n}{S^2} + 2(2 - \sigma_n - \sigma_t)\sin^2\alpha_a\right]\cos\alpha_a\,\mathrm{erf}\left(S\sin\alpha_a\right) \tag{21}$$

$$C_D = \left[\frac{2(2 - \sigma_n)}{\sqrt{\pi}\,S}\sin^2\alpha_a + \frac{2\sigma_t}{\sqrt{\pi}\,S}\cos^2\alpha_a\right]\exp\left(-S^2\sin^2\alpha_a\right) + 2\left[(2 - \sigma_n)\left(\sin^2\alpha_a + \frac{1}{2S^2}\right) + \sigma_t\cos^2\alpha_a\right]\sin\alpha_a\,\mathrm{erf}\left(S\sin\alpha_a\right) + 2\sigma_n\left(\frac{V_b}{V_R}\right)\sin^2\alpha_a \tag{22}$$

Equations (21) and (22) have been plotted by [Schaaf and Talbot].

(f) Show that (21) and (22) reduce to (19) and (20) in the limit as $S \to \infty$.

8.13 Verify the following integrals of the Maxwell distribution function, given by (8.2, 32), used in the text derivations:

$$\text{(i)}\ m\int_{-\infty}^{\infty}\int_{-\infty}^{\infty}\int_{-\infty}^{\infty} f\,dv_1\,dv_2\,dv_3 = \rho_a \tag{23}$$

$$\text{(ii)}\ m\int_{-\infty}^{\infty} dv_3\int_{-\infty}^{\infty} dv_2\int_{0}^{\infty} v_1 f\,dv_1 = \rho_a V_a \Gamma_1(S\cos\alpha) \tag{24}$$

$$\text{(iii)}\ m\int_{-\infty}^{\infty} dv_3\int_{-\infty}^{\infty} dv_2\int_{0}^{\infty} v_1 v_2 f\,dv_1 = \rho_a V_a^2 S\sin\alpha\,\Gamma_1(S\cos\alpha) \tag{25}$$

$$\text{(iv)}\ m\int_{-\infty}^{\infty} dv_3\int_{-\infty}^{\infty} dv_2\int_{0}^{\infty} v_1^2 f\,dv_1 = \rho_a V_a^2 \Gamma_2(S\cos\alpha) \tag{26}$$

where Γ_1 and Γ_2 are defined in (8.2, 35) and (8.2, 38). In this verification it is helpful to be aware of the following two definite integrals:

$$\text{(i)}\ \int_{a}^{\infty} \exp(-t^2)\,dt = \tfrac{1}{2}\sqrt{\pi}\,(1 - \operatorname{erf} a)$$

$$\text{(ii)}\ \int_{a}^{\infty} t^2 \exp(-t^2)\,dt = \tfrac{1}{4}\sqrt{\pi}\,(1 - \operatorname{erf} a) + \tfrac{1}{2} a \exp(-a^2)$$

Both are valid for any value of a, $-\infty < a < \infty$.

8.14 (a) Show that

$$\lim_{S\to\infty}\,[1 + \operatorname{erf}(S\cos\alpha)] = 2H(\cos\alpha) \tag{27}$$

(b) Using part (a), show that

$$\lim_{S\to\infty}\frac{\Gamma_1(S\cos\alpha)}{S} = H(\cos\alpha)\cos\alpha \tag{28}$$

(c) Show from the definitions of Γ_1 and Γ_2 that

$$\Gamma_2(x) = x\Gamma_1(x) + \tfrac{1}{4}(1 + \operatorname{erf} x) \tag{29}$$

(d) Show that

$$\lim_{S\to\infty}\frac{\Gamma_2(S\cos\alpha)}{S^2} = H(\cos\alpha)\cos^2\alpha \tag{30}$$

(e) Using these results, show that the expression for force and torque, (8.2, 42) and (8.2, 43), reduce when $S \to \infty$ to the hyperthermal expressions (8.2, 24) and (8.2, 26).

8.15 Consider (Fig. 8.18) a sphere of radius a and accommodation coefficients $\{\sigma_t, \sigma_n\}$ in a free-molecular flow of density ρ_a and speed V_R.

(a) Show using (8.2, 24) that the hyperthermal drag coefficient (based on the projected area πa^2) is given by

$$C_D = 2 - \sigma_n + \sigma_t + \left(\frac{4V_b}{3V_R}\right)\sigma_n \tag{31}$$

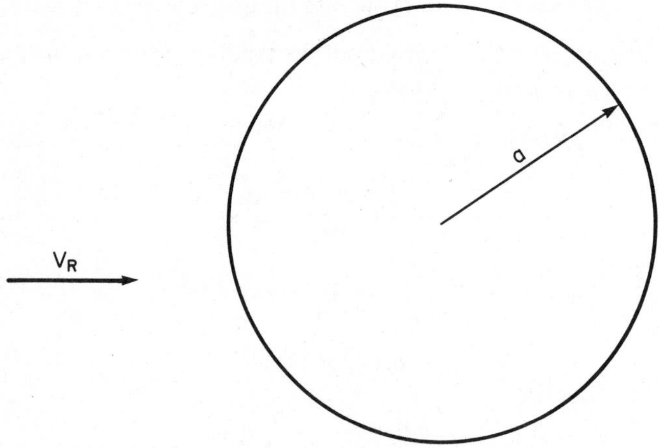

FIGURE 8.18 Sphere in a free-molecular flow. (See Problem 8.15.)

(*b*) Removing the hyperthermal assumption, show from (8.2, 42) that

$$C_D = \left(\frac{2 - \sigma_n + \sigma_t}{2S^2}\right)\left[\left(\frac{4S^4 + 4S^2 - 1}{2S}\right)\operatorname{erf} S + \left(\frac{2S^2 + 1}{\sqrt{\pi}}\right)\exp(-S^2)\right] + \left(\frac{4V_b}{3V_R}\right)\sigma_n \tag{32}$$

This equation has been plotted by [Schaaf and Talbot].

(*c*) Show that (32) reduces to (31) when $S \to \infty$.

8.16 A sphere of radius a rotates in a free-molecular flow $\{\rho_a, V_{cR}\}$. Assuming hyperthermal flow, total surface momentum accommodation, and negligible surface temperature, use (8.2, 12) to show that if the angular velocity of the sphere is

$$\boldsymbol{\omega} = \omega_{\|}\hat{\mathbf{V}}_R + \omega_{\perp}\mathbf{u}_{\perp} \qquad \left(\mathbf{u}_{\perp}^T\hat{\mathbf{V}}_R = 0\right)$$

then the consequent aerodynamic torque is

$$\mathbf{g}_c = -\tfrac{1}{4}\rho_a V_{cR} a^4\left(\omega_{\|}\hat{\mathbf{V}}_R + 3\pi\omega_{\perp}\mathbf{u}_{\perp}\right) \tag{33}$$

Deduce that the rotational energy of the sphere is being dissipated at the rate

$$\dot{T}_{\text{rot}} = -\tfrac{1}{4}\rho_a V_{cR} a^4\left(\omega_{\|}^2 + 3\pi\omega_{\perp}^2\right) \tag{34}$$

8.17 Consider a flat plate of area A exposed to a source of radiation pressure p at an angle of attack $\alpha_a \triangleq (\pi/2) - \alpha$. The surface properties are $\{\sigma_a, \sigma_{rs}, \sigma_{rd}\}$. Show that the forces perpendicular and parallel to the direction of radiation are given ($\sin\alpha_a \geq 0$) by

$$F_{\perp} = 2pA\left(\tfrac{1}{3}\sigma_{rd} + \sigma_{rs}\sin\alpha_a\right)\sin\alpha_a\cos\alpha_a \tag{35}$$

$$F_{\|} = pA\left(\sigma_n + \sigma_{rd} + \tfrac{2}{3}\sigma_{rd}\sin\alpha_a + 2\sigma_{rs}\sin^2\alpha_a\right)\sin\alpha_a \tag{36}$$

For a calculation of p in near-Earth elliptic orbits, see [Flanagan and Modi].

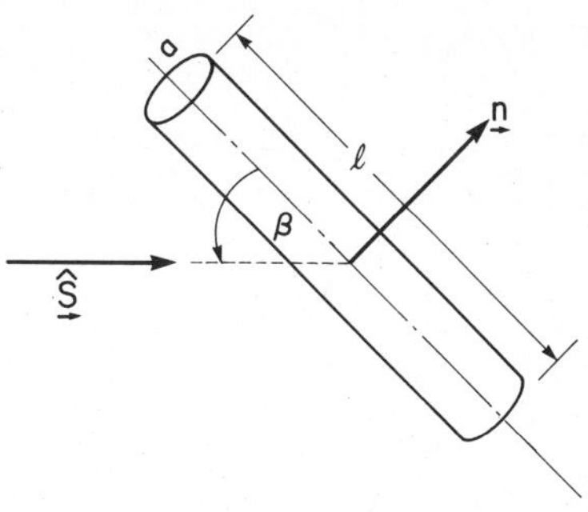

FIGURE 8.19 Solar-radiation pressure incident on a long slender cylinder. (See Problem 8.19.)

8.18 Show that the radiation force on a sphere of radius a whose surface properties are $\{\sigma_a, \sigma_{rs}, \sigma_{rd}\}$ is

$$F = \pi a^2 p\left(\sigma_{rs} + \tfrac{4}{9}\sigma_{rd}\right) \tag{37}$$

8.19 Calculate the radiation force on a cylinder of length l and radius a (Fig. 8.19). Assume that the axis of the cylinder makes an angle β with the radiation direction, that end effects can be ignored, and that the surface properties are $\{\sigma_a, \sigma_{rs}, \sigma_{rd}\}$. Show that

$$\mathbf{f} = 2pal\left[(\sigma_a + \sigma_{rd})\sin\beta\,\hat{\mathbf{S}} + \left(\frac{\pi}{6}\sigma_{rd} + \frac{4}{3}\sigma_{rs}\sin\beta\right)\sin\beta\,\mathbf{n}\right] \tag{38}$$

where $\mathbf{n}$ is a unit normal to the cylinder's axis, in the plane formed by $\hat{\mathbf{S}}$ and the cylinder axis.

8.20 Assume that in Earth's penumbra the gradient of solar radiation pressure (perpendicular to the sun's direction) is $dp/dy \doteq 10^{-12}$ N/m³. See Fig. 8.20. Motivated by large solar power satellites, consider a 20 km × 10 km planar rectangular surface normal to the sun's radiation. Show that the torque about the center of area is 8×10^3 N · m. See also [Sincarsin and Hughes, 2].

8.21 Denote by $\hat{\underset{\to}{\nabla}}$ a unit vector in the direction of $-\underset{\to}{\nabla}\phi_m$, where ϕ_m is the magnetic field potential external to a spacecraft.

(*a*) Show that the magnetic flux density $\underset{\to}{\mathbf{B}}$ for an Earth satellite can be written as

$$\underset{\to}{\mathbf{B}} = \frac{\mu_m}{R^3}(1 + 3\sin^2\lambda_m)\hat{\underset{\to}{\nabla}} \tag{39}$$

where λ_m is the latitude of the vehicle with respect to the geomagnetic equatorial plane.

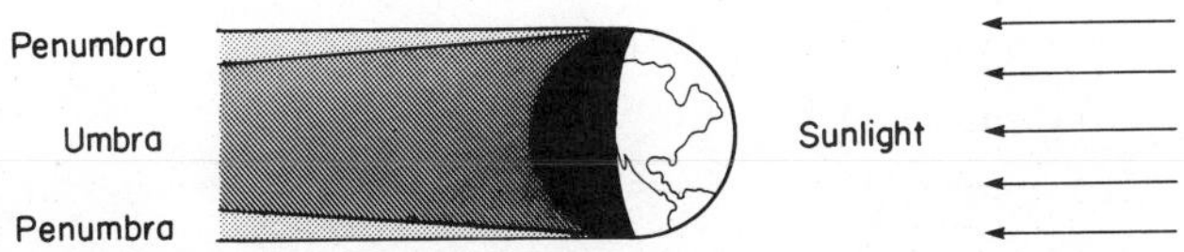

FIGURE 8.20 Gradient in solar-radiation pressure in penumbra. (See Problem 8.20.)

(*b*) Assume that the direction cosines of $\hat{\underset{\rightarrow}{\nabla}}$ and the magnetic moment $\underset{\rightarrow}{\mathbf{m}}_m$ in body axes are $\mathscr{F}_b \cdot \hat{\underset{\rightarrow}{\nabla}} \triangleq [c_{1\nabla} \quad c_{2\nabla} \quad c_{3\nabla}]^T$ and $\mathscr{F}_b \cdot \underset{\rightarrow}{\mathbf{m}}_m \triangleq m_m [c_{1m} \quad c_{2m} \quad c_{3m}]^T$. Show that the magnetic torque produced by the interaction of $\underset{\rightarrow}{\mathbf{m}}_m$ with $\underset{\rightarrow}{\mathbf{B}}$ can be expressed in body axes as follows:

$$\mathbf{g}_m \triangleq \mathscr{F}_b \cdot \underset{\rightarrow}{\mathbf{g}}_m = \frac{\mu_m m_m}{R^3}\left(1 + 3\sin^2\lambda_m\right)^{1/2} \begin{bmatrix} c_{2m}c_{3\nabla} - c_{3m}c_{2\nabla} \\ c_{3m}c_{1\nabla} - c_{1m}c_{3\nabla} \\ c_{1m}c_{2\nabla} - c_{2m}c_{1\nabla} \end{bmatrix} \tag{40}$$

(*c*) Compare (40) with the expression for gravity gradient torque, (8.1, 23), and comment.

(*d*) Assume that the attitude of the spacecraft is given relative to a set of orbiting axes $\mathscr{F}_o$ (see Fig. 9.1) and that the orientation of $\mathscr{F}_o$ is given in terms of the true in terms of the true anomaly η and the orbital elements Ω, i, and ω, with respect to a conventional geocentric coordinate system. Find expressions for $\{c_{1\nabla}, c_{2\nabla}, c_{3\nabla}\}$ in terms of the mentioned variables.

8.22 Use the same notation as in Problem 8.21 and denote the spacecraft angular velocity components in $\mathscr{F}_b$ by, as usual, $[\omega_1 \quad \omega_2 \quad \omega_3] \triangleq \mathscr{F}_b^T \cdot \underset{\rightarrow}{\boldsymbol{\omega}}$. Show from (8.4, 6) that the eddy current torque, expressed in $\mathscr{F}_b$, is given by

$$\mathbf{g} \triangleq \mathscr{F}_b \cdot \underset{\rightarrow}{\mathbf{g}} = -\frac{k\mu_m^2}{R^6}\left(1 + 3\sin^2\lambda_m\right)^2 \times \begin{bmatrix} 1 - c_{1\nabla}^2 & -c_{1\nabla}c_{2\nabla} & -c_{1\nabla}c_{3\nabla} \\ -c_{2\nabla}c_{1\nabla} & 1 - c_{2\nabla}^2 & -c_{2\nabla}c_{3\nabla} \\ -c_{3\nabla}c_{1\nabla} & -c_{3\nabla}c_{2\nabla} & 1 - c_{3\nabla}^2 \end{bmatrix} \begin{bmatrix} \omega_1 \\ \omega_2 \\ \omega_3 \end{bmatrix} \tag{41}$$

CHAPTER 9

GRAVITATIONAL STABILIZATION

The preceding chapters provide a foundation for understanding and predicting the attitude motion of most types of spacecraft. Even more important, they lay the groundwork for designing appropriate attitude behavior into the vehicle. One of the distinctions between *natural* and *artificial* satellites is that the motion (both attitude and orbital) of the latter is in principle controllable. It is sometimes desirable, for example, to maintain the whole spacecraft in a particular orientation. In other applications, it may be adequate to point one axis (e.g., an antenna) in a preferred direction. Indeed, it is quite common to have part of the spacecraft aimed in one direction (at Earth, for example) and another part oriented in a second direction (e.g., toward the sun).

In many modern applications, the attitude errors permitted are so small that only a fully automatic (fully "active") control system can meet mission specifications. Typically, an on-board attitude control system (ACS) includes sensing hardware for measuring attitude, attitude rate, or both ([Wertz]). This information is continuously presented to a computer whose software implements a preprogrammed control policy by issuing commands to appropriate hardware elements called effectors, or actuators, which in turn apply a corrective torque to the vehicle. The immense diversity of ACS specifications, sensors, actuators, and control policies makes "active" spacecraft attitude stabilization and control a voluminous study in itself, and this subject is for this reason beyond the scope of this book.

Many spacecraft stabilization techniques, however, depend for their success on the intrinsic vehicle dynamics, either by including in the configurational design appropriate dynamical elements (discrete dampers, for example), or by exploiting some naturally occurring force field to provide the desired stabilizing torque. These are called "passive stabilization" techniques since, in contradistinction to "active" systems, they consume no power, require no hardware for sensing or actuation, and make no demands on software. In other words, a spacecraft whose attitude is stabilized entirely by passive means would be said by a control systems engineer to have no attitude control system. Very few modern spacecraft employ totally passive, or totally active, attitude stabilization. Most are either predominantly passive, sometimes called "semipassive," or predominantly active ("semiactive"). Several examples of passive, semipassive, semiactive, and active attitude control have been given by [Sabroff, 2].

9.1 CONTEXT

This chapter deals with passive gravity gradient stabilization—ideally a totally passive scheme based on the gravitational torques discussed in Section 8.1. It can be used for one-axis (pointing axis) or three-axis control. It will be recalled from Section 8.1 that differential gravity acting over a spacecraft can produce a torque. The fundamental formula for this torque, obtained for a body immersed in the inverse-square gravitational field of a large primary, was given by (8.1, 23) and is repeated here for convenience:

$$\mathbf{g} = 3\left(\frac{\mu}{R_c^3}\right)\begin{bmatrix}(I_3 - I_2)c_{23}c_{33}\\(I_1 - I_3)c_{33}c_{13}\\(I_2 - I_1)c_{13}c_{23}\end{bmatrix} \tag{1}$$

The gravity gradient torque $\mathbf{g}$ has been expressed in the principal-axis frame for the body, $\mathscr{F}_p$; μ is the gravitational constant of the primary [for Earth, see (8.1, 10)]; R_c is the distance between the primary and spacecraft mass centers; I_i are, as usual, centroidal principal moments of inertia of the spacecraft; and c_{i3} are the direction cosines with respect to $\mathscr{F}_p$ of the local vertical at the spacecraft mass center. This local vertical is denoted by the unit vector $\underset{\rightarrow}{\hat{\mathbf{o}}}_3$, one of the three unit vectors in the *orbiting frame* $\mathscr{F}_o$ to be defined presently. Thus,

$$[c_{13} \quad c_{23} \quad c_{33}]^T = \mathscr{F}_p \cdot \underset{\rightarrow}{\hat{\mathbf{o}}}_3 \tag{2}$$

It is straightforward to show from (1) that $\underset{\rightarrow}{\mathbf{g}} \cdot \underset{\rightarrow}{\hat{\mathbf{o}}}_3 = 0$; there is no torque component about the local vertical.

The orbiting frame $\mathscr{F}_o$ is shown in Fig. 9.1. Its origin is at the spacecraft mass center; $\underset{\rightarrow}{\hat{\mathbf{o}}}_3$ points to the center of the primary, and $\underset{\rightarrow}{\hat{\mathbf{o}}}_2$ is perpendicular to the orbit,

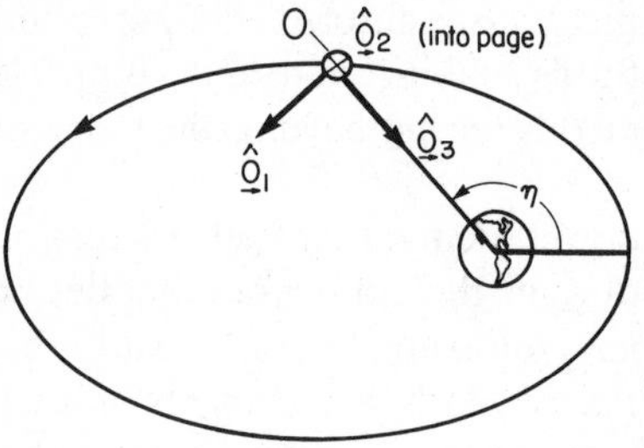

FIGURE 9.1 The orbiting frame $\mathscr{F}_o$.

its sense determined so that the third vector, $\underset{\rightarrow}{\hat{\mathbf{o}}}_1$, makes an acute angle with the orbital velocity vector. Extending aircraft terminology to spacecraft, $\{\underset{\rightarrow}{\hat{\mathbf{o}}}_1, \underset{\rightarrow}{\hat{\mathbf{o}}}_2, \underset{\rightarrow}{\hat{\mathbf{o}}}_3\}$ will be called the {roll, pitch, yaw} axes. Also shown in Fig. 9.1 is the angle η, the *true anomaly*. By convention, the roll, pitch, and yaw angles form the Euler angle sequence given by (2.3, 18), where $\theta_1 \to \alpha_1 \equiv$ roll, $\theta_2 \to \alpha_2 \equiv$ pitch, and $\theta_3 \to \alpha_3 \equiv$ yaw. Thus,

$$\mathbf{C}_{po} = \mathbf{C}_1(\alpha_1)\,\mathbf{C}_2(\alpha_2)\,\mathbf{C}_3(\alpha_3) \tag{3}$$

and it is evident from (2.2, 19) that

$$c_{13} = -\sin\alpha_2; \qquad c_{23} = \sin\alpha_1\cos\alpha_2; \qquad c_{33} = \cos\alpha_1\cos\alpha_2$$

In terms of Euler angles, then, the gravity gradient torque components tabulated in (1) become

$$\mathbf{g} = 3\left(\frac{\mu}{R_c^3}\right)\begin{bmatrix} (I_3 - I_2)\sin\alpha_1\cos\alpha_1\cos^2\alpha_2 \\ (I_3 - I_1)\cos\alpha_1\sin\alpha_2\cos\alpha_2 \\ (I_1 - I_2)\sin\alpha_1\sin\alpha_2\cos\alpha_2 \end{bmatrix} \tag{4}$$

However, as explained in Chapter 2, Euler angles are not recommended for numerical work with arbitrary attitude motions; they are most effective for infinitesimal analyses or when the motion is about a single axis so that only one angle is needed.

It is evident from (4) that gravity torque does not depend on α_3 (yaw). However, if the roll and pitch angles are infinitesimally small, then to first order,

$$\mathbf{g} = 3\left(\frac{\mu}{R_c^3}\right)\begin{bmatrix} (I_3 - I_2)\alpha_1 \\ (I_3 - I_1)\alpha_2 \\ 0 \end{bmatrix} \tag{5}$$

which demonstrates that there will be a stabilizing gravity gradient torque about the roll and pitch axes, provided I_3 is less than I_1 and I_2. (Remember, a positive rotational spring with constant $k > 0$ contributes a torque $-k\alpha$.) Thus, if the minor inertia axis of a satellite is pointing nearly vertically, the satellite will experience small gravity gradient torques about the roll and pitch axes that tend to move the minor axis toward the vertical. This implies that a spacecraft located at a fixed position above the Earth would have its minor axis stabilized to the local vertical (assuming all other environmental torques can be ignored). Further insight into this conclusion is available from the visualization discussed in Problem 9.1.

One must not be too hasty in applying this idea to Earth satellites, which are not located at fixed positions above Earth as they follow their orbital paths. The "local vertical" is a constantly changing direction. If a satellite is to be stabilized to the local vertical, it must continuously rotate about the orbit normal at the rate $\dot{\eta}$ (see Fig. 9.1). Therefore, the static analysis above, while showing promise for gravity gradient stabilization, must await a proper dynamical analysis for confirmation. This analysis is the subject of this and the next section. Then, in

Section 9.3, practical gravity gradient stabilization systems are described. In Section 9.4, the adverse influence of other environmental torques is admitted and some examples of flight experience are cited. An evaluation of the merits of gravity gradient stabilization concludes the chapter.

Attitude Motion of a Rigid Body in Orbit

The rotational motion of a single rigid body with no external torque acting upon it was the subject of Chapter 4. The chief objective of the present section is to extend this study to embrace the motion of a rigid body which experiences a gravity gradient torque as it orbits a gravitational primary. We shall henceforth refer to this primary as Earth, although any of the celestial primaries is of course equally appropriate.

The coupled attitude-orbital motion of a rigid body in a gravitational field is an interesting dynamical problem and accordingly has prompted a considerable body of literature. We shall now examine this problem briefly while at the same time realizing that it is in some respects somewhat esoteric from a satellite stabilization standpoint. Satellites are not rigid bodies. They are at best quasi-rigid. Chapter 5 gives ample evidence that quasi-rigidity can have a major impact on attitude motion. Indeed, as we shall see in Section 9.3, an adequate damping capability is an intrinsic feature of all practical gravity gradient stabilization designs. Furthermore, gravity gradient torques are inherently weak, and the disturbing influence of other environmental torques must eventually be considered. Also, the weak coupling from the attitude motion to the orbital motion will almost never be the principal orbital perturbation. Nevertheless, the attitude motion of a rigid body in orbit, with all but inverse-square gravity excluded, is still the best starting point from which to begin a study of gravity gradient stabilization.

Orbital-Attitude Coupling

Several analyses have been made of a rigid body in orbit with only gravity acting (e.g., [Doolin]). These tend to be rather long and would consume more space here than can be afforded. A general impression of the more interesting aspects can be conveyed somewhat more briefly perhaps. Let us begin with the equations for attitude motion. From Chapter 4,

$$\mathbf{I}\dot{\boldsymbol{\omega}} + \boldsymbol{\omega}^{\times}\mathbf{I}\boldsymbol{\omega} = \mathbf{g} \tag{6}$$

where $\boldsymbol{\omega}$ is the angular velocity of the principal-axis frame $\mathcal{F}_p$ with respect to an inertial frame $\mathcal{F}_i$ (i.e., the absolute angular velocity) and $\mathbf{g}$ is the external torque. We temporarily restrict $\mathbf{g}$ to arise solely from an inverse-square gravitational field. When stabilization with respect to $\mathcal{F}_i$ is the objective, $\boldsymbol{\omega}$ is of direct interest. For Earth-pointing applications (stabilization with respect to $\mathcal{F}_o$), one is concerned instead with the angular velocity of the satellite with respect to $\mathcal{F}_o$, denoted by $\underset{\rightarrow}{\boldsymbol{\omega}}_{\alpha}$.

Now,

$$\underset{\rightarrow}{\boldsymbol{\omega}} = \underset{\rightarrow\alpha}{\boldsymbol{\omega}} + \underset{\rightarrow o}{\boldsymbol{\omega}} \tag{7}$$

where $\underset{\rightarrow o}{\boldsymbol{\omega}}$ is the angular velocity of $\mathscr{F}_o$ with respect to an inertial frame $\mathscr{F}_i$. We can agree to express $\underset{\rightarrow\alpha}{\boldsymbol{\omega}}$ in $\mathscr{F}_p$ and $\underset{\rightarrow o}{\boldsymbol{\omega}}$ in $\mathscr{F}_o$:

$$\boldsymbol{\omega}_\alpha \triangleq \mathscr{F}_p \cdot \underset{\rightarrow\alpha}{\boldsymbol{\omega}}; \qquad \boldsymbol{\omega}_o \triangleq \mathscr{F}_o \cdot \underset{\rightarrow o}{\boldsymbol{\omega}} \tag{8}$$

whereupon, from (2.3, 9),

$$\boldsymbol{\omega} = \boldsymbol{\omega}_\alpha + \mathbf{C}_{po}\boldsymbol{\omega}_o \tag{9}$$

With ideal Earth-pointing stabilization, $\boldsymbol{\omega}_\alpha \equiv \mathbf{0}$ and $\mathbf{C}_{po} \equiv \mathbf{1}$.

It is now important to clarify a subtle point concerning the "orbit" and the "orbiting frame" shown in Fig. 9.1. As Newton first showed, a *point mass* moving solely under the influence of an inverse-square primary follows a conic-section path with respect to the primary. If the point mass has insufficient energy to escape, the conic section is an ellipse—a closed path, or *orbit*. However, if slight disturbing forces are present, the trajectory in general ceases to be an ellipse and ceases even to be a closed path (orbit). There is still merit however in visualizing an "instantaneous," or *osculating*, ellipse at each point in the trajectory (Fig. 9.2). Mathematically, the osculating ellipse at a point in the trajectory is defined to be the path the point mass would follow if all disturbing forces ceased at that point. To be precise, then, the ellipse shown in Fig. 9.1 is the osculating ellipse at the point O. In our present problem, we have proscribed all fields except the inverse-square gravitational field. Therefore, if the satellite were a point mass, the osculating ellipse would be constant (i.e., the path would revert to being simply an ellipse). With the satellite modeled as a rigid body of small but nonzero size, it turns out (as we shall shortly see) that the attitude and orbital motions are coupled: attitude motion can produce (small) perturbing forces on the satellite, deflecting it slightly from its purely elliptic path. It is this dependence of orbital motion on attitude motion that intrigues dynamicists.

On the other hand, the dependence of the attitude motion on the orbital motion is more readily seen. From Section 8.1 in general and (8.1, 21) in particular, we insert the expression for the gravitational torque on the right side

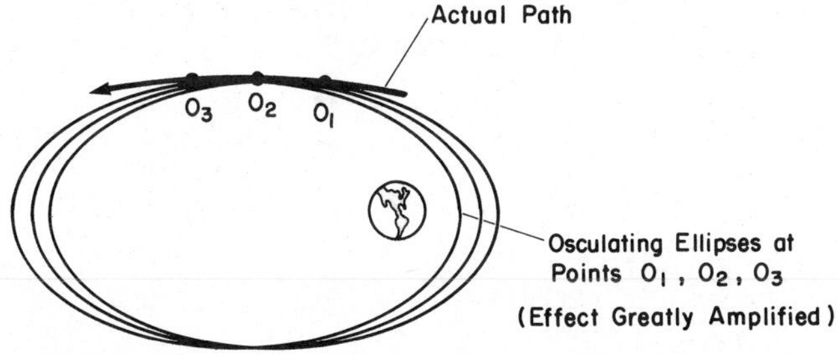

FIGURE 9.2 The osculating (instantaneous) ellipse.

of (6) to obtain

$$\mathbf{I}\dot{\boldsymbol{\omega}} + \boldsymbol{\omega}^{\times}\mathbf{I}\boldsymbol{\omega} = 3\left(\frac{\mu}{R_c^3}\right)\mathbf{c}_3^{\times}\mathbf{I}\mathbf{c}_3 + O(\varepsilon^2) \tag{10}$$

where, recall,

$$\mathbf{c}_3 = [c_{13} \quad c_{23} \quad c_{33}]^T \tag{11}$$

and $\varepsilon = l/R_c$ is a typical satellite length, nondimensionalized by R_c. For a discussion of the magnitude of ε, consult Section 8.1. It is immediately obvious that if R_c is not constant (i.e., if the orbit is not circular), the magnitude of the gravity gradient torque will vary. Also, $\mathbf{c}_3$ depends on the location of the local vertical, which varies throughout the orbit.

To find the effect of attitude motion on the orbit, we must carry the expansion of the gravitational field $\underset{\rightarrow}{\mathbf{f}}$ out to $O(\varepsilon^2)$. Since, on a mass element dm,

$$d\underset{\rightarrow}{\mathbf{f}} = -\frac{\mu\underset{\rightarrow}{\mathbf{R}}}{R^3}\,dm \qquad (\underset{\rightarrow}{\mathbf{R}} = \underset{\rightarrow}{\mathbf{R}}_c + \underset{\rightarrow}{\mathbf{r}}) \tag{12}$$

we have (see Problem 9.2)

$$\begin{aligned} d\underset{\rightarrow}{\mathbf{f}} = &\left(\frac{\mu\,dm}{R_c^2}\right)\underset{\rightarrow}{\hat{\mathbf{o}}}_3 \\ &+\left(\frac{\mu\,dm}{R_c^3}\right)(3\underset{\rightarrow}{\hat{\mathbf{o}}}_3\underset{\rightarrow}{\hat{\mathbf{o}}}_3 - \underset{\rightarrow}{\mathbf{1}})\cdot\underset{\rightarrow}{\mathbf{r}} \\ &+\left(\frac{3\mu\,dm}{2R_c^4}\right)\left[5(\underset{\rightarrow}{\hat{\mathbf{o}}}_3\cdot\underset{\rightarrow}{\mathbf{r}})^2\underset{\rightarrow}{\mathbf{1}} - r^2\underset{\rightarrow}{\mathbf{1}} - 2\underset{\rightarrow}{\mathbf{r}}\underset{\rightarrow}{\mathbf{r}}\right]\cdot\underset{\rightarrow}{\hat{\mathbf{o}}}_3 \\ &+O(\varepsilon^3) \end{aligned} \tag{13}$$

The first term is the familiar inverse-square force which leads to an elliptic orbit. The second line in (13) contains terms of $O(\varepsilon)$; these integrate to zero over the satellite because the origin of $\underset{\rightarrow}{\mathbf{r}}$ was chosen to be the spacecraft mass center. The third line in (13) contains the terms of $O(\varepsilon^2)$; these are the attitude-dependent disturbing forces $d\underset{\rightarrow}{\mathbf{f}}_d$ which cause the trajectory to deviate from an ellipse. Integrating (13) over the satellite, we find (as in Problem 9.2)

$$\begin{aligned} \underset{\rightarrow}{\mathbf{f}} = &\left(\frac{\mu m}{R_c^2}\right)\underset{\rightarrow}{\hat{\mathbf{o}}}_3 \\ &+\left(\frac{3\mu}{2R_c^4}\right)\left\{\left[(I_1+I_2+I_3)\underset{\rightarrow}{\mathbf{1}} + 2\underset{\rightarrow}{\mathbf{I}}\right]\cdot\underset{\rightarrow}{\hat{\mathbf{o}}}_3 - 5(\underset{\rightarrow}{\hat{\mathbf{o}}}_3\cdot\underset{\rightarrow}{\mathbf{I}}\cdot\underset{\rightarrow}{\hat{\mathbf{o}}}_3)\underset{\rightarrow}{\hat{\mathbf{o}}}_3\right\} \\ &+O(\varepsilon^3) \end{aligned} \tag{14}$$

To second order in ε, then, the disturbing force, expressed in $\mathscr{F}_o$, is

$$\mathbf{f}_d = \mathscr{F}_o \cdot \underset{\rightarrow}{\mathbf{f}}_d = \frac{3\mu}{2R_c^4}\begin{bmatrix} 2\underset{\rightarrow}{\hat{\mathbf{o}}}_1 \cdot \underset{\rightarrow}{\mathbf{I}} \cdot \underset{\rightarrow}{\hat{\mathbf{o}}}_3 \\ 2\underset{\rightarrow}{\hat{\mathbf{o}}}_2 \cdot \underset{\rightarrow}{\mathbf{I}} \cdot \underset{\rightarrow}{\hat{\mathbf{o}}}_3 \\ (I_1 + I_2 + I_3) - 3\underset{\rightarrow}{\hat{\mathbf{o}}}_3 \cdot \underset{\rightarrow}{\mathbf{I}} \cdot \underset{\rightarrow}{\hat{\mathbf{o}}}_3 \end{bmatrix} \tag{15}$$

Although the inertia dyadic $\underset{\rightarrow}{\mathbf{I}}$ is constant when expressed in any body-fixed frame (including $\mathscr{F}_p$), it is not constant in $\mathscr{F}_o$ but is attitude dependent. This attitude dependence is the object of the present derivation. Using vectrix notation (Appendix B), one calculates as follows:

$$\begin{aligned} \underset{\rightarrow}{\hat{\mathbf{o}}}_i \cdot \underset{\rightarrow}{\mathbf{I}} \cdot \underset{\rightarrow}{\hat{\mathbf{o}}}_3 &= \mathbf{1}_i^T \mathscr{F}_o \cdot \underset{\rightarrow}{\mathbf{I}} \cdot \mathscr{F}_o^T \mathbf{1}_3 \\ &= \mathbf{1}_i^T \mathbf{C}_{op} \mathscr{F}_p \cdot \underset{\rightarrow}{\mathbf{I}} \cdot \mathscr{F}_p^T \mathbf{C}_{po} \mathbf{1}_3 \\ &= \mathbf{1}_i^T \mathbf{C}_{op} \mathbf{I} \mathbf{C}_{po} \mathbf{1}_3 = \mathbf{c}_i^T \mathbf{I} \mathbf{c}_3 \qquad (i = 1, 2, 3) \end{aligned} \tag{16}$$

where $I = \operatorname{diag}\{I_1, I_2, I_3\}$ is the inertia dyadic expressed in $\mathscr{F}_p$ and $\mathbf{c}_i$ is the ith column of $\mathbf{C}_{po} \equiv \{c_{ij}\}$. With the aid of the scalar products in (16), the attitude-dependent gravitational force perturbations are, in $\mathscr{F}_o$,

$$\begin{aligned} f_{d1} &= \left(\frac{3\mu}{R_c^4}\right)(I_1 c_{11} c_{13} + I_2 c_{21} c_{23} + I_3 c_{31} c_{33}) \\ f_{d2} &= \left(\frac{3\mu}{R_c^4}\right)(I_1 c_{12} c_{13} + I_2 c_{22} c_{23} + I_3 c_{32} c_{33}) \\ f_{d3} &= \left(\frac{3\mu}{2R_c^4}\right)\left[(I_1 + I_2 + I_3) - 3\left(I_1 c_{13}^2 + I_2 c_{23}^2 + I_3 c_{33}^2\right)\right] \end{aligned} \tag{17}$$

correct to $O(\varepsilon^2)$. These results indicate that even if the spacecraft is in a purely inverse-square gravitational field, the mass center does not move as a point mass would move. This is due to the weak, attitude-dependent disturbance given to $O(\varepsilon^2)$ by (17), which deflects the mass center from its elliptic path. We conclude that the orbital motion affects the attitude motion, and vice versa.

Summary of the Coupled Orbital-Attitude Equations

We now assemble the coupled orbital-attitude motion equations. From now on, we drop the subscript c on R for simplicity. Looking first at the translational (orbital) equations, we have, from (14),

$$m\ddot{\underset{\rightarrow}{\mathbf{R}}} = \left(\frac{\mu m}{R^2}\right)\underset{\rightarrow}{\hat{\mathbf{o}}}_3 + \underset{\rightarrow}{\mathscr{F}}_o^T \mathbf{f}_d\left(R, \mathbf{C}_{po}\right) \tag{18}$$

The explicit form of $\mathbf{f}_d(R, \mathbf{C}_{po})$ was given in (17). The attitude dependence of $\mathbf{f}_d$ through $\mathbf{C}_{po}(t)$ is clear. A convenient frame in which to express (18) is $\mathscr{F}_o$. In $\mathscr{F}_o$,

$$\ddot{\underset{\rightarrow}{\mathbf{R}}} = \mathscr{F}_o^T \begin{bmatrix} -2\omega_{o2}\dot{R} - \dot{\omega}_{o2} R \\ -\omega_{o2}\omega_{o3} R \\ -\ddot{R} + \omega_{o2}^2 R \end{bmatrix} \tag{19}$$

where

$$[\omega_{o1} \quad \omega_{o2} \quad \omega_{o3}] \triangleq \mathscr{F}_o^T \cdot \underset{\rightarrow}{\omega}_o \tag{20}$$

and it has been noted that $\omega_{o1} \equiv 0$ from the definition of $\mathscr{F}_o$ (see Problem 9.3). Therefore, the orbital equations to be integrated are

$$\ddot{R} - \omega_{o2}^2 R = -\frac{\mu}{R^2} - \frac{f_{d3}(R, \mathbf{C}_{po})}{m} \tag{A}$$

$$(R^2\omega_{o2})^{\cdot} = \frac{-Rf_{d1}(R, \mathbf{C}_{po})}{m} \tag{B}$$

$$\omega_{o3} = \frac{-f_{d2}(R, \mathbf{C}_{po})}{m\omega_{o2}R} \tag{C}$$

Note that equation (C) is an algebraic relationship.

The rotational set is based on (10):

$$\mathbf{I}\dot{\boldsymbol{\omega}} + \boldsymbol{\omega}^\times \mathbf{I}\boldsymbol{\omega} = 3\left(\frac{\mu}{R^3}\right)\mathbf{c}_3^\times \mathbf{I}\mathbf{c}_3 \tag{D}$$

where $\mathbf{c}_3$ is the third column of $\mathbf{C}_{po}$. In turn, $\mathbf{C}_{po}$ is found by integrating the following equation [see (2.3, 5) in Chapter 2]:

$$\begin{aligned}\dot{\mathbf{C}}_{po} &= -\boldsymbol{\omega}_\alpha^\times \mathbf{C}_{po} \\ &= -(\boldsymbol{\omega} - \mathbf{C}_{po}\boldsymbol{\omega}_o)^\times \mathbf{C}_{po} \\ &= -\boldsymbol{\omega}^\times \mathbf{C}_{po} + \mathbf{C}_{po}\boldsymbol{\omega}_o^\times \end{aligned} \tag{E}$$

[The identity $(\mathbf{C}_{po}\boldsymbol{\omega}_o)^\times = \mathbf{C}_{po}\boldsymbol{\omega}_o^\times \mathbf{C}_{op}$ has been employed; see also (9).] The matrix $\boldsymbol{\omega}_o$ is of course available from integration of (B) and (C), with

$$\boldsymbol{\omega}_o = [0 \quad \omega_{o2} \quad \omega_{o3}]^T \tag{E$'$}$$

Next, the absolute attitude of the spacecraft can be determined from

$$\dot{\mathbf{C}}_{pi} = -\boldsymbol{\omega}^\times \mathbf{C}_{pi} \tag{F}$$

or, if desired, the attitude with respect to $\mathscr{F}_o$ is already available from equation (E). Finally, the absolute orientation of $\mathscr{F}_o$ can be found from

$$\mathbf{C}_{oi} = \mathbf{C}_{po}^T \mathbf{C}_{pi} \tag{F$'$}$$

Or, alternatively, instead of integrating (F) and using (F)′, we may directly integrate

$$\dot{\mathbf{C}}_{oi} = -\boldsymbol{\omega}_o^\times \mathbf{C}_{oi} \tag{F$''$}$$

The system of equations (A) through (F)″ is a complete set from which the attitude and orbital motion can be determined. As always, the kinematical equations (E), (F), and (F)″ can be taken literally (direction cosines used to represent attitude), or they can be taken indicatively if another parametrization is selected. (See Chapter 2 for a discussion of the alternatives.)

Also of interest are expressions for the system angular momentum and energy. It is shown in Problem 9.4 that these are indeed integrals of the motion, with the orbital terms orders of magnitude larger than the attitude terms.

Perturbation Ideas

The lightly coupled, nonlinear system (A) through (F)″ must in general be solved numerically. Further analytical progress can be made however by realizing that the orbital part of the solution is, after all, almost known. Because the coupling is small, the orbit must be almost a Keplerian ellipse. It would seem wise to explore how this knowledge can be used. One possibility (see Fig. 9.2) is to apply Lagrange's planetary equations. [Musen] has given a modern derivation of these equations. For example, the orbital semimajor axis a and eccentricity e change (slowly) as follows:

$$\dot{a} = \frac{2a^2}{(\mu p)^{1/2}}\left[(1 + e\cos\eta)f_{d1} - (e\sin\eta)f_{d3}\right]$$

$$\dot{e} = \left(\frac{p}{\mu}\right)^{1/2}\left[(\cos\eta + \cos E)f_{d1} - (\sin\eta)f_{d3}\right] \tag{21}$$

where the *parameter* p is $a(1 - e^2)$ and E, the *eccentric anomaly*, is related to time by Kepler's famous equation

$$E - e\sin E = \left(\frac{\mu}{a^3}\right)^{1/2}(t - t_p) \tag{22}$$

and t_p is the time at which the spacecraft passes through perigee (closest approach to Earth). In terms of E, the rapidly changing variables are

$$R = a(1 - e\cos E)$$

$$\eta = \tan^{-1}\left\{(1 - e^2)^{1/2}\sin E, \cos E - e\right\} \tag{23}$$

and the slowly changing variables $a(t)$ and $e(t)$ are found by integrating (21). Equations similar to (21) can be produced (see [Musen]) for the other four orbital elements, but we shall not pursue this idea any further here except to note that the distinction between rapidly and slowly changing variables has also led to the successful application of the method of averaging to this problem [Mohan, Breakwell, and Lange].

A different technique is the following: whereas Lagrange's planetary equations deal with an ellipse that is slowly evolving with time—the osculating ellipse—it is also possible to describe the motion with respect to a *fixed* ellipse. The fixed ellipse corresponds to $\mathbf{f}_d \equiv \mathbf{0}$. This approach is particularly attractive when one also linearizes with respect to position deviation. Thus, the difference between where the spacecraft actually is and where it would have been on the fixed reference ellipse without perturbations is taken to be a first-order quantity. We shall use the subscript e to denote quantities that refer to the reference ellipse.

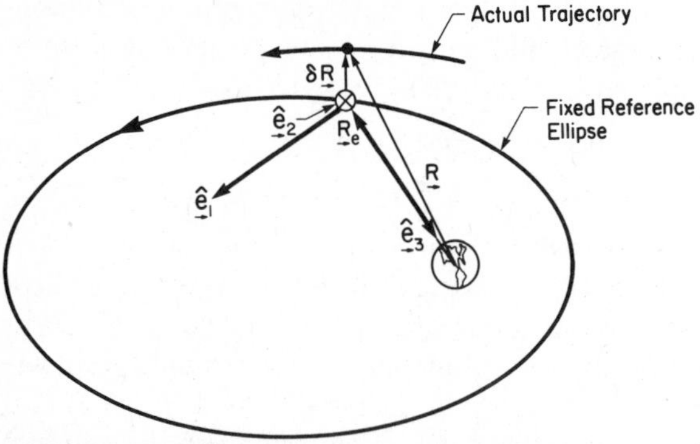

FIGURE 9.3 Fixed reference ellipse.

Thus, as in Fig. 9.3,

$$\underset{\rightarrow}{\mathbf{R}} = \underset{\rightarrow}{\mathbf{R}}_e + \delta\underset{\rightarrow}{\mathbf{R}} \tag{24}$$

indicating that the spacecraft is displaced by the vector $\delta\underset{\rightarrow}{\mathbf{R}}$ from its "reference ellipse" position. We shall treat $\delta\underset{\rightarrow}{\mathbf{R}}$ as first order and neglect second-order terms. The reference position $\underset{\rightarrow}{\mathbf{R}}_e(t)$ satisfies

$$\ddot{\underset{\rightarrow}{\mathbf{R}}}_e + \left(\frac{\mu}{R_e^3}\right)\underset{\rightarrow}{\mathbf{R}}_e = \underset{\rightarrow}{\mathbf{0}} \tag{25}$$

Also helpful is a frame $\mathscr{F}_e$, which is the orbiting frame for the reference ellipse. It is assumed that $\mathscr{F}_o$ differs only infinitesimally from $\mathscr{F}_e$.

Now, insert (24) into the translational motion equation (18):

$$m\,\delta\ddot{\underset{\rightarrow}{\mathbf{R}}} = \left(\frac{\mu m}{R^2}\right)\underset{\rightarrow}{\hat{\mathbf{o}}}_3 - \left(\frac{\mu m}{R_e^2}\right)\underset{\rightarrow}{\hat{\mathbf{e}}}_3 + \mathscr{F}_o^T\mathbf{f}_d(R, \mathbf{C}_{po}) \tag{26}$$

where $\underset{\rightarrow}{\hat{\mathbf{e}}}_3 \triangleq -\underset{\rightarrow}{\mathbf{R}}_e/R_e$. It can be shown (Problem 9.5) that

$$\underset{\rightarrow}{\hat{\mathbf{o}}}_3 - \underset{\rightarrow}{\hat{\mathbf{e}}}_3 = (\underset{\rightarrow}{\hat{\mathbf{e}}}_3\underset{\rightarrow}{\hat{\mathbf{e}}}_3 - \underset{\rightarrow}{\mathbf{1}})\cdot\frac{\delta\underset{\rightarrow}{\mathbf{R}}}{R_e} \tag{27}$$

$$R^{-2} = R_e^{-2}\left(1 + 2\frac{\delta R_3}{R_e}\right) \tag{28}$$

$$\delta\ddot{\underset{\rightarrow}{\mathbf{R}}} = \mathscr{F}_e^T\begin{bmatrix}\delta\ddot{R}_1 - \dot{\eta}^2\delta R_1 - 2\dot{\eta}\,\delta\dot{R}_3 - \ddot{\eta}\,\delta R_3\\ \delta\ddot{R}_2\\ \delta\ddot{R}_3 - \dot{\eta}^2\delta R_3 + 2\dot{\eta}\,\delta\dot{R}_1 + \ddot{\eta}\,\delta R_1\end{bmatrix} \tag{29}$$

where the absolute angular velocity of $\mathscr{F}_e$ is simply

$$\underset{\rightarrow}{\boldsymbol{\omega}}_e = [0 \quad -\dot{\eta} \quad 0]\mathscr{F}_e \tag{30}$$

and the components of $\delta\underset{\rightarrow}{\mathbf{R}}$ in $\mathscr{F}_e$ have been denoted as follows:

$$\delta\underset{\rightarrow}{\mathbf{R}} = [\delta R_1 \quad \delta R_2 \quad \delta R_3]\mathscr{F}_e \tag{31}$$

The final equations for $\delta\underset{\rightarrow}{\mathbf{R}}(t)$ are found by substitution of (27) through (29) into (26) and again discarding second-order terms. The result, in scalar form, is

$$m\begin{bmatrix} \delta\ddot{R}_1 - 2\dot{\eta}\,\delta\dot{R}_3 - \dot{\eta}^2\delta R_1 - \ddot{\eta}\,\delta R_3 + \left(\dfrac{\mu}{R_e^3}\right)\delta R_1 \\ \delta\ddot{R}_2 \qquad\qquad\qquad\qquad + \left(\dfrac{\mu}{R_e^3}\right)\delta R_2 \\ \delta\ddot{R}_3 + 2\dot{\eta}\,\delta\dot{R}_1 - \dot{\eta}^2\delta R_3 + \ddot{\eta}\,\delta R_1 - 2\left(\dfrac{\mu}{R_e^3}\right)\delta R_3 \end{bmatrix} = \mathbf{C}_{eo}\mathbf{f}_d(R, \mathbf{C}_{po}) \tag{32}$$

This is a set of linear differential equations with time-variable coefficients. The dependences $\eta(t)$ and $R_e(t)$ are known *a priori* from Newton's solution to the two-body problem:

$$\begin{aligned} R_e &= a(1 - e\cos E) \\ \dot{\eta} &= \frac{(\mu p)^{1/2}}{R_e^2} \\ \ddot{\eta} &= -2\left(\frac{\mu}{R_e^4}\right)ae(1 - e^2)^{1/2}\sin E \end{aligned} \tag{33}$$

and $E(t)$ was given implicitly earlier by (22).

As for the right side of (32), $\mathbf{C}_{eo}$ is only infinitesimally different from the unit matrix, the difference caused by the orbital perturbations $\delta\underset{\rightarrow}{\mathbf{R}}$. It can be shown (Problem 6.6) that

$$\mathbf{C}_{eo} = \begin{bmatrix} 1 & -\beta_3 & -\dfrac{\delta R_1}{R_e} \\ \beta_3 & 1 & -\dfrac{\delta R_2}{R_e} \\ \dfrac{\delta R_1}{R_e} & \dfrac{\delta R_2}{R_e} & 1 \end{bmatrix} \tag{34}$$

where $\dot{\eta}\beta_3 = (\delta R_2/R_e)^{\bullet}$. The possibility for deleting further second-order terms from the right side of (32) should also be noted, based on the observation that

$$R^{-4} = R_e^{-4}\left(1 + 4\frac{\delta R_3}{R_e}\right) \tag{35}$$

to first order. Of course, equations (32) are valid for any disturbance $\mathbf{f}_d$.

Nominally Circular Orbit

In general, the equations for orbital perturbations, (32), must be solved numerically in conjunction with the attitude motion equations. If the reference orbit is circular however, the left sides of the equations are much simplified. For a circular reference orbit,

$$p = a = R_e = \text{constant}; \qquad e = 0; \qquad \eta \equiv E; \qquad \dot{\eta} \equiv \omega_c \triangleq \left(\frac{\mu}{R_e^3}\right)^{1/2} \tag{36}$$

and (32) becomes

$$m\begin{bmatrix} \delta\ddot{R}_1 - 2\omega_c\delta\dot{R}_3 \\ \delta\ddot{R}_2 + \omega_c^2\delta R_2 \\ \delta\ddot{R}_3 + 2\omega_c\delta\dot{R}_1 - 3\omega_c^2\delta R_3 \end{bmatrix} = \mathbf{C}_{eo}\mathbf{f}_d(R, \mathbf{C}_{po}) \tag{37}$$

These are sometimes called the Euler–Hill equations. They permit the state transition matrix to be known in closed form, and superposition also applies. In general, $\mathbf{f}_d$ will be known only numerically, by solving the rotational motion equations numerically prior to using (17).

Attitude–Orbit Gravitational Coupling

The influence of attitude on the orbit must generally be ascertained numerically from the above, or similar, equations. Further analytical progress can be made however when additional assumptions are made. For example, [Moran] has studied the in-plane coupling for a dumbbell-shaped satellite of length $2l$ in a circular orbit. The possibility is demonstrated of a buildup in the orbital perturbation when the pitch librational period is equal to the orbital period. A radial change of $\sim 30l^2/R$ is possible after three orbits if the satellite librates with a amplitude of 90°; this represents a perturbation of 50 m for a solar power satellite (SPS) at geostationary altitude. Planar librations of a dumbbell satellite are studied in Problem 9.7.

[Lange] has examined attitude–orbital coupling; he linearizes about a circular reference orbit but includes both in-plane and out-of-plane motion. The pitch–orbital resonance phenomenon (when the pitch and orbital frequencies are approximately equal) is further illuminated. It is demonstrated that at resonance, for which $I_2 = 3(I_1 - I_3)$, a small initial eccentricity e in the orbit can eventually cause a buildup of pitch amplitude to $\sim (mR^2/I_2)^{1/2}e$ before energy begins to be transmitted from pitch back to the orbit. This pitch amplitude is $O(2500e)$ for an SPS in geostationary orbit.

The in-plane situation has been discussed further by [Mohan, Breakwell, and Lange] using the method of averaging. See also Chapter 4 of [Beletskii]. A recent in-depth numerical study has been undertaken by [Sincarsin and Hughes, 1, 2], who looked at attitude-orbital coupling for an SPS-type satellite, both with and without solar radiation pressure as a driving input.

It is emphasized once more that many of these results, although of theoretical interest, would no longer be valid if the effect of energy dissipation on pitch librations were included. The existence of many other orbital perturbations (luni-solar forces, solar pressure, atmospheric drag at low altitudes, and even control thruster firings) would also tend to mask the weak attitude-related gravitational disturbances discussed above. It should be pointed out as well that many of these more influential orbital perturbations are highly attitude dependent, thereby providing a relatively strong coupling of orbital motion to attitude motion. We shall in the sequel therefore neglect the gravitational coupling of attitude to orbit. In other words, we shall make the strong and usual assumption that, as far as gravity is concerned, the orbit is prescribed prior to solving for attitude. Attention is accordingly turned to the attitude dynamics of a satellite in an exactly circular or elliptic orbit.

9.2 EQUILIBRIA FOR A RIGID BODY IN A CIRCULAR ORBIT

The attitude motion equations for a rigid body in orbit were given earlier by (6.1, 10). These now take the form

$$\begin{aligned} I_1\dot{\omega}_1 &= (I_2 - I_3)\left(\omega_2\omega_3 - 3\omega_c^2 c_{23}c_{33}\right) \\ I_2\dot{\omega}_2 &= (I_3 - I_1)\left(\omega_3\omega_1 - 3\omega_c^2 c_{33}c_{13}\right) \\ I_3\dot{\omega}_3 &= (I_1 - I_2)\left(\omega_1\omega_2 - 3\omega_c^2 c_{13}c_{23}\right) \end{aligned} \tag{1}$$

since the orbital rate

$$\omega_c \triangleq \left(\frac{\mu}{R^3}\right)^{1/2} \tag{2}$$

is constant. To complete the system, we add the kinematical relations

$$\boldsymbol{\omega} = \boldsymbol{\omega}_\alpha + \mathbf{C}_{pc}\boldsymbol{\omega}_c \tag{3}$$

$$\dot{\mathbf{C}}_{pc} = -\boldsymbol{\omega}^\times\mathbf{C}_{pc} + \mathbf{C}_{pc}\boldsymbol{\omega}_c^\times \tag{4}$$

We use the subscript c in place of o as a reminder that the orbit is circular. In (3), $\boldsymbol{\omega}$ and $\boldsymbol{\omega}_\alpha$ are respectively the angular velocities, in $\mathscr{F}_p$, of $\mathscr{F}_p$ with respect to $\mathscr{F}_i$ and $\mathscr{F}_c$, and $\boldsymbol{\omega}_c$ is the angular velocity, in $\mathscr{F}_c$, of $\mathscr{F}_c$ with respect to $\mathscr{F}_i$. In fact, since

$$\boldsymbol{\omega}_c = [0 \quad -\omega_c \quad 0]^T \tag{5}$$

we may write (3) and (4) in the simpler form

$$\boldsymbol{\omega} = \boldsymbol{\omega}_\alpha - \omega_c\mathbf{c}_2 \tag{6}$$

$$\begin{aligned} \dot{\mathbf{c}}_1 &= -\boldsymbol{\omega}^\times\mathbf{c}_1 + \omega_c\mathbf{c}_3 \\ \dot{\mathbf{c}}_2 &= -\boldsymbol{\omega}^\times\mathbf{c}_2 \\ \dot{\mathbf{c}}_3 &= -\boldsymbol{\omega}^\times\mathbf{c}_3 - \omega_c\mathbf{c}_1 \end{aligned} \tag{7}$$

The scalar form of (7) is similar to (2.1, 13) in Chapter 2, but with the ω_c terms added. To find the attitude history, then, one must integrate (1) and two equations of (7), with the usual understanding that the analyst may use in place of direction cosines any rotational parametrization he or she wishes. Unfortunately, there is no general closed-form solution to this system, although many methods of approximation have been used in special cases.

Certain important properties can be deduced, however, without the trouble of complete integration. In addition to integrals of the motion associated with the orthonormality of $\mathbf{C}_{pc}$, there is the following nontrivial one (Problem 9.9):

$$H = \tfrac{1}{2}\boldsymbol{\omega}^T\mathbf{I}\boldsymbol{\omega} + \omega_c\boldsymbol{\omega}^T\mathbf{I}\mathbf{c}_2 + \tfrac{3}{2}\omega_c^2\mathbf{c}_3^T\mathbf{I}\mathbf{c}_3 \tag{8}$$

Another form for H, based on motions relative to $\mathscr{F}_c$, is obtained by inserting (6):

$$H = \tfrac{1}{2}\boldsymbol{\omega}_\alpha^T\mathbf{I}\boldsymbol{\omega}_\alpha + \tfrac{3}{2}\omega_c^2\mathbf{c}_3^T\mathbf{I}\mathbf{c}_3 - \tfrac{1}{2}\omega_c^2\mathbf{c}_2^T\mathbf{I}\mathbf{c}_2 \tag{9}$$

In this form, H is the extension, to a rigid body, of the H discussed for a point mass in Chapter 3, Problem 3.3. With the correspondences

$$m \to \int(\ )\,dm; \qquad \dot{\mathbf{r}} \to \boldsymbol{\omega}_\alpha^\times\mathbf{r}; \qquad v_o \to \omega_c R; \qquad \boldsymbol{\omega} \to -\omega_c\mathbf{c}_2 \tag{10}$$

it can be deduced from Problem 3.3 that

$$T_2 = \tfrac{1}{2}\boldsymbol{\omega}_\alpha^T\mathbf{I}\boldsymbol{\omega}_\alpha; \qquad T_o = \tfrac{1}{2}\omega_c^2\left(mR^2 + \mathbf{c}_2^T\mathbf{I}\mathbf{c}_2\right) \tag{11}$$

Likewise, the potential energy V for a rigid body in orbit is recorded in Problem 9.4, with $R_c \equiv R$. Therefore, apart from (constant) orbital contributions, the H of (9) has the form discussed in Problem 3.3:

$$H = T_2 + (V - T_o) \tag{12}$$

The combination $V - T_o$ is sometimes called the *dynamic potential*, with $-T_o$ thought of as the potential energy of centrifugal (inertial) forces.

It is important to note the existence of *equilibria* for which the relative motion is zero, $\boldsymbol{\omega}_\alpha \equiv \mathbf{0}$. These are also called *relative equilibria* because the satellite rotates uniformly as observed in $\mathscr{F}_i$ and is fixed in $\mathscr{F}_c$. If in (1) and (7) we set

$$\omega_1 \equiv 0; \qquad \omega_2 \equiv -\omega_c; \qquad \omega_3 \equiv 0; \qquad \mathbf{C}_{pc} \equiv \mathbf{1} \tag{13}$$

the motion equations are clearly satisfied. Thus, $\boldsymbol{\omega}_\alpha \equiv \mathbf{0}$, $\mathbf{C}_{pc} \equiv \mathbf{1}$ is indeed an equilibrium. We conclude that whenever one principal axis is vertical and a second principal axis is perpendicular to the orbit, an Earth-oriented equilibrium persists. Moreover, for a given satellite, there are 24 such equilibria: any one of the three principal axes, each with two senses, can point down; and, for each of these six possibilities, either of the two transverse axes, each with two senses, can point along the orbit normal. [Likins and Roberson] have proved that these are the *only* equilibria if the satellite is tri-inertial. In Problem 9.10, it is shown that these are also equilibria in the dynamical sense; the torque is identically zero for these 24, and only these 24, orientations.

Stability of Equilibria

If some of these equilibria were stable, this would augur well for the possibility of passive gravity gradient stabilization of Earth-pointing satellites. We shall now examine stability using both linearization (based on Sections A.3 and A.4) and Liapunov's method (Section A.5).

According to Section A.4, the motion equations can be linearized and, if infinitesimal asymptotic stability is demonstrated, the motion is asymptotically stable in some finite neighborhood of the equilibrium (see Theorem A.15). To linearize, set

$$\boldsymbol{\omega}_\alpha = \dot{\boldsymbol{\alpha}}; \qquad \mathbf{C}_{pc} = \mathbf{1} - \boldsymbol{\alpha}^\times \tag{14}$$

in (1) and (7). The resulting equations govern pitch, roll, and yaw librations in a circular orbit:

$$I_1\ddot{\alpha}_1 - (I_3 + I_1 - I_2)\omega_c\dot{\alpha}_3 + (I_2 - I_3)\omega_c^2\alpha_1 = 3\omega_c^2(I_3 - I_2)\alpha_1 \tag{15a}$$

$$I_2\ddot{\alpha}_2 = 3\omega_c^2(I_3 - I_1)\alpha_2 \tag{15b}$$

$$I_3\ddot{\alpha}_3 + (I_3 + I_1 - I_2)\omega_c\dot{\alpha}_1 + (I_2 - I_1)\omega_c^2\alpha_3 = 0 \tag{15c}$$

These important equations have been attributed to Lagrange; their left sides are in fact identical to equations (4.1, 12 and 13), derived earlier for the small attitude deviations of a rigid body from a state of uniform spin around a principal axis. The uniform spin rate is now $\nu \to -\omega_c$ since the uniform once-per-orbit absolute rotation is $-\omega_c\hat{\mathbf{p}}_2$. On the right sides of (15) are the gravity gradient torques mentioned at the beginning of this chapter, in (9.1, 5). It is characteristic of gravity gradient stabilization that the stabilizing torques are of the same form and magnitude as the dynamical terms in the motion equations, namely, the terms on the left in (15). This lends mathematical elegance but dims the prospect of making the stabilizing torques truly dominant.

It is also evident from (15) that pitch (α_2) and roll/yaw (α_1, α_3) are uncoupled in the linear approximation. For pitch stability, the condition

$$I_1 > I_3 \tag{16}$$

is necessary and sufficient. For roll/yaw, the form of the equations is

$$\mathscr{M}\ddot{\mathbf{q}} + \mathscr{G}\dot{\mathbf{q}} + \mathscr{K}\mathbf{q} = \mathbf{0} \tag{17}$$

where

$$\mathbf{q} \triangleq [\alpha_1 \quad \alpha_3]^T; \qquad \mathscr{M} \triangleq \operatorname{diag}\{I_1, I_3\}$$

$$\mathscr{G} \triangleq (I_3 + I_1 - I_2)\omega_c\begin{bmatrix} 0 & -1 \\ 1 & 0 \end{bmatrix}; \qquad \mathscr{K} \triangleq \omega_c^2\begin{bmatrix} 4(I_2 - I_3) & 0 \\ 0 & (I_2 - I_1) \end{bmatrix} \tag{18}$$

It is possible (see Section A.6) for $\mathscr{G}$ to be stabilizing even if $\mathscr{K}$ is not positive-definite, although, based on the evidence from Chapter 5, one might anticipate that such gyric stabilization is illusory. Certainly $\mathscr{K} > 0$ is sufficient for stability. Thus, for roll/yaw stability, it is sufficient that

$$I_2 > I_1; \qquad I_2 > I_3 \tag{19}$$

Combining this with (16), the condition for pitch stability, we have

$$I_2 > I_1 > I_3 \tag{20}$$

as sufficient conditions for three-axis stabilization. The interpretation is that a satellite is Liapunov attitude stable if its minor axis is vertical and its major axis is normal to the orbit.

From the variational equations (15) it can be seen that only the ratios $I_1 : I_2 : I_3$ are germane to the solution; as far as gravity gradient torque is concerned, a dressmaker's pin has the same motion as a lodge-pole. Also, ω_c plays no role in stability other than to define a characteristic time for the problem—the orbital period. Therefore, it is again appropriate to use the inertia ratio parameters k_1 and k_3, defined in Chapter 4, equation (4.4, 20), and explained in Fig. 4.14. In terms of k_1 and k_3, the condition (20) is written

$$k_1 > k_3 > 0 \tag{21}$$

If this seems at first to be more restrictive than the more liberal conditions ($k_1, k_3 > 0$) for a rigid body in torque-free motion, it should be borne in mind that the present conditions provide not only *directional* stability but *three-axis* stability. Although gravity stabilizes only pitch and roll directly, roll/yaw coupling extends roll stability to the yaw axis as well.

We now return to the possibility, considered by [DeBra and Delp], that $\mathscr{G}$ can stabilize even though $\mathscr{K} \not> 0$. The roll/yaw characteristic equation is

$$\det[\mathscr{M}s^2 + \mathscr{G}s + \mathscr{K}] = b_0 s^4 + b_1 s^2 + b_2 = 0 \tag{22}$$

where

$$\begin{aligned} b_0 &\triangleq I_1 I_3 \\ b_1 &\triangleq I_1 I_3 (1 + 3k_1 + k_1 k_3)\omega_c^2 \\ b_2 &\triangleq I_1 I_3 (4k_1 k_3)\omega_c^4 \end{aligned} \tag{23}$$

According to Table A.2, the motion is (infinitesimally) stable if and only if

$$\begin{aligned} &\text{(i)} \quad b_0 > 0; \qquad \text{(ii)} \quad b_1 > 0 \\ &\text{(iii)} \quad b_2 > 0; \qquad \text{(iv)} \quad b_1^2 - 4b_0 b_2 > 0 \end{aligned} \tag{24}$$

Therefore, the necessary and sufficient conditions for roll/yaw stability are

$$\begin{aligned} &\text{(i)} \quad 1 + 3k_1 + k_1 k_3 > 0 \\ &\text{(ii)} \quad k_1 k_3 > 0 \\ &\text{(iii)} \quad (1 + 3k_1 + k_1 k_3)^2 - 16 k_1 k_3 > 0 \end{aligned} \tag{25}$$

When these three conditions are combined with (16), which becomes $k_1 > k_3$, three-axis stability is available for the inertia ratios shown in Fig. 9.4. (Condition (i) above is not active but is subsumed by the other conditions.) The so-called Lagrange region, corresponding to the sufficient conditions (21), is identified; so also is the additional region permitted by the less restrictive necessary and sufficient conditions of [DeBra and Delp]. Since configurations in the DeBra–Delp region are statically unstable (in fact, $\mathscr{K}$ is negative-definite) and roll/yaw coupling is certainly present, we can anticipate from the stability theory of canonical matrix-second-order systems (Section A.6) that the slightest damping in

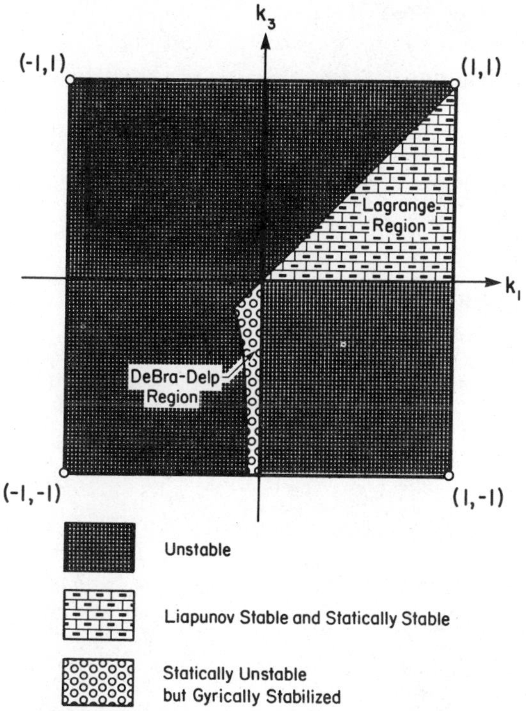

FIGURE 9.4 Infinitesimal stability diagram for the librations of a rigid body in a circular orbit. (Compare with Figs. 4.14 and 4.15.)

either roll or yaw will lead to instability in this region. The DeBra–Delp region is an interesting artifact of the zero damping assumption in the same way that minor-axis spin stability is predicted for torque-free motion.

The zero damping assumption creates another difficulty: the stable regions in Fig. 9.4 have been shown to be infinitesimally stable, but not asymptotically so. Thus, strictly speaking, stability is not guaranteed for finite motions. This difficulty can be overcome, for the Lagrange region at least, by returning to the nonlinear equations of motion and constructing a Liapunov function. The Hamiltonian H in (9) provides just such a function ([Beletskii]). The reference equilibrium is

$$\boldsymbol{\omega}_\alpha \equiv \mathbf{0}; \qquad \mathbf{C}_{pc} \equiv \mathbf{1} \tag{26}$$

for which

$$H \equiv H_o \triangleq \tfrac{1}{2}\omega_c^2(3I_3 - I_2) \tag{27}$$

Now, construct the following candidate Liapunov function:

$$\begin{aligned} v \triangleq H - H_o = {} & \tfrac{1}{2}\boldsymbol{\omega}_\alpha^T \mathbf{I} \boldsymbol{\omega}_\alpha \\ & + \tfrac{3}{2}\omega_c^2\left[(I_1 - I_3)c_{13}^2 + (I_2 - I_3)c_{23}^2\right] \\ & + \tfrac{1}{2}\omega_c^2\left[(I_2 - I_1)c_{12}^2 + (I_2 - I_3)c_{32}^2\right] \end{aligned} \tag{28}$$

The constraints $\mathbf{c}_2^T\mathbf{c}_2 = \mathbf{c}_3^T\mathbf{c}_3 = 1$ have been used. Because $v = 0$ for the reference equilibrium and $\dot{v} \equiv 0$ throughout the motion, we are able on the basis of Theorem A.18 to state that the equilibrium is stable if v is a positive-definite function, which it is (Problem 9.11) if

$$I_2 > I_1 > I_3 \tag{29}$$

This is precisely the Lagrange region in Fig. 9.4. There is now no remaining doubt about stability in the Lagrange region.

Pitch Motion in a Circular Orbit

If roll and yaw are initially quiescent, they are not excited by pitch motion. For a circular orbit, this is evident from the governing equations by substituting $\alpha_1 \equiv 0$, $\alpha_3 \equiv 0$. The rotation matrix $\mathbf{C}_{pc}$ and angular velocity $\boldsymbol{\omega}$ become, from (9.1, 3) and (9.1, 9),

$$\mathbf{C}_{pc} = \mathbf{C}_2(\alpha_2); \qquad \boldsymbol{\omega} = (\dot{\alpha}_2 - \omega_c)\mathbf{1}_2 \tag{30}$$

while the only nontrivial motion equation, for pitch, is

$$I_2\ddot{\alpha}_2 + 3\omega_c^2(I_1 - I_3)\sin\alpha_2\cos\alpha_2 = 0 \tag{31}$$

To simplify the notation, we rewrite this as

$$\ddot{\alpha}_2 + 3\omega_c^2 k_2 \sin\alpha_2\cos\alpha_2 = 0 \tag{32}$$

where the inertia ratio parameter for pitch is

$$k_2 \triangleq \frac{I_1 - I_3}{I_2} \equiv \frac{k_1 - k_3}{1 - k_1 k_3} \tag{33}$$

Note that k_2 is not independent of the earlier two inertia ratio parameters, k_1 and k_3.

An analogy between the pitch equation, (32), and the equation for a simple pendulum is immediate (Fig. 9.5). The only distinction is that while a

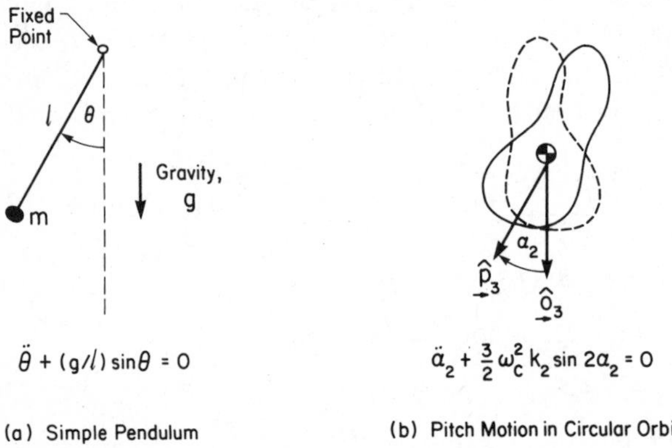

FIGURE 9.5 Analogy between pitch motion in a circular orbit and motion of a simple pendulum.

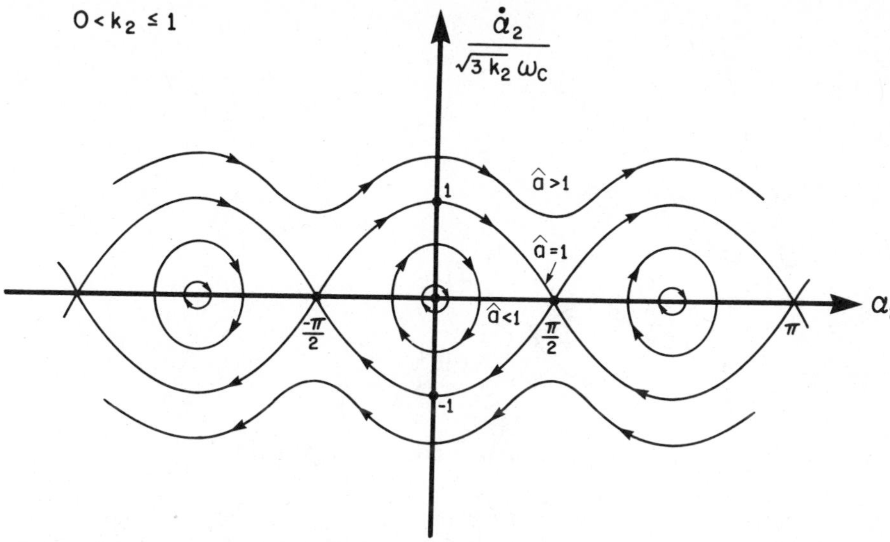

FIGURE 9.6 Phase plane for pitch in a circular orbit.

simple pendulum has stable equilibria at $\theta = \pm 2n\pi$ and unstable ones at $\theta = \pm(2n+1)\pi$, a rigid body in orbit has stable equilibria at $\alpha_2 = \pm n\pi$ and unstable ones at $\alpha_2 = \pm(2n+1)\pi/2$. Otherwise the motions are identical.

The energy integral associated with (32) is

$$\dot{\alpha}_2^2 + 3\omega_c^2 k_2 \sin^2\alpha_2 \triangleq a^2 \triangleq 3\omega_c^2 k_2 \hat{a}^2 = \text{constant} \tag{34}$$

where a (or $\hat{a}$) is calculated from initial conditions. A phase plane plot based on this integral appears in Fig. 9.6. It is evident that $0 < \hat{a} < 1$ produces librations, while $\hat{a} > 1$ results in tumbling. In fact, $\alpha_2(t)$ is available in closed form in terms of Jacobian elliptic functions, as is well known for the simple pendulum (see, for example, [Stern]):

$$\sin\alpha_2 = \begin{cases} \hat{a}\, sn\left[\dfrac{a(t-t_0)}{\hat{a}};\hat{a}\right] & (0 < \hat{a} < 1) \\ sn\left[a(t-t_0);\dfrac{1}{\hat{a}}\right] & (\hat{a} > 1) \end{cases} \tag{35}$$

Here, t_0 is the time at which $\alpha_2 = 0$, and the entry after the semicolon is the parameter k to be used with sn. The function sn is plotted for $k^2 = \frac{1}{2}$ in Fig. 4.7. We also note the asymptotic approximations

$$\alpha_2 \doteq \begin{cases} \hat{a}\sin\left[(3k_2)^{1/2}\omega_c(t-t_0)\right] & (\text{when } a \to 0) \\ a(t-t_0) & (\text{when } a \to \infty) \end{cases} \tag{36}$$

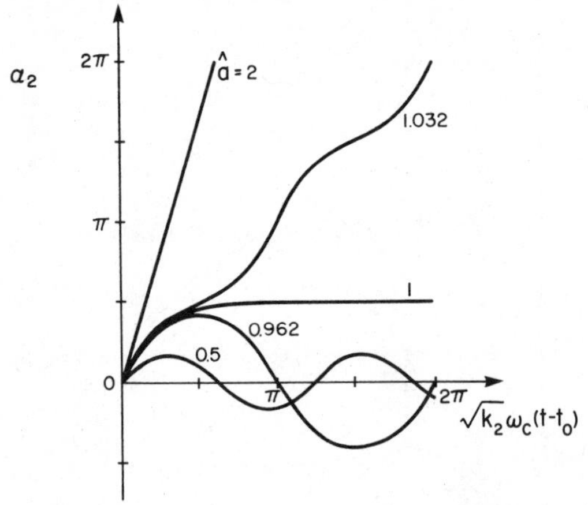

FIGURE 9.7 Pitch motion in a circular orbit. After [Moran].

These results are contained in early papers by [Schindler], [Klemperer], and [Moran] for dumbbell satellites ($k_2 = 1$). Figure 9.7, based on [Moran], graphs $\alpha_2(t)$ from (35) for several values of $\hat{a}$.

In addition to the critical value $\hat{a} = 1$ there are two other values of $\hat{a}$ of special interest. The period of the $sn\eta$ function is $\Delta\eta = 4K(k)$, $K(k)$ being Legendre's complete elliptic integral of the first kind. Therefore, in (35), when

$$\frac{a\,\Delta t}{\hat{a}} = 4K(\hat{a}) \tag{37}$$

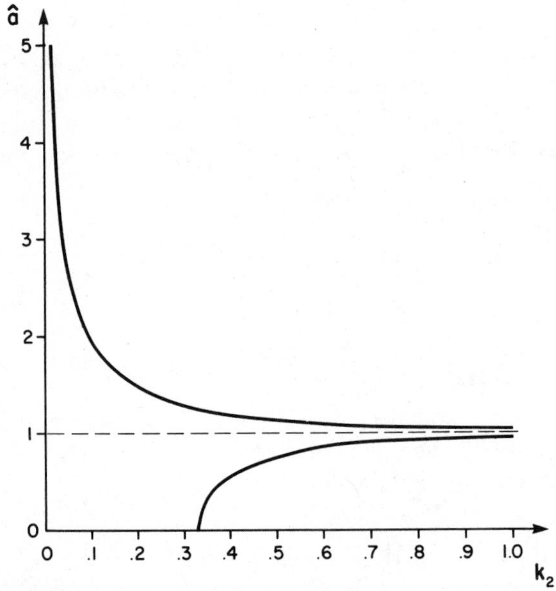

FIGURE 9.8 Condition for the periodic component of pitch motion in a circular orbit to have the orbital period.

α_2 has gone through a complete period. This period will coincide with the orbital period if $\Delta t = 2\pi/\omega_c$, that is, if

$$K(\hat{a}) = \tfrac{1}{2}\pi(3k_2)^{1/2} \tag{38}$$

Hence, a given spacecraft (a given k_2) will librate in pitch at the orbital frequency if its initial conditions produce an $\hat{a}$ such that (38) is satisfied. For $k_2 = 1$, the solution is $\hat{a} \doteq 0.962$, as shown in Fig. 9.7. The value of $\hat{a}$ that satisfies (38) in general is shown in Fig. 9.8. Similarly, if the periodic part of the tumbling motion for $\hat{a} > 1$ is to have a period equal to the orbital period, it is necessary that

$$K(\hat{a}^{-1}) = \tfrac{1}{2}\pi(3k_2)^{1/2}\hat{a} \tag{39}$$

For $k_2 = 1$, $\hat{a} \doteq 1.032$, and $\hat{a}$ is plotted in Fig. 9.8 in general. [Elrod] has pointed out that the condition (39) can be interpreted to mean that the satellite does not tumble as observed by an inertial observer; he also gives the practical details for such a "quasi-inertial" pointing system.

Bounds on the Librations of a Symmetrical Earth-Pointing Satellite

It has already been shown that a satellite in a circular orbit is in stable equilibrium if its minor axis is vertical and its major axis is normal to the orbit. Such equilibria populate the Lagrange region of Fig. 9.4. There are also cases where, in the absence of dissipation, a rigid body is in stable equilibrium when its minor axis is normal to the orbit and its major axis is tangential to the orbit; these cases populate the DeBra–Delp region in Fig. 9.4.

If a satellite has an axis of inertial symmetry, it can be oriented so that this axis coincides with the roll, pitch, or yaw axis. These three possibilities are now considered in turn. When the symmetry axis points *forward*, we set $I_1 = I_a$, $I_2 = I_3 = I_t$, $k_1 = 0$, and $k_3 = -k_t$, where $k_t \triangleq (I_a - I_t)/I_t$. These configurations are shown in Fig. 9.9. It can be shown (Problem 9.14) by analyzing the linearized equations for librations, (15), that there is (infinitesimal) $\{\alpha_2, \alpha_3\}$ stability, provided $0 < k_t \le 1$. This implies that the symmetry axis is directionally stable for $k_t > 0$. If however the symmetry axis is *normal to the orbit*, $I_2 = I_a$, $I_3 = I_1 = I_t$, and $k_1 = k_3 = k_t$. This family of configurations is also shown in Fig. 9.9. Pitch stability is neutral because there is no torque about the pitch axis. Again from (15), roll/yaw is stable in the Lagrange region for $0 < k_t \le 1$ and in the DeBra–Delp region for $k_o < k_t < 0$, where k_o is the first negative root of

$$(1 + 3k + k^2)^2 = 16k^2 \tag{40}$$

In fact, $k_o \doteq -0.146$. Thus, for $k_o < k_t \le 1$, the symmetry axis is directionally stable. Lastly, if the symmetry axis is nominally *vertical*, $I_3 = I_a$, $I_1 = I_2 = I_t$, $k_1 = -k_t$, and $k_3 = 0$. These configurations are also shown in Fig. 9.9. Stability analysis of (15) shows that although yaw stability is not present, pitch and roll are stable for $-1 \le k_t < 0$. Over this range, then, the axis of symmetry is directionally stable.

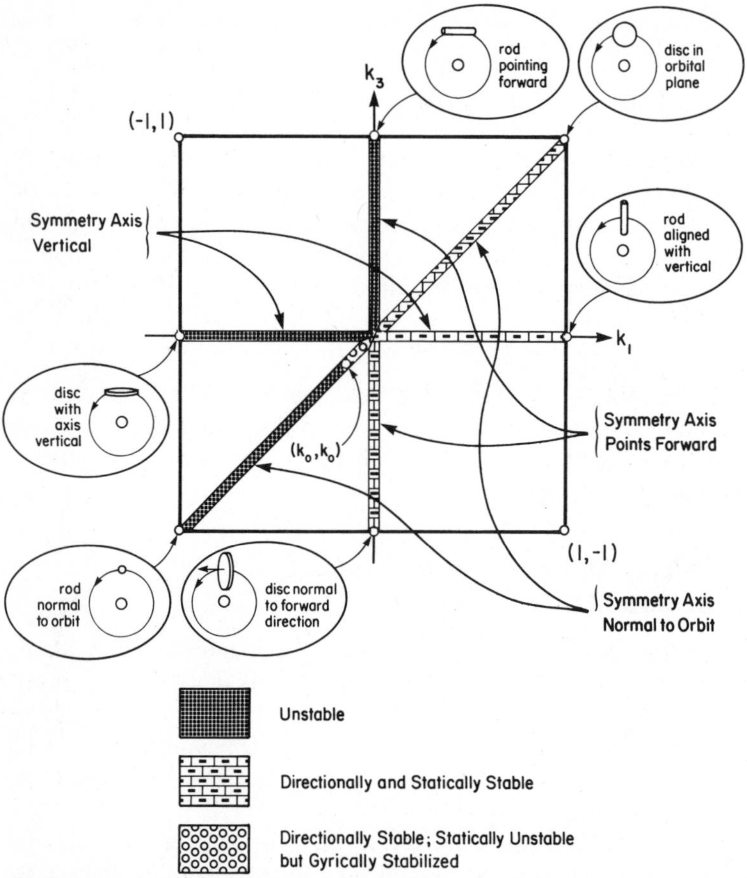

FIGURE 9.9 Stability diagram for librations of an inertially axisymmetrical satellite in a circular orbit. (Compare with Figs. 4.14 and 9.4.) Note: The "regions" are shown with finite width so that the legend can be used. The stability properties of these "regions" are to be interpreted as applying to the lines they surround.

These linear results can be confirmed by examining the system Hamiltonian, and specifically the function v defined in (28). More than that, v can be used to place a bound on the initial conditions (initial librational angles and rates) such that, if this bound is satisfied, the symmetry axis stays forever within $\frac{1}{2}\pi$ of its nominal direction. Such a motion is truly a libration; tumbling is excluded. Moreover, this knowledge is gained without complete integration of the motion equations. The following technique is similar to that used by Hill in the restricted three-body problem of celestial mechanics ([Szebehely]).

Assume first that the axis of symmetry points nominally forward. From (1), ω_1 = constant, a constant we shall set to zero. Also, from (6), $\omega_{\alpha 1} = \omega_c c_{12}$, where

$$[\omega_{\alpha 1} \quad \omega_{\alpha 2} \quad \omega_{\alpha 3}] \triangleq \boldsymbol{\omega}_\alpha^T \tag{41}$$

When v, known to be a constant of the motion, is reinterpreted in the present context, it can be written

$$\hat{v} \triangleq \frac{2v}{I_t \omega_c^2} = \hat{\omega}_{\alpha t}^2 + c_{12}^2 + 3k_t c_{13}^2 \tag{42}$$

where $\hat{\omega}_{\alpha t} \triangleq \omega_{\alpha t}/\omega_c$ and $\omega_{\alpha t} \triangleq (\omega_{\alpha 2}^2 + \omega_{\alpha 3}^2)^{1/2}$ is the transverse angular velocity component. It is clear from (42) that the state $c_{11} \equiv 1$, $\boldsymbol{\omega}_\alpha \equiv \mathbf{0}$ is Liapunov stable for $0 < k_t \leq 1$, confirming that the symmetry axis is directionally stable in the roll direction for $0 < k_t \leq 1$. Moreover, from

$$c_{12}^2 + 3k_t c_{13}^2 \leq \hat{v} \tag{43}$$

a condition can be found which ensures that the excursions of the symmetry axis

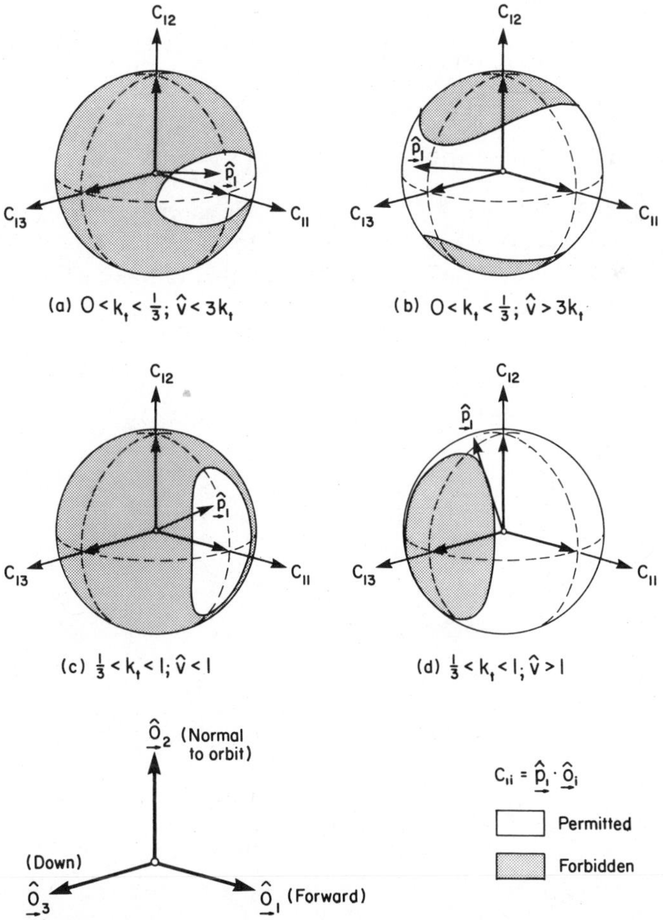

FIGURE 9.10 Libration bounds for a satellite whose axis of inertial symmetry points nominally forward. After [Auelmann].

will be bounded. No tumbling of this axis can occur if $c_{11} > 0$ always, that is, if

$$c_{12}^2 + c_{13}^2 < 1 \tag{44}$$

always. This condition is satisfied by (43) whenever the initial conditions in (42) are such that

$$\hat{v} < \begin{cases} 3k_t & \left(0 < k_t \le \frac{1}{3}\right) \\ 1 & \left(\frac{1}{3} \le k_t \le 1\right) \end{cases} \tag{45}$$

These results are shown geometrically in Fig. 9.10. The boundary between permitted and forbidden regions is defined by $\omega_{\alpha t} = 0$. In Figs. 9.10*a* and *c*, librational motion of the symmetry axis is guaranteed; tumbling is impossible.

Next, let the symmetry axis be nominally normal to the orbit. From (1), $\omega_2 =$ constant. We set this constant equal to $-\omega_c$ since we desire to have the pitch rate $\omega_{\alpha 2} \equiv 0$. With librations, the actual value of $\omega_{\alpha 2}$ is, from (6),

$$\omega_{\alpha 2} = (c_{22} - 1)\omega_c \tag{46}$$

Again v is the key, this time in the form

$$\hat{v} \triangleq \hat{\omega}_{\alpha t}^2 + (1 + k_t)(1 - c_{22})^2 + 3k_t c_{23}^2 + k_t\left(1 - c_{22}^2\right) \tag{47}$$

with the understanding that now $\omega_{\alpha t} \triangleq (\omega_{\alpha 1}^2 + \omega_{\alpha 3}^2)^{1/2}$. It is evident that the solution $c_{22} \equiv 1$, $\boldsymbol{\omega}_\alpha \equiv \mathbf{0}$ is Liapunov stable for $k_t > 0$. (The possibility $k_o < k_t < 0$ cannot be inferred, however.) Equation (47) may also be used to place an upper limit on $\hat{v}$ such that $c_{22} > 0$ always (Problem 9.15).

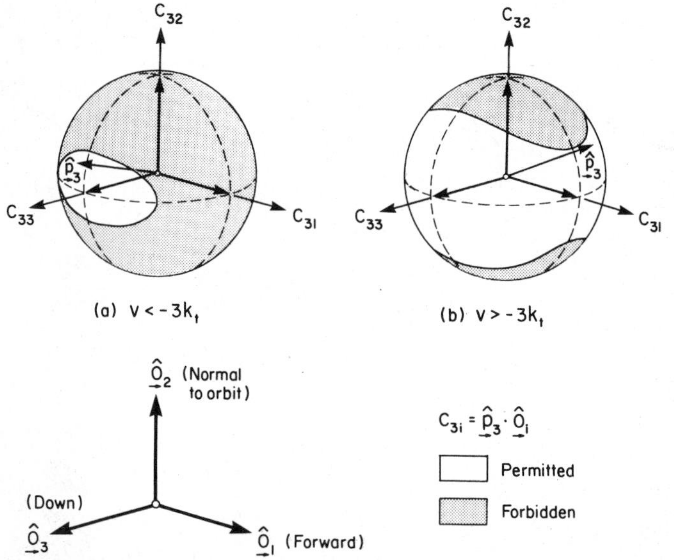

FIGURE 9.11 Libration bounds for a satellite whose axis of inertial symmetry is nominally vertical. After [Auelmann].

Lastly, let the symmetry axis point in a nominally vertical direction. Proceeding in a fashion analogous to the above derivations, we find

$$\hat{v} = \hat{\omega}_{\alpha t}^2 - 3k_t c_{31}^2 + (1 - 3k_t) c_{32}^2 \tag{48}$$

where now $\omega_{\alpha t} \triangleq (\omega_{\alpha 1}^2 + \omega_{\alpha 2}^2)^{1/2}$. It follows that the solution $c_{33} \equiv 1$, $\boldsymbol{\omega}_\alpha \equiv \mathbf{0}$ is Liapunov stable provided $k_t < 0$. Furthermore, as shown in Fig. 9.11, deviations of the symmetry axis from the vertical are bounded if $\hat{v} < -3k_t$.

Librations of an Earth-Pointing Satellite

We now remove the symmetry assumption and return to the more general case of a tri-inertial rigid body in a circular orbit, in an essentially Earth-pointing orientation. If gravity gradient is to be used for stabilization of such a satellite, it is important to know the magnitude of librations that may occur. This was done in the above paragraphs for symmetrical satellites (see, for example, Figs. 9.10 and 9.11). For tri-inertial satellites, the techniques are not as straightforward although useful information can still be deduced. It should also be remarked that the results obtained, while lending an understanding of the basic dynamics, are somewhat idealized inasmuch as (i) disturbance torques are not included, except vicariously through initial conditions, and (ii) the principal mechanism for absorbing disturbances, internal damping, is also absent. These objections will be removed in Sections 9.3 and 9.4.

Equations governing infinitesimal librations were given in (15). Basic stability conditions have already been derived (summarized in Fig. 9.4), and the Lagrange region has been confirmed as a region of stability for finite motions using the Liapunov function (28). Natural frequencies for small librations can also be calculated from (15), as tabulated graphically by [DeBra and Delp]. In principle, the Liapunov function v of (28) can be used to place bounds on initial conditions in such a manner that the motion of all three axes is entirely librational and tumbling is prevented. A sufficient condition for three-axis libration is

$$2v < \min\{3\omega_c^2(I_1 - I_3); \quad \omega_c^2(I_2 - I_1)\} \tag{49}$$

to be derived as Problem 9.11*c*. For example, if $\{I_1, I_2, I_3\} = \{4, 5, 3\}$ units, and if we take the small-libration approximation that $\mathbf{C}_{pc} = \mathbf{1} - \boldsymbol{\alpha}^\times$, then (49) shows that the librations will be bounded if

$$\frac{4\dot{\alpha}_1^2 + 5\dot{\alpha}_2^2 + 3\dot{\alpha}_3^2}{\omega_c^2} + 8\alpha_1^2 + 3\alpha_2^2 + \alpha_3^2 < 1 \tag{50}$$

initially. (Actually, it would be safer to replace 1 by 0.01 on the right side of this bound to ensure conformity with the assumption that $\|\boldsymbol{\alpha}\| \ll 1$.) See also [Zajac, 1].

Practical stabilization specifications require more than merely the absence of tumbling, of course, and a bound such as (49) has more to do with initial *attitude acquisition* (or *capture*, as it is also called) than stabilization tolerances. The function v can however also be used to establish narrower libration bounds, given a set of initial conditions. For example, consider a satellite whose inertia ratios are $\{k_1, k_3\} = \{0.9, 0.5\}$, and let the nonzero initial conditions be $\alpha_3(0) = -0.01$ rad and $\dot{\alpha}_2(0) = -0.13\omega_c$ rad/s. The satellite has a relatively large initial pitch rate and a relatively small initial yaw angle. An interesting question is: How much of this librational energy, initially in pitch, can eventually be transferred to roll/yaw? To answer accurately requires that the motion equations be integrated, but an upper bound is available more easily using v. From (28),

$$1.2c_{13}^2 + 1.35c_{23}^2 + 0.05c_{12}^2 + 0.45c_{32}^2 < 9.415 \times 10^{-3}$$

It follows that

$$0.05\left(c_{12}^2 + c_{32}^2\right) < 9.415 \times 10^{-3} \tag{51}$$

namely, that $c_{22} > 0.901$. According to this inequality, the angle between the pitch axis and the orbit normal cannot exceed $\cos^{-1} c_{22} = 25.7° = 0.45$ rad. Either this is a very loose bound or the initial pitch axis offset of 0.01 rad can grow to 45 times its initial value by drawing energy from the pitch libration.

This question has been taken up by [Kane, 1]. He shows that the pitch motion which occurs in the absence of roll and yaw, governed by (31), remains unchanged when roll and yaw are first-order infinitesimals. Although [Kane, 1] does not use yaw, pitch, and roll as defined by (9.1, 3), but angles $\{\theta_{1K}, \theta_{2K}, \theta_{3K}\}$, with

$$\mathbf{C}_{pc} = \mathbf{C}_2(-\theta_{3K})\mathbf{C}_1(\theta_{2K})\mathbf{C}_3(-\theta_{1K})$$

instead, retaining the present labeling of axes, it is not difficult to show that

$$\text{Roll} \equiv \alpha_1 \equiv \frac{\theta_{2K}}{\cos\theta_{3K}}$$

$$\text{Pitch} \equiv \alpha_2 \equiv -\theta_{3K}$$

$$\text{Yaw} \equiv \alpha_3 \equiv -\theta_{1K} - \theta_{2K}\tan\theta_{3K}$$

for first-order $\{\theta_{1K}, \theta_{2K}\}$, that is, for first-order $\{\alpha_1, \alpha_3\}$. Therefore, the conclusions Kane comes to for $\{\theta_{1K}, \theta_{2K}\}$ are immediately applicable to what we call "roll" and "yaw".

Roll and yaw, on the other hand, are influenced by finite pitch motion in the following manner:

$$\begin{bmatrix} \ddot{\alpha}_1 \\ \ddot{\alpha}_3 \end{bmatrix} = \omega_c^2 \mathbf{W}_P(\alpha_2, \dot{\alpha}_2)\begin{bmatrix} \alpha_1 \\ \alpha_3 \end{bmatrix} + \omega_c \mathbf{W}_D(\alpha_2, \dot{\alpha}_2)\begin{bmatrix} \dot{\alpha}_1 \\ \dot{\alpha}_3 \end{bmatrix} \tag{52}$$

We shall not take the space here to write out $\mathbf{W}_P$ and $\mathbf{W}_D$ explicitly. However, it is true that

$$\mathbf{W}_P(0,0) = \begin{bmatrix} -4k_1 & 0 \\ 0 & -k_3 \end{bmatrix}; \qquad \mathbf{W}_D(0,0) = \begin{bmatrix} 0 & (1-k_1) \\ (k_3 - 1) & 0 \end{bmatrix} \tag{53}$$

thus ensuring that when α_2 and $\dot{\alpha}_2$ are first-order infinitesimals also, (52) reduces,

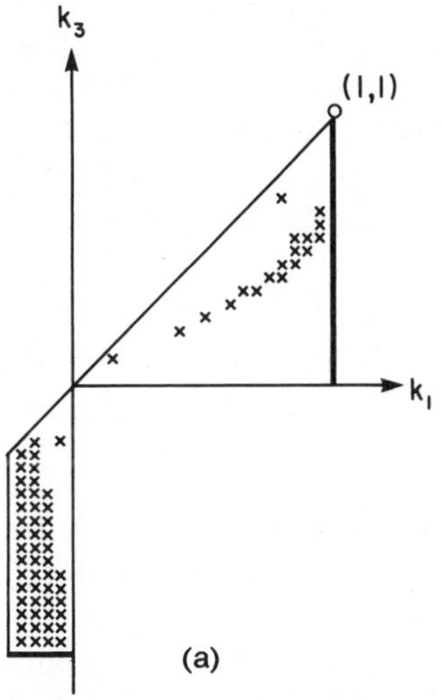

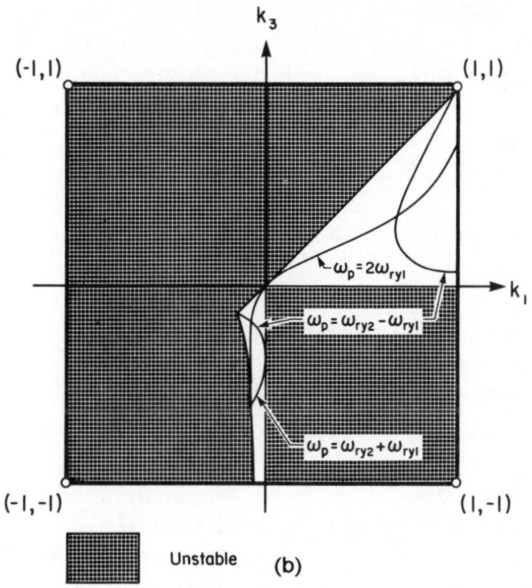

FIGURE 9.12 Librations of a rigid body in a circular orbit, with a reference pitch motion that is nonzero. (*a*) Infinitesimal stability diagram when reference pitch amplitude is 5°. (Compare with Fig. 9.4.) (*b*) Nonlinear internal resonance lines between pitch and roll / yaw.

as it should, to equations (15*a* and *c*). It should be recalled that the solution for $\alpha_2(t)$ is known, equation (35*a*), and this is to be used in $\mathbf{W}_P$ and $\mathbf{W}_D$. The roll/yaw equations (52) are therefore a set of linear differential equations with periodic coefficients. Floquet theory (Theorems A.9, A.10, and A.11) is available to infer stability properties.

[Kane, 1] has numerically integrated (52) and determined stability characteristics using three parameters: the satellite's inertia characteristics, $\{k_1, k_3\}$, and the pitch amplitude $\hat{a} \equiv \sin\alpha_{2,\max}$. A typical k_1–k_3 diagram is shown in Fig. 9.12*a* for a pitch amplitude $\alpha_{2,\max} = 5°$. This diagram is obtained in the following manner: once the initial conditions are chosen ($\alpha_2 = 5°$, $\dot{\alpha}_2 = 0$ in this case) and the satellite inertia distribution specified (by k_1 and k_3), the variational equations (52) are integrated over one pitch period, as required by Floquet theory, and the stability thereby determined. If found to be an unstable case, an × is plotted; otherwise, no symbol is plotted. Figure 9.12*a* should be compared with Fig. 9.4, which corresponds to infinitesimal pitch motion. Regions in the k_1–k_3 plane known to be unstable for infinitesimal pitching are not shown by [Kane, 1] in his presentation of Fig. 9.12*a*. Evidently, there are inertia distributions in the "stable" Lagrange region that are roll/yaw unstable when parametrically excited by a 5° pitch libration.

It is interesting to compare the approaches described above for finding the properties of the roll/yaw motion without actually integrating the nonlinear equations of motion numerically. The first considers the relative equilibrium $\alpha_1 = \alpha_2 = \alpha_3 = 0$, and linear variational equations are written for the small librations about this equilibrium. The resulting equations, (15), are linear with constant coefficients, and the inertia distributions that lead to instability are depicted in Fig. 9.4. Although instabilities for a linear constant-coefficient system are always characterized by exponential growth, this does not mean that the "actual" motion—the solution of the nonlinear equations—will be unbounded. This information is lost in the linearization. Similarly, when [Kane, 1] considers as his reference solution the periodic pitch oscillation governed by (34) and (35), still with $\alpha_1 = \alpha_3 = 0$, and proceeds by linearizing about this reference solution to arrive at the linear-periodic system (52), Floquet theory again predicts exponential growth for the unstable cases. But in "reality," that is, for the nonlinear motion equations, an unstable solution (a solution that cannot be bounded *arbitrarily* by choosing $\alpha_1^2(0) + \dot{\alpha}_1^2(0) + \alpha_3^2(0) + \dot{\alpha}_3^2(0)$ sufficiently small) may in fact be bounded. This is suggested by bounds of the type shown in Figs. 9.10 and 9.11, derived from an examination of the constant Hamiltonian. A specific example is studied in Problem 9.16.

Further light on this subject was cast by [Breakwell and Pringle], who pointed out that these new instabilities could be explained as *nonlinear internal resonances* of the form

$$\begin{aligned} \omega_p &= 2\omega_{ry1} \\ \omega_p &= \omega_{ry2} \pm \omega_{ry1} \end{aligned} \tag{54}$$

where ω_p is the frequency of pitch libration and ω_{ry1} and ω_{ry2} are the two frequencies associated with roll/yaw libration. We have already seen [e.g., from (32)] that

$$\omega_p = (3k_2)^{1/2}\omega_c \tag{55}$$

and ω_{ry1} and ω_{ry2} are the positive roots of

$$\omega^4 - (1 + 3k_1 + k_1k_3)\omega_c^2\omega^2 + 4k_1k_3\omega_c^4 = 0 \tag{56}$$

as indicated by (22) and (23). The internal resonance conditions (54) are shown plotted in Fig. 9.12*b*. A comparison with Fig. 9.12*a* indicates that internal resonance is indeed the correct explanation. This subject has also been studied further by [Likins and Wrout].

It bears observing once again that energy dissipation is not present in these theories. Indeed, damping is doubly missed, first because the parametric excitation (pitch) persists only in the absence of damping, and second because the slow transfer of energy to roll/yaw could in fact be continuously removed by suitable roll/yaw dissipation. Nevertheless, basic insight is obtained into the limiting case of very small damping.

Pitch Motion in an Elliptic Orbit

For a circular orbit, the local vertical rotates uniformly about the orbit normal. The absence of librational motion relative to the local vertical means that the satellite is rotating uniformly (with the orbital period) with respect to an inertial observer. If the orbit is elliptic, the local vertical does not rotate uniformly and pitch motion is excited. At the same time, gravity gradient torque about the pitch axis will tend to cause the minor inertia axis of the spacecraft to track the local vertical. Since no orbit can be perfectly circular, it is clearly worthwhile to examine the deleterious effect of orbital eccentricity on Earth-pointing attitude stabilization.

The governing equations of attitude motion for a rigid satellite are still given by (1), (6), and (7), repeated here for convenience and in a form more suitable for studying elliptic orbits:

$$\mathbf{I}\dot{\boldsymbol{\omega}} + \boldsymbol{\omega}^\times \mathbf{I}\boldsymbol{\omega} = 3\left(\frac{\mu}{R^3}\right)\mathbf{c}_3^\times \mathbf{I}\mathbf{c}_3 \tag{57}$$

$$\dot{\mathbf{c}}_1 + \boldsymbol{\omega}^\times \mathbf{c}_1 = \dot{\eta}\mathbf{c}_3 \tag{58}$$

$$\dot{\mathbf{c}}_3 + \boldsymbol{\omega}^\times \mathbf{c}_3 = -\dot{\eta}\mathbf{c}_1 \tag{59}$$

Here, $\boldsymbol{\omega}$ is the absolute angular velocity of the centroidal principal-axis frame $\mathcal{F}_p$, expressed in $\mathcal{F}_p$, and the direction cosines $\mathbf{c}_1$ and $\mathbf{c}_3$ are columns of the rotation matrix $\mathbf{C}_{po}$:

$$[\mathbf{c}_1 \quad \mathbf{c}_2 \quad \mathbf{c}_3] = \mathbf{C}_{po} = \mathcal{F}_p \cdot \mathcal{F}_o^T \tag{60}$$

The orbit is specified, and hence

$$R = \frac{p}{1 + e\cos\eta} \tag{61}$$

$$\dot{\eta} = \frac{(\mu p)^{1/2}}{R^2} \tag{62}$$

where p is the orbital parameter, $p = a(1 - e^2)$, and $\eta(t)$ is known implicitly from Kepler's equation. Thus, (57), (58), and (59) are a set of equations to be solved for $\boldsymbol{\omega}(t)$, $\mathbf{c}_1(t)$, and $\mathbf{c}_3(t)$, using any desired attitude parameters. Of course, $\mathbf{c}_2(t) = \mathbf{c}_3^\times(t)\mathbf{c}_1(t)$ completes the specification of $\mathbf{C}_{po}$, which shows the attitude of the spacecraft relative to the orbiting frame $\mathcal{F}_o$.

Even though the orbit is eccentric, roll and yaw will remain quiescent if they are initially so. Only pitch is excited. This can be seen from the motion equations (57), (58), and (59) which are satisfied by

$$\omega_1 = \omega_3 \equiv 0; \qquad \omega_2 = \dot{\alpha}_2 - \dot{\eta}$$
$$\mathbf{c}_1 = \begin{bmatrix} \cos\alpha_2 \\ 0 \\ \sin\alpha_2 \end{bmatrix}; \qquad \mathbf{c}_3 = \begin{bmatrix} -\sin\alpha_2 \\ 0 \\ \cos\alpha_2 \end{bmatrix} \tag{63}$$

and $\mathbf{c}_2 = [0 \quad 1 \quad 0]^T$. Only the pitch angle α_2 is nonzero; it satisfies

$$I_2\ddot{\alpha}_2 + 3\left(\frac{\mu}{R^3}\right)(I_1 - I_3)\sin\alpha_2\cos\alpha_2 = I_2\ddot{\eta} \tag{64}$$

showing clearly the excitation from the nonconstant $\dot{\eta}$. It is convenient to replace t by η as the independent variable and to denote derivatives with respect to η by primes. Straightforward application of the chain rule then produces the following pitch equation:

$$(1 + e\cos\eta)\alpha_2'' - (2e\sin\eta)\alpha_2' + 3k_2\sin\alpha_2\cos\alpha_2 = 2e\sin\eta \qquad (65)$$

The inertia parameter k_2 was given in (33). (Another possibility, considered in Problem 9.18, is to use the ecccentric anomaly as the independent variable.)

Equation (65) is a nonlinear, second-order, forced differential equation with variable coefficients. It and its related forms have been quite fully investigated in the literature. Keeping in mind that we are primarily interested in gravity gradient stabilization, which is practicable only for spacecraft configurations that include some form of energy dissipation and which are in nearly circular orbits, these investigations cannot here be given as extensive a treatment as their intrinsic interest warrants. However, a brief summary can be given. [Schechter] used the small eccentricity e as the basis for a perturbation expansion and was among the first to show that the effect of even small orbital eccentricities on the pitch motion can be significant.

Note that (65) is "exact" insofar as α_2 is concerned; it applies equally to tumbling and librational motion about the orbit normal. In studying librations, a linearized version is often adequate, on the grounds that a librational excursion large enough to violate the assumption of linearity would also be unacceptable as a technique for Earth-pointing stabilization. In addition to linearizing, we also follow [Beletskii] and [Modi and Brereton, 1] and use the transformation

$$z \triangleq (1 + e\cos\eta)\alpha_2 \qquad (66)$$

to obtain

$$z'' + \Omega^2(\eta)z = -2e\sin\eta \qquad (67)$$

where

$$\Omega^2 \triangleq \frac{3k_2 + e\cos\eta}{1 + e\cos\eta} \qquad (68)$$

When $e = 0$, this reduces, as it should, to the linearized pitch equation for circular orbits, namely, (15b). For small e, we can write the solution to (67) as a perturbation expansion:

$$z(\eta) = z_0(\eta) + ez_1(\eta) + e^2z_2(\eta) \qquad (69)$$

(It would be inconsistent to include terms of $O(e^3)$ in (69) because the simplification $\sin\alpha_2 \doteq \alpha_2$ has already been made.) We also observe from (68) that

$$\Omega^2 = \Omega_0^2 + e\Omega_1^2 + e^2\Omega_2^2 \qquad (70)$$

where

$$\Omega_0^2 = 3k_2; \qquad \Omega_1^2 = (1 - 3k_2)\cos\eta; \qquad \Omega_2^2 = (3k_2 - 1)\cos^2\eta \qquad (71)$$

Substituting (69) and (70) into (67) and equating like powers of e, we find

$$\begin{aligned} z_0'' + \Omega_0^2 z_0 &= 0 \\ z_1'' + \Omega_0^2 z_1 &= -2\sin\eta - \Omega_1^2 z_0 \\ z_2'' + \Omega_0^2 z_2 &= -\Omega_2^2 z_0 - \Omega_1^2 z_1 \end{aligned} \qquad (72)$$

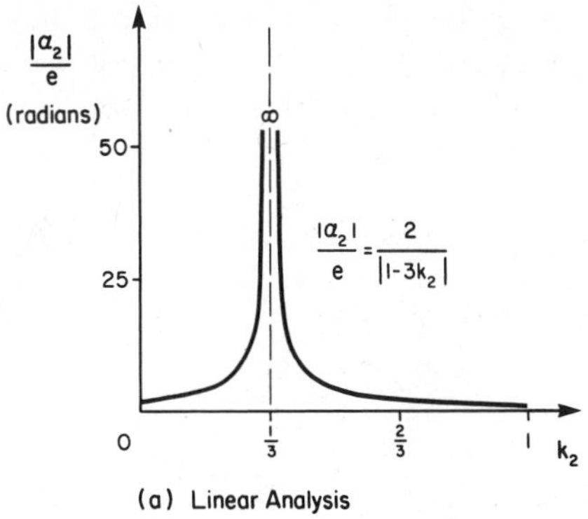

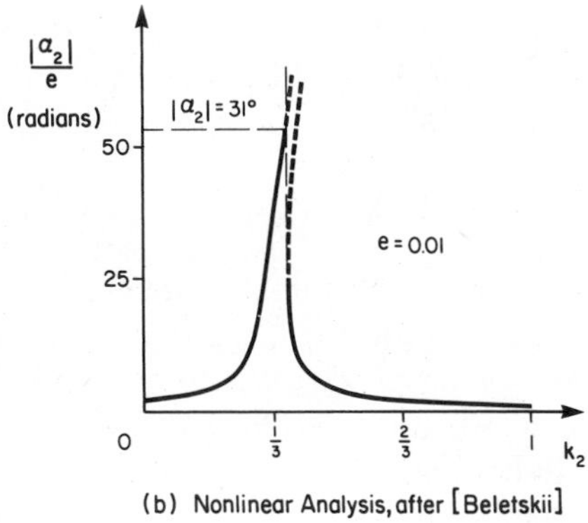

FIGURE 9.13 Amplitude of pitch librations forced by orbital eccentricity.

Since our interest is primarily in the response to orbital eccentricity, the natural librations at frequency Ω_0 will be suppressed: $z_0 \equiv 0$. The forced responses, z_1 and z_2, are given by

$$z_1 = \frac{2\sin\eta}{1 - 3k_2}; \qquad z_2 = \frac{\sin 2\eta}{4 - 3k_2} \tag{73}$$

It follows from (66) and (69) that

$$\alpha_2(\eta) = \left(\frac{2e}{1 - 3k_2}\right)\sin\eta - \left[\frac{3e^2}{(4 - 3k_2)(1 - 3k_2)}\right]\sin 2\eta \tag{74}$$

Although the case $3k_2 = 4$ is not possible physically, the condition $3k_2 = 1$ indicates a possible resonance in which the orbital frequency (the excitation frequency) is equal to the natural frequency of pitch libration. The amplitude of forced pitch libration is shown in Fig. 9.13*a*, based on the first term in the linear result (74).

To find the amplitude at resonance requires a nonlinear analysis. Following [Beletskii], we set

$$\alpha_2(\eta) = A\cos(\eta - \eta_o) \tag{75}$$

in the linear pitch equation (65). In other words, we seek a periodic solution at the frequency of excitation, and we take this periodic solution to be approximately sinusoidal. The amplitude A and phase η_o are to be determined. The further approximation

$$\sin\left[2A\cos(\eta-\eta_o)\right] \doteq 2J_1(2A)\cos(\eta-\eta_o) \tag{76}$$

is also useful, where J_1 is Bessel's function of the first kind of first order. The left side of (76) is thus replaced by its "fundamental" component (the first term in its Fourier series). After invoking harmonic balance, we find

$$eA\cos\eta_o = 0; \qquad \left[3k_2J_1(2A) - A\right]\cos\eta_o = 0$$

$$\left[3k_2J_1(2A) - A\right]\sin\eta_o = -2e \tag{77}$$

from which it follows that the amplitude of libration is given implicitly by

$$k_2 = \frac{A \pm 2e}{3J_1(2A)} \tag{78}$$

This is plotted in Fig. 9.13*b* for $e = 0.01$. Note that the most unfavorable inertia ratio is $3k_2 = 1.12$ (not 1.0), for which the amplitude is $\alpha_{2,\max} \doteq 31°$. For $3k_2 < 1.12$, there is only one periodic solution, which can be shown to be stable; for $3k_2 > 1.12$, three solutions are predicted, all still with the same period (the orbital period) but with three different amplitudes. The solution of intermediate amplitude (shown dashed in Fig. 9.13*b*) is unstable.

The only other troublesome inertia ratios for pitch librations in an elliptic orbit are those associated with so-called "parametric" excitation (see Appendix A, Section A.3). A glance at (67) and the approximation

$$\Omega^2 \doteq 3k_2 + (1 - 3k_2)e\cos\eta$$

shows clearly the analogy with a forced Mathieu equation. From the theory of Mathieu's equation, stability boundaries correspond to periodic solutions (Mathieu functions); these occur for particular combinations of k_2 and e. As $e \to 0$, the values of $3k_2$ for periodic solutions are

$$3k_2 = \frac{n^2}{4} \qquad (n = 0, 1, 2, \ldots) \tag{79}$$

Of course, $n > 3$ must be ruled out on physical grounds (since $k_2 \leq 1$). When $n = 0$, the gravity gradient torque is also zero, and when $n = 2$, the simple orbital resonance is obtained (see Fig. 9.13). We are left with the fractional-harmonic solutions corresponding to $n = 1$ and $n = 3$. The subharmonic $n = 1$, a libration with a period of two orbits, is studied in Problem 9.19*b*. [Beletskii] has calculated that the amplitude of this subharmonic librational response is 14° for $e = 0.01$.

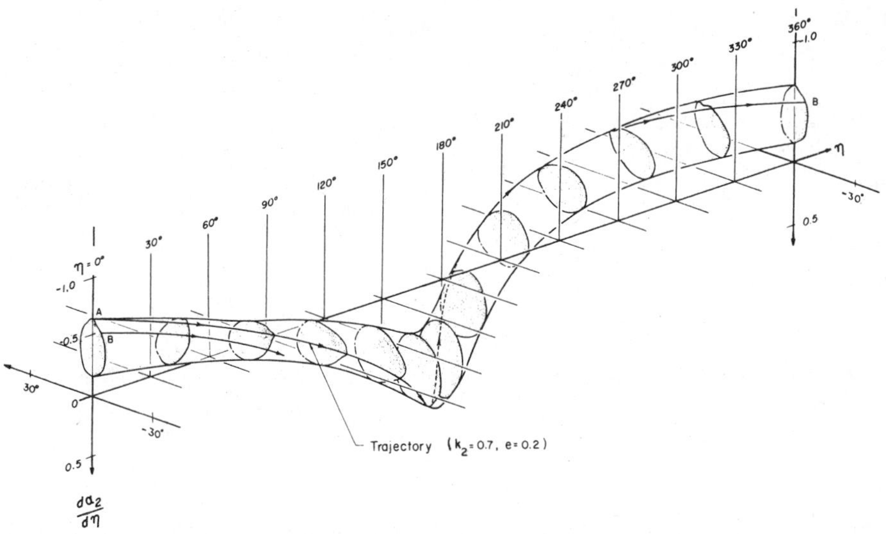

FIGURE 9.14 Pitch motion trajectory lying on invariant surface (for $k_2 = 0.7$, $e = 0.2$). After [Modi and Brereton, 3].

For large eccentricities the fundamental pitch equation (65) must be studied in its full nonlinear form. [Modi and Brereton, 3] have discovered periodic solutions whose periods are two and three orbital periods. Some of these are stable, others are not. Figure 9.14 shows a nearly periodic solution for $k_2 = 0.7$, $e = 0.2$, taken from [Modi and Brereton, 3]. The "trajectory" lies on the invariant surface that is discovered numerically to exist. Although the solution itself is not periodic, the invariant surface is periodic. This behavior, expected for *linear* dissipation-free differential equations with periodic coefficients ([Hughes, 1]), apparently also characterizes the solutions to the nonlinear equation (65). If the conditions at $\eta = 0$ were chosen to lie on the center of the "tube," the invariant surface for these initial conditions would degenerate to a curve, that is, the solution would be exactly periodic. The stability of this periodic solution is then evident from Fig. 9.14.

If finite roll and yaw angles are also present, one must deal with a sixth-order set of nonlinear differential equations with variable coefficients, and closed-form results are hard to come by. Numerical studies have been made by [Modi and Brereton, 2] and [Modi and Shrivastava, 1] for circular orbits, and by [Modi and Shrivastava, 2] for elliptic orbits. They found by numerical studies that invariant surfaces similar to the one shown in Fig. 9.14 for in-plane librations sometimes also exist for out-of-plane librations.

9.3 DESIGN OF GRAVITATIONALLY STABILIZED SATELLITES

The preceding analysis shows that it is possible, at least in principle, to stabilize the attitude of a satellite to an Earth-pointing orientation by exploiting the gravity gradient torque. For this stabilization to be effective, the chief requirement is a favorable inertia distribution: the parameters $\{k_1, k_2, k_3\}$ must be right. It appears, then, that by taking care to achieve the proper inertia distribution, stabilization is assured. But there are two important caveats associated with this optimistic conclusion. One is related to "inertia augmentation" and the other concerns effective damping. All serious proposals for gravity gradient stabilization, both paper studies and flight experiments, have dealt squarely with these two issues.

Inertia Augmentation

It can readily be seen from the linear equations of motion for the small librations of an Earth-pointing satellite, namely, (9.2, 15), that the "restoring spring stiffness" in pitch, roll, and yaw have the form

$$\text{“}k_{gg}\text{”} \sim \omega_c^2 \Delta I \quad \mathrm{N \cdot m/rad} \tag{1}$$

where ΔI is a typical difference between principal moments of inertia. Since $\omega_c^2 \sim R^{-3}$, the stabilizing torque drops off rapidly with altitude. The gravity gradient torque for near-Earth orbits, weak at best, is over 200 times weaker at geostationary altitude. Equation (1) also implies that ΔI should be made as large as possible. This leads to satellite configurations in which some moments of inertia are made as large as possible. The moment of inertia of a long slender rod about its mass center is $ml^2/12$. This suggests that a rod can be made to have a larger moment of inertia (for a given mass) by making it longer and more slender. Not surprisingly, many practical gravity gradient satellite designs include one or more long rods to bring about a mass-effective augmentation in ΔI.

As a simple illustration, consider the configuration shown in Fig. 9.15. Neglecting the size of the main body compared to the lengths of the rods, we have

$$\begin{aligned}
\text{Roll “stiffness”} &\sim (I_2 - I_3) = (I_{b2} - I_{b3}) + \tfrac{2}{3}(m_3 l_3^2 - m_2 l_2^2) \\
\text{Pitch “stiffness”} &\sim (I_1 - I_3) = (I_{b1} - I_{b3}) + \tfrac{2}{3}(m_3 l_3^2 - m_1 l_1^2) \\
\text{Yaw “stiffness”} &\sim (I_2 - I_1) = (I_{b2} - I_{b1}) + \tfrac{2}{3}(m_1 l_1^2 - m_2 l_2^2)
\end{aligned} \tag{2}$$

Whatever the inertias of the main body might be, it appears that these stiffnesses can be made as large as desired, even for a fixed expenditure of mass, merely by making the rods sufficiently long, according to the condition

$$m_3 l_3^2 \gg m_1 l_1^2 \gg m_2 l_2^2 \tag{3}$$

Unfortunately, this neglects two facts: (i) if the rods become too flimsy, they will be so structurally weak that flexible-body dynamics will deteriorate performance, and (ii) the external torques against which the gravity gradient stiffness is expected to act (e.g., aerodynamic or solar-radiation pressure) are themselves amplified as the rods get longer. Thus, the rods cannot be made arbitrarily long, and engineering compromises must be made.

Another basic point can also be made using the configuration in Fig. 9.15: Although it is clear from equation (2) that l_3 should be made as long as possible,

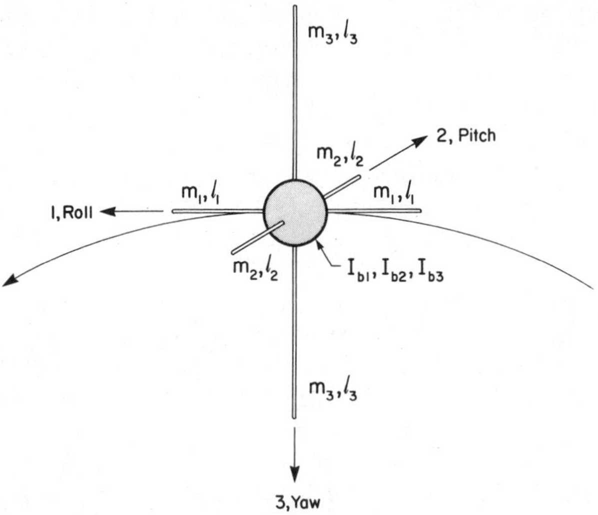

FIGURE 9.15 Typical scheme for inertia augmentation leading to gravity gradient stabilization.

the length l_1 requires a trade-off. Increasing l_1 increases yaw stiffness at the expense of pitch stiffness. As for l_2, its main benefit is to provide roll/yaw coupling, since the coupling coefficient in (9.2, 15) is

$$(I_3 + I_1 - I_2) = (I_{b3} + I_{b1} - I_{b2}) + \tfrac{4}{3} m_2 l_2^2 \tag{4}$$

Thus, l_2 can be made quite small, or even zero.

The need for long rods on gravity gradient satellites (together with antenna requirements on other types of satellites) has led to the development of long, light-weight metal tubes for space applications. These "booms" must be storable during launch and be unfurlable in orbit. Two reviews of the mechanical properties of the several types of booms have been given by [Staugaitis and Predmore] and by [NASA, 13].

Energy Dissipation—Effect of Quasi-Rigidity

It is amply clear from the developments in Chapters 5 and 7 that energy dissipation can be friend or foe in mechanical systems that include gyric coupling. It was demonstrated in Chapter 5 that the classical stability result for a spinning rigid body in torque-free motion, the major-or-minor-axis rule summarized in the k_1–k_3 stability diagrams of Figs. 4.14 and 4.15, is drastically altered by even a vanishingly small amount of damping, and becomes the major-axis rule represented by Fig. 5.3. To use the current vernacular: the bad news is that minor-axis spins are unstable; the good news is that major-axis spins are now (directionally) stable, *asymptotically to* $\underline{\vec{h}}$. We may anticipate similar conclusions when damping is introduced to the dissipation-free dynamical models of Section 9.2.

We begin by examining the general motion of a *quasi-rigid* body in a circular orbit. The notion of a "quasi-rigid" body, with its associated tiny "energy sink," was developed at the beginning of Chapter 5. The energy, normally constant in the absence of dissipation, slowly decreases. The leading term in the Hamiltonian H in (9.2, 8) is the kinetic energy. We might postulate an energy sink that reduces H slowly. Looking at (9.2, 8) alone, we might conclude that $\boldsymbol{\omega} = \mathbf{0}$ at minimum H. But according to the motion equations (9.2, 1), $\boldsymbol{\omega} \equiv \mathbf{0}$ cannot be a solution unless two of $\{c_{13}, c_{23}, c_{33}\}$ remain zero, in which case the third has unit magnitude, and according to (9.2, 7) this is not possible. We conclude that $\boldsymbol{\omega} \equiv \mathbf{0}$ is not a solution. Instead, we look at the alternate expression for H, equation (9.2, 9) and postulate an energy sink that slowly reduces the energy-like quantity H. At the minimum H, $\boldsymbol{\omega}_\alpha = \mathbf{0}$. In fact, $\boldsymbol{\omega}_\alpha = \mathbf{0}$ represents any one of the 24 possible equilibria (counting senses) studied in Problem 9.10. The issue is whether these equilibria are stable when vanishingly small damping is included. We already know from linear analysis of the librations about these equilibria that the only possibilities for equilibrium for a rigid body are (i) the major axis nominally normal to the orbit and the minor axis vertical (the Lagrange cases), and (ii) some inertia distributions in which the major axis is tangent to the orbit and the minor axis is normal to the orbit (the DeBra–Delp cases). These cases can be understood from studying Fig. 9.4 in conjunction with Fig. 4.14.

We now inquire whether these equilibria, infinitesimally stable for a rigid body, remain stable for a quasi-rigid body. The argument will proceed on the understanding that $I_2 > I_1 > I_3$ always; the two equilibria correspond to two different reference rotation matrices $\mathbf{C}_{pc}$. For the Lagrange cases, the equilibrium corresponds to $\mathbf{C}_{pc} = \mathbf{1}$, the unit matrix. The corresponding H for this relative equilibrium is $v = H - H_o = 0$, as indicated in (9.2, 27) and (9.2, 28). For the Debra–Delp cases, we have the reference orientation

$$\underset{\rightarrow}{\hat{\mathbf{p}}}_2 = \underset{\rightarrow}{\hat{\mathbf{c}}}_1; \qquad \underset{\rightarrow}{\hat{\mathbf{p}}}_3 = \underset{\rightarrow}{\hat{\mathbf{c}}}_2; \qquad \underset{\rightarrow}{\hat{\mathbf{p}}}_1 = \underset{\rightarrow}{\hat{\mathbf{c}}}_3 \tag{5}$$

for which

$$\mathbf{C}_{pc} = \begin{bmatrix} 0 & 0 & 1 \\ 1 & 0 & 0 \\ 0 & 1 & 0 \end{bmatrix} \tag{6}$$

Then, from (9.2, 28),

$$v = \tfrac{1}{2}\omega_c^2\left[3(I_1 - I_3) + (I_2 - I_1)\right] \tag{7}$$

for the DeBra–Delp equilibria. This value of v is positive for $I_2 > I_1 > I_3$. An energy sink that slowly reduces H, and therefore v, will therefore tend to bring the satellite from the DeBra–Delp orientation to the Lagrange orientation. While this argument is not strictly rigorous, inasmuch as it postulates an energy sink that drains H, it suggests that the DeBra–Delp equilibria are made unstable by dissipation.

As another variation on this theme, linear viscous damping terms can be added to the linear librational equations (9.2, 15). Since such terms represent direct librational damping, they would correspond to some interaction of the satellite with its external environment. Magnetic hysteresis rods (see damper mechanisms below) would provide such an interaction. [Likins and Columbus] have argued that such terms can also arise from eddy current damping. Alternatively, they can be thought of as heuristic damping terms. In any case, when terms $\{d_1\dot{\alpha}_2, d_2\dot{\alpha}_2, d_3\dot{\alpha}_3\}$ are added to each of (9.2, 15) respectively, the resulting system has the form

$$\mathscr{M}\ddot{\mathbf{q}} + \mathscr{D}\dot{\mathbf{q}} + \mathscr{G}\dot{\mathbf{q}} + \mathscr{K}\mathbf{q} = \mathbf{0} \tag{8}$$

where $\mathscr{M} = \mathbf{I}$, $\mathscr{D} = \operatorname{diag}\{d_i\}$, and

$$\mathbf{q} = [\alpha_1 \quad \alpha_2 \quad \alpha_3]^T$$

$$\mathscr{K} = \omega_c^2 \operatorname{diag}\{4(I_2 - I_3), 3(I_1 - I_3), (I_2 - I_1)\} \tag{9}$$

and the nonzero elements of the gyroscopic matrix $\mathscr{G}$ are

$$g_{31} = -g_{13} = g_{ry} \triangleq (I_3 + I_1 - I_2)\omega_c \tag{10}$$

Actually, the pitch equation is independent of the roll/yaw equations:

$$I_2\ddot{\alpha}_2 + d_2\dot{\alpha}_2 + 3\omega_c^2(I_1 - I_3)\alpha_2 = 0 \tag{11}$$

This shows that the condition for infinitesimal pitch stability, expressed in any of the equivalent forms

$$I_1 > I_3; \qquad k_1 > k_3; \qquad k_2 > 0 \tag{12}$$

now becomes a condition for asymptotic pitch stability.

The interesting question is whether *roll/yaw* librations are stable. These infinitesimal librations are governed by motion equations of the form

$$\mathscr{M}\ddot{\mathbf{q}} + \mathscr{D}\dot{\mathbf{q}} + \mathscr{G}\dot{\mathbf{q}} + \mathscr{K}\mathbf{q} = \mathbf{0} \tag{13}$$

where

$$\mathscr{M} = \operatorname{diag}\{I_1, I_3\}; \qquad \mathscr{D} = \operatorname{diag}\{d_1, d_3\}$$

$$\mathbf{q} = [\alpha_1 \quad \alpha_3]^T; \qquad \mathscr{K} = \omega_c^2 \operatorname{diag}\{4(I_2 - I_3), (I_2 - I_1)\}$$

$$\mathscr{G} = \begin{bmatrix} 0 & -g_{ry} \\ g_{ry} & 0 \end{bmatrix} \tag{14}$$

The system matrices in (13) satisfy the conditions required to use the Kelvin–Tait–Chetayev theorem (Appendix A). This theorem stipulates that asymptotic stability is possible only if $\mathscr{K} > 0$, that is, only if I_2 is the major axis. When combined with (12), it emerges that the librations of a quasi-rigid body in a circular orbit are stable only in the Lagrange region (Fig. 9.16). The tantalizing promise of the DeBra–Delp region cannot be a practical reality.

If Fig. 9.16 is compared with Fig. 5.3, it may at first appear that some of the stable cases have been lost. This is not a fair comparison. Whereas Fig. 5.3 is a stability diagram for *directional stability* of the spin axis, asymptotically to $\underrightarrow{\mathbf{h}}$, Fig. 9.16 is a diagram for asymptotic *three-axis stability*. On reflection, the disappearance of the DeBra–Delp region with the addition of damping (compare Fig.

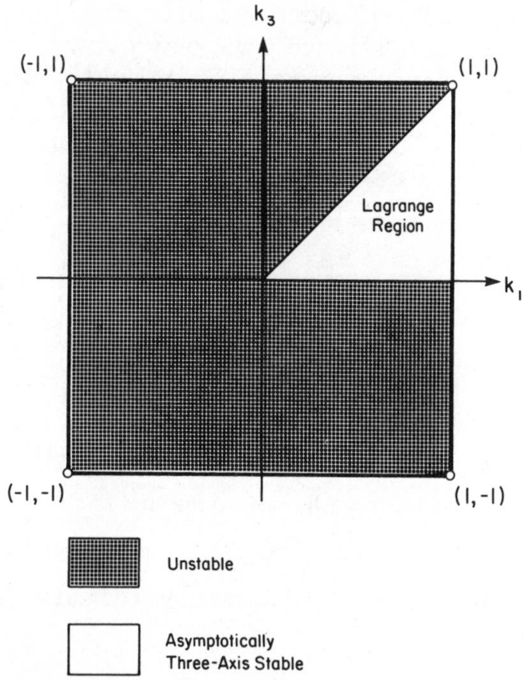

FIGURE 9.16 Stability diagram for the librations of a quasi-rigid body in a circular orbit. (Compare with Fig. 9.4.)

9.4 with Fig. 9.16) is not surprising. An Earth-pointing satellite is subject to two influences (nongravitational torques aside): (i) the basic spin stability that can be achieved by a once-per-orbit rotation (relative to $\mathscr{F}_i$), and (ii) a gravity gradient restoring torque. In the absence of damping, the former is stabilizing in roll/yaw if $I_2 > I_1$ and I_3, or if $I_2 < I_1$ and I_3 (Euler's rule). The latter is stabilizing if $I_3 < I_2$; see (9.1, 5). The Lagrange region, characterized by $I_2 > I_1 > I_3$, is thus simply a "major-axis spin" with a favorable gravity gradient torque. On the other hand, the DeBra–Delp region corresponds to a minor-axis spin with an unfavorable roll torque (though the pitch torque is favorable). This region (Fig. 9.4) corresponds to inertia distributions where the (supposedly) stabilizing minor-axis spin prevails over the destabiliziing gravity gradient roll torque. When damping is added, the minor-axis spin is no longer stabilizing, and the DeBra–Delp region has lost its *raison d'être*.

Rigid Body with Point Mass Damper

To demonstrate further the influence of damping on gravity gradient stability, we turn now to the system called $\mathscr{R} + \mathscr{P}$ in Chapter 3, Section 3.4. As shown in Figs. 3.6 and 3.7, it consists of a rigid body inside of which is a mass spring dashpot damper. The general (nonlinear) equations of motion were derived in

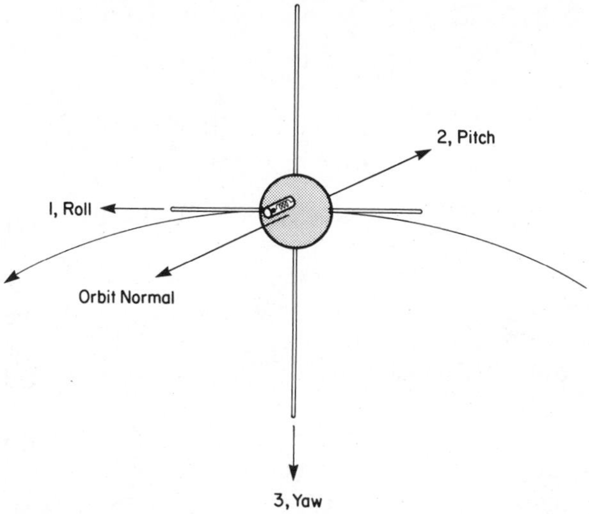

FIGURE 9.17 The system $\mathscr{R} + \mathscr{P}$ (i.e., rigid body with point mass damper), discussed in Section 3.4, as a gravity gradient satellite. (Compare with Figs. 3.6, 3.7, and 5.4.)

Section 3.4. Then, in Chapter 5, Section 5.2, this system was used to study the effect of internal damping on the directional stability of a spinning body. Linear motion equations were derived—linear in the deviations $\{\alpha_1, \alpha_2, \alpha_3\}$ from the reference simple spin and linear in the movement ξ of the damper mass $\mathscr{P}$ from its nominal position. These motion equations were summarized in equations (5.2, 11 through 14).

We can now place $\mathscr{R} + \mathscr{P}$ in a circular orbit and determine the influence of $\mathscr{P}$ on librational stability (Fig. 9.17). To do this, we need only interpret the spin rate ν in Chapter 5 as $-\omega_c$, the orbital rate, and add the gravity gradient force field. To the right of side of (5.2, 11), the gravity gradient torque must be added:

$$\mathbf{g}_{gg} = 3\omega_c^2 \begin{bmatrix} (I_3 - I_2)\alpha_1 \\ (I_3 - I_1)\alpha_2 \\ 0 \end{bmatrix} \tag{15}$$

as indicated in (9.1, 5). The damper motion $\xi(t)$ does not alter these torque components to first order. The net gravity gradient force on $\mathscr{R} + \mathscr{P}$ is zero, so equations (5.2, 10) are unchanged. But for $\mathscr{P}$ alone, we must add, in addition to the damper and spring force, $-c_d\dot{\xi} - d_d\xi$, a gravity gradient force $-m_d\omega_c^2\xi$ as indicated by the second term on the right side of (9.1, 13); see also Problem 9.1. Thus, in effect, the spring constant k_d is augmented as follows:

$$k_d \to k_d + m_d\omega_c^2 \tag{16}$$

When these modifications are made to (5.2, 11 and 12), the following set of motion equations for the small librations of $\mathscr{R} + \mathscr{P}$ in a circular orbit results:

$$I_2\ddot{\alpha}_2 + 3(I_1 - I_3)\omega_c^2\alpha_2 = 0$$
$$I_1\ddot{\alpha}_1 - (I_1 + I_3 - I_2)\omega_c\dot{\alpha}_3 + 4(I_2 - I_3)\omega_c^2\alpha_1 = 0$$
$$I_3\ddot{\alpha}_3 + (I_1 + I_3 - I_2)\omega_c\dot{\alpha}_1 + (I_2 - I_1)\omega_c^2\alpha_3 + m_d b(\ddot{\xi} + \omega_c^2\xi) = 0$$
$$\frac{m_d m_b}{m}\ddot{\xi} + c_d\dot{\xi} + (k_d + m_d\omega_c^2)\xi + m_d b(\ddot{\alpha}_3 + \nu^2\alpha_3) = 0 \quad (17)$$

These equations may be viewed as a combination of (5.2, 11 and 12) for $\mathscr{R} + \mathscr{P}$ with no gravity gradient field and (9.2, 15) for $\mathscr{R}$ in a gravity gradient field. The derivation just given of (17) has been rather brief to conserve space, and the diligent reader may wish to derive (17) carefully from first principles as suggested in Problem 9.20.

The system of equations (17) furnishes yet another opportunity to ascertain the influence of damping on the librations of a gravity gradient satellite. We have already used energy sink analysis (which has the benefit of being nonlinear, though not rigorous); and, in addition, the effect of *external* linear viscous damping has been found through the linear analysis represented by (14). In both cases, configurations in the Lagrange region are found to be asymptotically stable and those in the DeBra–Delp region to be unstable. With the $\mathscr{R} + \mathscr{P}$ system and (17), we can now determine the librational behavior with *internal* damping. The pitch equation in (17) is undamped; although another internal damper could readily be added to accomplish such damping, our chief interest is in the roll/yaw motion, for which stability is more problematic.

The development now will be parallel to the development in Section 5.2. The last three equations of (17) can be concisely written in matrix second-order canonical form:

$$\mathscr{M}\ddot{\mathbf{q}} + (\mathscr{D} + \mathscr{G})\dot{\mathbf{q}} + \mathscr{K}\mathbf{q} = \mathbf{0} \quad (18)$$

where

$$\mathbf{q} \triangleq \begin{bmatrix} \alpha_1 \\ \alpha_3 \\ \xi \end{bmatrix}; \qquad \mathscr{M} = \begin{bmatrix} I_1 & 0 & 0 \\ 0 & I_3 & m_d b \\ 0 & m_d b & \dfrac{m_d m_b}{m} \end{bmatrix}$$

$$\mathscr{K} = \begin{bmatrix} 4(I_2 - I_3)\omega_c^2 & 0 & \\ 0 & (I_2 - I_1)\omega_c^2 & m_d b\omega_c^2 \\ 0 & m_d b\omega_c^2 & k_d + m_d\omega_c^2 \end{bmatrix} \quad (19)$$

Also, $\mathscr{D} = \operatorname{diag}\{0, 0, c_d\}$, and the only nonzero elements of $\mathscr{G}$ are

$$g_{21} = -g_{12} = (I_1 + I_3 - I_2)\omega_c \quad (20)$$

The analogy with (5.2, 14) is immediate. The only changes in the present system of equations are gravity gradient adjustments to the diagonal elements of the "stiffness" matrix $\mathscr{K}$. Because $\mathscr{M} > 0$ (see Problem 5.3), $\mathscr{G}^T = -\mathscr{G}$, and $\mathscr{D}^T = \mathscr{D}$

≥ 0, it is sufficient for stability that $\mathscr{H} > 0$. Thus, if

$$\begin{aligned}&\textbf{(i)}\ I_2 - I_3 > 0\\ &\textbf{(ii)}\ I_2 - I_1 > 0\\ &\textbf{(iii)}\ (I_2 - I_1)(k_d + m_d\omega_c^2) > m_d^2 b^2 \omega_c^2\end{aligned} \tag{21}$$

stability is guaranteed. Note that (ii) is always weaker than (iii). According to this criterion, not all configurations in the Lagrange region are guaranteed to be stable. In presenting this result graphically, it is convenient to define the dimensionless parameters

$$\omega_d \triangleq \left(\frac{k_d}{m_d}\right)^{1/2}; \qquad \hat{\omega}_d \triangleq \frac{\omega_d}{\omega_c};$$

$$I_d \triangleq m_d b^2; \qquad \hat{I}_d \triangleq \frac{I_d}{I_3};$$

$$\hat{m}_b \triangleq \frac{m_b}{m} \tag{22}$$

together with the familiar inertia ratios k_1 and k_3. Then, the region guaranteed to be asymptotically stable is defined by

$$\begin{aligned}&\textbf{(i)}\ k_1 > 0\\ &\textbf{(ii)}\ k_3 > \frac{\hat{I}_d}{\hat{\omega}_d^2 + 1}\\ &\textbf{(iii)}\ k_1 > k_3\end{aligned} \tag{23}$$

Condition (iii) prevents instability.

To compare (23) more directly with the directional stability conditions (5.2, 18), an "effective" spring constant can be defined which combines the stiffness of the physical spring with gravity gradient stiffness:

$$k_{d,\text{eff}} \triangleq k_d + m_d\omega_c^2 \tag{24}$$

Then, with

$$\omega_{d,\text{eff}} \triangleq \left(\frac{k_{d,\text{eff}}}{m_d}\right)^{1/2}; \qquad \hat{\omega}_{d,\text{eff}} \triangleq \frac{\omega_{d,\text{eff}}}{\omega_c} \tag{25}$$

the stability conditions (23) become simply

$$k_1 > k_3 > \Xi_{d,\text{eff}} \tag{26}$$

where

$$\Xi_{d,\text{eff}} \triangleq \frac{\hat{I}_d}{\hat{\omega}_{d,\text{eff}}^2} \tag{27}$$

This tends to eliminate part of the Lagrange region, as illustrated in Fig. 9.18 for $\Xi_{d,\text{eff}} = 0.5$, an atypically large value. Again it is concluded that the benefits of damping are won at the expense of incurring the adverse effects of structural flexibility.

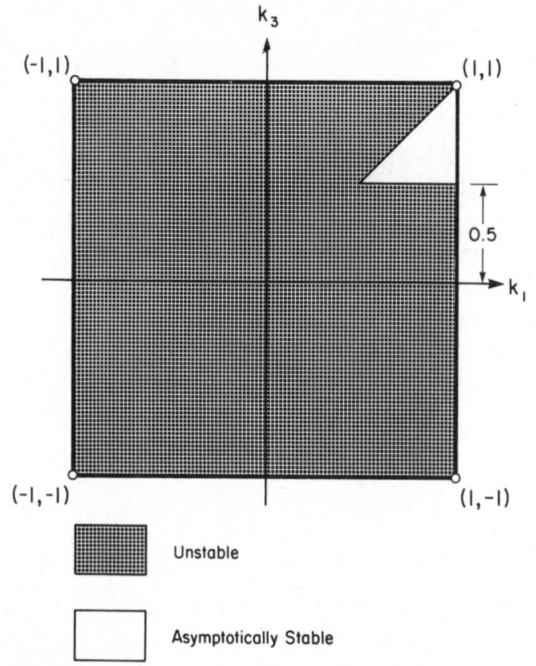

FIGURE 9.18 Stability diagram for librations of $\mathscr{R} + \mathscr{P}$ in a circular orbit when $\Xi_{d,\text{eff}} = 0.5$. (Compare with Figs. 9.4 and 9.16.)

It remains to show that $\mathscr{K} > 0$, and hence (26), is *necessary* for asymptotic stability. To accomplish this, we construct the characteristic equation corresponding to (18):

$$\det\left[\mathscr{M}s^2 + (\mathscr{D} + \mathscr{G})s + \mathscr{K}\right] = 0 \tag{28}$$

Written in terms of the dimensionless parameters defined above, this polmonimal reduces to

$$\left(\hat{m}_b s^2 + \hat{c}_d \omega_c s + \omega_c^2 \hat{\omega}_{d,\text{eff}}^2\right)\Delta(s^2) - \hat{I}_d\left(s^2 + \omega_c^2\right)^2\left(s^2 + 4k_1\omega_c^2\right) = 0 \tag{29}$$

where $\Delta(s)$ is the characteristic polynomial for roll/yaw librations with $k_d \to \infty$ (i.e., a single rigid body):

$$\Delta(s^2) \triangleq s^4 + (1 + 3k_1 + k_1 k_3)\omega_c^2 s^2 + 4k_1 k_3 \omega_c^4 \tag{30}$$

In studying (29), one possibility is to expand into a canonical sextic equation and find the roots numerically. This brute-force approach should be used only as a last resort because it can deal only with *specific* values of the parameters and cannot rigorously prove the elegant result that the sufficient conditions (26) are also necessary (except by engineering implication after a very large number of cases are computed). In an attempt to obtain literal stability conditions, the Routh–Hurwitz procedure could be undertaken, but this becomes unwieldy for a sextic polynomial. A more subtle alternative is to refer again to the matrix-second-order stability theory of Section A.6 in Appendix A and to realize that $\mathcal{K} > 0$ is a necessary condition for stability if the damping is *pervasive*. The damping is pervasive if there are no pure imaginary roots of the characteristic equation (29). Accordingly, we substitute $s = j\omega_c\beta$ into (29) to find out if such a root is possible. Equating the real and imaginary parts to zero gives

$$\hat{c}_d \beta \Delta(-\omega_c^2\beta^2) = 0 \tag{31}$$

$$(\hat{\omega}_{d,\text{eff}}^2 - \hat{m}_b\beta^2)\,\Delta(-\omega_c^2\beta^2) = \hat{I}_d(1-\beta^2)^2(4k_1 - \beta^2) \tag{32}$$

Now, $\beta \neq 0$; otherwise, from (32), $k_3 > \Xi_{d,\text{eff}}$ is violated. The only remaining possibility is $\Delta = 0$. From (32), this implies either $\beta^2 = 1$ or $\beta^2 = 4k_1$. However, from (30),

$$\Delta(-\omega_c^2) = -3\omega_c^4 k_1(1-k_3)$$

$$\Delta(-4\omega_c^2 k_1) = -4\omega_c^4 k_1(1-k_1)(1-k_3)$$

and neither of these can be zero for a tri-inertial, nonlaminar body. We conclude that there are no roots of the characteristic equation (29) on the imaginary axis. Therefore, the damping is pervasive, and therefore $\mathcal{K} > 0$ is a necessary condition for stability. In summary, the necessary and sufficient conditions for asymptotic stability of the librational motion of $\mathcal{R} + \mathcal{P}$ in a circular orbit as are given by (26) and as illustrated in Fig. 9.18 for $\Xi_{d,\text{eff}} = 0.5$.

Dampers Involving External Springs

There are several ways to classify the multitude of gravity gradient satellite designs that have been proposed. Some of these designs have aimed at stabilization only about the pitch and roll axes, thus maintaining the yaw axis nominally pointing toward Earth, while other designs attempted yaw stability as well. [Yaw stability tends to weaken pitch stability; see the remarks below equation (3).] Some designs, suited for low altitude, are not geometrically symmetrical, while the higher-altitude designs are symmetrical to balance solar radiation pressure. Perhaps the most distinguishing design feature, however, is the method for generating damping. In most cases, some form of internal relative motion is created that is coupled to the unwanted librations. By damping the former, one can ensure the decay of the latter. The basic idea of gravitational stabilization is as old as artificial satellites themselves. Roberson and Breakwell registered an early patent (1956), and [Kamm]'s "Vertistat" was one of the first published analyses.

One type of damper mechanism is a large, external spring. In the Rice–Wilberforce damper, described by [Buxton, Campbell, and Fosch], pitch and roll librations (and hence yaw, since it is coupled to roll) excite a "plunging" mode in the vertical spring (Fig. 9.19). This motion, in turn, causes a rotation of the disk immersed in damping fluid. A similar design has the dissipation in the spring itself (Fig. 9.20) and was the damping mechanism on the gravitationally stabilized satellites 1963-22A, 1963-38B, and 1963-49B. Figure 9.20 shows the

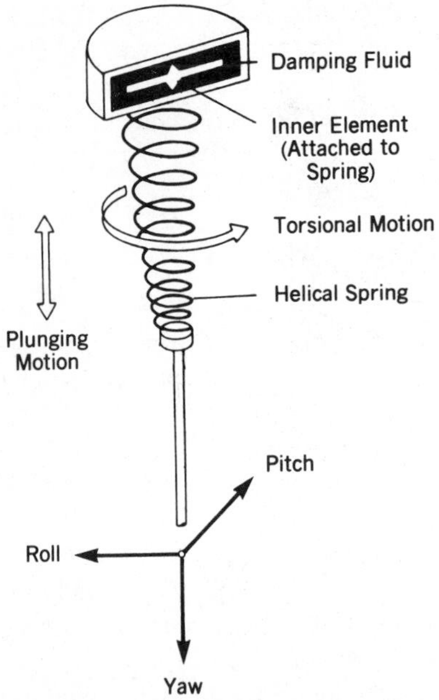

FIGURE 9.19 The Rice–Wilberforce damper. After [Buxton, Campbell, and Losch].

first of these satellites. Although these dampers are dynamically fascinating (e.g., see [Paul]), they have been supplanted by more effective alternatives.

Magnetic-Hysteresis Rods

If a slender rod of (magnetically) permeable material is attached to a satellite, rotation with respect to Earth's magnetic field will induce magnetic hysteretic losses. This idea is attractive in several respects: internally moving parts are not required, and the rods can be configured so as to contribute to inertia augmentation. For reasonable rod lengths, this effect is rather weak, made more so as altitude increases due to the R^{-3} dependence of the magnetic-field strength. It is also ineffective for librations about the magnetic-field lines.

Magnetic-hysteresis rods were used on satellite 1963-22A (Fig. 9.20) and on other satellites in the same series. Hysteresis rods are best suited to low-altitude, high-inclination orbits, especially for two-axis stabilization schemes (pitch and roll) and in conjunction with other damping mechanisms.

Spherical Tip Dampers

Two gravitationally stabilized satellites are exhibited in Fig. 9.21. It is obvious from their inertia distributions that they are to be two-axis stabilized. Further inspection reveals that the first satellite is suitable for low-altitude missions, while the second (symmetrical) configuration aspires to higher altitudes. Figure 9.21*a* shows a satellite launched in 1964 for the U.S. Naval Research Laboratory into a

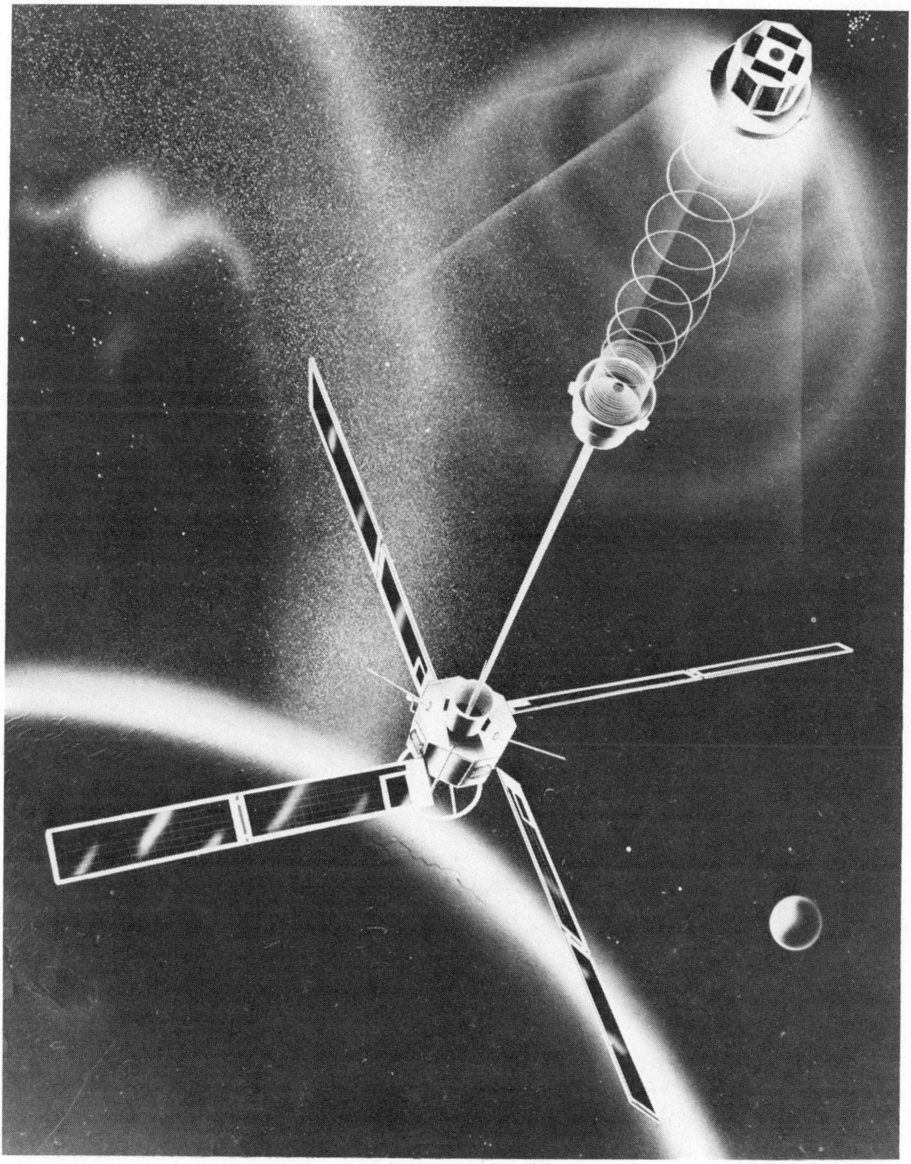

FIGURE 9.20 The "lossy spring" damper concept flown on satellite 1963-22A. From [Mobley and Fischell].

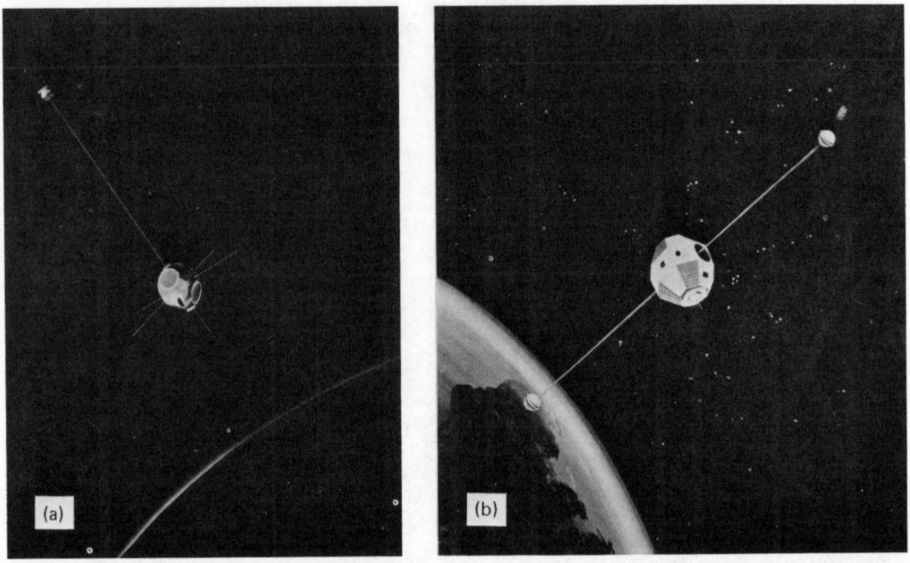

FIGURE 9.21 Two gravitationally stabilized satellites with viscous-fluid spherical dampers. From [Moyer and Katucki]. (*a*) NRL gravity gradient experiment (1964). (*b*) Gravity Gradient Test Satellite (1966).

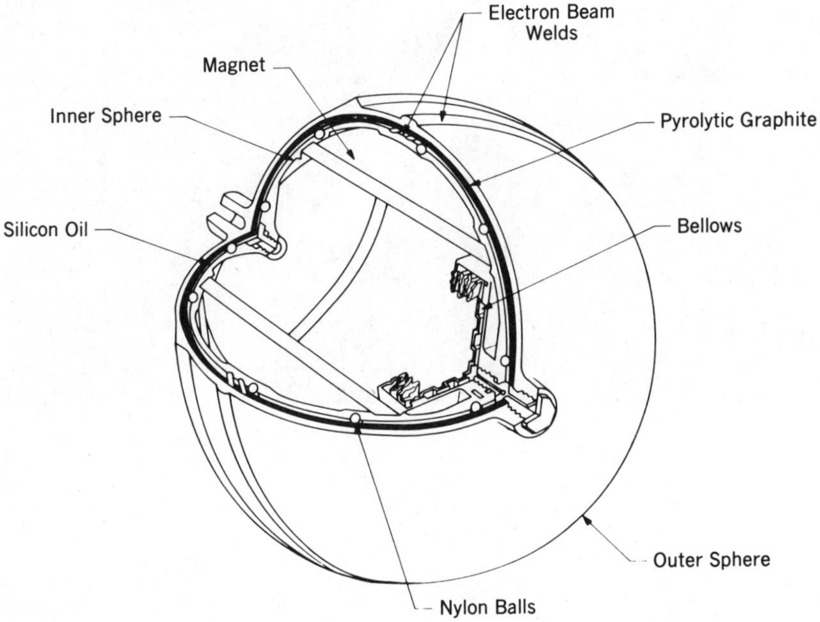

FIGURE 9.22 Viscous-fluid-filled, spherical, magnetically anchored damper. From [NASA, 14].

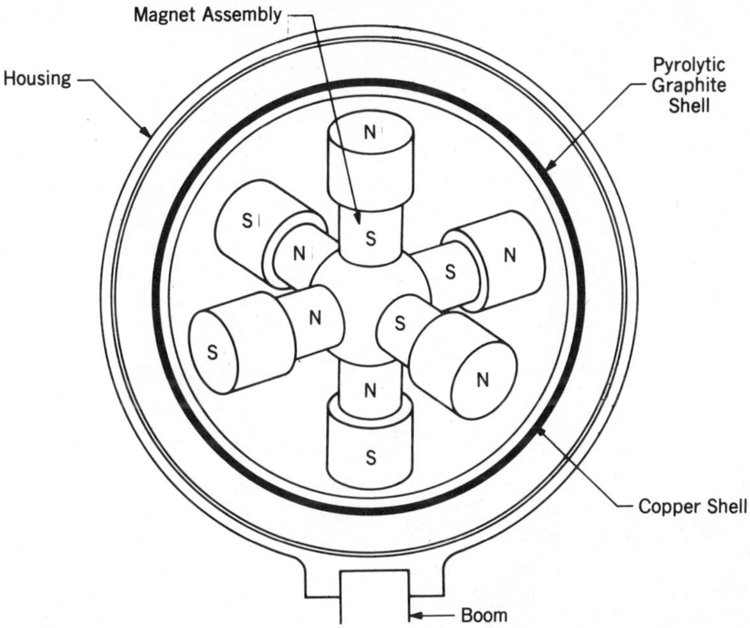

FIGURE 9.23 Eddy current spherical magnetically anchored damper. From [NASA, 14].

~ 900-km orbit for the purpose of gaining flight experience with gravity gradient stabilization ([Beal et al.]). Figure 9.21*b* shows the Gravity Gradient Test Satellite, launched in 1966 for the U.S. Air Force into a nearly geostationary orbit. For both of these satellites, the damping mechanism was a spherical ball damper designed by the General Electric Company. At the tip of the inertia augmenting booms is placed a sphere containing a second, concentric sphere anchored to Earth's magnetic field by a permanent magnet (Fig. 9.22). A viscous fluid between the inner and outer spheres provides the required damping. Design considerations for such satellites can be found in [Katucki and Moyer].

A similar idea is to replace the viscous fluid by an eddy-current damper (Fig. 9.23). This damper was also flown on an NRL satellite similar to that shown in Fig. 9.21*a* and on the GEOS (Geodetic Earth Orbiting Satellite) series. The two-axis-stabilized GEOS-II is shown in Fig. 9.24; it was launched in 1968 into a 1111 × 1574 km orbit.

Damping by Boom Articulation

A large number of the gravity gradient designs proposed feature an articulated damping boom: one or more of the inertia augmenting booms is hinged to the main body, and a suitable mechanism at this hinge provides the necessary damping. In the early 1960s, a number of analyses were published which

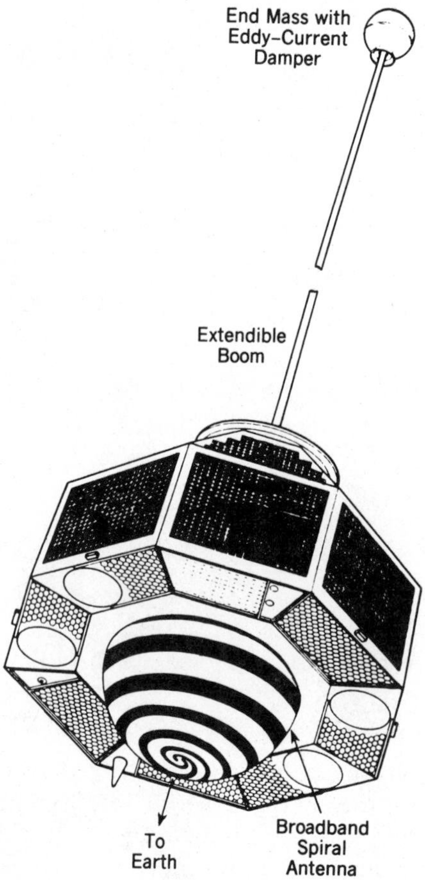

FIGURE 9.24 The GEOS-II space craft. After [Whisnant, Waszkiewicz, and Pisacane].

examined the feasibility of this concept (Fig. 9.25). A series of paper by [Paul, West, and Yu], [Fletcher, Rongved, and Yu], and [Yu] proposed a horizontally crossed pair of booms with two damping degrees of freedom (Fig. 9.25*a*). [Etkin, 1] showed encouraging transient performance for a configuration with two vertical booms, one above and one below the main body and each hinged with two damping degrees of freedom (Fig. 9.25*b*). [Tinling and Merrick] pointed out that all three librational degrees of freedom can be damped with only one extra degree of freedom (articulation) provided this rod and its axis of rotation are nominally in the horizontal plane (Fig. 9.25*c*). This damps pitch and roll directly, and yaw is damped indirectly through its coupling to roll. If geometrical symmetry is desirable (high-altitude orbits) and it is inconvenient to have an articulated rod pass through the center of the satellite, the skewed damper boom can be split into two parts ([Tinling, Merrick, and Watson]). [Sabroff, 1] studied a number of configurations, including the one shown in Fig. 9.25*d*, where the inertia augmentation booms along the pitch and roll axes are articulated for damping.

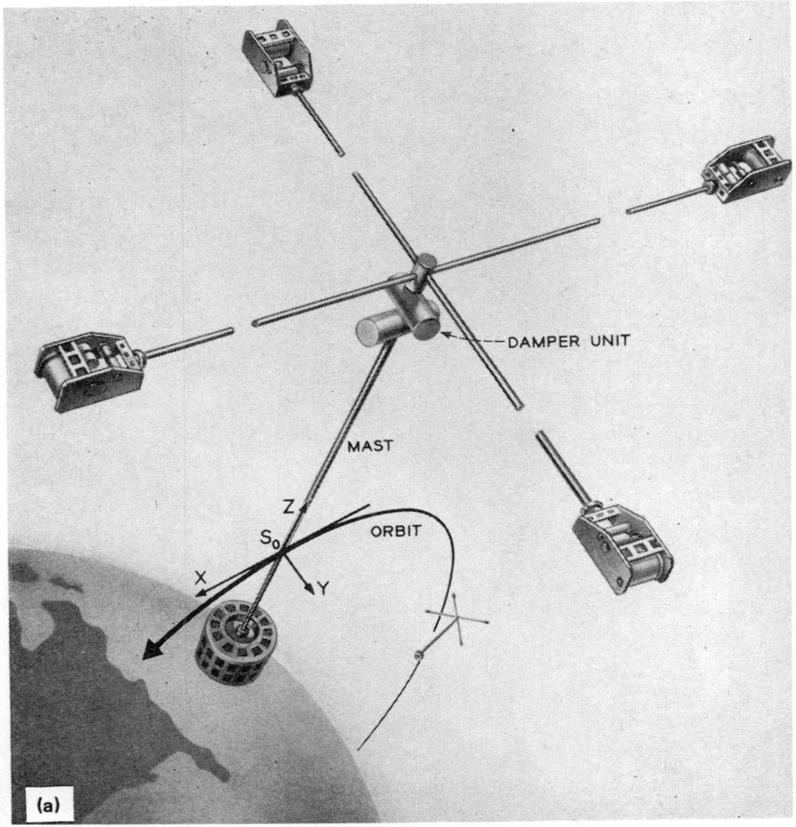

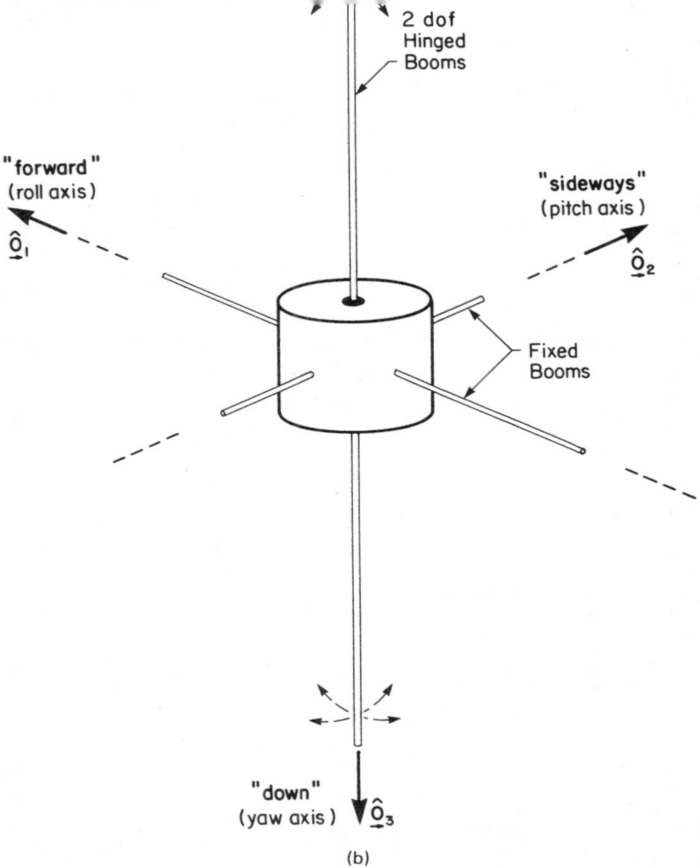

FIGURE 9.25 Articulated boom concepts. (*a*) Elevated horizontal damping booms. From [Paul, West, and Yu]. (*b*) Vertical damping booms. From [Etkin, 1]. (*c*) Single skewed boom. From [Tinling and Merrick]. (*d*) Horizontal damping booms. From [Sabroff, 1].

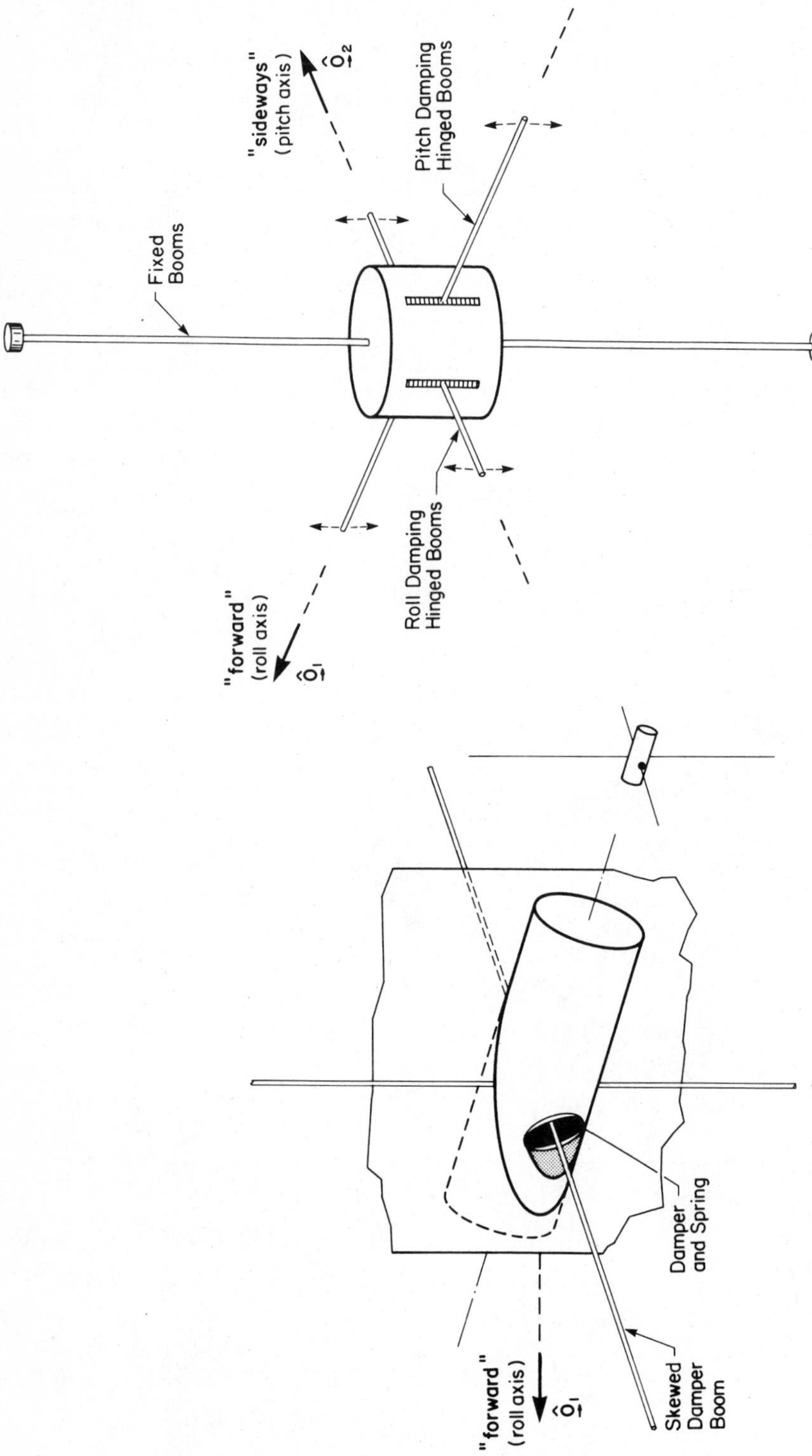

FIGURE 9.25 Continued

In all of the articulated designs, the equations of librational motion are of the form

$$\mathscr{M}\ddot{\mathbf{q}} + (\mathscr{D} + \mathscr{G})\dot{\mathbf{q}} + \mathscr{K}\mathbf{q} = \mathscr{f} \tag{33}$$

as might be expected. (Nongravitational disturbances are represented by $\mathscr{f}$). To understand the purpose for articulation however, it is more revealing to partition (33) into two parts, one corresponding to the roll/pitch/yaw motion characteristic of the main rigid body, and the other part corresponding to the nonrigid relative motion (the articulation). Thus, let

$$\mathbf{q} = \begin{bmatrix} \boldsymbol{\alpha}^T & \boldsymbol{\alpha}_a^T \end{bmatrix}^T \tag{34}$$

where $\boldsymbol{\alpha} = [\alpha_1 \quad \alpha_2 \quad \alpha_3]^T$ represents roll, pitch, and yaw $\boldsymbol{\alpha}_a$ represents the small angles through which the articulated booms rotate relative to the main body. The partitioned form of (33) is

$$\begin{bmatrix} \mathscr{M}_\alpha & \mathscr{M}_{\alpha a} \\ \mathscr{M}_{\alpha a}^T & \mathscr{M}_a \end{bmatrix} \begin{bmatrix} \ddot{\boldsymbol{\alpha}} \\ \ddot{\boldsymbol{\alpha}}_a \end{bmatrix} + \begin{bmatrix} \mathbf{O} & \mathbf{O} \\ \mathbf{O} & \mathscr{D}_a \end{bmatrix} \begin{bmatrix} \dot{\boldsymbol{\alpha}} \\ \dot{\boldsymbol{\alpha}}_a \end{bmatrix}$$
$$+ \begin{bmatrix} \mathscr{G}_\alpha & \mathbf{O} \\ \mathbf{O} & \mathbf{O} \end{bmatrix} \begin{bmatrix} \dot{\boldsymbol{\alpha}} \\ \dot{\boldsymbol{\alpha}}_a \end{bmatrix} + \begin{bmatrix} \mathscr{K}_\alpha & \mathscr{K}_{\alpha a} \\ \mathscr{K}_{\alpha a}^T & \mathscr{K}_a \end{bmatrix} \begin{bmatrix} \boldsymbol{\alpha} \\ \boldsymbol{\alpha}_a \end{bmatrix} = \begin{bmatrix} \mathscr{f}_\alpha \\ \mathscr{f}_a \end{bmatrix} \tag{35}$$

If the generalized forces $\mathscr{f}_a$ associated with articulation were such that relative motion was in fact prevented ($\boldsymbol{\alpha}_a \equiv \mathbf{0}$), that is, if

$$\mathscr{f}_a = \mathscr{M}_{\alpha a}^T \ddot{\boldsymbol{\alpha}} + \mathscr{K}_{\alpha a}^T \boldsymbol{\alpha} \tag{36}$$

in the steady state, then the remaining equations of motion, namely,

$$\mathscr{M}_\alpha \ddot{\boldsymbol{\alpha}} + \mathscr{G}_\alpha \dot{\boldsymbol{\alpha}} + \mathscr{K}_\alpha \boldsymbol{\alpha} = \mathscr{f}_\alpha$$

must revert to those describing the motion of a rigid body, as contained in (9.2, 15). This indicates that $\mathscr{M}_\alpha = \mathbf{I}$, with $\mathscr{G}_\alpha$ and $\mathscr{K}_\alpha$ as described in (10) and (11). The point is that $\boldsymbol{\alpha}$ is coupled to $\boldsymbol{\alpha}_a$ through $\mathscr{M}_{\alpha a}$ and $\mathscr{K}_{\alpha a}$, and $\boldsymbol{\alpha}_a$ is damped by the (usually diagonal) matrix $\mathscr{D}_\alpha$. Even though $\mathscr{D}$ is not positive-definite, it is positive-semidefinite and pervasive by virtue of $\mathscr{D}_a > 0$ through appropriate configurational design. In all cases, the proposers of the various designs attempt to optimize the configuration by choosing suitable boom lengths and damper constants. Of course, in these purely passive systems, even the best transient performance and response times cannot overcome the fact that the characteristic time in the dynamics is the orbital period. Moreover, one usually has to be content with a compromise between transient performance and steady-state response to disturbance torques. Usually, the transient performance can be enhanced by altering the length of a damper boom at the expense of weakening the basic gravity gradient stiffness needed to resist environmental disturbances. For further results on configurational optimization, the reader is also referred to [Zajac, 2], [Clark], [Connell, 2, 3] [Sarychev, Mirer, and Sazonov], and [Hartbaum et al.].

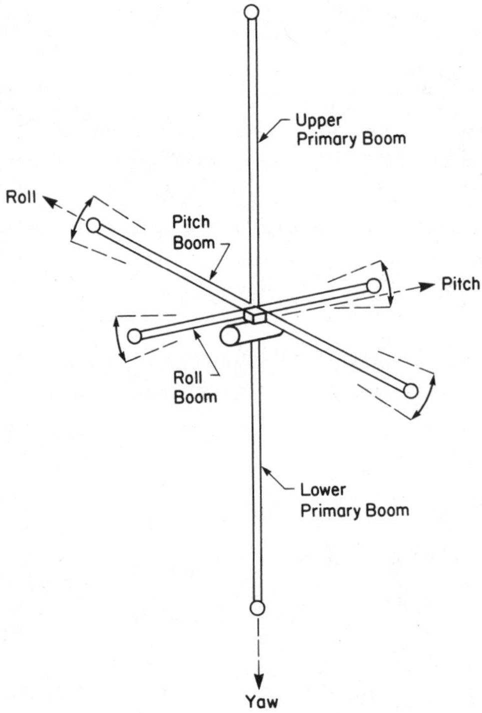

(a)

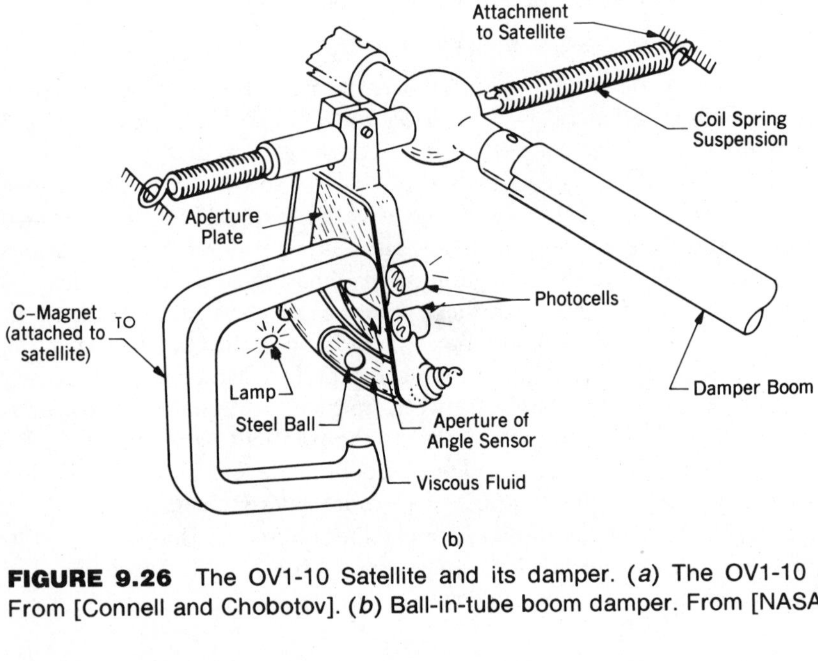

(b)

FIGURE 9.26 The OV1-10 Satellite and its damper. (*a*) The OV1-10 Satellite. From [Connell and Chobotov]. (*b*) Ball-in-tube boom damper. From [NASA, 14].

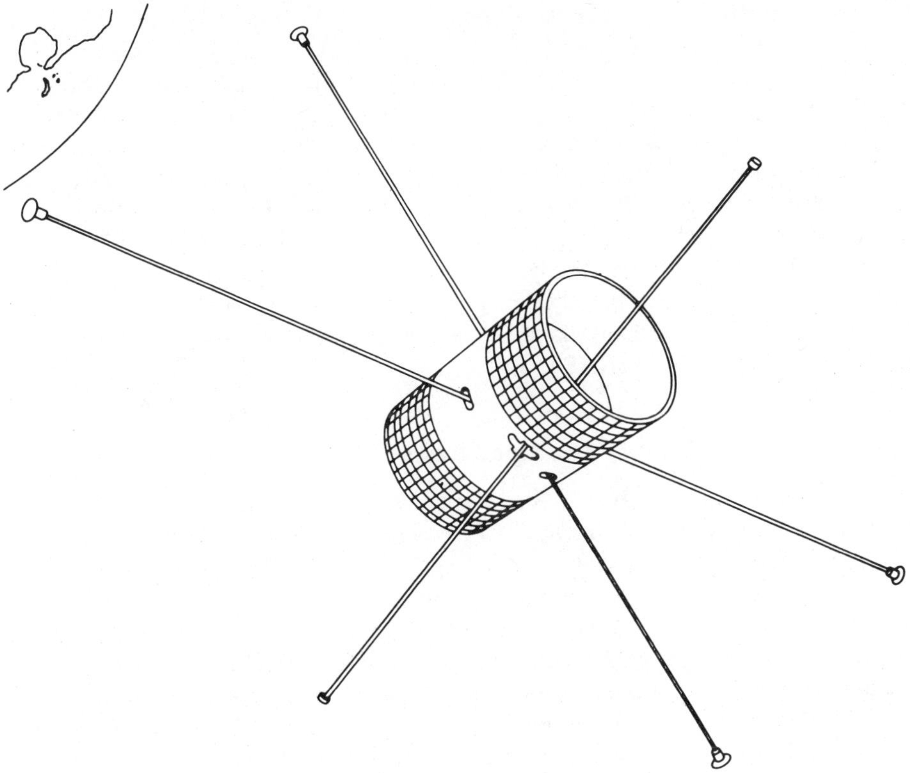

(a)

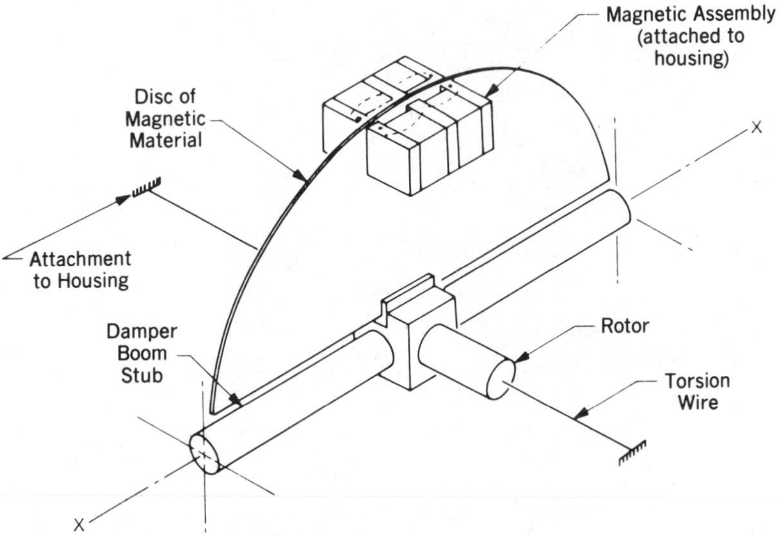

(b)

FIGURE 9.27 ATS Configuration and its damper. (*a*) The ATS configuration (failed to achieve circular orbit). From [Moyer and Katucki]. (*b*) Magnetic hysteresis boom damper. From [NASA, 14].

Two hardware representations of the "linear viscous dampers" at the base of the damper booms are shown in Figs. 9.26 and 9.27. The OV1-10 satellite (Fig. 9.26*a*) was inserted into a near-Earth orbit in 1966 and was gravitationally stabilized about all three axes. As shown in Fig. 9.26*b*, the articulated booms provide damping through a ball-in-tube damper. The motion of viscous fluid past the magnetically held steel ball leads to dissipation. The ATS spacecraft (Fig. 9.27*a*) used the skewed articulation boom developed by [Tinling and Merrick]. It was intended to be three-axis stabilized but never achieved circular orbit. Due to its high orbital eccentricity, gravity gradient stabilization was not achieved. The damper used on ATS was based on magnetic hysteresis (Fig. 9.27*b*).

Complete details on many damping devices used on gravitationally stabilized satellites can be found in [NASA, 14].

More Recent Gravity Gradient Applications

The Soviet Union's Salyut-6/Soyuz space station, shown on the cover of the January 1979 issue of *Astronautics and Aeronautics* and reproduced in Fig. 9.28, has spent much of its time in a gravity gradient-stabilized mode. Cosmonauts provide an interesting source of dynamical disturbance.

The U.S. Long Duration Exposure Facility (LDEF), shown in Fig. 9.29, is also three-axis stabilized gravitationally, as described by [Huckins, Breedlove, and Heinbockel]. Its normal inertia ratios are $\{k_1, k_3\} = \{0.055, 0.65\}$, and its librational frequencies are approximately $\{1.3, 1.7, 0.1\}\,\omega_c$ in {pitch, roll, yaw}. Damping is provided by a single viscous-fluid-filled, spherical magnetically anchored

FIGURE 9.28 The Salyut-6 Space Station (front cover of *Astronautics & Aeronautics*, January 1979).

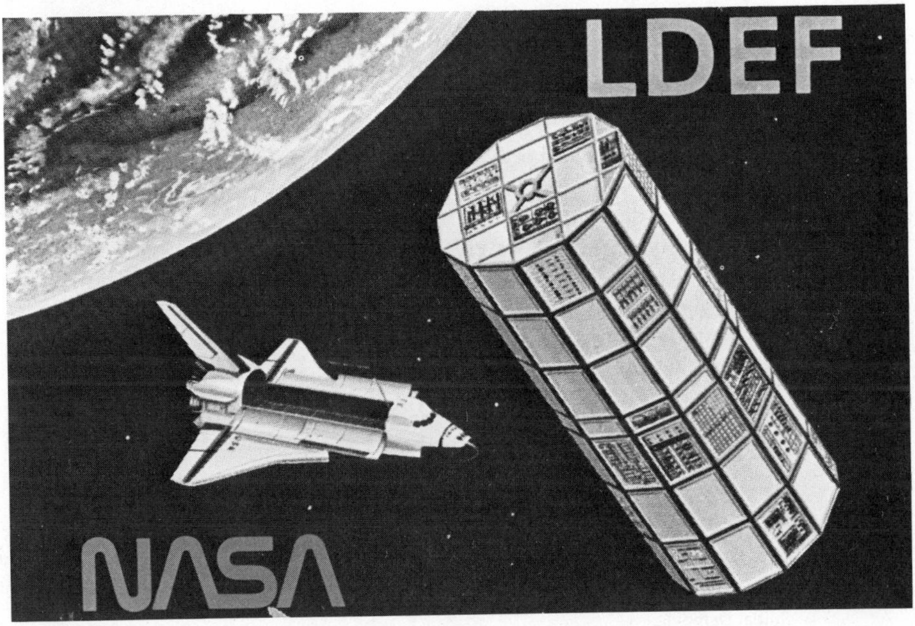

FIGURE 9.29 The Long Duration Exposure Facility.

damper of the type shown in Fig. 9.22. Located near the vehicle centroid, this damper is 20 cm in diameter and has a mass of 7 kg.

9.4 FLIGHT EXPERIENCE

Gravitational torques are quite weak, and satellites to be stabilized by the gravity gradient field must be carefully designed to withstand environmental disturbances. All of the disturbing torques described in Chapter 8 can be present to some degree, each causing unwanted librations. Several analyses of the response to specific torques can be found in the literature. [Nurre], [Modi and Shrivastava, 2, 3], and [Shrivastava and Modi], for example, have examined the librational consequences of aerodynamic torque at low altitudes. At high altitudes, solar radiation pressure becomes dominant; [Modi and Flanagan] and others have investigated this disturbance. Even meteoroidal disturbances can occasionally produce $\sim 0.5°$ attitude deviations ([Morgan and Yu]). Another subtle influence is the inertial nonstationarity of the orbit itself. Depending on the orbital radius and inclination, the orbit can precess at several degrees per day from Earth's oblateness. It should also be borne in mind that damping is as important in reducing response to external torques as it is in the decay of transients. Even a small, constant disturbing torque about the pitch axis can remove the Lagrange region stability of an undamped rigid body, as pointed out by [Garber]; see Problem 9.21. Space here does not permit a description of each of these analyses. For any real spacecraft, one can only predict librational motion through a detailed numerical simulation that includes all the relevant influences. We turn instead to a brief summary of flight experience with gravitationally stabilized satellites. Although the following discussion is not exhaustive, it is hoped that the examples selected will be interesting and instructive.

APL Satellites

The earliest "gravity-anchored" satellites were launched for the sole purpose of testing how well the theoretical predictions of gravity gradient stabilization could be attained in practice. (Parenthetically, one wonders why there have not been more attitude stabilization experiments in the past 25 years.) The first such satellite, 1963-22A, shown in Fig. 9.20, was launched in June 1963 into a near-Earth orbit. Also known as TRANSIT-5A, it was the first spacecraft to be successfully gravity oriented to the vertical (two-axis stabilization). As reported by [Mobley and Fischell], gravity capture and transient decay to a steady-state pointing error of ~ 6° was achieved within about a week (Fig. 9.30). Four additional satellites, also designed by the Applied Physics Laboratory at The

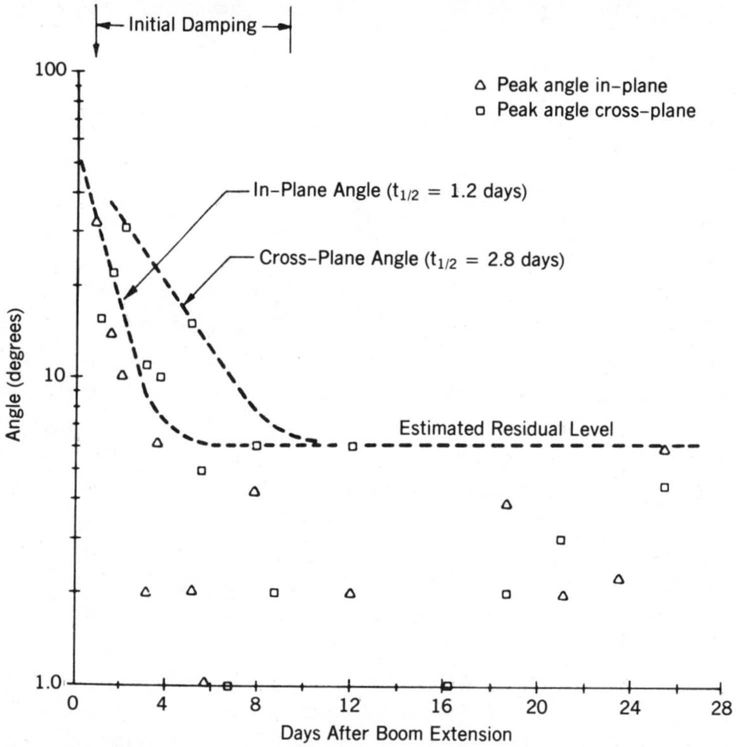

FIGURE 9.30 Capture and steady-state pointing error of the 1963-22A satellite. From [Mobley and Fischell].

Johns Hopkins University (JHU-APL), also achieved two-axis gravity gradient stabilization to essentially the same accuracy. In addition to the predicted disturbances, differential thermal heating of the stabilizing booms was detected to be a significant source of agitation on 1963-22A. The warped booms induced attitude errors. Moreover, in passing from Earth's shadow into sunlight, a dynamic version of this excitation was observed, descriptively named "thermal flutter"; this caused additional oscillations as high as 5°. Since the period of these boom-induced oscillations was observed to be ~ 15 s, this is a high-frequency disturbance when compared with the characteristic time of the other dynamics interactions—the orbital period. The source of thermal flutter was traced to the "open" cross section of the booms and to their radiative absorptivity. [Frisch, 1, 3] and [Kanning] have contributed detailed analyses that confirm this dynamic bending phenomenon, which not only excites librations directly through dynamic changes in the spacecraft inertia distribution but also excites them indirectly by altering the vehicle geometry as presented to aerodynamic or solar-radiation pressure. Static bending and thermal flutter were subsequently much reduced by boom seam interlocking, perforation, and silver plating. (A complete discussion of tubular booms is available in [NASA, 13].) [Garg and Hughes] have suggested that composite fibers could be used to replace much of the boom, thus reducing the exposed surface area to zero over the replaced length.

General Electric Satellites

In parallel with the five satellites designed by JHU-APL for the U.S. Navy, the General Electric Co. designed a series of gravity gradient satellite experiments for NRL (Naval Research Laboratory) using the GE magnetically anchored dampers shown in Figs. 9.22 and 9.23. A typical satellite in this series is shown in Fig. 9.21*a*. These were also low-altitude, two-axis-stabilized vehicles. [Beal et al.] reported pointing performance to be essentially the same as for the JHU-APL series.

Geodetic Earth Orbiting Satellites (GEOS)

The excitation from orbital eccentricity was significant for GEOS-I, similar in design to Fig. 9.24. In accordance with its mission to study geometrical and gravimetric geodesy, it was placed in a relatively near-Earth orbit whose eccentricity was $e = 0.072$. (The JHU-APL and GE designs described above were in orbits in the range $e = 0.002$–0.007.) Since it is not difficult to obtain (indeed, it is more difficult to avoid) pitch librations of amplitude $5e$ rad, pitch amplitudes up to 20° can be anticipated if orbital resonances are not avoided, as they were with GEOS-I. In spite of the relatively large eccentricity, GEOS-I was stabilized to the vertical to within 4° as shown in Fig. 9.31, extracted from the report by [Pisacane, Pardoe, and Hook].

GEOS-II (Fig. 9.24) was also inserted, in 1968, into what is for a gravity gradient satellite a quite elliptic orbit ($e = 0.032$). This by itself induces a 1.8°

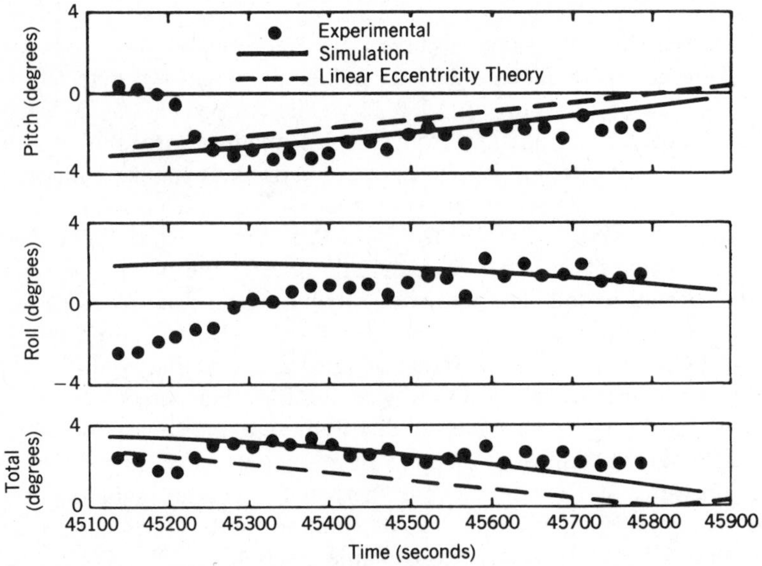

FIGURE 9.31 Earth pointing of GEOS-A. From [Pisacane, Pardoe, and Hook].

pitch oscillation. Nevertheless, two-axis stabilization (stabilization of the vertical) was achieved to within 5°. A comparison of flight data to computer simulation is shown in Fig. 9.32, taken from [Whisnant, Waszkiewicz, and Pisacane].

The OV1-10 Satellite

Three-axis stabilization was attempted for the OV1-10, shown in Fig. 9.26*a*. It was launched in December 1966 into a nearly circular orbit. Although basic gravity gradient satellite theory (Section 9.2) would indicate asymptotic stability with time constants of a few hours, OV1-10 suffered periodic inversions to an upside-down orientation and large, sometimes rather rapid, excursions in yaw ([Connell and Chobotov]). A large-scale analysis by [Connell, 1] isolated four possible sources for this anomalous behavior: (i) boom thermal flutter as the vehicle emerged from shadow into sunlight, (ii) a near-resonant interaction between the residual on-board magnetic dipole and Earth's magnetic field, (iii) small biases in the "springs" at the base of the damping booms, and (iv) slightly crooked damping booms. Since these causes were shown by numerical simulation to be sufficient to produce large yaw excursions and occasional tumbling, this OV1-10 experiment supplies valuable experience on the precautions necessary to achieve three-axis gravitational stabilization.

The Gravity Gradient Test Satellite (GGTS)

In anticipation of the many requirements for geostationary Earth-pointing communications satellites, the GGTS (Fig. 9.21*b*) was launched in June 1966 to

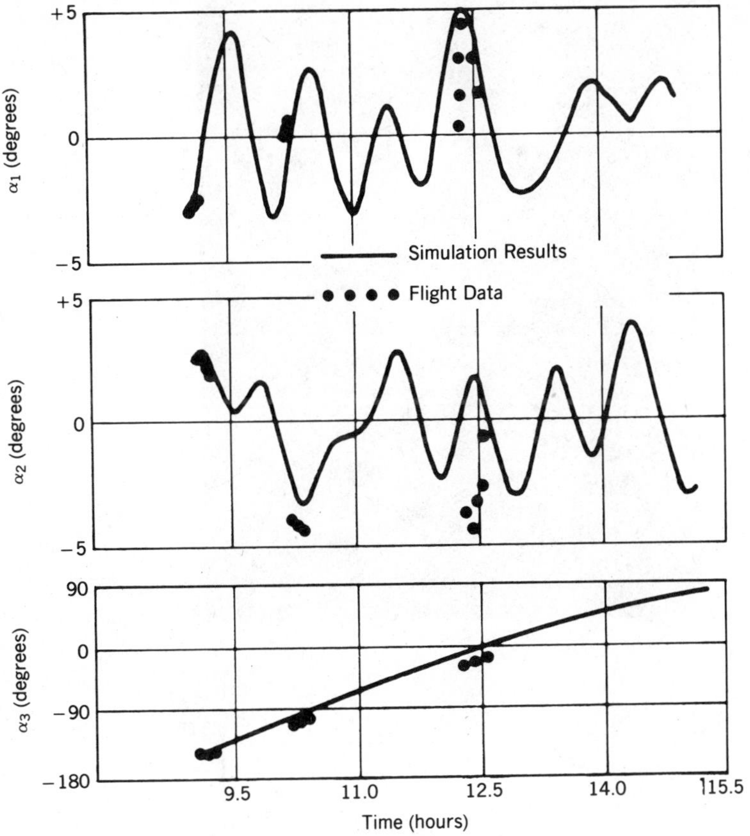

FIGURE 9.32 Earth-pointing performance of GEOS-II. From [Whisnant, Waszkiewicz, and Pisacane].

experiment with two-axis stabilization at near-synchronous altitude. GGTS was symmetrical, as dictated by the very weak gravity gradient field at these altitudes, and contained two of the GE spherical, magnetically anchored viscous dampers shown in Fig. 9.22, one at each tip. Although stabilization to within 8° was expected (within 60 days), librations of up to 15° continued after 60 days, according to [NASA, 14]. It should be mentioned that the direction of the magnetic field at 34,000 km altitude is somewhat unreliable, especially after periods of solar activity; and under these circumstances, a magnetically anchored damper can induce unwanted disturbances as it searches for the local field lines.

Department of Defense Gravity Experiment (DODGE Satellite)

The DODGE satellite is shown in Fig. 9.33. Launched in 1967, its aim was to assess the potential of three-axis stabilization in geostationary orbit. As shown in Figs. 9.33*b* and 9.33*c*, either of two configurations, both based on [Tinling and

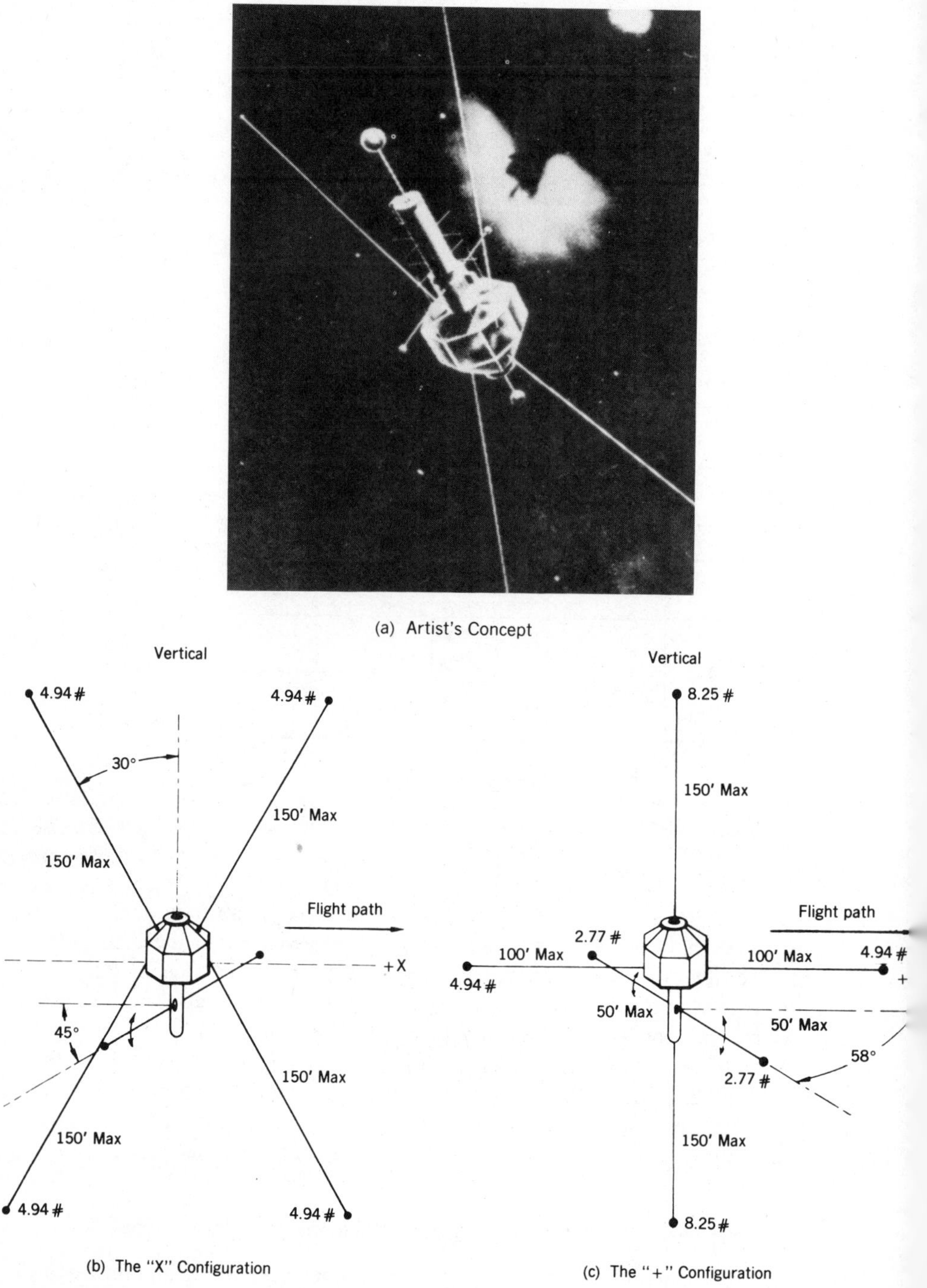

FIGURE 9.33 The DODGE Satellite. From [Fischell and Mobley]. (*a*) Artist's concept. (*b*) The "X" configuration. (*c*). The "+" configuration.

Merrick]'s single skewed damper boom (cf. Fig. 9.25*c*), could be studied. In addition to Tinling and Merrick's analysis, [Bainum and Mackison] have calculated that three-axis stabilization should be possible, even at geostationary altitude, for a satellite of this type. For redundancy, the boom damper was mounted on a torsion wire and used a combination of eddy current and magnetic-hysteresis dissipation.

Two types of active damping were also used on DODGE ([Fischell and Mobley]). In both cases a magnetometer signal was fed to appropriate electronics. In the "magnetic sample-and-hold" system, an amplifier produced a dipole moment over $t_1 \leq t \leq t_1 + \Delta t$ that was in the direction of Earth's magnetic field as measured at t_1. As the satellite wandered from its orientation at t_1, a restoring magnetic torque was thereby created which tended to damp librations. In the "enhanced magnetic hysteresis" system, the magnetometer reading was also amplified and fed to an electromagnet according to a different algorithm.

As illustrated by Fig. 9.34*a*, the torsion wire damper—the passive one—did not damp librations, in spite of the fact that the damper boom was known to be moving. The best results were obtained with the "X" configuration and active damping via the sample-and-hold technique ([Fischell and Mobley]). Figure 9.34*b* can be compared with Fig. 9.34*a*; pitch and roll were now within 10° and yaw within 25°.

Also flown on DODGE was a small momentum wheel aligned with the pitch axis to test, for the first time, the influence of enhanced roll/yaw coupling. (This is analyzed dynamically in Chapter 11.) When this wheel was turned on, the pitch librations were essentially unaffected, as expected; roll remained within 10° and yaw was brought to within 20°. Moreover, even better performance could be expected for an optimally designed wheel ([Fischell and Mobley]).

Radio Astronomy Explorer (RAE) Satellite

The RAE satellite (Fig. 9.35*a*) was launched in 1968 into a circular medium-altitude (6781 km) orbit—above the ionosphere. Consonant with its objectives in radio astronomy, it was largely an antenna made up of two long booms (460 m tip to tip) in the form of an "X." Based on [Tinling and Merrick]'s "inertia coupling" principle (Fig. 9.25*c*), a single damper boom (length 96 m) was incorporated, nominally in the roll/pitch plane, to oscillate in a plane at 65° to the roll/yaw plane. In this manner it was hoped to achieve three-axis gravity gradient stabilization. The successful "capture" of such a satellite to a nominal Earth-pointing attitude is itself a nontrivial dynamics problem ([Bowers and Williams]). Moreover, for booms this long, "flexible-body" dynamics must be included in preflight simulations. In particular, solar-induced thermal excitation was avoided by perforating the booms, silver plating their exteriors, and blackening their interiors.

As reported by [Blanchard], RAE successfully remained gravity gradient stabilized to within 5° in roll, 10° in pitch, and 35° in yaw. Typical flight data are exhibited in Fig. 9.35*b*.

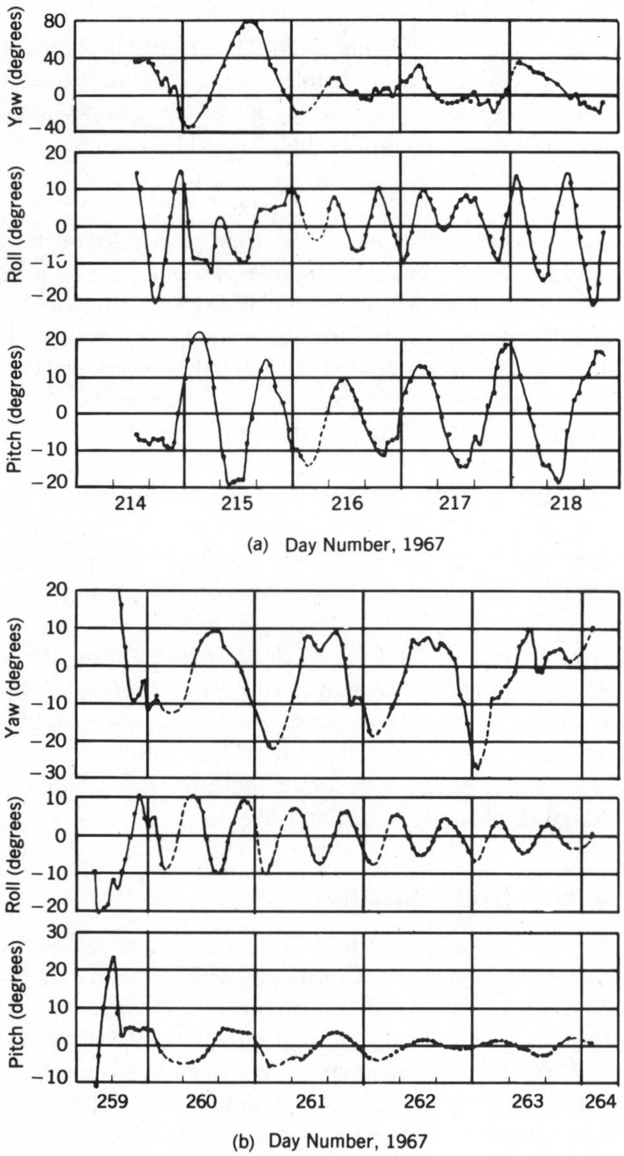

FIGURE 9.34 Flight results from the DODGE Satellite. From [Fischell and Mobley]. (*a*) Librations with a completely passive system; "X" configuration. (*b*) Librations with active damping; "X" configuration.

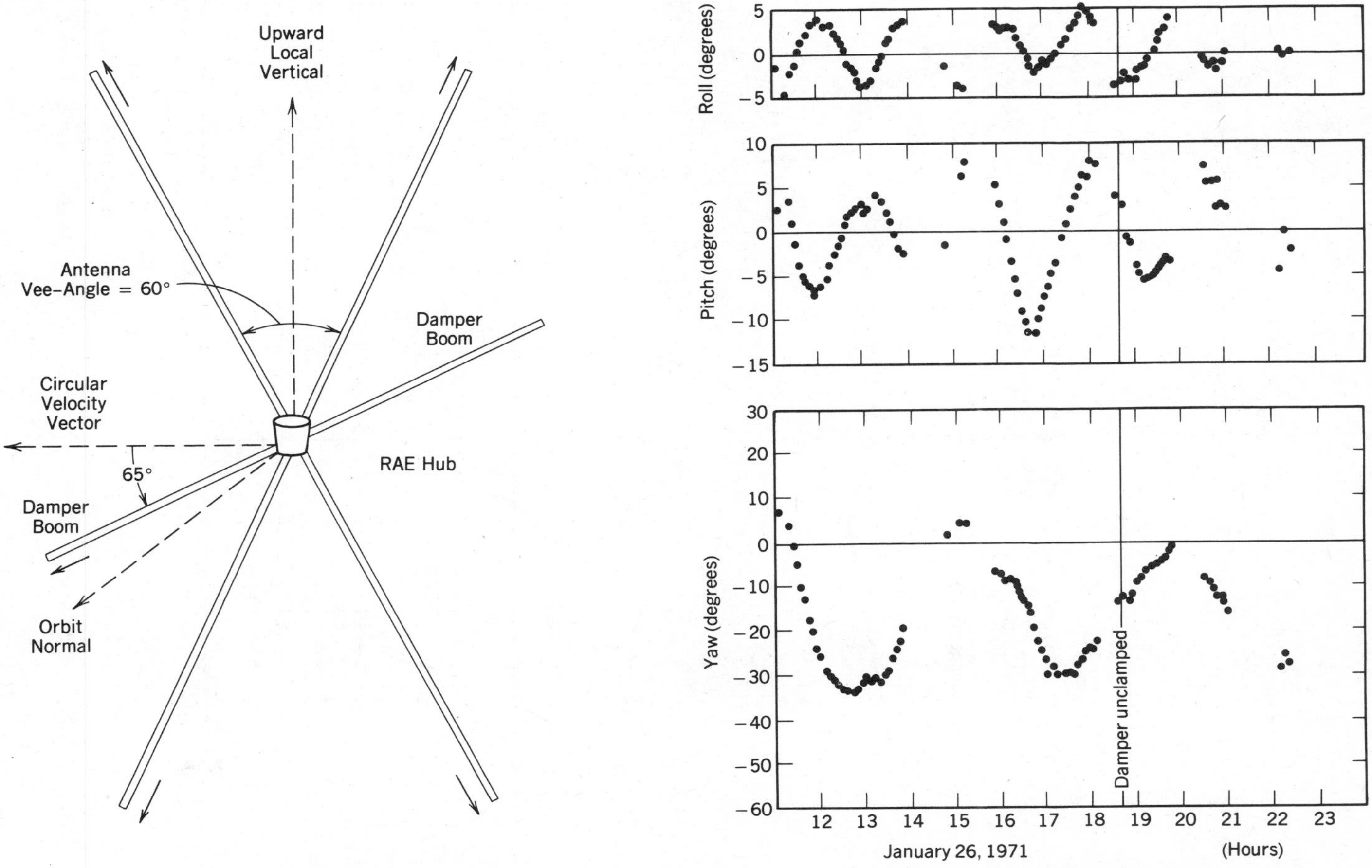

(a) The RAE satellite

(b) Typical flight data. From [Blanchard]

FIGURE 9.35 Gravity gradient stabilization of the RAE Satellite. (*a*) The RAE satellite. (*b*) Typical flight data. From [Blanchard].

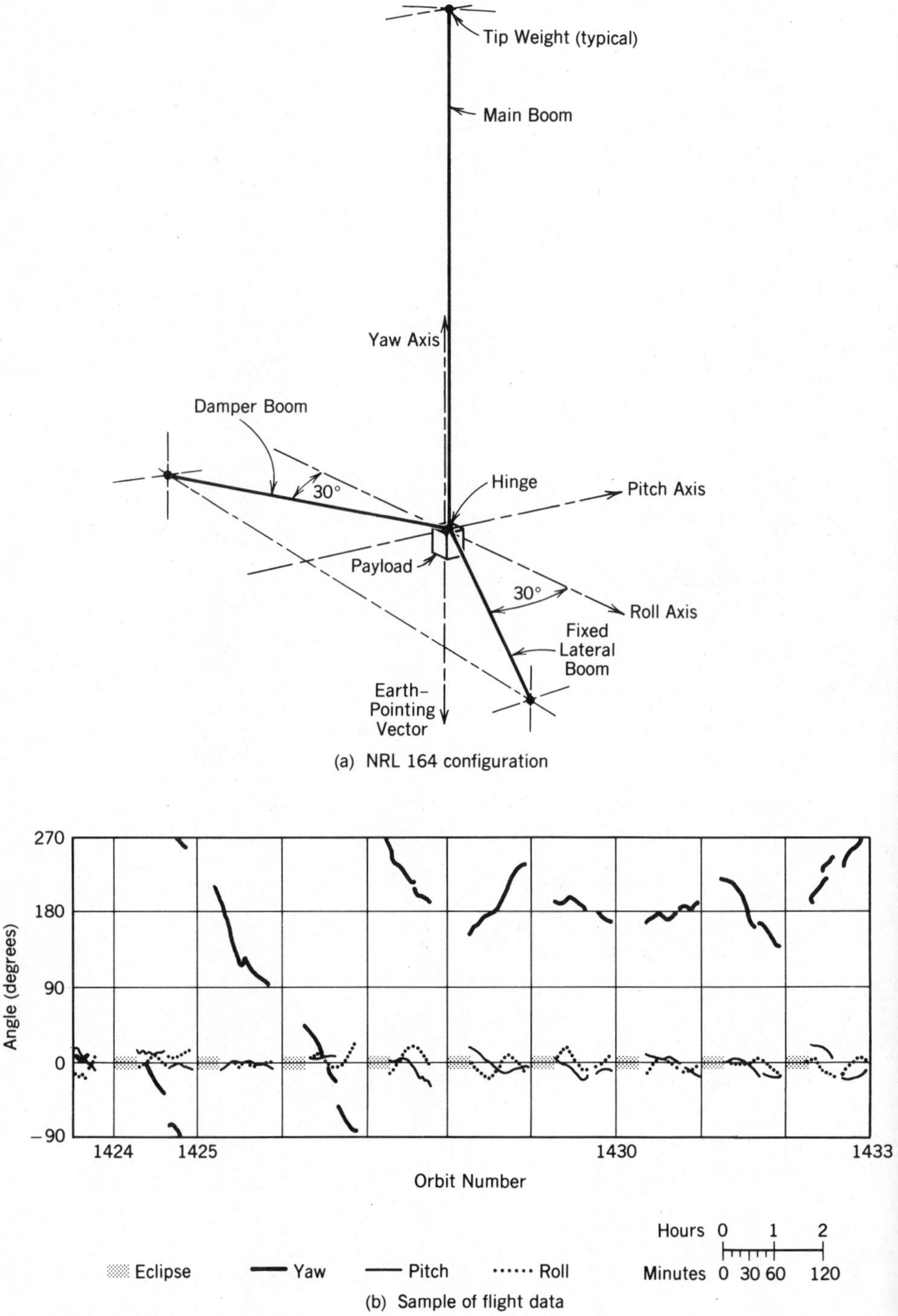

FIGURE 9.36 The NRL-164 Satellite. From [Goldman]. (*a*) NRL-164 configuration. (*b*) Sample of flight data.

The NRL-164 Satellite

For our last example of flight experience, we cite the case of the NRL-164 satellite, described by [Goldman] and shown in Fig. 9.36*a*. As its configuration suggests, it was intended to be three-axis stabilized in a near-Earth orbit in 1969. It had one vertical boom (unarticulated) and two booms in the horizontal (pitch/roll) plane, one of which was articulated to provide damping.

Flight data (Fig. 9.36*b*) showed that pitch and roll were only stabilized to within 40°, and yaw was not really stabilized at all. The explanation for this unacceptable performance centered on an interaction between the stabilizing booms and solar radiation. However, this was not the high-frequency "thermal flutter" analyzed by [Frisch, 1, 2] because the booms did not have an open cross section. Instead, it was a low-frequency disturbance most pronounced at low sun angles. Postflight dynamical analysis by [Goldman] was successful in establishing that such thermal distortions could indeed cause the observed instability.

The Role of Gravity Gradient Stabilization

It is clear from the preceding examples that the stabilizing effects of the gravity gradient force field predicted in theory can be demonstrated in flight. It is equally clear that this form of stabilization falls far short of sufficient for most modern applications. The best people in the business tried many configurations and many types of damping, yet their success was quite limited. The most basic fact about the gravity gradient torque is that it is very weak, and this simply could not be overcome in the face of numerous small environmental disturbances, some predicted and others unexpected. It is apparently possible to directionally stabilize the vertical (two-axis stabilization) to within ~ 20° at synchronous altitudes if symmetry is used. Three-axis stabilization almost never seems to function as intended, with large excursions (not to say inversions) commonly observed. Perhaps the DODGE satellite provides the best object lesson of all: the active damper performed better than the passive damper, and yaw librations were reduced when the small momentum wheel was employed. A passive stabilization scheme almost always evolves into semiactive one.

In the early 1960s, gravity gradient was heralded as "free" stabilization—no sensors, no power, no logic, and no actuators. Of course, it is not really free. It requires long booms, tip masses, and damping devices. And it doesn't work particularly well. Most modern spacecraft require pointing accuracies that are orders of magnitude better than can be derived from purely passive gravity-anchored stabilization; the weight budget for gravity gradient paraphernalia is better spent on sophisticated attitude sensors, microprocessors, and devices for managing internally stored angular momentum. Long-life geostationary satellites, for example, have a fuel requirement for north–south stationkeeping that is many times the amount needed for attitude control.

To summarize: for most applications, gravity gradient stabilization is inadequate. Nevertheless, for certain special situations, particularly in near-Earth orbits, it can profitably be used when accurate pointing is not required. The Salyut-6 space station (Fig. 9.28) and the Long Duration Exposure Facility (Fig. 9.29) furnish two current examples, and it is likely that the U.S. Space Station will also be stabilized with the assistance of the gravity gradient principles discussed in this chapter.

9.5 PROBLEMS

9.1 Consider the gravity gradient force field due to an inverse-square primary, about a point O at a distance R_c from the center of the primary. As derived in (9.1, 13), the vectorial expression for this force on a mass element dm displaced by $\underset{\rightarrow}{\mathbf{r}}$ from O is

$$d\underset{\rightarrow}{\mathbf{f}} = \frac{\mu\, dm}{R_c^3}(3\underset{\rightarrow}{\hat{\mathbf{o}}}_3\underset{\rightarrow}{\hat{\mathbf{o}}}_3 - \underset{\rightarrow}{\mathbf{1}}) \cdot \underset{\rightarrow}{\mathbf{r}} \tag{1}$$

where $\underset{\rightarrow}{\hat{\mathbf{o}}}_3$ is the unit vertical vector at O.

(*a*) Denote

$$\mathbf{r} \equiv [x \quad y \quad z]^T \triangleq \mathscr{F}_o \cdot \underset{\rightarrow}{\mathbf{r}} \tag{2}$$

and show that the gravity gradient force components in $\mathscr{F}_o$ are

$$d\mathbf{f} = \mathscr{F}_o \cdot d\underset{\rightarrow}{\mathbf{f}} = \frac{\mu\, dm}{R_c^3}[-x \quad -y \quad 2z]^T \tag{3}$$

(*b*) Discuss the geometrical interpretation of this result, shown in Fig. 9.37.

(*c*) Assume that the frame $\mathscr{F}_o$ rotates about $-\underset{\rightarrow}{\hat{\mathbf{o}}}_2$ at an absolute rate ω_o. Show that the additional "apparent force" thereby experienced by dm is

$$d\mathbf{f}_a = \mathscr{F}_o \cdot d\underset{\rightarrow}{\mathbf{f}}_a = dm \begin{bmatrix} \omega_o^2 x + 2\omega_o \dot{z} + \dot{\omega}_o z \\ 0 \\ \omega_o^2 x - 2\omega_o \dot{x} - \dot{\omega}_o x \end{bmatrix} \tag{4}$$

(For the meaning of "apparent force," see Problem 3.3.)

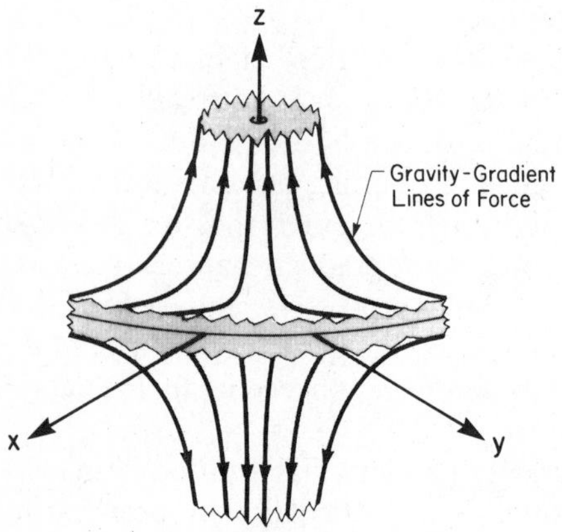

FIGURE 9.37 The gravity gradient force field. (See Problem 9.1.)

(*d*) How can the above results be applied to the attitude dynamics of a spacecraft orbiting Earth? See [Etkin, 1]. Also, compare (4) with (9.1, 29) in the text, and comment.

(*e*) Show that the gravity gradient force in (3) and the "apparent force" in (4) can be derived together from the "apparent" velocity-dependent potential

$$\frac{dV_a}{dm} = \frac{\mu}{2R_c^3}(x^2 + y^2 - 2z^2) + \omega_o(\dot{x}z - \dot{z}x) - \frac{1}{2}\omega_o^2(x^2 + z^2) \quad (5)$$

and discuss this result.

9.2 (*a*) Based on the inverse-square law given in (9.1, 12), show that the Taylor series expansion for the gravity field near a point $\underset{\rightarrow}{\mathbf{R}}_c$ is given by (9.1, 13), correct to $O(\varepsilon^2)$, where $\varepsilon \triangleq r/R_c$.

(*b*) Integrate this expression over a rigid body whose mass center is at $\underset{\rightarrow}{\mathbf{R}}_c$ to show that the gravitational force on the body is given by (9.1, 14), correct to $O(\varepsilon^2)$.

(*c*) Show that the attitude-dependent force, namely, the disturbance force of $O(\varepsilon^2)$, is given in the orbiting frame $\mathscr{F}_o$ by (9.1, 17).

9.3 A point in space has position $\underset{\rightarrow}{\mathbf{R}}$, velocity $\dot{\underset{\rightarrow}{\mathbf{R}}}$, and acceleration $\ddot{\underset{\rightarrow}{\mathbf{R}}}$, as measured with respect to an inertial frame $\mathscr{F}_i$. A reference frame $\mathscr{F}_o$ is set up such that $\underset{\rightarrow}{\hat{\mathbf{o}}}_3 = -\underset{\rightarrow}{\mathbf{R}}/R$ and $\underset{\rightarrow}{\hat{\mathbf{o}}}_2$ is in the direction $\dot{\underset{\rightarrow}{\mathbf{R}}} \times \underset{\rightarrow}{\mathbf{R}}$. Let the angular velocity of $\mathscr{F}_o$ with respect to $\mathscr{F}_i$ be $\underset{\rightarrow}{\boldsymbol{\omega}}_o$ and denote

$$\boldsymbol{\omega}_o \equiv [\omega_{o1} \quad \omega_{o2} \quad \omega_{o3}]^T \triangleq \mathscr{F}_o \cdot \underset{\rightarrow}{\boldsymbol{\omega}}_o$$

(*a*) Show that $\omega_{o1} \equiv 0$.

(*b*) Show that $\ddot{\underset{\rightarrow}{\mathbf{R}}}$, expressed in $\mathscr{F}_o$, is given by (9.1, 19).

9.4 A rigid body is in orbit around an inverse-square primary. Its mass center is located at $\underset{\rightarrow}{\mathbf{R}}$ and its absolute angular velocity is $\boldsymbol{\omega}$.

(*a*) Show that the total angular momentum of the satellite, *about the center of the primary*, is

$$\underset{\rightarrow}{\mathbf{h}} = m\underset{\rightarrow}{\mathbf{R}} \times \dot{\underset{\rightarrow}{\mathbf{R}}} + \underset{\rightarrow}{\mathbf{I}} \cdot \underset{\rightarrow}{\boldsymbol{\omega}} \quad (6)$$

where m = mass and $\underset{\rightarrow}{\mathbf{I}}$ = centroidal inertia dyadic.

(*b*) Show that the angular momentum expressed in principal axes is given by

$$\mathbf{h} \triangleq \mathscr{F}_p \cdot \underset{\rightarrow}{\mathbf{h}} = m\omega_{o2}R^2\mathbf{c}_2 + \mathbf{I}\boldsymbol{\omega} \quad (7)$$

where $\mathbf{c}_2$ is the second column of the rotation matrix $\mathbf{C}_{po} = \mathscr{F}_p \cdot \mathscr{F}_o^T$; the meanings of $\mathscr{F}_o$ and ω_{o2} were explained in the last problem.

(*c*) Show by substitution of the motion equations (A) through (E) in the text that

$$\dot{\mathbf{h}} + \boldsymbol{\omega}^{\times}\mathbf{h} = \mathbf{0}$$

namely, that $\underset{\rightarrow}{\mathbf{h}}$ is constant in $\mathscr{F}_i$.

(*d*) Show that the total kinetic energy of the satellite is

$$T = \tfrac{1}{2}m\dot{\underset{\rightarrow}{\mathbf{R}}} \cdot \dot{\underset{\rightarrow}{\mathbf{R}}} + \tfrac{1}{2}\underset{\rightarrow}{\boldsymbol{\omega}} \cdot \underset{\rightarrow}{\mathbf{I}} \cdot \underset{\rightarrow}{\boldsymbol{\omega}} \quad (8)$$

(*e*) Show that the kinetic energy expressed in terms of scalar quantities is

$$T = \tfrac{1}{2}m\left(\dot{R}^2 + \omega_{o2}^2R^2\right) + \tfrac{1}{2}\boldsymbol{\omega}^T\mathbf{I}\boldsymbol{\omega} \quad (9)$$

(*f*) Show by substitution of the motion equations (A) through (E) in the text that

$$(T + V)^{\cdot} = 0 \tag{10}$$

where V is the total gravitational potential energy of the satellite, given in Chapter 8, Equation (8.1, 24):

$$V = -\frac{\mu m}{R} - \frac{\mu}{2R^3}(I_1 + I_2 + I_3) + \frac{3\mu}{2R^3}\mathbf{c}_3^T \mathbf{I} \mathbf{c}_3 \tag{11}$$

(*g*) Examine the typical relative sizes of the terms in $\mathbf{h}$, T, and V, and comment.

9.5 Verify (9.1, 27) through (9.1, 29) in the text.

9.6 The position and velocity of a satellite are $\vec{\mathbf{R}}, \dot{\vec{\mathbf{R}}}$, and these are only slightly different from the values $\vec{\mathbf{R}}_e, \dot{\vec{\mathbf{R}}}_e$ on a reference ellipse, with $\delta\vec{\mathbf{R}} = \vec{\mathbf{R}} - \vec{\mathbf{R}}_e$. An orbiting frame $\mathcal{F}_o$ has $\hat{\vec{\mathbf{o}}}_3 = -\vec{\mathbf{R}}/R$ and $\hat{\vec{\mathbf{o}}}_2$ in the direction of $\dot{\vec{\mathbf{R}}} \times \vec{\mathbf{R}}$, while the orbiting frame $\mathcal{F}_e$ on the neighboring reference ellipse has $\hat{\vec{\mathbf{e}}}_3 = -\vec{\mathbf{R}}_e/R_e$ and $\hat{\vec{\mathbf{e}}}_2$ in the direction of $\dot{\vec{\mathbf{R}}}_e \times \vec{\mathbf{R}}_e$. Show that the rotation matrix $\mathbf{C}_{eo} = \vec{\mathcal{F}}_e \cdot \vec{\mathcal{F}}_o^T$ is given by (9.1, 34) in the text.

9.7 A dumbbell satellite undergoes pitching motion $\alpha(t)$ as shown in Fig. 9.38. Based on the general equations (A) through (F) in the text, show that the motion equations for this case reduce to

$$\ddot{R} - \dot{\eta}^2 R = -\frac{\mu}{R^2}\left[1 + \frac{3}{2}\left(\frac{l}{R}\right)^2 (3\cos^2\alpha - 1)\right]$$

$$\ddot{\alpha} - \ddot{\eta} + 3\left(\frac{\mu}{R^3}\right)\sin\alpha\cos\alpha = 0$$

$$l^2(\dot{\eta} - \dot{\alpha}) + R^2\dot{\eta} = h = \text{constant} \tag{12}$$

These agree with the equations given by [Moran], who also derives a perturbation solution using l/R as a small parameter.

9.8 A rigid satellite is in a circular orbit of radius R with one of its principal axes always pointing vertically and another principal axis always aligned with the orbit normal:

$$\mathbf{c}_3 \equiv [0 \quad 0 \quad 1]^T; \qquad \boldsymbol{\omega} \equiv [0 \quad -\omega_c \quad 0]^T$$

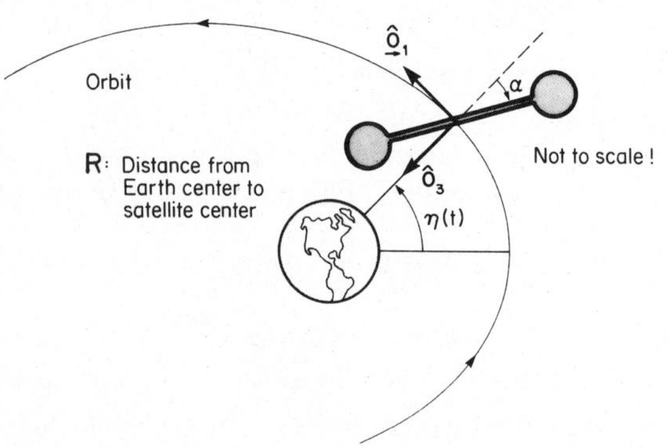

FIGURE 9.38 "Dumbbell" satellite. (See Problem 9.7.)

and the attitude motion equation (9.1, 10) is satisfied. Show that the orbital-disturbance equations (9.1, 17) and (9.1, 37) are also satisfied provided $\delta R_1 \equiv 0$, $\delta R_2 \equiv 0$, and

$$\frac{\delta R_3}{R} \equiv \left(\frac{2I_3 - I_1 - I_2}{2mR^2} \right) \tag{13}$$

Thus, the mass center is slightly displaced vertically from its reference circular orbit (about 5 m for an SPS satellite in geostationary orbit).

9.9 For a rigid body in a circular orbit, use the governing equations (9.2, 1) and (9.2, 7) to demonstrate that the energy-like quantity H is indeed conserved.

9.10 Prove the uniqueness of the 24 equilibria defined below equation (9.2, 13) in the text, starting with the motion equation

$$\mathbf{I}\dot{\boldsymbol{\omega}} + \boldsymbol{\omega}^{\times}\mathbf{I}\boldsymbol{\omega} = 3\omega_c^2 \mathbf{c}_3^{\times}\mathbf{I}\mathbf{c}_3 \tag{14}$$

given in component form by (9.2, 1). Proceed in the following steps:

(*a*) Assume that the definition of *dynamical equilibrium* is "external torques remain zero." Show from (14) that an orientation $\mathbf{C}_{pc}$ will have no torque acting at any instant for which $\mathbf{c}_3$ is an eigencolumn of

$$\mathbf{I}\mathbf{c}_3 = \lambda \mathbf{c}_3 \tag{15}$$

Explain why this means that one of the three principal axes must point vertically. Assume that the satellite is tri-inertial.

(*b*) Now assume that condition (15) persists as it would for a dynamical equilibrium. Explain why only rotation about the vertical is possible under this constraint. Let this rotation be $\beta(t)$. Show that $\boldsymbol{\omega}$ is of the form

$$\begin{aligned} \boldsymbol{\omega} &= [\ \dot{\beta} \quad -\omega_c \sin\beta \quad -\omega_c \cos\beta]^T \\ \boldsymbol{\omega} &= [-\omega_c \cos\beta \quad \dot{\beta} \quad -\omega_c \sin\beta]^T \\ \boldsymbol{\omega} &= [-\omega_c \sin\beta \quad -\omega_c \cos\beta \quad \dot{\beta}\ ]^T \end{aligned} \tag{16}$$

respectively for $\mathbf{c}_3 = \mathbf{1}_1, \mathbf{1}_2, \mathbf{1}_3$. Substitute into the motion equation (14) and show that the only possible "motion" satisfies

$$\sin\beta\cos\beta = 0 \tag{17}$$

namely, that the two horizontal axes must be in the orbital plane and normal to it, respectively.

(*c*) Since there are six possibilities for $\mathbf{c}_3$ in part (*a*) and, for each of these, four possibilities for $\mathbf{c}_1$ and $\mathbf{c}_2$ in part (*c*), what do you conclude about the number of possible dynamical equilibria?

9.11 As in the text, define $\{c_{ij}\}$ to be the rotation matrix that specifies the attitude of the satellite.

(*a*) Explain why $\{c_{13}(t), c_{23}(t), c_{12}(t), c_{32}(t)\}$ completely specify the attitude history of the satellite.

(*b*) Explain why it follows that the candidate Liapunov function v given by (9.2, 28) is positive-definite.

(*c*) Emulating [Beletskii], show that the attitude motion of a satellite in a circular orbit will be bounded (i.e., there will be no tumbling) provided that

$$v < \min\left\{\tfrac{3}{2}\omega_c^2(I_1 - I_3), \tfrac{1}{2}\omega_c^2(I_2 - I_1)\right\} \tag{18}$$

9.12 (*a*) Show by differentiating the solution for in-plane pitch motion in a circular orbit, (9.2, 35), that

$$\dot{\alpha}_2(t) = \begin{cases} a\,cn\left(\dfrac{a(t - t_o)}{\hat{a}}; \hat{a}\right) & (0 < \hat{a} < 1) \\ a\,dn\left[a(t - t_o); \hat{a}^{-1}\right] & (\hat{a} > 1) \end{cases} \tag{19}$$

Hint: The following identities for the Jacobian elliptic functions will prove useful:

(i) $\dfrac{d}{d\theta}\, sn\,\theta = cn\,\theta\, dn\,\theta$

(ii) $sn^2\theta + cn^2\theta = 1$

(iii) $dn^2\theta + k^2 sn^2\theta = 1$

(*b*) Hence, show that the solution (9.2, 35) satisfies the motion integral (9.2, 34).

(*c*) Derive the asymptotic approximations (9.2, 36) for $a \to 0$ and $a \to \infty$

(*d*) For $\hat{a} = 1$, show that the pitch motion obeys

$$\tan \tfrac{1}{2}\alpha_2 = \tanh\left[\tfrac{1}{2}(3k_2)^{1/2}\omega_c(t - t_o)\right] \tag{20}$$

and explain the physical meaning of this solution.

9.13 Show that the period of pitch libration in a circular orbit must be at least $T_c/\sqrt{3}$, where $T_c = 2\pi/\omega_c$ is the orbital period.

9.14 Consider a rigid body in a circular orbit. The body has an axis of inertial symmetry which points nominally forward. Thus, $I_1 = I_a$, $I_2 = I_3 = I_t$.

(*a*) By adapting the motion equations for small librations, (9.2, 15), determine stability conditions.

(*b*) Repeat this type of analysis, assuming this time that the symmetry axis nominally points normal to the orbit.

(*c*) Repeat again, assuming the symmetry axis points vertically.

9.15 A symmetrical rigid body is in a circular orbit with its symmetry axis pointing (nominally) normal to the orbit. A Liapunov function $\hat{v}$, based on the Hamiltonian, is given by (9.2, 47). Noting that

$$c_{22}^2 = 1 - c_{21}^2 - c_{23}^2$$

plot curves in the $\{c_{21}, c_{23}\}$ plane that correspond to $\hat{v}(c_{21}, c_{23}) = $ constant, with $\omega_{at} = 0$. Show how these may be used to place a bound on initial conditions such that no tumbling can occur.

9.16 A rigid body whose inertias are $\{I_1, I_2, I_3\} = \{4, 5, 6\}$ is in a circular orbit. Its initial state is

$$\alpha_1 = \dot{\alpha}_1 = \alpha_2 = \dot{\alpha}_3 = 0$$
$$\dot{\alpha}_2 = -0.13\omega_c; \qquad \alpha_3 = -0.01 \text{ rad}$$

where, as usual, $\{\alpha_1, \alpha_2, \alpha_3\}$ denote angles about the {roll, pitch, yaw} axes.

(*a*) Show that $\{k_1, k_2, k_3\} = \{0.9, 8/11, 0.5\}$.

(*b*) It was shown in (9.2, 51) et seq. that the conserved Hamiltonian, given by (9.2, 28), guarantees that $\cos^{-1} c_{22}$, the angle between the pitch axis and the orbit normal, cannot exceed 26°. Show from (9.2, 34 and 35) that the pitch libration which results from $\alpha_1 = \dot{\alpha}_1 = \alpha_2 = \alpha_3 = \dot{\alpha}_3 = 0$ and $\dot{\alpha}_2 = -0.13\omega_c$ has an amplitude $\alpha_{2,\max} \doteq 5°$. Find the point $\{k_1, k_3\} = \{0.9, 0.5\}$ on Kane's roll/yaw stability diagram (Fig. 9.12*a*) and discover whether Floquet theory indicates that the roll/yaw motion, excited parametrically by the periodic pitch motion, is unstable. How do you reconcile this with the fact that $\cos^{-1} c_{22} < 26°$? See also the nonlinear simulations in [Kane, 1].

9.17 The pitch equation of motion for a rigid body in an elliptic orbit (roll and yaw quiescent) is given by (9.2, 64). Use the chain rule of differentiation and the relations (9.2, 61) and (9.2, 62) to show that, when the true anomaly is used as the independent variable, the pitch equation is (9.2, 65).

9.18 The equation for motion about the pitch axis of a rigid spacecraft in an eccentric orbit is given by (9.2, 41). If the true anomaly η is used as the independent variable, the pitch equation is transformed to (9.2, 43). Show that if the *eccentric anomaly E* is used as the independent variable, the pitch equation is

$$(1 - e\cos E)\alpha_2'' - (e\sin E)\alpha_2' + 3k_2\sin\alpha_2\cos\alpha_2 = -\frac{2(1-e^2)^{1/2}}{1 - e\cos E} e\sin E \qquad (21)$$

where now the prime means d/dE.

9.19 (*a*) In connection with the pitch motion of a rigid body in an elliptic orbit, show that the motion equation (9.2, 65) is satisfied by $\alpha_2 = -\frac{1}{2}\eta$ provided the inertia parameter k_2 equals $2e$, where e is the eccentricity. Sketch this tumbling behavior.

(*b*) Find a parametrically resonant solution to (9.2, 65) in which the body librates in pitch at twice the orbital period. That is, find an approximate solution of the form

$$\alpha_2(\eta) = A\cos\left[\tfrac{1}{2}(\eta - \eta_o)\right] \qquad (22)$$

by following the same procedure initiated by (9.2, 75) in the text. Show that the amplitude of libration is given implicitly by

$$k^2 = \frac{A(2 \pm 3e)}{24J_1(2A)} \qquad (23)$$

(*c*) Show that the approximation $J_1(2A) \doteq A$ reduces the relation (9.2, 78) for the amplitude at orbital resonance to the one derived from linear theory and shown in Fig. 9.13*a*.

9.20 The system denoted $\mathscr{R} + \mathscr{P}$, a rigid body containing a mass spring dashpot damper, was illustrated in Fig. 3.6, and its general (nonlinear) equations of motion were derived in Section 3.4 of Chapter 3. By adding the gravitational force field expressed in (1) and linearizing about a reference Earth-pointing motion, show that the linearized motion equations for the librations of $\mathscr{R} + \mathscr{P}$ in a circular orbit are as shown in (9.3, 17) in the text. (*Hint*: The derivation of the motion equations for small deviations of $\mathscr{R} + \mathscr{P}$ from a simple spin, given in Section 5.2 of Chapter 5, should be helpful.)

9.21 Consider the attitude motion of a rigid body in a circular orbit, governed by equations (1), (6) and (7) in Section 9.2.

(*a*) Show that the motion

$$\boldsymbol{\omega} = \begin{bmatrix} 0 \\ \dot{\alpha}_2 - \omega_c \\ 0 \end{bmatrix}; \qquad \mathbf{C}_{pc} = \begin{bmatrix} \cos\alpha_2 & 0 & -\sin\alpha_2 \\ 0 & 1 & 0 \\ \sin\alpha_2 & 0 & \cos\alpha_2 \end{bmatrix} \tag{24}$$

satisfies these equations, that is, show that arbitrarily large pitch motion $\alpha_2(t)$ can exist without exciting roll or yaw.

(*b*) Assume that a constant disturbance g_d acts about the pitch axis. Show that the resulting steady-state pitch angle α_{2s} is given by

$$\sin 2\alpha_{2s} = \frac{2g_d}{2\omega_c^2(I_1 - I_3)} \tag{25}$$

Assume g_d is small enough that the right side is less than unity.

(*c*) Assume that the satellite attitude is given by the Euler angle sequence

$$\mathbf{C}_{pc} = \mathbf{C}_1(\alpha_1)\mathbf{C}_3(\alpha_3)\mathbf{C}_2(\alpha_2) \tag{26}$$

using the notation of (2.1, 28). Assume that α_1 and α_3 are first-order infinitesimals. Show that

$$\mathbf{C}_{pc} = \begin{bmatrix} c & \alpha_3 & -s \\ -\alpha_3 c + \alpha_1 s & 1 & \alpha_3 s + \alpha_1 c \\ s & -\alpha_1 & c \end{bmatrix}$$

$$\boldsymbol{\omega} = \begin{bmatrix} \dot{\alpha}_1 - \omega_c\alpha_3 \\ \dot{\alpha}_2 - \omega_c \\ \dot{\alpha}_3 + \omega_c\alpha_1 \end{bmatrix}$$

to first order, where $s = \sin\alpha_2$, $c = \cos\alpha_2$.

(*d*) Examine the librations about the steady-state $\alpha_2 \equiv \alpha_{2s}$ found in part (*b*) by deriving a set of motion equations linear in α_1 and α_3; show that these equations are

$$\ddot{\alpha}_1 - (1 - k_1)\omega_c\dot{\alpha}_3 + k_1\omega_c^2\left[(3c_s^2 + 1)\alpha_1 + 3s_s c_s \alpha_3\right] = 0$$

$$\ddot{\alpha}_3 + (1 - k_3)\omega_c\dot{\alpha}_1 + k_3\omega_c^2\left[3s_s c_s\alpha_1 + (3s_s^2 + 1)\alpha_3\right] = 0 \tag{27}$$

where $s_s = \sin\alpha_{2s}$, $c_s = \cos\alpha_{2s}$, and k_1 and k_3 have their usual meaning, (4.4, 20). As a check, show that when $\alpha_{2s} = 0$, these reduce to the roll/yaw equations (9.2, 15) for librations in torque-free motion.

(*e*) Show that the characteristic equation for the system (27) is

$$\begin{aligned}\phi_{ry}(s^2) &\triangleq s^4 + \Delta_1\omega_c^2 s^2 + \omega_c^4\Delta_2 \\ &= 3\omega_c^2(k_1 - k_3)(\omega_c\cos\alpha_{2s} + s\sin\alpha_{2s})s\sin\alpha_{2s}\end{aligned} \qquad (28)$$

where

$$\begin{aligned}\Delta_1 &\triangleq 1 + 3k_1 + k_1k_3 \\ \Delta_2 &\triangleq 4k_1k_3\end{aligned} \qquad (29)$$

(*f*) Note that $\phi_{ry}(s^2) = 0$ is the characteristic equation for roll/yaw motion when $\alpha_{2s} = 0$. Let the roots of $\phi_{ry} = 0$ be denoted $j\omega_{ry}$. Show that, for small α_{2s}, the roots of (28) are

$$s = j\omega_{ry} + \frac{3(k_1 - k_3)}{2(\Delta_1 - 2\omega_{ry}^2)}\alpha_{2s} \qquad (30)$$

What do you conclude about the stability of small librations about a small steady-state pitch angle? This situation was first studied by [Garber].

CHAPTER 10

SPIN STABILIZATION IN ORBIT

Chapters 4 and 5 were devoted to an exposition of how the spin axis of a spinning body in torque-free motion can be stabilized about an inertially fixed direction. To reiterate briefly, in Chapter 4 it was found that the spin axis of a rigid body (an abstract idealization) spinning about either its major or minor axis of inertia is infinitesimally directionally stable (Euler's result) and Liapunov stable for a major-axis spin. A quasi-rigid body, on the contrary, is unstable for a minor-axis spin, although the major-axis spin for such a body is asymptotically stable—the major-axis rule. Then, in Chapter 5, it was shown that the major-axis rule is, in turn, only an approximate stability condition and that when a physical damper of finite inertia is placed in an otherwise rigid body, the major-axis rule must be strengthened to ensure asymptotic stability. See, for example, Fig. 5.5 or equations (5.2, 16). In practice, such nutation dampers tend to be quite small, hence the major-axis rule is very nearly the correct stability condition.

The principle of spin stabilization has been used on a great many spacecraft and continues to find many applications. Properly used, it is an effective method of keeping a single vehicle axis (the major axis) pointing in an inertially fixed direction. Moreover, even spacecraft that have active attitude control systems in their operational orbits are frequently spin stabilized in an initial (transfer orbit) phase of their mission. Unlike the gravity gradient stabilization of the last section, in which a relatively weak environmental torque was to be exploited, the restoring torque for a spin-stabilized vehicle can in principle be made arbitrarily large merely by making the spin rate sufficiently rapid. Loosely speaking, the characteristic time for gravity gradient stabilization is the orbital period; for spin stabilization, it is the rotation period. From Euler's equations for the motion of a rigid body, it is known that the ratio $\|\mathbf{g}\|/\|\boldsymbol{\omega}^{\times}\mathbf{I}\boldsymbol{\omega}\|$—where $\|\mathbf{g}\|$ is the magnitude of the disturbing torque—can be made smaller by making the spin rate bigger. More precisely, from (4.4, 13),

$$|\alpha| \doteq \frac{|g|}{\nu^2 \Delta I} \tag{1}$$

where $|g|$ is the magnitude of the transverse torque, ν is the spin rate, and ΔI is the least of $I_2 - I_1$ and $I_2 - I_3$. (Here, I_2 is the maximum moment of inertia.) Thus, for a given spacecraft and a given $|g|$, wobble can be reduced to negligibly small proportions by a sufficiently rapid spin. How rapid is "sufficiently rapid"? A simple guideline is to view the gravity gradient torque as a typical disturbance. This disturbance is of the order of $\omega_o^2 \Delta I$, where ω_o is the orbital rate. Then, $|\alpha| \doteq (\omega_o/\nu)^2$. Even for a spin rate as low as 1 rpm, $(\omega_o/\nu)^2$ is $\sim 10^{-4}$ for a near-Earth orbit and $\sim 5 \times 10^{-7}$ at geostationary altitude. From this simple observation alone it can be seen that spin stabilization should be effective even in the vicinity of Earth if the spin rate is "several" rpm.

From a practical standpoint, two further observations should be made. First, some form of damping is essential to make the spin stabilization asymptotic. This guarantees that any slight wobble introduced by disturbances will tend to vanish. Second, it will be recalled from Chapters 4 and 5 that the directional stabilization is asymptotic to the angular-momentum vector. This vector is fixed inertially only in the absence of cross torques. In reality, the slight disturbing torques from the environment will cause $\underset{\rightarrow}{\mathbf{h}}$ to wander from its initial direction. It is elementary that it will wander an amount $\|\Delta \underset{\rightarrow}{\mathbf{h}}\| \doteq \|\underset{\rightarrow}{\mathbf{g}}\| \Delta t$ in time Δt, where $\|\underset{\rightarrow}{\mathbf{g}}\|$ is the magnitude of the cross torque. The angular deviation is then

$$\frac{\|\Delta \underset{\rightarrow}{\mathbf{h}}\|}{\|\underset{\rightarrow}{\mathbf{h}}\|} \doteq \frac{\|\underset{\rightarrow}{\mathbf{g}}\|}{\|\underset{\rightarrow}{\mathbf{h}}\|} \Delta t \tag{2}$$

so that, over a given period, with given cross torques, the angular deviation in the "directionally stable" direction can be reduced by increasing the spin rate. Unfortunately, Δt may be years long for an operational spin-stabilized satellite, and so, for many spacecraft, the angular momentum vector cannot be prevented from wandering over the vehicle lifetime. At each instant in this lifetime however, the satellite is directionally stabilized to its almost fixed, but slowly changing, $\underset{\rightarrow}{\mathbf{h}}$ vector.

Indeed, one of the interesting features of the dynamics of spin-stabilized satellites is that there are three characteristic times associated with the motion: the rotation period, the orbital period, and the lifetime. For a spacecraft spin stabilized at 10 rpm with a one-year lifetime, for example, these times are in the ratio $\nu : \omega_o : \omega_l = 5 \times 10^6 : 5 \times 10^3 : 1$ for a near-Earth orbit and $5 \times 10^6 : 5 \times 10^2 : 1$ for a geostationary orbit. These three widely separated characteristic times are the basis for the most basic assumption used in the study of practical spin-stabilized spacecraft: the dynamics of spin stabilization, the disturbances over one orbit, and the long-term motion of the angular momentum vector can be studied separately. The design is thus broken into three phases. In the first, the attitude motion equations are studied with a view to establishing the transient performance of the passive stabilization system; in this context, external torques are negligible, and the spacecraft is in effect situated at a point in inertial space remote from all environmental disturbances. The characteristic time in such studies is controlled by the spin rate ν. In the second phase, the spacecraft, successfully spin stabilized by the efforts in the first phase, is replaced in effect by an angular momentum vector, and the (small) $\Delta\underset{\rightarrow}{\mathbf{h}}$ caused by environmental disturbances over a single orbit is computed; this can often be done analytically. The third and final phase is to integrate, analytically or numerically, the small $\Delta\underset{\rightarrow}{\mathbf{h}}$ contributions for each orbit; this is done over the mission lifetime to establish how the spin axis orientation evolves over months and years. This last phase obviously is not relevant over short durations such as a transfer orbit application.

In this section, an understanding of the basic concept of directional stabilization by spin will be fashioned in a series of steps. First, a spin-stabilized rigid body subject to gravity gradient torque is discussed (Section 10.1). Then, in Section 10.2, nutation dampers are incorporated to make practical the idea of passive stabilization (asymptotic stability). Finally, in Section 10.3, the long-term influence of environmental torques is addressed, and some typical flight data are exhibited.

10.1 SPINNING RIGID BODY IN ORBIT

As usual, we shall begin with the "simplest satellite"—a single rigid body. Energy dissipation is therefore excluded, and so all conclusions reached must be viewed as tentative, subject to confirmation or contradiction with the inclusion of damping in Section 10.2. The only source of environmental torque considered explicitly in this section is gravity gradient torque. It can be argued that this is the torque most intrinsic to a satellite in orbit. Furthermore, for a wide class of vehicle configurations and over a considerable range of altitudes, this is the dominant torque.

Perhaps the most basic reason of all for examining the effect of gravity gradient torque is that elegant, general results can be obtained in closed form. Moreover, only the inertia distribution of the satellite enters the analysis (as represented by the ratios $I_1 : I_2 : I_3$), and this distribution is already assumed to be known in writing the "left side" of the motion equations. Thus, the influence of gravity gradient torque can be investigated without introducing additional spacecraft parameters. By contrast, solar pressure torque, while admittedly important at high altitudes, is very configuration dependent. The optical properties (absorptivity, specular reflectivity, diffuse reflectivity) of all surfaces must be specified, as must the details of the vehicle geometry. It is therefore difficult to perform an analysis that is of broad significance. To model accurately a *particular* satellite, of course, solar torque should often be included. But such an investigation usually leads to a computer simulation the results of which are of interest primarily to disciples of that specific configuration. The same can be said of aerodynamic torque, which becomes important at low altitudes. To achieve realistic results, the geometry of all spacecraft surfaces (and their surface accommodation characteristics) must be specified. In addition, one must have a realistic

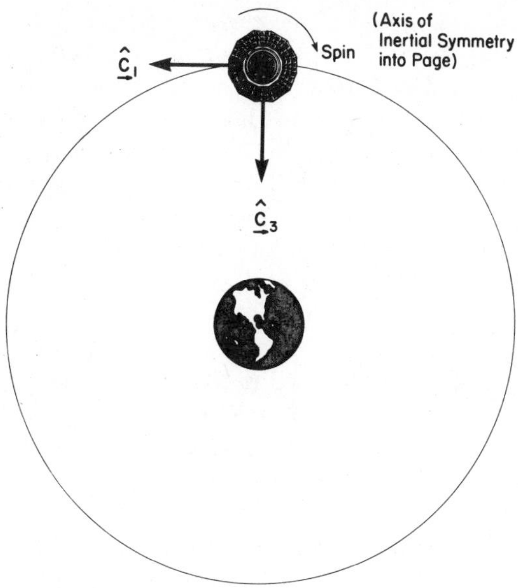

FIGURE 10.1 The Thomson equilibrium: a satellite spinning about its axis of symmetry, parallel to the orbit normal.

model for the atmosphere and of the orbit through that atmosphere. It is clear that the influence of aerodynamic torques is highly dependent on both configuration and mission. Therefore, while one is compelled to face these issues in preparing detailed simulations for low-altitude missions, it is exceedingly difficult to generalize the results obtained or to formulate closed-form conditions of wide application. For these reasons, attention will for the most part be directed to the consequences of gravity gradient torque on a spin-stabilized satellite.

Axisymmetrical Rigid Satellite in a Circular Orbit

Two further simplifications are now made. The orbit is assumed circular, and the satellite $\mathscr{R}$ possesses an axis of *inertial* symmetry (Fig. 10.1). (Note that the satellite need not be *geometrically* axisymmetrical, although $\mathscr{R}$ is usually drawn that way to imply inertial symmetry; for example, the body shown in Fig. 4.2 would qualify.) The axial and transverse inertias of $\mathscr{R}$ are I_a and I_t, respectively, and the orbital rate is $\omega_c = (\mu/R^3)^{1/2}$. The general case wherein the axis of spin is at an arbitrary orientation with respect to the inertial frame $\mathscr{F}_i$ is of interest, but in the simplest case the spin axis is nominally normal to the orbit. In this reference orientation there is no gravity gradient torque on the satellite. Using {roll, pitch, yaw} angles as conventionally defined—angles relating the body-fixed frame $\mathscr{F}_p$ to the orbiting frame $\mathscr{F}_c$—roll and yaw are nominally zero, although the pitch angle is arbitrarily large. The third column of $\mathbf{C}_{pc}$ for this reference motion has

$$c_{23} = 0 \tag{3}$$

This fact implies that the roll and yaw components of the gravity gradient torque are zero; see, for example, equation (9.1, 1). At the same time, the pitch component is also nominally zero because $I_1 - I_3 = 0$. The absence of torque in the reference motion means that a uniform spin normal to the orbit is a possible motion. To confirm this statement, one need only refer to the governing motion equations (9.2, 1 through 7) of Section 9.2. With $I_1 = I_3 = I_t$ and $I_2 = I_a$, and $\mathbf{C}_{pc} = \mathbf{C}_2[(\nu + \omega_c)t]$ [using the notation of (2.1, 28)], the motion equations reduce to $\omega_1 = \omega_3 = 0$, $\omega_2 = \text{constant} = \nu$.

Having established this simple reference motion, our next interest is to determine under what conditions it is stable. Using the by now familiar symbols $\{\alpha_1, \alpha_2, \alpha_3\}$ to denote small deviations of the body-fixed axes $\mathscr{F}_p$ from the state of uniform spin, we have the linearized equations derived in Section 4.4, namely, (4.4, 13), but with two modifications. First, we set $I_1 = I_3 = I_t$ and $I_2 = I_a$. Second, the gravity gradient torque components must be added to the right side:

$$I_a \ddot{\alpha}_2 = 0$$

$$I_t \ddot{\alpha}_1 + (2I_t - I_a)\nu\dot{\alpha}_3 + (I_a - I_t)\nu^2\alpha_1 = 3\omega_c^2(I_t - I_a)c_{23}c_{33}$$

$$I_t \ddot{\alpha}_3 - (2I_t - I_a)\nu\dot{\alpha}_1 + (I_a - I_t)\nu^2\alpha_3 = 3\omega_c^2(I_a - I_t)c_{13}c_{33} \tag{4}$$

where ν is the nominal spin rate with respect to $\mathscr{F}_i$ and $\{c_{13}, c_{23}, c_{33}\}$ are the direction cosines between $\mathscr{F}_p$ and the local vertical. These direction cosines must be derived with care because, unlike librations about the local vertical, the reference spin rate is no longer the orbital rate. That is, $\nu \neq -\omega_c$ in general.

A review of reference frames is in order. We are using four frames, as follows: $\mathscr{F}_p$, the body-fixed principal-axis frame; $\mathscr{F}_c$, the orbiting frame; $\mathscr{F}_i$, the inertial frame; and a fourth frame $\mathscr{F}_{\bar{p}}$, which we shall name the *reference spin frame*. $\mathscr{F}_{\bar{p}}$ rotates uniformly with respect to $\mathscr{F}_i$ about the 2-axis at rate ν:

$$\mathbf{C}_{\bar{p}i} = \mathbf{C}_2(\nu t) \tag{5}$$

For the reference motion, $\mathscr{F}_p \equiv \mathscr{F}_{\bar{p}}$, but with small excursions $\boldsymbol{\alpha}$:

$$\mathbf{C}_{p\bar{p}} = \mathbf{1} - \boldsymbol{\alpha}^{\times} \tag{6}$$

Finally, the relationship of $\mathscr{F}_c$ to $\mathscr{F}_i$ is

$$\mathbf{C}_{ci} = \mathbf{C}_2(-\omega_c t) \tag{7}$$

From these three relationships, any kinematical quantity of interest can be calculated. In particular, $\mathbf{c}_3$ is the third column of

$$\mathbf{C}_{pc} = \mathbf{C}_{p\bar{p}}\mathbf{C}_{\bar{p}i}\mathbf{C}_{ic} = \mathbf{C}_{p\bar{p}}\mathbf{C}_{\bar{p}i}\mathbf{C}_{ci}^T \tag{8}$$

Now,

$$\mathbf{C}_{\bar{p}i}\mathbf{C}_{ci}^T = \mathbf{C}_2(\nu t)\mathbf{C}_2(\omega_c t) = \mathbf{C}_2[(\nu + \omega_c)t] \tag{9}$$

Hence,

$$\mathbf{C}_{pc} = (\mathbf{1} - \boldsymbol{\alpha}^{\times})\mathbf{C}_2[(\nu + \omega_c)t] \tag{10}$$

The equations of motion, (4), therefore assume the form

$$\ddot{\alpha}_2 = 0$$
$$\ddot{\alpha}_1 + (1 - k_t)\nu\dot{\alpha}_3 + k_t\nu^2\alpha_1 + 3\omega_c^2 k_t(\alpha_1 c + \alpha_3 s)c = 0$$
$$\ddot{\alpha}_3 - (1 - k_t)\nu\dot{\alpha}_1 + k_t\nu^2\alpha_3 + 3\omega_c^2 k_t(\alpha_1 c + \alpha_3 s)s = 0 \qquad (11)$$

where $s = \sin(\nu + \omega_c)t$ and $c = \cos(\nu + \omega_c)t$, and

$$k_t = \frac{I_a - I_t}{I_t} \qquad (12)$$

is the relevant inertial parameter.

At first it might appear that Floquet theory is called for, inasmuch as (11) is a set of linear differential equations with periodic coefficients. (The roll gravity gradient torque has periodic components in body axes.) On closer examination however, it is possible to form the combinations

$$\gamma_1(t) \triangleq \alpha_1(t)\cos(\nu + \omega_c)t + \alpha_3(t)\sin(\nu + \omega_c)t$$
$$\gamma_2(t) \triangleq \alpha_2(t)$$
$$\gamma_3(t) \triangleq -\alpha_1(t)\sin(\nu + \omega_c)t + \alpha_3(t)\cos(\nu + \omega_c)t \qquad (13)$$

in terms of which the motion equations (11) become

$$\textbf{(i)}\quad \ddot{\gamma}_2 = 0$$
$$\textbf{(ii)}\quad \ddot{\gamma}_1 - (\nu + 2\omega_c + k_t\nu)\dot{\gamma}_3 + [k_t(3\omega_c^2 - \nu\omega_c) - \omega_c(\nu + \omega_c)]\gamma_1 = 0$$
$$\textbf{(iii)}\quad \ddot{\gamma}_3 + (\nu + 2\omega_c + k_t\nu)\dot{\gamma}_1 - [k_t\nu\omega_c + \omega_c(\nu + \omega_c)]\gamma_3 = 0 \qquad (14)$$

and the coefficients are now constant. In fact, the $\{\gamma_1, \gamma_3\}$ equations are in the canonical form

$$\mathscr{M}\ddot{\mathbf{q}} + \mathscr{G}\dot{\mathbf{q}} + \mathscr{K}\mathbf{q} = \mathbf{0}$$

with $\mathscr{M}^T = \mathscr{M} > 0$, $\mathscr{G}^T = -\mathscr{G}$, and $\mathscr{K}^T = \mathscr{K}$.

The variables $\{\gamma_1, \gamma_2, \gamma_3\}$ can be interpreted geometrically as three small angles in the following way. Consider a frame $\mathscr{F}_s$ that is identical with $\mathscr{F}_c$ for the reference motion but that is infinitesimally different from $\mathscr{F}_c$ in the perturbed motion by three small angles $\{\gamma_1, \gamma_2, \gamma_3\}$. That is,

$$\mathbf{C}_{sc} = (\mathbf{1} - \boldsymbol{\gamma}^{\times}) \qquad (15)$$

Frame $\mathscr{F}_s$ will be is called the *stroboscopic frame* because, although it is not fixed in $\mathscr{R}$, it coincides with $\mathscr{F}_p$ every time the spacecraft experiences one complete rotation with respect to the orbiting frame $\mathscr{F}_c$:

$$\mathbf{C}_{ps} = \mathbf{C}_2[(\nu + \omega_c)t] \qquad (16)$$

By comparing $\mathbf{C}_{ps}\mathbf{C}_{si}$ with $\mathbf{C}_{p\bar{p}}\mathbf{C}_{\bar{p}i}$ (they must be equal),

$$\mathbf{C}_2[(\nu + \omega_c)t](\mathbf{1} - \boldsymbol{\gamma}^{\times})\mathbf{C}_2(-\omega_c t) = (\mathbf{1} - \boldsymbol{\alpha}^{\times})\mathbf{C}_2(\nu t)$$

it can readily be shown that

$$\boldsymbol{\gamma}^{\times} = \mathbf{C}_2[-(\nu + \omega_c)t]\boldsymbol{\alpha}^{\times}\mathbf{C}_2[(\nu + \omega_c)t] \qquad (17)$$

In other words,

$$\boldsymbol{\gamma} = \mathbf{C}_2[-(\nu + \omega_c)t]\boldsymbol{\alpha} \tag{18}$$

which is just (13). The angles γ_1 and γ_3 are also conventionally called "roll" and "yaw," although they are not the same as the "true" roll and yaw angles α_1 and α_3.

The reason the transformation (13) successfully removes the periodic coefficients from (11) is because the inertia matrix, always constant in $\mathcal{F}_p$, is also constant in $\mathcal{F}_s$. In $\mathcal{F}_s$,

$$\begin{aligned}\mathcal{F}_s \cdot \underset{\rightarrow}{\mathbf{I}} \cdot \mathcal{F}_s^T &= \mathcal{F}_s \cdot \mathcal{F}_p^T[I_t\mathbf{1} + (I_a - I_t)\mathbf{1}_2\mathbf{1}_2^T]\mathcal{F}_p \cdot \mathcal{F}_s^T \\ &= \mathbf{C}_{sp}[I_t\mathbf{1} + (I_a - I_t)\mathbf{1}_2\mathbf{1}_2^T]\mathbf{C}_{ps} \\ &= I_t\mathbf{1} + (I_a - I_t)\mathbf{1}_2\mathbf{1}_2^T = \mathbf{I}\end{aligned}$$

from (16). This trick works only for axisymmetrical bodies and is the reason why they are studied prior to the more difficult case of tri-inertial bodies. Indeed, a more direct derivation of (14) is possible based on the property that $\underset{\rightarrow}{\mathbf{I}}$ is constant in $\mathcal{F}_s$. Following [Thomson, 2], the absolute velocity of $\mathcal{F}_s$ is, with the aid of (15) and (16),

$$\boldsymbol{\omega}_s = \begin{bmatrix} \dot{\gamma}_1 - \omega_c\gamma_3 \\ \dot{\gamma}_2 - \omega_c \\ \dot{\gamma}_3 + \omega_c\gamma_1 \end{bmatrix} \tag{19}$$

and the angular momentum components in $\mathcal{F}_s$ are

$$\mathbf{h}_s = \mathbf{I}\boldsymbol{\omega}_s + \mathbf{1}_2 I_a(\nu + \omega_c) \tag{20}$$

To write the equations of motion in $\mathcal{F}_s$, the components in $\mathcal{F}_s$ of the gravity torque are also required. These components are

$$\underset{\rightarrow}{\mathbf{g}} = [3\omega_c^2(I_t - I_a)\gamma_1 \quad 0 \quad 0]\mathcal{F}_s \equiv \mathbf{g}_s^T\mathcal{F}_s \tag{21}$$

Then, the motion equations are

$$\dot{\mathbf{h}}_s + \boldsymbol{\omega}_s^{\times}\mathbf{h}_s = \mathbf{g}_s \tag{22}$$

which produce (14), as before.

Directional Stability of the Thomson Equilibrium

The infinitesimal stability conditions for a rigid axisymmetrical satellite spinning nominally about the orbit normal can be inferred from (14). Obviously, γ_2 is unstable with respect to perturbations in $\dot{\gamma}_2$; this is usual for a spin-stabilized satellite. The best one can hope for is roll/yaw stability, that is, directional stability of the spin axis. Equations (14ii, iii) form a linear, undamped, gyric system, and so asymptotic stability is not possible; however, for the right combination of the dimensionless parameters k_t and $\hat{\nu} = \nu/\omega_c$, infinitesimal stability is in prospect. The characteristic equation corresponding to (14ii, iii) is

$$\left(\frac{s}{\omega_c}\right)^4 + b_1\left(\frac{s}{\omega_c}\right)^2 + b_2 = 0 \tag{23}$$

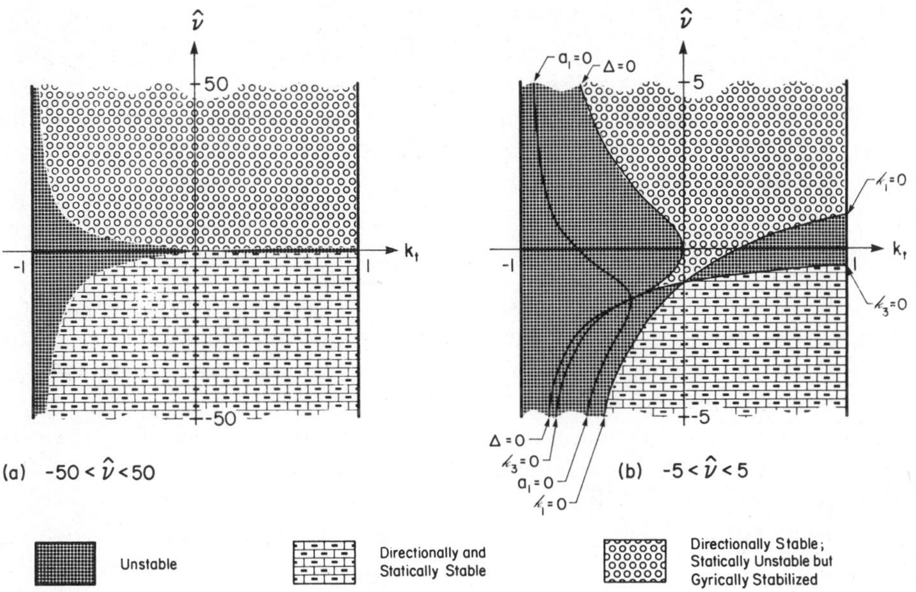

FIGURE 10.2 Stability diagrams for the Thomson equilibrium.

where

$$b_1 \triangleq (2 + 3k_t) + 2(1 + k_t)\hat{\nu} + (1 + k_t)^2\hat{\nu}^2$$
$$b_2 \triangleq \mathscr{k}_1\mathscr{k}_3 \tag{24}$$

where $\mathscr{k}_1$ and $\mathscr{k}_3$ are the diagonal elements of $\mathscr{K}$, after division by ω_c^2:

$$\mathscr{k}_1 \triangleq k_t(3 - \hat{\nu}) - 1 - \hat{\nu}$$
$$\mathscr{k}_3 \triangleq -k_t\hat{\nu} - 1 - \hat{\nu} \tag{25}$$

The conditions for infinitesimal directional stability are

(i) $b_1 > 0$
(ii) $b_2 > 0$
(iii) $\Delta \triangleq b_1^2 - 4b_2 > 0$ (26)

These conditions define the stability boundaries shown in Fig. 10.2; except for minor notational differences, this diagram was first published by [Kane, Marsh, and Wilson].

A distinction is made in Fig. 10.2 between the "statically stable" region corresponding to $\mathscr{K} > 0$ and the "statically unstable but gyrically stabilized" region corresponding to $\mathscr{K} < 0$. Figure 10.2*a* shows the stability diagram over the range of $\hat{\nu}$ likely to occur in practice. Figure 10.2*b* shows an expanded view of cases where the satellite is spinning so slowly that the centrifugal force field and the gravity gradient field are of similar magnitude. In particular, when $\hat{\nu} = -1$, the reference motion is zero relative to the orbiting frame, and the stability intervals along the line $\hat{\nu} = -1$ are seen to agree with the stability intervals along the line $k_1 = k_3 = k_t$ in Fig. 9.4, as they should. Figure 10.2 also indicates that the spin motion can be stabilized for any inertia ratio provided $|\nu|$ is made large enough. In interpreting these results however, two notes of caution should be sounded: (i) the stability proved is only infinitesimal stability (nonlinear terms have been neglected), and (ii) no damping has yet been introduced.

The Likins–Pringle Relative Equilibria

It has been pointed out by [Likins, 1] and [Pringle, 1] that a spinning symmetrical satellite in a circular orbit has other relative equilibria in addition to the Thomson equilibrium investigated above. By a "relative" equilibrium, we mean one in which the spin axis can remain fixed with respect to the orbiting frame $\mathscr{F}_c$. Unlike the Thomson equilibrium, where the spin axis is normal to the orbit and the gravitational torque vanishes, the Likins–Pringle equilibria will allow a nonzero torque if it causes the satellite to precess at a rate that matches the orbital rate. To demonstrate these equilibria and to study their stability properties, the stroboscopic frame $\mathscr{F}_s$ is of great value. It will be recalled from (16) that $\mathscr{F}_s$ is a frame intermediate to $\mathscr{F}_p$ and $\mathscr{F}_c$ in which $\mathscr{R}$ has a constant inertia matrix (because $\mathscr{R}$ is axisymmetrical). In fact, we define

$$\mathbf{C}_{pc} = \mathbf{C}_{ps}\mathbf{C}_{sc}; \qquad \mathbf{C}_{ps} = \mathbf{C}_2\left[(\nu + \omega_c)t\right] \tag{27}$$

in accordance with equation (16). Furthermore, we denote the direction cosine elements of $\mathbf{C}_{sc}$ by

$$\left\{c_{ij}^s\right\} \triangleq \mathbf{C}_{sc} \tag{28}$$

to distinguish them from the elements c_{ij} of $\mathbf{C}_{pc}$. For the Thomson equilibrium, $\mathbf{C}_{sc} = \mathbf{1}$; for the Likins–Pringle equilibria, $\mathbf{C}_{sc}$ is by definition a constant matrix, but different from $\mathbf{1}$.

The absolute angular velocity of $\mathscr{R}$ is

$$\boldsymbol{\omega} = \mathbf{1}_2(\nu + \omega_c) + \mathbf{C}_{ps}\boldsymbol{\omega}_s \tag{29}$$

where $\boldsymbol{\omega}_s$ is the absolute angular velocity of $\mathscr{F}_s$:

$$\boldsymbol{\omega}_s = \boldsymbol{\omega}_\gamma - \omega_c\mathbf{c}_2^s; \qquad \boldsymbol{\omega}_\gamma^\times \triangleq -\dot{\mathbf{C}}_{sc}\mathbf{C}_{sc}^T \tag{30}$$

and $\mathbf{c}_2^s$ is the second column of $\mathbf{C}_{sc}$. Equation (30) is similar to equation (9.2, 6), a comparison that suggests the analogy between the use of the stroboscopic frame $\mathscr{F}_s$ in the study of spinning symmetrical satellites and the use of the body-fixed frame $\mathscr{F}_p$ in the study of the librations of tri-inertial satellites. For an equilibrium relative to $\mathscr{F}_c$, $\boldsymbol{\omega}_\gamma \equiv \mathbf{0}$.

To formulate the motion equations in $\mathscr{F}_s$, we require the gravity gradient torque in that frame. For a symmetrical satellite,

$$\mathbf{g}_s = 3\omega_c^2\begin{bmatrix}(I_t - I_a)c_{23}^s c_{33}^s \\ 0 \\ (I_a - I_t)c_{13}^s c_{23}^s\end{bmatrix} \tag{31}$$

The equations of motion are then

$$\dot{\mathbf{h}}_s + \boldsymbol{\omega}_s^{\times}\mathbf{h}_s = \mathbf{g}_s \tag{32}$$

where

$$\mathbf{h}_s = \mathbf{I}\boldsymbol{\omega}_s + \mathbf{1}_2 I_a(\nu + \omega_c) \tag{33}$$

To establish equilibria relative to $\mathscr{F}_s$, one sets

$$\boldsymbol{\omega}_s = -\omega_c \mathbf{c}_2^s = \text{constant}$$

$$\mathbf{h}_s = -\omega_c \mathbf{I}\mathbf{c}_2^s + \mathbf{1}_2 I_a(\nu + \omega_c) = \text{constant} \tag{34}$$

and solves the algebraic equation $\boldsymbol{\omega}_s^{\times}\mathbf{h}_s = \mathbf{g}_s$. The nontrivial components of this equation in $\mathscr{F}_s$ are

$$\begin{aligned} &\text{(i)}\ \ k_t(3c_{23}^s c_{33}^2 - c_{22}^s c_{32}^s) + (1 + k_t)(1 + \hat{\nu})c_{32}^s = 0 \\ &\text{(ii)}\ \ k_t(3c_{23}^s c_{13}^s - c_{22}^s c_{12}^s) + (1 + k_t)(1 + \hat{\nu})c_{12}^s = 0 \end{aligned} \tag{35}$$

where, as before, $k_t \triangleq (I_a - I_t)/I_t$ and $\hat{\nu} \triangleq \nu/\omega_c$. The Thomson equilibrium, defined by $\{c_{ij}^s\} = \mathbf{1}$, satisfies these equations of course. The aim here is to find other solutions.

On multiplying (35i) by c_{12}^s and (35ii) by c_{32}^s and subtracting, we find

$$k_t c_{23}^s (c_{32}^s c_{13}^s - c_{12}^s c_{33}^s) = 0 \tag{36}$$

Note from equation (2.1, 13) that the expression in parentheses is just c_{21}^s. Therefore, apart from the uninteresting case of an inertially spherical satellite ($k_t = 0$), the Likins–Pringle equilibria are of two types:

$$c_{23}^s = 0; \qquad c_{21}^s \neq 0 \tag{37a}$$

$$c_{23}^s \neq 0; \qquad c_{21}^s = 0 \tag{37b}$$

In case (37*a*), the spin axis $\hat{\mathbf{p}}_2$ ($\equiv \underset{\rightarrow}{\hat{\mathbf{s}}}_2$) is normal to the vertical $\underset{\rightarrow}{\hat{\mathbf{c}}}_3$, but not normal to the forward direction $\underset{\rightarrow}{\hat{\mathbf{c}}}_1$; the spin axis generates a hyperboloid of revolution in inertial space (Fig. 10.3*b*). In case (37*b*), the spin axis is normal to the forward direction but not to the vertical; the spin axis generates a cone in inertial space (Fig. 10.3*c*). To find the dependence of c_{21}^s on k_t and $\hat{\nu}$ in the hyperbolic case, set $c_{23}^s = 0$ in (35) to obtain

$$\begin{aligned} c_{32}^s\left[k_t c_{22}^s - (1 + k_t)(1 + \hat{\nu})\right] &= 0 \\ c_{12}^s\left[k_t c_{22}^s - (1 + k_t)(1 + \hat{\nu})\right] &= 0 \end{aligned} \tag{38}$$

Now, it is not possible for both c_{12}^s and c_{32}^s to be zero, since then c_{21}^s would be zero, which it isn't [see (2.1, 13)]. Therefore, it is concluded for the hyperbolic case that

$$c_{23}^s = 0; \qquad c_{22}^s = \cos\bar{\gamma}_3 \triangleq \frac{(1 + k_t)(1 + \hat{\nu})}{k_t} \tag{39}$$

The right side of (39) cannot exceed unity in magnitude if the hyperbolic case is to be physically possible. As for the conical case, represented by (37*b*), it can be

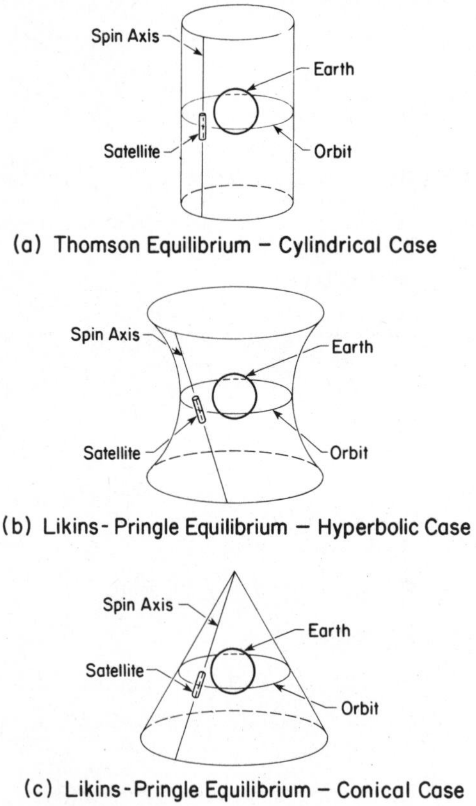

FIGURE 10.3 Types of relative equilibrium for a spinning symmetrical satellite in a circular orbit. After [Likins, 1].

shown (Problem 10.2) that

$$c_{21}^s = 0; \qquad c_{22}^s = \cos\bar{\gamma}_1 \triangleq \frac{(1 + k_t)(1 + \hat{\nu})}{4k_t} \tag{40}$$

Again, this relative equilibrium is not physically possible unless $|c_{22}^s| \le 1$.

A determination is now made of the infinitesimal stability conditions for the Likins–Pringle equilibria of a spinning axisymmetrical satellite in a circular orbit. First, the hyperbolic case shown in Fig. 10.3*b* is dealt with, for which we set

$$\mathbf{C}_{sc} = \mathbf{C}_2(\gamma_2)\mathbf{C}_1(\gamma_1)\mathbf{C}_3(\gamma_3 + \bar{\gamma}_3) \tag{41}$$

where $\bar{\gamma}_3$ is the finite tilt of the spin axis about the vertical. The wobble angles γ_1 and γ_3 are assumed to be infinitesimals, as in the less interesting angle γ_2. Thus,

$$\mathbf{C}_{sc} = (\mathbf{1} - \boldsymbol{\gamma}^{\times})\mathbf{C}_3(\bar{\gamma}_3) \tag{42}$$

where $\cos\bar{\gamma}_3$ is given by (39). From the definition (30) of $\boldsymbol{\omega}_\gamma$,

$$\boldsymbol{\omega}_\gamma = \dot{\boldsymbol{\gamma}} \tag{43}$$

to first order. Also, according to (42),

$$\mathbf{c}_2^s \triangleq \begin{bmatrix} c_{12}^s \\ c_{22}^s \\ c_{32}^s \end{bmatrix} = \begin{bmatrix} \sin\bar{\gamma}_3 + \gamma_3\cos\bar{\gamma}_3 \\ \cos\bar{\gamma}_3 - \gamma_3\sin\bar{\gamma}_3 \\ \gamma_2\sin\bar{\gamma}_3 - \gamma_1\cos\bar{\gamma}_3 \end{bmatrix} \tag{44}$$

$$\mathbf{c}_3^s \triangleq \begin{bmatrix} c_{13}^s \\ c_{23}^s \\ c_{33}^s \end{bmatrix} = \begin{bmatrix} -\gamma_2 \\ \gamma_1 \\ 1 \end{bmatrix} \tag{45}$$

Upon substitution of these expressions in (30) through (33), the linearized motion equations are found to be (Problem 10.3a)

$$\ddot{\gamma}_1 - (\omega_c\cos\bar{\gamma}_3)\dot{\gamma}_3 + 3k_t\omega_c^2\gamma_1 = 0$$

$$\ddot{\gamma}_2 + (\omega_c\sin\bar{\gamma}_3)\dot{\gamma}_3 = 0$$

$$\ddot{\gamma}_3 + (\omega_c\cos\bar{\gamma}_3)\dot{\gamma}_1 - (1 + k_t)(\omega_c\sin\bar{\gamma}_3)\dot{\gamma}_2 - k_t(\omega_c\sin\bar{\gamma}_3)^2\gamma_3 = 0 \tag{46}$$

Since we already know that stability about the spin axis is impossible, the second expression of (46) is not of direct interest; it will be integrated once, and $\dot{\gamma}_2$ will be substituted to obtain linear equations in the wobble angles γ_1 and γ_3:

$$\ddot{\gamma}_1 - (\omega_c\cos\bar{\gamma}_3)\dot{\gamma}_3 + 3k_t\omega_c^2\gamma_1 = 0$$

$$\ddot{\gamma}_3 + (\omega_c\cos\bar{\gamma}_3)\dot{\gamma}_1 + (\omega_c\sin\bar{\gamma}_3)^2\gamma_3 = 0 \tag{47}$$

The characteristic equation corresponding to (47) is

$$\left(\frac{s}{\omega_c}\right)^4 + b_1\left(\frac{s}{\omega_c}\right)^2 + b_2 = 0$$

$$b_1 = 3k_t + 1; \qquad b_2 = 3k_t\sin^2\bar{\gamma}_3$$

and the conditions for infinitesimal stability are

$$\textbf{(i)}\ b_1 > 0; \qquad \textbf{(ii)}\ b_2 > 0; \qquad \textbf{(iii)}\ b_1^2 - 4b_2 > 0 \tag{48}$$

The second of these, (48ii), implies that

$$k_t > 0 \tag{49}$$

and this also satisfies the other two. This is also the condition for "static" stability in (47). We therefore conclude ([Likins, 1]) that the hyperbolic Likins–Pringle spin equilibrium is infinitesimally directionally stable if $\mathscr{R}$ is spinning about its major axis (Fig. 10.4a).

For the conical case (Fig. 10.3c), set

$$\mathbf{C}_{sc} = (\mathbf{1} - \boldsymbol{\gamma}^{\times})\mathbf{C}_1(\bar{\gamma}_1) \tag{50}$$

where $\bar{\gamma}_1$ is the finite tilt of the spin axis about the forward direction and $\bar{\gamma}_1$ is given by (40). The three infinitesimal angles $\boldsymbol{\gamma}$ allow small attitude excursions about the nominal motion. From the definition of $\boldsymbol{\omega}_\gamma$ in (30), we still have $\boldsymbol{\omega}_\gamma = \dot{\boldsymbol{\gamma}}$,

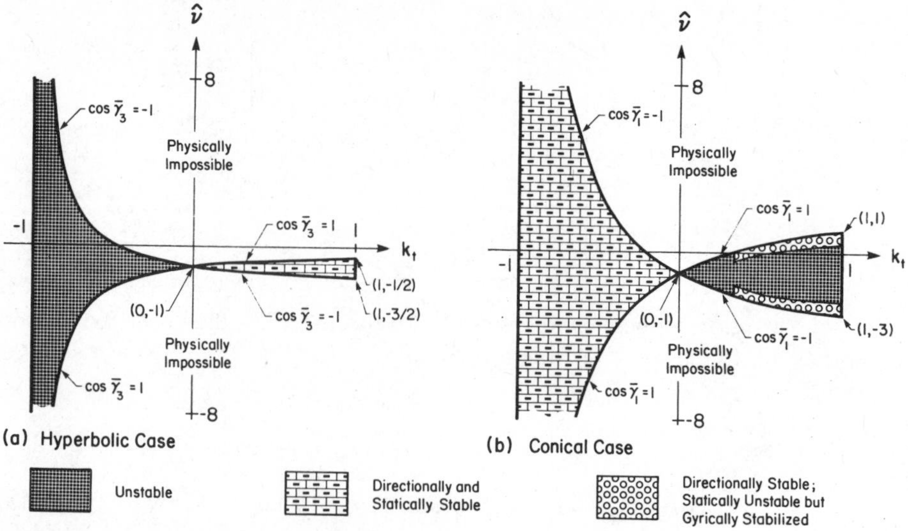

FIGURE 10.4 Stability diagrams for the Likins–Pringle equilibria.

but now

$$\mathbf{c}_2^s = \begin{bmatrix} \gamma_3 \cos\bar{\gamma}_1 + \gamma_2 \sin\bar{\gamma}_1 \\ \cos\bar{\gamma}_1 - \gamma_1 \sin\bar{\gamma}_1 \\ -\gamma_1 \cos\bar{\gamma}_1 - \sin\bar{\gamma}_1 \end{bmatrix} \tag{51}$$

$$\mathbf{c}_3^s = \begin{bmatrix} \gamma_3 \sin\bar{\gamma}_1 - \gamma_2 \cos\bar{\gamma}_1 \\ \sin\bar{\gamma}_1 + \gamma_1 \cos\bar{\gamma}_1 \\ -\gamma_1 \sin\bar{\gamma}_1 + \cos\bar{\gamma}_1 \end{bmatrix} \tag{52}$$

Upon substitution of these expressions in (30) through (33), the linearized motion equations become

$$\ddot{\gamma}_1 - (1 + k_t)(\omega_c \sin\bar{\gamma}_1)\dot{\gamma}_2 - (1 + 3k_t)(\omega_c \cos\bar{\gamma}_1)\dot{\gamma}_3 - 4k_t(\omega_c \sin\bar{\gamma}_1)^2 \gamma_1 = 0 \tag{53a}$$

$$\ddot{\gamma}_2 + (\omega_c \sin\bar{\gamma}_1)\dot{\gamma}_1 = 0 \tag{53b}$$

$$\ddot{\gamma}_3 + (1 + 3k_t)(\omega_c \cos\bar{\gamma}_1)\dot{\gamma}_1 - 3k_t \omega_c^2 \gamma_3 = 0 \tag{53c}$$

Integration of (53b) and substitution in (53a) yields the following equations for the wobble angles γ_1 and γ_3:

$$\ddot{\gamma}_1 - (1 + 3k_t)(\omega_c \cos\bar{\gamma}_1)\dot{\gamma}_3 + (1 - 3k_t)(\omega_c \sin\bar{\gamma}_1)^2 \gamma_1 = 0$$

$$\ddot{\gamma}_3 + (1 + 3k_t)(\omega_c \cos\bar{\gamma}_1)\dot{\gamma}_1 - 3k_t \omega_c^2 \gamma_3 = 0 \tag{54}$$

The characteristic equation corresponding to (54) is

$$\left(\frac{s}{\omega_c}\right)^4 + b_1\left(\frac{s}{\omega_c}\right)^2 + b_2 = 0$$

$$b_1 = 1 + 3k_t + 9k_t^2 - 9k_t(1 + k_t)\sin^2\bar{\gamma}_1$$

$$b_2 = -3k_t(1 - 3k_t)\sin^2\bar{\gamma}_1 \tag{55}$$

and the conditions for infinitesimal stability are still of the form (48). Also, it is recalled that $\cos\bar{\gamma}_1$ is given by (40). Figure 10.4*b* shows the $\{k_t, \hat{\nu}\}$ parameter plane for the conical case. Cases that are not possible physically are first excluded ($|k_t| > 1$ or $|c_{22}^s| > 1$). The remaining region is divided into three subregions: from (54), the spin is directionally and statically stable if and only if $k_t < 0$. If $k_t > 0$ but equations (48) are satisfied [with b_1 and b_2 given by (55)], the region is labeled "statically unstable but gyrically stabilized." If equations (48) are not satisfied, the motion is unstable.

Liapunov Stability

The stability diagrams shown in Figs. 10.2 and 10.4 show three types of regions: unstable, directionally stable (statically unstable but gyrically stabilized), and directionally and statically stable. It will now be shown that the last of these three regions is Liapunov stable. This will remove the uncertainty that accompanies the demonstration of infinitesimal, nonasymptotic stability. To this end, a Liapunov function is now constructed based on the Hamiltonian. The Hamiltonian for a rigid body in a circular orbit was given in (9.2, 9), repeated here for convenience:

$$H = \tfrac{1}{2}\boldsymbol{\omega}_\alpha^T\mathbf{I}\boldsymbol{\omega}_\alpha + \tfrac{3}{2}\omega_c^2\mathbf{c}_3^T\mathbf{I}\mathbf{c}_3 - \tfrac{1}{2}\omega_c^2\mathbf{c}_2^T\mathbf{I}\mathbf{c}_2 \tag{56}$$

To adapt this expression to present purposes, it is noted that

$$\boldsymbol{\omega}_\alpha = \mathbf{C}_{ps}\boldsymbol{\omega}_\gamma + (\nu + \omega_c)\mathbf{1}_2 \tag{57}$$

as follows from comparing (9.2, 6) with (29) and (30). Furthermore, the quadratic form $\mathbf{c}_3^T\mathbf{I}\mathbf{c}_3$ can be manipulated as follows:

$$\mathbf{c}_3^T\mathbf{I}\mathbf{c}_3 = \mathbf{c}_3^T\mathbf{C}_{ps}\mathbf{C}_{sp}\mathbf{I}\mathbf{C}_{ps}\mathbf{C}_{sp}\mathbf{c}_3 = \mathbf{c}_3^{sT}\mathbf{I}\mathbf{c}_3^s \tag{58}$$

because $\mathbf{C}_{sp}\mathbf{c}_3 = \mathbf{c}_3^s$ and

$$\mathbf{C}_{sp}\mathbf{I}\mathbf{C}_{ps} = \mathbf{I} = I_t\mathbf{1} + (I_a - I_t)\mathbf{1}_2\mathbf{1}_2^T \tag{59}$$

(The inertia matrix is $\mathbf{I}$ in either $\mathscr{F}_p$ or $\mathscr{F}_s$.) The same manipulation applies also to the last term in (56). When the above relations are inserted in (56), one finds (Problem 10.3) that

$$H = \tfrac{1}{2}I_a(\nu + \omega_c)^2 + I_t\omega_c^2 + \tfrac{1}{2}\boldsymbol{\omega}_\gamma^T\mathbf{I}\boldsymbol{\omega}_\gamma + I_a(\nu + \omega_c)\omega_{\gamma 2}$$
$$+ \tfrac{1}{2}(I_a - I_t)\omega_c^2\left[3(c_{23}^s)^2 - (c_{22}^s)^2\right] \tag{60}$$

This form of the Hamiltonian is suitable for dealing with the directional stability of a spinning rigid body in a circular orbit.

In the reference equilibrium (either the Thomson equilibrium or the hyperbolic or conical equilibria of Likins and Pringle), $\boldsymbol{\omega}_\gamma = \mathbf{0}$, and c_{23}^s and c_{22}^s have their reference values, indicated here by superscript o:

$$H_o = \tfrac{1}{2}I_a(\nu + \omega_c)^2 + I_t\omega_c^2 + \tfrac{1}{2}(I_a - I_t)\omega_c^2\left[3(c_{23}^{so})^2 - (c_{22}^{so})^2\right] \tag{61}$$

The appropriate Liapunov function is then

$$\begin{aligned} v \triangleq H - H_o &= \tfrac{1}{2}\boldsymbol{\omega}_\gamma^T\mathbf{I}\boldsymbol{\omega}_\gamma + I_a(\nu + \omega_c)\omega_{\gamma 2} \\ &\quad + \tfrac{1}{2}(I_a - I_t)\omega_c^2\left[3(c_{23}^s)^2 - 3(c_{23}^{so})^2 - (c_{22}^s)^2 + (c_{22}^{so})^2\right] \end{aligned} \tag{62}$$

Moreover, we also wish to incorporate in v the fact that there is no gravitational torque about $\underset{\rightarrow}{\hat{\mathbf{s}}}_2$, the symmetry axis of $\mathscr{R}$. This fact is demonstrated by the following calculation. From (8.1, 18) and using vectrices (Appendix B),

$$\underset{\rightarrow}{\hat{\mathbf{s}}}_2 \cdot \underset{\rightarrow}{\mathbf{g}} = 3\omega_c^2(\mathbf{1}_2^T\mathscr{F}_s)\cdot(\mathbf{1}_3^T\mathscr{F}_c)\times(\mathscr{F}_s^T\mathbf{I}\mathscr{F}_s)\cdot(\mathscr{F}_c^T\mathbf{1}_3) = 3\omega_c^2\mathbf{1}_2^T(\mathbf{c}_3^s)^\times\mathbf{I}\mathbf{c}_3^s = 0 \tag{63}$$

after using (59). Then, from (22),

$$\mathbf{1}_2^T(\dot{\mathbf{h}}_s + \boldsymbol{\omega}_s^\times\mathbf{h}_s) = 0 \tag{64}$$

But $\mathbf{1}_2^T\boldsymbol{\omega}_s^\times\mathbf{h}_s$ equals zero also, as can be seen by using (20). Therefore, $\mathbf{1}_2^T\dot{\mathbf{h}}_s = 0$. In other words, the component of the angular momentum along the symmetry axis is constant throughout the motion. Thus, from (20) and (30),

$$I_a(\omega_{\gamma 2} - \omega_c c_{22}^s) + I_a(\nu + \omega_c) = \text{constant}$$

We choose this constant based on its value in the reference equilibrium. Solving for $\omega_{\gamma 2}$, we find

$$\omega_{\gamma 2} = \omega_c(c_{22}^s - c_{22}^{so}) \tag{65}$$

This leads to the final form for the Liapunov function v:

$$\begin{aligned} v &= \tfrac{1}{2}I_t(\omega_{\gamma 1}^2 + \omega_{\gamma 3}^2) + \tfrac{1}{2}I_a\omega_c^2(c_{22}^s - c_{22}^{so})^2 + I_a\omega_c(\nu + \omega_c)(c_{22}^s - c_{22}^{so}) \\ &\quad + \tfrac{1}{2}(I_a - I_t)\omega_c^2\left[3(c_{23}^s)^2 - 3(c_{23}^{so})^2 - (c_{22}^s)^2 + (c_{22}^{so})^2\right] \end{aligned} \tag{66}$$

Liapunov stability can be proved if it can be demonstrated that v is positive-definite.

The three classes of relative equilibria have the following reference values:

Thomson:	$\boldsymbol{\omega}_\gamma = \mathbf{0}$;	$c_{22}^{so} = 1$; $c_{23}^{so} = 0$	(67*a*)
Likins–Pringle, hyperbolic:	$\boldsymbol{\omega}_\gamma = \mathbf{0}$;	$c_{22}^{so} = \cos\bar{\gamma}_3$; $c_{23}^{so} = 0$	(67*b*)
Likins–Pringle, conical:	$\boldsymbol{\omega}_\gamma = \mathbf{0}$;	$c_{22}^{so} = \cos\bar{\gamma}_1$; $c_{23}^{so} = \sin\bar{\gamma}_1$	(67*c*)

where $\cos\bar{\gamma}_1$ and $\cos\bar{\gamma}_3$ are given respectively by (40) and (39). To accommodate finite angular displacements with respect to these equilibria, we set, in the spirit of

(41),

$$\mathbf{C}_{sc} \equiv [\mathbf{c}_1^s \quad \mathbf{c}_2^s \quad \mathbf{c}_3^s] = \mathscr{C}\mathbf{C}_3(\bar{\gamma}_3) \tag{68a}$$

for the hyperbolic case and

$$\mathbf{C}_{sc} = \mathscr{C}\mathbf{C}_1(\bar{\gamma}_1) \tag{69a}$$

for the conical case, in the spirit of (50). (The rotation matrices $\mathscr{C}$ in (68a) and (69a) are not identical; no confusion should arise because the two cases are treated separately.) It follows from (68a) that, for the hyperbolic case,

$$c_{22}^s = c_{21}\sin\bar{\gamma}_3 + c_{22}\cos\bar{\gamma}_3 \tag{68b}$$

Hyperbolic:

$$c_{23}^s = c_{23} \tag{68c}$$

and from (69a) that, for the conical case,

$$c_{22}^s = c_{22}\cos\bar{\gamma}_1 - c_{23}\sin\bar{\gamma}_1$$

Conical:

$$c_{23}^s = c_{22}\sin\bar{\gamma}_1 + c_{23}\cos\bar{\gamma}_1 \tag{69b}$$

In both cases, c_{22} in nominally 1, and c_{23} is nominally 0.

For Thompson's equilibrium, (66) and (67a) yield the Liapunov function

$$v = \tfrac{1}{2}I_t\left(\omega_{\gamma1}^2 + \omega_{\gamma3}^2\right) - I_a\omega_c\nu\left(1 - c_{22}^s\right) + \tfrac{1}{2}\left[(3I_a - 4I_t)\left(c_{23}^s\right)^2 - I_t\left(c_{21}^s\right)^2\right]\omega_c^2 \tag{70}$$

By assigning specific values to v and setting $\omega_{\gamma1} = \omega_{\gamma3} = 0$ in (70), "zero velocity" curves on the unit sphere

$$\left(c_{21}^s\right)^2 + \left(c_{22}^s\right)^2 + \left(c_{23}^s\right)^2 = 1 \tag{71}$$

can be plotted in the same manner as for the torque-free motion of a rigid body (see Fig. 4.16) and for the librations of a symmetrical rigid body under gravitational torque in a circular orbit (see Figs. 9.10 and 9.11). These curves have been drawn by [Pringle, 1] and shown on the unit sphere by [Beletskii]. See Fig. 10.5.

It will be recalled that $\{c_{21}^s, c_{22}^s, c_{23}^s\}$ are the direction cosines of the symmetry axis $\vec{\mathbf{s}}_2$ with respect to the orbiting frame $\mathscr{F}_c$. Therefore, the zero velocity curves on the unit sphere can often be used to place bounds on the motion of the symmetry axis as viewed in $\mathscr{F}_c$. When the bound can be made arbitrarily close to the reference equilibrium by making the initial disturbances $\{\omega_{\gamma1}, \omega_{\gamma3}, c_{21}^s, c_{23}^s\}$ sufficiently small, this implies Liapunov stability. The conditions under which v is positive-definite are not obvious from (70). However, an expression for v accurate to second order is all that is required to show a finite region of stability. To second order,

$$v \doteq \tfrac{1}{2}I_t\left(\omega_{\gamma1}^2 + \omega_{\gamma3}^2\right) + \tfrac{1}{2}I_t\omega_c^2\left[k_3\left(c_{21}^s\right)^2 + k_1\left(c_{23}^s\right)^2\right] \tag{72}$$

using (71), where k_1 and k_3 were given in (25). Thus, for Liapunov stability, it is

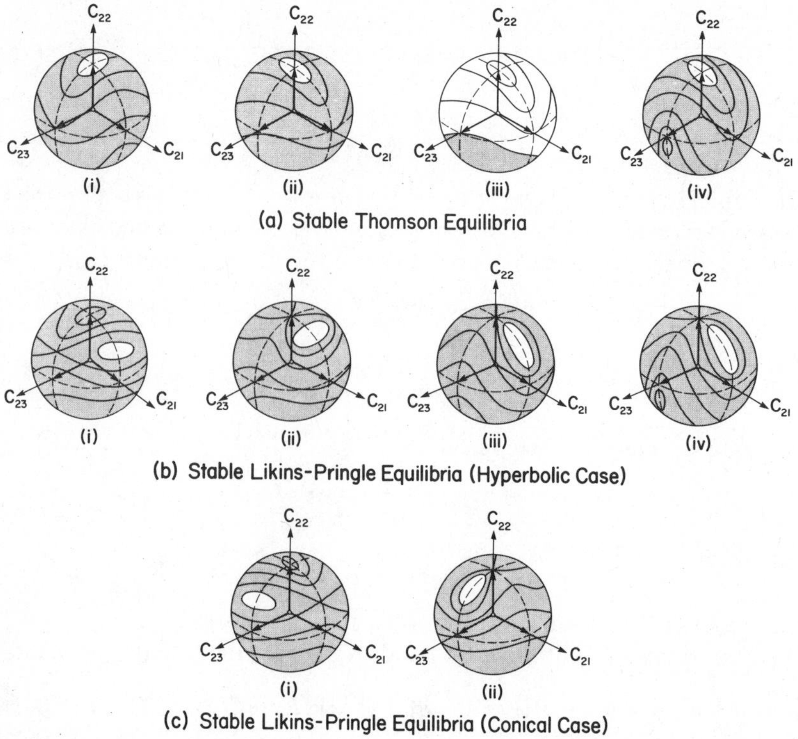

FIGURE 10.5 Bounds on finite deviations from reference spin equilibria. Based on [Pringle, 1] and [Beletskii].

sufficient that $\mathscr{k}_1 > 0$, $\mathscr{k}_3 > 0$; hence the region labeled "directionally and statically stable" in Fig. 10.2 is also Liapunov stable.

To turn to the hyperbolic relative equilibrium of Likins and Pringle (see Fig. 10.3*b*), the appropriate Liapunov function is still (66) with (67*b*) and (68*b*) inserted. The result (Problem 10.3) is

$$\begin{aligned} v = {} & \tfrac{1}{2}I_t\left(\omega_{\gamma 1}^2 + \omega_{\gamma 3}^2\right) + \tfrac{1}{2}I_a\omega_c^2\left(c_{21}\sin\bar{\gamma}_3 + c_{22}\cos\bar{\gamma}_3 - \cos\bar{\gamma}_3\right)^2 \\ & + \tfrac{1}{2}(I_a - I_t)\omega_c^2\left[\left(3 + \cos^2\bar{\gamma}_3\right)c_{23}^2 - \left(1 - 2\cos^2\bar{\gamma}_3\right)c_{21}^2\right. \\ & \left. + 2\cos\bar{\gamma}_3(1 - c_{22})(c_{21}\sin\bar{\gamma}_3 - \cos\bar{\gamma}_3)\right] \end{aligned} \tag{73}$$

Once again, zero velocity curves can be drawn on the unit sphere and bounds established on the deviations of the symmetry axis from its nominal orientation. As for positive-definiteness, v in (73) is, to second order,

$$v \doteq \tfrac{1}{2}I_t\left(\omega_{\gamma 1}^2 + \omega_{\gamma 3}^2\right) + \tfrac{1}{2}I_t\omega_c^2\left[c_{21}^2\sin^2\bar{\gamma}_3 + 3k_t c_{23}^2\right] \tag{74}$$

and v is clearly positive-definite for $k_t > 0$ (a major-axis spin). Therefore, the region labeled "directionally and statically stable" in Fig. 10.4*a* is in fact also Liapunov stable.

As for the conical case (Fig. 10.3c), the Liapunov function is found by using (67c) and (68c) in conjunction with (66). The result (Problem 10.3) is

$$\begin{aligned} v = {} & \tfrac{1}{2}I_t\left(\omega_{\gamma 1}^2 + \omega_{\gamma 3}^2\right) + \tfrac{1}{2}I_a\omega_c^2\left(c_{22}\cos\bar{\gamma}_1 - c_{23}\sin\bar{\gamma}_1 - \cos\bar{\gamma}_1\right)^2 \\ & + \tfrac{1}{2}(I_a - I_t)\omega_c^2\left[8(c_{22} - 1)\cos\bar{\gamma}_1(\cos\bar{\gamma}_1 + c_{23}\sin\bar{\gamma}_1)\right. \\ & \left. + \left(4\cos^2\bar{\gamma}_1 - 3\right)c_{21}^2 + 4\left(2\cos^2\bar{\gamma}_1 - 1\right)c_{23}^2\right] \end{aligned} \tag{75}$$

which reduces, to second order, to the following:

$$v \doteq \tfrac{1}{2}I_t\left(\omega_{\gamma 1}^2 + \omega_{\gamma 3}^2\right) + \tfrac{1}{2}I_t\omega_c^2\left[-3k_t c_{21}^2 + (1 - 3k_t)c_{23}^2\sin^2\bar{\gamma}_1\right] \tag{76}$$

When $k_t < 0$ (a minor-axis spin), v is positive-definite. Therefore, the region marked "directionally and statically stable" in Fig. 10.4b has been shown to be Liapunov stable.

Finite Deviations from Reference Spin Equilibria

To review briefly what has been accomplished thus far in the study of inertially symmetrical spinning satellites in a circular orbit, three types of relative equilibria have been found (Fig. 10.3). The most basic type—the Thomson equilibrium—exists for any spin rate ($\hat{\nu}$) and any inertia distribution (k_t). This equilibrium is moreover directionally stable for a wide range of $\hat{\nu}$ and k_t (Fig. 10.2), although the gyrically stabilized cases could not be shown to be Liapunov stable for finite deviations, and damping can be expected to remove this possibility (see Section 10.2). For the Likins–Pringle equilibria, the spin axis can in principle remain fixed in the orbiting reference frame $\mathscr{F}_c$ at angles other than 90° to the orbit normal; for a given orientation of the spin axis relative to $\mathscr{F}_c$, a constraint is placed on spin rate and inertia distribution: for example, given the orientation and k_t, $\hat{\nu}$ must have a certain value. Directionally stable regions in the $\{k_t, \hat{\nu}\}$ parameter plane have been calculated (Fig. 10.4), and the statically stable regions have also been shown to be Liapunov stable with respect to finite deviations from the reference (relative) equilibrium.

For large (noninfinitesimal) deviations from the reference equilibrium, the motion equations become nonlinear, and closed-form solution is impossible. The appropriate equations for numerical integration are now summarized. The motion equations, expressed in the "stroboscopic" frame $\mathscr{F}_s$, are given by (22), with the gravity gradient torque in $\mathscr{F}_s$ expressed for an axially symmetrical body by (31):

$$\begin{aligned} \dot{h}_{s1} + \omega_{s2}h_{s3} - \omega_{s3}h_{s2} &= 3\omega_c^2(I_t - I_a)c_{23}^s c_{33}^s \\ \dot{h}_{s3} + \omega_{s1}h_{s2} - \omega_{s2}h_{s1} &= 3\omega_c^2(I_a - I_t)c_{13}^s c_{23}^s \end{aligned} \tag{77a}$$

The axial component of the angular momentum is constant:

$$h_{s2} = I_a(\omega_{s2} + \omega_c + \nu) = \text{constant} \tag{77b}$$

the constant being determined from initial conditions. In turn, $\boldsymbol{\omega}_s$ is found from

(20):

$$\omega_{s1} = \frac{h_{s1}}{I_t}; \qquad \omega_{s3} = \frac{h_{s3}}{I_t}$$

$$\omega_{s2} = \frac{h_{s2}}{I_a} - \nu - \omega_c \tag{78}$$

The system is completed by the kinematical equations which are, from (28) and (30),

$$\dot{\mathbf{c}}_2^s = -\boldsymbol{\omega}_\gamma^\times \mathbf{c}_2^s; \qquad \dot{\mathbf{c}}_3^s = -\boldsymbol{\omega}_\gamma^\times \mathbf{c}_3^s \tag{79}$$

where, from (30),

$$\boldsymbol{\omega}_\gamma = \boldsymbol{\omega}_s + \omega_c \mathbf{c}_2^s \tag{80}$$

Thus, (77) through (80) represent 15 scalar equations (8 differential and 7 algebraic) for the 15 scalar quantities represented by $\{\mathbf{h}_s, \boldsymbol{\omega}_s, \boldsymbol{\omega}_\gamma, \mathbf{c}_2^s, \mathbf{c}_3^s\}$. In this book it is always implied that any appropriate set of variables can be used in place of the direction cosines (see Chapter 2). Equations (77) through (80) can be integrated with numerical efficiency since only elementary arithmetic operations are required. The use of Euler parameters instead might be more efficient since then four scalars would replace the six in $\mathbf{c}_2^s$ and $\mathbf{c}_3^s$. Euler angles are attractive from a visualization standpoint, but trigonometric functions tend to make the numerical integration less efficient.

The only hope of analytical results lies in the use of the Hamiltonian function; (70), (73), and (75) are such functions for the Thomson equilibrium and the Likins–Pringle equilibria, respectively. This approach has been taken by [Pringle, 1], who plots zero velocity contours (bounds) on the unit sphere associated with equation (71). Typical of such plots are the contours shown in Fig. 10.5, taken from [Beletskii]. Motion is confined to the unshaded regions. Further results can be found in the work of [Modi and Neilson, 1, 2]. The latter reference also investigates periodic solutions numerically.

For near-Earth orbits, the oblateness of Earth causes orbits inclined to the equatorial plane to precess slowly, and the "Thomson equilibrium" is no longer an equilibrium. [Pringle, 3] has analyzed the consequences of this orbital precession using perturbation methods and the method of averaging. He showed that as the spin rate $\hat{\nu}$ becomes larger, the deviations of the symmetry axis from the (precessing) orbit normal become large due to resonance with the precession rate. Indeed, for $\hat{\nu}$ greater than a critical value, the spin axis "ceases to remain near the orbit normal and describes a wandering path in inertial space." The motion has become that of a rigid symmetrical body in almost torque-free motion.

Tri-Inertial Rigid Satellite in a Circular Orbit

The stroboscopic frame $\mathcal{F}_s$, so useful for inertially axisymmetrical satellites, is no longer helpful for tri-inertial satellites, for which I_1, I_2, and I_3 are distinct. The only reference frame in which the inertia matrix is constant for a tri-inertial satellite is a body-fixed frame, and hence we return to $\mathscr{F}_p$, the centroidal principal-axis frame, to write motion equations. Unfortunately, the gravity gradient torque components are not constant in $\mathscr{F}_p$ as the satellite spins, even though the spin axis may be fixed in the orbiting frame $\mathscr{F}_c$. This leads to periodic coefficients in the linearized motion equations used for determined infinitesimal stability conditions. Thus, Floquet theory (see Section A.3 in Appendix A) and numerical integration must be used to establish stability boundaries.

The equations governing the motion of a triaxial rigid body in a circular orbit were given as (9.2, 1 through 7) in Chapter 9. These equations are summarized here for convenience:

$$\mathbf{I}\dot{\boldsymbol{\omega}} + \boldsymbol{\omega}^{\times}\mathbf{I}\boldsymbol{\omega} = 3\omega_c^2\mathbf{c}_3^{\times}\mathbf{I}\mathbf{c}_3 \tag{81}$$

$$\boldsymbol{\omega} = \boldsymbol{\omega}_\alpha - \omega_c\mathbf{c}_2 \tag{82}$$

$$\dot{\mathbf{c}}_2 = -\boldsymbol{\omega}_\alpha^{\times}\mathbf{c}_2; \qquad \dot{\mathbf{c}}_3 = -\boldsymbol{\omega}_\alpha^{\times}\mathbf{c}_3 \tag{83}$$

It will be recalled that $\boldsymbol{\omega}$ and $\boldsymbol{\omega}_\alpha$ are respectively the components (in $\mathscr{F}_p$) of the satellite's absolute angular velocity and its angular velocity with respect to the orbiting frame $\mathscr{F}_c$. The Hamiltonian for this motion (a constant) is given by (9.2, 8) or (9.2, 9). In (9.2, 13) the relative equilibrium $\boldsymbol{\omega}_\alpha \equiv \mathbf{0}$ was introduced, in which one principal axis of the satellite nominally points toward Earth and a second principal axis is nominally normal to the orbit. This is the so-called "Earth-pointing" case and suggests gravity gradient stabilization (Chapter 9). It is now shown that this equilibrium relative to $\mathscr{F}_c$ can, in a sense, be generalized. Suppose that one principal axis of the satellite initially points normal to the orbit (as in the Earth-pointing case) but that (unlike the Earth-pointing case) the vehicle is spinning about this principal axis (i.e., about the orbit normal) at an arbitrary rate $\dot{\alpha}_2$. As may be intuitively obvious, such a motion satisfies the motion equations. That is, the conditions

$$\boldsymbol{\omega}_\alpha = \dot{\alpha}_2\mathbf{1}_2; \qquad \boldsymbol{\omega} = (\dot{\alpha}_2 - \omega_c)\mathbf{1}_2$$

$$\mathbf{c}_2 \equiv \mathbf{1}_2; \qquad \mathbf{c}_3 \equiv [-\sin\alpha_2 \quad 0 \quad \cos\alpha_2]^T \tag{84}$$

do in fact satisfy (81) through (83), provided

$$I_2\ddot{\alpha}_2 + 3\omega_c^2(I_1 - I_3)\sin\alpha_2\cos\alpha_2 = 0 \tag{85}$$

This fact was mentioned earlier in the study of pitch motion in a circular orbit; see (9.2, 30) et seq. In particular, the pitch motion equation (85) was solved in terms of Jacobian elliptic functions. Figures 9.6 through 9.8 should be reviewed at this point. However, although the solution to (85) was obtained for arbitrary $\alpha_2(0)$ and $\dot{\alpha}_2(0)$, as recorded in (9.2, 35), the central interest in Chapter 9 was in librations about an Earth-pointing equilibrium, not (as now) in spinning satellites.

In a narrow sense, the solution (84) may be called a "relative equilibrium." It is not an equilibrium in the sense that no torque is acting, or in the sense that the satellite does not rotate. It is not even a relative equilibrium in the sense that motion with respect to the orbiting frame $\mathscr{F}_c$ is *uniform* (as are, for example, the Thompson and Likins–Pringle equilibria discussed on the preceding pages), because $\dot{\alpha}_2$ is not constant. Notwithstanding these observations, we shall nevertheless call (84) a relative equilibrium in the sense that *the spin axis is fixed in* $\mathscr{F}_c$ (and in $\mathscr{F}_i$). [Sperling] has shown that this relative equilibrium and the earlier ones of Thompson and Likins and Pringle for a symmetrical satellite constitute the only possible ones for a satellite in a circular orbit.

Stability by Floquet Theory

To ascertain the inertia distributions and spin rates for which this spin direction is stable, let

$$\mathbf{C}_{pc} = \mathbf{C}_1(\alpha_1)\mathbf{C}_3(\alpha_3)\mathbf{C}_2(\alpha_2) \tag{86}$$

using the principal rotations of (2.1, 28), and let α_1 and α_3 be first-order infinitesimals. The spin angle α_2 is not assumed to be small; indeed, it can be arbitrarily large as the spin evolves according to (85). The angular velocity of the satellite with respect to the orbiting frame is calculated from $\boldsymbol{\omega}_\alpha^\times = -\dot{\mathbf{C}}_{pc}\mathbf{C}_{pc}^T$ to be

$$\begin{aligned}\omega_{\alpha 1} &= \dot{\alpha}_1 + \dot{\alpha}_2\alpha_3\\ \omega_{\alpha 2} &= \dot{\alpha}_2\\ \omega_{\alpha 3} &= \dot{\alpha}_3 - \dot{\alpha}_2\alpha_1\end{aligned} \tag{87}$$

correct to first order in $\{\alpha_1, \alpha_3\}$. Furthermore,

$$\mathbf{c}_2 = \begin{bmatrix}\alpha_3\\ 1\\ -\alpha_1\end{bmatrix}; \qquad \mathbf{c}_3 = \begin{bmatrix}-\sin\alpha_2\\ \alpha_3\sin\alpha_2 + \alpha_1\cos\alpha_2\\ \cos\alpha_2\end{bmatrix} \tag{88}$$

whereupon the linearized motion equations for α_1 and α_3 become, from (81),

$$\begin{aligned}&\ddot{\alpha}_1 + [(1-k_1)\dot{\theta}_2]\dot{\alpha}_3 + k_1(\dot{\theta}_2^2 + 3\omega_c^2\cos^2\alpha_2)\alpha_1\\ &\qquad + [3\omega_c^2(k_1-k_2)\sin\alpha_2\cos\alpha_2]\alpha_3 = 0\\ &\ddot{\alpha}_3 - [(1-k_3)\dot{\theta}_2]\dot{\alpha}_1 + [3\omega_c^2(k_2+k_3)\sin\alpha_2\cos\alpha_2]\alpha_1\\ &\qquad + k_3(\dot{\theta}_2^2 + 3\omega_c^2\sin^2\alpha_2)\alpha_3 = 0\end{aligned} \tag{89}$$

where the inertia ratios $\{k_1, k_2, k_3\}$ have their usual meaning [see (4.4, 20) and (9.2, 33)]. The "absolute pitch rate" has been defined by

$$\dot{\theta}_2 \triangleq \dot{\alpha}_2 - \omega_c \tag{90}$$

The third motion equation (for α_2) is the pitch equation (85), rewritten as follows:

$$\ddot{\alpha}_2 + 3\omega_c^2 k_2\sin\alpha_2\cos\alpha_2 = 0 \tag{91}$$

In fact, (91) has been used in obtaining (89).

It has been recorded in (9.2, 35) that for spinning pitch motion,

$$\sin\alpha_2 = sn(at; \hat{a}^{-1}) \tag{92}$$

The elliptic function parameter $\hat{a}$ is defined by

$$\hat{a} \triangleq \frac{a}{(3k_2)^{1/2}\omega_c}; \qquad a \triangleq |\dot{\alpha}_2(0)| \tag{93}$$

assuming $\alpha_2(0) = 0$. Similarly,

$$\cos\alpha_2 = cn(at; \hat{a}^{-1}) \tag{94}$$

Thus, wherever $\sin\alpha_2$ and $\cos\alpha_2$ occur in (89), they are replaced by $sn(at)$ and $cn(at)$. Furthermore, from (9.2, 34),

$$\frac{\dot{\theta}_2}{\omega_c} = \pm\left[3k_2(\hat{a}^2 - sn^2 at)\right]^{1/2} - 1 \tag{95}$$

(The plus sign is used for spins opposite in sense to the orbit normal, that is, to the $+\underset{\rightarrow}{\mathbf{o}}_2$ direction, and the minus sign is used for spins in the same sense as the orbit normal.)

With relations (92), (94), and (95) in hand, all the coefficients in the motion equations (89) have been expressed in terms of $sn(at)$ and $cn(at)$. In particular, all coefficients are periodic with period equal to the period of the Jacobian elliptic functions:

$$T = \frac{4K(\hat{a}^{-1})}{a} \tag{96}$$

Therefore, according to Floquet theory (Section A.3), once the differential equations (89) have been integrated (numerically) over $0 \le t \le T$ for the four linearly independent sets of initial conditions

$$\begin{bmatrix} \alpha_1 \\ \alpha_3 \\ \dot{\alpha}_1 \\ \dot{\alpha}_3 \end{bmatrix} = \begin{bmatrix} 1 \\ 0 \\ 0 \\ 0 \end{bmatrix}, \begin{bmatrix} 0 \\ 1 \\ 0 \\ 0 \end{bmatrix}, \begin{bmatrix} 0 \\ 0 \\ 1 \\ 0 \end{bmatrix}, \begin{bmatrix} 0 \\ 0 \\ 0 \\ 1 \end{bmatrix} \tag{97}$$

all the information needed for a determination of infinitesimal stability is available (Problem 10.4).

The results of such a Floquet theory study have been reported by [Kane and Shippy]. Their results are exhibited in Fig. 10.6. The problem has three parameters: k_1, k_3, and a measure of spin rate. [The parameter k_2 is determined by k_1 and k_3; see (9.2, 33).] Kane and Shippy chose the average relative spin rate, made dimensionless by ω_c, as the measure of spin rate:

$$s \triangleq \frac{\bar{\dot{\alpha}}_2}{\omega_c} \equiv (\pm)\frac{2\pi}{\omega_c T} = \frac{\pi\dot{\alpha}_2(0)}{2\omega_c K(\hat{a}^{-1})} \tag{98}$$

Thus, infinitesimal stability is characterized by the three parameters $\{k_1, k_3, s\}$. Figure 10.6 shows three k_1–k_3 diagrams, for $s = 1, -1, -5$. The shaded regions were found to be unstable by examining the stability at a fine grid of points within each region.

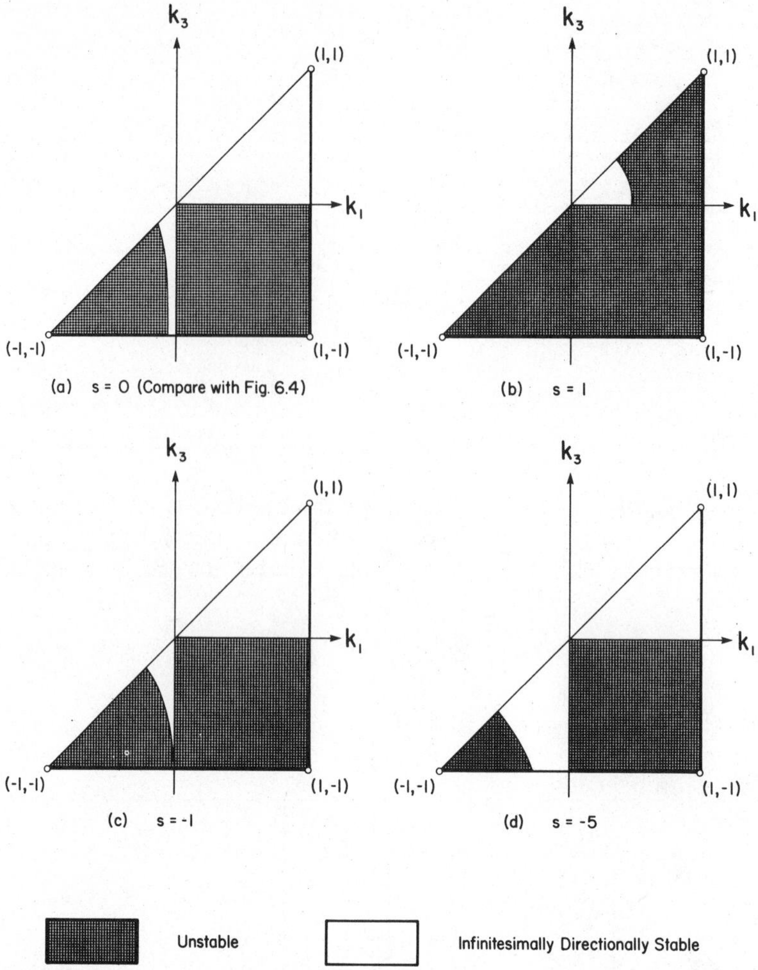

FIGURE 10.6 Stability diagrams for a tri-inertial satellite in a circular orbit, spinning about the orbit normal. Based on [Kane and Shippy].

It is instructive to compare the stability diagrams in Fig. 10.6 with other diagrams obtained earlier for dynamical situations to which the present one reduces in special cases. First, consider the symmetrical case $k_1 = k_3 = k_t$, $k_2 = 0$. The satellite is axially symmetrical, and its spin axis is normal to the orbit. Thus, we have Thomson's equilibrium for a symmetrical spinner, and $\hat{a} = \infty$, $K(\hat{a}^{-1}) = K(0) = \pi/2$. The relative spin parameter s equals $1 + \hat{\nu}$, and $\alpha_2 = (\nu + \omega_c)t$ and $\dot{\theta}_2 = \nu$. Under these circumstances, the motion equations (89) clearly reduce to (11) (which however can be transformed to (14) to eliminate the periodic coefficients), as they should. From Fig. 10.2 it is evident that, generally speaking, $\hat{\nu}$ must be negative for Liapunov stability, and this is borne out by Fig. 10.6 for the nonaxisymmetrical generalization. A second limiting case, also treated earlier, corresponds to infinitesimal α_2. Thus, we have simple librations about an Earth-pointing equilibrium. Then, $s = 0$, $\dot{\theta}_2 = -\omega_c$, and equations (89) reduce to (9.2, 15) for gravity gradient librations. The stability diagram for this case, Fig. 9.4, is consistent with Fig. 10.6*a*.

One interesting aspect to the comparison of Fig. 10.6 with Fig. 9.4 (gravity gradient librations) and with Fig. 10.2 (Thompson symmetrical spin librations) is that the important distinction between static and gyric stability appears to be lost. One can of course make educated guesses, but an ancillary criterion similar to "static stability" to accompany Floquet theory for dynamical systems would seem to be a fruitful area for further research.

[Meirovitch and Wallace] have shown that when the satellite is *almost* symmetrical, further analytical progress is possible by treating k_2 as a small parameter and expanding the motion as a power series in k_2. (This is a case of practical interest since many spin-stabilized satellites are of this type.) By dropping terms of $O(k_2^2)$, approximate stability boundaries and internal resonance lines can be found in closed form for $|k_2| \ll 1$.

Axisymmetrical Satellite Spinning in an Elliptic Orbit

When the orbit is not circular but elliptic, a new potential source for instabilities is present. The Likins–Pringle equilibria vanish altogether, although the Thomson equilibrium—spin axis normal to the orbit—remains. The question of interest is this: How is the stability diagram of Fig. 10.2 altered when a third parameter is introduced—the orbit eccentricity e? We shall not derive the modified motion equations here in detail, although all the elements required for this analysis have been given earlier in this chapter. Suffice it to say that ω_c is no longer constant (the absolute pitch rate $\dot{\theta}_2 = \dot{\alpha}_2 - \dot{\eta}$, where $\eta(t)$ is the true anomaly) and that the gravity gradient torque expression is, in $\mathscr{F}_p$,

$$\mathbf{g} = 3\left(\frac{\mu}{R^3}\right)\mathbf{c}_3^{\times}\mathbf{I}\mathbf{c}_3 \tag{99}$$

where $R = R(t)$. For an axisymmetrical satellite, (99) reduces to

$$\mathbf{g} = 3\left(\frac{\mu}{R^3}\right)\begin{bmatrix}(I_t - I_a)c_{23}c_{33}\\ 0\\ (I_a - I_t)c_{13}c_{23}\end{bmatrix} \tag{100}$$

The components of $\mathbf{g}$ in the stroboscopic frame $\mathscr{F}_s$ have the same form as in (100) but with $c_{i3} \to c_{i3}^s$, the direction cosines between $\mathscr{F}_s$ and the local vertical $\hat{\mathbf{o}}_3$. The linearized equations obtained earlier for a circular orbit, namely, (14), are then modified to incorporate $e \neq 0$, and the constant coefficients become periodic with period equal to the orbital period. Thus, Floquet theory (Section A.3) is in order. Figure 10.7*a* shows results obtained by [Kane and Barba] in 1965. These are superposed on Fig. 10.2*b* which corresponds to $e = 0$. They applied Floquet theory to a limited number of discrete points in the $\{k_t, \hat{\nu}\}$ plane. (Computing

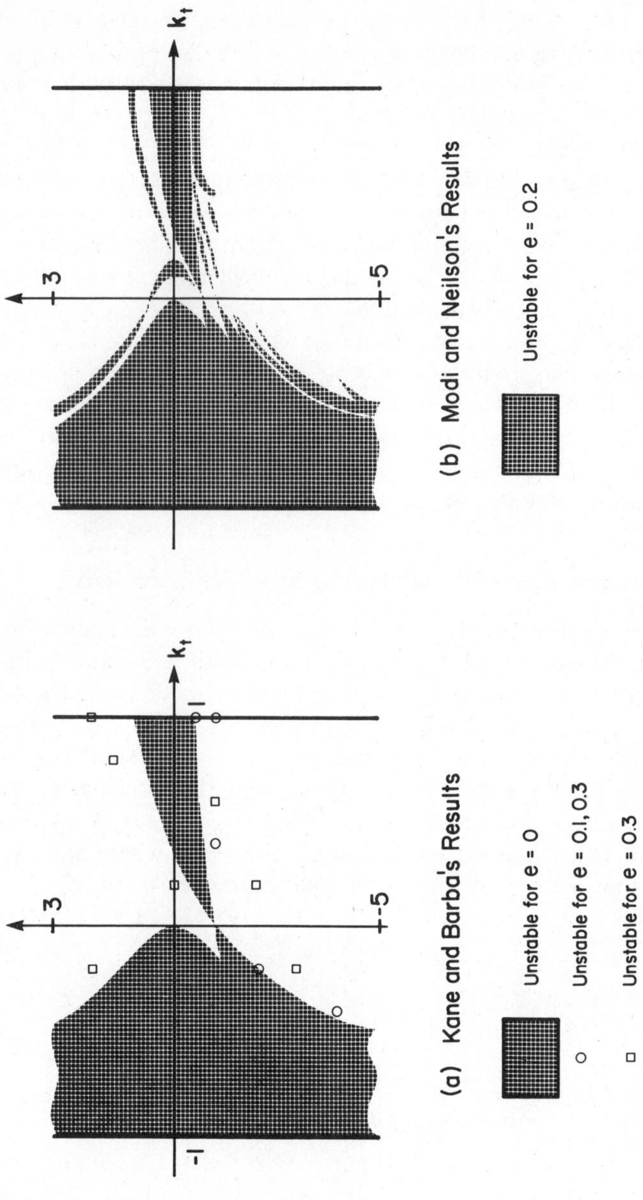

FIGURE 10.7 Effect of small orbital eccentricity on stability of an axisymmetrical rigid satellite spinning about the orbit normal. Based on [Kane and Barba] and [Modi and Neilson, 1].

was expensive in 1965.) Points indicated by small circles were found to be unstable at $e = 0.1$; these points, and also the points indicated by small squares, were unstable at $e = 0.2$. Many additional points were calculated in 1969 by [Modi and Neilson, 1], leading in effect to continuous stability boundaries; see Fig. 10.7*b* when $e = 0.2$.

Purely analytical results for elliptic orbits are not easily come by, even under "small wobble" assumptions. Wallace and Meirovitch show how to expand the motion in a power series in e, but their stability boundaries, although in closed form, are only approximate and do not disclose the intricacy evident in Fig. 10.7*b*. Different insights are available from the work of [Hitzl, 1] who used higher-order terms in the system Hamiltonian to show that *external nonlinear resonances* can occur when

$$\omega_o = 2\omega_{ry1}$$

$$\omega_o = \omega_{ry2} \pm \omega_{ry1} \qquad (101)$$

where ω_o is the mean orbital rate, $\omega_o = (\mu/a^3)^{1/2}$; $2a$ is the major diameter of the orbit; and ω_{ry1} and ω_{ry2} are the two frequencies associated with the "roll/yaw" wobbling of the spin axis about the Thomson equilibrium. In other words, ω_{ry1} and ω_{ry2} are the positive roots of

$$\left(\frac{\omega}{\omega_o}\right)^4 - b_1\left(\frac{\omega}{\omega_o}\right)^2 + b_2 = 0 \qquad (102)$$

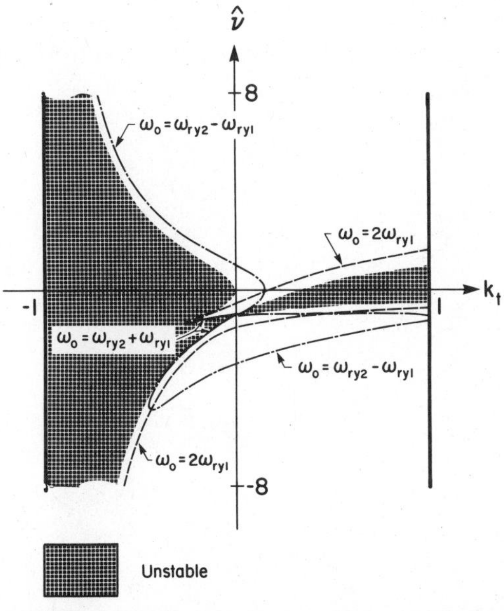

FIGURE 10.8 Nonlinear external resonance lines for a tri-inertial satellite in a slightly elliptic orbit, spinning about the orbit normal. After [Hitzl, 1].

where b_1 and b_2 are defined by (24). Conditions (101) are plotted in Fig. 10.8. It is evident from comparing Figs. 10.7 and 10.8 that the instabilities discovered numerically by [Kane and Barba] and [Modi and Neilson, 1] using Floquet theory do lie in the neighborhood of Hitzl's external resonance curves. In examining these nonlinear resonances, [Hitzl, 1] has done for the Thomson equilibrium of a symmetrical spinner in a slightly elliptic orbit what [Breakwell and Pringle] did for the pitch excitation of the roll/yaw motion of a tri-inertial Earth-pointing satellite (cf. Fig. 9.12)—the Floquet-derived instabilities are explained.

For the added complication of orbit nodal regression, see [Dvornychenko and Gerding].

Tri-Inertial Satellite Spinning in an Elliptic Orbit

If the satellite inertias are distinct, further mathematical difficulty ensues and closed-form results become even more rare. Various analytical approximations are required that make uncertain a precise interpretation of the results. For most purposes one must rely on computational simulation of the nonlinear equations of motion [e.g., (9.2, 1 through 7), adapted for elliptic orbits].

Nevertheless, some analytical progress has been made. [Chernous'ko] has applied the method of averaging to two situations. (This method requires the definition of a small parameter ε.) In one of these, ε represents the small deviation of the inertia distribution from spherical:

$$I_i = I_o + \varepsilon \Delta I_i \qquad (i = 1, 2, 3) \tag{103}$$

Chernous'ko shows that the motion is essentially of the Euler–Poinsot type (see Figs. 4.8 through 4.16) with a change in time scale, in effect. In the second situation, the method of averaging was applied to the "high-spin" case in which

$$\varepsilon \triangleq \frac{I_{\max}\omega_o}{\|\underset{\rightarrow}{\mathbf{h}}\|} \tag{104}$$

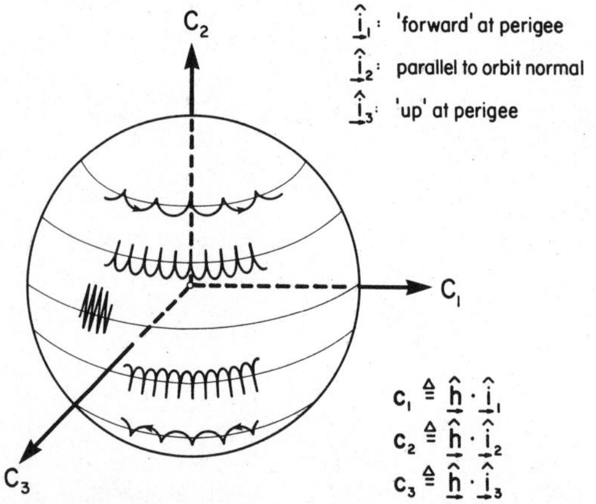

FIGURE 10.9 Gravity gradient effect on the spin vector of a rapidly spinning rigid body in a circular orbit. After [Chernous'ko].

Thus, ε is a measure of the importance of gravitational torque relative to the intrinsic gyric behavior of a spinning body. As $\varepsilon \to 0$, the motion must revert to Euler–Poinsot motion; for small ε, small deviations from this classical, torque-free motion can be calculated. Figure 10.9 shows the motion of the angular momentum vector $\underset{\rightarrow}{\mathbf{h}}$ with respect to an inertial frame $\mathscr{F}_i$. Evidently, $\underset{\rightarrow}{\mathbf{h}}$ tends to precess about the orbit normal, but there are in addition nutational wobbles of increasing severity as one selects spin directions increasingly different from the orbit normal.

[Cochran] has shown how to use Lie series to develop an approximate theory for a tri-inertial spinning satellite in an orbit that is not only elliptic but also evolving; both precession of the orbital plane due to nodal regression and rotation of the line of apsides are accommodated within his formulation. In spite of the complexity of his analysis, Cochran was able to show excellent agreement with Pegasus A flight data over a 20-day period.

10.2 DESIGN OF SPIN-STABILIZED SATELLITES

It must immediately be admitted that much of the preceding analysis (Section 10.1) is "theoretical" in the sense that many sources of torque have been neglected, even though for particular satellites one or more of the nongravitational torques may have a significant influence, particularly over the long term. Even more important, internal energy dissipation has been ignored in the foregoing discussion in spite of its pivotal implications for passive stabilization. However, the general case eludes closed-form analysis, and it is precisely such basic theory that must be relied upon to suggest possibilities and provide insights (see remarks in Chapter 1).

In this subsection, an overview is first given of the conclusions on the feasibility of spin stabilization that might reasonably be made on the basis of the various theoretical equilibria (and their stability) discussed in Section 10.1. The influence on stability of energy dissipation is then addressed, first heuristically via a "quasi-rigid body" analysis, and then more rigorously using nutation angle dampers. The long-term implications of environmental torques will be discussed in Section 10.3.

Possibilities for Spin Stabilization

The oft-quoted maxim, "Nature abhors a vacuum" seems not more true than the maxim "Nature abhors an equilibrium." Even under rather severe assumptions, very few equilibria for spinning satellites were uncovered in Section 10.1. The most robust is surely the Thomson equilibrium, in which the spacecraft spins about a principal axis aligned with the orbit normal. As first analyzed by [Thomson, 2], this equilibrium is associated with an inertially axisymmetrical rigid body in a circular orbit and is still possible in an elliptic orbit. Even if the satellite is not axisymmetrical, a principal-axis spin normal to the orbit is still an equilibrium as far as the spin axis is concerned (the spin *rate* undergoes some modulation from gravity gradient pitch torque). The basic stability diagram for the Thomson equilibrium of a rigid satellite was shown in Fig. 10.2. Typical modifications to this stability diagram to account for asymmetry and orbit eccentricity were presented in Figs. 10.6 and 10.7, respectively. The consequences of orbit evolution and nongravitational torques also tend to be adverse to stability. Nevertheless, spin about an axis normal to the orbit is a strong candidate for passive attitude stabilization.

By contrast, the Likins–Pringle equilibria are fragile. They are known to exist for axisymmetrical spinners in circular orbits subject to gravitational torque. But it is evident from the stability diagram in Fig. 10.4*a* that in the "hyperbolic" case the satellite has virtually no spin with respect to the orbiting frame ($\hat{\nu} + 1 \doteq 0$). Indeed, the hyperbolic case has the character more of an odd form of gravity gradient stabilization than of true "spin" stabilization. It is also not clear why anyone would wish to stabilize a satellite axis on a hyperboloid (Fig. 10.3*b*). In the conical case (Fig. 10.3*c*) on the other hand, the spin axis tends to remain aimed at an inertially fixed point, and this is of greater practical interest. Yet the stability diagram for conical equilibria (Fig. 10.4*b*) is again discouraging because high spin rates are possible only as $k_t \to -1$, that is, as $I_a \to 0$. Even a rapidly spinning symmetrical body has only a little angular momentum if it is pencil-shaped, and so we must conclude that the principal objective in spin stabilization —to obtain gyric stiffness—is not met in the conical case either. In both the hyperbolic and conical cases, then, we do not really have "true" spin stabilization but a delicate balance between a small gyric torque and a small gravity gradient torque. Moreover, these equilibria cease to exist if the satellite is asymmetrical or if the orbit is eccentric. And experience with attempts to use the "full" gravity gradient torque (Section 9.4), unweakened by gyric countertorque, does not make one sanguine about the prospects that this form of stabilization will be able to withstand environmental disturbance torques. It must therefore be concluded that the Likins–Pringle equilibria, in spite of their theoretical interest, are not altogether promising as practical spin stabilization techniques.

This leaves two strategies for spin stabilization: (i) the Thomson equilibrium which permits spins of arbitrary magnitude; and (ii) the "high-spin" strategy, in which the satellite spin vector $\underset{\rightarrow}{\mathbf{h}}$ is pointed in any direction of interest and is made so large that the relatively weak environmental torques take a long time to alter $\underset{\rightarrow}{\mathbf{h}}$ significantly. In both cases, the long-term effects of environmental torques and orbital evolution are encountered. These effects must either be tolerated or be compensated for by occasional active trimming of the angular-momentum vector.

Spinning Quasi-Rigid Satellite in Thomson Equilibrium

The heuristic notion of a quasi-rigid body was introduced in Chapter 5. Changes in the satellite's absolute angular-velocity vector $\underset{\rightarrow}{\boldsymbol{\omega}}$ excite internal relative motion, thereby producing dissipation. When $\underset{\rightarrow}{\boldsymbol{\omega}}$ ceases to change, the energy "sink" ceases to remove energy. It is not difficult to show that this condition is satisfied by the Thomson equilibrium for an inertially axisymmetrical spinning satellite in a circular orbit. Therefore, the associated Hamiltonian-based Liapunov function given by (10.1, 70), namely,

$$
\begin{aligned}
\upsilon = {} & \tfrac{1}{2} I_t \left(\omega_{\gamma 1}^2 + \omega_{\gamma 3}^2 \right) - I_a \omega_c \nu \left(1 - c_{22}^s\right) \\
& + \tfrac{1}{2} \left[(3 I_a - 4 I_t) \left(c_{23}^s\right)^2 - I_t \left(c_{21}^s\right)^2 \right] \omega_c^2
\end{aligned} \tag{1}
$$

is not only zero in the Thomson equilibrium but will tend to be reduced for perturbations about that equilibrium by the action of an energy sink. The question is, When is υ positive-definite? According to (10.1, 72), υ is positive-definite when $\mathscr{k}_1 > 0$, $\mathscr{k}_3 > 0$, with $\mathscr{k}_1$ and $\mathscr{k}_3$ defined by (10.1, 25). It can be concluded, at least within the strictures of energy sink analysis, that the region in Fig. 10.2 labeled "directionally and statically stable" is in fact asymptotically directionally stable when energy dissipation is taken into account. The region labeled "directionally stable; statically unstable but gyrically stabilized"—what might be referred to as the "DeBra–Delp stability region for the Thomson equilibrium"—cannot be investigated via this Liapunov function.

A more explicit treatment of the influence of damping on the stability of the Thomson equilibrium is obtained by adding damping terms to the linearized motion equations (10.1, 14) for γ_1 and γ_3, the small deviations from the reference spinning motion. With this done, the equations of motion of $\{\gamma_1, \gamma_3\}$ are of the form

$$
\mathscr{M} \ddot{\mathbf{q}} + (\mathscr{D} + \mathscr{G}) \dot{\mathbf{q}} + \mathscr{K} \mathbf{q} = \mathbf{0} \tag{2}
$$

where

$$
\mathbf{q} \triangleq \begin{bmatrix} \gamma_1 \\ \gamma_3 \end{bmatrix}; \qquad \mathscr{G} \triangleq \omega_c \begin{bmatrix} 0 & -\mathscr{g} \\ \mathscr{g} & 0 \end{bmatrix}; \qquad \mathscr{K} \triangleq \omega_c^2 \begin{bmatrix} \mathscr{k}_1 & 0 \\ 0 & \mathscr{k}_3 \end{bmatrix}
$$

$$
\begin{aligned}
\mathscr{g} &\triangleq k_t \hat{\nu} + \hat{\nu} + 2 \\
\mathscr{k}_1 &\triangleq k_t (3 - \hat{\nu}) - \hat{\nu} - 1 \\
\mathscr{k}_3 &\triangleq -k_t \hat{\nu} - \hat{\nu} - 1
\end{aligned}
$$

and $\mathscr{M} = \mathbf{1}$. The damping matrix $\mathscr{D}$ does not have to be specified beyond the stipulation that $\mathscr{D}^T = \mathscr{D} > 0$. Adding a damping matrix in this fashion was successful in Chapter 5 in connection with a spinning body in torque-free motion [see, for example, (5.1, 12) et seq.]. The same technique was also employed in the study of the librations of an Earth-pointing satellite [see, for example, (9.3, 8) et seq.]. [Likins and Columbus] have pointed out that such damping terms could arise physically from eddy currents in the conducting spacecraft structure rotating in Earth's magnetic field. (Such eddy currents could in principle also cause a very slow decay in the spin rate itself; see, for example, [Wilson], [Vigneron, 3], and [Dranovskii and Yanshin].) System (2) is of precisely the form addressed by the Kelvin–Tait–Chetayev theorem (Appendix A), which asserts that $\mathscr{K} > 0$ is required for stability. (Because $\mathscr{D} > 0$, the stability is in fact asymptotic.) Since $\mathscr{K} > 0$ implies that $\mathscr{k}_1 > 0$, $\mathscr{k}_3 > 0$, the region of asymptotic stability in the $\{k_t, \nu\}$ plane is precisely the region "directionally and statically stable" in Fig.

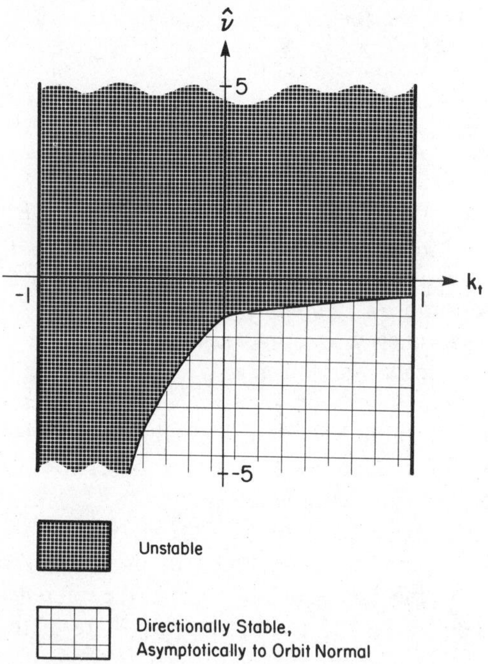

FIGURE 10.10 Stability diagram for the Thomson equilibrium with energy dissipation included. (Compare with Fig. 10.2.)

10.2. Therefore, it can be gleaned from both an energy sink analysis and a "damping matrix" treatment that, in reality, the stability diagram of Fig. 10.2, which was drawn without admission of energy dissipation, should be as shown in Fig. 10.10.

As a practical matter, one can show that if the spin rate is very high compared to the orbital rate (i.e., if $|\hat{\nu}| > 1$), the rate roll/yaw *decay* is very slow for $\hat{\nu} < 0$, and the rate of roll/yaw *growth* is very slow for $\hat{\nu} > 0$. (The limiting case $|\hat{\nu}| \gg 1$ is explored in Problem 10.5.) Therefore, for sufficiently high spin rates, the slight instability demonstrated for $\hat{\nu} \gg 1$ may be no more bothersome than other destabilizing effects (environmental torques, orbital evolution) not considered in the present analysis. If the latter are compensated for by occasional active intervention, the slight additional wobble from "spin instability" can be removed as part of the same operational corrections.

For the "high-spin" strategy, in which the satellite is spun about its major axis at a relatively rapid rate about an arbitrary inertial direction, the approach is similar. The motion is known to be asymptotically stable in the absence of environmental torques (Chapter 5). Environmental torques are generally destabilizing over the long term because, although $\underrightarrow{\boldsymbol{\omega}}$ remains essentially parallel to $\underrightarrow{\mathbf{h}}$, the direction of $\underrightarrow{\mathbf{h}}$ itself wanders slowly over the celestial sphere. For some applications, this evolution of $\underrightarrow{\mathbf{h}}$ may be acceptable; for others, periodic corrections to $\underrightarrow{\mathbf{h}}$ are required through occasional ground commands.

Rigid Body with Point Mass Damper, in Circular Orbit

The preceding two analyses (energy sink and subjoined damping matrix) have the virtue of simplicity but are not especially rigorous. (Exception: eddy current torques can result in a system of the form (2), as discussed above.) It is desirable to confirm these predictions, at least qualitatively, through the rigorous dynamical analysis of a specific energy dissipation mechanism. In this book, the point mass damper has repeatedly served this purpose. Thus, in Chapter 5, Section 5.2, the inference drawn for the (directional) stability of torque-free motion of a quasi-rigid spinning body $\mathscr{Q}$ (the major-axis rule) is verified through the analysis of $\mathscr{R} + \mathscr{P}$, a spinning rigid body containing a point mass damper (Fig. 5.4). It is found that the exact stability conditions, (5.2, 25), are more severe than indicated by the major-axis rule, although the exact conditions reduce to the major-axis rule in the limit as the dynamical significance of the damper, as measured by the dimensionless parameter Ξ_d, becomes vanishingly small (compare Figs. 5.3 and 5.5). In a similar fashion, the (directional) stability conditions for the torque-free motion of a quasi-rigid gyrostat $\mathscr{Q} + \mathscr{W}$ (Fig. 7.3), stated in (7.1, 37 and 38) and illustrated in Fig. 7.4, are substantiated by consideration of $\mathscr{R} + \mathscr{P} + \mathscr{W}$, a rigid gyrostat containing a point mass damper (Fig. 7.12). Again, it is found that the exact stability conditions, (7.2, 22), tend to be more severe than those predicted by energy sink arguments (compare Fig. 7.13 with Fig. 7.4) although, as argued in Problem 7.7, the exact stability conditions reduce to the energy sink stability conditions as $\Xi_d \to 0$. As the final example of this theme from earlier pages, consider a rigid body and point mass damper ($\mathscr{R} + \mathscr{P}$) librating about the local vertical in a circular orbit (Fig. 9.17). The exact conditions for three-axis stability given by (9.3, 26), namely, $k_1 > k_3 > \Xi_{d,\text{eff}}$, are again sharper than $k_1 > k_3 > 0$ as deduced for the librations of a quasi-rigid body $\mathscr{Q}$; and again the conditions for $\mathscr{R} + \mathscr{P}$ reduce to those for $\mathscr{Q}$ as $\Xi_{d,\text{eff}} \to 0$.

Based on these observations, one might expect a similar trend for an energy dissipating satellite spinning in the Thomson equilibrium. For vanishingly small mechanical dampers, the stability diagram in Fig. 10.10 should apply, with "flexible body" effects increasing in importance as the damper grows in size.

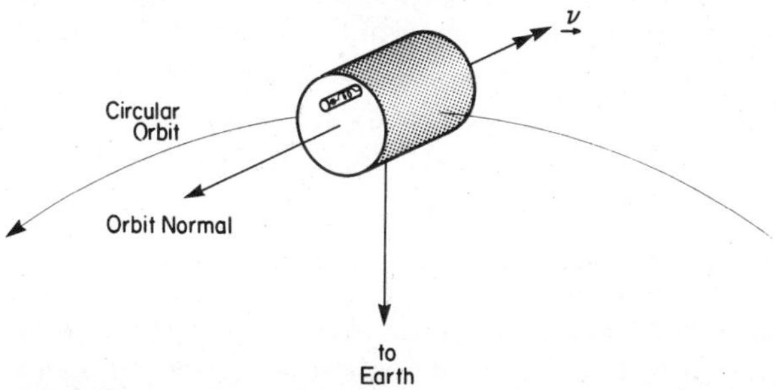

FIGURE 10.11 Inertially axisymmetrical rigid satellite, containing a point mass damper, in a circular orbit.

The physical system now to be considered is shown in Fig. 10.11: an inertially axisymmetrical rigid body $\mathscr{R}$ containing a point mass damper $\mathscr{P}$. The satellite $\mathscr{R} + \mathscr{P}$ is in a circular orbit and spins about its $\hat{\mathbf{p}}_2$ axis, the axis of inertial symmetry. The latter is nominally parallel to the orbital normal. Thus, we have the Thomson equilibrium. The equations of motion are those for the torque-free motion of $\mathscr{R} + \mathscr{P}$, as given by (5.2, 11 and 12), amended to include the gravity gradient torque on $\mathscr{R}$, the gravity gradient force on $\mathscr{P}$, and with $I_1 = I_3 = I_t$, $I_2 = I_a$. The resulting motion equations are

$$\ddot{\alpha}_2 = 0$$

$$\ddot{\alpha}_1 + (1 - k_t)\nu\dot{\alpha}_3 + k_t\nu^2\alpha_1 + 3\omega_c^2 k_t(\alpha_1 c + \alpha_3 s)c = 0$$

$$\ddot{\alpha}_3 - (1 - k_t)\nu\dot{\alpha}_1 + k_t\nu^2\alpha_3 + 3\omega_c^2 k_t(\alpha_1 c + \alpha_3 s)s + \frac{m_d b}{I_t}(\ddot{\xi} + \nu^2\xi) = 0$$

$$\frac{m_d m_b}{m}\ddot{\xi} + c_d\dot{\xi} + k_d\xi + m_d b(\ddot{\alpha}_3 + \nu^2\alpha_3) + m_d\omega_c^2\xi = 0 \quad (3)$$

where $s = \sin(\nu + \omega_c)t$, $c = \cos(\nu + \omega_c)t$, and $\{\alpha_1, \alpha_2, \alpha_3\}$ are the small librations of the body-fixed frame $\mathscr{F}_2$ that has an absolute spin ν about the $\vec{\mathbf{o}}_2$ axis. The small motion of the damper relative to $\mathscr{F}_p$ is $\xi(t)$. The motion equations of course reduce to (10.1, 11), the motion equations for $\mathscr{R}$ alone, when the damper becomes vanishingly small.

The periodic coefficients in the α_1 and α_3 motion equations present analytical difficulties. For $\mathscr{R}$ alone, these difficulties were removed by the transformation (10.1, 13), in which the new attitude variables $\boldsymbol{\gamma}$ represented deviations of the stroboscopic frame $\mathscr{F}_s$ from the orbital frame $\mathscr{F}_c$. This transformation successfully removed the periodic coefficients because $\mathscr{R}$ had, on account of symmetry, the same configuration in $\mathscr{F}_s$ as in $\mathscr{F}_p$. Unfortunately, this is no longer true for the $\mathscr{R} + \mathscr{P}$ satellite; the periodic coefficients are unavoidable. One possibility is to transform from $\boldsymbol{\alpha}$ to $\boldsymbol{\gamma}$ anyway and arrive at a set of equations that are like (10.1, 14), except for additional terms in ξ that have periodic coefficients. (There is, as well, an additional equation of motion for ξ.) For small m_d, the sinusoidal solutions for $m_d = 0$ can then be perturbed to first order in m_d to obtain insight into the limiting case $m_d \to 0$. This procedure would however be somewhat tedious; in an age of great "computer power," it may be as well to apply Floquet stability theory (Section A.3 of Appendix A) and perform the necessary numerical integrations. Unfortunately, this does not appear to have been done in the literature (to the author's knowledge). The reader is invited to fill this void in Problem 10.6.

An Example with Constant Coefficients

[Kane and Shippy, 2] have described a different system, one that avoids periodic coefficients and the consequent need for the numerical integrations called for by Floquet theory. Shown in Fig. 10.12 is one physical implementation of their model. Inside a hollow spinning cylinder $\mathscr{W}_b$ is a second, small cylinder $\mathscr{W}_d$

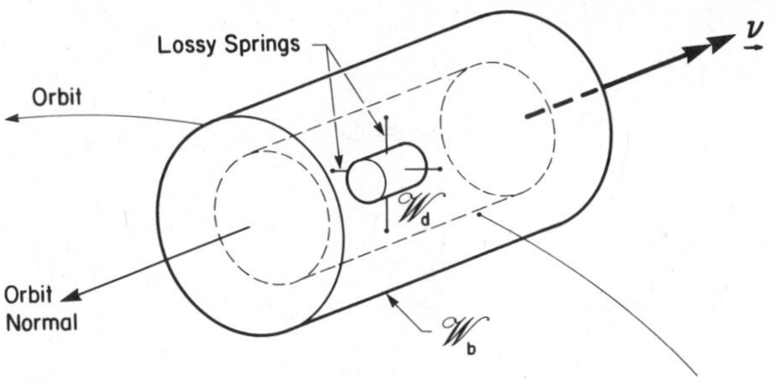

FIGURE 10.12 The system $\mathscr{W}_b + \mathscr{W}_d$ in a Thomson equilibrium.

attached to the outer cylinder by mechanisms (e.g., suitable wires) that provide both stiffness and damping. The two cylinders spin at the same rate, and their mass centers coincide. In the reference equilibrium, their symmetry axes also coincide. However, small attitude deviations $\{\alpha_1, \alpha_2, \alpha_3\}$ for the outer cylinder induce small attitude deviations $\{\alpha_1 + \beta_1, \alpha_2, \alpha_3 + \beta_3\}$ for the inner cylinder $\mathscr{W}_d$. Thus, β_1 and β_3 denote small attitude motions of $\mathscr{W}_d$ with respect to $\mathscr{W}_b$. The axial and transverse inertias of $\mathscr{W}_b$ and $\mathscr{W}_d$ are respectively I_{ab}, I_{tb}, I_{ad}, and I_{td}.

The relative motions $\beta_1(t)$ and $\beta_3(t)$ are counteracted by a stiffness and damping torque on $\mathscr{W}_d$:

$$\mathbf{g}_d = \mathscr{F}_p^T \begin{bmatrix} -k_d\beta_1 - c_d\dot{\beta}_1 \\ I_{ad}\ddot{\alpha}_2 \\ -k_d\beta_3 - c_d\dot{\beta}_3 \end{bmatrix} \tag{4}$$

Note that the spring constant k_d and the damping constant c_d are identical about the two transverse axes; otherwise, periodic coefficients cannot be avoided. The equations governing the small motion of the satellite $\mathscr{W}_b + \mathscr{W}_d$ about its Thomson equilibrium are found by adapting (10.1, 4) first to $\mathscr{W}_b$ and then to $\mathscr{W}_d$:

$$\begin{aligned} &I_{tb}\ddot{\alpha}_1 + (2I_{tb} - I_{ab})\nu\dot{\alpha}_3 + (I_{ab} - I_{tb})\nu^2\alpha_1 \\ &\qquad + 3\omega_c^2(I_{ab} - I_{tb})(\alpha_1 c + \alpha_3 s)c - c_d\dot{\beta}_1 - k_d\beta_1 = 0 \end{aligned} \tag{5}$$

$$\begin{aligned} &I_{tb}\ddot{\alpha}_3 - (2I_{tb} - I_{ab})\nu\dot{\alpha}_1 + (I_{ab} - I_{tb})\nu^2\alpha_3 \\ &\qquad + 3\omega_c^2(I_{ab} - I_{tb})(\alpha_1 c + \alpha_3 s)s - c_d\dot{\beta}_3 - k_d\beta_3 = 0 \end{aligned} \tag{6}$$

$$\begin{aligned} &I_{td}(\ddot{\alpha}_1 + \ddot{\beta}_1) + (2I_{td} - I_{ad})\nu(\dot{\alpha}_3 + \dot{\beta}_3) + (I_{ad} - I_{td})\nu^2(\alpha_1 + \beta_1) \\ &\qquad + 3\omega_c^2(I_{ad} - I_{td})[(\alpha_1 + \beta_1)c + (\alpha_3 + \beta_3)s]c + c_d\dot{\beta}_1 + k_d\beta_1 = 0 \end{aligned} \tag{7}$$

$$\begin{aligned} &I_{td}(\ddot{\alpha}_3 + \ddot{\beta}_3) - (2I_{td} - I_{ad})\nu(\dot{\alpha}_1 + \dot{\beta}_1) + (I_{ad} - I_{td})\nu^2(\alpha_3 + \beta_3) \\ &\qquad + 3\omega_c^2(I_{ad} - I_{td})[(\alpha_1 + \beta_1)c + (\alpha_3 + \beta_3)s]s + c_d\dot{\beta}_3 + k_d\beta_3 = 0 \end{aligned} \tag{8}$$

where, as before, $s = \sin(\nu + \omega_c)t$ and $c = \cos(\nu + \omega_c)t$. To manipulate these equations into their constant-coefficient form, we assemble the following simple linear combinations of (5) through (8):

$$
\begin{aligned}
[(5)+(7)]c+[(6)+(8)]s&=0\\
-[(5)+(7)]s+[(6)+(8)]c&=0\\
(7)c+(8)s&=0\\
-(7)s+(8)c&=0
\end{aligned}
\tag{9}
$$

After introduction of the transformations

$$
\begin{aligned}
\gamma_1 &= \alpha_1 c + \alpha_3 s; &\quad \delta_1 &= \beta_1 c + \beta_3 s\\
\gamma_3 &= -\alpha_1 s + \alpha_3 c; &\quad \delta_3 &= -\beta_1 s + \beta_3 c
\end{aligned}
\tag{10}
$$

and the parameters

$$
I_a \triangleq I_{ab} + I_{ad}; \qquad I_t \triangleq I_{tb} + I_{td} \tag{11}
$$

$$
k_t \triangleq \frac{I_a - I_t}{I_t}; \qquad k_{td} \triangleq \frac{I_{ad} - I_{td}}{I_{td}} \tag{12}
$$

$$
\hat{I}_{td} \triangleq \frac{I_{td}}{I_t}; \qquad \hat{c}_d \triangleq \frac{c_d}{I_t \omega_c}; \qquad \hat{k}_d \triangleq \frac{k_d}{I_t \omega_c^2} \tag{13}
$$

the set of equations (9) becomes

$$
\mathcal{M}\mathbf{q}'' + (\mathcal{D} + \mathcal{G})\mathbf{q}' + (\mathcal{K} + \mathcal{A})\mathbf{q} = \mathbf{0} \tag{14}
$$

where $\eta \triangleq \omega_c t$, primes denote differentiation with respect to η, and

$$
\mathbf{q} \triangleq [\gamma_1 \quad \gamma_3 \quad \delta_1 \quad \delta_3]^T \tag{15}
$$

The nonzero elements of the system inertia matrix $\mathcal{M}$ are

$$
\begin{aligned}
&m_{11} = m_{22} = 1\\
&m_{13} = m_{24} = m_{31} = m_{42} = m_{33} = m_{44} = \hat{I}_{td}
\end{aligned}
\tag{16}
$$

The nonzero elements of the damping matrix $\mathcal{D}$ and of the gyroscopic matrix $\mathcal{G}$ are

$$
\begin{aligned}
&d_{33} = d_{44} = \hat{c}_d\\
&g_{21} = -g_{12} = 2 + (1 + k_t)\hat{\nu}\\
&g_{23} = -g_{32} = g_{41} = -g_{14} = g_{43} = -g_{34} = \hat{I}_{td}[2 + (1 + k_{td})\hat{\nu}]
\end{aligned}
\tag{17}
$$

As for the stiffness matrix $\mathcal{K}$, it has the nonzero elements

$$
\begin{aligned}
k_{11} &= -1 - \hat{\nu} + k_t(3 - \hat{\nu})\\
k_{22} &= -1 - \hat{\nu} - k_t\hat{\nu}\\
k_{13} = k_{31} &= \hat{I}_{td}[-1 - \hat{\nu} + k_{td}(3 - \hat{\nu})]\\
k_{24} = k_{42} &= \hat{I}_{td}(-1 - \hat{\nu} - k_{td}\hat{\nu})\\
k_{33} &= k_{13} + \hat{k}_d\\
k_{44} &= k_{24} + \hat{k}_d
\end{aligned}
\tag{18}
$$

The matrix $\mathcal{A}$ will be discussed shortly.

It is evident from (16) that

$$\mathscr{M}^T = \mathscr{M} > 0 \tag{19}$$

as expected. The symmetry properties of $\mathscr{D}$ and $\mathscr{G}$, from (17), are

$$\mathscr{D}^T = \mathscr{D} \geq 0; \qquad \mathscr{G}^T = -\mathscr{G} \tag{20}$$

again as expected. As for $\mathscr{K}$, though it is evidently symmetric, it may not be positive-definite. This brings us to the matrix $\mathscr{A}$. Its nonzero elements are

$$a_{43} = -a_{34} = (1 + \hat{\nu})\hat{c}_d \tag{21}$$

and it is thus skew-symmetric:

$$\mathscr{A}^T = -\mathscr{A} \tag{22}$$

A term of the form $\mathscr{A}\mathbf{q}$, with $\mathscr{A}^T = -\mathscr{A}$, can be expected in the motion equations whenever there are rotational dampers that operate within a spinning reference frame. Such is the case for the present system $\mathscr{W}_b + \mathscr{W}_d$. The presence of this term precludes the use of the main theorems available for matrix-second-order systems. For example, the Kelvin–Trait–Chetayev theorem (Appendix A) does not apply. Some discussion of the effect of $\mathscr{A}$ on stability has been given by [Huseyin].

Two procedures for determining the stability properties associated with (14) are available:

(a) Find the system eigenvalues (numerically) and ascertain whether they all lie in the left half-plane.

(b) Find the characteristic equations, that is to say, find the coefficients of the polynomial equation

$$\det\left[\mathscr{M}s^2 + (\mathscr{D} + \mathscr{G})s + (\mathscr{K} + \mathscr{A})\right] = 0 \tag{23}$$

and use the Routh–Hurwitz stability criteria listed in Appendix A, Table A.1.

(A third possibility—numerical determination of the roots of the characteristic equation (23)—is not recommended because the process of finding the eigenvalues of a matrix by first finding the characteristic equation numerically and then solving for the roots numerically is an inferior method of finding the eigenvalues of a matrix.) In choosing between possibilities (a) and (b) above, it should be recognized that the chief attraction of the Routh–Hurwitz criteria is their potential for producing literal stability conditions (closed-form inequalities). Finding formulas for the nine coefficients of the eighth-degree polynomial in the characteristic equation (23) in terms of the system six parameters $\{k_t, k_{td}, \hat{\nu}, \hat{I}_{td}, \hat{c}_d, \hat{k}_d\}$ and subsequently applying the eight Routh–Hurwitz stability criteria unfortunately constitutes a task that verges on the intractable. Therefore, one would be forced to implement procedure (b) numerically. Having thus abandoned hope of literal stability conditions, and with only numerical results in prospect, one may as well revert to procedure (a) and ascertain not only

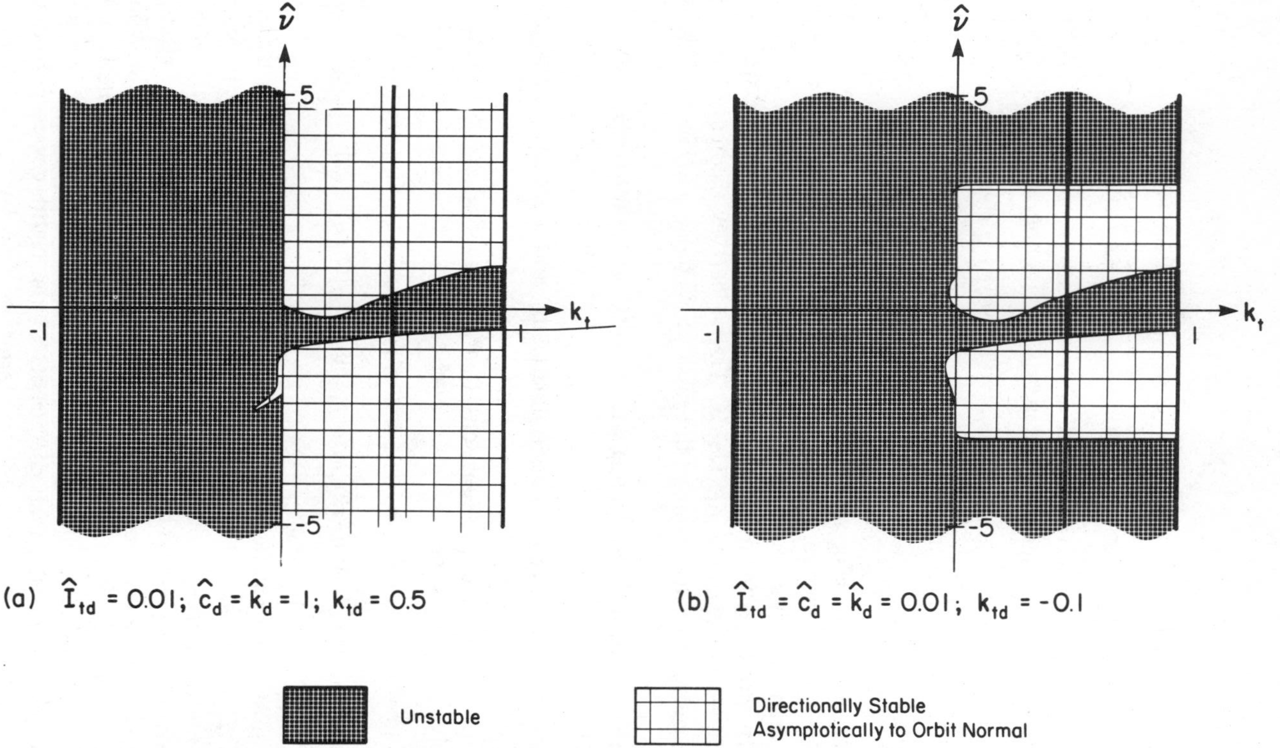

FIGURE 10.13 Stability diagrams for the system shown in Fig. 10.12. (Stability on the line $k_t = k_{td} = 0.5$ is not asymptotic; see Problem 10.7*b*.)

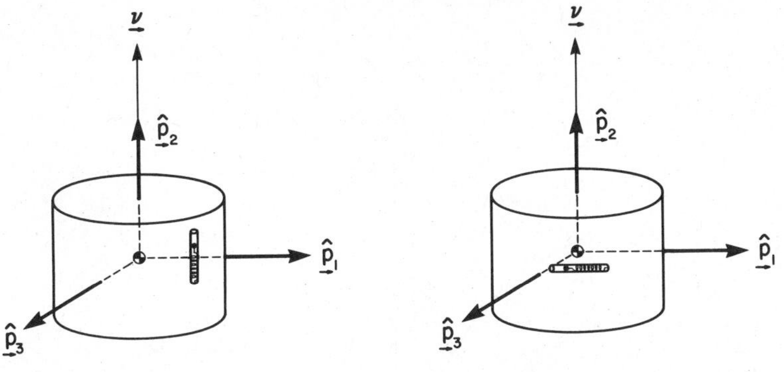

FIGURE 10.14 Two damper alternatives for spinners.

whether the eigenvalues are somewhere in the left half-plane but exactly where in that half-plane they are.

The latter procedure has been employed to find the typical stability diagrams for $\mathscr{W}_b + \mathscr{W}_d$ shown in Fig. 10.13. Through comparison of these diagrams with an earlier one in Fig. 10.10, we learn that quite different damping mechanisms can lead to quite different stability diagrams for spinning satellites in the Thomson equilibrium.

Wobble Dampers (Nutation Dampers)

It is by now apparent that one principal axis of a spacecraft (the "pointing axis") can often be passively stabilized with respect to an inertially fixed direction by spinning the vehicle about this axis. The stability is made asymptotic by including an appropriate on-board damping device. Reliable stabilization requires careful preflight dynamical modeling and stability analysis because precise conditions for stability depend on the configuration details of the damper.

Although most of the foregoing development has been focused on the primary issue—stability—it is in practice necessary to achieve more than mere stability. Performance specifications (e.g., time to half-amplitude) are normally also imposed, and damper parameters must often be chosen to maximize the rate of dissipation for a given damper mass. For point mass dampers connected to the main body via a spring and dashpot, this optimization is accomplished by *tuning* the damper so that its resonant frequency occurs at the frequency of excitation. For example, consider again the spinning satellite in essentially torque-free motion shown in Fig. 5.4. Assuming axisymmetry and a small damper mass, it can be shown (Problem 10.8) that the damper is tuned if

$$k_d \doteq m_d k_t^2 \nu^2 \tag{24}$$

where the damper mass and spring constant are respectively m_d and k_d, the spin rate is ν, and the inertia parameter is $k_t \triangleq (I_a/I_t) - 1 > 0$. The formula (24) is

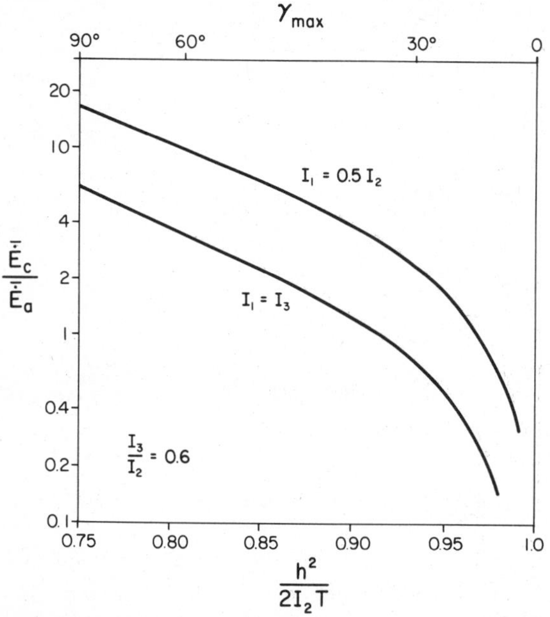

FIGURE 10.15 Relative efficiency of the two dampers shown in Fig. 10.14. From [Cochrane and Thompson].

based on linearized theory (small nutation angle γ), and detuning will occur if the initial value of γ is large. It should also be observed that the tuning condition is satisfied for only one spin rate, ν.

The possibility of a "circumferential" damper (instead of an "axial" damper) has been studied by [Schneider and Likins] and [Cochran and Thompson]. These two alternatives are shown schematically in Fig. 10.14. Our discussion to this point has centered on axial dampers because they are the appropriate choice for removing wobbling of *small* amplitude. Indeed, the circumferential damper does not damp wobble at all to first order in the wobble angles α_1 and α_3. For large initial nutation angles however, where nonlinear theory must be employed, the circumferential damper is coupled to wobbling and can even become more efficient than the axial damper.

Figure 10.15 compares the relative efficiencies of these two dampers as a function of maximum nutation angle $\gamma_{\max}$ (which can be calculated from the formulas in Chapter 4). Using the convention (as in Chapter 4) that $I_1 > I_2 > I_3$, and assuming that the wobble is about a major-axis spin, a minimization similar to (4.3, 7) gives

$$\omega_{1,\min} = h\left[\frac{I - I_2}{II_1(I_1 - I_2)}\right]^{1/2} \tag{25}$$

where $I \triangleq \frac{1}{2}h^2/T$. Then, since $h\cos\gamma = h_1 = I_1\omega_1$, we have

$$\cos\gamma_{max} = \left(\frac{I_1}{h}\right)\omega_{1,min} \tag{26}$$

In the present chapter however, we have adopted the convention that the spin is about $\underset{\rightarrow}{\hat{\mathbf{p}}}_2$. With $I_2 > I_3 > I_1$, (26) becomes

$$\cos\gamma_{max} = \left[\frac{I_2(I - I_3)}{I(I_2 - I_3)}\right]^{1/2} \tag{27}$$

In Fig. 10.15 it is assumed that each damper has been tuned to resonate at its frequency of excitation, and $\dot{E}_c$ and $\dot{E}_a$ are the average rates of energy extraction by the circumferential and axial dampers, respectively. The relative efficiency of these two dampers is plotted vs. $h^2/(2I_2T)$ and, using (27), vs. γ_{max}. It is evident that unless the initial nutation angle is very large, or unless the satellite is very asymmetrical, the axial damper is to be preferred. If damping is required over a wide range of nutation angles, it may be beneficial to incorporate both types of damper in the configuration.

A comment may be in order on the meaning of the term "nutation damper." The *nutation angle* (Fig. 4.6) is the angle between the angular momentum vector (inertially fixed over the short term) and the "spin axis" which, according to the major-axis rule, must be the axis of greatest inertia. For an inertially axisymmetrical satellite in torque-free motion, the nutation angle γ is fixed (see Section 4.2); for a nearly axisymmetrical satellite, the nutation angle γ varies slightly about a fixed mean value. *Nutation* is said to occur when the nutation angle changes. However, it is almost invariably the nutation angle itself that is objectionable in a spin-stabilized satellite. For example, a spin-stabilized axisymmetrical satellite that is not nutating but that displays a (constant) nutation angle of 20° is an embarrassment to its designers. Thus, what is customarily called a "nutation damper" might better be thought of as a "nutation angle inhibitor." A good nutation angle inhibitor would not *damp* nutation for a satellite whose nutation angle is 10°; it would *cause* nutation as it brought the nutation angle rapidly from 20° to 0°. For this reason the more pedestrian name "wobble damper" is arguably more precise than the more pedantic name "nutation damper." Despite this contention however, one must recognize the futility of rebelling against convention and accept the colloquial meaning of the term "nutation damper."

Nutation Dampers with Springs

The nutation dampers mentioned in the foregoing discussion are of the mass spring dashpot type (see Figs. 5.4, 10.11, 10.12, and 10.14). Such dampers are particularly amenable to analysis and are in principle closely analogous to the pendulum-type dampers planted in many spin-stabilized satellites. An early example is SYNCOM (Fig. 10.16*a*), the first geosynchronous satellite for active communications. Launched in 1963, its ~ 160-rpm spin provided stabilization

(a) SYNCOM 3 Satellite (b) Explorer 18 (on test stand), after [Corliss, 1]

FIGURE 10.16 Two spin-stabilized satellites.

over a six-year lifetime. A second example is Explorer 18 (Fig. 10.16*b*), also launched in 1963. Spin stabilized at 20 rpm, it probed the magnetosphere as it traveled on a slightly eccentric orbit (perigee altitude 192 km, apogee altitude 197 km); refer to [Butler] for details.

The mass spring dashpot concept can be engineered in many ways. For example, consider the "blade mass" design sketched in Fig. 10.17. In an empty case, the blade and ball would vibrate at their natural frequency, which can be tuned to the principal frequency of excitation. When the case is filled with silicone oil, an effective nutation damper is created. Incidentally, experience with two-degree-of-freedom nutation dampers indicates that better performance can usually be achieved for a given weight by using instead two one-degree-of-freedom dampers.

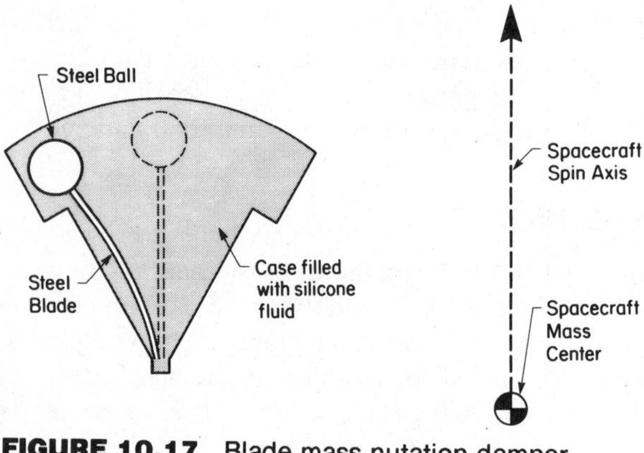

FIGURE 10.17 Blade mass nutation damper.

FIGURE 10.18 ISIS Satellite.

For some spacecraft, wobble energy is naturally dissipated in existing structural components. An example is ISIS (International Satellites for Ionospheric Studies), launched in 1969 under a joint Canada–United States program. The long dipole antenna booms (Fig. 10.18) are excited whenever the nutation angle is nonzero, and the consequent structural energy dissipation provides a natural source of wobble damping. This satellite also illustrates the point that it is the spin *momentum*, not the spin *rate*, that produces directional stability. Although ISIS spun at only ~ 2.9 rpm, the relatively large spin axis inertia contributed by the antenna booms produces an adequate spin axis momentum.

All dampers with springs have one common characteristic which is, paradoxically, both their greatest strength and their greatest weakness: they are tuned to a specific excitation frequency. At this frequency, they perform very well. Unfortunately, if the vehicle parameters are not ideal (perhaps the spin rate is not the nominal one, or the inertial properties have evolved over the satellite lifetime due to fuel consumption), spring-based dampers may perform badly. The designer is thus presented with an interesting quandary.

Pendulum Nutation Dampers

The idea of a pendulum damper is portrayed in Fig. 10.19. Several versions of this damper have been proposed, but the one shown is in a radially outward configuration, with an implied source of damping on the bob motion. The major point to be made is that the centrifugal "force" on the damper bob $\mathscr{P}$ produces a restoring moment equivalent to a spring. (In some designs an actual spring is used as well.) By an appropriate choice of χ, b, l_1, l_2, m_d, and the damping constant, an effective wobble damper can be designed.

We shall sketch an analysis of this damper that provides reasonably accurate design data. This analysis is also instructive in that a similar approach can be

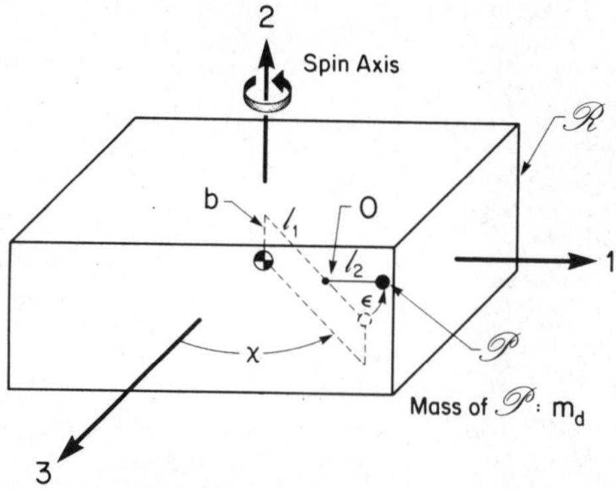

FIGURE 10.19 The pendulum damper concept.

used with several other types of damper. We begin with the assumption that it takes many precession periods for the damper to substantially affect the (small) wobbling. Thus, the spinning motion of the satellite is virtually the same as the motion of a tri-inertial rigid body, described by (4.3.1). This solution is, in fact,

$$\omega_2 \equiv \nu = \text{constant}$$

$$\omega_1 = \omega_{1o} \cos \Omega t; \qquad \omega_3 = -\omega_{3o} \sin \Omega t \tag{28}$$

where ω_{1o} and ω_{3o} are (small) constants and Ω is given by (4.4, 26). It is noted in passing that these constants are related as follows:

$$I_3 \Omega \omega_{3o} = (I_2 - I_1) \nu \omega_{1o} \tag{29}$$

The next step is to calculate the acceleration on the damper bob $\mathscr{P}$. Expressed in the principal-axis frame $\mathscr{F}_p$, this acceleration is

$$\mathbf{a}_m = \ddot{\mathbf{r}}_m + 2\boldsymbol{\omega}^{\times}\dot{\mathbf{r}}_m + (\dot{\boldsymbol{\omega}}^{\times} + \boldsymbol{\omega}^{\times}\boldsymbol{\omega}^{\times})\mathbf{r}_m \tag{30}$$

where $\boldsymbol{\omega}$ is available from (28) and $\mathbf{r}_m$ is (see Fig. 10.19) the position of $\mathscr{P}$ in $\mathscr{F}_p$:

$$\mathbf{r}_m = \begin{bmatrix} (l_1 + l_2)\sin\chi + l_2 \varepsilon \cos\chi \\ b \\ (l_1 + l_2)\cos\chi - l_2 \varepsilon \sin\chi \end{bmatrix} \tag{31}$$

Note that ε, like ω_1 and ω_3, is to be treated as a first-order infinitesimal. The inertial force on $\mathscr{P}$ is $-m_d \mathbf{a}_m$, and the moment of this force about O has the moment arm $\mathbf{r}_{Om} = \mathbf{r}_m - \mathbf{r}_O$ where

$$\mathbf{r}_O = [l_1 \sin\chi \quad b \quad l_1 \cos\chi]^T \tag{32}$$

Finally, we need the component of this inertial moment about the spin axis, namely, $-m_d \mathbf{1}_2^T \mathbf{r}_{Om}^{\times} \mathbf{a}_m$. The sum of this moment, the damping moment $-c_d \dot{\varepsilon}$ and the spring moment (if present) $-k_d \varepsilon$, must be zero. The resulting motion equation for $\mathscr{P}$ is (after some calculation)

$$\ddot{\varepsilon} + 2\zeta_d \omega_d \dot{\varepsilon} + \omega_d^2 \varepsilon = -\left(\frac{b}{l_2}\right)(\Omega + \nu)\omega_w \cos(\Omega t - \phi) \tag{33}$$

with

$$\omega_d^2 \triangleq \frac{k_d}{m_d l_2^2} + \frac{l_1}{l_2}\nu^2; \qquad \zeta_d \triangleq \frac{c_d}{2\omega_d m_d l_2^2} \tag{34}$$

and ω_w and ϕ defined by the following pair of equations:

$$\omega_w \sin\phi \triangleq \left(\frac{\Omega\omega_{1o} + \nu\omega_{3o}}{\Omega + \nu}\right)\sin\chi$$

$$\omega_w \cos\phi \triangleq \left(\frac{\Omega\omega_{3o} + \nu\omega_{1o}}{\Omega + \nu}\right)\cos\chi \tag{35}$$

The angle ϕ indicates the phase of the input, and ω_w is a good measure of "wobble."

The solution of (33) for the damper motion is

$$\varepsilon = E\sin(\Omega t - \phi - \psi) \tag{36}$$

with

$$E = -\frac{b}{l_2}\frac{(\Omega + \nu)\omega_w}{\Delta}$$

$$\Delta^2 = \left(\Omega^2 - \omega_d^2\right)^2 + 4\zeta_d^2\omega_d^2\Omega^2$$

$$\psi = \tan^{-1}\left(\Omega^2 - \omega_d^2, 2\zeta_d\omega_d\Omega\right) \tag{37}$$

Finally, the energy flux from the satellite, averaged over one cycle, is

$$\bar{\dot{T}} = -\frac{\Omega}{2\pi}\int_0^{2\pi/\Omega}[\text{Right-Hand Side of }(137)]\, m_d l_2^2 \dot{\varepsilon}\, dt$$

$$= \frac{-c_d(b/l_2)^2\Omega^2(\Omega + \nu)^2}{2\left[\left(\Omega^2 - \omega_d^2\right)^2 + 4\zeta_d^2\omega_d^2\Omega^2\right]}\omega_w^2 \tag{38}$$

because $m_d l_2^2$[R.H.S. of (33)] is the moment on $\mathscr{P}$ about O. It is equation (38) that provides guidelines for damper design (the proper choice of χ, b, l_1, l_2, m_d, c_d, and k_d). The object is to maximize the magnitude of $\bar{\dot{T}}$, consistent with good engineering practice (for example, the spring k_d may be deemed too complicated or too expensive), and to do so over the range of spin rates ν that are of concern. This pendulum damper is analyzed further in Problem 10.10.

One implementation of this type of pendulum damper was designed for the aeronomy satellite AEROS, a joint venture between West Germany and the United States. Figure 10.20 shows this vehicle, launched in 1972 and spin stabilized at ~ 10 rpm. The nutation damper is indicated in Fig. 10.20*b*. As described by [Zago and Leiss], AEROS included two such pendulum dampers, each swinging in a separate gas-filled case perpendicular to the spin axis. Thus, in this case, a gas, not oil, was selected as the dissipative medium. Designs based on eddy current induction or magnetic hysteresis (cf. Fig. 9.27*b*) are also possible.

(a) AEROS in Orbit

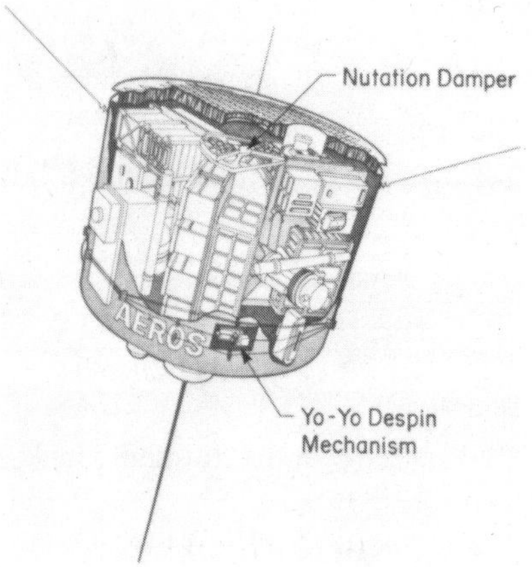

(b) View Showing Pendulum Nutation Damper, based on [Zago and Leiss]

FIGURE 10.20 The AEROS Satellite.

Ball-In-Tube Nutation Dampers

Two variations on the ball-in-tube theme will now be alluded to. The first is the "impact damper," one design of which is shown in Fig. 10.21*a*. Beryllium–copper balls roll inside the tube, where rolling friction and inelastic collisions at the tube ends consume mechanical energy. Two such impact dampers (tube axes parallel to the spin axis) are just barely visible at the tip of the radial boom on the left of the picture of Pioneer in Fig. 10.21*b*. The Pioneers formed a series of interplanetary

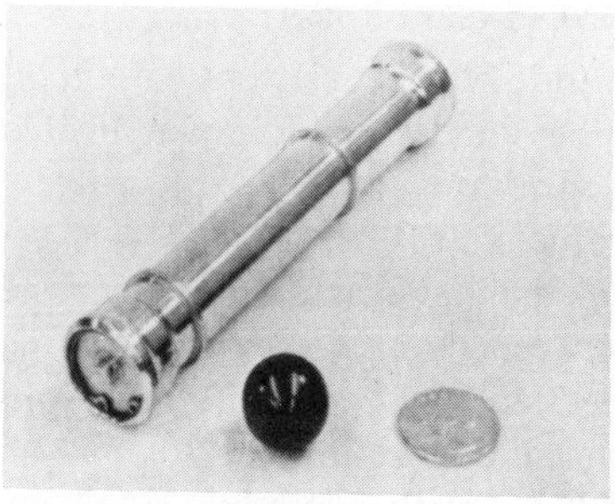

(a) TRW Impact Damper, from [De Bra]

(b) Pioneer 6, from [Corliss, 2]

FIGURE 10.21 Pioneer and its wobble damper.

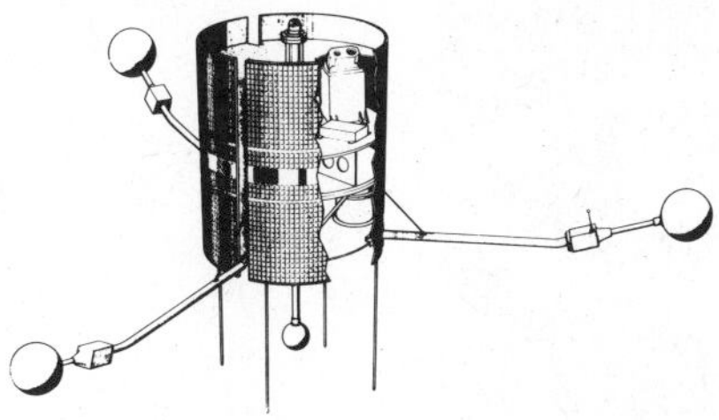

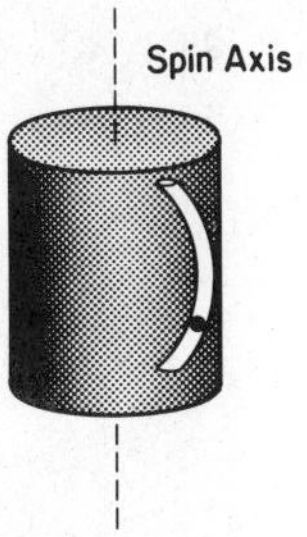

(a) ESRO 4, from [Willems] (b) Ball-in-Tube Pendulum Damper

FIGURE 10.22 ESRO 4 and its wobble damper.

spacecraft launched by NASA beginning in 1965, and the early members of this family were spin stabilized at ~ 60 rpm. This impact tube design of wobble damper is simple and reliable, although it cannot be expected to perform well for very small nutation angles.

A second type of ball-in-tube damper (Fig. 10.22) employs a curved tube, possibly filled with a liquid or a gas; centrifugal force causes the ball to roll toward the center of the tube. This is therefore, a type of centrifugal pendulum, and, as such, it has certain properties in common with the pendulum damper described above. In particular, since its natural frequency (which can be designed to a desired value by proper tube location and curvature) is proportional to the spin rate, once this damper is tuned, it is tuned at all spin rates (cf. Problem 10.10*c*). This type of damper was flown on several satellites, including ESRO-IV. (ESRO-IV launched in 1972 and directionally stabilized using a spin of ~ 40 rpm.)

Viscous-Ring Dampers

In some designs, the inertia and viscosity are provided by the same material, often mercury. Some tubes are partially filled, others completely so. Some tubes are rings or annuli. Some rings are centered on the spin axis, others are offset. The number of combinations is quite large. One such design is shown in Fig. 10.23*a*. A similar damper was flown on HELIOS, a solar-probe spacecraft that traveled in the mid-1970s to within 0.29 AU of the sun. HELIOS was spin stabilized at ~ 60 rpm ([Kutzer]).

Choice of Wobble Damper

It is clear that there is a very large number of wobble damper designs. The choice is a difficult one and must be based on characteristics such as large-angle vs. small-angle performance, reliability, importance of tuning, fatigue, weight, and electromagnetic cleanliness. Even concerns such as survivability during launch

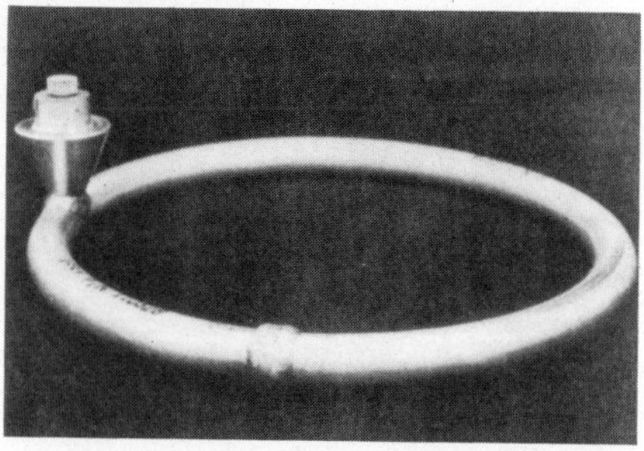

(a) Ring Damper with Expansion Chamber from [De Bra]

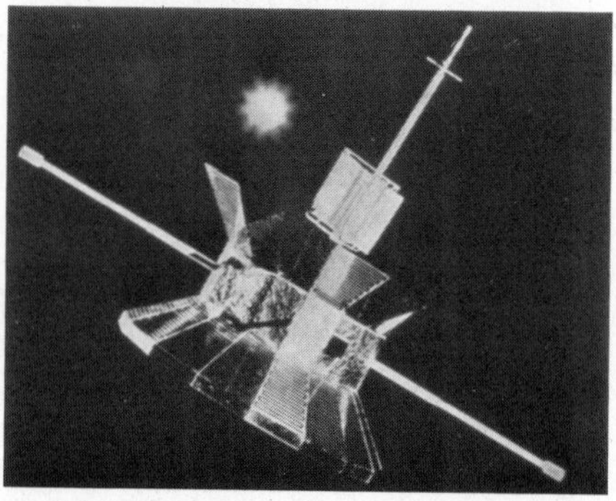

(b) HELIOS

FIGURE 10.23 A viscous-ring damper and a satellite application.

and ease of testing in Earth gravity may become deciding factors. A review and assessment of many designs has been carried out by [Ancher, Brink, and Pouw].

10.3 LONG-TERM EFFECTS OF ENVIRONMENTAL TORQUES, AND FLIGHT DATA

It would be helpful at this point to identify three contexts in which spin stabilization is an attractive strategy. First, many space vehicles are spin stabilized in *transfer orbit*—the trajectory between launch vehicle burnout and insertion into permanent orbit. Often a more sophisticated control system will be used once the permanent orbit is established, whereas temporary spin stabilization is attractive during the transfer orbit because of its simplicity. (As a further permutation, an active control system can also be used on a spinning satellite during transfer orbit when the spin axis is not the major axis of inertia ([Hrastar]).) Heretofore, the necessary spin has been imparted quite naturally by the spinning "third stage." In the era of the space shuttle, a special spin-up procedure will be used.

The second context for spin stabilization is the "high-spin" strategy, in which the spin rate is sufficiently high that environmental torques can be resisted;

attitude digressions of the spin axis become prominent only after a considerable period of time. The idea is that the spin axis is to be fixed in inertial space. The transfer orbit context explained in the last paragraph might be thought of as a special (short-term) case of the high-spin strategy.

In the third context, we have the Thomson equilibrium, for which the spin rate need not be so "high" but for which the spin axis must be perpendicular to the orbit. In exchange for this limitation (which may not in fact be a limitation for some missions), one has the assurance that with proper design the principal source of environmental disturbance—Earth's gravity gradient torque—is not destabilizing. For example, all spin-stabilized geostationary communications satellites to date have used the Thomson equilibrium; their spin axes have been nominally perpendicular to the equatorial plane. Over sufficiently long periods however, even this "equilibrium" proves impossible to maintain, in the face of other torques and orbit evolution, without some form of active intervention.

Spin stabilization often has advantages that go beyond attitude control. There are benefits also for thermal control. Constant rotation of the exposed surface can eliminate some "hot spots" and "cold spots." On the other hand, solar cells on this surface lose efficiency for precisely the same reason.

For the remainder of this section we shall be concerned with the long-term influences of environmental torques on satellites that are spin stabilized: the angular velocity vector, the spin vector, and the angular momentum vector are persistently codirectional. After some introductory analysis of the major origins of disturbing torque, flight data will be displayed to illustrate long term effects.

Gravitational Torque: Average Over One Spin Period

In general, a spinning satellite will experience, at each instant, a gravitational torque (Section 8.1). For rapidly spinning satellites, the principle of spin stabilization (see introductory remarks at the beginning of this chapter) is based on the requirement that a disturbance torque will take a long period of time to tilt the angular momentum vector $\underset{\rightarrow}{\mathbf{h}}$ away from its preselected direction. The purpose of this section is to lay the groundwork for a calculation of these long-term effects.

The instantaneous gravity gradient torque on a rigid body was given by (8.1, 16) in Chapter 8:

$$\underset{\rightarrow}{\mathbf{g}}_g = 3\left(\frac{\mu}{R^3}\right)\underset{\rightarrow}{\hat{\mathbf{o}}}_3 \times \underset{\rightarrow}{\mathbf{I}} \cdot \underset{\rightarrow}{\hat{\mathbf{o}}}_3 \tag{1}$$

It should be noted that the torque is calculated about the satellite mass center, μ is Earth's gravitational constant, R is the distance from Earth's center to the satellite, $\underset{\rightarrow}{\mathbf{I}}$ is the satellite inertia dyadic, and $\underset{\rightarrow}{\hat{\mathbf{o}}}_3$ is a unit vector from the satellite to Earth's center. If the rigid body possesses an axis of inertial symmetry defined by the unit vector $\underset{\rightarrow}{\hat{\mathbf{p}}}_2$, the specialized form of (1) is

$$\underset{\rightarrow}{\mathbf{g}}_g = 3\left(\frac{\mu}{R^3}\right)(I_a - I_t)\underset{\rightarrow}{\hat{\mathbf{o}}}_3 \times \underset{\rightarrow}{\hat{\mathbf{p}}}_2\underset{\rightarrow}{\hat{\mathbf{p}}}_2 \cdot \underset{\rightarrow}{\hat{\mathbf{o}}}_3 \tag{2}$$

This formula is derived in Problem 10.11.

Since the direction of the spin axis is essentially fixed, it is not the instantaneous torque that is of interest but the average torque over one satellite rotation. This average torque, impressed over a great many rotations, must eventually be compensated for if the spacecraft is to have its spin axis maintained in an inertially fixed direction. To calculate the average torque over one rotation, given by

$$\bar{\underset{\to}{\mathbf{g}}}_g \triangleq \frac{1}{2\pi}\int_0^{2\pi} \underset{\to}{\mathbf{g}}_g(\phi)\, d\phi \tag{3}$$

we require the integration of the instantaneous torque over one spacecraft rotation, $0 \leq \phi < 2\pi$. To carry out the integration in (3), we need the average of the inertia dyadic $\underset{\to}{\mathbf{I}}$ over one rotation. At each instant,

$$\underset{\to}{\mathbf{I}} = \sum_{i=1}^{3} I_i \underset{\to}{\hat{\mathbf{p}}}_i \underset{\to}{\hat{\mathbf{p}}}_i \tag{4}$$

with $\underset{\to}{\hat{\mathbf{p}}}_i$ being as usual the unit vectors along the principal axes. We assume that the spin is about $\underset{\to}{\hat{\mathbf{p}}}_2$; hence, $\underset{\to}{\hat{\mathbf{p}}}_2$ is fixed, and

$$\underset{\to}{\hat{\mathbf{p}}}_1 = \underset{\to}{\hat{\mathbf{p}}}_{1o} \cos\phi + \underset{\to}{\hat{\mathbf{p}}}_{3o} \sin\phi$$

$$\underset{\to}{\hat{\mathbf{p}}}_3 = \underset{\to}{\hat{\mathbf{p}}}_{3o} \cos\phi - \underset{\to}{\hat{\mathbf{p}}}_{1o} \sin\phi$$

The subscript o indicates the reference value when $\phi = 0$.

It follows that

$$\frac{1}{2\pi}\int_0^{2\pi} \underset{\to}{\mathbf{I}}\, d\phi = \frac{1}{2}(I_1 + I_3)\left(\underset{\to}{\hat{\mathbf{p}}}_{1o}\underset{\to}{\hat{\mathbf{p}}}_{1o} + \underset{\to}{\hat{\mathbf{p}}}_{3o}\underset{\to}{\hat{\mathbf{p}}}_{3o}\right) + I_2 \underset{\to}{\hat{\mathbf{p}}}_2 \underset{\to}{\hat{\mathbf{p}}}_2 \tag{5}$$

since the averages of $\sin\phi$, $\cos\phi$, and $\sin\phi\cos\phi$ are zero and the averages of $\sin^2\phi$ and $\cos^2\phi$ are $\frac{1}{2}$. Now, in view of (1) and (5), we need the following vector product:

$$\begin{aligned}\underset{\to}{\hat{\mathbf{o}}}_3 \times \left(\underset{\to}{\hat{\mathbf{p}}}_{1o}\underset{\to}{\hat{\mathbf{p}}}_{1o} + \underset{\to}{\hat{\mathbf{p}}}_{3o}\underset{\to}{\hat{\mathbf{p}}}_{3o}\right) \cdot \underset{\to}{\hat{\mathbf{o}}}_3 &= \underset{\to}{\hat{\mathbf{o}}}_3 \times \left[\underset{\to}{\hat{\mathbf{p}}}_{1o}\left(\underset{\to}{\hat{\mathbf{p}}}_{1o} \cdot \underset{\to}{\hat{\mathbf{o}}}_3\right) + \underset{\to}{\hat{\mathbf{p}}}_{3o}\left(\underset{\to}{\hat{\mathbf{p}}}_{3o} \cdot \underset{\to}{\hat{\mathbf{o}}}_3\right)\right] \\ &= \underset{\to}{\hat{\mathbf{o}}}_3 \times \left[\underset{\to}{\hat{\mathbf{o}}}_3 - \underset{\to}{\hat{\mathbf{p}}}_2\left(\underset{\to}{\hat{\mathbf{p}}}_2 \cdot \underset{\to}{\hat{\mathbf{o}}}_3\right)\right] = -\underset{\to}{\hat{\mathbf{o}}}_3 \times \underset{\to}{\hat{\mathbf{p}}}_2\underset{\to}{\hat{\mathbf{p}}}_2 \cdot \underset{\to}{\hat{\mathbf{o}}}_3\end{aligned}$$

Thus,

$$\bar{\underset{\to}{\mathbf{g}}}_g = 3\left(\frac{\mu}{R^3}\right)\left[I_2 - \frac{1}{2}(I_1 + I_3)\right]\underset{\to}{\hat{\mathbf{o}}}_3 \times \underset{\to}{\hat{\mathbf{p}}}_2\underset{\to}{\hat{\mathbf{p}}}_2 \cdot \underset{\to}{\hat{\mathbf{o}}}_3 \tag{6}$$

In particular, if the spacecraft is inertially axisymmetrical and is spinning about this axis of symmetry, the average torque reduces to

$$\bar{\underset{\to}{\mathbf{g}}}_g = 3\left(\frac{\mu}{R^3}\right)(I_a - I_t)\underset{\to}{\hat{\mathbf{o}}}_3 \times \underset{\to}{\hat{\mathbf{p}}}_2\underset{\to}{\hat{\mathbf{p}}}_2 \cdot \underset{\to}{\hat{\mathbf{o}}}_3 \tag{7}$$

which it should, because, since both $\underset{\to}{\hat{\mathbf{o}}}_3$ and $\underset{\to}{\hat{\mathbf{p}}}_2$ are constant in this calculation, so too is the instantaneous torque. Thus, (7) is identical to (2).

It is observed that the average torque is zero whenever the spin axis is perpendicular to the local vertical. This condition includes, in particular,

Thomson-type situations, for which the spin axis remains aligned with the orbit normal. Moreover, at points in the orbit where the spin axis points along the local vertical, the average torque is again zero.

Gravitational Torque: Average Over One Orbital Period

The preceding analysis in essence uses the *method of averaging*, in which "rapidly changing" variables are integrated out and attention then focuses on the behavior of the "slowly changing" variables. For a spinning satellite, the rapidly changing variable is the spin angle ϕ, and the slowly changing variables describe the long-term wandering of the spin axis under the influence of environmental torques. The averaging in the preceding paragraphs was based on the assumption that the direction and magnitude of the spin vector are essentially unchanged during one satellite rotation.

We now make a further assumption: the angular momentum of the spacecraft is essentially unchanged during one orbit. This assumption permits us to perform a further stage of averaging—over the orbital period. (Opportunities to use the method of averaging twice in succession are rare in engineering analysis.) During the orbit, both the local vertical ($\underset{\rightarrow}{\hat{\mathbf{o}}}_3$) and the orbital radius (R) vary. (For a circular orbit of course, R is constant.)

The needed geometrical ideas are presented in Fig. 10.24. An inertial reference frame $\mathscr{F}_i$ is associated with the orbit, as shown. The local vertical can be expressed in terms of $\mathscr{F}_i$ as follows:

$$\underset{\rightarrow}{\hat{\mathbf{o}}}_3 = -\underset{\rightarrow}{\hat{\mathbf{i}}}_1 \cos\eta - \underset{\rightarrow}{\hat{\mathbf{i}}}_2 \sin\eta \tag{8}$$

where η is the "true anomaly." We shall also need to know that

$$R = \frac{a(1 - e^2)}{1 + e\cos\eta} \tag{9}$$

where e and a are, as usual, the eccentricity and the semimajor diameter of the orbit. The last two pieces of information needed from orbit theory are the

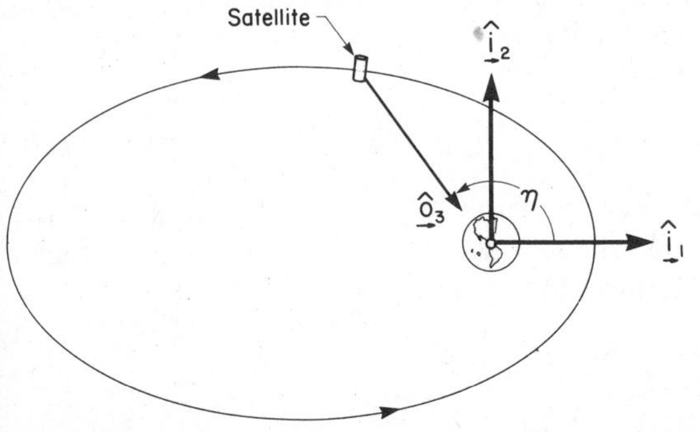

FIGURE 10.24 Geometry for orbital torque averaging.

following:

$$R^2\dot{\eta} = \left[\mu a(1-e^2)\right]^{1/2} \tag{10}$$

$$T_o = 2\pi\left(\frac{a^3}{\mu}\right)^{1/2} \tag{11}$$

The first expresses the constancy of the orbital angular momentum and the second is a formula for the orbital period; μ is the gravitational constant of the attracting body.

To average over one orbit, we form

$$\bar{\bar{\underset{\rightarrow}{\mathbf{g}}}}_g = \frac{1}{T_o}\int_0^{T_o} \bar{\underset{\rightarrow}{\mathbf{g}}}_g(t)\,dt \tag{12}$$

$$= \frac{1}{a^2(1-e^2)^{1/2}}\frac{1}{2\pi}\int_0^{2\pi} R^2(\eta)\bar{\underset{\rightarrow}{\mathbf{g}}}_g(\eta)\,d\eta \tag{13}$$

Upon substitution of (6), (8), and (9), we arrive at the expression sought:

$$\bar{\bar{\underset{\rightarrow}{\mathbf{g}}}}_g = \frac{3\mu}{2a^3(1-e^2)^{3/2}}\left[I_2 - \frac{1}{2}(I_1 + I_3)\right]\hat{\underset{\rightarrow}{\mathbf{p}}}_2 \times \hat{\underset{\rightarrow}{\mathbf{i}}}_3\hat{\underset{\rightarrow}{\mathbf{i}}}_3 \cdot \hat{\underset{\rightarrow}{\mathbf{p}}}_2 \tag{14}$$

where $\hat{\underset{\rightarrow}{\mathbf{i}}}_3$ is the orbit normal. Because all the periodically varying torque components have been averaged out, $\bar{\bar{\underset{\rightarrow}{\mathbf{g}}}}_g$ is called the *secular* gravitational torque. If the spin axis lies in the orbital plane or is perpendicular to the orbit, the secular gravitational torque is zero. As a practical design criterion, the value of I_2 is often chosen 5% to 15% higher than I_t so that secular gravitational torque is reduced as far as one dares without jeopardizing the major-axis spin ([NASA, 9]).

From the motion equation

$$\dot{\underset{\rightarrow}{\mathbf{h}}} = \bar{\bar{\underset{\rightarrow}{\mathbf{g}}}}_g \tag{15}$$

with $\underset{\rightarrow}{\mathbf{h}} = \hat{\underset{\rightarrow}{\mathbf{p}}}_2 h$ and $\bar{\bar{\underset{\rightarrow}{\mathbf{g}}}}_g = k_g \hat{\underset{\rightarrow}{\mathbf{p}}}_2 \times \hat{\underset{\rightarrow}{\mathbf{i}}}_3\hat{\underset{\rightarrow}{\mathbf{i}}}_3 \cdot \hat{\underset{\rightarrow}{\mathbf{p}}}_2$ (where k_g can be inferred from (14) above) one can determine the properties of the long-term motion. The component in the $\hat{\underset{\rightarrow}{\mathbf{p}}}_2$ direction is

$$\dot{h} = 0$$

showing that the spin rate remains constant, even over the long term. This might have been expected inasmuch as gravity is a conservative force field. The remainder of (15) is

$$\underset{\rightarrow}{\boldsymbol{\omega}}_p \times \hat{\underset{\rightarrow}{\mathbf{p}}}_2 h = k_g\left(\hat{\underset{\rightarrow}{\mathbf{p}}}_2 \times \hat{\underset{\rightarrow}{\mathbf{i}}}_3\hat{\underset{\rightarrow}{\mathbf{i}}}_3 \cdot \hat{\underset{\rightarrow}{\mathbf{p}}}_2\right) \tag{16}$$

where $\underset{\rightarrow}{\boldsymbol{\omega}}_p$ is the long-term angular velocity of the spin axis $\hat{\underset{\rightarrow}{\mathbf{p}}}_2$ with respect to the inertial frame $\mathscr{F}_i$. By convention, we take $\underset{\rightarrow}{\boldsymbol{\omega}}_p \cdot \hat{\underset{\rightarrow}{\mathbf{p}}}_2 = 0$, that is, all angular velocity along the spin axis is already included in the spin rate. Then, from (16),

$$\underset{\rightarrow}{\boldsymbol{\omega}}_p = -\left(\frac{k_g}{h}\right)\left(\hat{\underset{\rightarrow}{\mathbf{i}}}_3 \cdot \hat{\underset{\rightarrow}{\mathbf{p}}}_2\right)\hat{\underset{\rightarrow}{\mathbf{i}}}_3 \tag{17}$$

showing that the satellite spin vector precesses uniformly about the orbit normal.

Actually, the motion is not quite this simple because the orbit does not itself remain inertially fixed and there are usually other, nongravitational torques on the satellite.

Average Magnetic Torque

It was stated in Chapter 8, equation (8.4, 1), that the instantaneous torque on a satellite due to Earth's magnetic field is given by

$$\underset{\rightarrow}{\mathbf{g}}_m = \underset{\rightarrow}{\mathbf{m}}_m \times \underset{\rightarrow}{\mathbf{B}} \tag{18}$$

where $\underset{\rightarrow}{\mathbf{m}}_m$ is the net magnetic moment resident in the vehicle and $\underset{\rightarrow}{\mathbf{B}}$ is the local magnetic flux density of Earth's magnetic field. The average torque over one satellite spin rotation is

$$\underset{\rightarrow}{\bar{\mathbf{g}}}_m \triangleq \frac{1}{2\pi}\int_0^{2\pi} \underset{\rightarrow}{\mathbf{g}}_m(\phi)\, d\phi = m_{ms}\underset{\rightarrow}{\hat{\mathbf{p}}}_2 \times \underset{\rightarrow}{\mathbf{B}} \tag{19}$$

with m_{ms} the component of magnetic moment along the spin axis.

The second level of averaging—over the orbit—becomes more complicated than it was in the gravitational torque calculation because of the variation of $\underset{\rightarrow}{\mathbf{B}}$ over the orbit. As in Section 8.4.1, we write

$$\underset{\rightarrow}{\mathbf{B}} = \frac{\mu_m}{R^3}\left(\underset{\rightarrow}{\mathbf{1}} - 3\underset{\rightarrow}{\hat{\mathbf{o}}}_3\underset{\rightarrow}{\hat{\mathbf{o}}}_3\right)\cdot \underset{\rightarrow}{\hat{\mathbf{m}}}_3 \tag{20}$$

where μ_m is the strength of Earth's magnetic dipole and $\underset{\rightarrow}{\hat{\mathbf{m}}}_3$ is a unit vector in the direction of geomagnetic north. For another way of writing (20), see (8.4.4). We define

$$\underset{\rightarrow}{\bar{\bar{\mathbf{g}}}}_m \triangleq \frac{1}{T_o}\int_0^{T_o} \underset{\rightarrow}{\bar{\mathbf{g}}}_m(t)\, dt \tag{21a}$$

$$= 2k_m\underset{\rightarrow}{\hat{\mathbf{p}}}_2 \times \frac{1}{2\pi}\int_0^{2\pi}(1 + e\cos\eta)\left(\underset{\rightarrow}{\mathbf{1}} - 3\underset{\rightarrow}{\hat{\mathbf{o}}}_3\underset{\rightarrow}{\hat{\mathbf{o}}}_3\right)\cdot \underset{\rightarrow}{\hat{\mathbf{m}}}_3\, d\eta \tag{21b}$$

and

$$k_m \triangleq \frac{\mu_m m_{ms}}{2a^3(1 - e^2)^{3/2}} \tag{22}$$

The second form of (21) is arrived at after the substitution of (9), (13), and (19).

It is unfortunate that Earth is not constructed so that the north magnetic pole coincides with the north geographic pole; if it were, $\underset{\rightarrow}{\hat{\mathbf{m}}}_3$ would be inertially fixed and could be taken outside the integral in (21*b*). The dependence of $\underset{\rightarrow}{\hat{\mathbf{m}}}_3$ on t is not too complicated, but its dependence on η forces the analyst to handle Kepler's equation, (9.1, 22), among other unpleasantries. For a discussion of the various approximations that can be made and the corresponding expressions that are eventually derived, the reader is referred to [Thomas and Cappellari]. The approximation we shall make is the simplest one, namely, that $\underset{\rightarrow}{\hat{\mathbf{m}}}_3$ is almost fixed inertially. (This approximation should be best for low orbits; Earth, and $\underset{\rightarrow}{\hat{\mathbf{m}}}_3$, do not rotate much in 100 minutes.)

With $\hat{\underset{\rightarrow}{\mathbf{m}}}_3$ safely outside the integral, and $\hat{\underset{\rightarrow}{\mathbf{o}}}_3$ expressed as in (8)—see Fig. 10.24—the integration proceeds in a straightforward fashion. With the aid of the vector identity

$$\hat{\underset{\rightarrow}{\mathbf{i}}}_1\hat{\underset{\rightarrow}{\mathbf{i}}}_1 \cdot \hat{\underset{\rightarrow}{\mathbf{m}}}_3 + \hat{\underset{\rightarrow}{\mathbf{i}}}_2\hat{\underset{\rightarrow}{\mathbf{i}}}_2 \cdot \hat{\underset{\rightarrow}{\mathbf{m}}}_3 = \hat{\underset{\rightarrow}{\mathbf{m}}}_3 - \hat{\underset{\rightarrow}{\mathbf{i}}}_3\hat{\underset{\rightarrow}{\mathbf{i}}}_3 \cdot \hat{\underset{\rightarrow}{\mathbf{m}}}_3 \tag{23}$$

the result can be expressed as

$$\bar{\bar{\underset{\rightarrow}{\mathbf{g}}}}_m = k_m \hat{\underset{\rightarrow}{\mathbf{p}}}_2 \times \left(3\hat{\underset{\rightarrow}{\mathbf{i}}}_3\hat{\underset{\rightarrow}{\mathbf{i}}}_3 - \underset{\rightarrow}{\mathbf{1}}\right) \cdot \hat{\underset{\rightarrow}{\mathbf{m}}}_3 \tag{24}$$

Expressions such as (24) and the earlier similar expression for average gravitational torque are written in a compact vector notation for brevity. However, in the application of these expressions, it is usually necessary to express the various direction cosines in terms of all the angular parameters of the problem. For example, if an equatorial geocentric coordinate system is being used, the specification of the north magnetic pole ($\hat{\underset{\rightarrow}{\mathbf{m}}}_3$) requires two angles, the orientation of the spin axis ($\hat{\underset{\rightarrow}{\mathbf{p}}}_2$) with respect to the orbit requires two angles, and the orbit itself ($\hat{\underset{\rightarrow}{\mathbf{i}}}_1$ and $\hat{\underset{\rightarrow}{\mathbf{i}}}_2$) requires three angles. Therefore, (24) could be replaced by a very long trigonometric expression involving these seven angles. Here, we shall forbear to do so.

If magnetic torque were the only influence on the satellite attitude, we would have

$$\dot{\underset{\rightarrow}{\mathbf{h}}} = \bar{\bar{\underset{\rightarrow}{\mathbf{g}}}}_m \tag{25}$$

Because $\bar{\bar{\underset{\rightarrow}{\mathbf{g}}}}_m \cdot \hat{\underset{\rightarrow}{\mathbf{p}}}_2 = 0$, the *magnitude of* $\underset{\rightarrow}{\mathbf{h}}$ is unchanging. On the other hand, the long-term angular velocity of the spin axis would be

$$\underset{\rightarrow}{\boldsymbol{\omega}}_p = \frac{k_m}{h}\left(\underset{\rightarrow}{\mathbf{1}} - 3\hat{\underset{\rightarrow}{\mathbf{i}}}_3\hat{\underset{\rightarrow}{\mathbf{i}}}_3\right) \cdot \hat{\underset{\rightarrow}{\mathbf{m}}}_3 \tag{26}$$

When both gravitational and magnetic torques are considered, the long-term motion of the spin axis can be arrived at by superposing (17) and (26).

Eddy Current Torque

The two torques just studied—gravitational and magnetic—have been shown to make the spin vector evolve in direction only; the spin rate remains unchanged. A brief examination will now be made of two environmental influences that do affect spin rate.

The first of these is eddy current torque, which can be important for low-altitude spacecraft when the shell is conducting and permeable. The instantaneous value of this torque was given earlier in Chapter 8, equation (8.4, 6):

$$\underset{\rightarrow}{\mathbf{g}}_{ec} = -k\underset{\rightarrow}{\mathbf{B}} \times \left(\underset{\rightarrow}{\boldsymbol{\nu}} \times \underset{\rightarrow}{\mathbf{B}}\right) \tag{27}$$

It is helpful now to break up $\underset{\rightarrow}{\mathbf{B}}$ into its component $\underset{\rightarrow}{\mathbf{B}}_a$ along the spin axis and the transverse component $\underset{\rightarrow}{\mathbf{B}}_t$, whereupon (27) becomes (Problem 10.12*a*)

$$\underset{\rightarrow}{\mathbf{g}}_{ec} = -kB_t^2\underset{\rightarrow}{\boldsymbol{\nu}} + k\nu B_a \underset{\rightarrow}{\mathbf{B}}_t \tag{28}$$

The first term is clearly a despin torque, while the second term tends to be a precession torque. However, the second term averages out over one satellite rotation:

$$\bar{\mathbf{g}}_{ec} \triangleq \frac{1}{2\pi}\int_0^{2\pi} \mathbf{g}_{ec}(\phi)\, d\phi = -kB_t^2 \boldsymbol{\nu} \tag{29}$$

Therefore, the only consequence of an eddy current torque that matters is its tendency to reduce spin according to

$$\nu(t) = \nu(0)\exp\left\{-\frac{k}{I_a}\int_0^t B_t^2\, dt\right\} \tag{30}$$

The simple derivation of this roughly exponential decrease in spin rate with time is left for Problem 10.12*b*.

Eddy current torque is evidently very strongly dependent on altitude ($\sim R^{-6}$) and is therefore of potential significance only for low-altitude satellites. We could proceed to develop a theory for the orbit-averaged eddy current torque $\bar{\bar{\mathbf{g}}}_{ec}$, but the range of application is so narrow that we shall not devote to it the necessary space.

Solar-Motoring Torque

The second example of a torque that affects spin rate is again applicable to a specialized group of spinning satellites, those that undergo significant thermally induced distortions from solar radiation heating. The thermal distortions are periodic; in fact, they are sinusoidal at the spin frequency. Surface-dependent forces, chiefly solar radiation pressure (Section 8.3) and aerodynamic pressure (Section 8.2), are therefore also periodic. There is also a phase lag introduced because the structure takes a finite time to respond to the periodic heat input.

Although the solar-motoring phenomenon could occur in principle for any thermally responsive spinning satellite, the configuration for which it has been found in practice to be significant is the "crossed dipole" configuration (see, for example, the ISIS satellite in Fig. 10.18). Figure 10.25 shows such a satellite. Two pairs of dipole boom antennas are shown, and the spin vector is normal to the page. The solar vector is also assumed (temporarily) to lie in the plane of the page; one pair of booms is shown deflected (in the plane of the page) as a consequence of solar heating. It is visually evident that the torque about the mass center due to solar radiation pressure on the right-hand boom is less than if it were undeflected. At the same time, the torque on the left-hand boom is greater on account of its deflection. The implication is that a despin torque is produced.

To calculate the magnitude of this solar-motoring torque requires an analysis of the dynamics of a vibrating boom in a centrifugal force field, its response to a periodically varying thermal input, and the integrated effect of solar pressure on the deformed configuration. The interested reader can consult [Etkin and Hughes] for the details. This theory has been extended by [Hughes and Cherchas, 1] to cover arbitrary solar incidence angles.

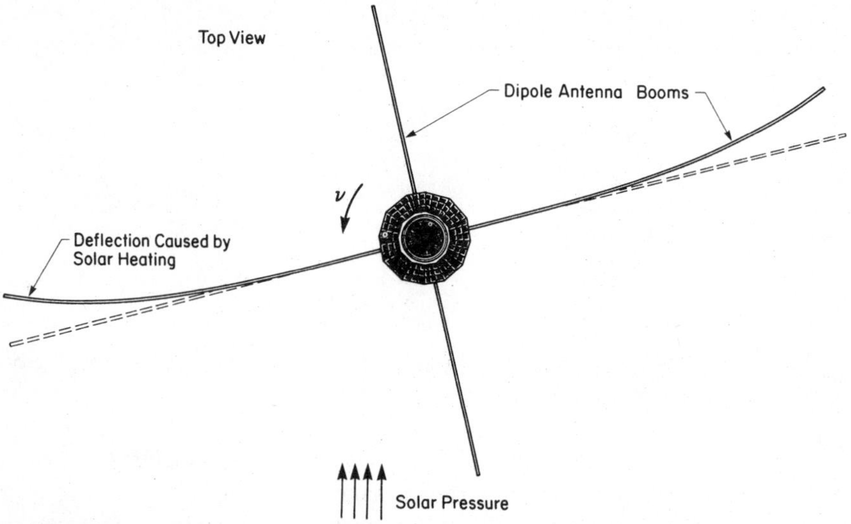

FIGURE 10.25 The solar motoring mechanism for a spinning satellite with long slender rods in the spin plane.

Two last comments on solar motoring: [Etkin and Hughes] have shown that the solar motoring torque on a dipole boom pair will be *pro*-spin if the spin rate is above the first natural frequency of in-plane vibration of the booms (which itself depends on the spin rate), although at such higher spin rates centrifugal straightening tends in any case to reduce the significance of the solar-motoring mechanism. And [Vigneron and Garrett] have pointed out that aerodynamic pressure also could combine with solar heating to produce "aero-solar" motoring as well as "solar-solar" motoring.

Other Torque Analyses

For each type of spin-stabilized satellite there are many torque mechanisms that should be considered quantitatively. For example, in connection with crossed-dipole spinners (Fig. 10.18), [Vigneron, 3] has identified seven torque mechanisms that tilt the spin vector and eight that affect the spin rate! It is obviously not feasible to include all possible torques for all possible satellites in this book. However, the general discussion of spacecraft torques in Chapter 8 and the more specific analyses above for four of the most important torques on spinners, go well beyond a superficial treatment of the subject.

Many detailed analyses of the effects of some or all of the major environmental torques have been published. A sampling of these is classified in Table 10.1. As is always the case, a highly accurate analysis of all the dynamical interactions that are likely to be important for a specific satellite leads inevitably to computer simulation, a process that produces large quantities of data for the particular satellite in question but that does not often give general insights.

TABLE 10.1
Additional Published Analyses of Environmental Torque Effects on Spin-Stabilized Satellites

Reference	Year	Gravity-gradient	Magnetic Field	Solar Radiation	Aerodynamics	Orbit Regression
		SOURCE OF TORQUE				
[Grasshoff]	1960	✓				
[Hultquist]	1961	✓				
[Clancy and Mitchell]	1964			✓		
[Beletskii, Chapter 7]	1965				✓	
[Beletskii, Chapter 8]	1965	✓			✓	✓
[Beletskii, Chapter 9]	1965		✓	✓		
[van den Brock]	1967	✓				✓
[Johnson]	1968				✓	
[Hodge]	1972	✓	✓	✓	✓	
[Modi and Pande]	1973	✓		✓		
[Hara]	1973	✓	✓			

Flight Data — Early Lessons

The first U.S. satellite, Explorer 1, shown in Fig. 10.26 was spin stabilized. At least, that was the plan. As it turned out, Explorer 1 taught a lesson so profound that the view of "rigid body" dynamics has not been the same since.

It is evident from Fig. 10.26 that a spin about the axis of symmetry is a minor-axis spin and that the small lateral whip antennas provide a source of

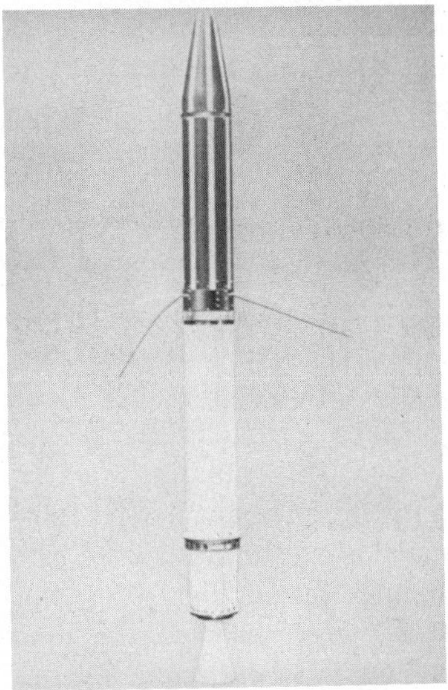

FIGURE 10.26 Explorer 1.

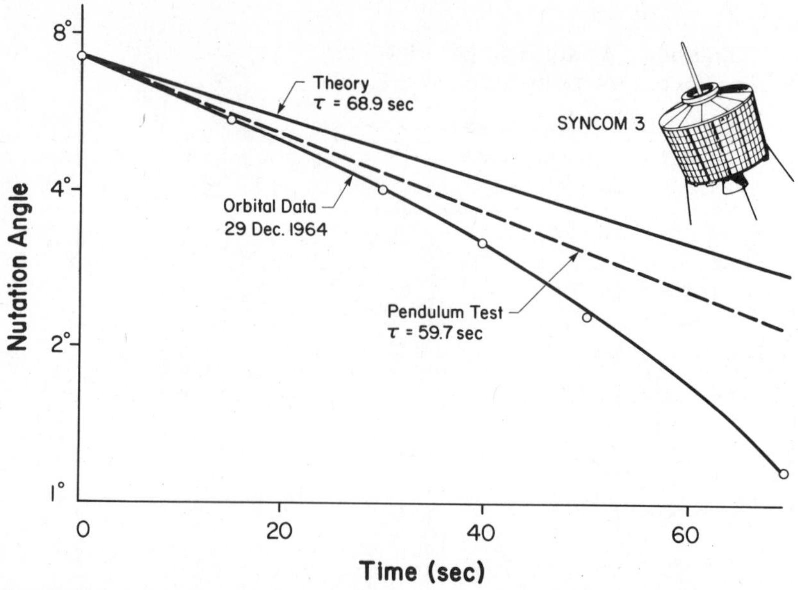

FIGURE 10.27 Nutation damper performance for SYNCOM 3.

structural energy dissipation. Prior to Explorer 1, the second inference would not have been made, because the question would not have been asked. The major-axis rule, described in detail in Section 5.1, states that the spin of a real physical body will not be stabilizing unless the spin is arranged to occur about the major axis of inertia. Otherwise, energy dissipation is destabilizing. Explorer 1 was spun about its minor axis and was consequently unstable.

[Bracewell and Garriott] appear to have been the first to enunciate the major-axis rule. (They appear also to have been the first, and perhaps the last, to describe the slowly degrading precession/nutation of a spinning rod-like body as "waltzing.") Their paper in *Nature*, published less than eight months after the Explorer 1 launch, speculated that it was the minor-axis spin stabilization concept that had failed.

Once the significance of energy dissipation was properly understood, dissipators were included as part of spacecraft design, as extensively discussed in Section 10.2. (A rare exception was Explorer 45, whose nutation damper was a casualty of a weight-reduction program; see [Flatley, 2].) For example, Fig. 10.27 shows the effectiveness of the damper on SYNCOM 3, which was spin stabilized at ~ 160 rpm; the data are from [DeBra]. (SYNCOM 3 has appeared earlier, in Fig. 10.16). The mission phase shown is an early, temporary one. If a nutation damper functions properly, the distinction between the spin axis and the angular-momentum vector quickly disappears. Once the nutation angle has shrunk to zero, the characteristic time for dynamical effects is not minutes (as in Fig. 10.27) but weeks or months. The question of spin stability having been resolved, the issue becomes the effect of environmental torques on the angular momentum vector.

Flight Data — Spin Rate

The first artificial Earth satellite, Sputnik 1, was observed to experience a decay in spin rate. Data from [Beletskii] have been used as the basis for Fig. 10.28. He attributes the despin to aerodynamic friction on the antennas. (Sputnik's mean height was 587 km.)

For the next example we choose a satellite launched one-half decade later into an almost circular orbit (1014 km altitude), namely, Alouette 1. [Etkin and

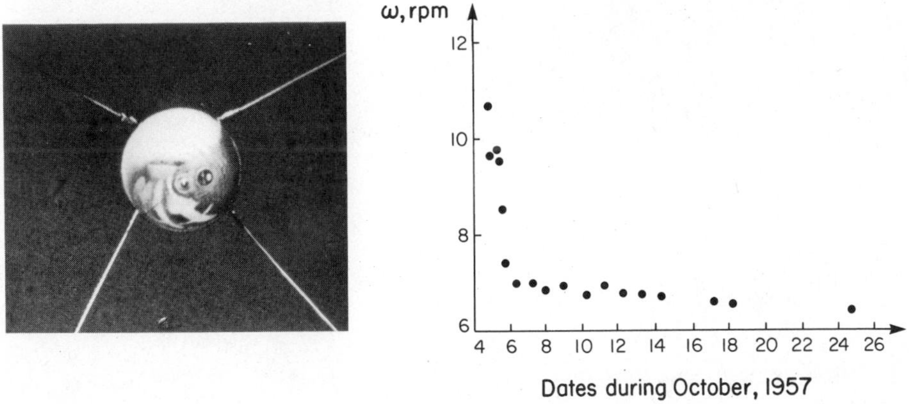

(a) Sputnik 1 (b) Spin Rate Data, based on [Beletskii]

FIGURE 10.28 Spin decay of Sputnik 1.

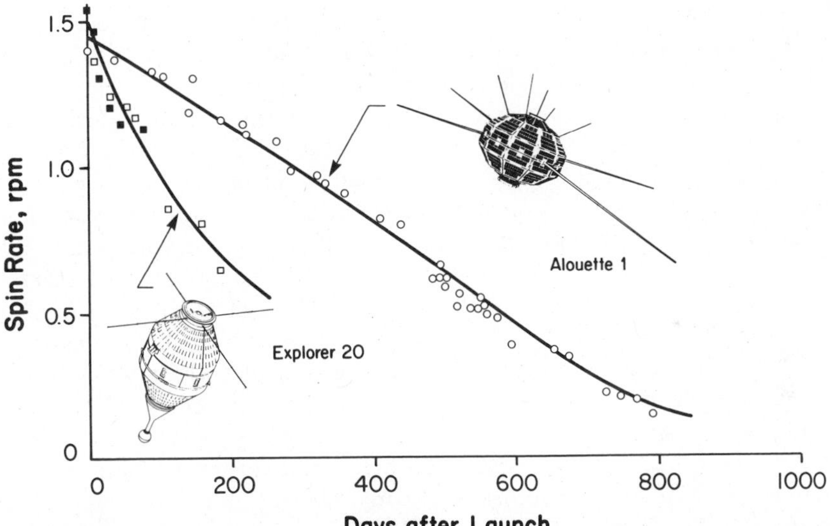

FIGURE 10.29 Spin decay caused by solar motoring for Alouette 1 and Explorer 20.

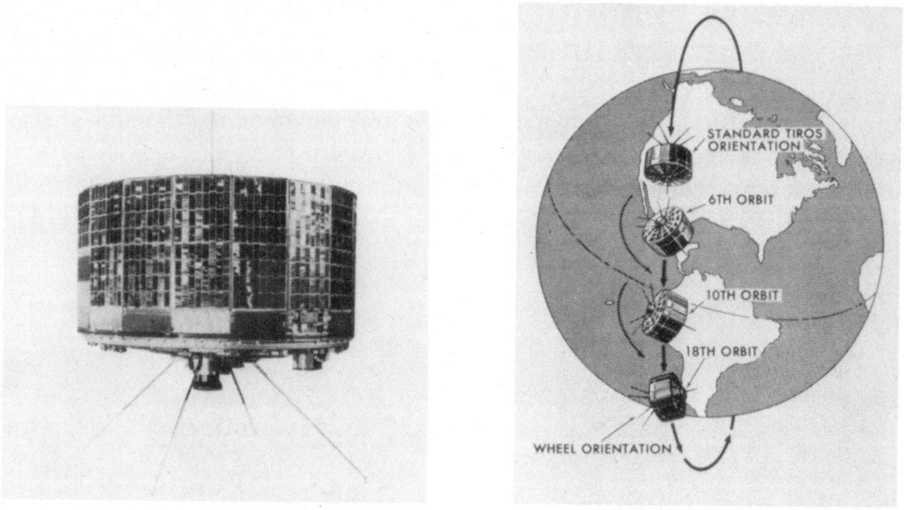

(a) Satellite Configuration [Schnapf]

(b) The TIROS IX Spin Vector Maneuver [NASA, 16]

FIGURE 10.30 The TIROS Satellite.

Hughes] have shown that the observed spin decay (Fig. 10.29) can be explained by the "solar motoring" mechanism. According to [Hughes and Cherchas, 2], the same hypothesis also explains the spin-down of another satellite with long, thermally responsive dipole antennas, Explorer 20. Solar motoring may occasionally be important also for spinning satellites whose thermally responsive elements are not long slender booms; for example, [Patterson and Kissell] have compared flight data with theory for PAGEOS, an aluminized balloon-like satellite.

The TIROS series of meterological satellites were spin stabilized in low Earth orbits (Fig. 10.30). An interesting feature of this series is that its early members (TIROS 1-8) retained initially the spin vector imparted by the launch vehicle. Thus, their spin vectors were nominally in the orbital plane; they followed what we have earlier called the "high spin" strategy. TIROS 9, on the other hand, was maneuvered in orbit (Fig. 10.30*b*), so that its spin axis was normal to the orbital plane (Thomson equilibrium). Figure 10.31*a* shows a typical spin rate history for TIROS 9 over 120 orbits. The diagram features a series of segments in each of which the spin rate decays, primarily from the action of eddy current torque ([Schnapf]). After each dozen orbits or so, a magnetic spin control system was energized to boost the spin rate.

Instances have been cited above for three mechanisms of spin decay: aerodynamic friction, solar motoring, and eddy currents. Further examples of flight data are offered in Table 10.2. We turn now to look at some flight experience with how the *direction* of the spin vector, ideally fixed, can evolve over many orbits.

TABLE 10.2
Additional Published Analyses of Flight Data for Spin-Stabilized Satellites

Reference	Year	Satellite	TYPE OF DATA: Spin Rate	Spin Direction	PRINCIPAL TORQUES: Eddy Currents	Aerodynamic Friction	Solar Motoring	Gravity Gradient	Magnetic Moment	Orbit Evolution	Magnetic Hysteresis	Solar Pressure
[Wilson]	1960	Vanguards 1,2	✓		✓							
[Naumann]	1962	Explorers 4,7,8,11		✓				✓	✓	✓		
[Thomas and Cappellari]	1964	Telstars 1,2		✓				✓	✓	✓		
[Beletskii], Chapter 10	1965	Sputnik 3	✓	✓	✓	✓		✓	✓	✓		
[Vigneron, 3]	1973	Alouette 2	✓	✓			✓	✓	✓	✓	✓	
[Vigneron, 3]	1973	ISIS 1	✓		(not conclusively identified)							
[Perry and Slater]	1973	Prospero	✓		✓							
[Davison and Merson]	1973	SKYNET (geostationary)		✓								✓
[Flatley, 2]	1975	Explorer 45 (no damper)	✓				✓					
[Patterson and Kissell]	1975	PAGEOS	✓				✓					

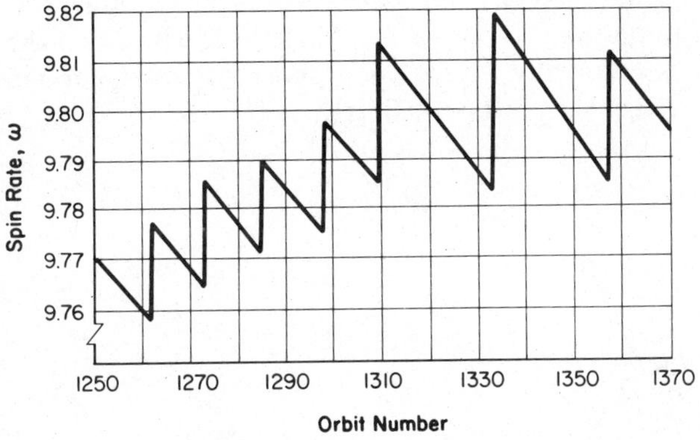

(a) Typical Spin Rate History for TIROS IX, [Schrapf]

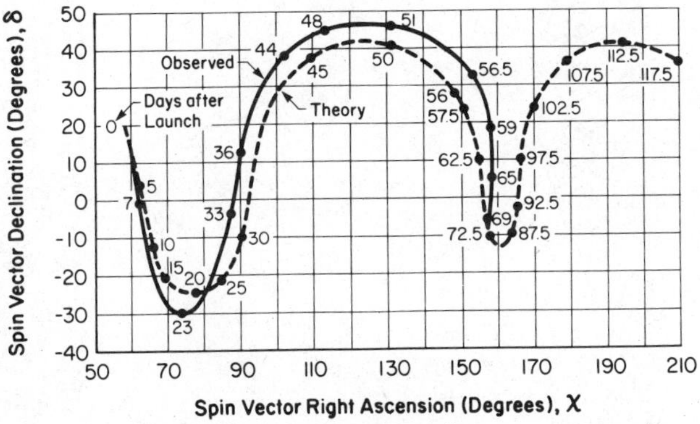

(b) Spin Vector History for TIROS I, [Bandeen and Manger]

FIGURE 10.31 TIROS flight data.

Flight Data — Spin Vector Direction

The chief causes of spin vector precession have tended to be gravitational and magnetic torques. One early example is TIROS 1, whose spin vector history over a postlaunch period of some four months is shown in Fig. 10.31*b*. (The *celestial coordinates* used here are explained in Fig. 10.32. The *declination* is positive north of the celestial equator; the *right ascension* is positive east of the vernal equinox.) To understand these data, one must realize that the satellite is in a nearly polar orbit. Therefore, the orbit normal is essentially parallel to the equator and so, as shown in earlier analysis, the precession due to gravity gradient will be about this orbit normal. The torque from residual axial magnetic moment is also expected, from the analysis above, to cause precession of the spin vector. [Bandeen and Manger] combined these two torques and demonstrated the good agreement with flight data shown in Fig. 10.31*b*.

For a second example of gravitational and magnetic torquing of the spin vector, we turn again to Alouette 2. Even though this satellite looks quite different from TIROS (Fig. 10.29), the basic dynamical actions are the same. [Graham] modeled these two torques for Alouette 1 and achieved the agreement with flight data shown in Fig. 10.33*a*. Unlike TIROS 1, for which the magnetic torque was dominant, the gravitational torque was strongest for Alouette 1. Therefore, the nodding of the spin vector is essentially a precession about the orbit normal, combined with a slow precession of the orbit normal itself.

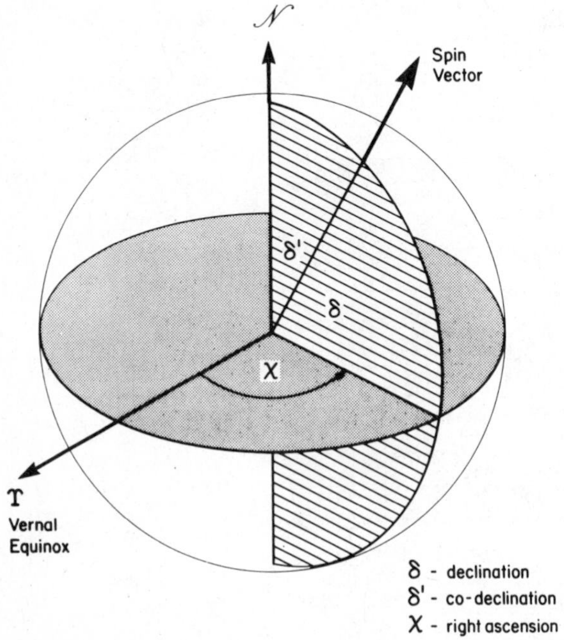

FIGURE 10.32 Spin vector orientation on the celestial sphere.

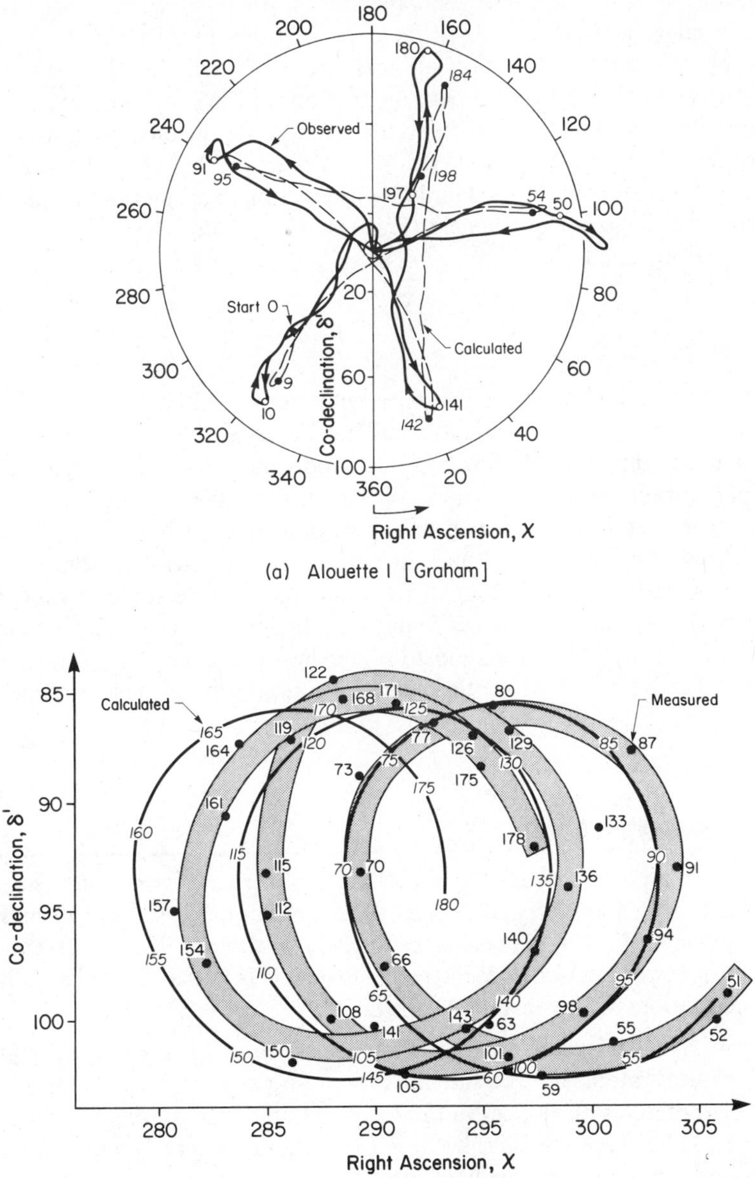

(a) Alouette I [Graham]

(b) ISIS-I [Vigneron,3]

FIGURE 10.33 Nodding of spin vector caused by gravitational and magnetic-moment torques.

Another view of this same phenomenon is provided by ISIS 1, shown in Fig. 10.18. [Vigneron, 3] has published the data shown in Fig. 10.33*b*. The path of the spin vector is the superposition of three factors. First, the gravitational torque (alone) would cause a circular path (coning about the orbit normal). Second, the center of this "circle" slowly drifts to the left because the orbit normal, which is at the center of the "circle," itself drifts slowly to the left. Finally, the spin axis trajectory is further modified slightly by interaction between the satellite magnetic moment and the geomagnetic field. This last effect is not large for ISIS 1 because its average altitude is 2000 km.

Other Flight Data

Much additional flight experience with spin-stabilized satellites has been published; some sources are tabulated in Table 10.2. From the data already presented in this section however, an understanding can be gained of the principal torque mechanisms and of when they are important for spinning satellites. Spin stabilization has been the workhorse of attitude control during the 1960s. Now reliable and well understood, it is usually used in conjunction with episodic, open-loop, active control to adjust the spin vector direction and rate. However, during the 1970s it was found that the highly successful characteristic of gyric "stiffness," created in its most basic form by spinning the entire spacecraft, can be achieved by more subtle means. We now turn to a discussion of these ideas.

10.4 PROBLEMS

10.1 In connection with the wobbling of a rigid, axisymmetrical satellite spinning about an axis normal to its circular orbit, show that the transformation of (10.1, 13) converts (10.1, 11), which are linear differential equations with periodic coefficients, to (10.1, 14), that is, linear differential equations with constant coefficients.

10.2 A rigid axisymmetrical satellite is spinning in a circular orbit whose orbital rate is ω_c. The satellite is spinning at rate $\nu + \omega_c$ about its axis of symmetry with respect to the "stroboscopic frame" $\mathscr{F}_s$ defined in Section 10.1. Let $\hat{\nu} = \nu/\omega_c$ and let $k_t = (I_a - I_t)/I_t$, where I_a and I_t are the axial and transverse moments of inertia of the satellite.

(*a*) Show that the spin axis can persist in an orientation normal to the forward direction but inclined to the vertical at an angle γ_3 given by

$$\cos\gamma_3 = \frac{(1 + k_t)(1 + \hat{\nu})}{4k_t} \tag{1}$$

provided the right side is less than unity in magnitude (the "conical" Likins–Pringle equilibrium).

(*b*) Show that the magnitude of the absolute angular velocity of the satellite is

$$\omega = \nu[1 + 2(1 + \hat{\nu})(1 - c_{22}^{s})] \tag{2}$$

(*c*) Show that $\omega = \nu$ for the Thomson equilibrium.

10.3 (*a*) Show that the Hamiltonian for the motion of a spinning symmetrical rigid body in a circular orbit is given by (10.1, 60).

(*b*) For Thompson's equilibrium (spin axis normal to orbit), show that the Hamiltonian-based Liapunov function is given by (10.1, 70) and, to second order, by (10.1, 72).

(*c*) For the hyperbolic Likins–Pringle equilibrium (spin axis normal to the vertical but inclined to the orbit normal), show that the corresponding Liapunov function is given by (10.1, 73) and, to second order, by (10.1, 74). [*Hint*: The definition of $\cos\gamma_3$, (10.1, 39) will be needed.]

(*d*) For the conical Likins–Pringle equilibrium (spin axis normal to the forward direction but inclined to the orbit normal), show that the corresponding Liapunov function is given by (10.1, 75) and, to second order, by (10.1, 76).

10.4 A tri-inertial rigid body is in a circular orbit and is spinning about the orbit normal. The infinitesimal angular deviations $\{\alpha_1, \alpha_3\}$ of the spin axis from the orbit normal are governed by the linear-periodic differential equations (10.1, 89).

(*a*) By defining the state vector

$$\mathbf{x} \triangleq \begin{bmatrix} \alpha_1 & \alpha_3 & \dfrac{\dot{\alpha}_1}{\omega_c} & \dfrac{\dot{\alpha}_3}{\omega_c} \end{bmatrix}^T \tag{3}$$

show that the state equations are of the form

$$\mathbf{x}' = \mathbf{A}(\eta)\mathbf{x} \tag{4}$$

where $\eta \triangleq \omega_c t$, $(\cdot)' \triangleq d/d\eta$, and $\mathbf{A}$ has the partitioned form

$$\mathbf{A} \triangleq \begin{bmatrix} \mathbf{O}_{2\times2} & \mathbf{1}_{2\times2} \\ \mathbf{A}_{21}(t) & \mathbf{A}_{22}(t) \end{bmatrix}$$

$$\mathbf{A}_{21}(t) \triangleq \begin{bmatrix} -k_1\left[(\theta_2')^2 + 3cn^2\,at\right] & 3(k_2 - k_1)\,sn\,at\,cn\,at \\ -3(k_2 + k_3)\,sn\,at\,cn\,at & -k_3\left[(\theta_2')^2 + 3sn^2\,at\right] \end{bmatrix}$$

$$\mathbf{A}_{22}(t) \triangleq \begin{bmatrix} 0 & (k_1 - 1)\theta_2' \\ (1 - k_3)\theta_2' & 0 \end{bmatrix}$$

The symbols are defined in (10.1, 89) et seq.

(*b*) Integrate (4) over the period $\Delta t = T$ given by (10.1, 96) four times, once for each of the elementary linearly independent initial conditions (10.1, 97) and verify the stability diagrams of [Kane and Shippy] shown in Fig. 10.6.

10.5 Consider the directional stability of a spinning, inertially axisymmetrical satellite in a circular orbit. The reference motion is the Thomson equilibrium whose stability, in the absence of damping, was considered in (10.1, 23) et seq., resulting in the stability diagram of Fig. 10.2. In this problem, the effect on stability of a small damping term is studied. For small deviations $\{\gamma_1, \gamma_3\}$ from the reference spin, the equations of motion are

$$\gamma_1'' + d\gamma_1' - g\gamma_3' + k_1\gamma_1 = 0$$
$$\gamma_3'' + g\gamma_1' + k_3\gamma_3 = 0 \tag{5}$$

where $(\cdot)' \triangleq d/d\eta$, $\eta = \omega_c t$, and g, k_1, and k_3 are given below equation (10.2, 2) in the text. The small damping parameter is d, with $d \ll 1$.

(*a*) Show that the damping is pervasive if $gk_3 \neq 0$.

(*b*) In the absence of damping ($d = 0$), show that the natural frequencies corresponding to high-spin motion ($|\hat{\nu}| \gg 1$) are

$$\omega_{ry1} \doteq 1; \qquad \omega_{ry2} \doteq |\hat{\nu}|(1 + k_t) \tag{6}$$

Note that the motion is always stable in this high-spin case. Note also that one mode is a "slow mode" while the other is a "fast mode."

(*c*) By substitution of

$$s = j\omega_{ry} - \delta$$

in the characteristic equation associated with (5), and with $O(\delta) = O(d)$, show that the two pure imaginary roots $j\omega_{ry1}$ and $j\omega_{ry2}$ with damping absent shift to the left when $d > 0$ by the following amounts for $|\hat{\nu}| \gg 1$:

$$\delta_1 = -\frac{d}{2(1 + k_t)\hat{\nu}}; \qquad \delta_2 = \frac{1}{2}d \tag{7}$$

Note that δ_2 is always asymptotically stabilizing whereas δ_1 is asymptotically stabilizing only if $\hat{\nu} < 0$; compare with Fig. 10.10.

(*d*) In the unstable situation where $\hat{\nu} \gg 1$, show that the number of orbits to double amplitude for the unstable mode is

$$N = \frac{(1 + k_t)\hat{\nu} \ln 2}{\pi d}$$

For $d = 10^{-4}$ and $\hat{\nu} = 10^3$, show that more than a million orbits are required to reach double amplitude.

10.6 (*a*) For a satellite modeled as shown in Fig. 10.11, derive the motion equations (10.2, 3).

(*b*) Using Floquet theory (Section A.3 of Appendix A), prepare diagrams in the $\{k_t, \hat{\nu}\}$ plane for directional stability (asymptotic to the orbit normal) corresponding to the following dimensionless parameters:

$$\hat{I}_{td} = 0.01; \qquad \hat{m}_d = 0.01; \qquad \hat{c}_d = 1; \qquad \hat{k}_d = 1$$

where

$$\hat{I}_{td} \triangleq \frac{m_d b^2}{I_t}; \qquad \hat{m}_d \triangleq \frac{m_d}{m}$$

$$\hat{c}_d \triangleq \frac{c_d}{m_d \omega_c}; \qquad \hat{k}_d \triangleq \frac{k_d}{m_d \omega_c^2} \tag{8}$$

(*c*) Can you offer any comments on the difficulties encountered when $\nu \doteq -\omega_c$?

10.7 (*a*) For a satellite modeled as shown in Fig. 10.12, derive the motion equations (10.2, 14).

(*b*) Suppose the inertia ratios for the inner and outer body are identical ($k_{td} = k_t$). Show that the characteristic equation (10.2, 23) reduces to

$$(s^4 + b_1 s^2 + b_2) p_4(s) = 0 \tag{9}$$

where b_1 and b_2 are defined by (10.1, 24) and the fourth-degree polynomial $p_4(s)$ is as follows:

$$p_4(s) \triangleq (s^2 + \delta s + \mathscr{k}_1 + \kappa)(s^2 + \delta s + \mathscr{k}_3 + \kappa) + [\mathscr{g} s + (1 + \hat{\nu})\delta]^2 \tag{10}$$

The symbols δ, κ, and $\mathscr{g}$ are defined as

$$\delta \triangleq \frac{\hat{c}_d}{\hat{I}_{td}(1 - \hat{I}_{td})}; \qquad \kappa \triangleq \frac{\hat{k}_d}{\hat{I}_{td}(1 - \hat{I}_{td})}$$

$$\mathscr{g} \triangleq 2 + (1 + k_t)\hat{\nu} \tag{11}$$

and $\mathscr{k}_1$ and $\mathscr{k}_3$ were defined earlier by (10.1, 25).

(*c*) Note that four of the roots now correspond to undamped sinusoidal oscillations provided the stability conditions (10.1, 27) are satisfied. Explain this result physically.

(*d*) Now remove the assumption that $k_{td} = k_t$, used in parts (*b*) and (*c*), and assume instead that $\hat{I}_{td}$, $\hat{c}_d$, and $\hat{k}_d$ are small:

$$\hat{I}_{td} = \varepsilon; \qquad \hat{c}_d = \varepsilon\delta; \qquad \hat{k}_d = \varepsilon\kappa \tag{12}$$

where $0 < \varepsilon \ll 1$. (*Note*: It is not implied that δ is necessarily small.) Show that, to $O(\varepsilon^2)$, the characteristic equation (10.2, 23) reduces to

$$(s^4 + b_1 s^2 + b_2) p_4(s) = 0 \tag{13}$$

where b_1 and b_2 are defined by (10.1, 24) and the fourth-degree polynomial $p_4(s)$ is

$$p_4(s) \triangleq (s^2 + \delta s + \mathscr{k}_{1d} + \kappa)(s^2 + \delta s + \mathscr{k}_{3d} + \kappa) + [\mathscr{g}_d s + (1 + \hat{\nu})\delta]^2 \tag{14}$$

where

$$\mathscr{k}_{1d} \triangleq k_{td}(3 - \hat{\nu}) - 1 - \hat{\nu}$$

$$\mathscr{k}_{3d} \triangleq -k_{td}\hat{\nu} - 1 - \hat{\nu}$$

$$\mathscr{g}_d \triangleq 2 + (1 + k_{td})\hat{\nu} \tag{15}$$

(*e*) Show that the stability diagram corresponding to the eighth-degree polynomial (13) is as shown in Fig. 10.34 for $k_{td} = -0.1$ and

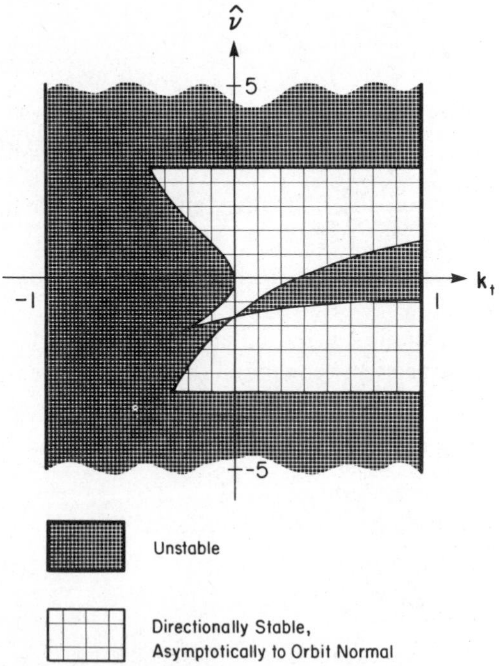

FIGURE 10.34 Stability diagram for Problem 10.7*e*.

$\delta = \kappa = 1$. [*Hint*: Use the conditions (10.1, 26) for b_1 and b_2 united with the Routh–Hurwitz conditions associated with $p_4(s)$.] Compare the $O(\varepsilon^2)$ diagram in Fig. 10.34 to the "exact" diagram in Fig. 10.13*b* for $\varepsilon = 0.01$.

10.8 Consider a spinning satellite containing a point mass damper as sketched in Fig. 5.4 and make the following assumptions: (i) the damper mass is very small, (ii) therefore the attitude motion is instantaneously very similar to undamped motion, (iii) external torques are weak and can be neglected over the short term, and (iv) the satellite is axisymmetrical. Show from the motion equations (5.2, 11 and 12) that the damper is "tuned" (i.e., has its resonance at the frequency of excitation) when

$$k_d \doteq m_d k_t^2 \nu^2 \tag{16}$$

where k_d is the damper spring constant, m_d is the damper mass, $k_t = I_a/I_t - 1$, and ν is the spin rate.

10.9 Consider an inertially axisymmetric rigid satellite, in a circular orbit, spinning about the orbit normal. That is, consider the Thomson equilibrium (Fig. 10.1). An infinitesimal analysis indicates that a substantial region in the $\{k_t, \hat{\nu}\}$ plane is "gyrically directionally stabilized," even though "statically unstable" (see Fig. 10.2). [Hitzl, 2] has shown that finite attitude excursions may produce instabilities in this region, particularly in

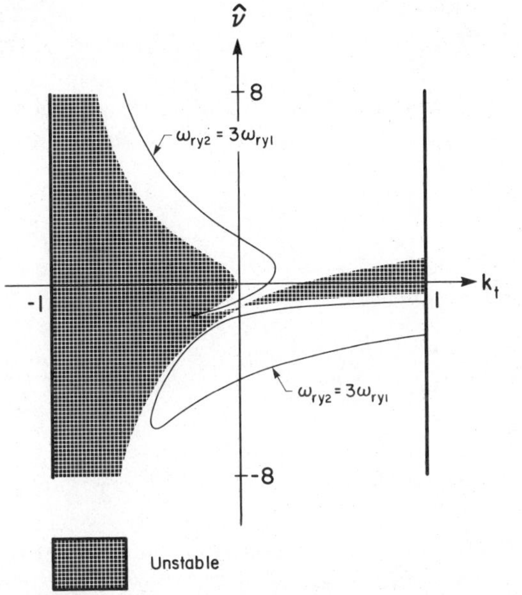

FIGURE 10.35 The nonlinear resonance lines $\omega_{ry2} = 3\omega_{ry1}$ for an axisymmetrical rigid satellite spinning normal to a circular orbit. (See Problem 10.9.)

the neighborhood of the *nonlinear internal resonance*

$$\omega_{ry2} = 3\omega_{ry1} \tag{17}$$

where ω_{ry1} and ω_{ry2} are the two natural frequencies associated with the "roll/yaw" wobbling of the spin axis; they are the positive roots of

$$\left(\frac{\omega}{\omega_c}\right)^4 - b_1\left(\frac{\omega}{\omega_c}\right)^2 + b_2 = 0 \tag{18}$$

where ω_c is the circular orbital rate and b_1 and b_2 are defined by (10.1, 24). Show that the curve in the $\{k_t, \hat{\nu}\}$ plane corresponding to (17) is as shown in Fig. 10.35.

10.10 This problem is concerned with the pendulum damper depicted in Fig. 10.19 and analyzed in Section 10.2.

(*a*) Show that, assuming the satellite's small wobbling motion can be approximated over short periods by the undamped expressions (10.2, 28), the motion equation for the pendulum damper is as indicated in (10.2, 33).

(*b*) Verify (10.2, 38), the equation giving the rate of energy loss for the satellite.

(*c*) If a spring is not allowed as part of the design, show that the tuning condition ($\Omega = \omega_d$) reduces to the following simple condition on the lengths l_1 and l_2:

$$(I_2 - I_1)(I_2 - I_3)l_2 = I_1 I_3 l_1 \tag{19}$$

Is it significant that this condition is independent of ν? Can you also conclude that l_1 is always less than l_2 for a tuned pendulum damper (without spring)?

(*d*) Show that the measure of wobble, ω_w, defined in (10.2, 35), is "good" in the sense that for an inertially axisymmetrical satellite, ω_w reduces to (see Section 4.2)

$$\omega_w = \omega_t$$

Show also that $\phi = \chi$.

(*e*) Still for an inertially axisymmetrical satellite, show that the rate of energy extraction to aid in suppressing wobble is

$$\dot{T} = -\frac{m_d b^2}{4\zeta_d}\left[\frac{(1+k_t)^2}{k_t}\right]\nu\omega_t^2 \tag{20}$$

(*f*) Noting from (7.1, 58) that

$$\dot{\gamma} = \frac{\dot{T}}{h_t k_t \nu} \tag{21}$$

[*Caution*: the sign convention for Ω in (7.1, 58) is opposite to that in the present context], and from (4.2, 12 through 15) that

$$\omega_t = (1 + k_t)\nu\gamma \tag{22}$$

show that

$$\dot{\gamma} = -\Lambda\nu\dot{\gamma} \tag{23}$$

where

$$\Lambda = \frac{1}{4\zeta_d}\frac{m_d b^2}{I_a}\left[\frac{(1+k_t)^4}{k_t^2}\right] \tag{24}$$

and hence that the pendulum damper reduces the nutation angle γ according to

$$\gamma = \gamma_o e^{-\Lambda\nu t} \tag{25}$$

with $\gamma_o = \gamma(0)$.

10.11 The gravity gradient torque on a rigid body $\mathscr{R}$ whose inertia dyadic is $\underset{\rightarrow}{\mathbf{I}}$ and which is at position $-\underset{\rightarrow}{\hat{\mathbf{o}}}_3 R_c$ with respect to the center of a planet whose gravitational constant is μ was derived in Chapter 8 to be as given by (8.1, 16). If the principal axis $\hat{\mathbf{p}}_2$ of $\mathscr{R}$ is an axis of inertial symmetry so that the axial and transverse inertias of $\mathscr{R}$ and I_a and I_t, show that the gravity gradient torque on $\mathscr{R}$ is given by

$$\underset{\rightarrow}{\mathbf{g}}_c = 3\left(\frac{\mu}{R_c^3}\right)(I_a - I_t)\underset{\rightarrow}{\hat{\mathbf{o}}}_3 \times \underset{\rightarrow}{\hat{\mathbf{p}}}_2\underset{\rightarrow}{\hat{\mathbf{p}}}_2 \cdot \underset{\rightarrow}{\hat{\mathbf{o}}}_3 \tag{26}$$

[*Hint*: The formula in Problem 3.7*b* should be useful.]

10.12 (*a*) Derive the formula (10.3, 28) for the eddy current torque on a spinning satellite. To begin, use (10.3, 27).

(*b*) Derive the consequent exponential decrease in spin rate, as stated in (10.3, 30).

CHAPTER 11

DUAL-SPIN STABILIZATION IN ORBIT: GYROSTATS AND BIAS MOMENTUM SATELLITES

The fundamental idea behind spin stabilization of spacecraft is that the direction about which the spin occurs tends to remain fixed. This strategy can be carried out either in concordance with gravitational torque (as when the spin axis is normal to the orbit) or in defiance of it (the "high spin" option). The ideas germane to torque-free motion of spinners are thoroughly discussed in Chapters 4 and 5, and the application to real spacecraft has been the subject of the previous chapter.

The present chapter traces the further evolution of these concepts and the formation of new species—*dual-spin* spacecraft—in which the vehicle comprises *two* bodies with a common axis of rotation. Thus, if the two bodies are labeled $\mathscr{B}_1$ and $\mathscr{B}_2$ and the (absolute) angular velocity of $\mathscr{B}_1$ is $\underset{\rightarrow}{\boldsymbol{\omega}}_1$, the (absolute) angular velocity of $\mathscr{B}_2$ must be $\underset{\rightarrow}{\boldsymbol{\omega}}_2 = \underset{\rightarrow}{\boldsymbol{\omega}}_1 + \omega_s \underset{\rightarrow}{\mathbf{a}}$, where ω_s is the spin of $\mathscr{B}_2$ relative to $\mathscr{B}_1$ and $\underset{\rightarrow}{\mathbf{a}}$ is the unit vector along their common axis. The reader may care to glance back at some figures in Chapters 6 and 7 for examples (in torque-free motion): Figs. 6.1, 7.2, 7.3, 7.5, 7.8, 7.11, and 7.14.

Virtually all dual-spin applications to date have had two special characteristics. First, one of the bodies is intended either to be inertially fixed or to be Earth-pointing; second, the other body is intended to be inertially axisymmetrical. For practical purposes, then, one of the bodies is "spinning" either not at all or very slowly, and the truly spinning body is a wheel or rotor. Sometimes this wheel is small, internal, and spins quite rapidly; sometimes it is large, external, and spins rather more slowly. By way of illustration, look at Anik C (Fig. 11.15) and Hermes (Fig. 11.24). Both are "Earth pointing" communications satellites in geostationary orbit, but a casual regard would suggest them to be members of two quite different dynamical families. Anik C consists largely of its rotor, which spins at ~ 100 rpm; its antenna assembly, however, is essentially nonspinning

(1 revolution per day). By contrast, every visible part of Hermes is Earth pointing (and thus essentially nonspinning); one requires "inside" information to become aware that a "momentum wheel," spinning at ~ 3000 rpm, resides within the central structure of the vehicle. Satellites like Anik C tend to be called "gyrostats," motivated by the classical gyrostat and encouraged by the terminology of Hughes Aircraft, the foremost commercial developer of this genre. Satellites like Hermes tend to be called "bias momentum satellites" in recognition of the fact that the internal momentum wheel is aboard precisely because it dominates the (uncontrolled) attitude dynamics.

From this brief discussion it will be evident that spacecraft of the dual-spin type have two quite different origins. One origin is the spin-stabilized satellite, for which it is a nuisance that everything on board is spinning and for which the major-axis rule has historically posed geometrical constraints difficult to reconcile with long, slender launch vehicles. It is better to have only *most* of the vehicle spinning, thus retaining gyric stiffness while providing at the same time a platform on which to mount devices to receive or transmit information. Furthermore, an important lesson of Chapter 7 (see, for example, Problem 7.3) is that a dual-spin design can readily be allowed to violate the major-axis rule. Thus, dual-spin stabilization is a natural extension of spin stabilization.

The second origin of dual-spin spacecraft is from quite the opposite point of view. One begins with a single-bodied configuration that is either inertially stabilized or Earth pointing. In either case the vehicle is essentially nonspinning and will require an active attitude control system ("active" in the sense of [Sabroff, 2]). As the designer reflects on the many sensors, actuators, fuel tanks, and so on, that an active control system typically entails, he is willing to consider the possibility that one or more devices for storing internal angular momentum (wheels) may simplify the control problem through their beneficial modification of the attitude dynamics.

This chapter begins (Section 11.1) by an examination of the neoclassical problem of a rigid gyrostat in a circular orbit, subject only to gravity gradient torque. Though this model is something of an idealization, analytical results are available that lay the foundation for understanding the numerical simulations of more sophisticated models. Gyrostats form the subject of Section 11.2, and bias momentum satellites are discussed in Section 11.3.

11.1 THE GYROSTAT IN ORBIT

The term "gyrostat" as used in this section is meant to have its classical meaning: a rigid body containing one or more rigid, spinning wheels. As explained in Section 6.1, if there are several wheels, they can for dynamics analysis purposes be replaced by a single equivalent wheel; only the total angular momentum stored in the wheels is of importance. Thus, our system is the $\mathscr{R} + \mathscr{W}$ gyrostat of Chapter 6. The immediate purpose here is to extend the discussion of torque-free gyrostats given in Chapter 6 to include the gravity gradient torque experienced by such gyrostats when in a circular orbit. An equally valid interpre-

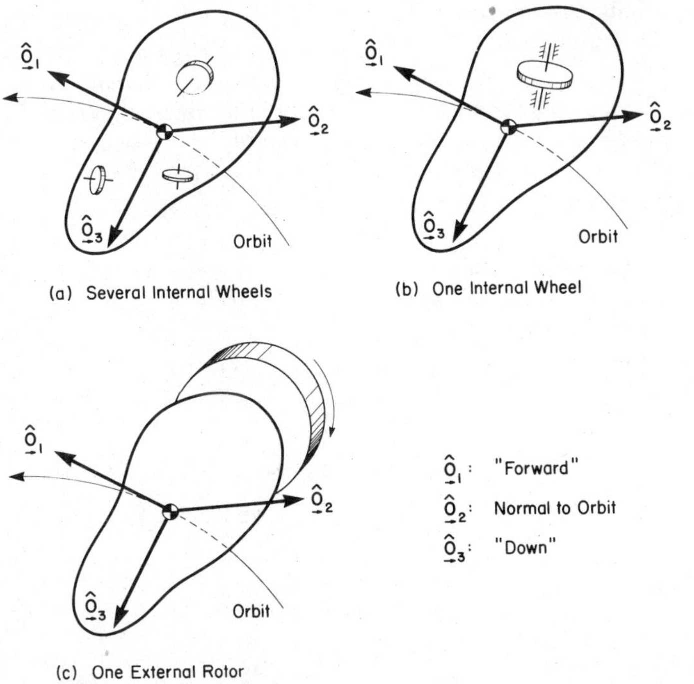

FIGURE 11.1 Three equivalent possibilities for a rigid gyrostat in a circular orbit.

tation of the analysis to follow is that it extends to gyrostats the analysis of relative equilibria made in Section 10.1 for rigid spinners in a circular orbit. A third possible interpretation is that a wheel is added to the gravitationally stabilized satellites of Chapter 9.

The appropriate motion equations for $\mathscr{R} + \mathscr{W}$ in orbit (Fig. 11.1) are available from (6.1, 6) and (8.1, 21). Expressed in matrix notation, they are

$$\mathbf{I}\dot{\boldsymbol{\omega}} + \boldsymbol{\omega}^{\times}(\mathbf{I}\boldsymbol{\omega} + h_s\mathbf{a}) = 3\omega_c^2\mathbf{c}_3^{\times}\mathbf{I}\mathbf{c}_3 \tag{1}$$

These equations are written in a body-fixed reference frame $\mathscr{F}_b$. The elements of $\mathbf{I}$ and $\boldsymbol{\omega}(t)$ are components in $\mathscr{F}_b$ of the inertia dyadic $\underset{\rightarrow}{\mathbf{I}}$ and the absolute angular velocity of $\mathscr{F}_b$, $\underset{\rightarrow}{\boldsymbol{\omega}}$. Similarly, $\mathbf{c}_3$ contains the components (i.e., direction cosines) of $\underset{\rightarrow}{\hat{\mathbf{c}}}_3$ in $\mathscr{F}_b$. The orbital frequency is ω_c, and the stored angular momentum in the wheel, relative to $\mathscr{F}_b$, has magnitude h_s and is in the direction specified by the unit vector $\underset{\rightarrow}{\mathbf{a}}$, whose components in $\mathscr{F}_b$ are contained in $\mathbf{a}$. With no loss in generality, we choose the body frame to be the principal-axis frame: $\mathscr{F}_b \equiv \mathscr{F}_p$.

Dynamical Equilibria

A *dynamical equilibrium* will be said to occur when the net external torque is zero and the rotational motion is uniform. In the present context, we would require the gravity gradient torque to remain zero, and $\boldsymbol{\omega} \equiv \boldsymbol{\nu} =$ constant. The former condition is familiar from earlier discussions: $\mathbf{c}_3^\times \mathbf{I}\mathbf{c}_3$ can be zero if and only if either (i) $\mathbf{c}_3 = \mathbf{0}$ (impossible), (ii) $\mathbf{I}\mathbf{c}_3 = \mathbf{0}$ (impossible), or (iii) $\mathbf{I}\mathbf{c}_3$ remains parallel to $\mathbf{c}_3$. Thus, one of the principal axes of the gyrostat must remain aligned with the local vertical. With no loss in generality, we let this principal axis be $\hat{\mathbf{p}}_3$. Then, $\mathbf{c}_3 \equiv \mathbf{1}_3$, $\mathbf{C}_{pc} = \mathbf{C}_3(\alpha_3)$, and $\boldsymbol{\omega}_\alpha = \mathbf{1}_3\dot{\alpha}_3$. Now, from (9.2, 6), $\boldsymbol{\nu} = \boldsymbol{\omega}_\alpha - \omega_c \vec{\mathbf{c}}_2$. It follows that $\boldsymbol{\omega}_\alpha \equiv \mathbf{0}$, $\alpha_3 \equiv \chi_3 =$ constant, and $\boldsymbol{\nu} = -\omega_c \mathbf{c}_2$. Hence, $\mathscr{F}_p$ is fixed with respect to $\mathscr{F}_c$; in other words, the orientation of the gyrostat is fixed in the orbiting frame.

It is at this point that our reasoning for a gyrostat differs from that for a rigid body. For a rigid body alone (no $\mathscr{W}$), (1) requires that $\boldsymbol{\nu}^\times \mathbf{I}\boldsymbol{\nu} = \mathbf{0}$; that is, $\mathbf{c}_2^\times \mathbf{I}\mathbf{c}_2 = \mathbf{0}$, and a second principal axis must be normal to the orbit. Thus, for a rigid body to achieve dynamical equilibrium in a circular orbit, the principal axes and orbiting axes must coincide. Not so for $\mathscr{R} + \mathscr{W}$, however. In fact, (1) becomes

$$\mathbf{I}\boldsymbol{\nu} + h_s \mathbf{a} = \lambda \boldsymbol{\nu} \tag{2}$$

Or, since $\boldsymbol{\nu} = -\omega_c \mathbf{c}_2$ and $\mathbf{c}_2 = [\sin\chi_3 \quad \cos\chi_3 \quad 0]^T$, we learn that

$$\frac{a_1^2}{(I_1 - \lambda)^2} + \frac{a_2^2}{(I_2 - \lambda)^2} = \frac{1}{J_s^2}$$
$$a_3 = 0 \tag{3}$$

where

$$J_s \triangleq \frac{h_s}{\omega_c} \tag{4}$$

In particular, no dynamical equilibrium is possible if a component of the angular momentum stored in the wheels exists in the vertical direction (i.e., if $a_3 \neq 0$).

The solution for λ from (3) is similar to the solution to (6.3, 4) in Chapter 6 (see also Fig. 6.2). When the stored angular momentum is slight, $J_s \ll I_{\min}$, there are four solutions for λ; on the other hand, when $J_s \gg I_{\max}$, there are only two solutions for λ. In both cases, and for each λ, the associated (constant) spin vector has the components

$$\nu_i = \frac{h_s a_i}{\lambda - I_i} \qquad (i = 1, 2) \tag{5}$$

and $\nu_3 = 0$. And the angle χ_3 between $\underset{\rightarrow}{\hat{\mathbf{p}}}_1$ and $\underset{\rightarrow}{\hat{\mathbf{c}}}_1$ (also between $\underset{\rightarrow}{\hat{\mathbf{p}}}_2$ and $\underset{\rightarrow}{\hat{\mathbf{c}}}_2$) is found from

$$\sin\chi_3 = \frac{J_s}{I_1 - \lambda} a_1; \qquad \cos\chi_3 = \frac{J_s}{I_2 - \lambda} a_2 \tag{6}$$

for each equilibrium (each λ).

Geometrically, the picture is as shown in Fig. 11.2. We have established that one of the principal axes of $\mathscr{R} + \mathscr{W}$ must persistently point "down" (so that there is no gravity gradient torque) and the axis of $\mathscr{W}$ must lie in the $\underset{\rightarrow}{\hat{\mathbf{p}}}_1$–$\underset{\rightarrow}{\hat{\mathbf{p}}}_3$ plane. If the axis of $\mathscr{W}$ makes an angle β_3 with $\underset{\rightarrow}{\hat{\mathbf{p}}}_2$ ($a_1 = -\sin\beta_3$, $a_2 = \cos\beta_3$), then $\underset{\rightarrow}{\hat{\mathbf{p}}}_1$

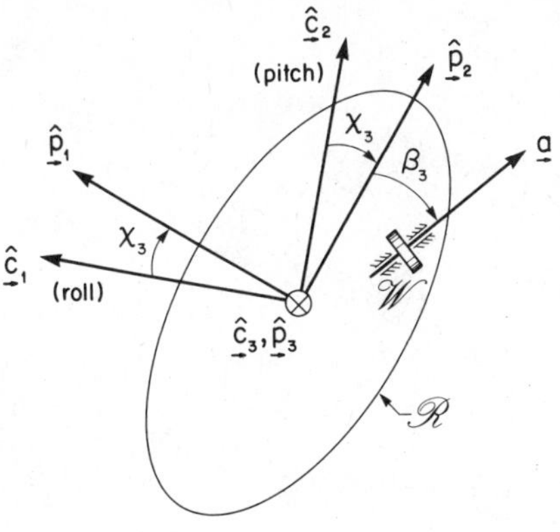

FIGURE 11.2 Geometry for dynamical equilibrium of $\mathscr{R} + \mathscr{W}$ in a circular orbit (see Fig. 11.5*a*.)

must, for dynamical equilibrium, make an angle χ_3, given by (6), with the "forward" direction $\hat{\mathbf{c}}_1$. In this way, the total absolute angular momentum of $\mathscr{R} + \mathscr{W}$ remains normal to the orbit. As a final result, the angle between the axis of $\mathscr{W}$ and the forward direction (the angle between $\mathbf{a}$ and $\hat{\mathbf{c}}_2$) is not difficult to calculate from the following relations:

$$\cos(\chi_3 + \beta_3) = J_s\left(\frac{a_1^2}{I_1 - \lambda} + \frac{a_2^2}{I_2 - \lambda}\right)$$

$$\sin(\chi_3 + \beta_3) = J_s\frac{a_1a_2(I_2 - I_1)}{(I_1 - \lambda)(I_2 - \lambda)} \tag{7}$$

This completes the identification of dynamical equilibria for the $\mathscr{R} + \mathscr{W}$ gyrostat in a circular orbit.

Simplifications When $\mathscr{R} + \mathscr{W}$ Is Axisymmetrical

We now inquire what simplifications are available if the gyrostat is inertially axisymmetrical. Let the axial and transverse moments of inertia of $\mathscr{R} + \mathscr{W}$ be I_a and I_t. One of the principal axes must point downward for a dynamical equilibrium. Let us first suppose that this axis is the symmetry axis, so that $I_3 = I_a$, and $I_1 = I_2 = I_t$. Then, (3) becomes

$$\lambda = I_t \pm J_s \tag{8}$$

since $a_1^2 + a_2^2 = 1$; and from (7) we learn that $\mathscr{W}$ must have its axis normal to the orbit, regardless of the value of h_s.

Next, suppose that a transverse axis points downward, so that the axis of symmetry points "horizontally" (i.e., in the $\hat{\mathbf{c}}_1$–$\hat{\mathbf{c}}_2$ plane). Then Fig. 11.2 still applies, and there is no special simplification.

One particular subcase should be mentioned, however. If the symmetry axis of $\mathscr{R} + \mathscr{W}$ is nominally normal to the orbit, as in Fig 11.3, we have in effect an extension of the "Thomson problem" (Fig. 10.1). As shown in Problem 11.2, the

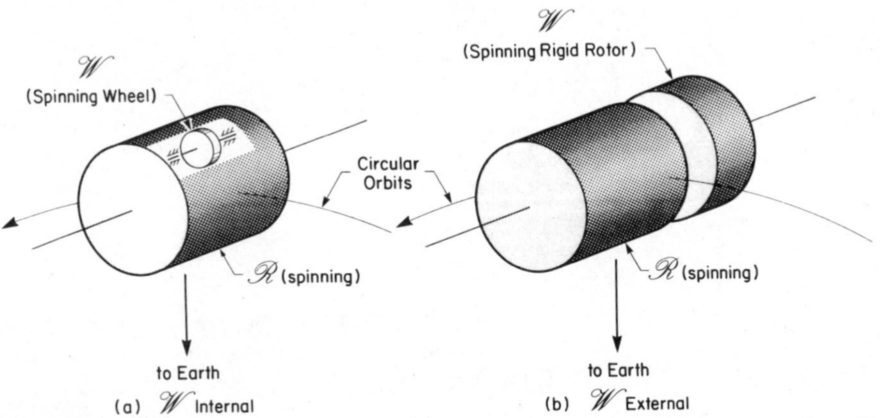

FIGURE 11.3 Two physical interpretations of the symmetrical gyrostat $\mathscr{R} + \mathscr{W}$ in circular orbit, with symmetry axis normal to the orbit. The stability diagram is the same as Fig. 10.2, with $\hat{\nu}$ replaced by $\hat{\nu}_{\text{eq}}$, given by equation (11.1, 9).

stability diagram for this situation is identical to Fig. 10.2 for an axisymmetrical rigid body, provided $\hat{\nu}$ is replaced by an equivalent, dimensionless spin rate $\hat{\nu}_{\text{eq}}$:

$$\nu_{\text{eq}} \triangleq \nu + \frac{h_s}{I_a}; \qquad \hat{\nu}_{\text{eq}} \triangleq \frac{\nu_{\text{eq}}}{\omega_c} \tag{9}$$

The physical meaning of (9) is simple: one replaces the axisymmetrical $\mathscr{R} + \mathscr{W}$ by an equivalent axisymmetrical rigid body. This equivalent body has the same inertia dyadic as $\mathscr{R} + \mathscr{W}$ and is spinning at a rate such that its angular momentum (normal to the orbit) is the same as the angular momentum of $\mathscr{R} + \mathscr{W}$. The equivalent spin rate is in fact ν_{eq}, given in (9) above. Figure 10.2 was applied by [Kane and Mingori] to an axisymmetrical $\mathscr{R} + \mathscr{W}$ spinning normally to the orbit. [White and Likins] have pointed out that the gyrostat-spinner equivalence represented by (9) is true for all orientations.

The Roberson Relative Equilibria

A *relative equilibrium* will be said to occur when the $\mathscr{R} + \mathscr{W}$ gyrostat is fixed in the orbiting frame $\mathscr{F}_c$. Note in particular that the gravity gradient torque is not necessarily zero; if it is nonzero however, it must precess the system about the orbit normal at a uniform rate (the orbital rate) so that the system appears fixed in $\mathscr{F}_c$.

A glance back at Fig. 10.3 will be helpful. This diagram shows three possible types of relative equilibrium for $\mathscr{W}$ alone in a circular orbit. The first type (the "cylindrical" case) is also a *dynamical* equilibrium (as we have defined it here). So, too, is the "hyperbolic" relative equilibrium: the gravity gradient torque is nominally zero, and the total angular momentum is normal to the orbit. For the "conical" case on the other hand, the gravity gradient torque is nonzero, but the resulting precession is about the orbit normal, at the orbital rate.

Our present task is to do for the $\mathscr{R}+\mathscr{W}$ gyrostat what Fig. 10.3 does for $\mathscr{W}$ alone. (Note that $\mathscr{R}+\mathscr{W}$ is not assumed to be axisymmetrical.) From (1), with $\boldsymbol{\omega} \equiv \boldsymbol{\nu} =$ constant and with $\boldsymbol{\nu} = -\omega_c \mathbf{c}_2$ so that the spin seen in $\mathscr{F}_c$ is zero, we deduce that

$$\mathbf{c}_2^{\times}(\mathbf{I}\mathbf{c}_2 - J_s\mathbf{a}) = 3\mathbf{c}_3^{\times}\mathbf{I}\mathbf{c}_3 \tag{10}$$

is the condition to be satisfied by all relative equilibria ([Roberson and Hooker]). Furthermore, define

$$J_{si} \triangleq J_s\mathbf{c}_i^T\mathbf{a} \qquad (i = 1,2,3) \tag{11}$$

Now, (10) is the vector condition for relative equilibrium, expressed in $\mathscr{F}_p$. It will be more convenient however to express this condition in $\mathscr{F}_c$ instead. Accordingly, we form these inner products: $\mathbf{c}_1^T(10)$, $\mathbf{c}_2^T(10)$, and $\mathbf{c}_3^T(10)$. Recognizing that

$$\mathbf{c}_1^T\mathbf{c}_2^{\times} = -(\mathbf{c}_2^{\times}\mathbf{c}_1)^T = -(-\mathbf{c}_3)^T = \mathbf{c}_3^T$$

along with five other similar identities, we arrive at the following three scalar conditions:

$$\mathbf{c}_1^T\mathbf{I}\mathbf{c}_2 = J_{s1}; \qquad \mathbf{c}_3^T\mathbf{I}\mathbf{c}_1 = 0; \qquad 4\mathbf{c}_2^T\mathbf{I}\mathbf{c}_3 = J_{s3} \tag{12}$$

Evidently, J_{s2} is of no consequence in the enumeration of equilibria. Physically, a relative equilibrium is not affected by the amount of angular momentum stored about the orbit normal.

Conditions (12) suggest that inertial quantities in the orbital frame $\mathscr{F}_c$, not in the principal axis frame $\mathscr{F}_p$, are most directly involved. Because $\mathscr{R}+\mathscr{W}$ is fixed in $\mathscr{F}_c$, its inertia matrix, which we shall denote in $\mathscr{F}_c$ by $\mathscr{I}$, in constant. In fact,

$$\mathscr{I} = \mathbf{C}_{cp}\mathbf{I}\mathbf{C}_{pc} \tag{13}$$

where, recall,

$$\mathbf{C}_{pc} = [\mathbf{c}_1 \quad \mathbf{c}_2 \quad \mathbf{c}_3] \tag{14}$$

(see Appendix B). Another way of writing (13) is

$$\mathscr{I}_{ij} = \mathbf{c}_i^T\mathbf{I}\mathbf{c}_j \qquad (i, j = 1,2,3) \tag{15}$$

where $\{\mathscr{I}_{ij}\} \equiv \mathscr{I}$. From this new point of view, conditions (12) for relative equilibrium can be rewritten

$$\mathscr{I}_{12} = J_{s1}; \qquad \mathscr{I}_{13} = 0; \qquad \mathscr{I}_{23} = \tfrac{1}{4}J_{s3} \tag{16}$$

Although a somewhat different approach has been followed here than by [Roberson and Hooker], the same conclusions are arrived at, namely:

- **(a)** For a principal axis of $\mathscr{R}+\mathscr{W}$ to point forward, it is necessary and sufficient that $J_{s1} = 0$.
- **(b)** For a principal axis of $\mathscr{R}+\mathscr{W}$ to point down, it is necessary and sufficient that $J_{s3} = 0$.
- **(c)** For a principal axis of $\mathscr{R}+\mathscr{W}$ to point normal to the orbit, it is necessary and sufficient that $J_{s1} = J_{s3} = 0$.

It is left to Problem 11.3 to demonstrate in detail the truth of these assertions.

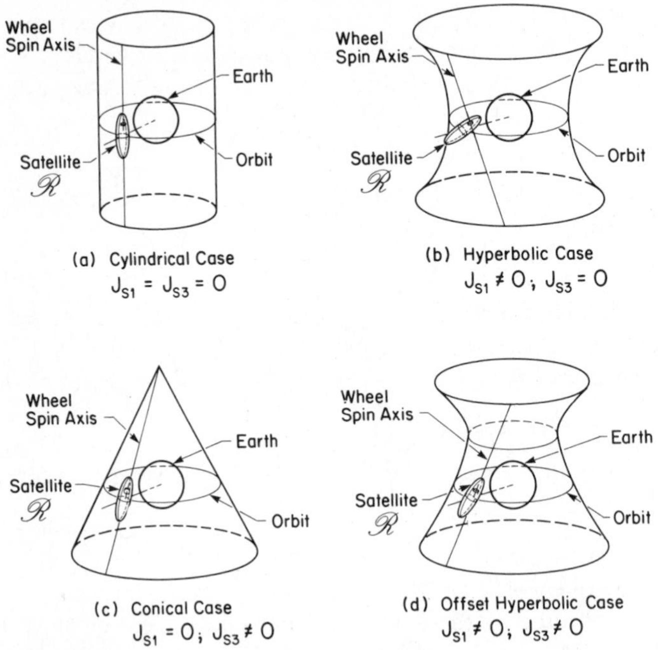

FIGURE 11.4 The Roberson equilibria for a gyrostat in a circular orbit, based on [Roberson, 6].

The above three statements provide a useful framework within which to classify the relative equilibria of $\mathscr{R} + \mathscr{W}$. The simplest case is $J_{s1} = J_{s3} = 0$, for which $\mathscr{I}_{ij} = 0$ $(i \neq j)$, showing that the principal axes of $\mathscr{R} + \mathscr{W}$ coincide with the orbiting frame $\mathscr{F}_c$. We shall call this the *cylindrical case* in analogy with the nomenclature used by [Likins, 1] for $\mathscr{W}$ without $\mathscr{R}$ (see Fig. 10.3). In the cylindrical case, then, $\mathscr{F}_p$ is aligned with $\mathscr{F}_c$; and because $J_{s1} = J_{s3} = 0$, the axis of $\mathscr{W}$ lies on a cylinder as the satellite revolves in its orbit (Fig. 11.4*a*).

Next, consider $J_{s1} \neq 0$ and $J_{s3} = 0$. From (16), one of the principal axes of $\mathscr{R} + \mathscr{W}$ points vertically down, although the other two principal axes are in general skewed in the "local horizontal" plane. Moreover, it is clear from (11) that the spin axis of $\mathscr{W}$, while still perpendicular to the vertical, is generally also inclined in the local horizontal plane. Therefore, we refer to this as the *hyperbolic case* (cf. Fig. 11.4*b*). It is straightforward to show that the gravity gradient torque on $\mathscr{R} + \mathscr{W}$ is zero (this relative equilibrium is therefore also a dynamical equilibrium). The total angular momentum is constant and normal to the orbit.

If $J_{s1} = 0$ and $J_{s3} \neq 0$, the axis of $\mathscr{W}$ is perpendicular to the forward direction (although generally skewed in the $\underset{\rightarrow}{\hat{\mathbf{c}}}_2$–$\underset{\rightarrow}{\hat{\mathbf{c}}}_3$ plane). And from (16) we learn that one of the principal axes of $\mathscr{R} + \mathscr{W}$ points forward (although the other two principal axes are in general skewed in the $\underset{\rightarrow}{\hat{\mathbf{c}}}_2$–$\underset{\rightarrow}{\hat{\mathbf{c}}}_3$ plane). Therefore, as shown in Fig. 11.4*c*, this case is designated the *conical case*.

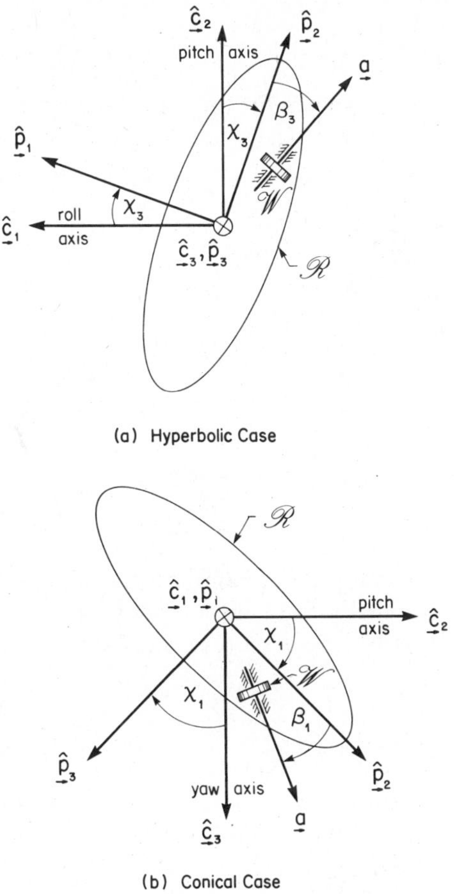

FIGURE 11.5 Orientation of relative equilibria.

Finally, when $J_{s1} \neq 0$ and $J_{s3} \neq 0$, the axis of $\mathscr{W}$ is inclined to all axes of $\mathscr{F}_c$; however, it lies again on a hyperboloid of one sheet, but now the center of the hyperboloid is offset along the orbit normal (Problem 11.4). This is shown in Fig. 11.4*d*. We shall call it the *offset-hyperbolic case*. In fact, the conical and hyperbolic cases are subcases of the offset-hyperbolic case, and the cylindrical case is, in turn, a limiting case for both the conical and hyperbolic cases.

Orientation of Gyrostats in Relative Equilibrium

The necessary relationships have now been given to permit a determination of the fixed orientation of the gyrostat $\mathscr{R} + \mathscr{W}$ with respect to the orbiting frame $\mathscr{F}_c$. In the hyperbolic case (Fig. 11.4*b*), one of the principal axes points vertically, whence (see Fig. 11.5*a*)

$$\mathbf{C}_{pc} = \mathbf{C}_3(\chi_3) \tag{17}$$

It follows from (13) that

$$\mathscr{I}_{12} = (I_1 - I_2)\sin\chi_3\cos\chi_3 \tag{18}$$

and, as a check, $\mathscr{I}_{23} = \mathscr{I}_{13} = 0$. Moreover, let the inclination of the $\mathscr{W}$-axis to the principal axis $\underset{\rightarrow}{\hat{\mathbf{p}}}_2$ be β_3 (in the local horizontal plane). Then,

$$\mathbf{a} = [-\sin\beta_3 \quad \cos\beta_3 \quad 0]^T \tag{19}$$

$$J_{s1} = J_s\mathbf{c}_1^T\mathbf{a} = -J_s\sin(\chi_3 + \beta_3)$$

$$J_{s3} = J_s\mathbf{c}_3^T\mathbf{a} = 0 \qquad \text{(a check)} \tag{20}$$

and so, from (16),

$$\sin(\chi_3 + \beta_3) = \kappa_H \sin 2\chi_3 \tag{21}$$

where the constant κ_H is defined by

$$\kappa_H \triangleq \frac{I_2 - I_1}{2J_s} \tag{22}$$

The subscript H refers to the hyperbolic case.

The transcendental equation for equilibrium attitude, namely, (21), was derived by [Roberson and Hooker], and a detailed plot of its solutions has been given by [Roberson, 6]. This plot is the basis for Fig. 11.6. One may specify β_3

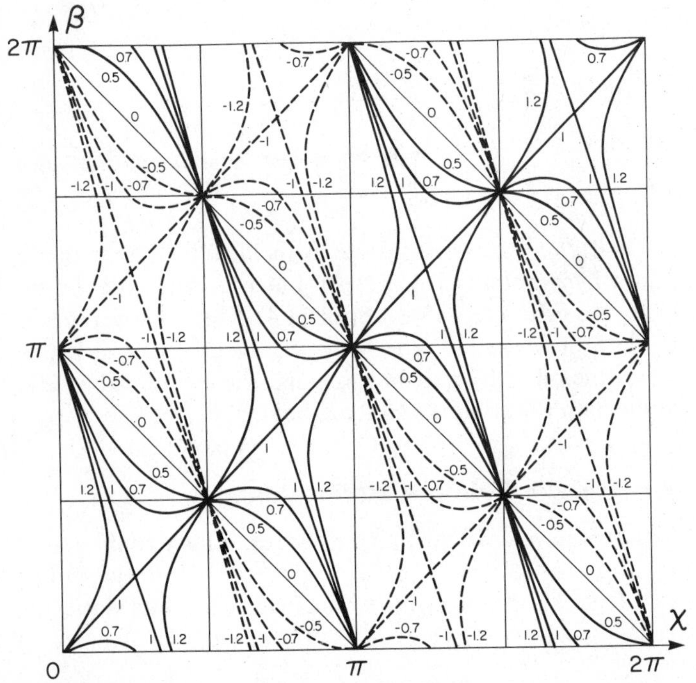

FIGURE 11.6 Solutions to $\sin(\chi + \beta) = \kappa \sin 2\chi$, based on [Roberson, 6]. (The curves are labeled according to the value of κ.)

(wheel direction) and ask for the orientation of the gyrostat (χ_3), or one may have a desired orientation (χ_3) in mind and wish to calculate the corresponding wheel inclination (β_3). For relatively low wheel momenta (specifically for $|\kappa_H| > 1$), there are four solutions for χ_3 for each β_3. However, for relatively high wheel momenta ($|\kappa_H| < \frac{1}{2}$), there are only two orientations of the gyrostat; and as the wheel momentum increases further, the orientation of the gyrostat is increasingly such as to make the axis of $\mathscr{W}$ normal to the orbit (Problem 11.5*a*).

It may often be preferable to solve a certain quartic equation (derived in Problem 11.4) than to solve the transcendental equation (21) for χ_3. This quartic was also given earlier as (3), in the study of dynamical equilibria.

Turning now to the conical case (Fig. 11.4*c*), we must replace (17) by

$$\mathbf{C}_{pc} = \mathbf{C}_1(\chi_1) \tag{23}$$

because one of the principal axes points forward. Then, from (13),

$$\mathscr{I}_{23} = (I_2 - I_3)\sin\chi_1\cos\chi_1 \tag{24}$$

Also, as shown in Fig. 11.5*b*, the axis of $\mathscr{W}$ is perpendicular to the forward direction and is inclined to $\underset{\rightarrow}{\hat{\mathbf{p}}}_2$ at the angle β_1, so that

$$\mathbf{a} = [0 \quad \cos\beta_1 \quad \sin\beta_1]^T \tag{25}$$

$$J_{s1} = 0 \qquad \text{(a check)}$$

$$J_{s3} = J_s \sin(\chi_1 + \beta_1) \tag{26}$$

The final equation for orientation is now available from (16):

$$\sin(\chi_1 + \beta_1) = \kappa_C \sin 2\kappa_1 \tag{27}$$

where

$$\kappa_C \triangleq 2\frac{I_2 - I_3}{J_s} \tag{28}$$

Happily, (27) is of exactly the same form as (21). Therefore, Fig. 11.6 applies to the conical case as well, as do all the above remarks about this figure. An alternative to solving the transcendental equation (27) for χ_1 is to solve a certain quartic (Problem 11.14).

Finally, we turn to the offset-hyperbolic case illustrated in Fig. 11.4*d*. We shall assume however that one of the principal axes of $\mathscr{R} + \mathscr{W}$ is in the orbital plane (the $\underset{\rightarrow}{\hat{\mathbf{c}}}_1$–$\underset{\rightarrow}{\hat{\mathbf{c}}}_3$ plane). [It can be shown (Problem 11.5*b*) that it is not possible for a principal axis to lie in the $\underset{\rightarrow}{\hat{\mathbf{c}}}_2$–$\underset{\rightarrow}{\hat{\mathbf{c}}}_3$ plane, or the $\underset{\rightarrow}{\hat{\mathbf{c}}}_1$–$\underset{\rightarrow}{\hat{\mathbf{c}}}_2$ plane.] This is a sufficient condition for an offset-hyperbolic equilibrium to exist; it is not, however, a necessary one. The most general case in which none of the principal axes lies in any of the three principal planes associated with the orbit is treated in [Longman and Roberson].

We choose $\underset{\rightarrow}{\hat{\mathbf{p}}}_1$ as the principal axis that is to lie in the orbital plane and choose the rotation matrix $\mathbf{C}_{pc}$ to be

$$\mathbf{C}_{pc} = \mathbf{C}_1(\chi_1)\mathbf{C}_2(\chi_2) \tag{29}$$

which does indeed have 0 as its (1, 2) element. The cross products of inertia in $\mathscr{F}_c$ are, from (13) and (29),

$$\begin{aligned}\mathscr{I}_{13} &= \left(I_2 \sin^2 \chi_1 + I_3 \cos^2 \chi_1 - I_1\right) \sin \chi_2 \cos \chi_2 \\ \mathscr{I}_{12} &= \left(I_2 - I_3\right) \sin \chi_1 \cos \chi_1 \sin \chi_2 \\ \mathscr{I}_{23} &= \left(I_2 - I_3\right) \sin \chi_1 \cos \chi_1 \cos \chi_2\end{aligned} \tag{30}$$

Now, according to (16), $\mathscr{I}_{13}$ is to vanish. However, if $\sin \chi_2 = 0$, $\mathscr{I}_{12} = 0$ also. But $\mathscr{I}_{12} = J_{s1}$. Thus, $\sin \chi_2 \neq 0$. (Otherwise, our case degenerates back to the conical one.) Similarly, $\cos \chi_2 \neq 0$. It follows that

$$I_2 \sin^2 \chi_1 + I_3 \cos^2 \chi_1 = I_1 \tag{31}$$

and hence that

$$\begin{aligned}\sin \chi_1 &= (\pm)_1 \left(\frac{I_1 - I_3}{I_2 - I_3}\right)^{1/2} \\ \cos \chi_1 &= (\pm)_2 \left(\frac{I_2 - I_1}{I_2 - I_3}\right)^{1/2}\end{aligned} \tag{32}$$

The notations $(\pm)_1$ and $(\pm)_2$ are used to keep track of independent sign choices. It is interesting that the four solutions (32) for χ_1 do not depend on the wheel momentum $h_s\mathbf{a}$. It is also evident from (32) that I_1 must be the intermediate moment of inertia for $\mathscr{R} + \mathscr{W}$; this avoids square roots of negative quantities and sines and cosines greater than unity.

It is now time to use the other two facts stated in (16). Together with the second and third parts of (30), these show that

$$\{\sin \chi_2, \cos \chi_2\} = \frac{\{2J_{s1}, \tfrac{1}{2}J_{s3}\}}{(I_2 - I_3) \sin 2\chi_1} \tag{33}$$

Assuming that the desired attitude of the gyrostat is known (i.e., χ_1 and χ_2), (32) and (33) show how to design $\mathscr{R} + \mathscr{W}$ to achieve this orientation. In conjunction with (33), one must note from the definition of J_{si} in (11) that

$$\begin{aligned}J_{s1} &= J_s[a_1 \cos \chi_2 + a_2 \sin \chi_1 \sin \chi_2 + a_3 \cos \chi_1 \sin \chi_2] \\ J_{s3} &= J_s[-a_1 \sin \chi_2 + a_2 \sin \chi_1 \cos \chi_2 + a_3 \cos \chi_1 \cos \chi_2]\end{aligned} \tag{34}$$

where $\{a_1, a_2, a_3\}$ are the direction cosines of the $\mathscr{W}$-axis expressed in $\mathscr{F}_p$. On the other hand, if $\mathscr{R} + \mathscr{W}$ is given *a priori*, so that $\{\mathbf{I}, J_s, \mathbf{a}\}$ are known, and it is required to determine the equilibrium orientation $\{\chi_1, \chi_2\}$, one can first find χ_1 from (32) and then iterate using (33) and (34) to find χ_2. Because of the restriction that the axis of intermediate inertia must lie in the orbital plane, there are constraints on the allowed values of $\{\mathbf{I}, J_s, \mathbf{a}\}$ (see Problem 11.7).

Much more detail on general offset-hyperbolic equilibria is available in [Longman, 1, 2]. One interesting result is that if $\mathscr{W}$ is aligned with a principal axis, say, $\vec{\hat{\mathbf{p}}}_2$, and the momentum of $\mathscr{W}$ has the critical value

$$J_s = 2[(I_2 - I_3)(I_2 - I_1)]^{1/2} \tag{35}$$

there exists in configuration space not just a few isolated equilibrium points but a continuous curve of equilibria. This can lead in turn to a slow tumbling motion of $\mathscr{R} + \mathscr{W}$. The stability of this tumbling has been investigated by [Longman, 3]. Another interesting possibility is analyzed by [Anchev]: if $\mathscr{W}$ is physically implemented as three independent wheels along the principal axes of $\mathscr{R} + \mathscr{W}$, it is possible to transfer from the basic gravity-gradient-stabilized equilibrium (wheels off) to any desired gyrostat equilibrium through appropriate spin-up histories for the three wheels.

Librations about Equilibrium

As usual, the question that arises immediately on having found equilibria is: For what system parameters are these equilibria stable? We accordingly now look at a linearized version of the motion equations in which infinitesimal variations are taken about the reference equilibrium solution. The appropriate motion equations were given earlier by (1). The angular velocity $\boldsymbol{\omega}$ can be expressed as

$$\boldsymbol{\omega} = \dot{\boldsymbol{\alpha}} + (\mathbf{1} - \boldsymbol{\alpha}^{\times})\boldsymbol{\nu} \tag{36}$$

Here, $\boldsymbol{\alpha}$ represents the three small rotational deviations of $\mathscr{R} + \mathscr{W}$ from its equilibrium orientation [see (2.3, 33) in Chapter 2], and $\boldsymbol{\nu}$ is, as before, the equilibrium spin rate,

$$\boldsymbol{\nu} = -\omega_c \bar{\mathbf{c}}_2 \tag{37}$$

The overbar on $\bar{\mathbf{c}}_2$ is a reminder that it is the equilibrium value of $\mathbf{c}_2$ that is being referred to. This leaves us free to retain the symbol $\mathbf{c}_2$ for the direction cosines of the orbit normal $\underset{\rightarrow}{\hat{c}}_2$ expressed in the (slightly perturbed) $\mathscr{F}_p$. In fact,

$$\mathbf{c}_2 = (\mathbf{1} - \boldsymbol{\alpha}^{\times})\bar{\mathbf{c}}_2 \tag{38a}$$

Similarly,

$$\mathbf{c}_3 = (\mathbf{1} - \boldsymbol{\alpha}^{\times})\bar{\mathbf{c}}_3 \tag{38b}$$

After insertion of (36) through (38) in (1) and deletion of second-order terms, we arrive at the linear equations for librations about equilibrium:

$$\mathscr{M}\boldsymbol{\alpha}'' + \mathscr{G}\boldsymbol{\alpha}' + \mathscr{K}\boldsymbol{\alpha} = \mathbf{0} \tag{39}$$

where the primes denote differentiation with respect to the orbital angle η ($\eta = \omega_c t$).

The coefficient matrices in (39) are found to be (Problem 11.8a)

$$\mathscr{M} = \mathbf{I} \tag{40a}$$

$$\mathscr{G} = (\mathbf{I}\bar{\mathbf{c}}_2)^{\times} - \mathbf{I}\bar{\mathbf{c}}_2^{\times} - \bar{\mathbf{c}}_2^{\times}\mathbf{I} - J_s\mathbf{a}^{\times} \tag{40b}$$

$$\mathscr{K} = \bar{\mathbf{c}}_2^{\times}\mathbf{I}\bar{\mathbf{c}}_2^{\times} - (\mathbf{I}\bar{\mathbf{c}}_2)^{\times}\bar{\mathbf{c}}_2^{\times} + J_s\mathbf{a}^{\times}\bar{\mathbf{c}}_2^{\times} + 3(\mathbf{I}\bar{\mathbf{c}}_3)^{\times}\bar{\mathbf{c}}_3^{\times} - 3\bar{\mathbf{c}}_3^{\times}\mathbf{I}\bar{\mathbf{c}}_3^{\times} \tag{40c}$$

Obviously, $\mathscr{M}^T = \mathscr{M} > 0$; and, almost as obviously, $\mathscr{G}^T = -\mathscr{G}$. It is however far from obvious that $\mathscr{K}^T = \mathscr{K}$. In fact, the last property cannot be demonstrated from (40c) alone; it is necessary to incorporate the equilibrium conditions, namely, (10). Instead of doing this directly, it is somewhat easier and more instructive to rewrite the variational equations (39) in the orbital frame $\mathscr{F}_c$. To this end, define

$$\boldsymbol{\alpha}_c(t) \triangleq \mathbf{C}_{c\bar{p}}\boldsymbol{\alpha}(t) \tag{41}$$

with $\mathbf{C}_{c\bar{p}}$ the rotation matrix from $\mathscr{F}_{\bar{p}}$ to $\mathscr{F}_c$ and $\mathscr{F}_{\bar{p}}$ the reference frame $\mathscr{F}_p$ when $\mathscr{R} + \mathscr{W}$ is in its reference equilibrium orientation. Then, $\boldsymbol{\alpha} = \mathbf{C}_{\bar{p}c}\boldsymbol{\alpha}_c$. This substitution into (39), and premultiplication by $\mathbf{C}_{c\bar{p}}$, leads to the following motion equations in the new coordinates:

$$\mathscr{M}_c\boldsymbol{\alpha}_c'' + \mathscr{G}_c\boldsymbol{\alpha}_c' + \mathscr{K}_c\boldsymbol{\alpha}_c = \mathbf{0} \tag{42}$$

where

$$\mathscr{M}_c \triangleq \mathbf{C}_{c\bar{p}}\mathbf{I}\mathbf{C}_{\bar{p}c} \equiv \mathscr{I} \tag{43a}$$

$$\mathscr{G}_c \triangleq \mathbf{C}_{c\bar{p}}\mathscr{G}\mathbf{C}_{\bar{p}c} = \mathscr{g}_c^{\times} \tag{43b}$$

$$\mathscr{K}_c = \mathbf{C}_{c\bar{p}}\mathscr{K}\mathbf{C}_{\bar{p}c} = \begin{bmatrix} k_{c11} & k_{c12} & 0 \\ k_{c12} & k_{c22} & k_{c23} \\ 0 & k_{c23} & k_{c33} \end{bmatrix} \tag{43c}$$

where $\mathscr{I}$ is as defined earlier by (13) and (15), and

$$\begin{aligned}
g_{c1} &= J_{s1} \\
g_{c2} &= (\mathscr{I}_{22} - \mathscr{I}_{11} - \mathscr{I}_{33}) - J_{s2} \\
g_{c3} &= -\tfrac{1}{2}J_{s3} \\
k_{c11} &= 4(\mathscr{I}_{22} - \mathscr{I}_{33}) - J_{s2} \\
k_{c22} &= 3(\mathscr{I}_{11} - \mathscr{I}_{33}) \\
k_{c33} &= (\mathscr{I}_{22} - \mathscr{I}_{11}) - J_{s2} \\
k_{c12} &= -3J_{s1} \\
k_{c23} &= \tfrac{3}{4}J_{s3}
\end{aligned} \tag{44}$$

Some stamina is needed to work out the details of this derivation (Problem 11.8*b*). The fact that $\{\bar{\mathbf{c}}_1, \bar{\mathbf{c}}_2, \bar{\mathbf{c}}_3\}$ form a dextral orthonormal triad is used repeatedly, and the equilibrium conditions (16) are also involved. One consequence of (43*c*) is that $\mathscr{K}_c$ (and therefore $\mathscr{K}$) is symmetric. Note also that although the conditions for equilibrium do not depend on the component of the $\mathscr{W}$ momentum normal to the orbit (i.e., on J_{s2}), the *stability* of these equilibria *does* depend on J_{s2}.

The characteristic equation associated with (42) is

$$\begin{aligned} b_0 s^6 + b_1 s^4 + b_2 s^2 + b_3 &\triangleq \det\left(\mathscr{M}_c s^2 + \mathscr{G}_c s + \mathscr{K}_c\right) \\ &\equiv \det\left(\mathscr{M} s^2 + \mathscr{G} s + \mathscr{K}\right) = 0 \end{aligned} \tag{45}$$

where

$$b_0 = \det \mathcal{M}_c = \det \mathcal{M} = I_1 I_2 I_3 \tag{46a}$$

$$\begin{aligned} b_1 = &\mathcal{I}_{11}\mathcal{I}_{22}k_{c33} + \mathcal{I}_{22}\mathcal{I}_{33}k_{c11} + \mathcal{I}_{33}\mathcal{I}_{11}k_{c22} \\ &- \tfrac{1}{2}\mathcal{I}_{11}J_{s3}k_{c23} - 2\mathcal{I}_{33}J_{s1}k_{c12} - J_{s1}^2 k_{c33} - \tfrac{1}{16}J_{s3}^2 k_{c11} \\ &+ \mathcal{I}_{11}g_{c1}^2 + \mathcal{I}_{22}g_{c2}^2 + \mathcal{I}_{33}g_{c3}^2 + 2J_{s1}g_{c1}g_{c2} + \tfrac{1}{2}J_{s3}g_{c3}g_{c2} \end{aligned} \tag{46b}$$

$$\begin{aligned} b_2 = &\mathcal{I}_{11}k_{c22}k_{c33} + \mathcal{I}_{22}k_{c33}k_{c11} + \mathcal{I}_{33}k_{c11}k_{c22} \\ &- \mathcal{I}_{11}k_{c23}^2 - \mathcal{I}_{33}k_{c12}^2 - \tfrac{1}{2}J_{s3}k_{c11}k_{c23} - 2J_{s1}k_{c12}k_{c33} \\ &+ g_{c1}^2 k_{c11} + g_{c2}^2 k_{c22} + g_{c3}^2 k_{c33} + 2g_{c2}g_{c3}k_{c23} + 2g_{c1}g_{c2}k_{c12} \end{aligned} \tag{46c}$$

$$b_3 = \det \mathcal{K}_c = \det \mathcal{K} = k_{c11}k_{c22}k_{c33} - k_{c12}^2 k_{c33} - k_{c23}^2 k_{c11} \tag{46d}$$

We are now in a position to deal with the issue of stability.

The "librations" about an equilibrium are in fact really librations only if the equilibrium in question is stable. For motion equations of the kind described by (39) or (42), it is known (Section A.6 in Appendix A) that $\mathcal{K}_c > 0$ (or, equivalently, $\mathcal{K} > 0$) is sufficient for stability, though this is not a necessary condition. By analogy with the Lagrange region shown in Fig. 6.4 for the librations of $\mathcal{R}$ (alone) in a circular orbit, we shall refer to the regions in parameter space where $\mathcal{K}_c > 0$ as the L-region. Regions of stability where $\mathcal{K}_c \not> 0$ will for brevity be called the DD-region (in analogy with the DeBra–Delp region in Fig. 9.4).

Necessary and sufficient conditions for the stability of (42), that is, necessary and sufficient conditions for the six roots of the characteristic equation (45) to lie on the imaginary axis, are given in Appendix A, Table A.2. One of these conditions is $b_3 > 0$. Thus, $\det \mathcal{K}_c > 0$. To have stability when $\det \mathcal{K}_c > 0$ but $\mathcal{K}_c \not> 0$, precisely two of the three eigenvalues of $\mathcal{K}_c$ must be negative. Thus, in the DD-region, this "statically unstable" pair is "gyrically stabilized."

Stability of Cylindrical Equilibrium

The simplest (and most important) type of equilibrium is the so-called "cylindrical" case, in which the principal axes (at equilibrium) coincide with the orbital axes and the spin axis of $\mathcal{W}$ is also nominally perpendicular to the orbit. In this cylindrical case, shown schematically in Fig. 11.4a, we know that

$$J_{s1} = J_{s3} = 0; \qquad J_{s2} = J_s \tag{47}$$

There is no need to distinguish between $\mathcal{F}_{\bar{p}}$ and $\mathcal{F}_c$, or between $\{\mathcal{M}_c, \mathcal{G}_c, \mathcal{K}_c\}$ and $\{\mathcal{M}, \mathcal{G}, \mathcal{K}\}$; also, $\boldsymbol{\alpha}_c(t) \equiv \boldsymbol{\alpha}(t)$. Therefore, the librational equations are of the conical form (39), with

$$\mathcal{M} = \mathbf{1} \tag{48a}$$

$$\mathcal{G} = g^{\times}; \qquad g = (I_2 - I_1 - I_3 - J_s)[0 \quad 1 \quad 0]^T \tag{48b}$$

$$\mathcal{K} = \operatorname{diag}\{4(I_2 - I_3) - J_s, 3(I_1 - I_3), (I_2 - I_1) - J_s\} \tag{48c}$$

As we should expect, pitch motion (α_2) and roll/yaw motion (α_1 and α_3) are mutually uncoupled.

For pitch stability we have the familiar condition

$$I_1 > I_3 \tag{49}$$

and the roll/yaw stability conditions can be expressed via the inertia parameters k_1 and k_3 used throughout this book. Recall that $k_1 \triangleq (I_2 - I_3)/I_1$ and $k_3 \triangleq (I_2 - I_1)/I_3$. Then, the characteristic equation for roll/yaw is, from (48),

$$s^4 + b_1 s^2 + b_2 = 0 \tag{50}$$

where

$$b_1 = \mathscr{k}_1 + \mathscr{k}_3 + (\mathscr{k}_1 - 3k_1 - 1)(\mathscr{k}_3 - 1) \tag{50a}$$

$$b_2 = \mathscr{k}_1 \mathscr{k}_3 \tag{50b}$$

$$\mathscr{k}_1 = 4k_1 - \hat{J}_s \frac{1 - k_1 k_3}{1 - k_3} \tag{50c}$$

$$\mathscr{k}_3 = k_3 - \hat{J}_s \frac{1 - k_1 k_3}{1 - k_1} \tag{50d}$$

$$\hat{J}_s \triangleq \frac{J_s}{I_2} \tag{51}$$

The necessary and sufficient conditions for stability are

(i) $b_1 > 0$
(ii) $b_2 > 0$
(iii) $b_1^2 - 4b_2 > 0$
(iv) $k_1 - k_3 > 0$ (52)

The last of these requirements is a restatement of the pitch stability condition (49). On the other hand, the conditions that correspond to static stability ($\mathscr{K} > 0$) are

$$\mathscr{k}_1 > 0; \qquad \mathscr{k}_3 > 0; \qquad k_1 - k_3 > 0 \tag{53}$$

Moreover, these last three conditions are also sufficient for Liapunov stability (Problem 11.9*d*). In the k_1–k_3 plane, points that satisfy (53) are in the L-region; points that satisfy (52) but not (53) are in the DD-region.

[Crespo da Silva, 1] has published stability diagrams in the k_1–k_3 plane, with $\hat{J}_s$ as a parameter. Similar plots for other values of $\hat{J}_s$ are available in [Fujii, 1] and [Longman, Hagedorn, and Beck]. A sampling of four such diagrams is offered here in Fig. 11.7. Figure 9.4, for $\hat{J}_s = 0$, can be added to this group. As $\hat{J}_s$ becomes more positive, the DD-region evidently grows at the expense of the L-region. Indeed, it is not difficult to show (Problem 11.10) that as $\hat{J}_s \to \infty$, the DD-region occupies the whole lower-right half-square, while for $\hat{J}_s \to -\infty$ this same area is occupied by the L-region. This behavior is consistent with Fig. 6.7, drawn for $\mathscr{R} + \mathscr{W}$ in torque-free motion: as $\hat{\Omega}_{po} \to \infty$, all inertia distributions

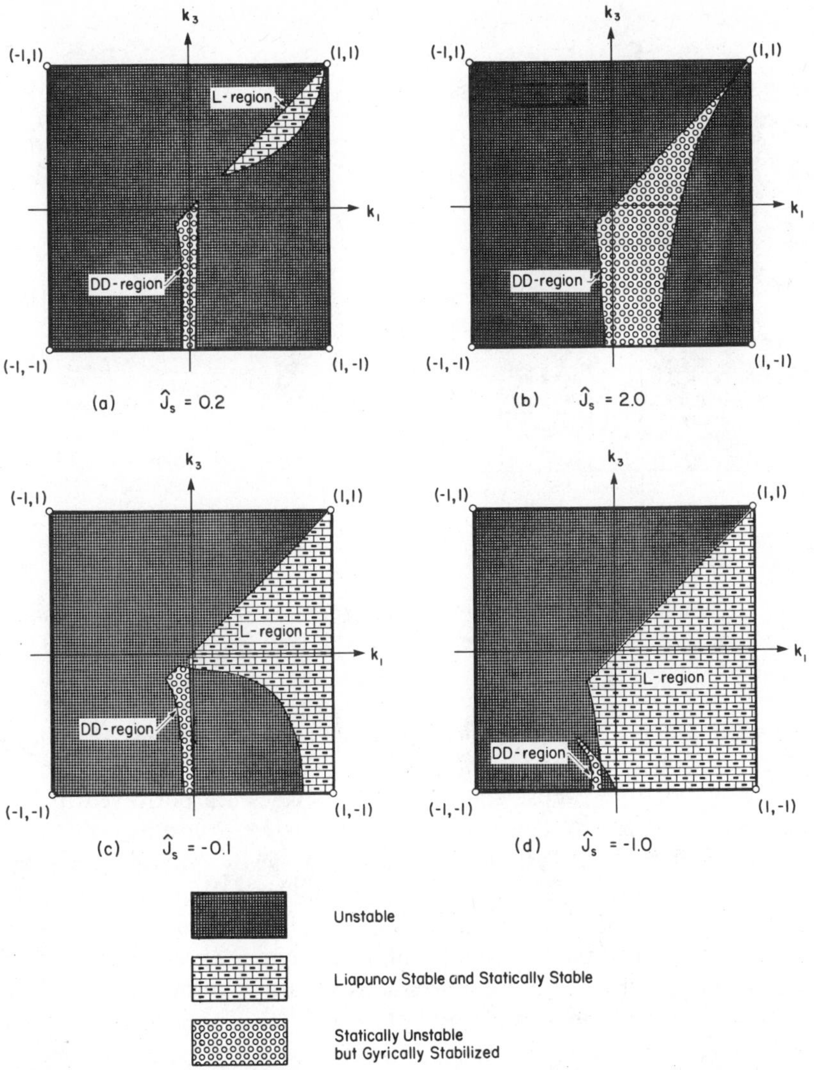

FIGURE 11.7 Stability diagrams for a gyrostat $\mathscr{R} + \mathscr{W}$ in a circular orbit (cylindrical case: principal axis aligned with orbital axes, and stored momentum normal to orbit). (Compare with Fig. 9.4.)

are stable; and as $\hat{\Omega}_{po} \to -\infty$, all are gyrically stabilized. (The symbol definitions (51) and (6.4, 27) imply that

$$\hat{J}_s \leftrightarrow -\frac{(I_1 I_3)^{1/2}}{I_2}\hat{\Omega}_{po} \tag{54}$$

is the relationship between $\hat{J}_s$ and $\hat{\Omega}_{po}$.) The reason for the whole square being stabilized in Fig. 6.7 and only half the square in Fig. 11.7 is that only directional

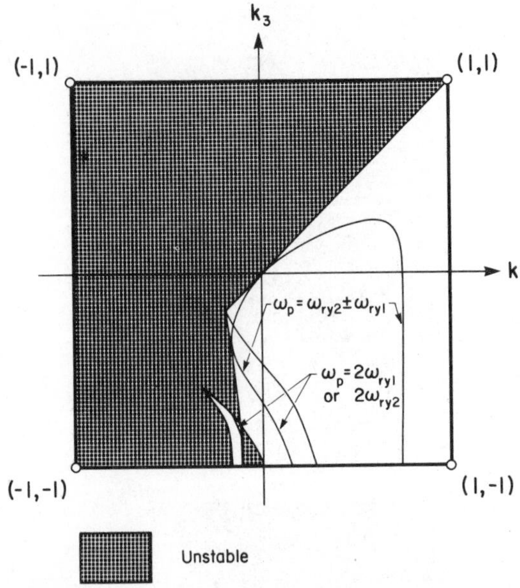

FIGURE 11.8 Nonlinear internal resonance lines for $\mathscr{R} + \mathscr{W}$ gravitationally stabilized in a circular orbit ($J_s = -1.0$). After [Crespo da Silva, 2]. (Compare with Figs. 9.12*b* and 11.7*d*.)

(one-axis) stability is available in the former, while attitude (three-axis) stability can be achieved in the latter for $k_1 > k_3$.

The reference "motion" whose stability characteristics are portrayed in Fig. 11.7 is the equilibrium $\boldsymbol{\alpha} \equiv \mathbf{0}$. However, it is also possible to use as a reference motion $\alpha_1 = \alpha_3 = 0$, $\alpha_2 \not\equiv 0$. Just as for $\mathscr{R}$ alone, a steady pitch libration can exist even with roll/yaw quiescent. In fact, this pitch libration is unaffected by the presence of $\mathscr{W}$ because $\mathscr{W}$ is spinning normal to the orbit. The pitch motion is still governed by (9.2, 31)—the equation for pitch motion in a circular orbit—and the solution is still given by (9.2, 35); for librations, we are interested in $0 < \hat{a} < 1$. If small (first-order) roll and yaw perturbations now occur, the periodic pitch motion can cause parametric excitation through nonlinear resonances. For $\mathscr{R}$ alone, this additional potential for instability can be seen from diagrams such as Fig. 9.12*a* and can be explained with the aid of the nonlinear internal resonance conditions depicted in Fig. 9.12*b*. The same principles apply to $\mathscr{R} + \mathscr{W}$. [Crespo de Silva, 2] has calculated the internal resonance curves for $\mathscr{R} + \mathscr{W}$ (Fig. 11.8 is typical) and has confirmed by numerical integration of the nonlinear equations of attitude motion, (1), that large-angle excursions do indeed occur in the neighborhood of these curves.

Stability of Noncylindrical Equilibria

Some typical stability results are now given for some of the "noncylindrical" equilibria depicted in Fig. 11.4. Consider first the hyperbolic case. Following [Longman, Hagedorn, and Beck], we shall adopt the simplest subcase in which the axis of $\mathscr{W}$ lies parallel to a principal axis. Thus, with reference to Fig. 11.5*a*, it is to be noted that $\beta_3 = 0$; and the solution to the transcendental equation (21) is obtained quite simply:

$$\cos \chi_3 = \frac{1}{2\kappa_H} \tag{55}$$

(Note that $\sin \chi_3 = 0$ is not allowed.) Evidently, the condition

$$|\kappa_H| \geq \tfrac{1}{2} \tag{56}$$

must be satisfied; if the wheel speed is too high ($|\kappa_H|$ too small), no equilibrium of the class under consideration can exist.

The stability analysis proceeds along the general lines laid down by (42) through (46). As with all hyperbolic equilibria, the fact that $J_{s3} = 0$ is helpful. Figure 11.9 gives the flavor of the stability results. The first diagram, Fig. 11.9*a*, seems at first glance to be in conflict with Fig. 9.4. In both cases, a rigid body in circular orbit librates about equilibrium; since $J_s \rightarrow 0$, $\mathscr{W}$ becomes part of $\mathscr{R}$. The explanation for this paradox can be given with the aid of Fig. 11.5*a*. The *cylindrical* equilibrium ($\sin \chi_3 = 0$) has already been eliminated from the present discussion, and only the *hyperbolic* equilibrium, with $\cos \chi_3 = (2\kappa_H)^{-1}$, remains. As $J_s \rightarrow 0$, we have $\kappa_H \rightarrow \infty$, and $\chi_3 \rightarrow \pm 90°$. In other words, as $J_s \rightarrow 0$ for the hyperbolic equilibrium, the principal axes do eventually coincide with the orbiting axes, but with $\{\hat{\mathbf{p}}_1, \hat{\mathbf{p}}_2, \hat{\mathbf{p}}_3\} = \{\underset{\rightarrow}{\hat{\mathbf{c}}}_2, -\underset{\rightarrow}{\hat{\mathbf{c}}}_1, \underset{\rightarrow}{\hat{\mathbf{c}}}_3\}$. Thus, $\{I_1, I_2, I_3\} \rightarrow \{I_2, I_1, I_3\}$, and the parameters $\vec{k}_1$ and $\vec{k}_3$ change accordingly. With this transformation it is not difficult to show that the regions in Fig. 9.4 map precisely into the regions shown in Fig. 11.9*a*. The remaining three diagrams in Fig. 11.9 illustrate how the various regions evolve as the wheel spin is increased in either direction. (Note that (56) is violated in the regions marked "physically impossible.") It is also interesting to compare these stability diagrams with those obtained in Fig. 10.4*a* for $\mathscr{W}$ alone; this is done in Problem 11.12.

Now consider the conical case (Figs. 11.4*c*). In general, the principal axis $\hat{\mathbf{p}}_2$ and the $\mathscr{W}$-axis will be inclined in the $\underset{\rightarrow}{\hat{\mathbf{c}}}_2$–$\underset{\rightarrow}{\hat{\mathbf{c}}}_3$ plane. Again following [Longman, Hagedorn and Beck], we simplify by restricting β_1 to zero. Thus, the axis of $\mathscr{W}$ is parallel to $\hat{\mathbf{p}}_2$ and is inclined at χ_1 to $\underset{\rightarrow}{\hat{\mathbf{c}}}_2$. The transcendental equation that implies the condition for relative equilibrium, namely, (27), now is simply solved:

$$\cos \chi_1 = \frac{1}{2\kappa_C} \tag{57}$$

For physically possible equilibria of this class to occur, the inertia distribution and wheel momentum must be such that

$$|\kappa_C| \geq \tfrac{1}{2} \tag{58}$$

The stability of such equilibria can be ascertained on the basis of (42) through (46), noting that $J_{s1} = 0$. Typical stability diagrams are shown in Fig. 11.10 (see also Problem 11.13). The stability diagram for $J_s = 0$ will transform to the earlier results of Fig. 9.4 provided the correspondence $\{I_1, I_2, I_3\} \rightarrow \{I_1, I_3, I_2\}$ is used.

A great many additional stability diagrams pertaining to a rigid gyrostat in a circular orbit have been prepared by [Hughes and Golla]. In addition, the stability

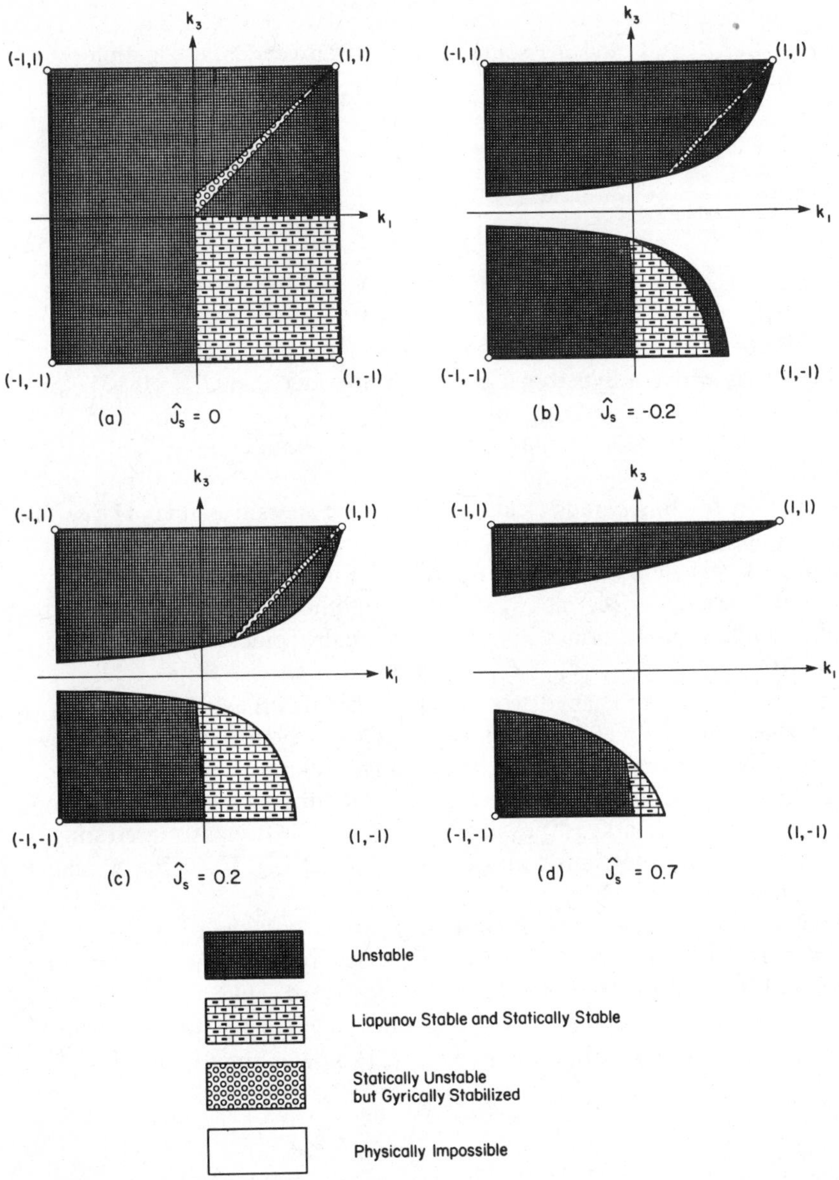

FIGURE 11.9 Stability diagrams for a gyrostat $\mathscr{R} + \mathscr{W}$ in a circular orbit—hyperbolic cases (see Fig. 11.5*a*) with $\beta_3 = 0$.

relationship between "Kelvin's gyrostat" and the "apparent gyrostat" (see Section 6.1) is further clarified by [Li and Longman]. In the former, a wheel speed control system that keeps h_s constant is implied, whereas in the latter, the absolute spin rate of the rotor or wheel is constant. In view of the reality of bearing friction, the "apparent gyrostat" will in practice also require a wheel speed control system to maintain the desired absolute spin rate.

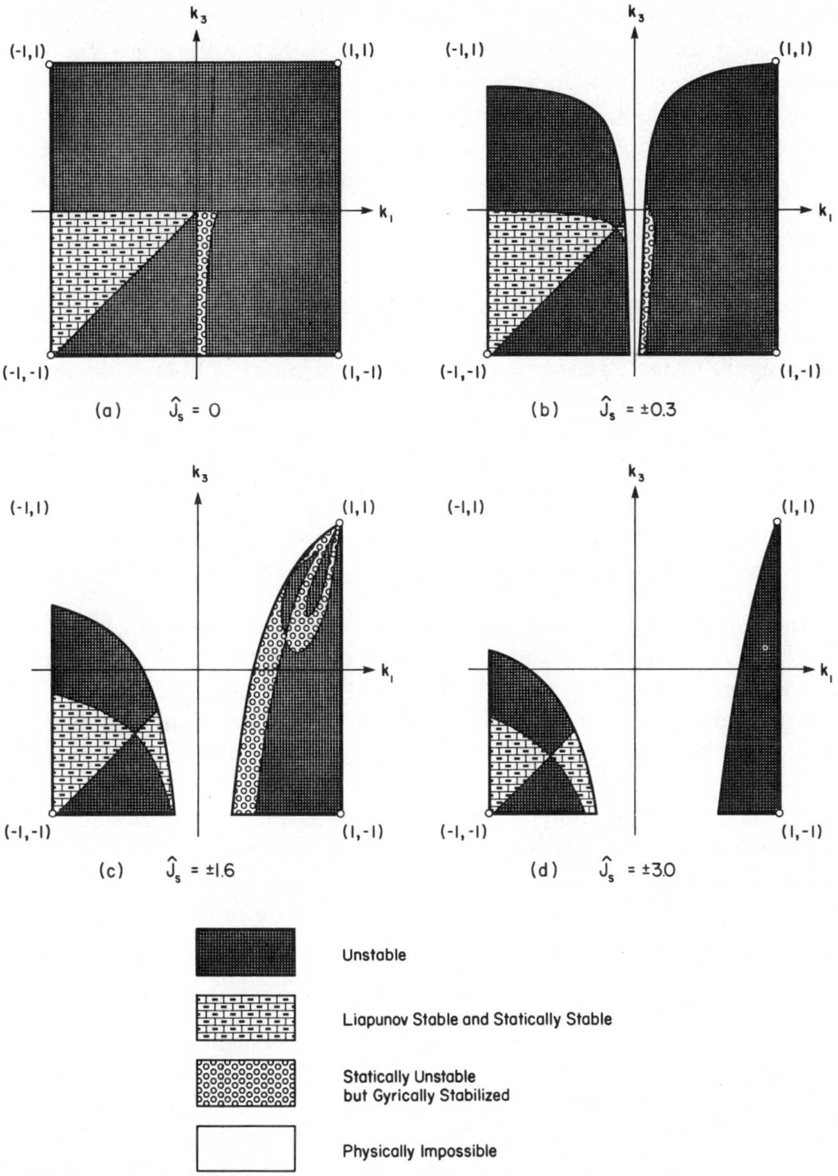

FIGURE 11.10 Stability diagrams for a gyrostat $\mathscr{R} + \mathscr{W}$ in a circular orbit—conical cases (see Fig. 11.5*b*) with $\beta_1 = 0$.

Before closing this section, we remark that many of the techniques explained earlier in this book can also be applied to gain further information concerning the characteristics of $\mathscr{R} + \mathscr{W}$ equilibria. To be specific, much of the methodology used in Section 9.1 for a rigid body $\mathscr{R}$ in orbit (i.e., under the influence of gravity gradient torque) can be extended to $\mathscr{R} + \mathscr{W}$ by making modifications appropriate to the addition of $\mathscr{W}$. For example, if $\hat{\underset{\rightarrow}{\mathbf{p}}}_1$ or $\hat{\underset{\rightarrow}{\mathbf{p}}}_3$ is an axis of inertial symmetry, libration bounds similar to those of Figs. 9.10, 9.11, and 10.5 can be used to aid visualization and to understand finite attitude excursions. These are again based on the Hamiltonian function (see Problems 11.6, 11.9, and 11.11). Analytical approximations for slight inertial assymmetry, Floquet analyses for slight orbital eccentricity, invariant surfaces (cf. Fig. 9.14) that characterize the nonlinear equations of motion—these ideas, and others, can in principle be applied to study the $\mathscr{R} + \mathscr{W}$ system in orbit. To the author's knowledge, such investigations have not been published. Moreover, we must not lose sight of the fact that the $\mathscr{R} + \mathscr{W}$ system is something of an abstract object, inasmuch as it ignores energy dissipation on either $\mathscr{R}$ or $\mathscr{W}$. As seen repeatedly in developments earlier in this book, no dynamical analysis, however elegant mathematically, can be said to apply to real spacecraft unless the implications of energy dissipation are also understood. In view of these remarks, we turn now instead to an examination, from an engineering viewpoint, of the two types of $\mathscr{R} + \mathscr{W}$ (gyrostat) spacecraft that have reached maturity over the past decade.

11.2 GYROSTATS WITH EXTERNAL ROTORS

The basic idea of influencing the attitude dynamics of a satellite by incorporating one or more spinning wheels into the design has been exploited in an almost bewildering number of ways. Indeed, the present state of the art tends to make an ordered exposition very difficult, but very necessary. A rational introduction to the many diverse gyrostat-type ideas in the literature requires that the subject be developed in a sequence of logical steps. To this end, a dissipation-free, torque-free rigid body containing a wheel or rotor was analyzed in Chapter 6; the importance of internal energy dissipation was then recognized in Chapter 7; and the influence of gravitational torque, which must be acknowledged for any orbiting satellite, has been the topic for discussion thus far in the present chapter.

To develop our ideas further, it becomes necessary to make a distinction that is not fundamentally dynamical in character. In this section, we shall consider gyrostats for which the wheel is *external*; this wheel is, for such gyrostats, universally called a *rotor*. Then, in the Section 11.3, we shall consider gyrostats for which the wheel is *internal*; this wheel is universally called, simply a "wheel" (although more descriptive labels such as "momentum wheel" and "flywheel" are often used). This external–internal distinction is motivated primarily by the two design channels into which basic gyrostat ideas have flowed in engineering practice.

In the external-rotor gyrostat, most of the mass of the satellite is resident in the rotor. To a large extent, the satellite *is* the rotor, with additional parts (antennas, solar arrays) despun as dictated by mission requirements. Such designs are the direct descendents of spinning satellites (Chapter 10) and indeed are often referred to loosely as "spinners" even though important parts of the satellite are not spinning. This class of satellite has undergone much development by, among others, the Hughes Aircraft Company, who have applied the term *Gyrostat* to their type of design. Thus, although classically the wheel of a gyrostat is *internal* to its container (as explained in Section 3.5), in modern satellite usage a Gyrostat has an *external* rotor. For the remainder of our discussion, the word Gyrostat, when capitalized, will indicate this special, more modern, meaning.

Other satellite attitude stabilization systems also use the dynamical principle of the gyrostat, although their configurations are visually quite different from Gyrostats. These satellites contain, internally, one or more relatively small but rapidly spinning wheels; the sole function of these wheels is to lend such gyricity as is beneficial to attitude stabilization. It may be helpful to look upon these satellites as being the descendents of the gravity gradient satellites discussed at length in Chapter 9; they will be called *bias momentum satellites*, in accordance with current parlance. It should also be noted that whereas the Gyrostat rotor usually contains much of the rotational inertia of the satellite and spins relatively slowly (~ 50 rpm), the momentum wheel in a bias momentum satellite contributes almost negligibly to the satellite's rotational inertia but spins relatively rapidly (~ 3000 rpm).

With this helpful distinction understood, another important point can be made. Although both Gyrostats and bias momentum satellites are very similar dynamically (in that they are both orbiting gyrostats), nevertheless, in practice, the rather Gyrostat rotor tends also to act as a container for all sorts of satellite paraphernalia, including, for example, thruster fuel tanks; Gyrostat rotors must therefore be modeled as energy dissipative. The compact, rapidly spinning wheels on bias momentum satellites, on the contrary, are about as close as one can get in reality to spinning, symmetrical *rigid* bodies. In this respect, the developments in Chapter 7 relating to dissipative rotors (see, for example, the idealizations of Figs. 7.8, 7.9, and 7.11) is more relevant to Gyrostats that to bias momentum satellites. Attention should also be given to the several conclusions reached in Problem 7.3.

The Basic Concept

The basic guiding principle behind the successful functioning of a passive Gyrostat is found from the derivations in Chapter 7. The mildly approximate (but very revealing) major-axis rule for Gyrostats was displayed in (7.1, 118). That rule is repeated here with slightly altered symbols. For stability,

$$\frac{\dot{T}_{sp}}{I_{ar}} + \frac{\dot{T}_{sr}}{I_{ar} - (I_1 I_3)^{1/2}} < 0 \qquad (1)$$

according to energy sink analysis (very small damper). For external-rotor Gyrostats, the "carrier" tends to be called the *platform*, and the "wheel," as mentioned, is called the "rotor"; hence the changes of subscripts, $e \to p$ and $w \to r$. The physical model used was shown in Fig. 7.11, adapted here as Fig. 11.11. In equation (1), I_{ar} is the rotor's axial inertia and I_1 and I_3 are the transverse

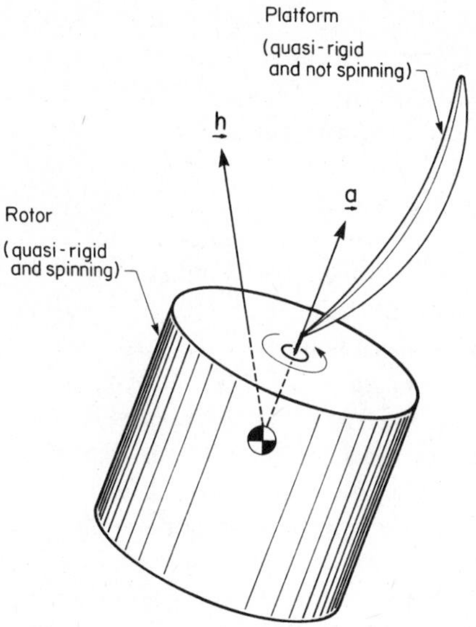

FIGURE 11.11 The Gyrostat concept. (See Fig. 7.11.)

inertias of the entire vehicle. The rates of energy dissipation in the platform and rotor are denoted $\dot{T}_{sp}$ and $\dot{T}_{sr}$, respectively, and these quantities must of course be negative for a passive design. The Gyrostat will be said to be *oblate* (*prolate*) according to whether $I_{ar}/(I_1 I_3)^{1/2} > 1$ (< 1). Note that the platform inertia along the spin axis is not included in this discrimination.

Equation (1) indicates that all oblate Gyrostats are asymptotically stable. For prolate Gyrostats however, the stabilizing influence of damping in the platform must prevail over the destabilizing influence of damping in the rotor. Thus, for passive directional stabilization, the Gyrostat must either be oblate or have a wobble damper within the platform. *Active* wobble control is often used nowadays however, as will be discussed briefly below.

Wobble Dampers

Many devices have been proposed to damp out Gyrostat wobble, and several have been used in orbit. (The reader may wish to review the wobble damper concepts and designs discussed in Section 9.3 for gravity-gradient-stabilized satellites, as well as those described in Section 10.2 for spin stabilized satellites.) In designing such dampers for prolate Gyrostats, it must always be borne in mind that they must be large enough to offset the destabilizing effect of damping in the rotor. Rotor damping is most often caused by fuel sloshing. Although the dynamics of fuel sloshing is beyond our intended scope here, the experimental results reported by [Martin] in connection with INTELSAT IV are of interest. A recent colloquium, [INTELSAT/ESA], summarizes the state of the art as of 1984. The review of [Ancher, Brink, and Pouw] should also again be referred to.

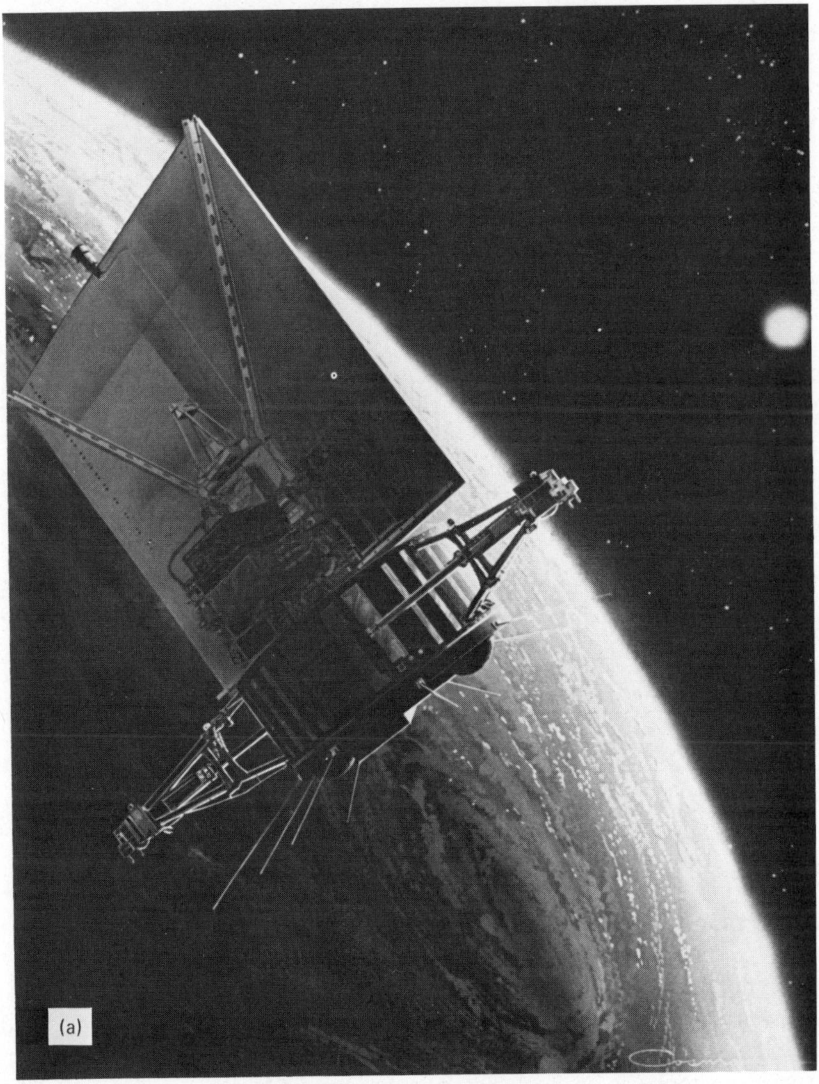

FIGURE 11.12 OSO-1 Gyrostat and its wobble damper. (*a*) OSO-1. (*b*) Pendulous viscous-fluid wobble damper used in OSO-1.

Three early designs developed especially for gyrostats were described by [Taylor and Conway], [Auelmann and Lane], and [Spencer, 4]. These were, respectively, a viscous-ring damper (similar to the one shown in Fig. 10.23*a*), a ball-in-tube damper (not unlike the one shown in Fig. 10.22*b*), and the cantilevered-mass (or tuned-pendulum) damper shown in Fig. 11.12*b*. The latter was flown on OSO-1 in 1962—the first Gyrostat in orbit (Fig. 11.12*a*). Complete dynamical analyses can be found in the cited references.

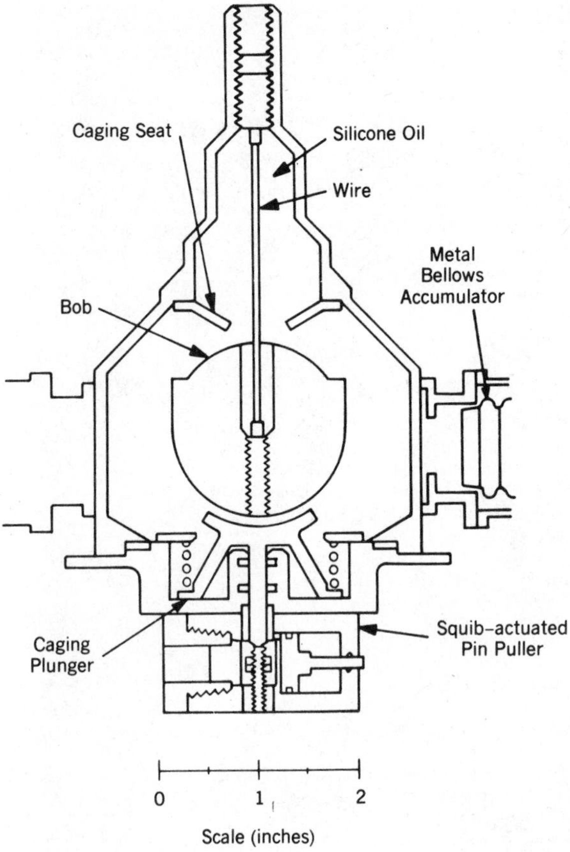

(b)

FIGURE 11.12 (Continued)

Analytical investigations of basic design principles for Gyrostat dampers have been given by [Cloutier, 1, 2] and by [Haines and Leondes]. Cloutier considered mechanical dampers of the mass-spring, cantilevered-beam, and pendulum types, with special emphasis on assuring their performance over a wide range of body spin rates, while Haines and Leondes considered eddy current dampers. In all cases, the aim is to choose the physical parameters of the damper to achieve mass-efficient dissipation.

The TACSAT satellite (Fig. 11.13) is of great interest, being the first prolate Gyrostat. Launched in February 1969, it demonstrated the principle that sufficient damping in the platform can be made to stabilize an otherwise unstable vehicle. Its wobble damper (Fig. 13*b*) was a two-degree-of-freedom pendulum suspended in a case filled with viscous fluid. The flight data reported by [C. R. Johnson] show intervals of $\sim 1°$ wobble interspersed between much tighter accuracies. Although the 1° wobble was of concern to the satellite's designers

FIGURE 11.13 TACSAT Gyrostat and its wobble damper. (*a*) TACSAT. (*b*) Pendulum wobble damper used in TACSAT.

(using postflight analysis, it was traced to certain dynamical characteristics of the rotor bearing) the basic correctness of the theory underlying prolate Gyrostats was amply demonstrated.

A "second generation" prolate Gyrostat—INTELSAT IV—is shown in Fig. 11.14. Its damper again utilized a pendulum, but the damping action was provided by eddy current dissipation. According to [Neer, 2], INTELSAT IV flight data demonstrate a residual nutation angle less than 3 arcseconds and nutation damping time constants as short as 18 seconds.

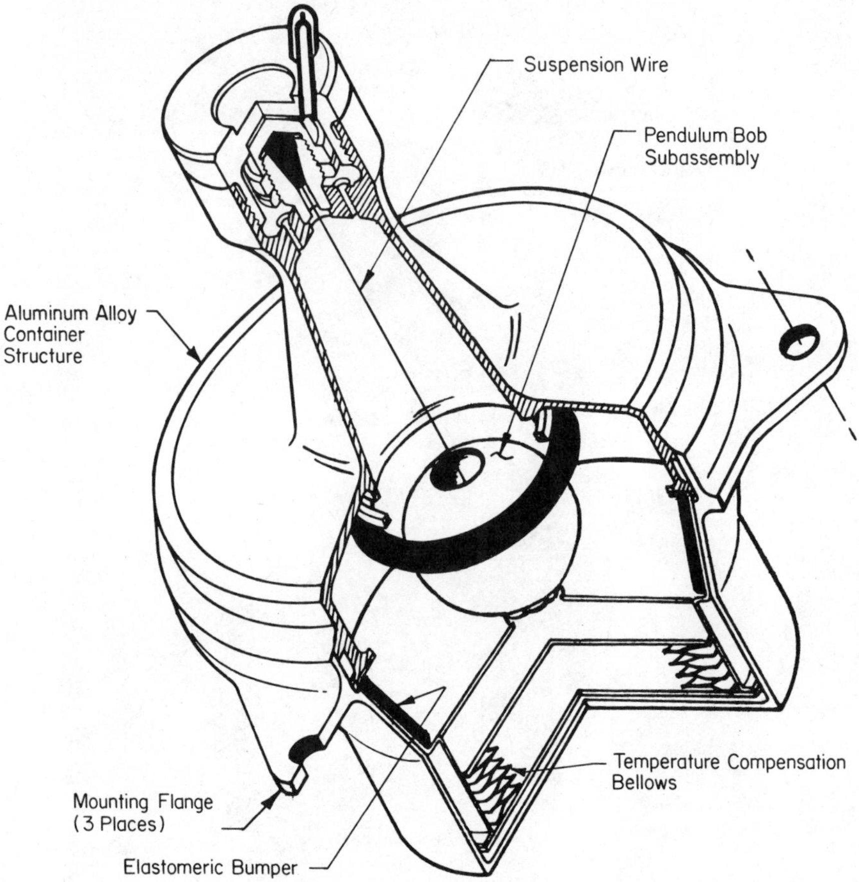

(b)

FIGURE 11.13 (Continued)

Bearing Alignment and Flexibility

Fundamental to the Gyrostat concept is a bearing assembly about which the rotor rotates relative to the despun platform. The basic theory assumes an ideal bearing, but small bearing imperfections lead in practice to small pointing perturbations. The rotor is never perfectly axisymmetrical, the bearing axis never passes precisely through the rotor mass center, and the bearing axis is never perfectly aligned with the rotor's principal axis of inertia. In consequence, there are small static and dynamic imbalances that lead to slight wobbling. These effects have been quantified by several dynamical analyses; see, for example, [McIntyre and Gianelli, 1].

Nor can a bearing housing be completely rigid. [Cronin's] paper contains some typical bearing stiffness data which indicate the magnitude of the flexibility and also show a considerable degree of nonlinearity in the torque deflection characteristic. Transverse bearing deflections are also accompanied by transverse bearing damping—a factor of concern in detailed stability analysis ([Scher]).

Attitude Acquisition and Reacquisition

Immediately after launch, Gyrostats usually spin as a single body, namely, platform and rotor together. The attitude acquisition process includes procedures for despinning the platform and energizing the despin control system (see below). These procedures are designed to be as dynamically benign as possible, to avoid large attitude deviations and transverse momentum buildup.

If a failure eventually occurs in the despin control system, bearing friction will tend to reduce the relative spin between platform and rotor until the satellite ultimately spins as a single body. This can be especially unsettling for prolate vehicles because such a spin is of course unstable (Chapter 5), leading to a "flat spin" in which the rotor symmetry axis rotates in a plane perpendicular to $\underset{\rightarrow}{\mathbf{h}}$. Assuming the original failure can be corrected, recovery from a flat spin requires reorientation through 90° the analysis of which presents a challenging problem in

FIGURE 11.14 INTELSAT IV Gyrostat and its wobble damper. (*a*) INTELSAT IV. (*b*) Pendulous eddy current wobble damper used on INTELSAT IV.

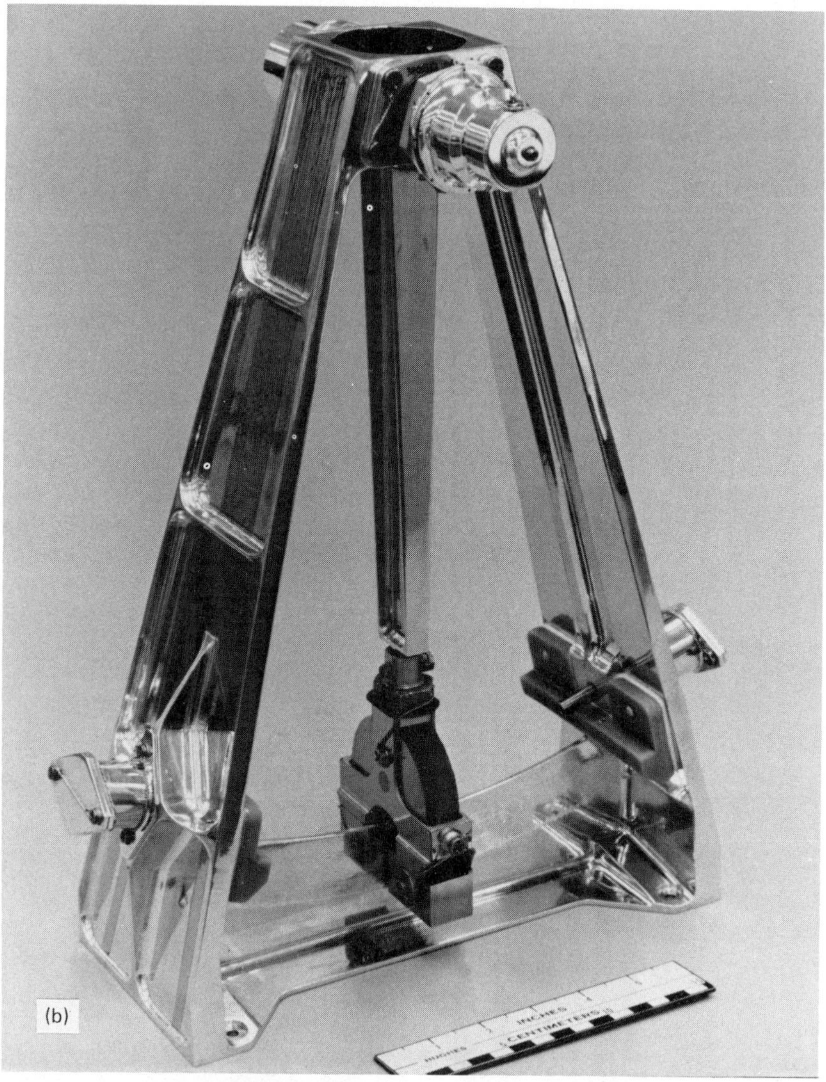

FIGURE 11.14 (Continued)

nonlinear attitude dynamics. Indeed, some form of approximation is inevitable for analytical work. [Gebman and Mingori] have worked out a perturbation solution which indicates recovery duration and residual nutation angle for one class of recovery procedure. [Hubert] has indicated that a wide range of initial orientation states, including complete inversion, can be transferred to the "acquired" state by first spinning up the rotor to a desired rate and then allowing passive damping to operate during a transitional phase in which the rotor's relative spin is kept constant. Further analysis of platform despin dynamics has been given by [Adams], based on in-orbit experience.

Nonlinear Phenomena

When the attitude motion of a Gyrostat consists of more than small deviations from the reference equilibrium, several phenomena can exhibit themselves which are fascinating dynamically and potentially troublesome operationally. Two of the most important nonlinear phenomena for Gyrostats are *nonlinear resonances* and *trap states*.

The idea of nonlinear resonance was introduced earlier, in Chapter 9, in connection with the librations of a rigid body in a circular orbit (see, for example, Fig. 9.12). It was shown there that even though infinitesimal roll/yaw librations are mutually uncoupled from pitch librations, as indicated by linear analysis, finite pitch motion can excite roll/yaw librations through the higher-order terms whose effects become increasingly important as the pitch amplitude grows. The same phenomenon can occur for Gyrostats as well. [Fujii, 2] has analyzed such resonances using the method of averaging; his analysis is aimed at explicating how best to damp such resonances using the passive platform damper already on board. [Cochran] has used the method of averaging also, as a tool to uncover the contribution of slight rotor asymmetry and platform dynamic imbalance (small products of inertia). [Cochran and Beaty] have show that the solution for the averaged variables can be written using Jacobian elliptic functions (recall Fig. 4.7). In connection with all these analyses, it should be remarked that during the platform despin maneuver (attitude acquisition), a wide range of dynamical states is traversed, thus making very real the possibility that such resonances could be encountered.

A different sort of nonlinear phenomenon is the "trap state" (an appellation that, like many others in engineering, makes up in descriptive impact what it lacks in syntactical finesse). The attitude and attitude rate variables may enter regions of state space (possibly "point" regions) from which they will never leave; in particular, they will never converge to their intended values, the values corresponding to a directionally stabilized equilibrium. As explained by [Scher and Farrenkopf], two trap states can be identified for Gyrostats having asymmetrical or unbalanced rotors. The first is a minimum energy state, to which the spacecraft settles when despin control is deactivated. The second, encountered while despinning the platform, is characterized by a stalling of the despin process and a violent increase in nutation. Fortunately, it has also been shown that the state can be extricated from either trap by active torquing of the rotor at an appropriate natural frequency. The analysis of [Tsuchiya] is also relevant.

Remarks on Active Control

Our discussion of Gyrostats has not emphasized those aspects of spacecraft stabilization that are active ("active" in Sabroff's sense); the aim of this book is to widen understanding of the dynamical, not the control, characteristics of these vehicles. However, lest a distorted view of Gyrostats as complete attitude stabilization systems be created, a few brief remarks will be made on active control components. One such component has already been alluded to several times: the despin control system, without which bearing friction would slowly eliminate the relative spin between the rotor and the platform. In other words, the rotor would spin down and the platform would spin up. This process would not change the system angular momentum, of course, but the system energy would be reduced. However, one can track the object (usually Earth) at which the platform is supposed to be pointing using an active sensor and feed the measured pointing error to an appropriate control electronics unit. When, in turn, this unit governs the despin control motor, the perfect-bearing assumption made in the basic theory can be closely approximated. Having thus approximated the theoretical ideal, directional stability can be achieved.

For prolate Gyrostats, some form of nutation damping is also required, as described above. As sensors and microelectronics have become better developed, "active" options have tended to gain the ascendancy at the expense of "passive" (or purely dynamical) options. Indeed, many modern Gyrostats rely for their wobble damping not so much on passive dampers as on active control systems that exploit platform asymmetry. Active wobble control of *single*-body, *minor*-axis spinners had been developed during the 1960s (see, for example, [Grasshoff, 2]). Such control often became especially necessary in transfer orbit, before the ultimate stabilization system became operational. The critical wobble sensor was an accelerometer whose signal, after suitable compensation (and with due regard for phase on a spinning vehicle), triggered thrusters to produce the desired transverse torques. The task was made easier by axial asymmetry.

Following this tradition, it was recognized that a Gyrostat by its very nature possesses an active control component not available on a spinner—a despin control system—and that the thrusters necessarily used for spinners might be replaced somehow by this already available active loop. [Phillips] and [Slafer and Marbach] showed how to exploit "inertial coupling" (available through the platform's products of inertia) as a measure of wobble. They showed furthermore how this measurement, when suitably fed to the already present despin control system, could be made to produce the desired wobble damping, again via inertial coupling. The most successful sensing element has been the accelerometer; the papers by [Smay and Slafer] and [Slafer and Smay] can be consulted for much further detail. The general conclusion is that this method of wobble damping often proves to be more weight efficient than do passive dampers.

For some special applications, a part of the platform—a communications antenna, for example—must be continuously articulated. This additional active control requirement would exist for these applications even if the directional stabilization of the satellite itself were perfect. Reaction torques caused by this articulation inevitably affect the overall wobble situation. [McIntyre and Gianelli] have shown that, for many applications, such autotracking can in fact *increase* the directional stability margin. Thus, in addition to performing its primary function of payload pointing, the articulation control system may produce the side benefit of helping to stabilize the spacecraft. [Slafer] has described just such a

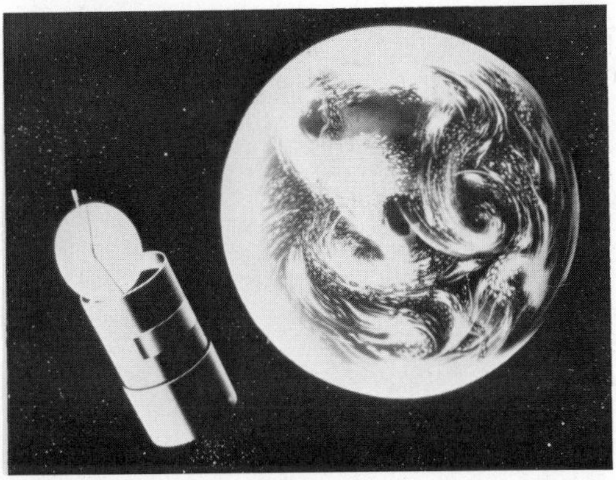

FIGURE 11.15 Anik C Gyrostat.

situation with the OSO-8, whose control system design criteria, developed with nutation stability in mind, were verified through a series of in-orbit tests.

Finally, just as with single-body spinners, it is necessary to compensate periodically for the external torques on a Gyrostat, torques that tilt the angular momentum vector $\underset{\rightarrow}{\mathbf{h}}$ (the vector to which the wobble control system, whether active or passive, strains to align the vehicle). Small changes in the orbital plane pose geometrical misalignments of an essentially similar kind. These realignments are brought about in practice by control procedures that are neither active nor passive, but ground controlled. As an example, consider Anik C, shown in Fig. 11.15. Launched in 1982 as a geostationary communications satellite, Anik C was a prolate Gyrostat with active wobble control. Figure 11.16 shows flight data during a five-month period in 1983. The top two diagrams indicate the satellite's position with respect to a geocentric equatorial coordinate system. (Note in Fig. 11.16*a* the daily variations in longitude due to slight orbital eccentricity.) Of even greater interest in the present context is the variation of attitude codeclination (defined in Fig. 10.32). Ground-commanded thruster firings are required every 10 days or so, to keep the spin vector within 0.15° of geographic North.

11.3 BIAS MOMENTUM SATELLITES

Remarks have already been made at the outset of the last section highlighting the difference between Gyrostats and bias momentum satellites. Although they are in many respects dynamically indistinguishable (both are based on the classical gyrostat) the details of their implementation are sufficiently distinct that they should be thought of to some degree as two separate categories. They are also visibly distinct in that a Gyrostat possesses a large external rotor, whereas a bias momentum satellite has a relatively small internal momentum wheel. Nonetheless, there is no doubt that the angular momentum of a Gyrostat is "biased," just as there is no doubt that what we shall call a "bias momentum" satellite is

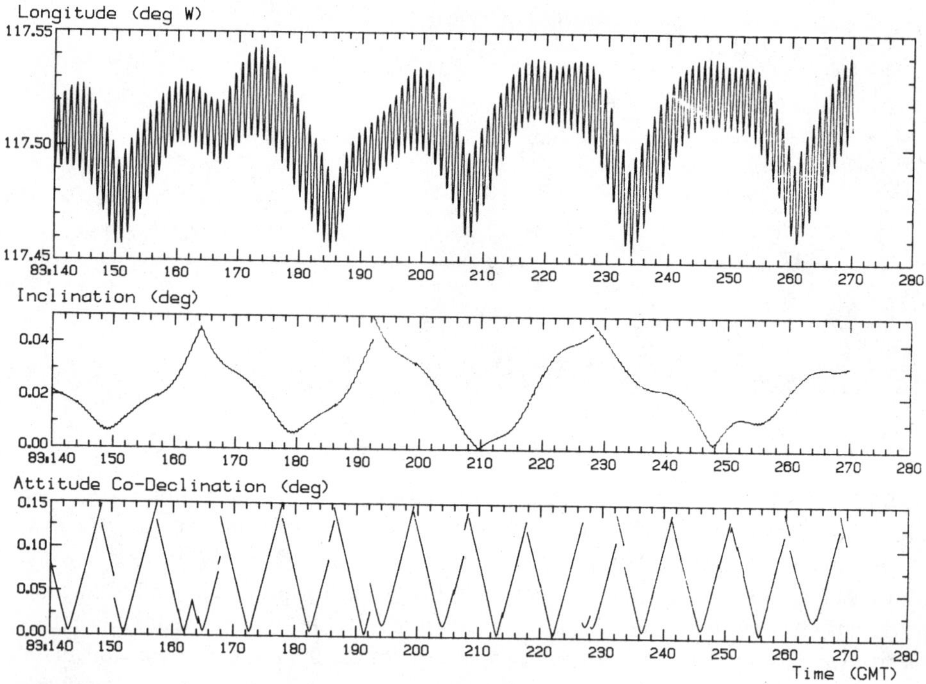

FIGURE 11.16 Spin vector history for Anik C.

identical to the classical gyrostat; however, the terminology used in the sequel conforms to modern usage in the aerospace literature.

Bias momentum satellites tend to have more active control elements in their designs, although both Gyrostats and bias momentum satellites come in a wide range of designs on the passive–active scale. Both must be actively controlled to the extent that a motor, informed by measurement, keeps the rotor (or wheel) spinning, else bearing friction would soon produce, for both designs, a single spinning body. Distinctions begin to occur with respect to the remaining three fundamental degrees of rotational freedom, those that characterize the attitude of the main body (called "the platform" in the case of Gyrostats).

There are of course design considerations other than dynamics and control that are important in deciding whether a Gyrostat or a bias momentum configuration is most appropriate for a particular mission. For example, thermal control may be made easier by a configuration in which most of the satellite is spinning, but extensive electrical connections across a major rotary joint may prove unnecessarily expensive. Although discussion of the thermal and electrical subsystems is beyond the scope of this book, these examples serve to illustrate why the Gyrostat and bias momentum alternatives, although dynamically quite similar, are distinct in practice. It is also interesting to observe that the requisite technologies for both designs are now so advanced that for many missions there is little to choose between the two competing alternatives.

Augmented Gravity Gradient Stabilization

It has been recognized for some time in the history of gravity gradient stabilization that a small momentum wheel nominally oriented along the pitch axis (i.e., the orbit normal) enhances roll/yaw coupling and thereby provides greater latitude in arranging the inertia distribution, as is so all-important for gravity gradient designs. Experience with the DODGE satellite (1967) has already been mentioned (Section 9.4). Typical analyses in the literature of that time are those of [Mingori and Kane] and [Palmer, Blackiston, and Farrenkopf]. In both cases, the small momentum wheel is suspended in gimbals via spring damper connections to provide roll/yaw coupling and passive damping. The motion equations for this situation are of the form

$$\mathscr{M}\ddot{\mathbf{q}} + (\mathscr{D} + \mathscr{G})\dot{\mathbf{q}} + \mathscr{K}\mathbf{q} = \mathscr{f} \tag{1}$$

and the wheel momentum and gimbal spring and damping constants, which enter into $\mathscr{D}$, $\mathscr{G}$, and $\mathscr{K}$, provide design freedom (details are left to Problem 11.16). The optimization of such "gyrodampers" has been explored by [Sarychev and Mirer].

GEOS-C is an example of a satellite whose attitude stabilization was based on the gravity gradient, augmented by a small pitch momentum wheel (Fig. 11.17). Launched in 1975 to perform experiments in Earth physics and oceanography, GEOS-C was three-axis stabilized to within 3.6° in pitch and 1.5° in roll and yaw (Fig. 11.17*b*).

The Basic Concept

Although a small pitch momentum wheel can sometimes be used to augment gravity gradient stabilization, the term "bias momentum" really refers to a larger momentum wheel, used on a satellite that does not rely on the gravity gradient for stabilization. The basic idea in its simplest form is shown in Fig. 11.18. The satellite is intended to be stabilized to the orbiting axes (in other words, to be Earth pointing). An active attitude sensor (indicated by the "eye" in Fig. 11.18) focuses on Earth, and any deviations from the vertical are measured. That is, pitch and roll attitude errors are sensed and fed to appropriate control systems which in turn cause corrective pitch and roll torques to be applied.

Thus, pitch and roll are stabilized—but what about yaw? If yaw error were also sensed, a control loop could be added for yaw as well. Unfortunately, yaw sensing is a complex proposition, and the development of space-proved yaw sensing hardware has lagged behind Earth sensors. Even though now available, these devices are very expensive, and consequently yaw sensing is avoided unless absolutely necessary. It is often preferable to use a momentum wheel $\mathscr{W}$, as in Fig. 11.18. This couples roll and yaw together, so that stabilization for roll becomes stabilization for yaw also.

To discuss this idea precisely and quantitatively, we examine the motion equations pertaining to the system shown in Fig. 11.18. These equations were

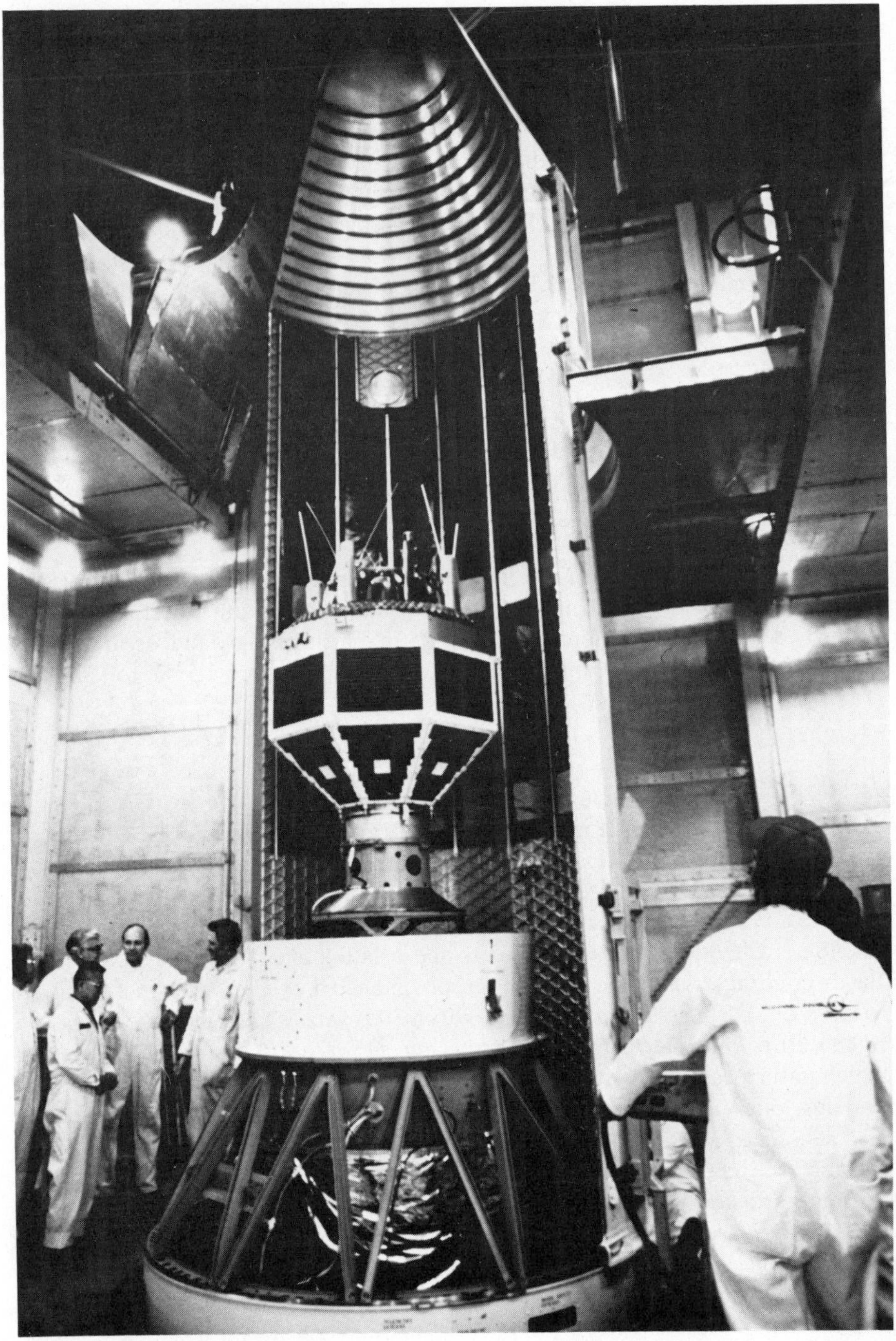

FIGURE 11.17 GEOS-C spacecraft and some typical flight data. After [Lerner and Coriell]. (*a*) GEOS-C spacecraft. (*b*) Typical flight data.

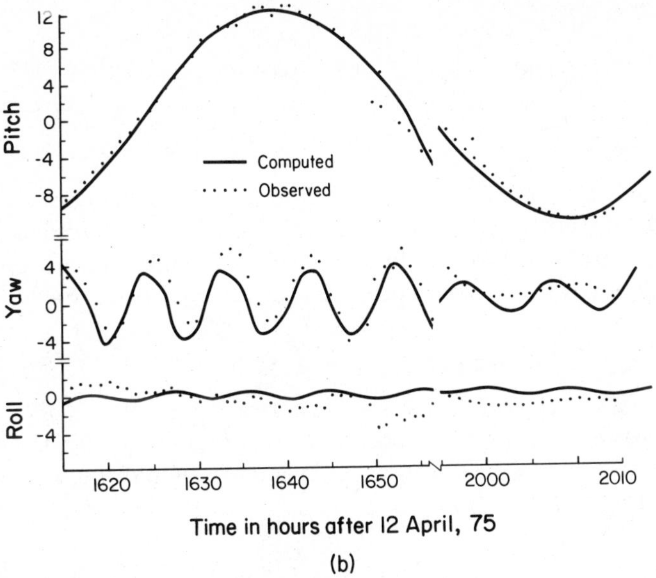

FIGURE 11.17 (continued)

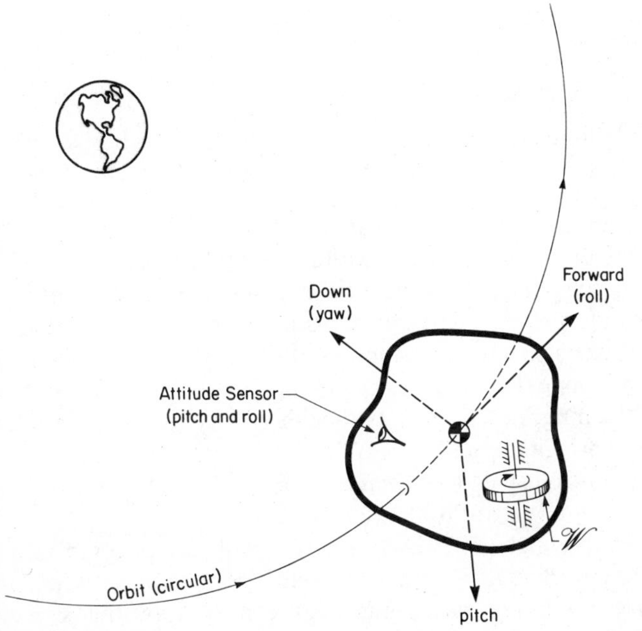

FIGURE 11.18 The basic bias momentum satellite concept.

given earlier by (11.1, 47 and 48):

$$I_1\ddot{\alpha}_1 + (I_2 - I_1 - I_3 - J_s)\omega_c\dot{\alpha}_3 + [4(I_2 - I_3) - J_s]\omega_c^2\alpha_1 = g_{1c} + g_{1d} \quad (2)$$

$$I_3\ddot{\alpha}_3 - (I_2 - I_1 - I_3 - J_s)\omega_c\dot{\alpha}_1 + (I_2 - I_1 - J_s)\omega_c^2\alpha_3 = g_{3c} + g_{3d} \quad (3)$$

$$I_2\ddot{\alpha}_2 + 3(I_1 - I_3)\omega_c^2\alpha_2 = g_{2c} + g_{cd} \quad (4)$$

Here, $J_s \triangleq h_s/\omega_c$, as usual, and the subscripts c and d on the torque components connote "control" and "disturbance" torques. The nature of the closed-loop control torques $\{g_{1c}, g_{2c}, g_{3c}\}$ will be discussed briefly presently. For bias momentum systems as the term is normally used, and as distinct from gravity gradient augmentation, one has

$$J_s \gg \max\{I_1, I_2, I_3\} \quad (5)$$

In other words, the angular momentum of $\mathscr{W}$ is much greater than the angular momentum of $\mathscr{R} + \mathscr{W}$ arising from its once-per-orbit inertial rotation. Under this assumption, the motion equations for roll and yaw, (2) and (3), simplify to

$$\begin{aligned} I_1\ddot{\alpha}_1 - h_s\dot{\alpha}_3 - h_s\omega_c\alpha_1 &= g_{1c} + g_{1d} \\ I_3\ddot{\alpha}_3 + h_s\dot{\alpha}_1 - h_s\omega_c\alpha_3 &= g_{3c} + g_{3d} \end{aligned} \quad (6)$$

The characteristic frequencies (without feedback control) are found from the characteristic equation

$$I_1I_3s^4 + h_s^2s^2 + h_s^2\omega_c^2 = 0 \quad (7)$$

Recalling from (6.2, 7) that the precession frequency in torque-free motion is

$$\Omega_p = \frac{h_s}{(I_1I_3)^{1/2}} \quad (8)$$

we have

$$s^4 + \Omega_p^2(s^2 + \omega_c^2) = 0 \quad (9)$$

Making assumption (5), we find the two approximate roll/yaw frequencies:

$$\omega_{ry} \doteq \omega_c, \Omega_p \quad (10)$$

Thus, in the limiting case indicated by (5), the satellite has two roll/yaw modes.

In the first mode, the satellite does not rotate with respect to inertial space, but the pitch axis has a slight roll (or yaw) error. This error, as the satellite orbits Earth, is interpreted alternately every quarter-orbit as purely roll error and then as purely yaw error. Obviously, the frequency of this mode is the orbital frequency ω_c. In the second mode, the satellite does in fact precess with respect to inertial space in a (small) coning motion, just as though gravity torque were not acting [because, with the approximation (5), it isn't]. The precession frequency of $\mathscr{R} + \mathscr{W}$ in torque-free motion is just Ω_p (see Section 6.2). Small corrections to the frequency expressions (10) are given in Problem 11.17.

A typical momentum wheel is shown in Fig. 11.19. The basic idea of spinning a wheel on bearings is simple enough, but a complete design which extracts maximum momentum for minimum mass, and that can spin in the space environment with virtually no chance of failure, requires a major feat of engineer-

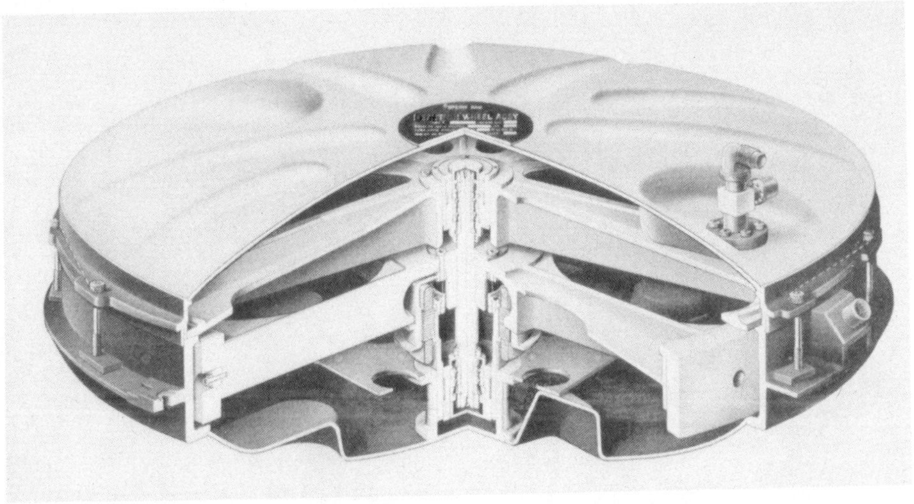

FIGURE 11.19 Typical momentum wheel.

ing. We also note from (3) that, even after stabilization, a quasi-steady yaw torque g_{3d} will lead to a quasi-steady yaw error

$$\alpha_3 = -\frac{g_{3d}}{J_s\omega_c^2} \equiv -\frac{g_{3d}}{h_s\omega_c} \tag{11}$$

This yaw error gradually becomes a roll error over the next quarter-orbit (a long time compared to the desired response time of the control system) and cannot be eliminated. For bias momentum systems of the type shown in Fig. 11.18, then, (11) provides a simple but crucial design equation for the size of the wheel. Once a permissible yaw error (α_3) has been decided upon for the mission, and once the maximum yaw torque (g_{3d}) has been estimated, (11) gives the minimum wheel momentum (h_s) that will produce the desired gyric yaw "stiffness."

There are important variations to the basic bias momentum theme shown in Fig. 11.18. One can, for example, mount $\mathscr{W}$ in *gimbals*, as shown in Fig. 11.20. The principle here is that the wheel can be tipped slightly to provide storage of small angular momentum components in the plane of the orbit without tipping the entire satellite, as must be done for *fixed-wheel* configurations. The strategy works particularly well if the in-plane disturbance torques are periodic (viewed from an inertial reference frame) so that the gimbal angles required to absorb the angular impulse are not required to keep on growing.

The motion equations [under assumption (5)] for a double-gimbal momentum wheel are

$$\begin{aligned}
I_1\ddot{\alpha}_1 + I_{tw}\ddot{\beta}_1 - h_s(\dot{\alpha}_3 + \dot{\beta}_3) - h_s\omega_c(\alpha_1 + \beta_1) &= g_{1c} + g_{1d} \\
I_3\ddot{\alpha}_3 + I_{tw}\ddot{\beta}_3 + h_s(\dot{\alpha}_1 + \dot{\beta}_1) - h_s\omega_c(\alpha_3 + \beta_3) &= g_{3c} + g_{3d} \\
I_{tw}(\ddot{\alpha}_1 + \ddot{\beta}_1) - h_s(\dot{\alpha}_3 + \dot{\beta}_3) - h_s\omega_c(\alpha_1 + \beta_1) &= g_{\beta 1} \\
I_{tw}(\ddot{\alpha}_3 + \ddot{\beta}_3) + h_s(\dot{\alpha}_1 + \dot{\beta}_1) - h_s\omega_c(\alpha_3 + \beta_3) &= g_{\beta 3}
\end{aligned} \tag{12}$$

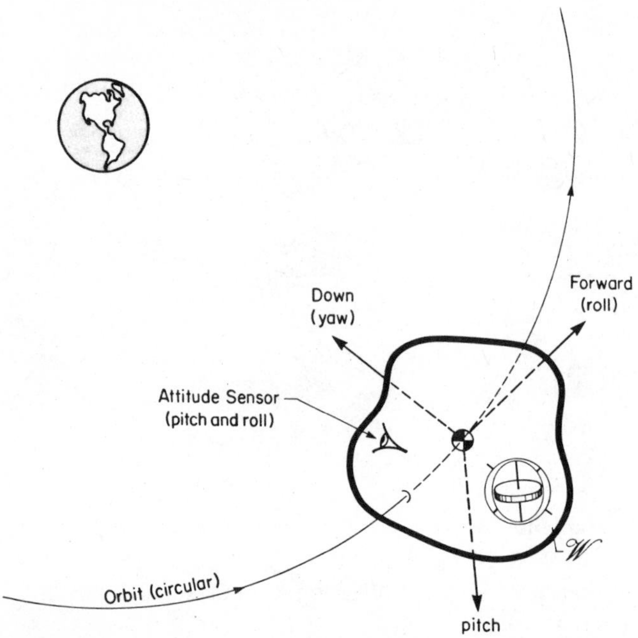

FIGURE 11.20 A double-gimballed bias momentum system.

where $\{\beta_1, \beta_3\}$ are the gimbal angles, $\{g_{\beta 1}, g_{\beta 3}\}$ are the gimbal torques and I_{tw} is the transverse wheel inertia. These equations follow from those derived in Problem 11.16 after the gimbal springs and dampers have been dropped and assumption (5) imposed. The natural frequencies follow from the characteristic equation

$$s^4\left[s^4 + \Omega_{pw}^2\left(s^2 + \omega_c^2\right)\right] = 0 \tag{13}$$

where

$$\Omega_{pw} \triangleq \frac{h_s}{I_{tw}} \tag{14}$$

is the free precession frequency for the wheel alone. In a fashion similar to (9) and (10), approximate frequencies are found to be

$$\omega \doteq \omega_c, \Omega_{pw} \tag{15}$$

(plus the zero frequencies). However, unlike (10), which gives roll/yaw frequencies for the satellite, (15) refers to the wheel alone. In roll/yaw, the wheel is in effect not connected to the satellite until the control system is turned on.

Remarks on Active Control

The discussion has now reached the point where a look at active control features would begin. Although we shall not do that here, a brief indication will be given of the interface between active control and the dynamics. Let us return to the bias momentum satellite of Fig. 11.18, namely, to momentum wheel design without gimbals. It has already been indicated in Fig. 11.18 that Earth sensing is necessary for both pitch (α_2) and roll (α_1). This information is available to the roll and yaw controllers, so that (6) becomes

$$g_{1c} = g_{1c}(\alpha_1); \qquad g_{3c} = g_{3c}(\alpha_1) \tag{16}$$

and to the pitch controller, so that (4) becomes

$$g_{2c} = g_{2c}(\alpha_2) \tag{17}$$

We assume throughout this section that the principal axes of the vehicle are nominally aligned with the orbiting axes; therefore the pitch control system is independent of the roll/yaw control system. Given the feedback laws indicated in their simplest form by (16) and (17), appropriate actuators must be chosen. For roll/yaw, these are usually thrusters, but for pitch a different possibility exists: using $\mathscr{W}$ as a *reaction wheel*. Normally, a reaction wheel is much smaller than a momentum wheel and is intended not to contribute a momentum bias but to provide an object against which the rest of the satellite can torque in its attempt to point properly. The angular impulse contributed by external disturbance torques can thus be passed on the reaction wheel for storage. Changes to the angular momentum of $\mathscr{R} + \mathscr{W}$ caused by external torque are passed on, via the control system, to $\mathscr{W}$. A fixed momentum wheel, then, serves two functions: it

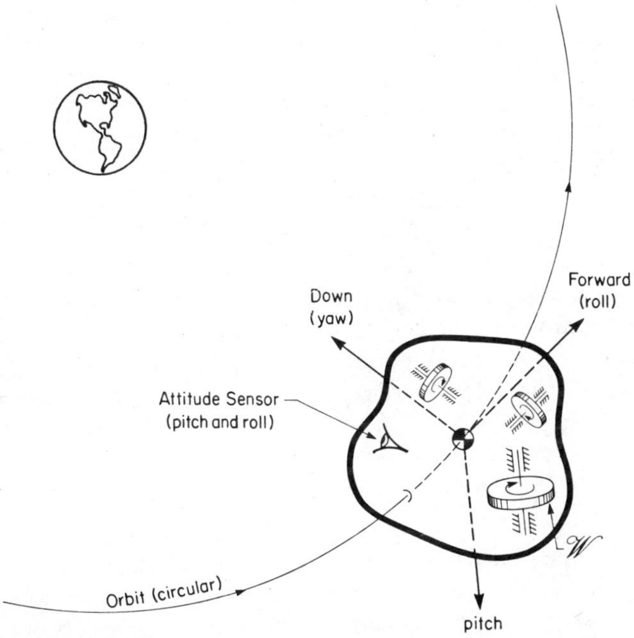

FIGURE 11.21 Bias momentum wheel used with roll and yaw reaction wheels.

contributes a momentum bias, coupling roll and yaw; and it serves as a reaction wheel for the pitch control system.

When using a gimballed momentum wheel, by contrast, angular momentum can be stored also along the roll and yaw axes, without tipping $\mathscr{R}$ (see Fig. 11.20). Referring to the relevant motion equations, (12), four control torque terms have been mentioned: $\{g_{1c}, g_{3c}, g_{\beta 1}, g_{\beta 3}\}$. The first pair, g_{1c} and g_{3c}, represent torques external to $\mathscr{R} + \mathscr{W}$; these would still have to be provided, if desired, by thrusters. The new possibility (now that $\mathscr{W}$ is not fixed in $\mathscr{R}$) is to use the gimbal control torques, as follows:

$$g_{\beta 1} = g_{\beta 1}(\alpha_1, \beta_1, \beta_3); \qquad g_{\beta 3} = g_{\beta 3}(\alpha_1, \beta_1, \beta_3) \tag{18}$$

The feedback provided by (18) can in principle be used to provide better attitude stabilization of $\mathscr{R}$ than for fixed-wheel designs, although at the expense of added complexity. The pitch control system for a gimballed wheel is essentially the same as for a fixed wheel.

Yet another possibility, sketched in Fig. 11.21, is to keep the momentum wheel fixed and use either a roll reaction wheel, a yaw reaction wheel, or both. In this concept, instead of tilting the main wheel, small components of angular momentum in the orbital plane can be stored in the small roll and yaw wheels.

Two Examples: MAGSAT and SEASAT

MAGSAT is a small scientific satellite launched in 1979 to study near-Earth magnetic field; it is shown in Fig. 11.22. Weighing only 184 kg, it is stabilized by a small, ungimballed momentum wheel—small compared to larger satellites. Even so, $J_s/I_{\max} > 100$, thus justifying the assumption often made in the preceding discussion. Figure 11.22*b* shows some typical flight data over an 11-day period. The centers of the small circles pinpoint the location of the vehicle angular momentum vector $\underset{\rightarrow}{\mathbf{h}}$, and the circles themselves represent precession of the pitch axis with respect to $\underset{\rightarrow}{\mathbf{h}}$. Note that $\underset{\rightarrow}{\mathbf{h}}$ itself precesses during this period due to environmental torques.

SEASAT is an ocean survey satellite (as its name implies). Its purpose is to map Earth's geoid. Shown pictorially in Fig. 11.23, its attitude control system comprises a pitch momentum wheel together with a rather smaller roll reaction wheel for absorbing roll torques. The fixed momentum wheel has a strength of 20 N · m · s, which, based on the principle underlying (11), restrains the yaw error to 0.75°.

Bearing Flexibility

Just as with gyrostat designs, bias momentum stabilization systems rely on ideal bearings, although the small size of a momentum wheel (relative to a Gyrostat rotor) implies less drastic consequences from bearing imperfections.

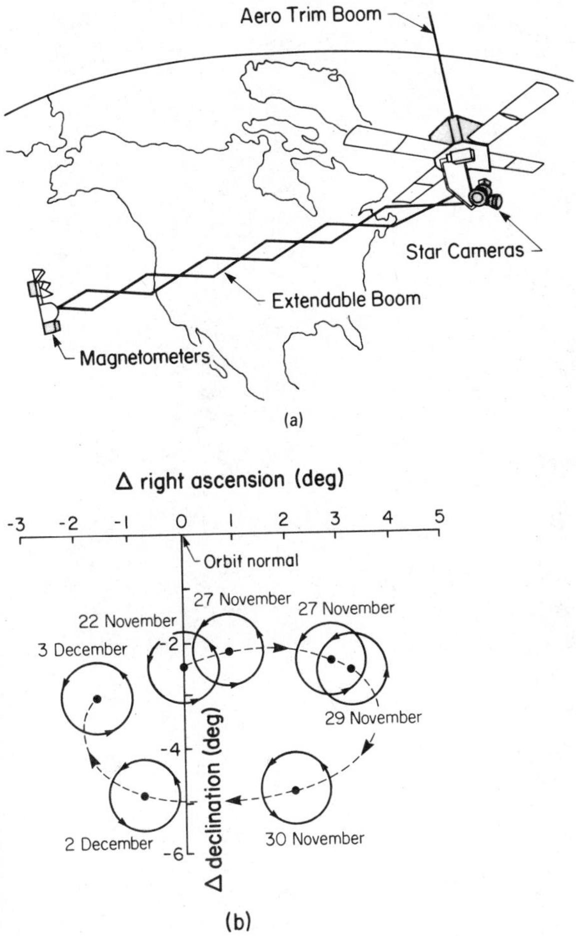

FIGURE 11.22 MAGSAT, with typical flight data. (*a*) MAGSAT. (*b*) Typical flight data.

[Bainum, Fuechsel, and Fedor] have analyzed a momentum wheel with two flexing degrees of freedom with respect to the main body, having in mind the SAS (Small Astronomy Satellite). Bearing flexibility was shown, as might be expected, to be a potential source of attitude instability. However, for SAS, their analysis confirmed that the degradation in stability margin was not serious.

[Tonkin's] analysis emphasizes a further aspect of the problem that is physically meaningful. It will be recalled from several analyses earlier in this book that the question of whether the spin vector of a particular spinning body is parallel or antiparallel to the precession vector of the larger system of which it forms a part can be critical. A general example occurs in the discussion of a system of coaxial quasi-rigid wheels in Section 7.1 (see also Fig. 7.10). According to Tonkin, when

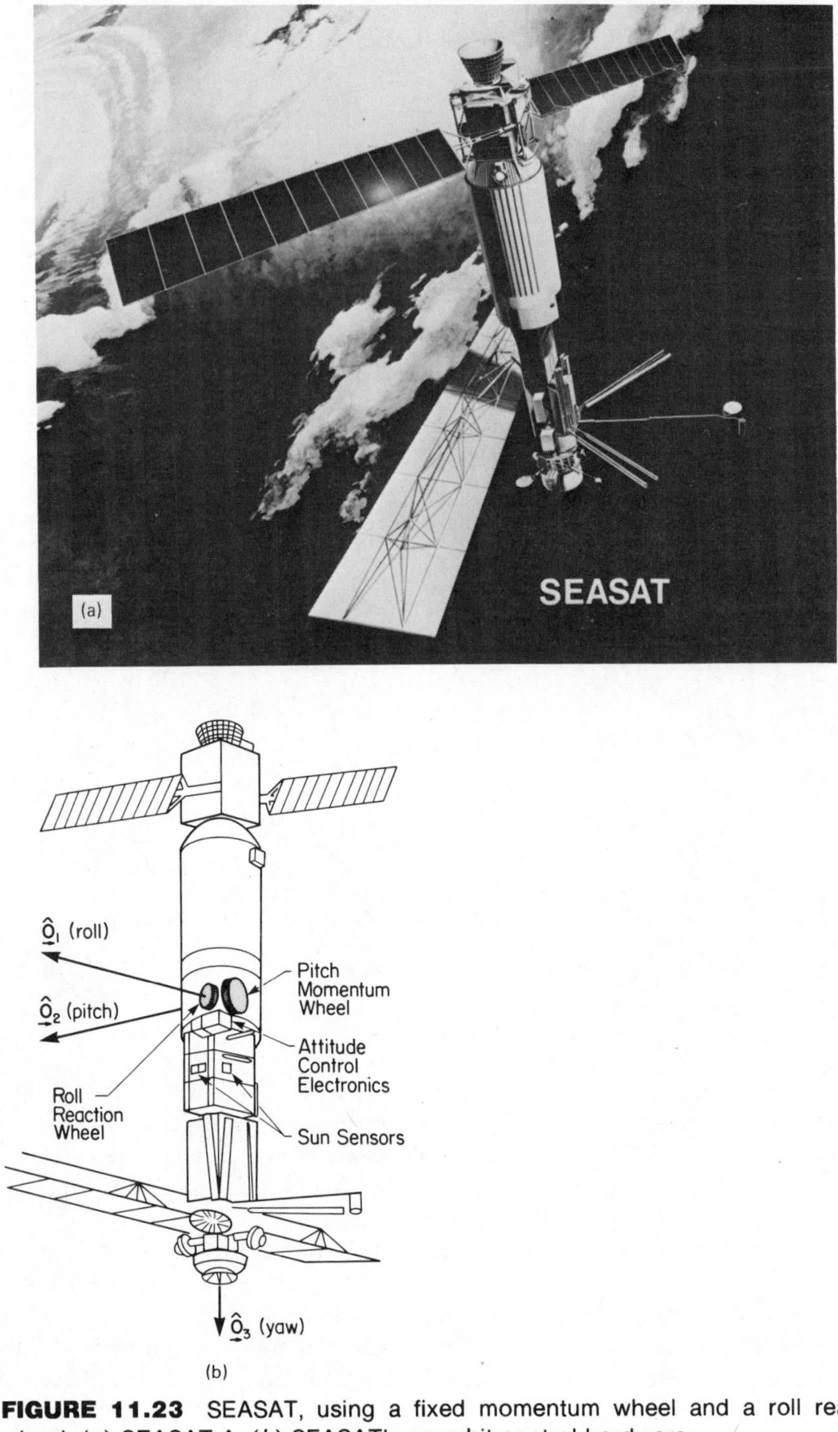

(b)

FIGURE 11.23 SEASAT, using a fixed momentum wheel and a roll reaction wheel. (*a*) SEASAT-A. (*b*) SEASAT's on-orbit control hardware.

the wheel spin vector has the same sense as the precession vector, the effect is more destabilizing than when it is of opposite sense.

Attitude Acquisition and Reacquisition

The momentum wheel spin-up maneuver for OSO was typical of the attitude acquisition procedure initiated after launch for a bias-momentum satellite. (The discussion of this topic for gyrostats earlier in this section can be referred to with profit.) [Sen and Bainum] have analyzed several variations of this procedure, including the question of whether a nutation damper on the main body is helpful. They found that such a damper is essentially irrelevant for symmetrical spacecraft, but that when there is misalignment and the spacecraft is asymmetrical, the behavior is more problematical.

The more drastic prospect of recovery from a flat spin (reacquisition after a failure) is addressed for prolate monospinners by [Barba, Furumoto, and Leliakov], and for bias momentum dual spinners by [Kaplan and Patterson]. In the latter analysis, an attempt is made simultaneously to reorient the satellite and to spin up the wheel; physical interpretation is emphasized.

The key acquisition problem for bias momentum satellites—transition between (perhaps active) spin stabilization immediately after orbit injection and three-axis control—has been dealt with by [Barba and Aubrun]. The problem is dynamically similar to that of recovery of a Gyrostat from a flat spin (see earlier discussion) because initially the entire spacecraft, including the nonspinning wheel, is spinning about the (eventual) yaw axis, whereas after the maneuver the wheel is to spin about the pitch axis with the remainder of the spacecraft nonspinning (except for its small once-per-orbit Earth-pointing rotation rate). The wheel speed control algorithm is open-loop: a constant wheel torque is applied. This adds energy to the system, although the total angular momentum $\underrightarrow{\mathbf{h}}$ remains constant. Barba and Aubrun present a dynamically meaningful interpretation of the maneuver using a "momentum sphere," which is identical to the unit attitude sphere introduced earlier in Fig. 4.13, apart from the magnitude applification h (where h is the magnitude of $\underrightarrow{\mathbf{h}}$, a constant for this maneuver). They make use of the intersections of the two fundamental quadric surfaces which are associated with the dynamics, just as in Fig. 6.6. However, just as energy can be extracted from the system by an energy sink (see Fig. 7.1), the same idea is used to ascertain the impact of a wheel spin-up motor (which may extract or contribute energy to the system, depending on polarity). The overall conclusion is that the simple constant-torque control strategy can be made to produce the desired final result: a properly oriented bias momentum spacecraft, ready for the baseline attitude control system to be turned on.

Further dividends are paid if, instead of applying constant wheel torque, a variable torque is used. The torque can be applied in three stages ([Vigneron and Staley]). By taking more advantage of the intrinsic dual-spin dynamics, acquisition can be achieved for appropriate inertia distributions in less time, with less wobble damping required, or, equivalently, with less final nutation angle.

Hermes in Its Death Throes: A Study in Attitude Dynamics

Hermes (known before launch as the Communications Technology Satellite) had a fixed momentum wheel and offset roll/yaw thrusters (Fig. 11.24). The flight performance of its attitude control system has been described by [Vigneron and Millar]. The solar arrays extended along the pitch axis so that, viewed as a dual-spin satellite, it was prolate; the solar arrays were part of the "despun body," or "platform," and provided ample damping. The Hermes pitch : roll : yaw inertia ratios were approximately 1 : 10 : 10, so that it was in fact *very* prolate.

One can review much about spacecraft attitude dynamics by following developments as they transpired after an Earth sensor failed near the end of 1979, four years after launch. This caused the automatic control system to go into a constant-wheel-spin mode to protect against the failure; at this stage, the bias momentum Hermes had become in effect a prolate Gyrostat. Thruster firings then produced a slow pitch rotation and a roll/yaw precession with a nutation angle of 20°; this stage corresponded to a prolate, truly dual-spin satellite. With time-degraded batteries and solar arrays no longer tracking the sun, this constant-wheel-speed mode could be sustained for only a few hours. As deciphered by [Vigneron and Krag] using ground-based optical measurements, the relative speed between the wheel and the main body gradually subsided over the next half-hour, transmuting Hermes into a prolate monospinner. As such, it was of course as unstable as Explorer 1.

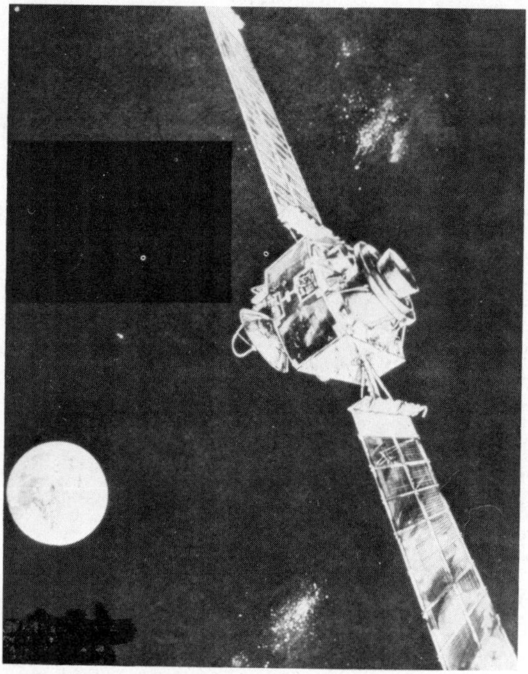

FIGURE 11.24 Hermes.

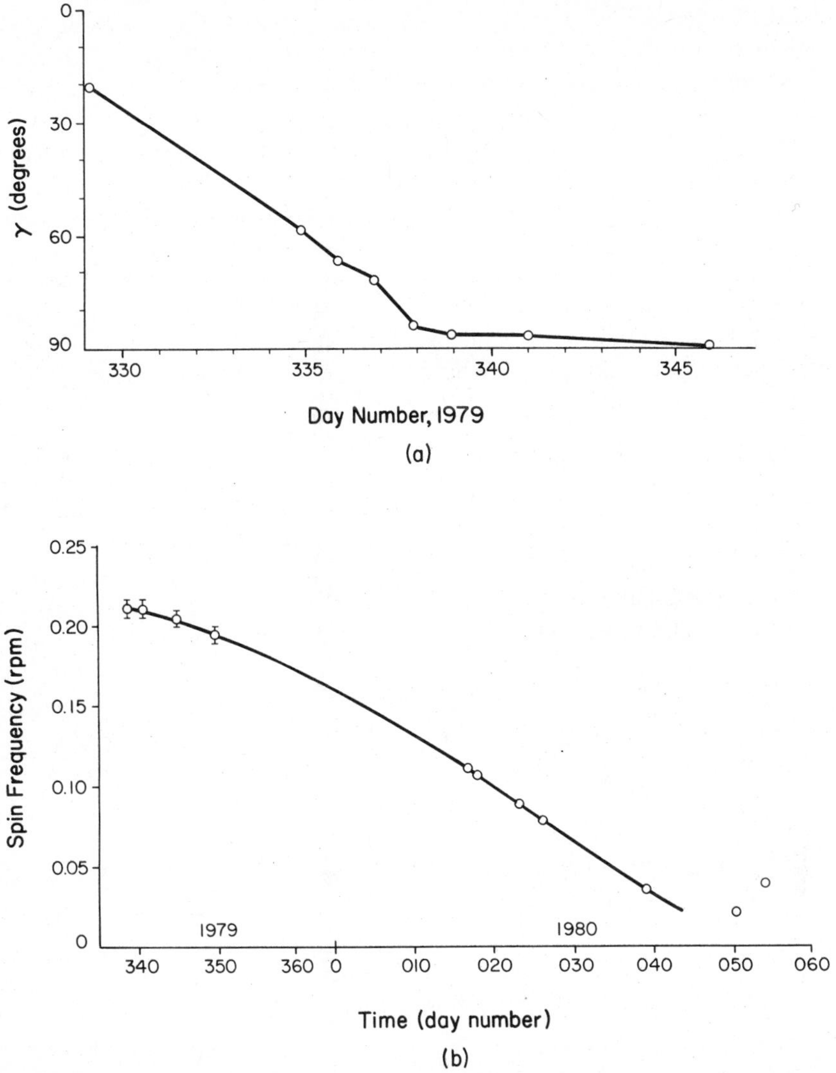

FIGURE 11.25 Hermes at the end. (*a*) Transition from prolate to oblate monospinner. (*b*) Transition from oblate monospinner to feeble gravity gradient satellite.

Over the next two weeks, Hermes gradually transferred into a flat spin (Fig. 11.25); in other words, it became an oblate monospinner, spinning about what had previously been its yaw axis. (The process is shown in Fig. 5.2.) In the absence of external torques, this state of major-axis spin would have persisted indefinitely. However, continued observation revealed that the spin rate, initially about 0.2 rpm, gradually lessened. [Vigneron and Krag] have cited indirect evidence that this was due to a "propeller" effect in which the two solar panels interacted with solar radiation pressure: a relative rotational offset of as little as 2.5° between the two panels would be sufficient to explain the rate of despin observed. In any case, the spin declined over the next two months to the point where Hermes became captured, at least intermittently, by Earth's gravity gradient field (Fig. 11.25*b*). It became in effect a poorly designed gravity gradient satellite.

In summary, over a two-month period, Hermes exhibited the characteristics of six of the cases considered in this book: it progressed from bias momentum to prolate Gyrostat, to true dual spinner, to prolate monospinner, to oblate monospinner, and finally to the point where its dynamics was dominated by the gravity gradient force field. Would that a camera crew could have filmed this multiple metamorphosis. What an excellent teaching resource it would have made!

11.4 PROBLEMS

11.1 The conditions that must obtain for a rigid gyrostat $\mathscr{R} + \mathscr{W}$ to be in a state of *dynamical equilibrium* in a circular orbit are given in the text as (11.1, 2) through (11.1, 6). This problem considers two limiting cases: very small stored angular momentum and very high stored angular momentum.

(*a*) If the relative angular momentum stored in $\mathscr{W}$ is very small, in the sense that

$$0 < h_s \ll \omega_c \min\{I_1, I_2\} \tag{1}$$

show that the equilibria revert, as they should, to those for a single rigid body (see Section 9.2 and Problem 9.10). That is, the three principal axes of $\mathscr{R} + \mathscr{W}$ coincide with the forward, sideways, and downward directions, that is, with the orbiting frame $\mathscr{F}_c$.

(*b*) If the relative angular momentum stored in $\mathscr{W}$ is very large, in the sense that

$$h_s \gg \omega_c \max\{I_1, I_2\} \tag{2}$$

show that equilibria require that the axis of $\mathscr{W}$ be normal to the orbit, regardless of its orientation in $\mathscr{R}$.

11.2 Consider the symmetrical $\mathscr{R} + \mathscr{W}$ gyrostat in a circular orbit, with the symmetry axis normal to the orbit, as shown in Fig. 11.3.

(*a*) By examining the derivation of motion equations for (symmetrical) $\mathscr{R}$ alone in this situation—specifically, the discussion surrounding (10.1, 22) in Chapter 10—explain why the nominal angular momentum is the operative quantity, whether resident in $\mathscr{R}$ alone or resident in $\mathscr{R} + \mathscr{W}$. [*Caution*: The symbol $\mathbf{h}_s$ in (10.1, 22) denotes the components of the total angular momentum, expressed in the stroboscopic frame $\mathscr{F}_s$, whereas for $\mathscr{R} + \mathscr{W}$ we have used $\mathbf{h}_s$ to denote the angular momentum of $\mathscr{W}$ relative to $\mathscr{R}$, expressed in the principal axes of $\mathscr{R} + \mathscr{W}$.] Hence, verify (11.1, 8).

(*b*) Suppose the bearings supporting the rotation of $\mathscr{W}$ relative to $\mathscr{R}$ have some friction, so that their relative rotation decreases over time. What change will this make to pointing stability?

11.3 Prove the three assertions made on page 429 regarding the relative equilibria of the gyrostat $\mathscr{R} + \mathscr{W}$ in a circular orbit.

11.4 The geometrical interpretations of the relative equilibria of the gyrostat $\mathscr{R} + \mathscr{W}$ in a circular orbit are examined in this problem (see Section 11.1 and Fig. 11.4). The most general case is the offset-hyperbolic equilibrium, in which the axis of $\mathscr{W}$ is inclined to all three of the basis vectors in the orbiting frame $\mathscr{F}_c$. To be specific, let the axis of $\mathscr{W}$ be inclined, first at an angle α about the vertical, then at an angle β about the (new) forward direction. Denote by y a coordinate perpendicular to the orbital plane and by x and z the coordinates in the orbital plane (the origin of the xyz-system is thus the center of the orbit). The orbital radius is R.

(*a*) Show that as the axis of $\mathscr{W}$ travels around the circular orbit, it lies always on the one-sheet hyperboloid whose equation is

$$(x^2 + z^2)\cos^2\alpha\cos^2\beta = y^2\sin^2\alpha + (R\cos\beta + y\sin\beta)^2\cos^2\alpha \quad (3)$$

(*b*) Show that the center of this hyperboloid is offset in the y-direction by an amount

$$\Delta y = \frac{R\cos^2\alpha\sin\beta\cos\beta}{\sin^2\alpha + \cos^2\alpha\sin^2\beta} \quad (4)$$

(*c*) As a particular case, consider the (central) hyperbolic case for which $\beta = 0$. Show from (3) that the equation for the hyperboloid is

$$(x^2 + z^2)\cos^2\alpha = y^2\sin^2\alpha + R^2\cos^2\alpha \quad (5)$$

(*d*) As a different particular case, consider the conical case $\alpha = 0$, for which the hyperboloid degenerates to a cone. Show that (3) reduces to

$$x^2 + z^2 = (R + y\tan\beta)^2 \quad (6)$$

Why is this the equation for a cone? What is the distance between the orbit center and the apex of the cone? Does this agree with (4) above?

(*e*) What happens when α and β are both zero?

11.5 This problem also deals with the relative equilibria for a gyrostat $\mathscr{R} + \mathscr{W}$ in a circular orbit (Section 11.1 and Fig. 11.4).

(*a*) Show from (11.1, 1) that as the relative angular momentum of $\mathscr{W}$ becomes large, the equilibrium orientation of a gyrostat tends to be such that the axis of $\mathscr{W}$ is normal to the orbit. Can you explain this physically? Is this finding consistent with the small-κ solutions to (11.1, 21) and (11.1, 27)?

(*b*) For the offset-hyperbolic case, show that no relative equilibrium exists in which a principal axis of the gyrostat is perpendicular to either the "forward" or "vertical" directions. (Forward is along $\underset{\rightarrow}{\hat{\mathbf{c}}}_1$; vertical is along $\underset{\rightarrow}{\hat{\mathbf{c}}}_3$.)

11.6 This problem deals with the gyrostat $\mathscr{R} + \mathscr{W}$ in a circular orbit from the viewpoint of analytical mechanics (in the sense of [Lanczos]).

(a) Explain, using (8.1, 19), why the rotational potential energy of $\mathscr{R} + \mathscr{W}$ is given by

$$V = \tfrac{3}{2}\omega_c^2 \mathbf{c}_3^T \mathbf{I} \mathbf{c}_3 \tag{7}$$

assuming (as we have in this chapter) that the attitude motion does not affect the circular orbit. (The orbital rate is ω_c, $\mathbf{I}$ is the expression of the inertia dyadic for $\mathscr{R} + \mathscr{W}$ in centroidal principal-axis frame $\mathscr{F}_p$, and the elements of the column matrix $\mathbf{c}_3$ are the direction cosines in $\mathscr{F}_p$ of the local vertical.)

(b) Show, using (3.5, 28), that the rotational kinetic energy of $\mathscr{R} + \mathscr{W}$ is given by

$$T = \tfrac{1}{2}\boldsymbol{\omega}^T \mathbf{I} \boldsymbol{\omega} + h_s \mathbf{a}^T \boldsymbol{\omega} + \tfrac{1}{2} I_s \omega_s^2 \tag{8}$$

where $\boldsymbol{\omega} = \mathscr{F}_p \cdot \underset{\rightarrow}{\boldsymbol{\omega}}$, $\underset{\rightarrow}{\boldsymbol{\omega}}$ is the absolute angular velocity of $\mathscr{R}$, $\mathbf{a} = \mathscr{F}_p \cdot \underset{\rightarrow}{\mathbf{a}}$, $\underset{\rightarrow}{\mathbf{a}}$ is the orientation of the axis of $\mathscr{W}$, $h_s \mathbf{a}$ is the angular momentum of $\mathscr{W}$ relative to $\mathscr{R}$, and $h_s = I_s \omega_s$.

(c) Denote by $\boldsymbol{\omega}_\alpha$ the components in $\mathscr{F}_p$ of the angular velocity of $\mathscr{R}$ with respect to the orbiting frame $\mathscr{F}_c$. That is, let $\boldsymbol{\omega} = \boldsymbol{\omega}_\alpha - \omega_c \mathbf{c}_2$ where the elements of $\mathbf{c}_2$ are the direction cosines in $\mathscr{F}_p$ of the orbit normal. Furthermore, expand $T = T_2 + T_1 + T_0$, with T_2 comprising terms quadratic in $\boldsymbol{\omega}_\alpha$, and so on. Show that the (apparent) Hamiltonian is

$$H = \tfrac{1}{2}\boldsymbol{\omega}^T \mathbf{I} \boldsymbol{\omega} + \tfrac{3}{2}\omega_c^2 \mathbf{c}_3^T \mathbf{I} \mathbf{c}_3 + \omega_c \mathbf{c}_2^T (\mathbf{I}\boldsymbol{\omega} + h_s \mathbf{a}) + \tfrac{1}{2} I_s \omega_s^2 \tag{9}$$

[*Hint*: Compare with Problem 3.3.] This result agrees with [Longman, 2], apart from the dynamically unimportant last term.

11.7 Consider an offset-hyperbolic equilibrium for a gyrostat in a circular orbit for which the intermediate axis of inertia $\hat{\mathbf{p}}_1$ lies in the orbital plane. Show that the following two conditions hold:

$$\text{(i)} \qquad 16 J_{s1}^2 + J_{s3}^2 = 16(I_1 - I_3)(I_2 - I_1)$$

$$\text{(ii)} \qquad 4\Gamma^2 J_s^2 (\sigma^2 + a_1^2) - 5\Gamma J_s \sigma + 1 = 0 \tag{10}$$

where J_s is defined in (11.1, 4), J_{s1} and J_{s3} are defined in (11.1, 11), and $\{a_1, a_2, a_3\}$ are the direction cosines of the wheel $\mathscr{W}$ with respect to $\mathscr{F}_p$. The two new symbols in (10) are

$$\Gamma \triangleq \frac{1}{2(I_2 - I_3)\sin 2\chi_1}$$

$$\sigma \triangleq a_2 \sin \chi_1 + a_3 \cos \chi_1 \tag{11}$$

with χ_1 given by (11.1, 32).

11.8 (a) Show that the linearized motion equations for the librations of an $\mathscr{R} + \mathscr{W}$ gyrostat about one of its relative equilibria in a circular orbit are as given by (11.1, 39 and 40).

(b) By transforming these equations to components in the orbital frame, show that they may also be written as in (11.1, 42) through (11.1, 44).

(c) Can you reconcile these latter equations with the librational equations for $\mathscr{R}$ (alone) in a circular orbit, namely, (9.2, 15) through (9.2, 18)?

11.9 Consider the librations $\boldsymbol{\alpha}(t)$ of a gyrostat $\mathscr{R} + \mathscr{W}$ about a relative equilibrium. The gyrostat is in a circular orbit.

(*a*) Explain, using help from Problem 2.25 in Chapter 2, why the angular velocity $\boldsymbol{\omega}$ is given by

$$\boldsymbol{\omega} = \boldsymbol{\nu} + (\dot{\boldsymbol{\alpha}} + \boldsymbol{\nu}^{\times}\boldsymbol{\alpha}) - \tfrac{1}{2}\boldsymbol{\alpha}^{\times}\dot{\boldsymbol{\alpha}} + \tfrac{1}{2}(\boldsymbol{\alpha}\boldsymbol{\alpha}^T - \alpha^2\mathbf{1})\boldsymbol{\nu} \tag{12}$$

to second order in $\boldsymbol{\alpha}$ and $\dot{\boldsymbol{\alpha}}$. (*Note*: $\alpha^2 \equiv \boldsymbol{\alpha}^T\boldsymbol{\alpha}$.)

(*b*) In a similar fashion, derive the following second-order relations:

$$\mathbf{c}_2 = \left[\mathbf{1} - \boldsymbol{\alpha}^{\times} + \tfrac{1}{2}(\boldsymbol{\alpha}\boldsymbol{\alpha}^T - \alpha^2\mathbf{1})\right]\bar{\mathbf{c}}_2 \tag{13}$$

and similarly for $\mathbf{c}_3$. [*Note*: $\bar{\mathbf{c}}_2$ is the reference equilibrium value of $\mathbf{c}_2$.]

(*c*) Use the above results in conjunction with the Hamiltonian recorded above as (9) to show that, to second order,

$$H = H_0 + H_1 + H_2 \tag{14}$$

$$H_0 = \frac{1}{2}\omega_c^2\left[3\bar{\mathbf{c}}_3^T\mathbf{I}\bar{\mathbf{c}}_3 - \bar{\mathbf{c}}_2^T\mathbf{I}\bar{\mathbf{c}}_2 + J_{s2} + I_s\left(\frac{\omega_s}{\omega_c}\right)^2\right] \tag{15}$$

$$H_1 = 0 \tag{16}$$

$$H_2 = \tfrac{1}{2}\dot{\boldsymbol{\alpha}}_c^T\boldsymbol{\mathscr{I}}\dot{\boldsymbol{\alpha}}_c + \tfrac{1}{2}\omega_c^2\boldsymbol{\alpha}_c^T\boldsymbol{\mathscr{K}}_c\boldsymbol{\alpha}_c \equiv \tfrac{1}{2}\dot{\boldsymbol{\alpha}}^T\mathbf{I}\dot{\boldsymbol{\alpha}} + \tfrac{1}{2}\omega_c^2\boldsymbol{\alpha}^T\boldsymbol{\mathscr{K}}\boldsymbol{\alpha} \tag{17}$$

where the meanings of the symbols are clear from (11.1, 39) through (11.1, 44).

(*d*) Construct a Liapunov function to show that an equilibrium is Liapunov stable if $\boldsymbol{\mathscr{K}}$ evaluated at that equilibrium is positive-definite.

11.10 With reference to Fig. 11.7 and equations (11.1, 48 through 53), show that as $\mathscr{W}$ spins very rapidly about the orbit normal ($|\hat{J}_s| \to \infty$), all $\mathscr{R} + \mathscr{W}$ configurations with $I_1 > I_3$ have stable attitude librations.

11.11 Suppose $\mathscr{R} + \mathscr{W}$ possesses an axis of inertial symmetry and let it be $\vec{\hat{\mathbf{p}}}_3$. Thus, the symmetry axis always points nominally down, although the $\vec{\text{axis}}$ of $\mathscr{W}$ continues to be nominally normal to the orbit.

(*a*) Show that the Liapunov testing function $\hat{v}$, given in (9.2, 48) for $\mathscr{R}$ alone, requires the following extra term on account of $\mathscr{W}$:

$$\Delta\hat{v} = 2\left(\frac{J_s}{I_t}\right)(c_{22} - 1) \tag{18}$$

(*b*) How are the libration bounds affected by J_s?

(*c*) Plot diagrams similar to those shown in Fig. 9.11 to exhibit your findings.

(*d*) Repeat parts (*a*) through (*c*) assuming the symmetry axis is "forward," namely, it is $\vec{\hat{\mathbf{p}}}_1$ instead of $\vec{\hat{\mathbf{p}}}_3$.

11.12 Following [Longman, Hagedorn, and Beck], consider *hyperbolic* equilibria for the gyrostat $\mathscr{R} + \mathscr{W}$ in a circular orbit, wherein, as a special subcase, the axis of $\mathscr{W}$ is parallel to $\hat{\mathbf{p}}_2$. (For the meaning of "hyperbolic equilibrium," consult Figs. 11.4*b* $\vec{\text{and}}$ 11.5*a*.) Thus, $\beta_3 = 0$. Some stability diagrams for these equilibria are shown in Fig. 11.9. In this problem, stability results for the hyperbolic equilibrium of $\mathscr{R} + \mathscr{W}$ (Section 11.1 and Fig. 11.4*b*) are compared to those obtained earlier (Section 10.1 and Fig. 10.3*b*) for the hyperbolic equilibrium of a symmetrical spinning rigid body in a circular orbit. To make this comparison, the rigid body $\mathscr{R}$ must also be inertially symmetrical: $\mathscr{R} + \mathscr{W} \to \mathscr{W}_1 + \mathscr{W}_2$.

(a) Explain why the comparison between $\mathscr{W}_1 + \mathscr{W}_2$ and an "equivalent wheel" $\mathscr{W}_{eq}$ should require that

(i) Their axial inertias be equal:

$$I_a(\mathscr{W}_{eq}) = I_a(\mathscr{W}_1 + \mathscr{W}_2) \tag{19}$$

(ii) Their transverse inertias be equal:

$$I_t(\mathscr{W}_{eq}) = I_t(\mathscr{W}_1 + \mathscr{W}_2) \tag{20}$$

(iii) Their absolute angular momenta in their reference condition (i.e., prior to perturbations) be equal:

$$\underset{\rightarrow}{\mathbf{h}}(\mathscr{W}_{eq}) = \underset{\rightarrow}{\mathbf{h}}(\mathscr{W}_1 + \mathscr{W}_2) \tag{21}$$

(*b*) Show that the requirements of part (*a*) are fulfilled if

$$\hat{\nu}_{eq} = \hat{J}_s - 1 \tag{22}$$

where $\nu_{eq} = \omega_c \hat{\nu}_{eq}$ is the absolute spin rate of $\mathscr{W}_{eq}$, and where $\hat{J}_s \triangleq J_s/I_a$, with $I_a = I_a(\mathscr{W}_{eq})$ and $J_s \triangleq h_s/\omega_c$. Here, h_s is the magnitude of the angular momentum of $\mathscr{W}_2$ relative to $\mathscr{W}_1$.

(*c*) By comparing the motion equations for $\mathscr{W}_1 + \mathscr{W}_2$, namely, (11.1, 39 and 40), under present assumptions to the motion equations for $\mathscr{W}_{eq}$, namely, (10.1, 46), deduce that

$$\bar{\gamma}_{3,eq} = \chi_3; \qquad \boldsymbol{\gamma}_{eq} \equiv \boldsymbol{\alpha} \tag{23}$$

and that (22) is also true.

(*d*) Based on the results in parts (*a*), (*b*), and (*c*), give an interpretation of the stability diagram shown in Fig. 10.4*a* that is applicable to the gyrostat $\mathscr{W}_1 + \mathscr{W}_2$ in a circular orbit.

(*e*) What is the connection between the developments above and equation (11.1, 9)?

11.13 In parallel with the previous problem, consider now the *conical* equilibria for a gyrostat $\mathscr{R} + \mathscr{W}$ in a circular orbit, with the axis of $\mathscr{W}$ parallel to the principal axis $\hat{\mathbf{p}}_2$ of $\mathscr{R} + \mathscr{W}$. (The conical equilibria are visualized in Figs. 11.4*c* and 11.5*b*, and now $\beta_1 = 0$.) Some stability diagrams for these equilibria are shown in Fig. 11.10. The specialization in this problem is to cases where $\mathscr{R}$ is also axially symmetrical (same axis as $\mathscr{W}$), so that $\mathscr{R} + \mathscr{W} \to \mathscr{W}_1 + \mathscr{W}_2$.

(*a*) After reviewing the points raised by parts (*a*) and (*b*) of the previous problem (which are equally applicable to the present problem), compare the motion equations for $\mathscr{W}_1 + \mathscr{W}_2$, namely, (11.1, 39 and 40) under present assumptions, to the motion equations for $\mathscr{W}_{eq}$ in conical equilibrium, namely, (10.1, 53), and hence deduce that

$$\bar{\gamma}_{1,eq} = \chi_1; \qquad \gamma_{eq} \equiv \alpha \tag{24}$$

along with (22).

(*b*) Based on part (*a*), give an interpretation of the stability diagram shown in Fig. 10.4*b* that is applicable to the gyrostat $\mathscr{W}_1 + \mathscr{W}_2$ in a circular orbit.

11.14 In this problem, the hyperbolic and conical cases of relative equilibrium for $\mathscr{R} + \mathscr{W}$ in a circular orbit are reexamined from a different viewpoint. The algebraic condition for a relative equilibrium was shown in the text to be

$$\mathbf{c}_2^\times(\mathbf{I}\mathbf{c}_2 - J_s\mathbf{a}) = 3\mathbf{c}_3^\times\mathbf{I}\mathbf{c}_3 \tag{25}$$

(*a*) Consider first the hyperbolic case, for which the gravity gradient torque is zero (i.e., the right side of the above equation is zero). Explain why $\mathbf{c}_3 = \mathbf{1}_3$ and why

$$\mathbf{I}\mathbf{c}_2 - J_s\mathbf{a} = \lambda\mathbf{c}_2 \tag{26}$$

Deduce the quartic equation (11.1, 3) for λ and the consequent expressions (11.1, 6) for χ_3, the angle through which the gyrostat is tilted in orbit.

(*b*) Consider next the conical case, for which the principal axes are tilted around the forward (roll) direction. Explain why $\mathbf{c}_1 = \mathbf{1}_1$ and why, as a consequence,

$$\mathbf{c}_2^\times\mathbf{I}\mathbf{c}_2 + \mathbf{c}_3^\times\mathbf{I}\mathbf{c}_3 = \mathbf{0} \tag{27}$$

(*c*) Still for the conical case, use (25) and (27) in combination to show that

$$4\mathbf{I}\mathbf{c}_2 - J_s\mathbf{a} = 4\lambda\mathbf{c}_2 \tag{28}$$

Thence, deduce the quartic equation

$$\frac{a_2^2}{(I_2 - \lambda)^2} + \frac{a_3^2}{(I_3 - \lambda)^2} = \frac{16}{J_s^2} \tag{29}$$

for λ, and that $a_1 = 0$.

(*d*) Finally, having thus determined λ, show that the tilt about the roll axis is given by

$$\sin\chi_1 = \frac{-J_s}{4(I_3 - \lambda)}a_3; \qquad \cos\chi_1 = \frac{J_s}{4(I_2 - \lambda)}a_2 \tag{30}$$

11.15 Consider the $\mathscr{R} + \mathscr{W} + \mathscr{P}$ system of Fig. 7.12 in a circular orbit, and therefore subject to a gravity gradient force field. In addition, $\mathscr{R}$ is taken to be Earth-pointing, so that the system is subject to a once-per-orbit rotation. Assume that the axis of $\mathscr{W}$ is parallel to a system principal axis. The context is shown in Fig. 11.26.

(*a*) Derive motion equations for the small librations $\boldsymbol{\alpha}(t)$ of $\mathscr{R}$ with respect to orbiting axes. [*Note*: This is an important working example of a dual-spin satellite with a discrete damper. Although to conserve

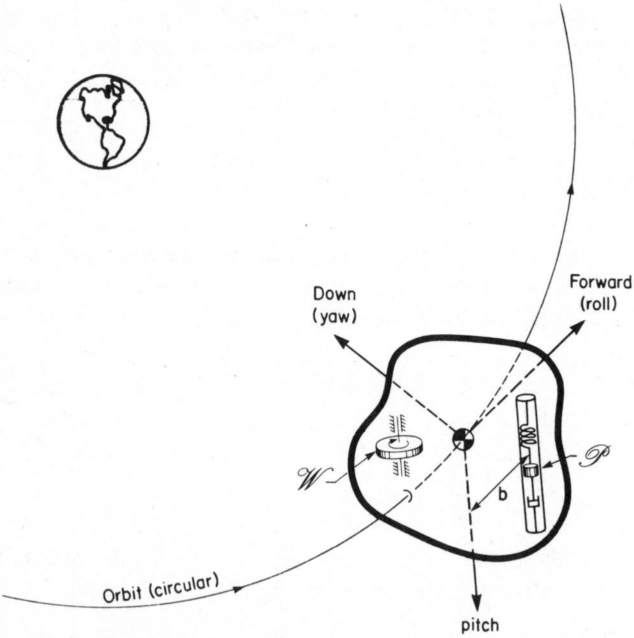

FIGURE 11.26 The system $\mathscr{R} + \mathscr{P} + \mathscr{W}$ in a circular orbit. The main body $\mathscr{R}$ is Earth-pointing. (See Problem 11.15.)

space, these equations are not derived in the text, all the necessary elements have been discussed in detail earlier in this book: the motion equations must reduce to (7.2, 13 and 18) for $\mathscr{R} + \mathscr{W} + \mathscr{P}$ in torque-free motion when the gravity gradient torque terms are dropped, they must reduce to (9.3, 17) for $\mathscr{R} + \mathscr{P}$ in a circular orbit when the wheel's relative angular momentum $\underset{\rightarrow}{\mathbf{h}}_s$ is set to zero, and they must reduce to (11.1, 39 and 48) for $\mathscr{R} + \mathscr{W}$ in a circular orbit when the damper mass m_d vanishes.] See [Bainum, Fuechsel, and Mackison] for an analogous analysis appropriate to the SAS-A satellite.

(*b*) Prepare stability diagrams in the k_1–k_3 plane, using system parameters as appropriate, and compare with Figs. 5.3, 6.7, 7.4, 9.18, and 11.7.

11.16 Consider a rigid body $\mathscr{R}$ which contains a momentum wheel $\mathscr{W}$ and which experiences gravitational torque in a circular orbit. The axis of $\mathscr{W}$ is parallel to the pitch axis of the system, and the angular momentum of $\mathscr{W}$ relative to $\mathscr{R}$ is h_s. As usual, we denote the principal moments of inertia of $\mathscr{R} + \mathscr{W}$ by $\{I_1, I_2, I_3\}$ and the orbital frequency by ω_c. The transverse moment of inertia of $\mathscr{W}$ is I_{tw}. The momentum wheel is suspended within $\mathscr{R}$ on two orthogonal gimbals, restrained by springs and dampers whose parameters are $\{c_1, c_3, k_1, k_3\}$. (The gimbal axes are parallel to roll and yaw.) Let roll and yaw be denoted $\{\alpha_1, \alpha_3\}$, and let the corresponding gimbal angles be $\{\beta_1, \beta_3\}$.

(*a*) Derive attitude motion equations for the roll/yaw motion, using $\{\alpha_1, \alpha_3, \beta_1, \beta_3\}$ as coordinates. In particular, neglecting the inertia of the gimbals themselves, show that these equations are of the form

$$\mathscr{M}\mathbf{q}'' + (\mathscr{D} + \mathscr{G})\mathbf{q}' + \mathscr{K}\mathbf{q} = \not{f} \tag{31}$$

where

$$\mathscr{M} \triangleq \begin{bmatrix} I_1 & 0 & I_{tw} & 0 \\ 0 & I_2 & 0 & I_{tw} \\ I_{tw} & 0 & I_{tw} & 0 \\ 0 & I_{tw} & 0 & I_{tw} \end{bmatrix} \tag{32}$$

$$\mathscr{D} \triangleq \operatorname{diag}\{0, 0, \hat{c}_1, \hat{c}_3\} \tag{33}$$

$$\mathscr{G} \triangleq \begin{bmatrix} 0 & I_2 - I_1 - I_3 - J_s & 0 & -J_s \\ & 0 & J_s & 0 \\ \text{(skew)} & & 0 & -J_s \\ & & & 0 \end{bmatrix} \tag{34}$$

$$\mathscr{K} \triangleq \begin{bmatrix} 4(I_2 - I_3) - J_s & 0 & -J_s & 0 \\ & I_2 - I_1 - J_s & 0 & -J_s \\ \text{(symmetric)} & & -J_s + \hat{k}_1 & 0 \\ & & & -J_s + \hat{k}_3 \end{bmatrix} \tag{35}$$

$$\not{f} \triangleq \begin{bmatrix} \dfrac{g_1}{\omega_c^2} & \dfrac{g_3}{\omega_c^2} & 0 & 0 \end{bmatrix}^T \tag{36}$$

where primes denote $d/d(\omega_c t)$; g_1 and g_3 are the roll and yaw torques (excluding gravity gradient); $J_s = h_s/\omega_c$; $\hat{c}_1 = c_1/\omega_c$, $\hat{k}_1 = k_1/\omega_c^2$, and similarly for $\hat{c}_3$ and $\hat{k}_3$.

(*b*) Derive the motion equation for pitch $\alpha_2(t)$.

(*c*) Based on your motion equations from parts (*a*) and (*b*), derive conditions for asymptotic attitude stability.

(*d*) Plot a few typical k_1–k_3 stability diagrams of the type found throughout this book, using reasonable values for the other dimensionless parameters.

(*e*) Compare and contrast your diagrams in part (*d*) with part (*b*) of Problem 11.15.

11.17 Consider the bias momentum satellite $\mathscr{R} + \mathscr{W}$ visualized in Fig. 11.18 and whose equations of motion were given by (11.3, 2 through 4). This problem considers the open-loop system (i.e, the dynamics of the spacecraft before feedback control is applied) as represented by the left side of the motion equations.

(*a*) Show that the frequency of pitch libration is

$$\omega_p = \sqrt{\frac{3(I_1 - I_3)}{I_2}}\,\omega_c \tag{37}$$

Explain physically why the relative angular momentum of $\mathscr{W}$, namely, h_s, does not enter into this expression.

(*b*) Under the assumption that

$$J_s \ll \max\{I_1, I_2, I_3\} \tag{38}$$

show that the roll/yaw frequencies are

$$\omega_{ry} = \omega_c, \Omega_p \tag{39}$$

where Ω_p, the free-body precession frequency, is given by $\Omega_p = h_s/(I_1 I_3)^{1/2}$.

(*c*) Show that the small corrections to (39) when first-order quantities I_i/J_s are included are indicated by

$$\omega_{ry} = \omega_c\left[1 - \frac{3(I_2 - I_3)}{2J_s}\right] \tag{40}$$

$$\omega_{ry} = \Omega_p\left[1 - \frac{2I_2 - I_1 - I_3}{2J_s}\right] \tag{41}$$

11.18 (*a*) Derive the motion equations (11.3, 12) for the free (uncontrolled) motions of a double-gimballed momentum wheel, under the assumption that the wheel momentum h_s is much larger than $I_i\omega_c$ ($i = 1, 2, 3$).

(*b*) Show further that the characteristic equation is approximately given by (11.3, 13), leading to the frequencies in (11.3, 15).

(*c*) To what motion do the roots $s^4 = 0$ correspond?

11.19 The bias momentum satellites in Section 11.3 have all had their momentum wheels aligned with the pitch axis (i.e., nominally with the orbit normal). They have also had their principal axes of inertia nominally aligned with the orbiting axes. This problem considers a spacecraft that has two of its principal axes tilted with respect to the pitch and yaw axes, to avoid obscuring its communications antenna (Fig. 11.27*a*). This reference alignment produces a steady torque due to Earth's gravity gradient field. Also shown in Fig. 11.27*a* is a yaw-axis bias momentum wheel whose intent is to cancel the gravity gradient roll torque.

(*a*) Explain, using Figs. 11.27*b* and 11.4, why this situation is a particular case of a gyrostat in "conical" equilibrium.

(*a*) Show that the nominal offset of the pitch/yaw principal axes with respect to principal axes is given by χ_1, where

$$\chi_1 = \tfrac{1}{2}\sin^{-1}\kappa_C^{-1}$$

and κ_C is given by (11.1, 28).

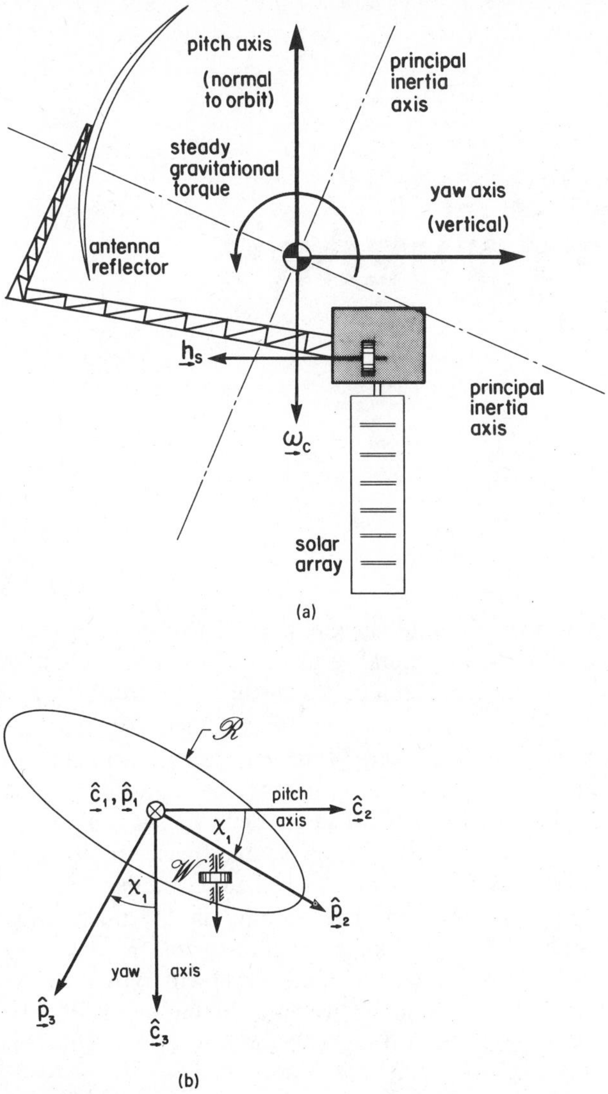

FIGURE 11.27 A system requiring bias momentum along the yaw axis. (See Problem 11.19.) (*a*) Bias momentum along the yaw axis. (*b*) Dynamical abstraction: a particular case of the "conical equilibrium."

(*c*) Perform a stability analysis of this system using k_1–k_3 diagrams (as elsewhere in this book) using χ_1 or κ_C as a parameter (compare with [Hughes, 3]).

(*d*) Add a pitch momentum wheel to the satellite (which does not affect the conditions for equilibrium, but affects their stability) and compare with [Hughes, 4] and [Hughes and Golla].

APPENDIX A

ELEMENTS OF STABILITY THEORY

The notion of stability is a relatively familiar one, particularly to mathematicians, scientists, and engineers. Stated in general terms, the behavior of a system is said to be stable if it is not adversely affected by the disturbances to which it is inevitably subjected. This general notion is rather vague and therefore does not qualify as a mathematically precise definition. Moreover, as one attempts to fashion a more precise definition, it becomes apparent that the concept of stability is a complicated one indeed. In fact, no single definition is universally appropriate. The author encountered over 35 distinguishable species of stability while preparing this appendix, and many more specialized varieties are undoubtedly available. [Bellman] has remarked that stability is a "much overburdened word with an unstabilized definition." Fortunately, the applications discussed elsewhere in this book can be adequately understood with only a few of these definitions, and attention will be confined to these in this appendix. In addition, certain definitions specialized to attitude dynamics will also be introduced.

Part of the reason for the proliferation of stability definitions is the diversity of applications where stability is important. As with so many techniques and concepts in applied mathematics, the earliest definition of stability arose in connection with celestial mechanics. In studying dynamics of systems of point masses moving under the influence of their mutual gravitational fields, Lagrange called a motion "stable" if none of the point masses escaped (i.e., arrived at an infinite distance from another point mass). Thus, a system is *Lagrange stable* if all its motions are *bounded*. In more modern times, celestial mechanicians have found several other specialized definitions of stability to be convenient, as discussed, for example, by [Szebehely]. The question of stability arises in many other branches of the natural sciences as well. For instance, the characteristics of stellar evolution are determined in part by the stability of certain nuclear processes, and the transition from laminar to turbulent flow was only theoretically explicable when it was viewed as a stability problem (see [Schlichting]). Even cancer in living cells can be usefully modeled as an instability.

Stability concepts are also important in the applied sciences, where stable behavior is almost always a performance requirement; if the system is not stable, the design criteria will specify that it be stabilized. A first example relates to an ultimate source of energy: the achievement of controlled fusion power depends critically on the solution of plasma stability problems. As a second example, the flight of aircraft is feasible only because stabilization techniques are well understood ([Etkin, 2]); often the pilot plays a role in this stabilization. In a similar vein, aeroelasticity is concerned ([Bisplinghoff, Ashley, and Halfman]) with the avoidance of divergence (static instability) and flutter (dynamic instability). Furthermore, the present book is founded on the fact that attitude stabilization is a necessity for a properly functioning spacecraft.

Several excellent books are available on stability theory, among them those of [Hahn, 1, 2], [Cesari], [Bellman], [Chetayev], [LaSalle and Lefschetz], and [Leipholz]. Consequently, no attempt is made here to present an exhaustive or even representative picture of the subject. The objective instead is to select from the existing literature those items which are most relevant to spacecraft attitude dynamics. In keeping with this policy, such theorems as may occur will only rarely be accompanied by proofs.

A.1 STABILITY DEFINITIONS

To be viable in physical applications, a mathematical definition of stability must characterize those properties which are inherent in the notion of "stability" for that particular application and, at the same time, it must be amenable to analytical treatment. Consider, for example, the Lagrange definition of stability alluded to a moment ago (all motions bounded). This definition satisfies the second requirement since boundedness is a property with which the analyst is equipped to deal. Unfortunately, it is a definition which is not, for our purposes, sufficiently stringent.

By far the most successful definition of stability is due to Liapunov, who was, interestingly enough, endeavoring to develop criteria for physical motions. His definition can be approached in the following way: The weakness in Lagrange's definition is that the motion may indeed be bounded but altogether too large. Therefore, one wishes to be able to specify the upper bound, and only if the motion remains always within this specified bound does it qualify as a "stable" motion. Furthermore, since the bound that is actually acceptable will vary from one application to another, one is naturally led to specify that the disturbed motion shall always stay within an *arbitrarily* low bound provided the troublesome disturbances are small enough.

Liapunov Stability

In order to state a precise definition, consider the differential system

$$\dot{\mathbf{y}} = \mathbf{g}(\mathbf{y}, t) \tag{1}$$

Attention is focused on the particular solution of (1) which proceeds from the initial condition

$$\mathbf{y} = \mathbf{y}_o \quad \text{at} \quad t = t_o \tag{2}$$

The resulting trajectory will be denoted $\mathbf{y}(t; \mathbf{y}_o, t_o)$, the functional dependence of $\mathbf{y}$ on $\mathbf{y}_o$ and t_o to be included when helpful. It is assumed that the function $\mathbf{g}(\mathbf{y}, t)$ is of such a nature that there is in fact one and only one solution. That is, we assume existence and uniqueness. The conditions which $\mathbf{g}(\mathbf{y}, t)$ must satisfy to have such a nature may be found in any standard text on ordinary differential equations. Next, consider *all* solutions starting at t_o, that is, solutions which at $t = t_o$ are different from $\mathbf{y}_o$ by an amount $\Delta\mathbf{y}_o$. Then, the resulting trajectory $\mathbf{y}(t; \mathbf{y}_o + \Delta\mathbf{y}_o, t_o)$ will be different from the original trajectory $\mathbf{y}(t; \mathbf{y}_o, t_o)$ by an amount

$$\Delta\mathbf{y}(t; \mathbf{y}_o, \Delta\mathbf{y}_o, t_o) \triangleq \mathbf{y}(t; \mathbf{y}_o + \Delta\mathbf{y}_o, t_o) - \mathbf{y}(t; \mathbf{y}_o, t_o) \tag{3}$$

Here, $\Delta\mathbf{y}_o$ plays the role of a "disturbance" whose effect $\Delta\mathbf{y}$ is propagated onward in time. We shall refer to $\Delta\mathbf{y}$ as the *perturbation* and to $\mathbf{y} + \Delta\mathbf{y}$ as the *perturbed* solution or trajectory. The question to be answered is: Will the "effect" $\Delta\mathbf{y}$ be arbitrarily small for all subsequent time $t > t_o$ if the "cause" $\Delta\mathbf{y}_o$ is sufficiently small (but not, of course, zero)? If the answer is in the affirmative, then the solution $\mathbf{y}(t; \mathbf{y}_o, t_o)$ is said to be stable in the sense of Liapunov or, simply, L-stable.

Definition A.1: (L-Stability) *The solution* $\mathbf{y}(t; \mathbf{y}_o, t_o)$ *is said to be Liapunov stable (L-stable) if there exists a number* $\delta > 0$ *such that, for any preassigned* $\varepsilon > 0$, *one can maintain* $\|\Delta\mathbf{y}\| < \varepsilon$ *for all* $t \geq t_o$, *by choosing any* $\Delta\mathbf{y}_o$ *subject to the constraint that* $\|\Delta\mathbf{y}_o\| < \delta$.

This definition employs the norm $\|\mathbf{v}\|$ of the column $\mathbf{v}$. Therefore, we shall pause briefly and review the relevant properties of norms. To qualify as a norm, $\|\mathbf{v}\|$ must possess the following properties:

(a) $\|\mathbf{v}\| > 0$ for all $\mathbf{v} \neq \mathbf{0}$

(b) $\|\mathbf{0}\| = 0$

(c) The "triangle" inequality must hold:

$$\|\mathbf{v}_1 + \mathbf{v}_2\| \leq \|\mathbf{v}_1\| + \|\mathbf{v}_2\| \text{ for all } \mathbf{v}_1, \mathbf{v}_2 \tag{4}$$

(d) $\|\kappa\mathbf{v}\| = |\kappa|\,\|\mathbf{v}\|$ for all $\mathbf{v}$, where κ is any complex constant

Thus, the norm is intended to be a measure of the size of $\mathbf{v}$ or, in other words, the distance from the origin of the point $(v_1, \ldots, v_n)$. There are many acceptable definitions of $\|\mathbf{v}\|$ which possess these properties. The most geometrically appeal-

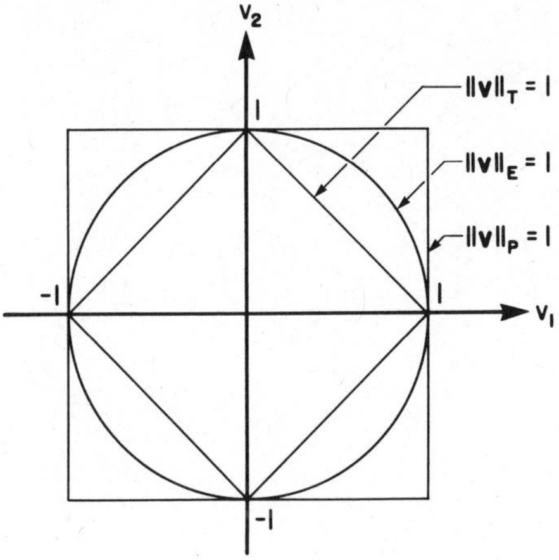

FIGURE A.1 Illustration comparing three norms.

ing is the Euclidean norm defined by

$$\textit{Euclidean norm:} \qquad \|\mathbf{v}\|_E \triangleq \left(v_1^2 + \cdots + v_n^2\right)^{1/2} \tag{5}$$

However, the definition

$$\textit{Taxicab norm:} \qquad \|\mathbf{v}\|_T \triangleq |v_1| + \cdots + |v_n| \tag{6}$$

is often more useful in proofs and can be shown to satisfy the required conditions (4). So also can the definition

$$\|\mathbf{v}\|_P \triangleq \max_{i=1}^{n} |v_i| \tag{7}$$

These three alternatives are illustrated in Fig. A.1, where the loci of $\|\mathbf{v}\|_E = 1$, $\|\mathbf{v}\|_T = 1$, and $\|\mathbf{v}\|_P = 1$ are shown.

Norms of square matrices will also be used in the sequel. The norm of $\mathbf{M}$ is defined by

$$\|\mathbf{M}\| = \text{minimum value of } k \text{ such that } \|\mathbf{Mv}\| \leq k\|\mathbf{v}\| \tag{8}$$

Thus, the actual value of $\|\mathbf{M}\|$ will depend on the norm chosen for $\|\mathbf{v}\|$. Corresponding to a Euclidean norm, we have

$$\|\mathbf{M}\|_E^2 = \max_{\mathbf{v}} \left\{ \frac{\mathbf{v}^T\mathbf{M}^T\mathbf{M}\mathbf{v}}{\mathbf{v}^T\mathbf{v}} \right\} \tag{9}$$

Other possibilities include

$$\|\mathbf{M}\| \triangleq \sum_{i,j} |a_{ij}| \tag{10}$$

and

$$\|\mathbf{M}\| \triangleq \left(\sum_{i,j} a_{ij}^2 \right)^{1/2} \tag{11}$$

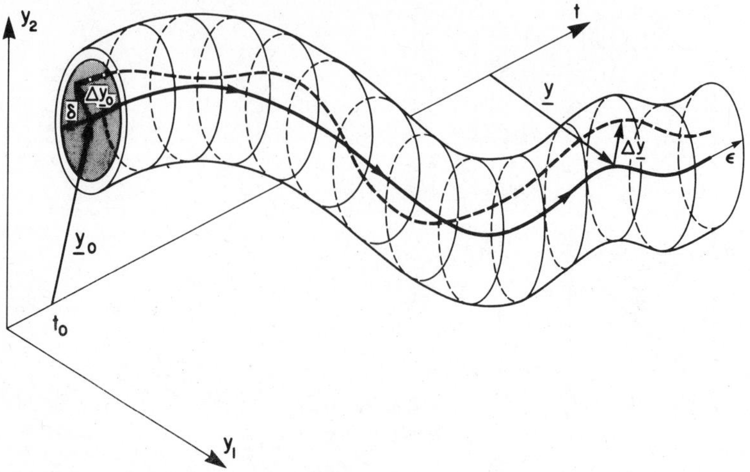

FIGURE A.2 Liapunov's definition of stability.

Visualization of L-Stability

To return to Definition A.1, L-stability is clarified with the aid of Fig. A.2. Here, the trajectory $\mathbf{y}(t, \mathbf{y}_o, t_o)$ is shown for $t \geq t_o$, and the bound $\|\Delta \mathbf{y}\| < \varepsilon$ is represented by a "tube" of radius ε surrounding this reference trajectory. The reference trajectory $\mathbf{y}(t, \mathbf{y}_o, t_o)$ is L-stable if it is possible to find a circle with $\mathbf{y}_o$ as center and of radius δ, such that all trajectories starting on this circular disk remain within the tube.

A very simple example is the scalar equation

$$\dot{y} = 1 \tag{12}$$

with the stability of the solution corresponding to $y(t_o) = y_o$ to be considered. In this example, matters are greatly simplified by the availability of a simple closed-form solution

$$y(t; y_o, t_o) = y_o + t - t_o \tag{13}$$

The same is true of the perturbation Δy due to an initial disturbance Δy_o:

$$\Delta y \equiv \Delta y_o \tag{14}$$

So in this case the initial disturbance remains at its initial value, and the solution (13) is evidently L-stable, since to ensure that $|\Delta y| < \varepsilon$ for all $t \geq t_o$, we merely choose $|\Delta y_o| < \varepsilon$. For this example, then, $\delta = \varepsilon$. Observe that δ does not depend on t_o; this important question is discussed further below. (Observe also that the solution (13) is not Lagrange stable since it is unbounded.)

A second simple example is the scalar equation

$$\dot{y} = y \tag{15}$$

We consider the stability of the solution corresponding to $y(t_o) = y_o$. Again, a closed-form solution is available:

$$y(t; y_o, t_o) = y_o \exp(t - t_o) \tag{16}$$

The perturbation Δy due to an initial disturbance Δy_o is clearly

$$\Delta y = \Delta y_o \exp(t - t_o) \tag{17}$$

and the presence of the exponential precludes the possibility of finding a δ such that, if $|\Delta y_o| < \delta$, then $|\Delta y| < \varepsilon$ for all $t - t_o \geq 0$. Therefore, solution (13) is not L-stable. In fact, it is not possible to find *any* bound on $|\Delta y|$, much less an arbitrarily small one. On the other hand, all solutions of the scalar equation

$$\dot{y} = -y \tag{18}$$

can be shown in a similar fashion to be L-stable. This is because

$$\Delta y = \Delta y_o \exp(t_o - t) \tag{19}$$

The second-order example

$$\ddot{y} + y = 0 \tag{20a}$$

can be rewritten in the required form (1) by setting $y = y_1$, $\dot{y} = y_2$:

$$\dot{y}_1 = y_2; \qquad \dot{y}_2 = -y_1 \tag{20b}$$

Consider the stability of the solution corresponding to $\mathbf{y}(t_o) = \mathbf{y}_o$, which we know is

$$\mathbf{y} = \mathbf{T}(t - t_o)\mathbf{y}_o \tag{21}$$

with

$$\mathbf{T}(t) = \begin{bmatrix} \cos t & \sin t \\ -\sin t & \cos t \end{bmatrix} \tag{22}$$

The perturbation due to an initial disturbance $\Delta\mathbf{y}_o$ is

$$\Delta\mathbf{y} = \mathbf{T}(t - t_o)\,\Delta\mathbf{y}_o \tag{23}$$

Since the norm of $\mathbf{T}$ is certainly bounded, $\|\mathbf{T}\| < \overline{T}$, then

$$\|\Delta\mathbf{y}\| < \|\mathbf{T}(t - t_o)\|\,\|\Delta\mathbf{y}_o\| < \overline{T}\|\Delta\mathbf{y}_o\| \tag{24}$$

and so, for any preassigned ε, we can have $\|\Delta\mathbf{y}\| < \varepsilon$ for all $t - t_o \geq 0$ and for all $\|\Delta\mathbf{y}_o\| < \delta$, provided we choose $\delta(\varepsilon) = \varepsilon/\overline{T}$.

The previous four examples have been of a particularly simple variety, namely, linear stationary systems (i.e., linear differential equations with constant coefficients). To introduce the aspect of time dependence, consider the stability of the solution of

$$\dot{y} + \frac{y}{t} = 0 \tag{25}$$

corresponding to $y(t_o) = y_o$; this solution is given by

$$y = \frac{y_o t_o}{t} \tag{26}$$

The perturbation Δy due to an initial disturbance Δy_o is

$$\Delta y = \frac{\Delta y_o t_o}{t} \tag{27}$$

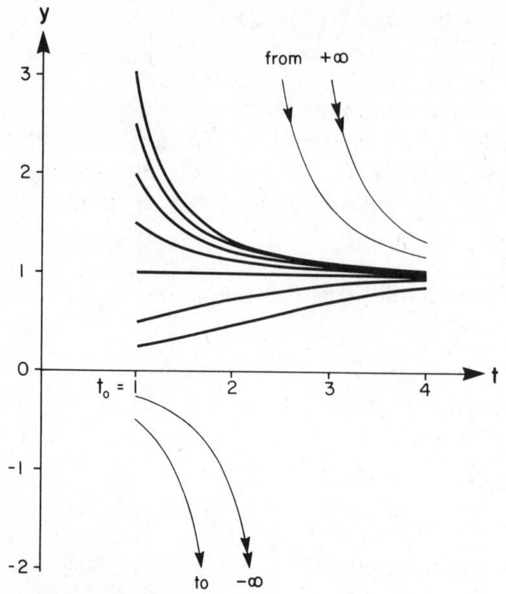

FIGURE A.3 Solutions to $\dot{y} = y(1 - y)$ for $t > 1$.

An immediate complication is seen to be that, if $t_o < 0$, then Δy goes to infinity as $t \to 0$. This behavior is described by saying that the solution has a *finite escape time*. To avoid finite escape times, we shall restrict $t_o > 0$, in which case $|\Delta y|$ can always be made $< \varepsilon$ for all $t \geq t_o$ by choosing $|\Delta y_o| < \varepsilon$, that is, by choosing $\delta = \varepsilon$.

As an example of a nonlinear system, consider the stability of the solution of

$$\dot{y} = y(1 - y) \tag{28}$$

that corresponds to $y(t_o) = y_o$. This solution is

$$y = \frac{y_o \exp(t - t_o)}{1 - y_o[1 - \exp(t - t_o)]} \tag{29}$$

We shall examine the stability of the two solutions of greatest interest, namely, $y \equiv 0$ and $y \equiv 1$, the *equilibrium* solutions. For the solution $y \equiv 0$, the perturbation Δy due to an initial disturbance Δy_o has the closed-form solution

$$\Delta y = \frac{\Delta y_o \exp(t - t_o)}{1 - \Delta y_o[1 - \exp(t - t_o)]} \tag{30}$$

since in this case $y_o = \Delta y_o$ and we may use (29). It is clear from (30) that $|\Delta y| \to 1$ as $t \to \infty$, regardless of how $|\Delta y_o|$ is restricted. Therefore, the solution $y \equiv 0$ is not L-stable. In fact, as shown in Fig. A.3, the perturbed solutions are particularly ill behaved for $\Delta y_o < 0$; they become infinite as $t \to t_o + \ln[1 - (\Delta y_o)^{-1}]$. For the solution $y \equiv 1$ on the other hand, the perturbation Δy due to an initial disturbance Δy_o has the closed-form solution

$$\Delta y = \frac{\Delta y_o}{(1 + \Delta y_o)\exp(t - t_o) - \Delta y_o} \tag{31}$$

This perturbation is well behaved (provided $\Delta y_o > -1$), and the denominator is always greater than unity for $t > t_o$. Therefore, $|\Delta y|$ can always be kept less than any preassigned $\varepsilon > 0$ merely by choosing $|\Delta y_o| < \varepsilon$ if $\varepsilon \leq 1$, or choosing $|\Delta y_o| < 1$ if $\varepsilon > 1$. That is, $\delta(\varepsilon) = \min\{\varepsilon, 1\}$.

Related Definitions

The above examples suggest that in studying the stability of solutions $\mathbf{y}(t; \mathbf{y}_o, t_o)$ to system (1), some further definitions will prove useful. First, we give the rather self-evident definition of instability:

Definition A.2: (Instability) *The solution* $\mathbf{y}(t; \mathbf{y}_o, t_o)$ *is said to be unstable if it is not L-stable.*

Moreover, the δ in Definition A.1 will generally depend on both ε and t_o. It is important to know whether $\delta(\varepsilon, t_o)$ depends on t_o or not. It always depends on ε: at the very least we must have $\delta \leq \varepsilon$. However, it is desirable that δ not depend on t_o since if it does not, a bound on initial disturbances exists such that $\|\Delta y\| < \varepsilon$ can be satisfied regardless of the instant at which the initial disturbance occurs. Therefore, we have

Definition A.3: (Uniform L-Stability) *If, in Definition* A.1, *a* $\delta(\varepsilon, t_o)$ *can be found that is independent of* t_o, *then the L-stability is said to be uniform.*

The possible dependence of δ on t_o can be traced to the dependence of $\mathbf{g}$ on t in (1). If $\mathbf{g}$ does not depend on t, the system is said to be *stationary*, or *time invariant*. Since this also precludes the possibility of a prescribed control function $\mathbf{u}(t)$, as in $\dot{\mathbf{y}} = \mathbf{g}(\mathbf{y}, \mathbf{u})$, the system is also said to be *autonomous*. It follows that the L-stability of solutions of autonomous systems will always be uniform in t.

For reasons that will become apparent as we proceed, L-stability is usually not a strong enough property in practice. In the definition of L-stability, the disturbance was represented by $\Delta\mathbf{y}_o$ at $t = t_o$ and no further disturbances were contemplated for $t > t_o$; the trajectory was still governed by the original equation, (1). In reality, however, disturbances are constantly impressing themselves on the system; and even though these disturbances may be bounded, their cumulative effect throughout $t_o \leq t < \infty$ may not be sufficiently bounded. This is unacceptable from a practical standpoint, and so an additional condition is placed on the asymptotic behavior of the perturbations:

Definition A.4: (Attractive Solution) *The solution* $\mathbf{y}(t; \mathbf{y}_o, t_o)$ *is said to be attractive if there exists a number* $\delta_A > 0$ *such that* $\|\Delta\mathbf{y}\| \to 0$ *as* $t \to \infty$ *for all* $\|\Delta\mathbf{y}_o\| < \delta_A$.

This definition now allows us to define the type of stability best suited to engineering applications:

Definition A.5: (Asymptotic Stability) *The solution* $\mathbf{y}(t; \mathbf{y}_o, t_o)$ *is said to be asymptotically stable if it is both L-stable and attractive.*

Again, the question of uniformity with respect to t_o arises, but we shall not belabor this aspect since, for all applications considered in this book, asymptotic stability is in fact uniform in t_o.

Thus far, all the definitions of stability have reflected *local* properties of the trajectories $\mathbf{y}(t; \mathbf{y}_o, t_o)$. L-Stability itself is a local property since, for an ε of interest, δ may have to be extremely small; similarly, in the definition of asymptotic stability, the largest possible value of δ_A may be small. However, if δ_A can be arbitrarily large and yet still $\|\Delta\mathbf{y}\| \to 0$ as $t \to \infty$ for all $\|\Delta\mathbf{y}\| < \delta_A$, attractiveness is no longer local. We have the following definition:

Definition A.6: (Global Asymptotic Stability) *The solution* $\mathbf{y}(t; \mathbf{y}_o, t_o)$ *is said to be globally asymptotically stable if it is L-stable and every trajectory converges to it as* $t \to \infty$.

A physical system modeled according to (1) will in reality have numerous disturbing influences acting upon it which are not explicitly modeled. These may be taken into account by examining the associated system

$$\dot{\mathbf{y}}_\Delta = \mathbf{g}(\mathbf{y}_\Delta, t) + \Delta\mathbf{g}(t) \tag{32}$$

where $\Delta\mathbf{g}(t)$ represents unmodeled physical disturbances. It is of great importance to know how the solution $\mathbf{y}_\Delta(t; \mathbf{y}_o + \Delta\mathbf{y}_o, t_o)$ of (32) relates to the reference solution $\mathbf{y}(t; \mathbf{y}_o, t_o)$ of (1). The desirable relationship is expressed in the following definition:

Definition A.7: (Total Stability) *The solution* $\mathbf{y}(t; \mathbf{y}_o, t_o)$ *is said to be totally stable if there exist numbers* $\delta > 0$ *and* $\bar{g} > 0$ *such that, for any preassigned* $\varepsilon > 0$, *one can maintain*

$$\|\mathbf{y}_\Delta(t; \mathbf{y}_o + \Delta\mathbf{y}_o, t_o) - \mathbf{y}(t; \mathbf{y}_o, t_o)\| < \varepsilon$$

for all $t \geq t_o$ *by choosing any* $\Delta\mathbf{y}_o$ *subject to the constraint* $\|\Delta\mathbf{y}_o\| < \delta$ *and any* $\Delta\mathbf{g}$ *subject to constraint:* $\|\Delta\mathbf{g}(t)\| < \bar{g}$, *for all* $\|\Delta\mathbf{y}\| < \varepsilon$ *and all* $t \geq t_o$.

Thus, for total stability, it is required that the perturbation in the reference trajectory can be arbitrarily bounded by sufficiently limiting both the initial and persistent disturbances.

A further origin of disturbance lies in the parameters of the system. Physically, they will not be exactly equal to their mathematically assumed values, for a variety of reasons. This observation is emphasized by writing (1) in the form

$$\dot{\mathbf{y}} = \mathbf{g}(\mathbf{y}, t; \mathbf{p}) \tag{33}$$

where $\mathbf{p}$ contains the (constant) parameters of the system. The question arises as to the effects of parameter errors $\Delta\mathbf{p}$. We should wish that no qualitative change occur in the solutions of (33). This wish is expressed in the following definition of robustness:

Definition A.8: (Robustness) *The system* $\dot{\mathbf{y}} = \mathbf{g}(\mathbf{y}, t; \mathbf{p})$ *is said to be robust if there exists a number* $\bar{p} > 0$ *such that if* $\|\Delta\mathbf{p}\| < \bar{p}$, *the topological behavior of the system trajectories is unchanged.*

This definition is from [Hahn, 1, 2], who calls the property "structural stability." We prefer the term "robustness" here since it is less confusing (mathematical structures vs. physical structures) and conforms more nearly to current engineering usage.

An elementary but important example of the foregoing definitions is the damped linear oscillator described by the equation

$$\ddot{y} + 2\zeta\omega\dot{y} + \omega^2 y = 0 \tag{34}$$

The stability of the equilibrium solution $y(t) \equiv 0$ is of interest, and ζ and ω are parameters. Since this equation has an explicit and well-known solution in terms of elementary functions, stability questions can be resolved simply by examining the properties of the closed-form solution. The equilibrium is (i) stable if $\zeta \geq 0$, (ii) asymptotically stable if $\zeta > 0$, and (iii) unstable if $\zeta < 0$. These conclusions are independent of the initial conditions y_o and $\dot{y}_o$, and so the asymptotic stability is global. Note also that the system (34) ceases to be robust at the *branch value* $\zeta = 0$.

A further sort of stability now will be introduced to complete the definitions we shall need. It is concerned with the particular instance of equation (1) where the system is autonomous, $\mathbf{g}$($\mathbf{y}$ only), and where there is a periodic solution $\mathbf{y}_T(t)$, that is, $\mathbf{y}_T(t + T) \equiv \mathbf{y}_T(t)$. Such periodic solutions (oscillations) are important both in engineering applications and in the general study of nonlinear differential equations. The books by [Stern], [Mickens], [Hagadorn, 1], and [Nayfeh and Mook] may be consulted for the main results. The key point to be made is that it may be quite acceptable if the disturbed trajectory remains arbitrarily close to the reference (periodic) trajectory, and it may *not* be necessary to insist that points on the original and neighboring trajectories *which correspond to the same instant of time* be arbitrarily close. In other words, *phase* information may not be important. To cover this situation, Poincaré introduced the concept of *orbital stability*, defined as follows:

Definition A.9: (Orbital Stability) *The oscillation* $\mathbf{y}_T(t)$ *is said to be orbitally stable if there exist a number* $\delta > 0$ *and a time* $t_1 > 0$ *such that, for any preassigned* $\varepsilon > 0$, *one can maintain*

$$\|\mathbf{y}(t; \mathbf{y}_o) - \mathbf{y}_T(t_1)\| < \varepsilon$$

for all $t \geq 0$ *by choosing any* $\mathbf{y}_o$ *subject to the constraint* $\|\mathbf{y}_o - \mathbf{y}_T(0)\| < \delta$.

Since the system is autonomous, there has been no loss in generality in setting $t_o = 0$.

The asymptotic behavior of the perturbed motion is also of interest, and one is again led to a version of asymptotic stability whose definition is tailored to oscillations:

Definition A.10: (Asymptotic Orbital Stability) *The oscillation* $\mathbf{y}_T(t)$ *is said to be asymptotically orbitally stable if it is orbitally stable and*

$$\|\mathbf{y}(t;\mathbf{y}_o) - \mathbf{y}_T(t_1)\| \to 0$$

as $t \to \infty$.

One of the simplest and most studied equations with which to demonstrate orbital stability is van der Pol's equation:

$$\ddot{y} - \alpha^2(1 - y^2)\dot{y} + y = 0 \tag{35}$$

It is nonlinear and autonomous and certainly possesses the quiescent solution $y \equiv 0$. However, it also possesses a periodic solution (*limit cycle*) as suggested by the following observations. When $0 < y \ll 1$, the equation is approximately $\ddot{y} - \alpha^2\dot{y} + y = 0$, which may be thought of as negative damping in the neighborhood of the origin; hence the origin is likely to be unstable. On the other hand, when $y \gg 1$, the equation is approximately $\ddot{y} + \alpha^2 y^2 \dot{y} + y = 0$, which may be thought of as positive nonlinear damping for large excursions from the origin. The compromise between these two tendencies is an asymptotically stable limit cycle of the type shown in Fig. A.4*a*.

An important point can be made with an equation similar to van der Pol's:

$$\ddot{y} + \alpha^2(1 - y^2)\dot{y} + y = 0 \tag{36}$$

By a similar line of reasoning, the origin ($y \equiv 0$) is predicted to be asymptotically stable for small initial values of $(\Delta y, \Delta \dot{y})$, but large perturbations can be expected to increase. This behavior is illustrated in Fig. A.4*b*. The origin is thus only *locally* asymptotically stable, and the limit cycle is unstable.

Reflections on the Definitions

There are several ways in which engineering realities are reflected in the above stability definitions:

- Uncertainty in physical laws, $\to \Delta\mathbf{g}(t)$
- Uncertainty in physical constants, $\to \Delta\mathbf{p}$
- Uncertainty in measured parameters, $\to \Delta\mathbf{p}$
- Quasi-steady variation in parameters, $\to \Delta\mathbf{p}$
- Manufacturing tolerances, $\to \Delta\mathbf{p}$
- Mathematical approximations, $\to \Delta\mathbf{g}(t)$
- Unmodeled continuous disturbances, $\to \Delta\mathbf{g}(t)$
- Effects of all past disturbances, $\to \Delta\mathbf{y}(t)$

The essence of stability is that the *effects* from all these disturbances can be made

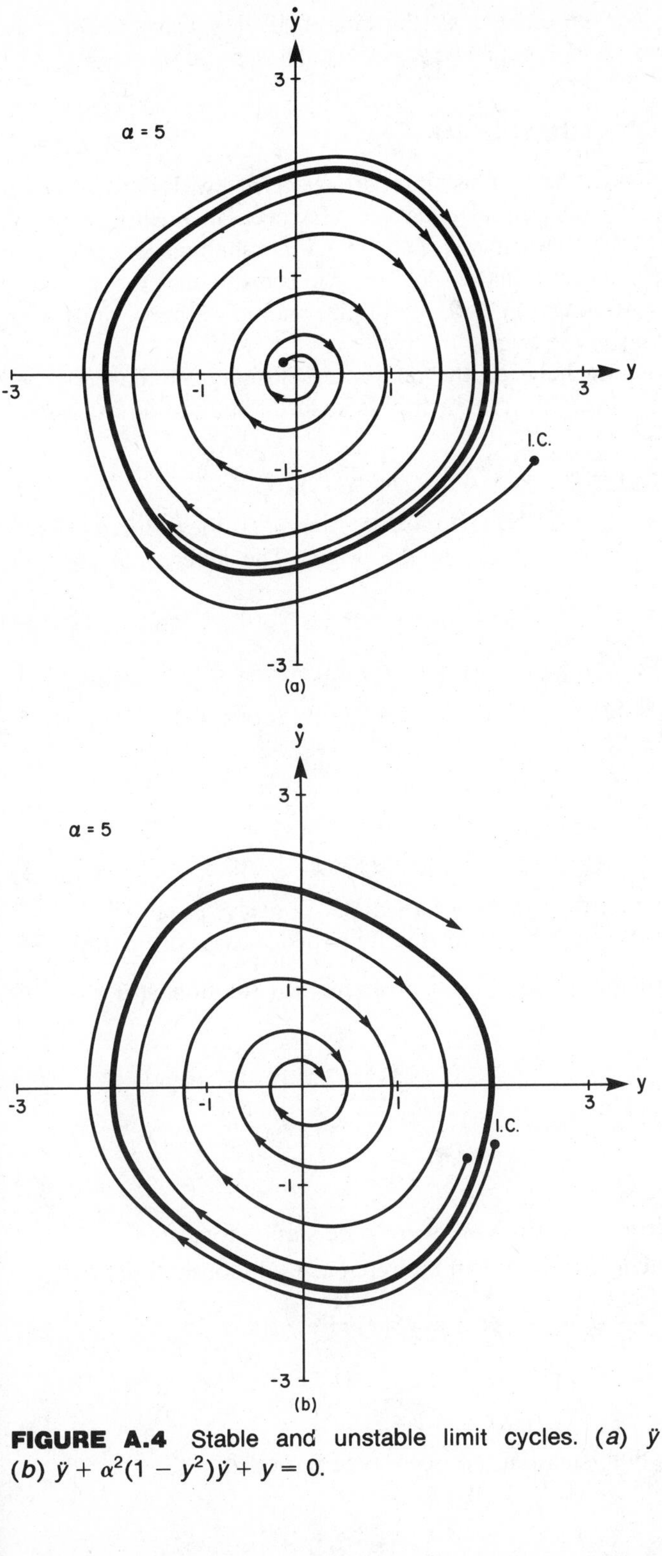

FIGURE A.4 Stable and unstable limit cycles. (*a*) $\ddot{y} - \alpha^2(1 - y^2)\dot{y} + y = 0$. (*b*) $\ddot{y} + \alpha^2(1 - y^2)\dot{y} + y = 0$.

as small as necessary (as represented by the bound $\|\Delta \mathbf{y}\| < \varepsilon$) by making the disturbances themselves small enough ($\|\Delta \mathbf{g}\| < \bar{g}$; $\|\Delta \mathbf{p}\| < \bar{p}$; $\|\Delta \mathbf{y}_o\| < \delta$).

A.2 STABILITY OF THE ORIGIN

It should be noted that the definitions above are concerned with the stability of a particular solution of a differential equation. However, in dealing with the stability of a particular reference solution $\mathbf{y}(t; \mathbf{y}_o, t_o)$, it is usually most convenient to do so in terms of the coordinates $\Delta \mathbf{y}$. In $\Delta \mathbf{y}$ coordinates, the reference (unperturbed) solution is always $\Delta \mathbf{y} \equiv \mathbf{0}$, the origin, and so we may with this in mind refer to the *stability of the origin*.

Now that the distinction between the reference solution and the perturbed solution $\mathbf{y}(t) + \Delta \mathbf{y}(t)$ is amply clear, a slightly less cluttered notation can henceforth be used for $\Delta \mathbf{y}$:

$$\mathbf{x}(t) \triangleq \mathbf{y}(t; \mathbf{y}_o + \Delta \mathbf{y}_o, t_o) - \mathbf{y}(t; \mathbf{y}_o, t_o) \tag{1}$$

Thus, $\mathbf{x}(t)$ is the perturbation in the reference solution $\mathbf{y}(t)$. Viewed in $\mathbf{x}$-coordinates, we consider always the stability of the origin. The differential equation governing the perturbations to

$$\dot{\mathbf{y}} = \mathbf{g}(\mathbf{y}, t) \tag{2}$$

is readily derived. For, defining $\tilde{\mathbf{y}}(t) \triangleq \mathbf{y}(t; \mathbf{y}_o, t_o)$ for brevity,

$$\dot{\mathbf{x}} = \dot{\mathbf{y}}(t; \mathbf{y}_o + \mathbf{x}_o, t_o) - \dot{\mathbf{y}}(t; \mathbf{y}_o, t_o) = \mathbf{g}(\tilde{\mathbf{y}} + \mathbf{x}, t) - \mathbf{g}(\tilde{\mathbf{y}}, t) \tag{3}$$

So, if we define

$$\mathbf{f}(\mathbf{x}, t) \triangleq \mathbf{g}[\tilde{\mathbf{y}}(t) + \mathbf{x}, t] - \mathbf{g}[\tilde{\mathbf{y}}(t), t] \tag{4}$$

we have

$$\dot{\mathbf{x}} = \mathbf{f}(\mathbf{x}, t); \qquad \mathbf{f}(\mathbf{0}, t) \equiv \mathbf{0} \tag{5}$$

The advantage of (5) is that the "reference solution" is always the origin.

An examination of (4) reveals the following four special cases of interest:

(A) The original system stationary and the reference solution $\tilde{\mathbf{y}}(t)$ an equilibrium solution:

$$\frac{\partial \mathbf{g}}{\partial t} \equiv \mathbf{0}; \qquad \tilde{\mathbf{y}}(t) \equiv \mathbf{y}_e \tag{6}$$

Therefore, from (4),

$$\frac{\partial \mathbf{f}}{\partial t} \equiv \mathbf{0} \tag{7}$$

and the perturbation equations are seen to be stationary also.

(B) The original system stationary and the reference solution $\tilde{\mathbf{y}}(t)$ periodic:

$$\frac{\partial \mathbf{g}}{\partial t} \equiv \mathbf{0}; \qquad \tilde{\mathbf{y}}(t) \equiv \mathbf{y}_T(t) \equiv \mathbf{y}_T(t + T) \tag{8}$$

Therefore, from (4),

$$\mathbf{f}(\mathbf{x}, t + T) \equiv \mathbf{f}(\mathbf{x}, t) \tag{9}$$

and the perturbation equations are seen to be periodic.

(C) The original system periodic and the reference solution $\tilde{\mathbf{y}}(t)$ an equilibrium:

$$\mathbf{g}(\mathbf{y}, t+T) \equiv \mathbf{g}(\mathbf{y}, t); \qquad \tilde{\mathbf{y}}(t) \equiv \mathbf{y}_e \tag{10}$$

Therefore, from (4),

$$\mathbf{f}(\mathbf{x}, t+T) \equiv \mathbf{f}(\mathbf{x}, t) \tag{11}$$

and the perturbation equations are seen to be periodic in this case also.

(D) The original system periodic (period T_1) and the reference solution $\tilde{\mathbf{y}}(t)$ also periodic (period T_2), with T_1 and T_2 commensurable:

$$\mathbf{g}(\mathbf{y}, t+T_1) \equiv \mathbf{g}(\mathbf{y}, t)$$

$$\tilde{\mathbf{y}}(t) \equiv \mathbf{y}_{T_2}(t) \equiv \mathbf{y}_{T_2}(t+T_2)$$

$$pT_1 = qT_2; \ p, q \text{ integers} \tag{12}$$

Therefore, from (4),

$$\mathbf{f}(\mathbf{x}, t+T) \equiv \mathbf{f}(\mathbf{x}, t) \tag{13}$$

and the perturbation equations are given again periodic.

A glance at equations (7), (9), (11), and (13) shows that, as far as the perturbation equations are concerned, the above four cases reduce to two possibilities: stationary perturbation equations (case A), or periodic perturbation equations (cases B, C, and D). Moreover, it is clear that the stability definitions of Section A.1 are most readily applied when the perturbation equations can be solved in closed form. Unfortunately, this is not generally possible. It is then natural to inquire whether system (5) can be approximated in some manner near the origin and, if so, whether the implications for stability of such an approximation can be determined. These questions lead directly to the linear approximation.

A.3 THE LINEAR APPROXIMATION

We now consider the linear approximation to the system

$$\dot{\mathbf{x}} = \mathbf{f}(\mathbf{x}, t); \qquad \mathbf{f}(\mathbf{0}, t) \equiv \mathbf{0} \tag{1}$$

Usually, $\mathbf{f}$ is either independent of t (the *stationary* case) or periodic in t (the *periodic* case). The basis for the approximation is a Taylor expansion of (1), which gives

$$\dot{\mathbf{x}} = \mathbf{A}(t)\mathbf{x} + \mathbf{n}(\mathbf{x}, t) \tag{2}$$

where

$$\mathbf{A}(t) \triangleq \left[\frac{\partial \mathbf{f}(\mathbf{x}, \mathbf{t})}{\partial \mathbf{x}^T}\right]_{\mathbf{x}=\mathbf{0}} \tag{3a}$$

$$\mathbf{n}(\mathbf{0}, t) = \mathbf{0} \tag{3b}$$

The notation on the right-hand side of (3a) is a short form:

$$\frac{\partial \mathbf{f}}{\partial \mathbf{x}^T} \triangleq \begin{bmatrix} \frac{\partial f_1}{\partial x_1} & \cdots & \frac{\partial f_1}{\partial x_n} \\ \vdots & & \vdots \\ \frac{\partial f_n}{\partial x_1} & \cdots & \frac{\partial f_n}{\partial x_n} \end{bmatrix} \tag{4}$$

It is clear that if $\mathbf{f}$ is stationary, $\mathbf{A}$ is also, whereas if $\mathbf{f}$ is periodic (period T), $\mathbf{A}$ is also periodic with period T. Not all functions $\mathbf{f}$ can be differentiated as required by (3), but all the functions of interest to us can be.

Since the linear part of $\mathbf{f}$ has been extracted in (2), the remaining term, $\mathbf{n}(\mathbf{x}, t)$, must be nonlinear, with the property that

$$\|\mathbf{n}(\mathbf{x}, t)\| = o(\|\mathbf{x}\|) \tag{5}$$

uniformly in t. This notation implies that

$$\lim_{\|\mathbf{x}\| \to 0} \frac{o(\|\mathbf{x}\|)}{\|\mathbf{x}\|} = 0 \tag{6}$$

Thus, the closer the *state* $\mathbf{x}(t)$ comes to the origin, the less important the term $\mathbf{n}(\mathbf{x}, t)$ becomes.

The object of the linear approximation is to permit us to study the stability properties of (2) via the linear approximation

$$\mathbf{x}(t) \doteq \delta\mathbf{x}(t) \tag{7}$$

where

$$\delta\dot{\mathbf{x}} = \mathbf{A}(t)\,\delta\mathbf{x} \tag{8}$$

As we shall see, some of the stability properties of (2) are retained by (8), although some stability information is lost. The advantage of (8) over (2) of course is that (8) can be treated more easily analytically. In addition to "linear approximation," (8) is sometimes called the *infinitesimal approximation*, or the *variational approximation* to (2). If for example the origin is stable for (8), we say that the origin is *infinitesimally stable* (or *variationally stable*) for (2).

We proceed by examining the stability of (8), first when $\mathbf{A}(t)$ is constant, and then when $\mathbf{A}(t)$ is periodic. Then, in Section A.4, we shall consider how the stability properties associated with (8) are related to those associated with (2).

Stability of Linear Stationary Systems

The linear system

$$\delta\dot{\mathbf{x}} = \mathbf{A}\,\delta\mathbf{x} \tag{9}$$

is said to be stationary if $\mathbf{A}$ is constant. We are particularly interested in the conditions under which the origin is asymptotically stable; mere L-stability, in which the origin is not attractive, is of mathematical interest only and is unacceptable for engineering applications.

The solution of (9) is well known:

$$\delta\mathbf{x}(t) = \exp(\mathbf{A}t)\,\delta\mathbf{x}_o \tag{10}$$

where $\delta\mathbf{x}(0) = \delta\mathbf{x}_o$ is a specified initial condition. This closed-form solution permits analytical stability properties to be derived relatively easily (although, for very high-order systems, numerical problems can intrude). Let the eigenvalues of **A** be

$$\lambda_i\{\mathbf{A}\} = \sigma_i + j\omega_i \qquad (i = 1, \ldots, n) \tag{11}$$

Furthermore, let

$$\bar{\sigma} \triangleq \max_{i=1}^{n}\{\sigma_i\} \tag{12}$$

Then, we have the following theorems:

Theorem A.1: *When the eigenvalues λ_i of* **A** *are distinct, the solution $\delta\mathbf{x} \equiv \mathbf{0}$ of $\delta\dot{\mathbf{x}} = \mathbf{A}\,\delta\mathbf{x}$ is L-stable if $\bar{\sigma} \leq 0$.*

Theorem A.2: *The solution $\delta\mathbf{x} \equiv \mathbf{0}$ of $\delta\dot{\mathbf{x}} = \mathbf{A}\,\delta\mathbf{x}$ is asymptotically stable if $\bar{\sigma} < 0$.*

Theorem A.3: *The solution $\delta\mathbf{x} \equiv \mathbf{0}$ of $\delta\dot{\mathbf{x}} = \mathbf{A}\,\delta\mathbf{x}$ is unstable if $\bar{\sigma} > 0$.*

Not all cases are covered by these theorems. In particular, the situation where $\bar{\sigma} = 0$ and the eigenvalues of **A** are not distinct is not covered.

On the basis of these theorems, one may determine stability properties by examining the eigenvalues of **A**. This must be done numerically for all but very low-order systems ($n = 2, 3, 4$). There is however an intermediate-order group of systems ($n = 4, \ldots, 8$, say) where literal stability criteria can still often be derived. To these criteria we now turn.

The Routh–Hurwitz Problem

The Routh–Hurwitz problem addresses the question of whether the eigenvalues of **A** are in the (open) left half-plane (LHP) based on the coefficients of the *characteristic polynomial*

$$p_n(\lambda) \triangleq \det(\lambda\mathbf{1} - \mathbf{A}) \tag{13}$$

For intermediate-order systems, which frequently occur in this book, the determinant in (13) can often be expanded literally. Let

$$\lambda^n + a_1\lambda^{n-1} + \cdots + a_{n-1}\lambda + a_n \triangleq p_n(\lambda) \tag{14}$$

Then, the *characteristic equation*, namely,

$$p_n(\lambda) = 0 \tag{15}$$

has the eigenvalues of **A** as its roots. Although the *exact* location of the λ_i may not be available in closed form from the $\{a_1, \ldots, a_n\}$, a lesser question—whether the λ_i are all somewhere in the LHP—is answerable using a finite number of arithmetic operations on the a_i.

An often useful theorem is the following:

Theorem A.4: *If any one of the a_i is negative, the solution $\delta \mathbf{x} \equiv \mathbf{0}$ of $\delta \dot{\mathbf{x}} = \mathbf{A}\,\delta \mathbf{x}$ is unstable.*

This theorem gives necessary conditions for stability. It frequently allows the analyst to detect instability at a glance.

However, a set of necessary and sufficient conditions for asymptotic stability is even more desirable. This problem has been solved independently by Routh and Hurwitz. We prefer Hurwitz's criteria here because fractions are avoided. His criteria are based on the *Hurwitz matrix*:

$$\mathbf{H} \triangleq \begin{bmatrix} a_1 & a_3 & a_5 & \cdots & 0 \\ 1 & a_2 & a_4 & \cdots & 0 \\ 0 & a_1 & a_3 & \cdots & 0 \\ \vdots & \vdots & \vdots & \vdots & \vdots \\ 0 & 0 & 0 & \cdots & 0 \\ 0 & 0 & 0 & \cdots & a_n \end{bmatrix} \tag{16}$$

(we take $a_k = 0$ for $k > n$.) Next, the principal minors of $\mathbf{H}$ are formed:

$$\Delta_1 \triangleq a_1$$

$$\Delta_2 \triangleq \det \begin{bmatrix} a_1 & a_3 \\ 1 & a_2 \end{bmatrix}$$

$$\vdots$$

$$\Delta_n \triangleq \det \mathbf{H}$$

Then, we have

Theorem A.5: *Necessary and sufficient conditions for asymptotic stability of the solution $\delta \mathbf{x} \equiv \mathbf{0}$ of $\delta \dot{\mathbf{x}} = \mathbf{A}\,\delta \mathbf{x}$ are*

$$\Delta_i > 0 \qquad (i = 1, \ldots, n)$$

It is helpful also to note that $\Delta_n = a_n \Delta_{n-1}$, so that the condition $\Delta_n > 0$ can be replaced by $a_n > 0$. Several other similar relationships can also be demonstrated, leading to the most economical statement of these criteria being expressed as in Table A.1. For system orders greater than $n = 8$, the criteria tend to become too unwieldy, and, even worse, the determinant in (13) becomes intractable.

Stability Boundaries

In practice, the system matrix $\mathbf{A}$ depends on a set of parameters, $\mathbf{p}$: $\mathbf{A} = \mathbf{A}(\mathbf{p})$. It follows from (13) and (14) that the coefficients of the characteristic equation also depend on these parameters: $a_i = a_i(\mathbf{p})$. The Routh–Hurwitz criteria of Theorem A.5 thus define those regions in parameter space that correspond to

TABLE A.1
The Routh–Hurwitz Criteria

System Order	Criteria
$n = 1$	$a_1 > 0$
$n = 2$	$a_1, a_2 > 0$
$n = 3$	$a_1, a_3, \Delta_2 > 0$, or $a_2, a_3, \Delta_2 > 0$
$n = 4$	$a_1, a_2, a_4, \Delta_3 > 0$, or $a_1, a_3, a_4, \Delta_3 > 0$
$n = 5$	$a_1, a_3, a_5, \Delta_2, \Delta_4 > 0$, or $a_2, a_4, a_5, \Delta_2, \Delta_4 > 0$
$n = 6$	$a_1, a_2, a_4, a_6, \Delta_3, \Delta_5 > 0$, or $a_1, a_3, a_5, a_6, \Delta_3, \Delta_5 > 0$
$n = 7$	$a_1, a_3, a_5, a_7, \Delta_2, \Delta_4, \Delta_6 > 0$, or $a_2, a_4, a_6, a_7, \Delta_2, \Delta_4, \Delta_6 > 0$
$n = 8$	$a_1, a_2, a_4, a_6, a_8, \Delta_3, \Delta_5, \Delta_7 > 0$, or $a_1, a_3, a_5, a_7, a_8, \Delta_3, \Delta_5, \Delta_7 > 0$

asymptotic stability. These regions are separated from the unstable regions by *stability boundaries.* As a point $\mathbf{p}$ traverses this boundary, from asymptotic stability to instability, at least one eigenvalue in the complex λ-plane must cross the $j\omega$-axis.

If the eigenvalue that crosses the $j\omega$-axis is real, then

$$a_n(\mathbf{p}) = 0 \tag{17}$$

(i.e., the crossing occurs at the origin). This follows from the fact that

$$a_n = \lambda_1\lambda_2 \cdots \lambda_n \tag{18}$$

On the other hand, if a pair of complex conjugate eigenvalues passes into the right half-plane (RHP), as $\mathbf{p}$ passes through the stability boundary, then

$$\Delta_{n-1}(\mathbf{p}) = 0 \tag{19}$$

This follows from Orlando's formula:

$$\Delta_{n-1} = (-1)^{n(n-1)/2} \prod_{\substack{i,k=1 \\ i<k}}^{n} (\lambda_i + \lambda_k) \tag{20}$$

This examination of possibilities leads to the following theorem:

Theorem A.6: *The stability boundaries in parameter space for the system* $\delta\dot{\mathbf{x}} = \mathbf{A}(\mathbf{p})\,\delta\mathbf{x}$ *must satisfy one of the following conditions*:

$$\begin{aligned} a_n(\mathbf{p}) &= 0 \\ \Delta_{n-1}(\mathbf{p}) &= 0 \end{aligned} \tag{21}$$

An example for two parameters p_1 and p_2 is shown in Fig. A.5.

Stability of Linear Time-Variable Systems

The definitions of stability in force for our present discussion (introduced in Section A.1) require an examination of *all* initial conditions $\delta\mathbf{x}_o$ within the neighborhood $\|\delta\mathbf{x}_o\| < \delta$, and they also require that the resulting perturbed

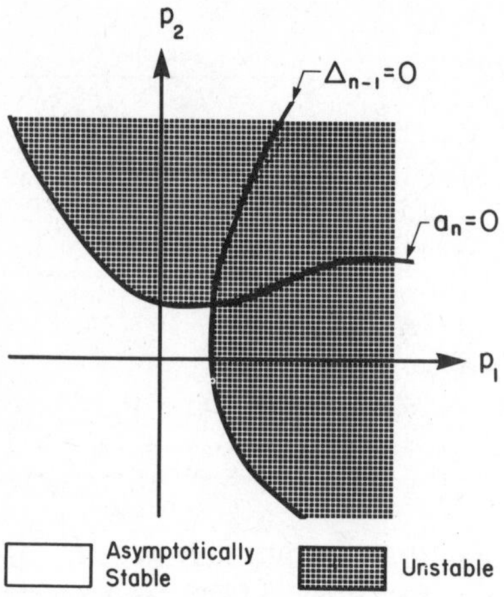

FIGURE A.5 Stability boundaries in the parameter plane.

solutions be examined for *all* $t > t_o$. Therefore, a strictly numerical investigation of stability requires in principle the examination of an infinite number of solutions, each over an infinite interval of time. The advantage of the linear approximation is that the number of solutions to be examined is no longer infinite but merely n—quite an improvement! This reduction follows from the fact that the solution to

$$\delta\dot{\mathbf{x}} = \mathbf{A}(t)\,\delta\mathbf{x}; \qquad \delta\mathbf{x}(t_o) = \delta\mathbf{x}_o \tag{22}$$

is

$$\delta\mathbf{x}(t) = \mathbf{\Phi}(t; t_o)\,\delta\mathbf{x}_o \tag{23}$$

where the $n \times n$ matrix $\mathbf{\Phi}$ satisfies

$$\dot{\mathbf{\Phi}}(t; t_o) = \mathbf{A}(t)\,\mathbf{\Phi}(t; t_o) \tag{24a}$$

$$\mathbf{\Phi}(t_o; t_o) = \mathbf{1} \tag{24b}$$

In other words, the n columns of $\mathbf{\Phi}$ represent n solutions to $\delta\dot{\mathbf{x}} = \mathbf{A}(t)\,\delta\mathbf{x}$ in terms of which any other solution can be written, as in (23). These n solutions are initially linearly independent; in fact, they are initially orthonormal—see (24b). Moreover, they *remain* linearly independent, as can be demonstrated from the fact that, from (24a),

$$\det\mathbf{\Phi}(t_2; t_o) = \det\mathbf{\Phi}(t_1; t_o)\exp\left\{\int_{t_1}^{t_2} \operatorname{trace}\mathbf{A}(t')\,dt'\right\} \tag{25}$$

after some application of linear algebra. In particular, with $t_2 = t$ and $t_1 = t_o$, (25) becomes

$$\det\mathbf{\Phi}(t; t_o) = \exp\left\{\int_{t_o}^{t} \operatorname{trace}\mathbf{A}(t')\,dt'\right\} \tag{26}$$

Now, the right-hand side of (26) is never zero for any finite time. Therefore, det $\mathbf{\Phi}$ is never zero, and the columns of $\mathbf{\Phi}$ remain always linearly independent.

Thus, we see the great virtue of linearity: it replaces the infinite requirement "all solutions with $\|\mathbf{x}_o\| < \delta$" by the finite requirement "n linearly independent solutions with $\|\mathbf{x}_o\| < \delta$." Indeed, from (23), it can be seen that

$$\|\delta\mathbf{x}(t)\| = \|\mathbf{\Phi}(t; t_o)\,\delta\mathbf{x}_o\| \leq \|\mathbf{\Phi}(t; t_o)\|\,\|\delta\mathbf{x}_o\| \tag{27}$$

Thus, if $\delta\mathbf{x}_o$ is chosen inside a sphere a radius δ, and if $\mathbf{\Phi}$ is bounded,

$$\|\mathbf{\Phi}(t; t_o)\| \leq \overline{\Phi} \tag{28}$$

we have

$$\|\delta\mathbf{x}(t)\| \leq \overline{\Phi}\delta \tag{29}$$

A requirement that $\delta\mathbf{x}$ must always lie inside a sphere of radius ε can clearly be met by choosing $\delta \leq \varepsilon/\overline{\Phi}$. For linear systems, then, Lagrange stability (boundedness) implies L-stability. However, as seen from earlier examples (such as Fig. A.4a), this connection disappears for nonlinear systems.

Asymptotic stability for (22) can be investigated by ascertaining whether, in addition to L-stability, it is true that

$$\lim_{t\to\infty} \|\delta\mathbf{x}(t)\| = 0 \tag{30}$$

As indicated by (27), the issue reduces to whether or not

$$\lim_{t\to\infty} \|\mathbf{\Phi}(t; t_o)\| = 0 \tag{31}$$

If so, then the origin is an asymptotically stable solution.

We summarize our findings in the form of the following two theorems:

Theorem A.7: *The solution $\delta\mathbf{x} \equiv \mathbf{0}$ of $\delta\dot{\mathbf{x}} = \mathbf{A}(t)\,\delta\mathbf{x}$ is L-stable if $\mathbf{\Phi}$ is bounded.*

Theorem A.8: *The solution $\delta\mathbf{x} \equiv \mathbf{0}$ of $\delta\dot{\mathbf{x}} = \mathbf{A}(t)\,\delta\mathbf{x}$ is asymptotically stable if it is L-stable and if, in addition, $\|\mathbf{\Phi}\| \to 0$ as $t \to \infty$.*

With respect to Theorem A.8, note that it has not been stated that $\|\delta\mathbf{x}\| \to 0$ *exponentially*, only *asymptotically*. The stronger condition of exponential asymptotic stability, taken for granted in stationary systems, is connected with the idea of *uniformity* (see Definition A.3).

Stability of Linear Periodic Systems

We have seen that linearity confers the benefit that only n solutions must be examined, instead of an infinite number. However in general, this examination must continue for all $t > t_o$. For linear periodic systems however, the infinite-time interval can be replaced by a T-interval, as will now be shown.

As a mnemonic, we shall use the symbol $\mathbf{P}$ instead of $\mathbf{A}$ when referring to periodic systems. Thus, the system under discussion is

$$\delta\dot{\mathbf{x}} = \mathbf{P}(t)\,\delta\mathbf{x} \tag{32a}$$

$$\mathbf{P}(t + T) \equiv \mathbf{P}(t) \tag{32b}$$

The several origins of periodic systems were identified in Section A.2.

The first point to recognize is that in general there is no closed-form solution to (32) for $n > 1$. (Several approximation techniques can be used however when the time dependence of $\mathbf{P}(t)$ has certain marked characteristics.) Nevertheless, periodicity is still a great boon (relative to general time variability) because, if n linearly independent solutions are known (even numerically) over one period, any solution for all subsequent time can be expressed in terms of them. Moreover, periodicity often occurs in the attitude dynamics of spinning systems; hence the study of linear periodic systems is of particular interest in the present context.

The key to the idea, due to Floquet, is to realize from (23) not only that $\mathbf{\Phi}(t; t_o)$ is a matrix solution to (32) but that $\mathbf{\Phi}(t + T; t_o)$ is as well. By definition,

$$\dot{\mathbf{\Phi}}(t + T; t_o) = \mathbf{P}(t + T)\mathbf{\Phi}(t + T; t_o) \tag{33}$$

(It matters not whether the overdot means differentiation with respect to $t + T$, or with respect to t.) From the periodicity of $\mathbf{P}$,

$$\dot{\mathbf{\Phi}}(t + T; t_o) = \mathbf{P}(t)\,\mathbf{\Phi}(t + T; t_o) \tag{34}$$

In other words, the columns of $\mathbf{\Phi}(t + T; t_o)$ satisfy (32a), and they must therefore be expressible as linear combinations of the columns of $\mathbf{\Phi}(t; t_o)$. Denoting the ith column of $\mathbf{\Phi}$ by $\boldsymbol{\phi}_i$, we have

$$\begin{aligned}\boldsymbol{\phi}_i(t + T; t_o) &= \sum_{j=1}^{n} m_{ji}\boldsymbol{\phi}_j(t; t_o)\\ &= [\boldsymbol{\phi}_1(t; t_o) \quad \cdots \quad \boldsymbol{\phi}_n(t; t_o)]\begin{bmatrix} m_{1i} \\ \vdots \\ m_{ni}\end{bmatrix}\\ &= \mathbf{\Phi}(t; t_o)\mathbf{m}_i\end{aligned} \tag{35}$$

(The definition of $\mathbf{m}_i$ is evident.) Collecting all n relations of the type (35), we have

$$\mathbf{\Phi}(t + T; t_o) = \mathbf{\Phi}(t; t_o)\mathbf{M} \tag{36}$$

with

$$\mathbf{M} \triangleq [\mathbf{m}_1 \quad \cdots \quad \mathbf{m}_n] \tag{37}$$

The role of the multiplicative matrix $\mathbf{M}$ is that the fundamental solutions (columns of $\mathbf{\Phi}$) over any given period can be found, from (36), in terms of the fundamental solutions over the previous period.

This idea can be extended indefinitely—to $t \to \infty$, in fact, as required for stability determinations. To do so, let τ be a time variable confined to the first period:

$$t_o \le \tau < t_o + T \tag{38}$$

Any $t > t_o$ can be expressed as

$$t = \tau + \nu T \tag{39}$$

where the integer ν is chosen in accordance with (38). Then,

$$\delta\mathbf{x}(t) = \mathbf{\Phi}(t; t_o)\,\delta\mathbf{x}_o = \mathbf{\Phi}(\tau; t_o)\mathbf{M}^{\nu}\,\delta\mathbf{x}_o \tag{40}$$

In this form, it is evident that the conditions for stability depend on the boundedness of $\mathbf{\Phi}$ over the first period $(0 \le \tau - t_o < T)$ and on whether the matrix factor $\mathbf{M}^\nu$ amplifies or diminishes the solution in subsequent periods.

Boundedness over the first period is never in doubt for the attitude dynamics applications in this book. The real issue is the growth or decay of $\mathbf{M}^\nu$. This depends on the eigenvalues of $\mathbf{M}$, which we denote by μ. Since $\mathbf{M}$ is a general real matrix, the μ_i $(i = 1, \ldots, n)$ are usually complex, and the growth of $\mathbf{M}^\nu$ depends on the values of $|\mu_i|$. Let

$$\bar{\mu} \triangleq \max_{i=1}^{n} |\mu_i| \tag{41}$$

Then, we have the following theorems:

Theorem A.9: *When the eigenvalues μ_i of $\mathbf{M}$ are distinct, the solution $\delta\mathbf{x} \equiv \mathbf{0}$ of $\delta\dot{\mathbf{x}} = \mathbf{P}(t)\,\delta\mathbf{x}$ is L-stable if $\bar{\mu} \le 1$.*

Theorem A.10: *The solution $\delta\mathbf{x} \equiv \mathbf{0}$ of $\delta\dot{\mathbf{x}} = \mathbf{P}(t)\,\delta\mathbf{x}$ is asymptotically stable if $\bar{\mu} < 1$.*

Theorem A.11: *The solution $\delta\mathbf{x} \equiv \mathbf{0}$ of $\delta\dot{\mathbf{x}} = \mathbf{P}(t)\,\delta\mathbf{x}$ is unstable if $\bar{\mu} > 1$.*

For the modification to Theorem A.9 when the μ_i are indistinct, see [Cesari]. Theorems A.9, A.10, and A.11 for linear periodic systems are analogous to Theorems A.1, A.2, and A.3 for linear stationary systems.

Further Results

Significant deductions concerning the stability of linear periodic systems can sometimes be made without even solving for $\mathbf{\Phi}$ over the first period. An important instance arises in connection with (26). Since

$$\mathbf{M} = \mathbf{\Phi}(t_o + T; t_o) \tag{42}$$

we have

$$\det \mathbf{M} \equiv \mu_1 \mu_2 \cdots \mu_n = \exp\left\{ \int_0^T \operatorname{trace} \mathbf{P}(\tau)\, d\tau \right\} \tag{43}$$

The following theorem can therefore be deduced:

Theorem A.12: *If* trace $\mathbf{P}(t)$ *averages to zero over one period, the solution $\delta\mathbf{x} \equiv \mathbf{0}$ of $\delta\dot{\mathbf{x}} = \mathbf{P}(t)\,\delta\mathbf{x}$ cannot be asymptotically stable.*

The proof rests on the observation that since the product of the eigenvalue magnitudes $|\mu_i|$ is unity, one cannot be less than unity unless another is greater than unity.

A second instance in which it is not necessary to solve for the eigenvalues of $\mathbf{M}$ occurs when the original (nonlinear) system is autonomous but has a periodic solution (see Section A.2):

$$\dot{\mathbf{y}}_T = \mathbf{g}(\mathbf{y}_T); \qquad \mathbf{y}_T(t + T) \equiv \mathbf{y}_T(t) \tag{44}$$

In this case,

$$\mathbf{P}(t) = \left[\frac{\partial \mathbf{g}}{\partial \mathbf{y}^T}\right]_{\mathbf{y}=\mathbf{y}_T(t)} \tag{45}$$

Differentiating the first part of (44), we see that

$$\ddot{\mathbf{y}}_T = \mathbf{P}(t)\dot{\mathbf{y}}_T \tag{46}$$

Furthermore, since $\mathbf{y}_T$ is periodic, $\dot{\mathbf{y}}_T$ is periodic also. This shows that at least one solution of $\delta\dot{\mathbf{x}} = \mathbf{P}(t)\,\delta\mathbf{x}$, namely, $\delta\mathbf{x} \equiv \dot{\mathbf{y}}_T$, is periodic. Asymptotic stability is thereby precluded. Thus, we have

Theorem A.13: *A periodic solution $\mathbf{y}_T(t)$ of the autonomous system $\dot{\mathbf{y}} = \mathbf{g}(\mathbf{y})$ cannot be asymptotically stable.*

It was in fact for just such solutions that the concept of orbital stability (Definitions A.9 and A.10) was introduced:

Theorem A.14: *If $n - 1$ of the characteristic multipliers $|\mu_i|$ associated with the periodic solution $\mathbf{y}_T(t)$ of the autonomous system $\dot{\mathbf{y}} = \mathbf{g}(\mathbf{y})$ are less than unity, then $\mathbf{y}_T(t)$ is asymptotically orbitally stable.*

For a proof, see [Coddington and Levinson].

An understanding of linear periodic systems is enhanced through a study of Mathieu's equation, which can be written in the form

$$\ddot{\theta} + (k_0 + k_1 \cos t)\theta = 0 \tag{47}$$

We shall not take the space to study this equation in detail here; its properties are discussed in many texts. Suffice it to say that the stability of the solution $\theta \equiv 0$ is not without some surprises. For example, the solution is unstable for arbitrarily small values of k_1 if k_0 takes on the critical values $m^2/4$, $m = 1, 2, 3, \ldots$. Even more surprising perhaps is that the solution can be stable for $k_0 < 0$, provided k_1 is selected in an appropriate range of values.

A.4 NONLINEAR INFERENCES FROM INFINITESIMAL STABILITY PROPERTIES

To recapitulate, the nonlinear system

$$\dot{\mathbf{x}} = \mathbf{A}(t)\mathbf{x} + \mathbf{n}(\mathbf{x}, t); \qquad \mathbf{n}(\mathbf{0}, t) = \mathbf{0} \tag{1}$$

with

$$\|\mathbf{n}(\mathbf{x}, t)\| = o(\|\mathbf{x}\|) \qquad \text{(uniformly in } t\text{)} \tag{2}$$

has been approximated in Section A.3 by the linear system

$$\delta\dot{\mathbf{x}} = \mathbf{A}(t)\,\delta\mathbf{x} \tag{3}$$

the latter being increasingly appropriate as $\delta\mathbf{x}$ becomes vanishingly small, $\delta\mathbf{x} \to \mathbf{0}$. As a matter of semantics, the stability properties of (3) are said to be the *infinitesimal* stability properties of (1). At issue in this section is the question of how the stability properties of (1) and (3) are related. Certainly, L-stability of the origin for (3) does not guarantee L-stability of the origin for (1), as can be seen from the simple example

$$\begin{aligned} \dot{x}_1 &= x_2 + \beta x_1 \left(x_1^2 + x_2^2 \right) \\ \dot{x}_2 &= -x_1 + \beta x_2 \left(x_1^2 + x_2^2 \right) \end{aligned} \tag{4}$$

In the linear approximation ($\dot{x}_1 = x_2, \dot{x}_2 = -x_1$), the origin is readily shown to be L-stable; but the origin for the nonlinear system (4) is unstable if $\beta > 0$, as seen from the fact that

$$\frac{d}{dt} \|\mathbf{x}\|_E = \beta \|\mathbf{x}\|_E^3 \tag{5}$$

This example also shows why a demonstration of infinitesimal stability for a nonlinear system, while much preferable to instability, is really only a first step in proving stability. The solution $\mathbf{x} \equiv \mathbf{0}$ of (4) is asymptotically stable if $\beta < 0$, but this cannot be seen from the linear approximation.

If a linear stationary approximation indicates *asymptotic* stability however, this property is retained by the nonlinear system:

Theorem A.15: *If the solution $\delta\mathbf{x} \equiv \mathbf{0}$ of the system $\delta\dot{\mathbf{x}} = \mathbf{A}\,\delta\mathbf{x}$ is asymptotically stable, so, too, is the solution $\mathbf{x} \equiv \mathbf{0}$ of the system $\dot{\mathbf{x}} = \mathbf{A}\mathbf{x} + \mathbf{n}(\mathbf{x})$, $\|\mathbf{n}(\mathbf{x})\| = o(\|\mathbf{x}\|)$.*

For a proof, see [Bellman]. Upon this theorem rests the justification for the ubiquitous use of the linear approximation in stability investigations of equilibria for stationary systems.

A similar result holds for time-variable systems:

Theorem A.16: *If the solution $\delta\mathbf{x} \equiv \mathbf{0}$ of the system $\delta\dot{\mathbf{x}} = \mathbf{A}(t)\,\delta\mathbf{x}$ is uniformly asymptotically stable, so, too, is the solution $\mathbf{x} \equiv \mathbf{0}$ of the system $\dot{\mathbf{x}} = \mathbf{A}(t)\mathbf{x} + \mathbf{n}(\mathbf{x}, t)$, where $\|\mathbf{n}(\mathbf{x}, t)\| = o(\|\mathbf{x}\|)$ uniformly in t.*

A proof for this theorem can be found in [Coddington and Levinson]. The uniformity conditions insisted upon by this theorem are not just mathematical frills. In particular, it is *not* true that if $\delta\mathbf{x} \equiv \mathbf{0}$ is asymptotically stable for $\delta\dot{\mathbf{x}} = \mathbf{A}(t)\,\delta\mathbf{x}$, then $\mathbf{x} \equiv \mathbf{0}$ is asymptotically stable for $\dot{\mathbf{x}} = \mathbf{A}(t)\mathbf{x} + \mathbf{n}(\mathbf{x}, t)$, as [Bellman] has shown by counterexample. Fortunately, for the specific class of time-variable linear systems of chief interest in this book—linear periodic systems—the periodicity guarantees uniformity.

The linearized system is also helpful in detecting instability.

Theorem A.17: *If the solution $\delta\mathbf{x} \equiv \mathbf{0}$ of the system $\delta\dot{\mathbf{x}} = \mathbf{A}(t)\,\delta\mathbf{x}$ is unstable, so, too, is the solution $\mathbf{x} \equiv \mathbf{0}$ of the system $\dot{\mathbf{x}} = \mathbf{A}(t)\mathbf{x} + \mathbf{n}(\mathbf{x}, t)$, where $\|\mathbf{n}(\mathbf{x}, t)\| = o(\|\mathbf{x}\|)$ uniformly in t.*

A proof is given by [Coddington and Levinson]. This theorem applies in particular to the two classes of linear approximation in which we are most interested: linear stationary systems and linear periodic systems.

What, then, has been lost in the linear approximation? At least three things. First, infinitesimal L-stability, while encouraging, does not guarantee L-stability of the original (nonlinear) system. Second, even though asymptotic stability of linear stationary systems and linear periodic systems guarantees asymptotic stability for their nonlinear counterparts, the *domain* of attraction to the origin cannot be determined from the linear approximation (see, for example, Fig. A.4*b*). Third, even though instability for the linear approximation guarantees instability for the original nonlinear system, the digressions from the origin may not be objectionable for engineering purposes if they are sufficiently bounded (e.g., if the stable limit cycle in Fig. A.4*a* is sufficiently close to the origin).

Fundamental to these distinctions are the concepts of *local* and *global* stability properties. (The more cumbersome and grammatically weird usages "in the small" and "in the large" are also encountered in the literature respectively for "local" and "global.") These properties depend upon the *domain of attraction*, defined as the set of initial conditions $\mathbf{x}_o$ such that $\mathbf{x}(t) \to 0$ as $t \to \infty$ (see Definition A.4). The domain of attraction for linear systems (including stationary and periodic ones) is always infinite. Linear systems, by their very nature, cannot distinguish (qualitatively) between "large" and "small" perturbations. Thus, the property of asymptotic stability is, for linear systems, always global. This property may or may not be global however for the associated nonlinear system. This is precisely the information that is lost in the linearization process.

The dual remark for instability is that even if the linearized system is unstable, a combination of weak repulsion contributed by the linear approximation and strong attraction contributed by the nonlinear terms may confine the "exact" solution to within an acceptable neighborhood of the origin.

A.5 LIAPUNOV'S METHOD

As explained in Section A.3, the problem with the stability definitions in Section A.1 is that even though they have proved to be the most useful ones from both mathematical and engineering standpoints, they do unfortunately seem to require that an infinitude of neighboring trajectories be examined, each for an infinite period of time. Section A.3 demonstrated that these stiff requirements can be relaxed to require that a set of n neighboring trajectories be examined for, at most, a finite period of time, to ascertain infinitesimal stability properties. Nonetheless, as made clear in Section A.4, important questions cannot be answered on the basis of the linearized system alone. These remaining issues can often be addressed with the aid of a method due to Liapunov. The method now to be discussed is often called *Liapunov's second method*, in contradistinction to *Liapunov's first method*, the latter being the linearization process discussed in Section A.3, whose major consequences—some of which were studied by Liapunov—were briefly sketched in Section A.4.

The elegant feature of Liapunov's method is reminiscent of the Routh—Hurwitz approach to the stability of linear stationary systems. In both these methods, the analyst is prepared to sacrifice knowledge of the exact solution (which may be available only numerically) in return for literal stability conditions. Liapunov's method is "direct" in that it is based on the differential equations themselves, not on their solutions *per se*.

To motivate the discussion, consider again the example in Section A.4 represented by (A.4, 4). The function

$$v(\mathbf{x}) \triangleq x_1^2 + x_2^2 \tag{1a}$$

has two properties that recommend themselves:

(a) $v(\mathbf{x}) > 0$ for all $\mathbf{x}$ except $\mathbf{x} = \mathbf{0}$.

(b) v satisfies

$$\dot{v} = 2\beta v^2 \tag{1b}$$

as v follows any solution $\mathbf{x}(t)$ of (A.4, 4).

These two properties suggest that the trajectory $\mathbf{x}(t)$ tends to move toward (or away from) the origin according to whether $\beta < 0$ (or $\beta > 0$). Thus, the origin is attractive if $\beta < 0$ and repulsive if $\beta > 0$. Note that this conclusion follows from the interaction of the system differential equations (A.4, 4) with the test function (1) and that it was arrived at without actually solving (A.4, 4).

Sign-Definite Functions

The above example is especially simple in its geometrical interpretation because the function v is just the squared distance from the origin. However, the idea of "distance" must now be substantially generalized, and this requires the notion of *positive-definiteness*:

***Definition A.11:* (*Positive-Definite Function*)** *A (continuously differentiable) function $v(\mathbf{x})$ is said to be positive-definite if $v(\mathbf{0}) = 0$ and $v(\mathbf{x}) > 0$ at every point* $\mathbf{x} \neq \mathbf{0}$.

Associated with positive-definiteness are the following similar definitions:

***Definition A.12:* (*Positive-Semidefinite Function*)** *Same as for a positive-definite function, except that $v(\mathbf{x}) > 0$ becomes $v(\mathbf{x}) \leq 0$.*

***Definition A.13:* (*Negative-Definite Function*)** *Same as for a positive-definite function, except that $v(\mathbf{x}) > 0$ becomes $v(\mathbf{x}) < 0$.*

***Definition A.14:* (*Negative-Semidefinite Function*)** *Same as for a negative-definite function, except that $v(\mathbf{x}) < 0$ becomes $v(\mathbf{x}) \leq 0$.*

The geometrical interpretation of a positive-definite function is that $v(\mathbf{x}) = c > 0$ corresponds to a closed hypersurface enclosing the origin.

Moreover, this hypersurface degenerates to the origin as $v(\mathbf{x}) \to 0$. This is precisely the property asked of $\mathbf{x}(t)$ if it is to be asymptotically stable. The necessary connection between the solution $\mathbf{x}(t)$ of the object system

$$\dot{\mathbf{x}} = \mathbf{f}(\mathbf{x}, t); \qquad \mathbf{f}(\mathbf{0}, t) \equiv \mathbf{0} \tag{2}$$

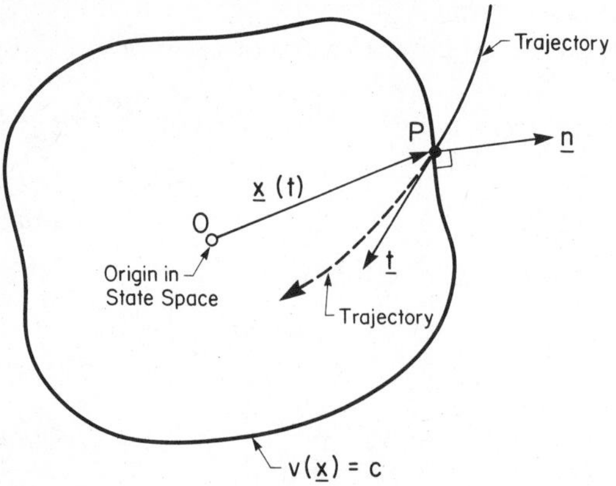

FIGURE A.6 Relationship between state trajectory and Liapunov surface.

and the $\mathbf{x}$ that can wander over the surface $v(\mathbf{x}) = c$ is clearly that $\mathbf{x}(t)$ should at all times lie on $v(\mathbf{x}) = c$. This connection imposes itself when the surface is forced to follow along as the solution $\mathbf{x}(t)$ evolves:

$$[\dot{v}(\mathbf{x}, t)]_{(2)} = \left[\left(\frac{\partial v}{\partial \mathbf{x}}\right)^T \dot{\mathbf{x}}\right]_{(2)} = \left(\frac{\partial v}{\partial \mathbf{x}}\right)^T \mathbf{f}(\mathbf{x}, t) \tag{3}$$

This idea is illustrated geometrically in Fig. A.6. At time t_1 the solution $\mathbf{x}(t_1)$ is shown, as is the particular surface of the family $v(\mathbf{x}) = c$ that contains $\mathbf{x}(t_1)$. The "trajectory" shown is the solution $\mathbf{x}(t)$. The unit tangent vector to this trajectory at the point P is $\mathbf{t} \triangleq \dot{\mathbf{x}}/\|\dot{\mathbf{x}}\| \equiv \mathbf{f}/\|\mathbf{f}\|$; the unit (outward) normal vector to the surface at P is $\mathbf{n} \triangleq (\partial v/\partial \mathbf{x})/\|\partial v/\partial \mathbf{x}\|$. The idea is that the solution trajectory must pierce the hypersurface in an inward direction, so that $\mathbf{n}^T\mathbf{t} < 0$. Comparison with (3) shows that this requirement is consistent with a shrinking value of v, that is $\dot{v} < 0$.

The Main Theorems

The geometrical ideas just developed are made precise in the following theorems, whose proofs may be found in [Hahn, 2].

Theorem A.18: (L-Stability) *The solution* $\mathbf{x} \equiv \mathbf{0}$ *of the system* $\dot{\mathbf{x}} = \mathbf{f}(\mathbf{x}), \mathbf{f}(\mathbf{0}) = \mathbf{0}$, *is L-stable if there exists a positive-definite function* $v(\mathbf{x})$ *such that* $(\partial v/\partial \mathbf{x})^T \mathbf{f}(\mathbf{x})$ *is negative-semidefinite or identically zero.*

Theorem A.19: (Asymptotic Stability) *The solution* $\mathbf{x} \equiv \mathbf{0}$ *of the system* $\dot{\mathbf{x}} = \mathbf{f}(\mathbf{x}), \mathbf{f}(\mathbf{0}) = \mathbf{0}$, *is asymptotically stable if there exists a positive-definite function* $v(\mathbf{x})$ *such that* $(\partial v/\partial \mathbf{x})^T \mathbf{f}(\mathbf{x})$ *is negative-definite.*

Theorem A.20: (Instability) *The solution* $\mathbf{x} \equiv \mathbf{0}$ *of the system* $\dot{\mathbf{x}} = \mathbf{f}(\mathbf{x}), \mathbf{f}(\mathbf{0}) = \mathbf{0}$, *is unstable if there exists a positive-definite or sign-indefinite function* $v(\mathbf{x})$ *such that* $(\partial v/\partial \mathbf{x})^T \mathbf{f}(\mathbf{x})$ *is positive-definite.*

Note that these theorems give *sufficient* conditions. Note also that when a chosen test function $v(\mathbf{x})$ doesn't satisfy the stated conditions, the stability question is still moot. If a test function $v(\mathbf{x})$ satisfies the conditions of one of these theorems, this successful test function is called a *Liapunov function*.

As a simple example of the technique, we revisit the example in the last section given by (A.4, 4), but now an additional term is added to the right-hand side of each equation:

$$\dot{x}_1 = \alpha x_1 + x_2 + \beta x_1\left(x_1^2 + x_2^2\right) \tag{4a}$$

$$\dot{x}_2 = -x_1 - \alpha x_2 + \beta x_2\left(x_1^2 + x_2^2\right)$$

The function

$$v(\mathbf{x}) = x_1^2 + 2\alpha x_1 x_2 + x_2^2 \tag{4b}$$

is selected as a candidate Liapunov function. Note that $\|\mathbf{x}\|_E$ does not suffice as a Liapunov function, as it did in (A.4, 5), so that the simple use of Euclidean distance as a metric is no longer successful; the new function $v(\mathbf{x})$ does not have this deficiency. The linear approximation to (4*a*) produces the characteristic equation $\lambda^2 + 1 - \alpha^2 = 0$, indicating instability if $|\alpha| > 1$ and infinitesimal stability if $|\alpha| < 1$. To resolve the stability question for finite perturbations when $|\alpha| < 1$, we use (4*b*), noting that v is positive-definite if $|\alpha| < 1$, and that the time derivative of v along any trajectory satisfying the governing differential equations is

$$\dot{v}|_{(4)} = 2\beta\left(x_1^2 + x_2^2\right)v \tag{5}$$

Obviously, this function is negative-definite for $\beta < 0$. Therefore, we conclude from Theorem A.19 that the origin is asymptotically stable if $|\alpha| < 1$ and $\beta < 0$. When this intelligence is combined with the earlier result (from the linear approximation) that $|\alpha| > 1$ corresponds to infinitesimal instability and Theorem A.17 is applied, the stability characteristics are known for all $\{\alpha, \beta\}$; these are summarized in Fig. A.7.

Obviously, the focus of the investigation is finding a Liapunov function. In the last example, this process was atypically simple; more often, it is very difficult. The search for Liapunov functions is often simplified for mechanical systems by appealing to integrals of the motion, usually an energy integral. For example, the system

$$\ddot{q} + k(q) = 0 \tag{6}$$

when rewritten in state coordinates ($x_1 = q$; $x_2 = \dot{q}$) becomes

$$\dot{x}_1 = x_2; \qquad \dot{x}_2 = -k(x_1) \tag{7}$$

The kinetic energy is $\frac{1}{2}x_2^2$, and the potential energy is

$$V(x_1) = \int_0^{x_1} k(q)\,dq \tag{8}$$

Therefore, the Liapunov function

$$v(\mathbf{x}) = V(x_1) + \tfrac{1}{2}x_2^2 \tag{9}$$

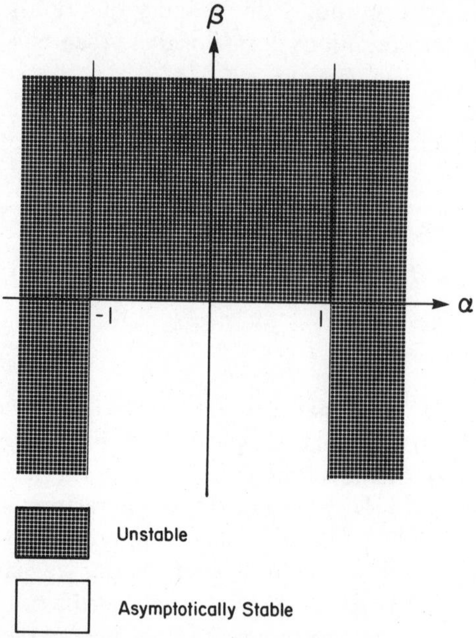

FIGURE A.7 Stability diagram for the system given by equation (A.5, 4).

can easily be shown to be constant for system (6). Looking at Theorem A.18, we can show L-stability if $v(\mathbf{x})$ is positive-definite. This will be the case in some neighborhood of the origin provided that, for some $a > 0$,

$$qk(q) > 0; \qquad 0 < |q| < a \tag{10}$$

The power of the method is exemplified by the generality of (6) and (10); note also that the absence of a general solution to (6) was no obstacle in the search for the stability condition (10). Many further examples of the use of an energy integral as a Liapunov function can be found in the main text of this book.

Extensions to the Main Theorems

Lest the reader suppose that the "main theorems" constitute a complete set of tools for treating differential systems, or for mechanical systems at least, the following elementary example will give pause:

$$\ddot{q} + 2\zeta\omega\dot{q} + \omega^2 q = 0 \tag{11}$$

The natural state variables for this damped harmonic oscillator are $x_1 \triangleq q$ and $x_2 \triangleq \dot{q}$, and the global asymptotic stability of this system for $\zeta > 0$, $\omega^2 > 0$ is one of the best-known facts in dynamics. Yet Theorem A.19 grinds to a halt when the energy integral

$$v(\mathbf{x}) = \tfrac{1}{2}\omega^2 x_1^2 + \tfrac{1}{2}x_2^2 \tag{12}$$

is used as a Liapunov function because, although v is positive-definite for $\omega^2 > 0$, its trajectory-fixed derivative

$$\dot{v} = -2\zeta\omega x_2^2$$

is not negative-definite in $\mathbf{x}$. Theorem A.18 guarantees L-stability, but one can hardly settle for that.

The problem is that Theorem A.19 requires $v(\mathbf{x})$ to *continuously* decrease to the origin, whereas the "attraction" property (Definition A.4) requires only that $v(\mathbf{x})$ *eventually* decrease to the origin. Therefore, an improvement to Theorem A.19 is the following:

Theorem A.21: *The solution* $\mathbf{x} \equiv \mathbf{0}$ *of the system* $\dot{\mathbf{x}} = \mathbf{f}(\mathbf{x})$, $\mathbf{f}(\mathbf{0}) = \mathbf{0}$, *is asymptotically stable if there exists a positive-definite function* $v(\mathbf{x})$ *such that* $(\partial v/\partial \mathbf{x})^T\mathbf{f}(\mathbf{x})$ *is negative-semidefinite and such that there is no non-null solution such that* $(\partial v/\partial \mathbf{x})^T\mathbf{f}(\mathbf{x})$ *is identically zero.*

A proof can be found in [Kalman and Bertram].

This theorem resolves the dilemma of why such a powerful method as Liapunov's apparently falters with such a simple system as a damped harmonic oscillator. Although $\dot{v}$ given by (12) is not negative-definite, as required by Theorem A.19, it is also true that $x_2 \triangleq \dot{q} \equiv 0$ is not a solution of (11) unless $x_1 \triangleq q \equiv 0$ also. Therefore, by Theorem A.21, a damped harmonic oscillator is asymptotically stable. This example is used of course for pedagogical purposes only: one does not need to call upon Liapunov's method for linear-stationary systems.

An improvement due to [Chetayev] can also be made to the main instability theorem, Theorem A.20:

Theorem A.22: *Let* $\mathcal{N}$ *be a neighborhood of the origin, and let* $\mathcal{R}$ *be a region in* $\mathcal{N}$ *such that the origin is a boundary point of* $\mathcal{R}$. *Then, the solution* $\mathbf{x} \equiv \mathbf{0}$ *of the system* $\dot{\mathbf{x}} = \mathbf{f}(\mathbf{x})$, $\mathbf{f}(\mathbf{0}) = \mathbf{0}$, *is unstable if there exists a function* $v(\mathbf{x})$ *that is positive-definite in* $\mathcal{R}$, *with* $v(\mathbf{x}) = 0$ *on the boundary of* $\mathcal{R}$, *such that* $(\partial v/\partial \mathbf{x})^T\mathbf{f}(\mathbf{x}) > 0$ *in* $\mathcal{R}$.

The theorem is geometrically self-evident (Fig. A.8). Any nonzero initial conditions in $\mathcal{R}$ will produce a trajectory that must exit through the outer boundary of

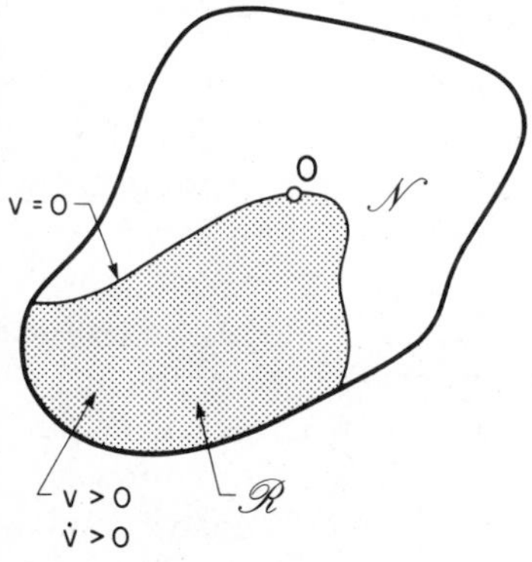

FIGURE A.8 Neighborhood in state space, used in Theorem A.22.

$\mathscr{R}$ and therefore cannot be confined to within an ε-sphere of the origin. Therefore, the origin is not L-stable.

Among the best examples of the use of Theorem A.22 is the demonstration that the intermediate-axis spin of a rigid body in torque-free motion is unstable (Section 4.4). Theorems similar to Theorems A.18 through A.22 are also known ([Hahn, 2]) for the nonautonomous case $\dot{\mathbf{x}} = \mathbf{f}(\mathbf{x}, t)$, $\mathbf{f}(\mathbf{0}, t) \equiv \mathbf{0}$; we shall not introduce them, however, because they are not needed in this book.

A.6 STABILITY OF LINEAR STATIONARY MECHANICAL SYSTEMS

It would be best to explain at once the two new ideas in the title of this section. First, we speak for the first time of the stability of a *system*, rather than the more restricted idea of the stability of a *particular solution* of a system (as called for in the Definitions A.1 through A.10). The reason for this terminology can be gleaned from the following theorem:

Theorem A.23: *All solutions of the linear system* $\delta\dot{\mathbf{x}} = \mathbf{A}(t)\,\delta\mathbf{x}$ *have the same stability properties as the null solution* $\delta\mathbf{x} \equiv \mathbf{0}$.

The proof of this important fact is elementary. That this property of linear systems does not hold for nonlinear systems can be seen, for example, in Fig. A.4.

Second, we wish to be precise about the term *linear mechanical system*. In fact, the term *linear dynamical system* is more appealing, but *dynamical system* has been appropriated by system theorists in reference to a much more general class of systems. By a linear mechanical system we shall in this section mean a stationary matrix-second-order system of the form

$$\mathscr{M}\ddot{\mathbf{q}} + (\mathscr{D} + \mathscr{G})\dot{\mathbf{q}} + (\mathscr{K} + \mathscr{A})\mathbf{q} = \mathscr{f}(t) \tag{1}$$

which possesses the following properties:

$$\mathscr{M}^T = \mathscr{M} > 0 \tag{2}$$

$$\mathscr{D}^T = \mathscr{D} \not< 0 \tag{3}$$

$$\mathscr{G}^T = -\mathscr{G} \tag{4}$$

$$\mathscr{K}^T = \mathscr{K}; \qquad \mathscr{A}^T = -\mathscr{A} \tag{5}$$

That is, $\mathscr{M}$ is symmetric and positive-definite, $\mathscr{D}$ is symmetric and non-negative-definite, $\mathscr{G}$ and $\mathscr{A}$ are skew-symmetric, and $\mathscr{K}$ is symmetric. The coefficient matrices $\{\mathscr{M}, \mathscr{D}, \mathscr{G}, \mathscr{K}, \mathscr{A}\}$ are respectively associated with inertial, damping, gyric, stiffness, and constraint damping forces, and $\mathscr{f}(t)$ represents any other generalized forces acting on the system. The N elements of $\mathbf{q}$ are physical displacements.

The reason that (1) qualifies as more than just a general matrix-second-order system, and as a "mechanical" system, lies in the properties (2) through (5). Certainly, (2) gives $\mathscr{M}$ two properties that distinguish it from a general real matrix. It may at first appear that $\mathscr{D} + \mathscr{G}$ is a general real matrix, inasmuch as any real matrix can be expressed as the sum of a symmetric matrix and a skew-symmetric matrix. The fact that $\mathscr{D}$ is nonnegative-definite ($\mathscr{D} \nless 0$), however, places a restriction on $\mathscr{D} + \mathscr{G}$. Moreover, the completely different physical origins of the terms $\mathscr{D}\dot{\mathbf{q}}$ and $\mathscr{G}\dot{\mathbf{q}}$ justify their separate identification, and this observation applies as well to $\mathscr{K}\mathbf{q}$ and $\mathscr{A}\mathbf{q}$. A general reference book on this class of system is that of [Huseyin].

It is possible of course to convert (1) into a general linear stationary system of the form $\dot{\mathbf{x}} = \mathbf{A}\mathbf{x}$ studied in the first part of Section A.3, by defining

$$\mathbf{x} \triangleq \begin{bmatrix} \mathbf{q} \\ \dot{\mathbf{q}} \end{bmatrix}; \qquad \mathbf{A} \triangleq \begin{bmatrix} \mathbf{O} & \mathbf{1} \\ -\mathscr{M}^{-1}(\mathscr{K} + \mathscr{A}) & -\mathscr{M}^{-1}(\mathscr{D} + \mathscr{G}) \end{bmatrix} \tag{6}$$

after which Theorems A.1 through A.6 are applicable. The intent here, in contrast, is to exploit the conditions (2) through (5) [or, equivalently, the special form of $\mathbf{A}$ in (6)] to arrive at stability results tailored to the special properties of linear mechanical systems.

"Rigid" Systems

The simplest instance of equation (1) has all matrices absent except $\mathscr{M}$. The resulting autonomous system, $\mathscr{M}\ddot{\mathbf{q}} = \mathbf{0}$, is so mathematically trivial it would not be worth mentioning were it not so fundamental to vehicle dynamics. Because its solution is of form $\mathbf{q}(t) = \mathbf{a} + \mathbf{b}t$, we deduce the following theorem:

Theorem A.24: *The system* $\mathscr{M}\ddot{\mathbf{q}} = \mathbf{0}$, $\mathscr{M}^T = \mathscr{M} > 0$, *is unstable.*

Rigid Gyric* Systems

A system of the form

$$\mathscr{M}\ddot{\mathbf{q}} + \mathscr{G}\dot{\mathbf{q}} = \mathbf{0} \tag{7}$$

with $\mathscr{M}^T = \mathscr{M} > 0$ and $\mathscr{G}^T = -\mathscr{G}$, occurs in connection with spinning rigid systems. In spite of its relative simplicity, it is not immediately obvious what the stability conditions for this system are. By means of a nonsingular transformation, $\mathbf{q} = \mathbf{T}_1\mathbf{q}_1$, however, (7) can be transformed ([Hadley]) to become

$$\ddot{\mathbf{q}}_1 + \mathscr{G}_1\mathbf{q}_1 = \mathbf{0} \tag{8}$$

with the property $\mathscr{G}_1^T = -\mathscr{G}_1$ left intact, since $\mathscr{G}_1 = \mathbf{T}_1^T\mathscr{G}\mathbf{T}_1$. Next, an orthonormal transformation, $\mathbf{q}_1 = \mathbf{T}_2\mathbf{q}_2$, is employed ([Greub]) to convert equation (8) to

$$\ddot{\mathbf{q}}_2 + \mathscr{G}_2\mathbf{q}_2 = \mathbf{0} \tag{9}$$

*The word *gyroscopic* is normally used here. Unfortunately, this term is quite defective, unless used in its proper reference to the noun *gyroscope*. The syllable *gyro-* is operative; the syllable *-scope* is specious. We coin the more accurate and appropriate term *gyric* to refer to the present type of torque. The associated noun is then *gyricity*.

where $\mathscr{G}_2 = \mathbf{T}_2^T \mathscr{G}_1 \mathbf{T}_2$ has the canonical form

$$\mathscr{G}_2 = \begin{bmatrix} \mathbf{O} & \mathbf{O} \\ \mathbf{O} & \mathscr{G}_{2P} \end{bmatrix} \tag{10}$$

$$\mathscr{G}_{2P} = \text{diag}\{\mathbf{S}_1, \dots, \mathbf{S}_P\} \tag{11}$$

$$\mathbf{S}_p = \begin{bmatrix} 0 & -\omega_p \\ \omega_p & 0 \end{bmatrix} \qquad (\omega_p \neq 0), (p = 1, \dots, P) \tag{12}$$

Obviously, $\mathbf{S}_p^T = -\mathbf{S}_p$, $p = 1, \dots, P$. Therefore, $\mathscr{G}_{2P}^T = -\mathscr{G}_{2P}$ and $\mathscr{G}_2^T = -\mathscr{G}_2$. Thus, (9) through (12) is the canonical form of (7). It is evident that (9) comprises two subsystems. One is a rigid nongyric system,

$$\ddot{\mathbf{q}}_{r2} = \mathbf{0} \tag{13}$$

The other subsystem is a set of P 2×2 uncoupled rigid gyric systems,

$$\ddot{\mathbf{q}}_{p2} + \mathbf{S}_p \dot{\mathbf{q}}_{p2} = \mathbf{0} \qquad (p = 1, \dots, P) \tag{14}$$

The former subsystem is unstable, while the latter is stable (this can readily be shown from its elementary solution). Therefore, the key to stability must be the absence of the $(N - 2P) \times (N - 2P)$ zero matrix in the upper left partition of (10). This absence can be simply tested by calculation of

$$\begin{aligned} \det \mathscr{G}_2 &= 0^{N-2P} \det \mathscr{G}_{2P} \\ &= 0^{N-2P} \prod_{p=1}^{P} \det \mathbf{S}_p \\ &= 0^{N-2P} \prod_{p=1}^{P} \omega_p^2 \end{aligned} \tag{15}$$

Thus, the necessary and sufficient condition that (7) be stable is that $\det \mathscr{G}_2 \neq 0$. However,

$$\begin{aligned} \det \mathscr{G}_2 &= \det \mathbf{T}_2^T \det \mathscr{G}_1 \det \mathbf{T}_2 \\ &= \det \mathscr{G}_1 \\ &= \det \mathbf{T}_1^T \det \mathscr{G} \det \mathbf{T}_1 \end{aligned} \tag{16}$$

Certainly, $\det \mathbf{T}_1^T = \det \mathbf{T}_1 \neq 0$. Therefore, the condition reduces to $\det \mathscr{G} \neq 0$ and we have the following theorem:

> ***Theorem A.25:*** *The system* $\mathscr{M}\ddot{\mathbf{q}} + \mathscr{G}\dot{\mathbf{q}} = \mathbf{0}$, $\mathscr{M}^T = \mathscr{M} > 0$, $\mathscr{G}^T = -\mathscr{G}$ *is stable if and only if* $\det \mathscr{G} \neq 0$.

As a corollary, all rigid gyric systems with an odd number of coordinates (N odd) are unstable.

Conservative Systems

A system of the form

$$\mathscr{M}\ddot{\mathbf{q}} + \mathscr{K}\mathbf{q} = \mathbf{0} \tag{17}$$

with $\mathcal{M}^T = \mathcal{M} > 0$ and $\mathcal{K}^T = \mathcal{K}$ will be called a *conservative* system. It is one of the most important and best understood systems in dynamics. Just as with the rigid system of Theorem A.24 and the rigid gyric system of Theorem A.25, asymptotic stability is out of the question.

Definition A.15: (Static Stability) *A linear mechanical system that includes a term* $\mathcal{K}\mathbf{q}$, $\mathcal{K}^T = \mathcal{K}$, *is said to be statically stable if* $\mathcal{K} > 0$.

Some writers choose to define static stability in terms of the more limited requirement $\det \mathcal{K} > 0$ because the condition $\det \mathcal{K} = 0$ can be interpreted as the *static stability boundary* (recall Theorem A.6). In any case, the conditions for stability of (17) are well known:

Theorem A.26: *The system* $\mathcal{M}\ddot{\mathbf{q}} + \mathcal{K}\mathbf{q} = \mathbf{0}$, $\mathcal{M}^T = \mathcal{M} > 0$, $\mathcal{K}^T = \mathcal{K}$, *is stable if and only if it is statically stable.*

Two proofs will be given for this important system, both of which are brief and instructive.

The first proof rests on the existence of a real nonsingular transformation $\mathcal{E}$ such that

$$\mathcal{E}^T\mathcal{M}\mathcal{E} = \mathbf{1}; \qquad \mathcal{E}^T\mathcal{K}\mathcal{E} = \hat{\mathcal{K}} \tag{18}$$

where $\hat{\mathcal{K}}$ is diagonal (see [Parlett]). The determination of $\mathcal{E}$ and $\hat{\mathcal{K}}$ is called the *general symmetric eigenvalue problem*. It is therefore evident that system (17) reduces to a set of uncoupled equations of the form

$$\ddot{\eta}_\alpha + \hat{k}_\alpha \eta_\alpha = 0 \qquad (\alpha = 1, \ldots, N) \tag{19}$$

where $\hat{k}_\alpha$ $(\alpha = 1, \ldots, N)$ are the diagonal elements of $\hat{\mathcal{K}}$. If one of the $\hat{k}_\alpha < 0$, the corresponding subsystem is exponentially unstable; system (17) is therefore also unstable in this case. If one of the $\hat{k}_\alpha = 0$, the corresponding subsystem is again unstable since it has a term $\sim t$; again (17) is unstable. If however $\omega_\alpha^2 \triangleq \hat{k}_\alpha > 0$ for all α, all the subsystems (19) are stable, and so, too, is system (17). The necessary and sufficient conditions that all $\hat{k}_\alpha > 0$ is that $\mathcal{K} > 0$.

For the second proof, we select the total system energy

$$v(\mathbf{q}, \dot{\mathbf{q}}) = \tfrac{1}{2}\dot{\mathbf{q}}^T\mathcal{M}\dot{\mathbf{q}} + \tfrac{1}{2}\mathbf{q}^T\mathcal{K}\mathbf{q} \tag{20}$$

as a Liapunov function and use Theorem A.18. Since $\dot{v} = 0$ for solutions that satisfy (17), and since v is positive-definite if $\mathcal{K} > 0$, stability can be inferred. (Note that $\mathcal{K} > 0$ has only been shown to be a sufficient condition for stability by this argument. However, Liapunov technique can be used also to show necessity, as well.)

Effect of Damping

If a damping term $\mathcal{D}\dot{\mathbf{q}}$ is added to (17), we have

$$\mathcal{M}\ddot{\mathbf{q}} + \mathcal{D}\dot{\mathbf{q}} + \mathcal{K}\mathbf{q} = \mathbf{0} \tag{21}$$

with $\mathcal{M}^T = \mathcal{M} > 0$, $\mathcal{K}^T = \mathcal{K}$, $\mathcal{D}^T = \mathcal{D} \nless 0$; asymptotic stability is now a possibility. The matter turns on the sign-definiteness properties $\mathcal{D}$ and $\mathcal{K}$.

To continue the discussion, the following definitions are required:

Definition A.16: (Complete Damping) *A linear mechanical system that includes a term* $\mathscr{D}\dot{\mathbf{q}}, \mathscr{D}^T = \mathscr{D}$, *is said to be completely damped if* $\mathscr{D} > 0$.

Definition A.17: (Pervasive Damping) *A linear mechanical system that includes a term* $\mathscr{D}\dot{\mathbf{q}}, \mathscr{D}^T = \mathscr{D}$, *is said to be pervasively damped if and only if, for all solutions* $\mathbf{q} \not\equiv \mathbf{0}$, $\dot{\mathbf{q}}^T\mathscr{D}\dot{\mathbf{q}} \geq 0$ *and* $\mathscr{D}\dot{\mathbf{q}} \not\equiv \mathbf{0}$.

Note that complete damping is a property of the matrix $\mathscr{D}$ alone, whereas pervasive damping is a property of the system as a whole. For (21) in particular, pervasive damping is a property of the $\{\mathscr{M}, \mathscr{D}, \mathscr{K}\}$ triplet.

With "complete" damping, even if any set of $N - 1$ coordinates among $\{q_1, \ldots, q_N\}$ were constrained not to move, damping would still take place. This leads to the following theorem:

Theorem A.27: *The system* (21), *with* $\mathscr{M}^T = \mathscr{M} > 0$, $\mathscr{D}^T = \mathscr{D}$, $\mathscr{K}^T = \mathscr{K}$, *is asymptotically stable if it is statically stable and completely damped.*

The proof rests on the total energy (20) being a positive-definite function $\{\mathbf{q}, \dot{\mathbf{q}}\}$. The energy changes according to

$$\dot{v} = -\dot{\mathbf{q}}^T\mathscr{D}\dot{\mathbf{q}} \tag{22}$$

Though not negative-definite in $\{\mathbf{q}, \dot{\mathbf{q}}\}$ as required by Theorem A.19, the complete damping condition $\mathscr{D} > 0$ ensures that $\dot{v} \equiv 0$ implies that $\dot{\mathbf{q}} \equiv 0$. This can happen only if $\mathbf{q} \equiv \mathbf{0}$ also, as is clear from (21). Therefore, by Theorem A.21, the system is asymptotically stable.

Although this theorem is elegant and has a relatively simple proof, it gives only sufficient conditions for asymptotic stability. In many instances involving space vehicles, these conditions ($\mathscr{D} > 0$, $\mathscr{K} > 0$) are not met. This is where pervasive damping (Definition A.17) is indispensable. Pervasive damping means that energy is dissipated in *all* possible motions. A system with 100 masses and 100 springs can be pervasively damped with only one dashpot if all characteristic motions of the system involve the damper. This requirement is stated mathematically by the following theorem:

Theorem A.28: *The system* (21) *with* $\mathscr{M}^T = \mathscr{M} > 0$, $\mathscr{D}^T = \mathscr{D}$, $\mathscr{K}^T = \mathscr{K}$, *is asymptotically stable if it is statically stable, with* $\mathscr{D} \not< 0$ *and*

$$\operatorname{rank}\left[\mathscr{D} \quad \mathscr{K}\mathscr{M}^{-1}\mathscr{D} \quad (\mathscr{K}\mathscr{M}^{-1})^2\mathscr{D} \quad \cdots \quad (\mathscr{K}\mathscr{M}^{-1})^{N-1}\mathscr{D}\right] = N \tag{23}$$

The proof ([Walker and Schmitendorf]) still rests on the energy (20) as a Liapunov function and on (22) as its rate of decrease. If the damping is complete of course, $\mathscr{D} > 0$ and $\operatorname{rank}\mathscr{D} = N$, which immediately satisfies (23). Theorem A.28 generalizes Theorem A.27 to cases where the damping is not complete but is nonetheless pervasive. In fact, (23) is the necessary and sufficient condition for pervasive damping. According to Theorem A.21, the issue is whether $\dot{\mathbf{q}}^T\mathscr{D}\dot{\mathbf{q}} \equiv 0$, apart from the quiescent state, $\mathbf{q} \equiv \mathbf{0}$. A critical observation is that $\dot{\mathbf{q}}^T\mathscr{D}\dot{\mathbf{q}} = 0$ implies $\mathscr{D}\dot{\mathbf{q}} = \mathbf{0}$. However, since $\mathscr{D}$ is no longer positive-definite, this does not in turn imply by itself that $\mathbf{q} \equiv \mathbf{0}$. Appeal must be made to the system itself. The question is whether

$$\mathcal{M}\ddot{\mathbf{q}} + \mathcal{D}\dot{\mathbf{q}} + \mathcal{K}\mathbf{q} = \mathbf{0}; \qquad \mathcal{D}\dot{\mathbf{q}} \equiv \mathbf{0} \tag{24}$$

implies that $\mathbf{q} \equiv \mathbf{0}$. An equivalent (and simpler) question is whether

$$\mathcal{M}\ddot{\mathbf{q}} + \mathcal{K}\mathbf{q} = \mathbf{0}; \qquad \mathcal{D}\dot{\mathbf{q}} \equiv \mathbf{0} \tag{25}$$

implies that $\mathbf{q} \equiv \mathbf{0}$. Readers familiar with multivariable control theory will recognize that this is equivalent to asking whether the system

$$\frac{d}{dt}\begin{bmatrix}\mathbf{q}\\ \dot{\mathbf{q}}\end{bmatrix} = \begin{bmatrix}\mathbf{O} & \mathbf{1}\\ -\mathcal{M}^{-1}\mathcal{K} & \mathbf{O}\end{bmatrix}\begin{bmatrix}\mathbf{q}\\ \dot{\mathbf{q}}\end{bmatrix} \tag{26}$$

is *observable* via the measurements $\mathcal{D}\dot{\mathbf{q}}$ (i.e., via measurements of all the generalized damping forces). At this point, then, the proof transfers to a proof of observability. Application of the rank test for observability produces (23).

A physical interpretation of (23) can be achieved by applying it to the transformed version of (21) in which (undamped) modal coordinates are employed. If one sets $\mathbf{q} = \mathcal{E}\boldsymbol{\eta}$, with $\mathcal{E}^T\mathcal{M}\mathcal{E} = \mathbf{1}$, $\mathcal{E}^T\mathcal{K}\mathcal{E} = \boldsymbol{\Omega}^2 > 0$, and $\hat{\mathcal{D}} = \mathcal{E}^T\mathcal{D}\mathcal{E}$, then Theorem A.28 still applies, with $\mathcal{M} \to \mathbf{1}$, $\mathcal{K} \to \boldsymbol{\Omega}^2$, $\mathcal{D} \to \hat{\mathcal{D}}$. Since

$$\operatorname{rank}\,[\hat{\mathcal{D}} \quad \boldsymbol{\Omega}^2\hat{\mathcal{D}} \quad \boldsymbol{\Omega}^4\hat{\mathcal{D}} \quad \cdots \quad \boldsymbol{\Omega}^{2N-2}\hat{\mathcal{D}}] = N \tag{27}$$

where

$$\boldsymbol{\Omega}^2 = \operatorname{diag}\{\omega_1^2, \ldots, \omega_N^2\} \tag{28}$$

is equivalent (for distinct frequencies) to requiring that no row of $\hat{\mathcal{D}}$ be zero, (27) requires in effect that there be no mode for which $\ddot{\eta}_\alpha + \omega_\alpha^2\eta_\alpha = 0$, that is, that there be no undamped modes.

Conservative Gyric Systems

A system of the form

$$\mathcal{M}\ddot{\mathbf{q}} + \mathcal{G}\dot{\mathbf{q}} + \mathcal{K}\mathbf{q} = \mathbf{0} \tag{29}$$

with $\mathcal{M}^T = \mathcal{M} > 0$, $\mathcal{G}^T = -\mathcal{G}$, and $\mathcal{K}^T = \mathcal{K}$ will be called a conservative gyric system. The terminology conservative is appropriate because the system energy, given by (20), is constant for (29). Therefore, we have, from Theorem A.18, the following result:

Theorem A.29: *The system* (29), *with* $\mathcal{M}^T = \mathcal{M} > 0$, $\mathcal{G}^T = -\mathcal{G}$, $\mathcal{K}^T = \mathcal{K}$, *is stable if it is statically stable.*

Note in comparison with Theorem A.26 however that when the gyric term $\mathcal{G}\dot{\mathbf{q}}$ is added, static stability, while still a sufficient condition for stability, is no longer claimed to be a necessary one.

It is quickly seen that asymptotic stability for system (29) is impossible because the characteristic equation

$$p_{2N}(s) \triangleq \det(\mathcal{M}s^2 + \mathcal{G}s + \mathcal{K}) = 0 \tag{30}$$

is in fact a polynomial in s^2:

$$\begin{aligned} p_{2N}(-s) &= \det(\mathscr{M}s^2 - \mathscr{G}s + \mathscr{K}) \\ &= \det(\mathscr{M}^T s^2 + \mathscr{G}^T s + \mathscr{K}^T) \\ &= \det(\mathscr{M}s^2 + \mathscr{G}s + \mathscr{K})^T \\ &= \det(\mathscr{M}s^2 + \mathscr{G}s + \mathscr{K}) = p_{2N}(s) \end{aligned} \tag{31}$$

so that, if s is a root, $-s$ is also a root. Note that this property rules out asymptotic stability, since if one root is in the LHP, another root must be in the RHP. For stability, all roots must be of the form $s = j\omega$.

A very simple system of this type is

$$\begin{aligned} \ddot{q}_1 - g\dot{q}_2 + k_1 q_1 &= 0 \\ \ddot{q}_2 + g\dot{q}_1 + k_2 q_2 &= 0 \end{aligned} \tag{32}$$

whose characteristic equation is

$$s^4 + (k_1 + k_2 + g^2)s^2 + k_1 k_2 = 0 \tag{33}$$

The Routh–Hurwitz tests (for asymptotic stability) are not applicable, but in this simple case the roots of the characteristic equation (33), written for brevity as

$$s^4 + b_1 s^2 + b_2 = 0 \tag{34}$$

can be directly found. Instead of requiring that the s-roots lie on the imaginary axis, it is easier to require, equivalently, that the s^2-roots lie on the negative real axis. The necessary and sufficient conditions for this to happen are

$$b_1 > 0 \tag{35a}$$

$$b_2 > 0 \tag{35b}$$

$$b_1^2 - 4b_2 > 0 \tag{35c}$$

as can be seen from the solution to (34). For (33), then,

$$k_1 + k_2 + g^2 > 0 \tag{36a}$$

$$k_1 k_2 > 0 \tag{36b}$$

$$\Xi \triangleq (k_1 - k_2)^2 + 2g^2(k_1 + k_2) + g^4 > 0 \tag{36c}$$

These conditions are plotted in the k_1–k_2 plane in Fig. A.9. The second and fourth quadrants are eliminated by (36b), and all points southwest of the line (36a) are also unstable. The most interesting condition is (36c), which translates into a parabolic region in the third quadrant, skewed at 45° to the axes, and touching the axes at $(-g^2, 0)$ and $(0, -g^2)$. All points within this parabola are also unstable. Note however the region next to the origin in the third quadrant which, although statically unstable, is stabilized by the gyric forces. With g sufficiently large, then, some statically unstable conservative systems can be gyrically stabilized.

This observation is now extended to the general system (29). It is necessary and sufficient that all zeros of the polynomial

$$f_N(\mu) \triangleq \det\left[\mathscr{M}\mu + \mathscr{G}\sqrt{\mu} + \mathscr{K}\right] \tag{37}$$

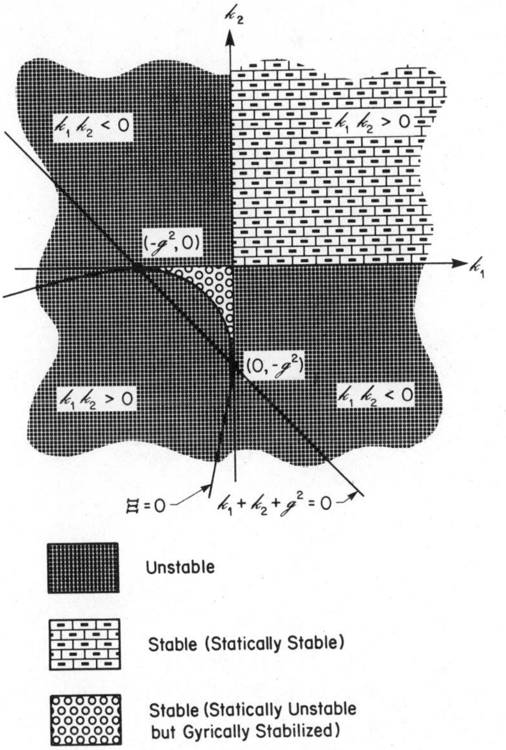

FIGURE A.9 Simple illustration of gyric stabilization.

lie on the negative real axis. According to Sturm's theorem, the number of real roots of $f_N(\mu) = 0$ in the interval $a < \mu < b$ is equal to $V(a) - V(b)$, where $V(\mu)$ is the number of sign changes in the sequence

$$\{f_N(\mu), f_N'(\mu), r_1(\mu), \ldots, r_\nu(\mu)\} \tag{38}$$

Here, $-r_1(\mu)$ is the remainder after the division $f_N(\mu)/f_N'(\mu)$, $-r_2(\mu)$ is the remainder after the division $f_N'(\mu)/r_1(\mu), \ldots,$ and $-r_\nu(\mu)$ is the remainder after the division $r_{\nu-2}(\mu)/r_{\nu-1}(\mu)$, such that $r_\nu(\mu)$ is a nonzero constant. This proves the following theorem:

Theorem A.30: *The system* (29), *with* $\mathcal{M}^T = \mathcal{M} > 0$, $\mathcal{G}^T = -\mathcal{G}$, $\mathcal{K}^T = \mathcal{K}$, *is stable if and only if*

$$V(-\infty) = N \tag{39}$$

$$V(0) = 0 \tag{40}$$

where $V(\mu)$ *is the number of sign changes in the sequence* (38) *and* $f_N(\mu)$ *is defined by* (37).

Literal stability conditions derived from this theorem will be presented shortly.

First, as a simple example, let $N = 2$, whence $f(\mu)$ is of the form

$$f_2(\mu) = b_0\mu^2 + b_1\mu + b_2 \tag{41}$$

TABLE A.2
Stability Criteria for Conservative Gyric Systems

System Order n	Criteria
$N = 2;\ n = 2N = 4$	$(b_0 > 0)\ b_1, b_2, \Xi_1 > 0$ $\Xi_1 \triangleq b_1^2 - 4b_0 b_2$
$N = 3;\ n = 2N = 6$	$(b_0 > 0)\ b_2, b_3, \Xi_2, \Xi_3, \Xi_4 > 0$ $\Xi_2 \triangleq b_1 b_2 - 9b_0 b_3$ $\Xi_2 \triangleq b_1^2 - 3b_0 b_2$ $\Xi_4 \triangleq b_1^2 b_2^2 - 4b_1^3 b_3 - 4b_0 b_2^2 - 27 b_0^2 b_3^2 + 18 b_0 b_1 b_2 b_3$
$N = 3;\ n = 2N = 6$	$(b_0 > 0)\ b_1, b_3, \Xi_3, \Xi_4, \Xi_5 > 0$ $\Xi_5 \triangleq b_2 b_1^2 - 4b_2^2 b_0 + 3b_0 b_2 b_3$
$N = 3;\ n = 2N = 6$	$(b_0 > 0)\ b_1, b_2, b_3, \Xi_4 > 0$

where $b_0 = \det \mathscr{M} > 0$. Then,

$$f_2'(\mu) = 2b_0\mu + b_1; \qquad r_1(\mu) = \frac{b_1^2}{4b_0} - b_2$$

(In this case, $\nu = 1$.) Then,

$$V(-\infty) = V\{+, -, r_1\}; \qquad V(0) = V\{b_2, b_1, r_1\}$$

and the conditions (39) and (40) require that $r_1 > 0$, $b_2 > 0$, $b_1 > 0$. With $b_0 = 1$, these are identical to (35).

The apparently simple form of conditions (39) and (40) belies the arithmetic complexity that ensues when they are expressed in terms of the coefficients for larger N. Conditions for $N = 2$ and $N = 3$ are given in Table A.2. In fact, three different sets of conditions are given for $N = 3$. The first set is derived from the Sturm sequence approach (Theorem A.30); the second set was derived by [Longman, Hagedorn, and Beck]; and the third set was derived by [Hughes and Golla]. It is not immediately obvious that these three sets of conditions are equivalent, although their theoretical bases prove in effect that they must be so. Golla has shown in an unpublished analysis that they are in fact equivalent. He has also shown that the condition $\Xi_3 > 0$ is redundant in both the first and second sets of criteria, which explains why these *five* criteria are equivalent to the *four* criteria given in the third set. Clearly, the third set is the most economical and should be used in practice for $N = 3$. Extensions to these results for $N > 3$ have not been made (to the author's knowledge). It should also be noted that the Routh–Hurwitz criteria for the coefficients b_i are subsumed by these sets of criteria because the requirement that all roots lie on the negative real axis requires, in particular, that the all roots lie in the open left half-plane.

To close our discussion of the stability of gyric conservative systems, an elegant theorem on instability due to [Hagadorn, 2] will be given:

Theorem A.31: *The system* (29), *with* $\mathscr{M}^T = \mathscr{M} > 0$, $\mathscr{G}^T = -\mathscr{G}$, $\mathscr{K}^T = \mathscr{K}$, *is unstable if* $4\mathscr{K} + \mathscr{G}^T\mathscr{M}^{-1}\mathscr{G}$ *is negative-definite.*

This theorem provides a relatively simple check for instability and indicates in a quantitative manner the struggle between static instability and gyric stabilization. Certainly, $\mathscr{G}^T\mathscr{M}^{-1}\mathscr{G}$ is nonnegative-definite (it will in fact be positive-definite if $\det\mathscr{G} \neq 0$); the question is whether the positivity characteristics of $\mathscr{G}^T\mathscr{M}^{-1}\mathscr{G}$ win out over the negativity characteristics of $\mathscr{K}$. For the simple example represented by (32), Theorem A.31 indicates that the region $k_1, k_2 < -g^2/4$ is unstable. This region is easily marked out on Fig. A.9, especially when it is realized that the vertex of the parabola defined by $\Xi = 0$ is the point $(-g^2/4, -g^2/4)$.

Effect of Damping

When a damping term $\mathscr{D}\dot{\mathbf{q}}$ is added to (29), we have

$$\mathscr{M}\ddot{\mathbf{q}} + (\mathscr{D} + \mathscr{G})\dot{\mathbf{q}} + \mathscr{K}\mathbf{q} = \mathbf{0} \tag{42}$$

with $\mathscr{M}^T = \mathscr{M} > 0$, $\mathscr{D}^T = \mathscr{D} \not< 0$, $\mathscr{G}^T = -\mathscr{G}$, and $\mathscr{K}^T = \mathscr{K}$. The preceding discussion has shown that, although a gyric term can never destablize a stable conservative system, it can often stabilize an unstable one. The further addition of energy dissipation has consequences that are sometimes beneficial, sometimes otherwise. Specifically, damping tends to convert statically stable systems into *asymptotically stable* ones and to convert gyrically stabilized, statically unstable systems into *unstable* ones. The first tendency is well known: damped oscillators are asymptotically stable. The second tendency is perhaps less well known but can be quickly illustrated by adding a damping term to the simple gyric conservative system (32):

$$\begin{aligned} \ddot{q}_1 - g\dot{q}_2 + k_1 q_1 &= 0 \\ \ddot{q}_2 + d\dot{q}_2 + g\dot{q}_1 + k_2 q_2 &= 0 \end{aligned} \tag{43}$$

The characteristic equation is now

$$s^4 + ds^3 + \left(k_1 + k_2 + g^2\right)s + dk_1 s + k_1 k_2 = 0 \tag{44}$$

and the Routh–Hurwitz criteria (Table A.1) apply; they reduce to

$$d > 0 \tag{45a}$$

$$k_1 + k_2 + g^2 > 0 \tag{45b}$$

$$k_1 k_2 > 0 \tag{45c}$$

$$d^2 g^2 k_1 > 0 \tag{45d}$$

With reference to Fig. A.9, (45*c*) eliminates the second and fourth quadrants of the k_1–k_2 plane, and (45*b*) eliminates points southwest of the line $k_1 + k_2 + g^2 = 0$. The important new condition is (45*d*), which requires that $k_1 > 0$. Together with (45*c*), this condition demands static stability. In summary, the first quadrant is now asymptotically stable, but the price paid is that the gyrically stable region in Fig. A.9 is now unstable regardless of how large g is or how small $d > 0$ is.

These ideas are expanded for the general system (42) by the Kelvin–Tait–Chetayev theorem. Slight variants of the KTC theorem exist in the literature. We shall consider the situation in two steps, one relating to asymptotic stability, the other to instability.

Theorem A.32: *(KTC-I) When the system* (42), *with* $\mathcal{M}^T = \mathcal{M} > 0$, $\mathcal{D}^T = \mathcal{D}$, $\mathcal{G}^T = -\mathcal{G}$, $\mathcal{K}^T = \mathcal{K}$, *is statically stable and completely damped, it is asymptotically stable.*

In effect, this theorem states that a set of coupled damped harmonic oscillators remain asymptotically stable even when gyric effects are present. The proof again uses the energy (20) as a Liapunov function. By virtue of static stability ($\mathcal{K} > 0$), $v(\mathbf{q}, \dot{\mathbf{q}})$ is positive-definite. Furthermore, $\dot{v} = -\dot{\mathbf{q}}^T\mathcal{D}\dot{\mathbf{q}}$, whence complete damping ($\mathcal{D} > 0$) ensures that $\dot{v} \equiv 0 \rightarrow \dot{\mathbf{q}} \equiv \mathbf{0} \rightarrow \mathbf{q} \equiv \mathbf{0}$, from (42). Thus, by Theorem A.21, Theorem A.32 is proved.

The second part of the KTC Theorem relates to instability:

Theorem A.33: *(KTC-II) When the system* (42) *is completely damped, it is unstable if it is statically unstable.*

By "statically unstable" we mean that $\mathcal{K}$ has at least one negative real eigenvalue. [Zajac, 1] has further shown that (42) has as many eigenvalues in the right half-plane as there are negative eigenvalues of $\mathcal{K}$.

Unfortunately, the conditions called for by the KTC theorems ($\mathcal{D} > 0$, $\mathcal{K} > 0$) do not always conform to the assumptions made in spacecraft attitude dynamics analysis. Once again, the essential requirements is not *complete* damping (Definition A.16) but *pervasive* damping (Definition A.17), as pointed out by [Zajac, 2]. The following theorem due to [Hughes and Gardner] provides necessary conditions for asymptotic stability that are much sharper than the KTC-I conditions of Theorem A.32:

Theorem A.33: *The system* (42) *with* $\mathcal{M}^T = \mathcal{M} > 0$, $\mathcal{D}^T = \mathcal{D}$, $\mathcal{G}^T = -\mathcal{G}$, $\mathcal{K}^T = \mathcal{K}$, *is asymptotically stable if it is statically stable, with* $\mathcal{D} \nless 0$ *and*

$$\operatorname{rank}\left[\mathbf{C}^T \quad \mathbf{A}^T\mathbf{C}^T \quad (\mathbf{A}^T)^2\mathbf{C}^T \quad \cdots \quad (\mathbf{A}^T)^{N-1}\mathbf{C}^T\right] = N \tag{46}$$

where $\mathbf{C} \triangleq [\mathbf{O} \quad \mathcal{D}]$ *and* $\mathbf{A}$ *is defined by* (6).

Once again, asymptotic stability rests on pervasive damping, the necessary and sufficient condition for which is (46). This condition guarantees that the system is completely observable using the damping forces as outputs. If a part of the system is not observable in this way, it is undamped and the stability is no longer asymptotic.

A.7 STABILITY IDEAS SPECIALIZED TO ATTITUDE DYNAMICS

Even this quite extensive appendix does not do justice to the vast subject of stability theory, and important questions have not been addressed. Nonetheless, the discussion is reasonably complete for the purposes of this book. Many further examples of the applications of these theorems can be found of course in the main text itself.

There are also several stability ideas that are associated exclusively with attitude dynamics. Although a fuller discussion of these ideas can be found in the main text, the idea of *directional* (or *pointing*) stability will be briefly mentioned here.

Consider a *unit vector* $\vec{\mathbf{u}}(t)$ such that, in the absence of perturbations, $\vec{\mathbf{u}}$ is fixed in direction: $\vec{\mathbf{u}}(t) \equiv \vec{\mathbf{u}}_o$. This corresponds to the "reference solution" in the definitions of stability. However, if, at time $t = t_o$, $\vec{\mathbf{u}}$ experiences a perturbation $\Delta\vec{\mathbf{u}}_o$, then, for $t \geq t_o$,

$$\vec{\mathbf{u}}(t) = \vec{\mathbf{u}}_o + \Delta\vec{\mathbf{u}}(t) \tag{1}$$

with of course $\Delta\vec{\mathbf{u}}(t_o) = \Delta\vec{\mathbf{u}}_o$. The stability questions arising are these: Can $\|\Delta\vec{\mathbf{u}}(t)\|$ be arbitrarily bounded for $t \geq t_o$ by sufficiently bounding $\|\Delta\vec{\mathbf{u}}_o\|$? Does $\|\Delta\vec{\mathbf{u}}\| \to 0$ as $t \to \infty$? If the answer to the former question is in the affirmative, it is natural, following Definition A.1, to say that $\vec{\mathbf{u}}(t) \equiv \vec{\mathbf{u}}_o$ is *directionally stable*. If both answers are positive, then, in accord with Definition A.5, we could say that $\vec{\mathbf{u}}(t) \equiv \vec{\mathbf{u}}_o$ is *asymptotically directionally stable*.

Next, consider an idea that is slightly more complicated but of essentially the same sort. Suppose that $\vec{\mathbf{u}} \equiv \vec{\mathbf{u}}_o$ is directionally stable, but that

$$\lim_{t \to \infty} \vec{\mathbf{u}}(t) = \vec{\mathbf{u}}_\infty \tag{2}$$

with $\vec{\mathbf{u}}_\infty \neq \vec{\mathbf{u}}_o$. In accordance with the premise of directional stability, the asymptotic direction $\vec{\mathbf{u}}_\infty$ can be made to differ as little as desired from $\vec{\mathbf{u}}_o$ merely by making $\Delta\vec{\mathbf{u}}_o$ sufficiently small. If $\vec{\mathbf{u}}_o$ is directionally stable, one can arbitrarily confine $\|\vec{\mathbf{u}}(t) - \vec{\mathbf{u}}_o\|$ for *any* $\vec{\mathbf{u}}(t)$, $0 \leq t < \infty$, including, in particular, $\vec{\mathbf{u}}_\infty$. And it is also clear that $\vec{\mathbf{u}}_\infty$ depends on $\Delta\vec{\mathbf{u}}_o$: as $\Delta\vec{\mathbf{u}}_o \to \vec{\mathbf{0}}$, $\vec{\mathbf{u}}_\infty \to \vec{\mathbf{u}}_o$. Nevertheless, the fact that $\vec{\mathbf{u}}(t)$ tends to $\vec{\mathbf{u}}_\infty(\Delta\vec{\mathbf{u}}_o)$ and not to $\vec{\mathbf{u}}_o$ as $t \to \infty$ precludes a statement that $\vec{\mathbf{u}}(t)$ is asymptotically directionally stable. We say instead that $\vec{\mathbf{u}}(t)$ is *directionally stable, asymptotically to* $\vec{\mathbf{u}}_\infty$.

These special-purpose stability ideas are particularly relevant to attitude dynamics. The $\vec{\mathbf{u}}$ vector in question is usually a unit vector fixed in a body whose attitude is of interest.

APPENDIX B

VECTRICES

From Chapter 2 it is plain that a spacecraft attitude dynamicist must be dextrous in the simultaneous use of many reference frames. Modern spacecraft typically are configured in a manner that suggests dynamical models in terms of several subbodies; each of these bodies is then assigned one or more reference frames. When extrinsic elements are added to the model—for example, orbital geometry, attitude sensing references, pointing targets, and the several torque sources described in Chapter 8—the number of reference frames multiplies still further. An appropriate notational device is needed to deal with these numerous reference frames. A *vectrix* (a word coined by the author) provides such a device.

Many of the differential equations of dynamics are in the first instance expressed in vectorial form, a form independent of reference frame. Even when Lagrangian or Hamiltonian formulations of mechanics are used ([Lanczos]), a vectorial framework is often the most expeditious way of formulating the kinetic energy, potential energy, and virtual-work expressions. Although it is frequently advisable to delay the scalar expression of vector quantities until certain initial steps have been completed, one goal of the analysis is usually to arrive eventually at a scalar system of motion equations. In converting vector equations to their scalar equivalents, one rule is inviolable: *every term in the vector equation must be expressed in the same frame*. For analyses that employ many reference frames, this rule necessitates a careful accounting of which vector has been expressed in which frame and requires a means of transforming components in one frame to components in another. Vectrices provide a straightforward and reliable formalism for keeping track of these expressions and transformations and for deducing reliably the correct scalar equations.

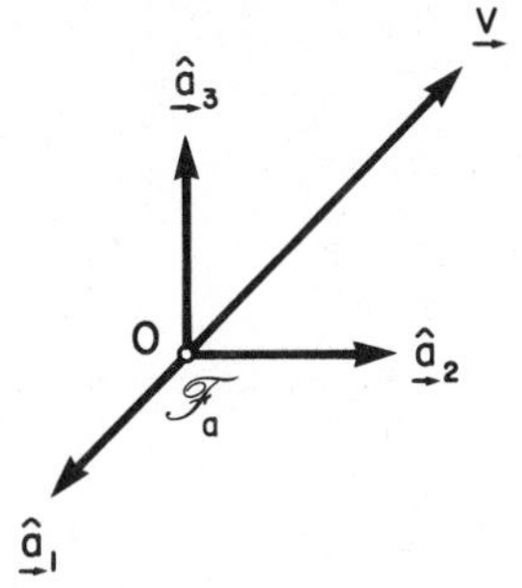

FIGURE B.1 A reference frame $\mathcal{F}_a$, and a vector $\underrightarrow{v}$.

B.1 REMARKS ON TERMINOLOGY

We must be clear on what is meant by a "vector." The term is used consistently in this book to mean a quantity possessing both magnitude and direction in three-dimensional space. Thus, we say, for example, that the velocity $\underrightarrow{v}$ is a vector. Such quantities have been called *Gibbsian vectors*, referring to the early expository textbook by [Gibbs]. Any vector may of course be expressed in any reference frame $\mathcal{F}$ of interest. In general, $\mathcal{F}$ consists of any three non-coplanar basis vectors, although we shall treat only frames whose basis vectors are of unit length, mutually perpendicular, and right-handed (Fig. B.1). This resolution of $\underrightarrow{v}$ in terms of its components $\{v_1, v_2, v_3\}$ in $\mathcal{F}_a$ may be written

$$\underrightarrow{v} = v_1\underrightarrow{\hat{\mathbf{a}}}_1 + v_2\underrightarrow{\hat{\mathbf{a}}}_2 + v_3\underrightarrow{\hat{\mathbf{a}}}_3 \tag{1}$$

To benefit from matrix notation, the components are often assembled as a 3×1 column matrix which we denote by $\mathbf{v}$:

$$\mathbf{v} \triangleq [v_1 \quad v_2 \quad v_3]^T \tag{2}$$

Although the column matrix $\mathbf{v}$ is frequently referred to as a vector—and understandably so from a system theory viewpoint—we shall not do so here, to avoid confusion. Only the symbol $\underrightarrow{v}$ (which is independent of reference frame) will be called a *vector*; the symbol $\mathbf{v}$ (which presupposes a vector $\underrightarrow{v}$ *and* a particular reference frame) will be called simply a *column matrix*.

A further source of possible confusion should be noted. Mathematicians refer to $\mathbf{v}$ as a vector in a more general sense: it is an element of a *linear vector space*. The set of all $n \times 1$ column matrices (and for that matter the set of all $n \times m$ rectangular matrices) is a linear vector space; a 5×7 matrix is, in this sense, a vector. We shall not require this terminology. Instead of saying "matrix vector notation" we shall say, simply, "matrix notation."

B.2 VECTRICES

Experience has shown that the dividends justify the investment in learning vectrix notation. For example, [Likins, 3] and [Wittenburg] have developed similar formalisms. The central element in vectrix notation is the analogy between the right side of (B.1, 1) and the inner product of two column matrices $\mathbf{u}$ and $\mathbf{v}$.

For if we have, in addition to $\mathbf{v}$, a matrix $\mathbf{u}$:

$$\mathbf{u} \triangleq [u_1 \quad u_2 \quad u_3]^T \tag{1}$$

the inner product of $\mathbf{u}$ and $\mathbf{v}$ is written

$$\mathbf{v}^T\mathbf{u} = v_1u_1 + v_2u_2 + v_3u_3 = \mathbf{u}^T\mathbf{v} \tag{2}$$

To construct the analogy, we define a *vectrix* $\mathcal{F}_a$ for frame $\mathcal{F}_a$ in terms of its basis vectors as follows:

$$\mathcal{F}_a \triangleq [\underset{\rightarrow}{\hat{\mathbf{a}}}_1 \quad \underset{\rightarrow}{\hat{\mathbf{a}}}_2 \quad \underset{\rightarrow}{\hat{\mathbf{a}}}_3]^T \tag{3}$$

As its name implies, a vectrix has a split personality, possessing simultaneously the characteristics of both a vector and a matrix. Thus, *the elements of the fundamental vectrix $\mathcal{F}_a$ are a set of basis vectors*. Using the vectrix $\mathcal{F}_a$, a vector $\underset{\rightarrow}{\mathbf{v}}$ is expressed in terms of its components $\mathbf{v}$ in $\mathcal{F}_a$:

$$\underset{\rightarrow}{\mathbf{v}} = \mathbf{v}^T\mathcal{F}_a \equiv \mathcal{F}_a^T\mathbf{v} \tag{4}$$

Equation (4) is the symbolic way of stating that "$\underset{\rightarrow}{\mathbf{v}}$ is the vector whose components in $\mathcal{F}_a$ are the elements of the column matrix $\mathbf{v}$."

By analog with matrix *outer product*, we define

$$\mathcal{F}_a \cdot \mathcal{F}_a^T \triangleq \begin{bmatrix} \underset{\rightarrow}{\hat{\mathbf{a}}}_1 \cdot \underset{\rightarrow}{\hat{\mathbf{a}}}_1 & \underset{\rightarrow}{\hat{\mathbf{a}}}_1 \cdot \underset{\rightarrow}{\hat{\mathbf{a}}}_2 & \underset{\rightarrow}{\hat{\mathbf{a}}}_1 \cdot \underset{\rightarrow}{\hat{\mathbf{a}}}_3 \\ \underset{\rightarrow}{\hat{\mathbf{a}}}_2 \cdot \underset{\rightarrow}{\hat{\mathbf{a}}}_1 & \underset{\rightarrow}{\hat{\mathbf{a}}}_2 \cdot \underset{\rightarrow}{\hat{\mathbf{a}}}_2 & \underset{\rightarrow}{\hat{\mathbf{a}}}_2 \cdot \underset{\rightarrow}{\hat{\mathbf{a}}}_3 \\ \underset{\rightarrow}{\hat{\mathbf{a}}}_3 \cdot \underset{\rightarrow}{\hat{\mathbf{a}}}_1 & \underset{\rightarrow}{\hat{\mathbf{a}}}_3 \cdot \underset{\rightarrow}{\hat{\mathbf{a}}}_2 & \underset{\rightarrow}{\hat{\mathbf{a}}}_3 \cdot \underset{\rightarrow}{\hat{\mathbf{a}}}_3 \end{bmatrix} = \begin{bmatrix} 1 & 0 & 0 \\ 0 & 1 & 0 \\ 0 & 0 & 1 \end{bmatrix} \equiv \mathbf{1} \tag{5}$$

This leads in turn to expressions for the relationship between $\underset{\rightarrow}{\mathbf{v}}$ and $\mathbf{v}$ that are the converse of (4):

$$\begin{aligned} \mathbf{v} &= \mathbf{1v} = \mathcal{F}_a \cdot \mathcal{F}_a^T\mathbf{v} = \mathcal{F}_a \cdot \underset{\rightarrow}{\mathbf{v}} \equiv \underset{\rightarrow}{\mathbf{v}} \cdot \mathcal{F}_a \\ \mathbf{v}^T &= \mathbf{v}^T\mathbf{1} = \mathbf{v}^T\mathcal{F}_a \cdot \mathcal{F}_a^T = \underset{\rightarrow}{\mathbf{v}} \cdot \mathcal{F}_a^T \equiv \mathcal{F}_a^T \cdot \underset{\rightarrow}{\mathbf{v}} \end{aligned} \tag{6}$$

Expressed in words, $\mathbf{v}$ is the column matrix whose elements are the components of $\underset{\rightarrow}{\mathbf{v}}$ in $\mathcal{F}_a$.

Note that vectrices permit the use of both vectors and column matrices in the same equation, whether that equation is itself a vector equation or a matrix equation. The most basic examples of this fact are (4), which incorporates both $\underset{\rightarrow}{\mathbf{v}}$ and $\mathbf{v}$ in a vector equation, and (6), which incorporates both $\underset{\rightarrow}{\mathbf{v}}$ and $\mathbf{v}$ in a matrix equation.

In an analogous manner, we define

$$\mathcal{F}_a \times \mathcal{F}_a^T \equiv \begin{bmatrix} \underset{\rightarrow}{\hat{\mathbf{a}}}_1 \\ \underset{\rightarrow}{\hat{\mathbf{a}}}_2 \\ \underset{\rightarrow}{\hat{\mathbf{a}}}_3 \end{bmatrix} \times [\underset{\rightarrow}{\hat{\mathbf{a}}}_1 \quad \underset{\rightarrow}{\hat{\mathbf{a}}}_2 \quad \underset{\rightarrow}{\hat{\mathbf{a}}}_3]$$

$$\triangleq \begin{bmatrix} \underset{\rightarrow}{\hat{\mathbf{a}}}_1 \times \underset{\rightarrow}{\hat{\mathbf{a}}}_1 & \underset{\rightarrow}{\hat{\mathbf{a}}}_1 \times \underset{\rightarrow}{\hat{\mathbf{a}}}_2 & \underset{\rightarrow}{\hat{\mathbf{a}}}_1 \times \underset{\rightarrow}{\hat{\mathbf{a}}}_3 \\ \underset{\rightarrow}{\hat{\mathbf{a}}}_2 \times \underset{\rightarrow}{\hat{\mathbf{a}}}_1 & \underset{\rightarrow}{\hat{\mathbf{a}}}_2 \times \underset{\rightarrow}{\hat{\mathbf{a}}}_2 & \underset{\rightarrow}{\hat{\mathbf{a}}}_2 \times \underset{\rightarrow}{\hat{\mathbf{a}}}_3 \\ \underset{\rightarrow}{\hat{\mathbf{a}}}_3 \times \underset{\rightarrow}{\hat{\mathbf{a}}}_1 & \underset{\rightarrow}{\hat{\mathbf{a}}}_3 \times \underset{\rightarrow}{\hat{\mathbf{a}}}_2 & \underset{\rightarrow}{\hat{\mathbf{a}}}_3 \times \underset{\rightarrow}{\hat{\mathbf{a}}}_3 \end{bmatrix} \equiv \begin{bmatrix} \underset{\rightarrow}{\mathbf{0}} & \underset{\rightarrow}{\hat{\mathbf{a}}}_3 & -\underset{\rightarrow}{\hat{\mathbf{a}}}_2 \\ -\underset{\rightarrow}{\hat{\mathbf{a}}}_3 & \underset{\rightarrow}{\mathbf{0}} & \underset{\rightarrow}{\hat{\mathbf{a}}}_1 \\ \underset{\rightarrow}{\hat{\mathbf{a}}}_2 & -\underset{\rightarrow}{\hat{\mathbf{a}}}_1 & \underset{\rightarrow}{\mathbf{0}} \end{bmatrix} \tag{7}$$

Thus, $\mathcal{F}_a \times \mathcal{F}_a^T$ is a 3×3 skew-symmetric vectrix whose elements are basis vectors, arranged as shown in (7).

The two operations, $\mathcal{F}_a \cdot \mathcal{F}_a^T$ defined by (5) and $\mathcal{F}_a \times \mathcal{F}_a^T$ defined by (7), are the only two operations between fundamental vectrices that we shall require.

Several Vectors

Consider now several vectors, $\underrightarrow{u}, \underrightarrow{v}, \underrightarrow{w}, \ldots,$ all expressed in the same frame, $\mathcal{F}_a$, and define the component matrices as follows:

$$
\begin{aligned}
[u_1 \quad u_2 \quad u_3]^T &\equiv \mathbf{u} \triangleq \mathcal{F}_a \cdot \underrightarrow{u} \\
[v_1 \quad v_2 \quad v_3]^T &\equiv \mathbf{v} \triangleq \mathcal{F}_a \cdot \underrightarrow{v} \\
[w_1 \quad w_2 \quad w_3]^T &\equiv \mathbf{w} \triangleq \mathcal{F}_a \cdot \underrightarrow{w} \\
&\vdots
\end{aligned}
\tag{8}
$$

Or, equivalently, $[\underrightarrow{u} \quad \underrightarrow{v} \quad \underrightarrow{w}] = \mathcal{F}_a^T[\mathbf{u} \quad \mathbf{v} \quad \mathbf{w}]$. Vector operations among these vectors can readily be expressed, in component form, in $\mathcal{F}_a$. If, for example,

$$\underrightarrow{u} = c_1\underrightarrow{v} + c_2\underrightarrow{w} + \cdots \tag{9}$$

then

$$\mathbf{u} = c_1\mathbf{v} + c_2\mathbf{w} + \cdots \tag{10}$$

as can be obtained by operating on (9) by $\mathcal{F}_a \cdot$. The scalar product between two vectors is also handled readily:

$$\underrightarrow{u} \cdot \underrightarrow{v} = \left(\mathbf{u}^T\mathcal{F}_a\right) \cdot \left(\mathcal{F}_a^T\mathbf{v}\right) = \mathbf{u}^T\mathcal{F}_a \cdot \mathcal{F}_a^T\mathbf{v} = \mathbf{u}^T\mathbf{1}\mathbf{v} = \mathbf{u}^T\mathbf{v} \equiv \mathbf{v}^T\mathbf{u} \tag{11}$$

The cross vector product is slightly more complicated. From (7),

$$
\begin{aligned}
\underrightarrow{u} \times \underrightarrow{v} &= \mathbf{u}^T\mathcal{F}_a \times \mathcal{F}_a^T\mathbf{v} \\
&= \mathbf{u}^T \begin{bmatrix} \underrightarrow{\mathbf{0}} & \underrightarrow{\hat{\mathbf{a}}}_3 & -\underrightarrow{\hat{\mathbf{a}}}_2 \\ -\underrightarrow{\hat{\mathbf{a}}}_3 & \underrightarrow{\mathbf{0}} & \underrightarrow{\hat{\mathbf{a}}}_1 \\ \underrightarrow{\hat{\mathbf{a}}}_2 & -\underrightarrow{\hat{\mathbf{a}}}_1 & \underrightarrow{\hat{\mathbf{0}}} \end{bmatrix} \mathbf{v} \\
&= [u_1 \quad u_2 \quad u_3] \begin{bmatrix} \underrightarrow{\hat{\mathbf{a}}}_3 v_2 - \underrightarrow{\hat{\mathbf{a}}}_2 v_3 \\ \underrightarrow{\hat{\mathbf{a}}}_1 v_3 - \underrightarrow{\hat{\mathbf{a}}}_3 v_1 \\ \underrightarrow{\hat{\mathbf{a}}}_2 v_1 - \underrightarrow{\hat{\mathbf{a}}}_1 v_2 \end{bmatrix} \\
&= \mathcal{F}_a^T \begin{bmatrix} u_2 v_3 - u_3 v_2 \\ u_3 v_1 - u_1 v_3 \\ u_1 v_2 - u_2 v_1 \end{bmatrix} \\
&= \mathcal{F}_a^T \mathbf{u}^\times \mathbf{v}
\end{aligned}
\tag{12}
$$

where

$$\mathbf{u}^{\times} \triangleq \begin{bmatrix} 0 & -u_3 & u_2 \\ u_3 & 0 & -u_1 \\ -u_2 & u_1 & 0 \end{bmatrix} \tag{13}$$

The notation $(\ )^{\times}$ requires the formation of a skew-symmetric matrix $\mathbf{u}^{\times}$ from the elements of $\mathbf{u}$ according to the pattern dictated by (13). Several notations for this operation are extant in the literature, including $\tilde{\mathbf{u}}$, $\mathbf{U}$, and $C_p(\mathbf{u})$. The last is rather space consuming.

An objection to $\mathbf{U}$ is that it demands a new letter of the alphabet, and these may be in short supply in a large-scale analysis. The most commonly used notation, $\tilde{\mathbf{u}}$, does not suffer from the disadvantages just mentioned; however the type-setting can become somewhat cluttered if one wishes to add other overscripts (as in the overdot designating time derivative, for example), or if one wishes to form the skew matrix corresponding to an expression such as $\mathbf{Ax} + \mathbf{By}$. The author proposes that the notation $(\ )^{\times}$ be used because no confusion can arise with raising a column matrix to a power, no new letters in the alphabet are required, it is concise, it does not interfere with other oversymbols, expressions such as $(\mathbf{Ax} + \mathbf{By})^{\times}$ can be handled simply, other notations calling for a re-arrangement of matrix elements (the transpose, for example) are similarly placed, and, perhaps most important, this notation is suggestive of its function as the matrix equivalent of a vector product. Note that

$$\mathbf{u}^{\times T} = -\mathbf{u}^{\times} \tag{14}$$

That is, $\mathbf{u}^{\times}$ is always skew-symmetric.

Multiple products are handled in the same manner. For example, using the vectrix $\mathscr{F}_a$ one readily obtains

$$\underrightarrow{\mathbf{u}} \cdot \underrightarrow{\mathbf{v}} \times \underrightarrow{\mathbf{w}} = \mathbf{u}^T\mathbf{v}^{\times}\mathbf{w} \tag{15}$$

$$\underrightarrow{\mathbf{u}} \times (\underrightarrow{\mathbf{v}} \times \underrightarrow{\mathbf{w}}) = \mathscr{F}_a^T\mathbf{u}^{\times}\mathbf{v}^{\times}\mathbf{w} \tag{16}$$

Well-known vector identities also have their easily recognized counterpart when expressed in terms of their components in $\mathscr{F}_a$. Thus,

(a) $\underrightarrow{\mathbf{u}} \times \underrightarrow{\mathbf{u}} \equiv \underrightarrow{\mathbf{0}} \rightarrow \mathbf{u}^{\times}\mathbf{u} \equiv \mathbf{0}$

(b) $\underrightarrow{\mathbf{u}} \times \underrightarrow{\mathbf{v}} \equiv -\underrightarrow{\mathbf{v}} \times \underrightarrow{\mathbf{u}} \rightarrow \mathbf{u}^{\times}\mathbf{v} \equiv -\mathbf{v}^{\times}\mathbf{u}$

(c) $\underrightarrow{\mathbf{u}} \times (\underrightarrow{\mathbf{v}} \times \underrightarrow{\mathbf{w}}) \equiv (\underrightarrow{\mathbf{u}} \cdot \underrightarrow{\mathbf{w}})\underrightarrow{\mathbf{v}} - (\underrightarrow{\mathbf{u}} \cdot \underrightarrow{\mathbf{v}})\underrightarrow{\mathbf{w}} \rightarrow \mathbf{u}^{\times}\mathbf{v}^{\times}\mathbf{w} \equiv (\mathbf{u}^T\mathbf{w})\mathbf{v} - (\mathbf{u}^T\mathbf{v})\mathbf{w} \equiv (\mathbf{v}\mathbf{u}^T - \mathbf{u}^T\mathbf{v}\mathbf{1})\mathbf{w}$

Dyadics

A *dyadic* has the property that scalar multiplication with a vector produces a vector. Thus, if $\underrightarrow{\mathbf{D}}$ is a dyadic and $\underrightarrow{\mathbf{v}}$ is a vector, $\underrightarrow{\mathbf{v}} \cdot \underrightarrow{\mathbf{D}}$ and $\underrightarrow{\mathbf{D}} \cdot \underrightarrow{\mathbf{v}}$ are both vectors—*different* vectors, in general. Two vectors placed side by side, with neither a dot nor a cross between them (such as $\underrightarrow{\mathbf{u}}\underrightarrow{\mathbf{v}}$) can be regarded as a dyadic, provided we define

$$\underset{\rightarrow}{\mathbf{v}} \cdot (\underset{\rightarrow}{\mathbf{u}}\underset{\rightarrow}{\mathbf{w}}) \triangleq (\underset{\rightarrow}{\mathbf{v}} \cdot \underset{\rightarrow}{\mathbf{u}})\underset{\rightarrow}{\mathbf{w}}$$
$$(\underset{\rightarrow}{\mathbf{u}}\underset{\rightarrow}{\mathbf{w}}) \cdot \underset{\rightarrow}{\mathbf{v}} \triangleq (\underset{\rightarrow}{\mathbf{w}} \cdot \underset{\rightarrow}{\mathbf{v}})\underset{\rightarrow}{\mathbf{u}} \tag{17}$$

This explains the origin of the term *dyad*, which means "a pair."

Just as a vector can be expressed in a particular frame as a linear combination of the basis vectors of the frame [see (1)], a dyadic can be expressed in a particular frame as a linear combination of the elemental dyadics of that frame:

$$\underset{\rightarrow}{\mathbf{D}} = \sum_{i=1}^{3} \sum_{j=1}^{3} d_{ij} \underset{\rightarrow}{\hat{\mathbf{a}}}_i \underset{\rightarrow}{\hat{\mathbf{a}}}_j \qquad (i, j = 1, 2, 3) \tag{18}$$

The nine terms on the right side of this equation can be concisely represented with the aid of vectrices:

$$\underset{\rightarrow}{\mathbf{D}} = \mathscr{F}_a^T \mathbf{D} \mathscr{F}_a \tag{19}$$

where $\mathbf{D}$ is the matrix $\{d_{ij}\}$ and (19) is interpreted as a matrix quadratic form. This result does for dyadics what (4) does for vectors. The converse of (19), namely,

$$\mathbf{D} = \mathscr{F}_a \cdot \underset{\rightarrow}{\mathbf{D}} \cdot \mathscr{F}_a^T \tag{20}$$

should similarly be compared to (6) for vectors.

Consider now a dyadic $\underset{\rightarrow}{\mathbf{D}}$ and a vector $\underset{\rightarrow}{\mathbf{v}}$, both expressed in $\mathscr{F}_a$; (4) and (20) represent the situation. By using vectrix operations, it is straightforward to show that

$$\underset{\rightarrow}{\mathbf{D}} \cdot \underset{\rightarrow}{\mathbf{v}} = \mathscr{F}_a^T \mathbf{D}\mathbf{v} \equiv \mathbf{v}^T \mathbf{D}^T \mathscr{F}_a$$
$$\underset{\rightarrow}{\mathbf{v}} \cdot \underset{\rightarrow}{\mathbf{D}} = \mathscr{F}_a^T \mathbf{D}^T \mathbf{v} \equiv \mathbf{v}^T \mathbf{D} \mathscr{F}_a \tag{21}$$

Similarly,

$$\underset{\rightarrow}{\mathbf{u}} \cdot \underset{\rightarrow}{\mathbf{D}} \cdot \underset{\rightarrow}{\mathbf{v}} = \mathbf{u}^T \mathbf{D} \mathbf{v} \equiv \mathbf{v}^T \mathbf{D}^T \mathbf{u} \tag{22}$$

It is instructive to note also the particular form that (21) and (22) take when $\underset{\rightarrow}{\mathbf{D}} = \underset{\rightarrow}{\mathbf{u}}\underset{\rightarrow}{\mathbf{v}}$. Note also that the unit dyadic, expressed in *any* reference frame, becomes the unit matrix.

B.3 SEVERAL REFERENCE FRAMES

It is evident that vectrix notation is useful in forming the bridge between vectors and dyadics and their matrix representations in a particular reference frame. It is in the context of several reference frames, however, that the benefits of vectrix notation are even more clearly visible.

Let $\mathscr{F}_a$ and $\mathscr{F}_b$ be the vectrices corresponding to frames $\mathscr{F}_a$ and $\mathscr{F}_b$ (Fig. B.2). Then, we define

$$\mathscr{F}_b \cdot \mathscr{F}_a^T \triangleq \begin{bmatrix} \underset{\rightarrow}{\hat{\mathbf{b}}}_1 \cdot \underset{\rightarrow}{\hat{\mathbf{a}}}_1 & \underset{\rightarrow}{\hat{\mathbf{b}}}_1 \cdot \underset{\rightarrow}{\hat{\mathbf{a}}}_2 & \underset{\rightarrow}{\hat{\mathbf{b}}}_1 \cdot \underset{\rightarrow}{\hat{\mathbf{a}}}_3 \\ \underset{\rightarrow}{\hat{\mathbf{b}}}_2 \cdot \underset{\rightarrow}{\hat{\mathbf{a}}}_1 & \underset{\rightarrow}{\hat{\mathbf{b}}}_2 \cdot \underset{\rightarrow}{\hat{\mathbf{a}}}_2 & \underset{\rightarrow}{\hat{\mathbf{b}}}_2 \cdot \underset{\rightarrow}{\hat{\mathbf{a}}}_3 \\ \underset{\rightarrow}{\hat{\mathbf{b}}}_3 \cdot \underset{\rightarrow}{\hat{\mathbf{a}}}_1 & \underset{\rightarrow}{\hat{\mathbf{b}}}_3 \cdot \underset{\rightarrow}{\hat{\mathbf{a}}}_2 & \underset{\rightarrow}{\hat{\mathbf{b}}}_3 \cdot \underset{\rightarrow}{\hat{\mathbf{a}}}_3 \end{bmatrix} \triangleq \mathbf{C}_{ba} \tag{1}$$

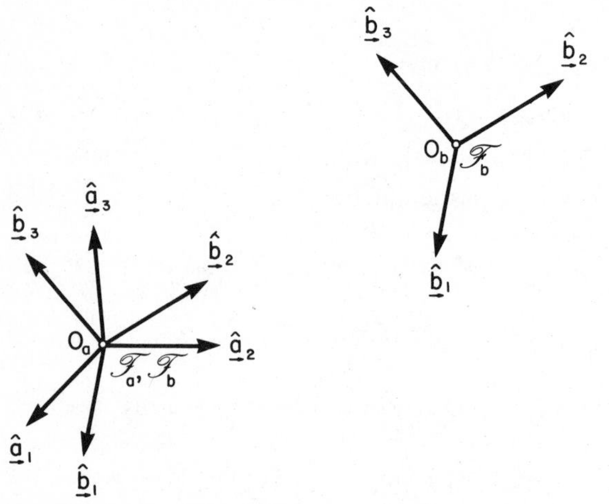

FIGURE B.2 Two reference frames, $\mathscr{F}_a$ and $\mathscr{F}_b$. (From the standpoint of *components*, origins of reference frames are unimportant.)

The matrix of direction cosines, $\mathbf{C}_{ba}$, is precisely the rotation matrix discussed at length in Chapter 2. The *ba* subscript is not strictly necessary for just two frames, but we wish to develop a notation suitable for many frames. The rotation $\mathbf{C}_{ba}$ transforms the components of a vector in $\mathscr{F}_a$ to the components of that same vector in $\mathscr{F}_b$. For, let

$$\mathbf{u}_a \triangleq \mathscr{F}_a \cdot \underset{\rightarrow}{u}; \qquad \mathbf{u}_b \triangleq \mathscr{F}_b \cdot \underset{\rightarrow}{u}$$

represent the components of the vector $\underset{\rightarrow}{u}$ expressed in $\mathscr{F}_a$ and $\mathscr{F}_b$. Then

$$\mathbf{u}_b = \mathscr{F}_b \cdot \mathscr{F}_a^T \mathbf{u}_a = \mathbf{C}_{ba}\mathbf{u}_a \tag{2}$$

We denote the inverse transformation $\mathbf{C}_{ba}^{-1}$ by $\mathbf{C}_{ab}$ for notational convenience, and so we have the companion relationship

$$\mathbf{u}_a = \mathbf{C}_{ab}\mathbf{u}_b \tag{3}$$

Since $\mathbf{C}_{ba}$ is an orthonormal matrix (see Chapter 2), $\mathbf{C}_{ab} = \mathbf{C}_{ba}^T$. In general, the transformation between any two vectrices $\mathscr{F}_a$ and $\mathscr{F}_b$ is written in either of the two equivalent forms:

$$\mathscr{F}_a = \mathbf{C}_{ab}\mathscr{F}_b \qquad \text{or} \qquad \mathscr{F}_a^T = \mathscr{F}_b^T\mathbf{C}_{ba} \tag{4}$$

This is verified by premultiplying the second form by $\mathscr{F}_b \cdot$ or postmultiplying the first by $\cdot \mathscr{F}_b^T$ and using (1).

If a third frame $\mathscr{F}_c$ is introduced, let $\mathbf{u}_c \triangleq \mathscr{F}_c \cdot \underset{\rightarrow}{u}$. Then,

$$\mathbf{u}_c = \mathbf{C}_{cb}\mathbf{u}_b = \mathbf{C}_{cb}\mathbf{C}_{ba}\mathbf{u}_a$$

Since $\mathbf{u}_c$ also equals $\mathbf{C}_{ca}\mathbf{u}_a$, however, it follows that

$$\mathbf{C}_{ca} \equiv \mathbf{C}_{cb}\mathbf{C}_{ba} \tag{5}$$

This result is illustrative of a general property of the double-subscript notation for rotation matrices. The relation

$$\mathbf{C}_{ca} \equiv \mathbf{C}_{cg}\mathbf{C}_{gb}\mathbf{C}_{bd}\mathbf{C}_{da} \tag{6}$$

is a more complicated example of the same principle.

Several Vectors in Several Reference Frames

Relations of the type (4) furnish the means to proceed at will between any two reference frames of interest. Consider several vectors, $\{\underset{\rightarrow}{\mathbf{u}}, \underset{\rightarrow}{\mathbf{v}}, \underset{\rightarrow}{\mathbf{w}}, \ldots\}$ and several frames, $\{\mathcal{F}_a, \mathcal{F}_b, \ldots\}$. Denote by $\mathbf{u}_i$ the components of $\underset{\rightarrow}{\mathbf{u}}$ in $\mathcal{F}_i$, and similarly for $\underset{\rightarrow}{\mathbf{v}}, \underset{\rightarrow}{\mathbf{w}}, \ldots$. Then, the following conversions are illustrative:

$$\underset{\rightarrow}{\mathbf{u}} + \underset{\rightarrow}{\mathbf{v}} = \mathcal{F}_a^T\mathbf{u}_a + \mathcal{F}_b^T\mathbf{v}_b = \mathcal{F}_a^T(\mathbf{u}_a + \mathbf{C}_{ab}\mathbf{v}_b) \tag{7a}$$

$$\underset{\rightarrow}{\mathbf{u}} \cdot \underset{\rightarrow}{\mathbf{v}} = \mathbf{u}_d^T\mathcal{F}_d \cdot \mathcal{F}_b^T\mathbf{u}_b = \mathbf{u}_d^T\mathbf{C}_{db}\mathbf{v}_b \tag{7b}$$

$$\underset{\rightarrow}{\mathbf{u}} \times \underset{\rightarrow}{\mathbf{v}} = \mathbf{u}_a^T\mathcal{F}_a \times \mathcal{F}_b^T\mathbf{v}_b = \mathbf{u}_a^T\mathcal{F}_a \times \mathcal{F}_a^T\mathbf{C}_{ab}\mathbf{v}_b = \mathcal{F}_a^T\mathbf{u}_a^{\times}\mathbf{C}_{ab}\mathbf{v}_b \tag{7c}$$

These examples also serve to demonstrate that with practice one often finds it unnecessary to carry out all the intermediate vectrix operations. Once the meaning of the symbols is clearly understood, one can usually proceed directly to the result.

Before leaving (7c), we use it to derive an important identity. Converting to $\mathcal{F}_b$, we have

$$\underset{\rightarrow}{\mathbf{u}} \times \underset{\rightarrow}{\mathbf{v}} = \mathcal{F}_b^T\mathbf{C}_{ba}\mathbf{u}_a^{\times}\mathbf{C}_{ab}\mathbf{v}_b \tag{8}$$

However, we could have proceeded as follows:

$$\underset{\rightarrow}{\mathbf{u}} \times \underset{\rightarrow}{\mathbf{v}} = \mathbf{u}_a^T\mathbf{C}_{ab}\mathcal{F}_b \times \mathcal{F}_b^T\mathbf{v}_b = \mathcal{F}_b^T(\mathbf{C}_{ba}\mathbf{u}_a)^{\times}\mathbf{v}_b \tag{9}$$

Comparing (8) and (9), we derive the following identity:

$$(\mathbf{C}_{ab}\mathbf{u}_b)^{\times} \equiv \mathbf{C}_{ab}\mathbf{u}_b^{\times}\mathbf{C}_{ba} \tag{10}$$

This frequently useful identity holds true for any two frames $\mathcal{F}_a$ and $\mathcal{F}_b$.

Some further examples of vectrix operations are now considered. Let

$$\underset{\rightarrow}{\mathbf{u}} = \mathcal{F}_a^T\mathbf{u}; \qquad \underset{\rightarrow}{\mathbf{v}} = \mathcal{F}_b^T\mathbf{v}; \qquad \underset{\rightarrow}{\mathbf{w}} = \mathcal{F}_c^T\mathbf{w}; \qquad \underset{\rightarrow}{\mathbf{r}} = \mathcal{F}_d^T\mathbf{r} \tag{11}$$

Then,

$$\underset{\rightarrow}{\mathbf{u}} \cdot \underset{\rightarrow}{\mathbf{v}} \times \underset{\rightarrow}{\mathbf{w}} = \mathbf{u}^T\mathbf{C}_{ab}\mathbf{v}^{\times}\mathbf{C}_{bc}\mathbf{w} \tag{12}$$

and

$$\underset{\rightarrow}{\mathbf{u}} \times (\underset{\rightarrow}{\mathbf{v}} \times \underset{\rightarrow}{\mathbf{w}}) = \mathcal{F}_a^T\mathbf{u}^{\times}\mathbf{C}_{ab}\mathbf{v}^{\times}\mathbf{C}_{bc}\mathbf{w} \tag{13}$$

when expressed in $\mathcal{F}_a$. Furthermore, the familiar vector identities

(a) $\underset{\rightarrow}{\mathbf{u}} \cdot \underset{\rightarrow}{\mathbf{v}} \times \underset{\rightarrow}{\mathbf{w}} \equiv \underset{\rightarrow}{\mathbf{v}} \cdot \underset{\rightarrow}{\mathbf{w}} \times \underset{\rightarrow}{\mathbf{u}} \equiv \underset{\rightarrow}{\mathbf{w}} \cdot \underset{\rightarrow}{\mathbf{u}} \times \underset{\rightarrow}{\mathbf{v}}$
(b) $\underset{\rightarrow}{\mathbf{u}} \times (\underset{\rightarrow}{\mathbf{v}} \times \underset{\rightarrow}{\mathbf{w}}) + \underset{\rightarrow}{\mathbf{v}} \times (\underset{\rightarrow}{\mathbf{w}} \times \underset{\rightarrow}{\mathbf{u}}) + \underset{\rightarrow}{\mathbf{w}} \times (\underset{\rightarrow}{\mathbf{u}} \times \underset{\rightarrow}{\mathbf{v}}) \equiv \underset{\rightarrow}{\mathbf{0}}$
(c) $\underset{\rightarrow}{\mathbf{u}} \times (\underset{\rightarrow}{\mathbf{v}} \times \underset{\rightarrow}{\mathbf{w}}) \equiv (\underset{\rightarrow}{\mathbf{u}} \cdot \underset{\rightarrow}{\mathbf{w}})\underset{\rightarrow}{\mathbf{v}} - (\underset{\rightarrow}{\mathbf{u}} \cdot \underset{\rightarrow}{\mathbf{v}})\underset{\rightarrow}{\mathbf{w}}$
(d) $(\underset{\rightarrow}{\mathbf{u}} \times \underset{\rightarrow}{\mathbf{v}}) \cdot (\underset{\rightarrow}{\mathbf{w}} \times \underset{\rightarrow}{\mathbf{r}}) \equiv (\underset{\rightarrow}{\mathbf{u}} \cdot \underset{\rightarrow}{\mathbf{w}})(\underset{\rightarrow}{\mathbf{v}} \cdot \underset{\rightarrow}{\mathbf{r}}) - (\underset{\rightarrow}{\mathbf{u}} \cdot \underset{\rightarrow}{\mathbf{r}})(\underset{\rightarrow}{\mathbf{v}} \cdot \underset{\rightarrow}{\mathbf{w}})$

which are often called upon in this book (particularly in Chapter 3), can be converted into the following component forms using vectrix operations:

(a′) $\mathbf{u}^T\mathbf{C}_{ab}\mathbf{v}^{\times}\mathbf{C}_{bc}\mathbf{w} \equiv \mathbf{v}^T\mathbf{C}_{bc}\mathbf{w}^{\times}\mathbf{C}_{ca}\mathbf{u} \equiv \mathbf{w}^T\mathbf{C}_{ca}\mathbf{u}^{\times}\mathbf{C}_{ab}\mathbf{v}$

(b′) $\mathbf{C}_{da}\mathbf{u}^{\times}\mathbf{C}_{ab}\mathbf{v}^{\times}\mathbf{C}_{bc}\mathbf{w} + \mathbf{C}_{db}\mathbf{v}^{\times}\mathbf{C}_{bc}\mathbf{w}^{\times}\mathbf{C}_{ca}\mathbf{u} + \mathbf{C}_{dc}\mathbf{w}^{\times}\mathbf{C}_{ca}\mathbf{u}^{\times}\mathbf{C}_{ab}\mathbf{v} \equiv \mathbf{0}$

(c′) $\mathbf{u}^{\times}\mathbf{C}_{ab}\mathbf{v}^{\times}\mathbf{C}_{bc}\mathbf{w} \equiv \mathbf{C}_{ab}\mathbf{v}\mathbf{u}^T\mathbf{C}_{ac}\mathbf{w} - \mathbf{C}_{ac}\mathbf{w}\mathbf{u}^T\mathbf{C}_{ab}\mathbf{v}$

(d′) $\mathbf{u}^T\mathbf{C}_{ab}\mathbf{v}^{\times}\mathbf{C}_{bc}\mathbf{w}^{\times}\mathbf{C}_{cd}\mathbf{r} \equiv \mathbf{u}^T\mathbf{C}_{ac}\mathbf{w}\mathbf{v}^T\mathbf{C}_{bd}\mathbf{r} - \mathbf{u}^T\mathbf{C}_{ad}\mathbf{r}\mathbf{v}^T\mathbf{C}_{bc}\mathbf{w}$

Each of the identities (b)–(d) has been expressed, as each must be, in a *particular* reference frame. However, by premultiplying by an appropriate rotation matrix, each can be converted to components in a different reference frame.

In a similar fashion, relationships involving dyadics can be dealt with. Suppose, for example, that a certain dyadic $\underrightarrow{\mathbf{D}}$ has components $\mathbf{D}_a$ in $\mathcal{F}_a$, and $\mathbf{D}_b$ in $\mathcal{F}_b$. That is,

$$\mathbf{D}_a \triangleq \mathcal{F}_a \cdot \underrightarrow{\mathbf{D}} \cdot \mathcal{F}_a^T; \qquad \mathbf{D}_b = \mathcal{F}_b \cdot \underrightarrow{\mathbf{D}} \cdot \mathcal{F}_b^T \tag{14}$$

Then, it is a simple exercise to show that

$$\mathbf{D}_b = \mathbf{C}_{ba}\mathbf{D}_a\mathbf{C}_{ab} \tag{15}$$

Or, consider a *symmetric* dyadic $\underrightarrow{\mathbf{D}}$, which has the property that

$$\underrightarrow{\mathbf{v}} \cdot \underrightarrow{\mathbf{D}} \equiv \underrightarrow{\mathbf{D}} \cdot \underrightarrow{\mathbf{v}} \tag{16}$$

for all vectors $\underrightarrow{\mathbf{v}}$. It follows from vectrix algebra that $\underrightarrow{\mathbf{D}}$, expressed in any reference frame, produces a symmetric 3×3 matrix. It can also be shown that $\underrightarrow{\mathbf{u}}\underrightarrow{\mathbf{v}} + \underrightarrow{\mathbf{v}}\underrightarrow{\mathbf{u}}$ is always a symmetric dyadic for any vectors $\underrightarrow{\mathbf{u}}$ and $\underrightarrow{\mathbf{v}}$.

B.4 KINEMATICS OF VECTRICES

If two reference frames $\mathcal{F}_a$ and $\mathcal{F}_b$ are rotating with respect to each other, the vectrix $\mathcal{F}_b$ has an angular velocity with respect to $\mathcal{F}_a$, and *vice versa*. We shall denote the former angular velocity by $\underrightarrow{\boldsymbol{\omega}}_{ba}$ and define $\underrightarrow{\boldsymbol{\omega}}_{ab}$ similarly. The order of the subscripts is important. By symmetry,

$$\underrightarrow{\boldsymbol{\omega}}_{ab} + \underrightarrow{\boldsymbol{\omega}}_{ba} = \underrightarrow{\mathbf{0}}$$

This is a special case of the law of addition of angular velocities proved in (2.3, 10).

Evidently, observers in $\mathcal{F}_a$ and $\mathcal{F}_b$ do not see the same motions, owing to their own relative motion. We denote vector time derivatives as seen in $\mathcal{F}_a$ and $\mathcal{F}_b$ by an overdot ($\dot{\ }$) and an overcircle ($\mathring{\ }$), respectively. By definition, then,

$$\dot{\mathcal{F}}_a \equiv 0; \qquad \mathring{\mathcal{F}}_b \equiv 0 \tag{1}$$

As for $\dot{\mathcal{F}}_b$, we can proceed as follows. We know from vector calculus that

$$\dot{\underrightarrow{\hat{\mathbf{b}}}}_1 = \underrightarrow{\boldsymbol{\omega}}_{ba} \times \underrightarrow{\hat{\mathbf{b}}}_1; \qquad \dot{\underrightarrow{\hat{\mathbf{b}}}}_2 = \underrightarrow{\boldsymbol{\omega}}_{ba} \times \underrightarrow{\hat{\mathbf{b}}}_2; \qquad \dot{\underrightarrow{\hat{\mathbf{b}}}}_3 = \underrightarrow{\boldsymbol{\omega}}_{ba} \times \underrightarrow{\hat{\mathbf{b}}}_3 \tag{2}$$

These may be combined to form a single vectrix relation:

$$\dot{\mathcal{F}}_b = \underrightarrow{\boldsymbol{\omega}}_{ba} \times \mathcal{F}_b \tag{3}$$

Now consider an arbitrary vector $\underrightarrow{\mathbf{v}}(t)$, and let the components of $\underrightarrow{\mathbf{v}}$ in $\mathcal{F}_a$ and $\mathcal{F}_b$ be $\mathbf{v}_a$ and $\mathbf{v}_b$, respectively:

$$\underrightarrow{\mathbf{v}} = \mathcal{F}_a^T \mathbf{v}_a = \mathcal{F}_b^T \mathbf{v}_b \tag{4}$$

Then,

$$\dot{\underrightarrow{\mathbf{v}}} = \mathcal{F}_a^T \dot{\mathbf{v}}_a + \dot{\mathcal{F}}_a^T \mathbf{v}_a = \mathcal{F}_a^T \dot{\mathbf{v}}_a \tag{5}$$

Similarly,

$$\mathring{\underrightarrow{\mathbf{v}}} = \mathcal{F}_b^T \dot{\mathbf{v}}_b \tag{6}$$

Note that the time derivative of a column matrix, unlike the time derivative of a vector, *can have only one meaning*, and so the simple overdot can be used without fear of ambiguity. Also, from (4), we find that

$$\begin{aligned}\dot{\underrightarrow{\mathbf{v}}} &= \mathcal{F}_b^T \dot{\mathbf{v}}_b + \dot{\mathcal{F}}_b^T \mathbf{v}_b \\ &= \mathcal{F}_b^T \dot{\mathbf{v}}_b + \underrightarrow{\boldsymbol{\omega}}_{ba} \times \mathcal{F}_b^T \mathbf{v}_b \\ &= \mathring{\underrightarrow{\mathbf{v}}} + \underrightarrow{\boldsymbol{\omega}}_{ba} \times \underrightarrow{\mathbf{v}} \end{aligned} \tag{7}$$

which is a familiar result. Less well known perhaps is the matrix equivalent of (7). To derive it, express the components of $\underrightarrow{\boldsymbol{\omega}}_{ba}$ in $\mathcal{F}_b$:

$$\underrightarrow{\boldsymbol{\omega}}_{ba} = \boldsymbol{\omega}_{ba}^T \mathcal{F}_b \tag{8}$$

Then, in a similar manner to the derivation of (B.2, 12), we have

$$\dot{\mathcal{F}}_b^T = \mathcal{F}_b^T \boldsymbol{\omega}_{ba}^{\times} \tag{9}$$

Comparing (5) and (7), we have

$$\mathcal{F}_a^T \dot{\mathbf{v}}_a = \mathcal{F}_b^T \left(\dot{\mathbf{v}}_b + \boldsymbol{\omega}_{ba}^{\times} \mathbf{v}_b\right) \tag{10}$$

Premultiplying by $\mathcal{F}_a \cdot$ and using (B.3, 1), we conclude that

$$\dot{\mathbf{v}}_a = \mathbf{C}_{ab}\left(\dot{\mathbf{v}}_b + \boldsymbol{\omega}_{ba}^{\times} \mathbf{v}_b\right) \tag{11}$$

This is the matrix equivalent of (7).

Where higher time derivatives are required, the above relations are used recursively. The second derivative of $\underrightarrow{\mathbf{v}}$ as seen in $\mathcal{F}_a$ is, from (7),

$$\begin{aligned}\ddot{\underrightarrow{\mathbf{v}}} &= (\dot{\underrightarrow{\mathbf{v}}})^{\circ} + \underrightarrow{\boldsymbol{\omega}}_{ba} \times \dot{\underrightarrow{\mathbf{v}}} \\ &= \left(\overset{\circ\circ}{\underrightarrow{\mathbf{v}}} + \underrightarrow{\boldsymbol{\omega}}_{ba} \times \mathring{\underrightarrow{\mathbf{v}}} + \mathring{\underrightarrow{\boldsymbol{\omega}}}_{ba} \times \underrightarrow{\mathbf{v}}\right) + \underrightarrow{\boldsymbol{\omega}}_{ba} \times \left(\mathring{\underrightarrow{\mathbf{v}}} + \underrightarrow{\boldsymbol{\omega}}_{ba} \times \underrightarrow{\mathbf{v}}\right) \\ &= \overset{\circ\circ}{\underrightarrow{\mathbf{v}}} + 2\underrightarrow{\boldsymbol{\omega}}_{ba} \times \mathring{\underrightarrow{\mathbf{v}}} + \mathring{\underrightarrow{\boldsymbol{\omega}}}_{ba} \times \underrightarrow{\mathbf{v}} + \underrightarrow{\boldsymbol{\omega}}_{ba} \times \left(\underrightarrow{\boldsymbol{\omega}}_{ba} \times \underrightarrow{\mathbf{v}}\right)\end{aligned} \tag{12}$$

One may write $\dot{\underrightarrow{\boldsymbol{\omega}}}_{ba}$ instead of $\mathring{\underrightarrow{\boldsymbol{\omega}}}_{ba}$ in this equation since they are equal. The matrix equivalent of (12) is found by inserting

$$\ddot{\underrightarrow{\mathbf{v}}} = \mathcal{F}_a^T \ddot{\mathbf{v}}_a; \qquad \overset{\circ\circ}{\underrightarrow{\mathbf{v}}} = \mathcal{F}_b^T \ddot{\mathbf{v}}_b; \qquad \mathring{\underrightarrow{\boldsymbol{\omega}}}_{ba} = \mathcal{F}_b^T \dot{\boldsymbol{\omega}}_{ba} \tag{13}$$

together with (4) and (6); the result is

$$\ddot{\mathbf{v}}_a = \mathbf{C}_{ab}\left[\ddot{\mathbf{v}}_b + 2\boldsymbol{\omega}_{ba}^{\times}\dot{\mathbf{v}}_b + \left(\dot{\boldsymbol{\omega}}_{ba}^{\times} + \boldsymbol{\omega}_{ba}^{\times}\boldsymbol{\omega}_{ba}^{\times}\right)\mathbf{v}_b\right] \tag{14}$$

Once the symbols are thoroughly understood, matrix relations such as (14) may be written directly from the vectorial form (12). Vectrices, however, make the derivation precise and should always be used whenever confusion might be occasioned by the simultaneous use of more than one reference frame. In general, time derivatives of vectors in different reference frames generally require that one of the following two identities

$$\overset{\circ}{\mathscr{F}}_a = -\underset{\rightarrow}{\boldsymbol{\omega}}_{ba} \times \mathscr{F}_a \tag{15}$$

$$\overset{\circ}{\mathscr{F}}_a = -\mathscr{F}_a^T\mathbf{C}_{ab}\boldsymbol{\omega}_{ba}^{\times}\mathbf{C}_{ba} \tag{16}$$

be used in the derivation.

Similar procedures hold for dyadics $\underset{\rightarrow}{\mathbf{D}}(t)$. Using the above notation for time derivatives measured in $\mathscr{F}_a$ and $\mathscr{F}_b$, we can derive the following results using vectrices:

$$\underset{\rightarrow}{\dot{\mathbf{D}}} = \mathscr{F}_a^T\dot{\mathbf{D}}_a\mathscr{F}_a = \mathscr{F}_b^T\left(\dot{\mathbf{D}}_b + \boldsymbol{\omega}_{ba}^{\times}\mathbf{D}_b - \mathbf{D}_b\boldsymbol{\omega}_{ba}^{\times}\right)\mathscr{F}_b \tag{17}$$

$$\dot{\mathbf{D}}_a = \mathbf{C}_{ab}\left(\dot{\mathbf{D}}_b + \boldsymbol{\omega}_{ba}^{\times}\mathbf{D}_b - \mathbf{D}_b\boldsymbol{\omega}_{ba}^{\times}\right)\mathbf{C}_{ba} \tag{18}$$

$$\begin{aligned}\ddot{\mathbf{D}}_a = \mathbf{C}_{ab}\big[&\ddot{\mathbf{D}}_b + 2\boldsymbol{\omega}_{ba}^{\times}\dot{\mathbf{D}}_b - 2\dot{\mathbf{D}}_b\boldsymbol{\omega}_{ba}^{\times} - 2\boldsymbol{\omega}_{ba}^{\times}\mathbf{D}_b\boldsymbol{\omega}_{ba}^{\times} \\ &+ \left(\dot{\boldsymbol{\omega}}_{ba}^{\times} + \boldsymbol{\omega}_{ba}^{\times}\boldsymbol{\omega}_{ba}^{\times}\right)\mathbf{D}_b - \mathbf{D}_b\left(\dot{\boldsymbol{\omega}}_{ba}^{\times} - \boldsymbol{\omega}_{ba}^{\times}\boldsymbol{\omega}_{ba}^{\times}\right)\big]\mathbf{C}_{ba}\end{aligned} \tag{19}$$

As a last example of dyadic operations, if $\underset{\rightarrow}{\mathbf{h}}(t) = \underset{\rightarrow}{\mathbf{I}}(t) \cdot \underset{\rightarrow}{\boldsymbol{\omega}}(t)$ and all three are expressed in $\mathscr{F}_b$, then

$$\underset{\rightarrow}{\dot{\mathbf{h}}} = \mathscr{F}_b^T\left(\mathbf{I}\dot{\boldsymbol{\omega}}_{ba} + \dot{\mathbf{I}}\boldsymbol{\omega}_{ba} + \boldsymbol{\omega}_{ba}^{\times}\mathbf{I}\boldsymbol{\omega}_{ba}\right) \tag{20}$$

This example has relevance to attitude dynamics, with $\underset{\rightarrow}{\mathbf{h}}$, $\underset{\rightarrow}{\mathbf{I}}$, and $\underset{\rightarrow}{\boldsymbol{\omega}}$ being respectively an angular momentum vector, an inertia dyadic, and an angular velocity.

B.5 DERIVATIVE WITH RESPECT TO A VECTOR

Consider a scalar function of $\underset{\rightarrow}{\mathbf{v}}$, $f(\underset{\rightarrow}{\mathbf{v}})$. It is sometimes useful (see Chapter 3) to define the derivative of f with respect to $\underset{\rightarrow}{\mathbf{v}}$. We are interested only in functions for which the Taylor approximation

$$f(\underset{\rightarrow}{\mathbf{v}} + d\underset{\rightarrow}{\mathbf{v}}) = f(\underset{\rightarrow}{\mathbf{v}}) + \left(\frac{df}{d\underset{\rightarrow}{\mathbf{v}}}\right) \cdot d\underset{\rightarrow}{\mathbf{v}} + o(\|d\underset{\rightarrow}{\mathbf{v}}\|) \tag{1}$$

is valid. Equation (1) has introduced the derivative $df/d\underset{\rightarrow}{\mathbf{v}}$ in a setting that indicates its intended meaning: it is itself a vector whose projection in any direction is readily found from (1). In particular, its projection in the direction defined by the unit vector $\underset{\rightarrow}{\mathbf{n}}$ is

$$\underset{\rightarrow}{\mathbf{n}} \cdot \left(\frac{df}{d\underset{\rightarrow}{\mathbf{v}}}\right) = \lim_{\varepsilon \to 0} \frac{f(\underset{\rightarrow}{\mathbf{v}} + \varepsilon\underset{\rightarrow}{\mathbf{n}}) - f(\underset{\rightarrow}{\mathbf{v}})}{\varepsilon} \tag{2}$$

When the limit in (2) exists, the directional derivative exists.

Since $df/d\underrightarrow{\mathbf{v}}$ is a vector, its components can be expressed in any reference frame. The most important such instance occurs when $\underrightarrow{\mathbf{v}}$ and $df/d\underrightarrow{\mathbf{v}}$ are both expressed in the same frame. Thus, if

$$\mathbf{v} \triangleq \mathscr{F}_a \cdot \underrightarrow{\mathbf{v}} \tag{3}$$

then, from (2),

$$\mathscr{F}_a \cdot \left(\frac{df}{d\underrightarrow{\mathbf{v}}}\right) = \lim_{\varepsilon \to 0} \begin{bmatrix} f(\underrightarrow{\mathbf{v}} + \varepsilon \underrightarrow{\hat{\mathbf{a}}}_1) - f(\underrightarrow{\mathbf{v}}) \\ f(\underrightarrow{\mathbf{v}} + \varepsilon \underrightarrow{\hat{\mathbf{a}}}_2) - f(\underrightarrow{\mathbf{v}}) \\ f(\underrightarrow{\mathbf{v}} + \varepsilon \underrightarrow{\hat{\mathbf{a}}}_3) - f(\underrightarrow{\mathbf{v}}) \end{bmatrix} \div \varepsilon \tag{4}$$

The entries in (4) are scalars, and they may therefore be calculated in any reference frame, including $\mathscr{F}_a$. Therefore,

$$\mathscr{F}_a \cdot \left(\frac{df}{d\underrightarrow{\mathbf{v}}}\right) = \lim_{\varepsilon \to 0} \begin{bmatrix} f(\mathbf{v} + \varepsilon \mathbf{1}_1) - f(\mathbf{v}) \\ f(\mathbf{v} + \varepsilon \mathbf{1}_2) - f(\mathbf{v}) \\ f(\mathbf{v} + \varepsilon \mathbf{1}_3) - f(\mathbf{v}) \end{bmatrix} \div \varepsilon \tag{5}$$

where $\mathbf{1}_1 \triangleq [1 \quad 0 \quad 0]^T$, and so on. The entries in (5) are evidently the partial derivatives of f with respect to the components of $\mathbf{v}$:

$$\mathscr{F}_a \cdot \left(\frac{\partial f}{\partial \underrightarrow{\mathbf{v}}}\right) \equiv \frac{\partial f}{\partial \mathbf{v}} \triangleq \begin{bmatrix} \dfrac{\partial f}{\partial v_1} \\ \dfrac{\partial f}{\partial v_2} \\ \dfrac{\partial f}{\partial v_3} \end{bmatrix} \tag{6}$$

If there are two reference frames involved, $\mathscr{F}_a$ and $\mathscr{F}_b$, then $\underrightarrow{\mathbf{v}}$ and $df/d\underrightarrow{\mathbf{v}}$ can be expressed in either $\mathscr{F}_a$ or $\mathscr{F}_b$. Let

$$\mathbf{v}_a \triangleq \mathscr{F}_a \cdot \underrightarrow{\mathbf{v}}; \qquad \mathbf{v}_b \triangleq \mathscr{F}_b \cdot \underrightarrow{\mathbf{v}}; \qquad \mathbf{v}_b = \mathbf{C}_{ba}\mathbf{v}_a \tag{7}$$

By applying (6) twice, we get

$$\mathscr{F}_a \cdot \left(\frac{df}{d\underrightarrow{\mathbf{v}}}\right) = \frac{\partial f}{\partial \mathbf{v}_a}; \qquad \mathscr{F}_b \cdot \left(\frac{df}{d\underrightarrow{\mathbf{v}}}\right) = \frac{\partial f}{\partial \mathbf{v}_b} \tag{8}$$

Recall from (B.3, 4) that $\mathscr{F}_b = \mathbf{C}_{ba}\mathscr{F}_a$, so the second of (8) can be converted to

$$\mathbf{C}_{ba}\mathscr{F}_a \cdot \left(\frac{df}{d\underrightarrow{\mathbf{v}}}\right) = \frac{\partial f}{\partial \mathbf{v}_b} \tag{9}$$

from which we extract

$$\frac{\partial f}{\partial \mathbf{v}_b} = \mathbf{C}_{ba}\frac{\partial f}{\partial \mathbf{v}_a} \tag{10}$$

This is the result desired.

In dynamics applications, the time dependence of $\underset{\rightarrow}{\mathbf{v}}$ is important and leads implicitly to $f(t)$. From (1),

$$\dot{f} = \dot{\underset{\rightarrow}{\mathbf{v}}} \cdot \left(\frac{df}{d\underset{\rightarrow}{\mathbf{v}}} \right) = (\overset{\circ}{\underset{\rightarrow}{\mathbf{v}}} + \underset{\rightarrow}{\boldsymbol{\omega}}_{ba} \times \underset{\rightarrow}{\mathbf{v}}) \cdot \left(\frac{df}{d\underset{\rightarrow}{\mathbf{v}}} \right)$$

$$= \dot{\mathbf{v}}_a^T \frac{\partial f}{\partial \mathbf{v}_a} = (\dot{\mathbf{v}}_b + \boldsymbol{\omega}_{ba}^{\times} \mathbf{v}_b)^T \frac{\partial f}{\partial \mathbf{v}_b} \tag{11}$$

(The conventions of Section B.4 are still in force.) Moreover, the vector $df/d\underset{\rightarrow}{\mathbf{v}}$ itself will generally change with time, and its time derivatives, as seen in $\mathscr{F}_a$ and $\mathscr{F}_b$, are respectively denoted by $(df/d\underset{\rightarrow}{\mathbf{v}})^{\bullet}$ and $(df/d\underset{\rightarrow}{\mathbf{v}})^{\circ}$. The relationship between them is, as always,

$$\left(\frac{df}{d\underset{\rightarrow}{\mathbf{v}}} \right)^{\bullet} = \left(\frac{df}{d\underset{\rightarrow}{\mathbf{v}}} \right)^{\circ} + \underset{\rightarrow}{\boldsymbol{\omega}}_{ba} \times \frac{df}{d\underset{\rightarrow}{\mathbf{v}}} \tag{12}$$

From (8),

$$\frac{df}{d\underset{\rightarrow}{\mathbf{v}}} = \mathscr{F}_a^T \frac{\partial f}{\partial \mathbf{v}_a} = \mathscr{F}_b^T \frac{\partial f}{\partial \mathbf{v}_b} \tag{13}$$

and so the three scalar equation equivalent to (13) are assembled as the matrix equation

$$\frac{d}{dt}\left(\frac{\partial f}{\partial \mathbf{v}_a} \right) = \mathbf{C}_{ab}\left[\frac{d}{dt}\left(\frac{\partial f}{\partial \mathbf{v}_b} \right) + \boldsymbol{\omega}_{ba}^{\times} \frac{\partial f}{\partial \mathbf{v}_b} \right] \tag{14}$$

This result should be compared with (B.4, 11).

APPENDIX C

LIST OF SYMBOLS

This appendix lists only principal symbols and notations; these are used as consistently as possible throughout the book. Many other symbols, including those listed below, have other meanings in local contexts; such symbols are defined as and when they occur.

C.1 LOWERCASE SYMBOLS

Note: for Greek symbols, see Section C.3.

$\underrightarrow{\mathbf{a}}$	unit vector defining rotation in Euler's theorem (Chapter 2)
$\mathbf{a}$	components of $\underrightarrow{\mathbf{a}}$ (above)
$\underrightarrow{\mathbf{a}}$	unit vector along wheel axis, or along rigid-body symmetry axis
$\mathbf{a}$	components in $\mathscr{F}_p$ of $\underrightarrow{\mathbf{a}}$ (above) (see Section C.2)
a	semimajor diameter of elliptic orbit
$\mathbf{c}$	first moment of inertia, $\triangleq \int \underrightarrow{\mathbf{r}}\, dm$
$\mathbf{c}$	components of $\underrightarrow{\mathbf{c}}$ in $\mathscr{F}_b$ or $\mathscr{F}_p$ (see Section C.2)
$\mathbf{c}_1, \mathbf{c}_2, \mathbf{c}_3$	columns of $\mathbf{C}_{bo} \triangleq \mathscr{F}_b \cdot \mathscr{F}_o^T$; often $\mathscr{F}_b = \mathscr{F}_p$, and often $\mathscr{F}_o = \mathscr{F}_c$ (see Section C.2 and Appendix B)
$\mathbf{c}_1^s, \mathbf{c}_2^s, \mathbf{c}_3^s$	columns of $\mathbf{C}_{sc} \triangleq \mathscr{F}_s \cdot \mathscr{F}_c^T$ (see Section C.2 and Appendix B)
c_1, c_2, c_3	direction cosines of $\underrightarrow{\mathbf{h}}$ in $\mathscr{F}_p$ (see Section C.2 and Appendix B)
c_{ij}	$(i, j = 1, 2, 3)$ elements of $\mathbf{C}_{bo} \triangleq \mathscr{F}_b \cdot \mathscr{F}_o^T$; often $\mathscr{F}_b = \mathscr{F}_p$, and $\mathscr{F}_o \equiv \mathscr{F}_c$ (see Section C.2 and Appendix B)
$\mathbf{c}_p$	center of aerodynamic pressure
c_d	damper constant
cn	Jacobian elliptic function, $cn(t; k)$
d_{ij}	$(i, j = 1, \ldots, N)$ elements of damping matrix $\mathscr{D}$
dn	Jacobian elliptic function, $dn(t; k)$
e	orbit eccentricity
$\underrightarrow{\mathbf{f}}$	total external force on vehicle
$\mathbf{f}$	components of $\underrightarrow{\mathbf{f}}$ in $\mathscr{F}_b$ or $\mathscr{F}_p$ (see Section C.2)

$\boldsymbol{f}$	$N \times 1$ matrix of generalized forces
$\underrightarrow{\mathbf{g}}$	total external torque on vehicle, about an agreed reference point in vehicle
$\mathbf{g}$	components of $\underrightarrow{\mathbf{g}}$ in $\mathscr{F}_b$ or $\mathscr{F}_p$ (see Section C.2)
$\mathscr{g}_{ij}$	$(i, j = 1, \ldots, N)$ element of gyricity matrix $\mathscr{G}$
$\underrightarrow{\mathbf{h}}$	angular momentum of vehicle, about an agreed reference point in vehicle
$\mathbf{h}$	components of $\underrightarrow{\mathbf{h}}$ in $\mathscr{F}_b$ or $\mathscr{F}_p$ (see Section C.2)
h	magnitude of $\underrightarrow{\mathbf{h}}$, $\triangleq \lVert\underrightarrow{\mathbf{h}}\rVert = \lVert\mathbf{h}\rVert$
$\underrightarrow{\mathbf{h}}_s$	angular momentum of wheel $\mathscr{W}$ relative to body $\mathscr{R}$
$\mathbf{h}_s$	angular momentum of vehicle in $\mathscr{F}_s$ (see Section C.2)
$\underrightarrow{\mathbf{h}}_w$	absolute angular momentum of the wheel $\mathscr{W}$
h_a	magnitude of component of absolute angular momentum along symmetry axis
h_t	magnitude of component of absolute angular momentum transverse to symmetry axis
j	"imaginary" operator: $j^2 = -1$
k_1, k_2, k_3	dimensionless inertia ratios: $k_1 = (I_2 - I_3)/I_1$; $k_2 = (I_1 - I_3)/I_2$; $k_3 = (I_2 - I_1)/I_3$
k_t	dimensionless inertia ratio for symmetrical vehicles: $k_t = (I_a - I_t)/I_t$
k_d	spring constant for damper
k	argument in $K(k)$, the Legendre complete elliptic integral (first kind)
$\mathscr{k}_{ij}$	$(i, j = 1, \ldots, N)$ elements of "stiffness" matrix $\mathscr{K}$
m	vehicle mass
m_b	mass of rigid body $\mathscr{R}$
m_d	mass of damper $\mathscr{P}$
m_p	mass of a gravitational primary
$\mathscr{m}_{ij}$	$(i, j = 1, \ldots, N)$ elements of inertia matrix $\mathscr{M}$
$\underrightarrow{\mathbf{n}}$	unit normal vector
$\mathbf{n}$	components of $\underrightarrow{\mathbf{n}}$ (above)
n	system order, as defined in the theory of ordinary differential equations
$\underrightarrow{\mathbf{p}}$	momentum of vehicle
$\mathbf{p}$	components of $\underrightarrow{\mathbf{p}}$ in $\mathscr{F}_b$ or $\mathscr{F}_p$ (see Section C.2)
$\boldsymbol{p}$	$N \times 1$ matrix of generalized momenta
p	orbital parameter; for elliptic orbits, $p \triangleq a(1 - e^2)$
p	solar radiation pressure per unit area of totally absorbing surface
$\mathbf{q}$	$N \times 1$ matrix of generalized coordinates
$\underrightarrow{\mathbf{r}}$	position vector within a body
$\mathbf{r}$	components of $\underrightarrow{\mathbf{r}}$ in $\mathscr{F}_b$ or $\mathscr{F}_c$ (see Section C.2); three independent scalar position variables
$\mathbf{r}_c$	position of mass center, $\mathbf{r}_c \triangleq \mathbf{c}/\mathrm{m} = \text{constant}$
s	Laplace transform variable
sn	Jacobian elliptic function, $sn(t; k)$
$\underrightarrow{\mathbf{v}}$	velocity
$\boldsymbol{v}$	$N \times 1$ matrix of generalized velocities
v	Liapunov function

C.2 UPPERCASE SYMBOLS

Note: for Greek symbols, see Section C.4.

The notation used for specific reference frames is summarized first, followed by other major uppercase notation.

Reference Frames (see also Appendix B)

$\mathscr{F}_a$ a generic dextral orthonormal reference frame represented by the vectrix $\mathscr{F}_a \triangleq [\hat{\underset{\to}{\mathbf{a}}}_1 \quad \hat{\underset{\to}{\mathbf{a}}}_2 \quad \hat{\underset{\to}{\mathbf{a}}}_3]^T$

$\mathscr{F}_b$ body-fixed reference frame represented by the vectrix $\mathscr{F}_b \triangleq [\hat{\underset{\to}{\mathbf{b}}}_1 \quad \hat{\underset{\to}{\mathbf{b}}}_2 \quad \hat{\underset{\to}{\mathbf{b}}}_3]^T$

$\mathscr{F}_c$ "orbiting" frame—the notation for $\mathscr{F}_o$ when the orbit is circular; represented by the vectrix $\mathscr{F}_c \triangleq [\hat{\underset{\to}{\mathbf{c}}}_1 \quad \hat{\underset{\to}{\mathbf{c}}}_2 \quad \hat{\underset{\to}{\mathbf{c}}}_3]^T$

$\mathscr{F}_i$ inertial reference frame (the frame in which Newton's laws are valid; for the purposes of this book, $\mathscr{F}_i$ is an Earth-centered frame, having axes fixed with respect to the Milky Way); represented by the vectrix $\mathscr{F}_i \triangleq [\hat{\underset{\to}{\mathbf{i}}}_1 \quad \hat{\underset{\to}{\mathbf{i}}}_2 \quad \hat{\underset{\to}{\mathbf{i}}}_3]^T$

$\mathscr{F}_o$ "orbiting" frame, represented by the vectrix $\mathscr{F}_o \triangleq [\hat{\underset{\to}{\mathbf{o}}}_1 \quad \hat{\underset{\to}{\mathbf{o}}}_2 \quad \hat{\underset{\to}{\mathbf{o}}}_3]^T$; here, $\hat{\underset{\to}{\mathbf{o}}}_3$ is a unit vector pointing to the center of Earth, $\hat{\underset{\to}{\mathbf{o}}}_1$ is also in the orbital plane, "forward" and perpendicular to $\hat{\underset{\to}{\mathbf{o}}}_3$; and $\hat{\underset{\to}{\mathbf{o}}}_2 = \hat{\underset{\to}{\mathbf{o}}}_3 \times \hat{\underset{\to}{\mathbf{o}}}_1$ (see Fig. 9.1)

$\mathscr{F}_p$ principal-axis reference frame—the notation for $\mathscr{F}_b$ when the inertia matrix is diagonal; represented by the vectrix $[\hat{\underset{\to}{\mathbf{p}}}_1 \quad \hat{\underset{\to}{\mathbf{p}}}_2 \quad \hat{\underset{\to}{\mathbf{p}}}_3]^T$

$\mathscr{F}_{\bar{p}}$ notation for $\mathscr{F}_p$ when the gyrostat $\mathscr{R} + \mathscr{W}$ is in its reference equilibrium orientation (see Section 11.1)

$\mathscr{F}_s$ "stroboscopic" reference frame—a frame whose $\hat{\underset{\to}{\mathbf{s}}}_2$ axis coincides with the spin of a symmetrical spinner in a circular orbit and whose $\hat{\underset{\to}{\mathbf{s}}}_1$ and $\hat{\underset{\to}{\mathbf{s}}}_3$ axes are found by (constant) rotations, $\bar{\gamma}_1$ and $\bar{\gamma}_3$, with respect to $\mathscr{F}_c$

$\mathscr{F}_w$ wheel-fixed principal-axis frame

Other Uppercase Symbols

$\mathscr{A}$ $N \times N$ constraint damping matrix

$\mathscr{B}$ a body, either rigid or flexible

$\underset{\to}{\mathbf{B}}$ magnetic flux density (usually Earth's field)

$\mathbf{C}_{ab}$ rotation matrix, defined as $\mathscr{F}_a \cdot \mathscr{F}_b^T$ (see Appendix B); elements are direction cosines

$\mathbf{C}$ shorthand for $\mathbf{C}_{bo} \triangleq \mathscr{F}_b \cdot \mathscr{F}_o^T$; often $\mathscr{F}_b = \mathscr{F}_p$ or $\mathscr{F}_o = \mathscr{F}_c$ (see definitions of reference frames above and Appendix B)

$\mathbf{C}_1, \mathbf{C}_2, \mathbf{C}_3$ principal rotations (see Section 2.1)

$\mathscr{D}$ $N \times N$ damping matrix

$\mathscr{G}$ $N \times N$ gyricity matrix (see footnote, p. 511)

G universal gravitational constant [see equation (8.1,2)]

$\mathbf{H}$ Hurwitz matrix [see equation (A.3,16)]

H Hamiltonian function

H Heaviside function: $H(x) = 1$ for $x \geq 0$; $H = 0$ for $x < 0$

$\mathscr{H}$ angular momentum ellipsoid

$\underset{\to}{\mathbf{I}}$ second-moment-of-inertia dyadic, taken about mass center

$\mathbf{I}$ second-moment-of-inertia matrix, taken about mass center: $\underset{\to}{\mathbf{I}}$ expressed in $\mathscr{F}_b$

I_{ij} $(i, j = 1,2,3)$ elements of $\mathbf{I}$

I_1, I_2, I_3 elements of $\mathbf{I}$ when $\mathscr{F}_b = \mathscr{F}_p$, that is, principal moments of inertia

$\mathscr{I}$ second-moment-of-inertia matrix, taken about mass center: $\underset{\to}{\mathbf{I}}$ expressed in $\mathscr{F}_c$

$\mathscr{I}_{ij}$ $(i, j = 1,2,3)$ elements of $\mathscr{I}$

I $h^2/2T$

I_a axial moment of inertia

I_s moment of inertia of wheel or rotor, $\mathscr{W}$, about spin axis

I_t transverse momentum of inertia

$\underset{\rightarrow}{\mathbf{J}}$	second-moment-of-inertia dyadic, taken about body-fixed point other than mass center
$\mathbf{J}$	second-moment-of-inertia matrix, taken about body-fixed point other than mass center: $\underset{\rightarrow}{\mathbf{J}}$ expressed in $\mathscr{F}_b$
J_{ij}	$(i, j = 1, 2, 3)$ elements of $\mathbf{J}$
J_s	h_s/ω_c
J_{si}	$J_s \mathbf{c}_i^T \mathbf{a}$ $(i = 1, 2, 3)$
$\mathscr{K}$	$N \times N$ "stiffness' matrix
$K(k)$	Legendre's complete elliptic integral (first kind)
$\mathscr{M}$	$N \times N$ system inertia matrix
N	number of generalized coordinates, that is, number of degrees of freedom
$\mathbf{O}$	rectangular (or square) matrix of zeros ($\mathbf{0} \equiv$ zero column; $\mathbf{0}^T \equiv$ zero row)
$\mathscr{P}$	a point mass, that is, a "body" possessing mass but which is of zero size
$\mathscr{Q}$	a quasi-rigid body (see Chapter 5)
$\mathscr{Q}_w$	a quasi-rigid wheel or rotor
$\underset{\rightarrow}{\mathbf{R}}$	position vector
R	distance from Earth center to satellite
$\mathscr{R}$	a rigid body
T	kinetic energy
$\mathscr{T}$	kinetic energy ellipsoid
$\underset{\rightarrow}{\mathbf{V}}$	local velocity of ideal (nonrotating) atmosphere relative to satellite
$\underset{\rightarrow}{\mathbf{V}}_R$	local velocity of atmosphere relative to satellite
V	potential energy
$\mathscr{V}$	vehicle
$\mathscr{W}$	wheel or rotor

C.3 LOWERCASE GREEK SYMBOLS

$\boldsymbol{\alpha}$	librations of $\mathscr{F}_b$ with respect to uniformly spinning reference motion; often $\mathscr{F}_b = \mathscr{F}_p$
$\alpha_1, \alpha_2, \alpha_3$	elements of $\boldsymbol{\alpha}$; if the uniformly spinning reference motion corresponds to an Earth-pointing, orbiting frame, $\mathscr{F}_o$ (or $\mathscr{F}_c$), then α_1, α_2, and α_3 are respectively called roll, pitch, and yaw.
α	angle of attack
$\boldsymbol{\beta}$	angles of internal wheel with respect to main body
$\boldsymbol{\gamma}$	librations of stroboscopic frame
γ	nutation angle
$\boldsymbol{\varepsilon}$	first three Euler parameters $\{\varepsilon_1, \varepsilon_2, \varepsilon_3\}$ [see equation (2.2, 9)]
$\boldsymbol{\theta}$	rotation angles giving absolute orientation of body, that is, $\mathscr{F}_b$ with respect to $\mathscr{F}_i$
κ_C	dimensionless parameter for conical equilibrium of gyrostat [see equation (11.1, 28)]
κ_H	dimensionless parameter for hyperbolic equilibrium of gyrostat; [see equation (11.1, 22)]
λ	eigenvalue
λ	proportionality factor in gyrostat equilibrium equation [see equation (6.1, 23)]
μ	Gm_p, the gravitational constant for an attraction primary
$\underset{\rightarrow}{\boldsymbol{\nu}}$	constant reference spin velocity vector

$\boldsymbol{\nu}$	components of $\underset{\rightarrow}{\boldsymbol{\nu}}$ in $\mathscr{F}_b$ or $\mathscr{F}_p$
ν	constant reference spin rate, $= \|\boldsymbol{\nu}\|$
$\hat{\nu}$	ν/ω_c
ξ	coordinate associated with damper (see Fig. 3.6)
$\boldsymbol{\eta}$	$N \times 1$ matrix of modal coordinates
η	fourth Euler parameter [see equation (2.2, 10)]
η	orbital anomaly (see Fig. 9.1)
$\underset{\rightarrow}{\boldsymbol{\rho}}$	position vector relative to mass center
ρ_a	atmospheric density
$\sigma(\underset{\rightarrow}{\mathbf{r}})$	mass density at $\underset{\rightarrow}{\mathbf{r}}$
ϕ	angle in Euler's theorem (see Fig. 2.3)
χ_1	gyrostat offset about roll axis in conical equilibrium (see Fig. 11.5)
χ_3	gyrostat offset about yaw axis in hyperbolic equilibrium (see Fig. 11.5)
$\underset{\rightarrow}{\boldsymbol{\omega}}$	absolute angular velocity, that is, angular velocity with respect to $\mathscr{F}_i$
$\underset{\rightarrow}{\boldsymbol{\omega}}_{ba}$	angular velocity of $\mathscr{F}_b$ with respect to $\mathscr{F}_a$
$\underset{\rightarrow}{\boldsymbol{\omega}}_s$	angular velocity of $\mathscr{W}$ relative to $\mathscr{R}$
$\underset{\rightarrow}{\boldsymbol{\omega}}_w$	angular velocity of $\mathscr{W}$ relative to $\mathscr{F}_i$
$\boldsymbol{\omega}_\alpha$	angular velocity of $\mathscr{F}_p$ with respect to $\mathscr{F}_o$ (usually, $\mathscr{F}_o = \mathscr{F}_c$)
$\boldsymbol{\omega}_\gamma$	angular velocity of $\mathscr{F}_s$ with respect to $\mathscr{F}_c$
ω_o	orbital circular frequency; that is, $\omega_o = 2\pi/P_o$, where P_o is orbital period
ω_c	value of ω_o when orbit is circular; $\omega_c = (\mu/R^3)^{1/2}$

C.4 UPPERCASE GREEK SYMBOLS

$\underset{\rightarrow}{\boldsymbol{\Omega}}$	spin rate of a precessing system, relative to the precessing frame
Ω	magnitude of $\underset{\rightarrow}{\boldsymbol{\Omega}}$
$\underset{\rightarrow}{\boldsymbol{\Omega}}_p$	angular velocity of precession
Ω_p	magnitude of $\underset{\rightarrow}{\boldsymbol{\Omega}}_p$
$\boldsymbol{\Omega}$	$N \times N$ diagonal matrix of natural frequencies; $\boldsymbol{\Omega} = \text{diag}\{\omega_1, \ldots, \omega_N\}$

C.5 OTHER NOTATIONAL CONVENTIONS

Subscripts

a	absorbed; or axial
b	main body
c	relating to mass center; circular orbit; or control
C	conical gyrostat equilibrium
d	diffuse; disturbance; or damper
eff	effective
g	gravity
H	hyperbolic gyrostat equilibrium
i	inertial
m	magnetic
n	normal
o	initial value
p	pitch

r	reflected
ry	roll/yaw
s	specular; or stroboscopic
t	transverse; or tangential
w	wheel

Other

$\underrightarrow{(\;)}$	a vector
$\mathbf{A}, \mathbf{a}, \ldots$	matrices
$\mathbf{A}^T$	transpose
$\mathbf{A}^H$	Hermitian transpose $\overline{\mathbf{A}}^T$
$\mathbf{v}^\times$	cross-product operation on 3×1 matrix $\mathbf{v}$ [see equation (B.2, 13)]
$\underrightarrow{\hat{(\;)}}$	unit vector
$\overline{(\;)}$	complex conjugate; or average value
$\dot{\underrightarrow{(\;)}}$	absolute time derivative (in $\mathscr{F}_i$)
$\overset{\circ}{\underrightarrow{(\;)}}$	relative time derivative (in a rotating frame)
$\doteq$	approximately equal
$\triangleq$	defined equal
$\equiv$	identically equal
$\lvert\;\rvert$	absolute value
$\lVert\;\rVert$	norm
$\mathbf{A} > 0$	symmetric matrix $\mathbf{A}$ is positive-definite (has all eigenvalues real and positive)
$\mathbf{A} \geq 0$	symmetric matrix $\mathbf{A}$ is positive-semidefinite (has all eigenvalues real, with at least one positive and at least one zero)
$\mathbf{A} \not< 0$	symmetric matrix $\mathbf{A}$ is nonnegative-definite ($\mathbf{A} \geq 0$ or $\mathbf{A} > 0$)
$\mathbf{A} < 0$	symmetric matrix $\mathbf{A}$ is negative definite ($-\mathbf{A} > 0$)
$\mathbf{A} \leq 0$	symmetric matrix $\mathbf{A}$ is negative-semidefinite ($-\mathbf{A} \geq 0$)
$\mathbf{A} \not> 0$	symmetric matrix $\mathbf{A}$ is nonpositive-definite ($-\mathbf{A} \not< 0$)
$\mathbf{O}$	rectangular zero matrix (possibly square)
$\mathbf{0}$	zero column matrix
$\mathbf{0}^T$	zero row matrix
$O(\varepsilon^2)$	terms of order ε^2 and higher
$o(\varepsilon)$	terms such that $o(\varepsilon)/\varepsilon \to 0$ as $\varepsilon \to 0$
$\partial f/\partial \underrightarrow{\mathbf{v}}$	see equation (B.5, 1) et seq.
$\partial f/\partial \mathbf{v}$	see equation (B.5, 8)
$\partial \mathbf{f}/\partial \mathbf{x}^T$	see equation (A.3, 4)
$\underrightarrow{\nabla}$	divergence operator
$\boldsymbol{\nabla}$	$[\partial/\partial r_1 \quad \partial/\partial r_2 \quad \partial/\partial r_3]^T$
$\oint$ or $\oiint$	surface integral
$\mathbf{1}$	unit matrix
$\mathbf{1}_1, \mathbf{1}_2, \mathbf{1}_3$	unit columns: $\begin{bmatrix} 1 \\ 0 \\ 0 \end{bmatrix}, \begin{bmatrix} 0 \\ 1 \\ 0 \end{bmatrix}, \begin{bmatrix} 0 \\ 0 \\ 1 \end{bmatrix}$
$\odot$	sun
$\oplus$	Earth
☾	moon

REFERENCES

Abzug, M. J.: "Active Satellite Attitude Control," in *Guidance and Control of Aerospace Vehicles* (Ed: C. T. Leondes), McGraw-Hill, New York, 1963, pp. 331–425.

Adams, G. N.: "Dual-Spin Spacecraft Dynamics During Platform Spinup," *J. Guid. Control*, **3** (No. 1), 29–36 (1980).

Alfriend, K. T., and Hubert, C. H.: "Stability of a Dual-Spin Satellite with Two Dampers," *J. Spacecr. Rockets*, **11** (No. 7), 469–474 (1974).

Ancher, L. J., Brink, H., and Pouw, A.: "Study on Passive Nutation Dampers" (in 3 volumes), ESA Contractor Report CR(P)-788, Noordwijk, Holland, December 1975.

Anchev, A. A.: "Equilibrium Attitude Transitions of a Three-Rotor Gyrostat in a Circular Orbit," *AIAA J.*, **11** (No. 4), 467–472 (1973).

Auelmann, R. R.: "Regions of Libration for a Symmetrical Satellite," *AIAA J.*, **1** (No. 6), 1445–1447 (1963).

Auelmann, R. R., and Lane, P. T.: "Design and Analysis of Ball-In-Tube Nutation Dampers," in *Proc. of the Symposium on Attitude Stabilization and Control of Dual-Spin Spacecraft*, USAF Rept. No. SAMSO-TR-68-191, 1968.

Bainum, P. M., Fuechsel, P. G., and Fedor, J. V.: "Stability of a Dual-Spin Spacecraft with a Flexible Momentum Wheel," *J. Spacecr. Rockets*, **9** (No. 9), 640–646 (1972).

Bainum, P. M., Fuechsel, P. G., and Mackison, D. L.: "Motion and Stability of a Dual-Spin Satellite with Nutation Damping," *J. Spacecr. Rockets*, **7** (No. 6), 690–696 (1970).

Bainum, P. M., and Mackison, D. L.: "Gravity-Gradient Stabilization of Synchronous Orbiting Satellites," *J. Br. Interplanet. Soc.*, **21**, 341–369 (1968).

Bandeen, W. R., and Manger, W. P.: "Angular Motion of the Spin Axis of the TIROS I Meteorological Satellite Due to Magnetic and Gravitational Torques," NASA TN D-571, Washington, D.C., April 1961.

Barba, P. M., and Aubrun, J. N.: "Satellite Attitude Acquisition by Momentum Transfer," *AIAA J.*, **14** (No. 10), 1382–1386 (1976).

Barba, P. M., Furumoto, N., and Leliakov, I. P.: "Techniques for Flat-Spin Recovery of Spinning Satellites," in *Proc. of the AIAA Guidance and Control Conference* (AIAA Paper 73-859), Key Biscayne, Fla., August 1973.

Beal, R. T., et al.: "System Performance and Attitude Sensing of Three Gravity-Gradient-Stabilized Satellites," NASA SP-107, Washington, D.C., 1966.

Beletskii, V. V.: *Motion of an Artificial Satellite About its Center of Mass* (translated from the Russian), NASA TT F-429, 1966.

Bellman, R.: *Stability Theory of Differential Equations*, McGraw-Hill, New York, 1953.

Beusch, J. U., and Smith, N. P.: "Stable Equilibria of Satellites Containing a Momentum Wheel in a Controlled Gimbal," *J. Spacecr. Rockets*, **8** (No. 7), 736–742 (1971).

Bisplinghoff, R. L., Ashley, H., and Halfman, R. L.: *Aeroelasticity*, Addison-Wesley, Reading, Mass., 1955.

Blanchard, D. E.: "Flight Results from the Gravity-Gradient Controlled RAE-1 Satellite," in *Proc. of the VI IFAC Automatic Control in Space Symposium*, Yerevan, U.S.S.R., 1974.

Bowers, E. J., Jr., and Williams, C. E.: "Optimization of RAE Satellite Boom Deployment Timing," *J. Spacecr. Rockets*, **7** (No. 9), 1057–1062 (1970).

Bracewell, R. N., and Garriott, O. K.: "Rotation of Artificial Earth Satellites," *Nature*, **182**, 760–762 (1958).

Breakwell, J. V., and Pringle, R., Jr.: "Nonlinear Resonances Affecting Gravity-Gradient Stability," in *Proc. of the XVI International Astronautical Congress*, Athens, September 1965, pp. 305–325.

Brereton, R. C., and Modi, V. J.: "Stability of the Planar Librational Motion of a Satellite in an Elliptic Orbit," in *Proc. of the XVII International Astronautical Congress*, Madrid, October 1966.

Butenin, N. V.: *Elements of the Theory of Nonlinear Oscillations*, Blaisdell, New York 1965.

Butler, P.: "The IMP-I (Explorer XVII) Satellite," in *Record of the IEEE International Space Electronics Symposium*, 1964.

Buxton, A. C., Campbell, D. E., and Losch, K.: "Rice/Wilberforce Gravity-Gradient Damping System," in NASA SP-107, 1966.

Caputo, M.: *The Gravity Field of the Earth from Classical and Modern Methods*, Academic, New York, 1967.

Cesari, L.: *Asymptotic Behavior and Stability Problems in Ordinary Differential Equations* (3rd edition), Springer-Verlag, New York, 1971.

Charlamov, P. V.: "On the Velocity Distribution in a Rigid Body" (in Russian), *Mech. Tverdovo Tela*, **1**, 77–81 (1969).

Cherchas, D. B., and Hughes, P.C.: "Attitude Stability of a Dual-Spin Satellite with a Large Flexible Solar Array," *J. Spacecr. Rockets*, **10** (No. 2), 126–132 (1973).

Chernous'ko, F. L.: "On the Motion of a Satellite about its Center of Mass Under the Action of Gravitational Moments," *PMM*, **27** (No. 3), 474–483 (1963).

Chetayev, N. G.: *Stability of Motion* (translated from the Russian), Pergamon, New York, 1961.

Coddington, E. A., and Levinson, N.: *Theory of Ordinary Differential Equations*, McGraw-Hill, New York, 1955.

Churchyard, J. N.: "A Universal Three-Angle Basis for Rotational Kinematic Analysis, Simulation, and Control," *J. Spacecr. Rockets*, **9** (No. 10), 779–781 (1972).

Clancy, T. F., and Mitchell, T. P.: "Effects of Radiation Forces on the Attitude of an Artificial Earth Satellite," *AIAA J.*, **2** (No. 3), 517–524 (1974).

Clark, J. P. C.: "Response of a Two-Body Gravity-Gradient System in a Slightly Eccentric Orbit," *J. Spacecr. Rockets*, **7** (No. 3), 294–298 (1970).

Cloutier, G. J. [1]: "Nutation Damper Design Principles for Dual-Spin Spacecraft," *J. Astronaut. Sci.*, **16** (No. 2), 79–87 (1969).

Cloutier, G. J. [2]: "Variable Spin Rate, Two-Degree-of-Freedom, Nutation Damper Dynamics," in *Astronautics 1975* (*Adv. Astronaut. Sci.*, **33**), Univelt, San Diego, Calif., 1975.

Cochran, J. E.: "Nonlinear Resonances in the Attitude Motion of Dual-Spin Spacecraft," *J. Spacecr. Rockets*, **14** (No. 9), 562–572 (1977).

Cochran, J. E., and Beaty, J. R.: "Near-Resonant and Transition Attitude Motion of a Class of Dual-Spin Spacecraft," *J. Astronaut. Sci.*, **26** (No. 1), 19–45 (1978).

Cochran, J. E., Jr., and Thompson, J. A.: "Nutation Dampers vs. Precession Dampers for Asymmetric Spacecraft," *J. Guid. Control*, **3**, 22–28 (1980).

Connell, G. M. [1]: "An Investigation of the Attitude Dynamics of the OV1-10 Satellites," Aerospace Corp. Rept. TR-0066(5306)-10, El Segundo, Calif., September 1969.

Connell, G. M. [2]: "Degree-of-Freedom Requirements for Hinged Multibody Satellites," *J. Spacecr. Rockets*, **6** (No. 4), 501–503 (1969).

Connell, G. M. [3]: "Optimal Configurations for Hinged, Two-Body Satellites," *J. Spacecr. Rockets*, **6** (No. 9), 1024–1030 (1969).

Connell, G. M., and Chobotov, V.: "Possible Effects of Boom Flutter on the Attitude Dynamics of the OV1-10 Satellite," *J. Spacecr. Rockets*, **6** (No. 1), 90–92 (1969).

Corliss, W. R. [1]: *Scientific Satellites*, NASA SP-133, 1967.

Corliss, W. R. [2]: *The Interplanetary Pioneers*, NASA SP-278, 1972.

Crespo da Silva, M. [1]: "Attitude Stability of a Gravity-Stabilized Gyrostat Satellite," *Celestial Mech.*, **2**, 147–165 (1970).

Crespo da Silva, M. [2]: "Non-Linear Resonant Attitude Motions in Gravity-Stabilized Gyrostat Satellites," *Int. J. Non-Linear Mech.*, **7**, 621–641 (1972).

Cronin, D. L.: "Influence of Nonlinear Rotor-Platform Interface on Dual-Spin Attitude Dynamics," *AIAA J.*, **14** (No. 10), 1395–1401 (1976).

Cupit, C. R.: "Rotation Matrix Generation," *Simulation*, **15**, 145–147 (1970).

Davidson, J. R., and Armstrong, R. L.: "Effect of Crew Motions on Spacecraft Orientation," *AIAA J.*, **9** (No. 2), 232–238 (1971).

Davison, G. J., and Merson, R. H.: "The Effect of Direct Solar Radiation on the Attitude of the SKYNET Spacecraft," *J. Br. Interplanet. Soc.*, **26**, 228–241 (1973).

DeBra, D. B.: "Principles and Developments in Passive Attitude Control," in *Recent Developments in Space Flight Mechanics*, (Ed.: E. Burgess), Univelt, San Diego, 1966.

DeBra, D. B., and Delp, R. H.: "Rigid Body Attitude Stability and Natural Frequencies in a Circular Orbit," *J. Astronaut. Sci.*, **8**, 14–17 (1961).

Doolin, B. F.: "Gravity Torque on an Orbiting Vehicle," NASA TN D-70, September 1959.

Dranovskii, V. I., and Yanshin, A. M.: "Influence of Dissipative Moments from Eddy Currents on Orientation of a Rotation-Stabilized Satellite," *Kosm. Issled.*, **13**, (No. 4), 487–494 (1975). (Translated by Plenum Publishing Corp.)

Duty, R. L., and Bean, J. T.: "An Attitude Reference System with Discrete-Correction Capability," *J. Spacecr. Rockets*, **9** (No. 9), 647–650 (1972).

Dvornychenko, V. N., and Gerding, R. B.: "The Effect or Orbital Eccentricity and Nodal Regression of Spin-Stabilized Satellites," *Celestial Mech.*, **16**, 263–274 (1977).

Eichelberger, R. J., and Gehring, J. W.: "Effects of Meteoroid Impacts on Space Vehicles," *ARS J.*, **32** (No. 10), 1583–1591 (1962).

Elrod, B. D.: "Quasi-Inertial Attitude Mode for Orbiting Spacecraft," *J. Spacecr. Rockets*, **9** (No. 12), 889–895 (1972).

Etkin, B. [1]: "Dynamics of Gravity-Oriented Orbiting Systems with Application to Passive Stabilization," *AIAA J.*, **2** (No. 6), 1008–1014 (1964).

Etkin, B. [2]: *Dynamics of Atmospheric Flight*, Wiley, New York, 1972.

Etkin, B., and Hughes, P. C.: "Explanation of the Anomalous Spin Behavior of Satellites with Long, Flexible Antennae," *J. Spacecr. Rockets*, **4** (No. 9), 1139–1145 (1967).

Evans, W. J.: "Aerodynamic and Radiation Disturbance Torques on Satellites Having Complex Geometry," in *Torques and Attitude Sensing in Earth Satellites* (Ed.: S. F. Singer), Academic, New York, 1964.

Fang, A. C., and Zimmerman, B. G.: "Digital Simulation of Rotational Kinematics," NASA TN D-5302, Washington, D.C., October 1969.

Fischell, R. E., and Mobley, F. F.: "Gravity-Gradient Stabilization Studies with the DODGE Satellite," AIAA Paper No. 70-69, AIAA Eighth Aerospace Sciences Meeting, New York, January 1970.

Flanagan, R. C., and Modi, V. J.: "Radiation Forces on a Flat Plate in Ecliptic Near-Earth Orbits," *CASI Trans.*, **3** (No. 2), 147–158 (1970).

Flatley, T. W. [1]: "Equilibrium States for a Class of Dual-Spin Spacecraft," NASA TR R-362, Washington, D.C., March 1971.

Flatley, T. W. [2]: "Nutational Behavior of Explorer-45," in *Significant Achievements in Science and Technology*, NASA SP-361, Washington, D.C., 1975.

Fletcher, H. J., Rongved, L., and Yu, E. Y.: "Dynamics Analysis of a Gravitationally Oriented Satellite," *Bell System Tech. J.*, **42** (No. 5), 2239–2266 (1963).

Frisch, H. P. [1]: "Thermal Bending Plus Twist of a Thin Walled Cylinder of Open Section with Application to Gravity-Gradient Booms," NASA TN D-4069, Washington, D. C., August 1967.

Frisch, H. P. [2]: "Thermally Induced Vibrations of Long Thin-Walled Cylinders of Open Section," in *Proc. of the Tenth AIAA Structures, Structural Dynamics, and Materials Conference*, pp 405–420 (1969).

Frisch, H. P. [3]: "A Vector-Dyadic Development of the Equations of Motion for N Coupled Rigid Bodies and Point Masses," NASA TN D-7767, Washington, D.C., 1974.

Fujii, H. [1]: "Attitude Stability of a Rigid Dual-Spin Satellite," *Mem. Metr. Col. Tech.* (Tokyo), **No. 2**, 33–42 (1974).

Fujii, H. [2]: "Librations of a Parametrically Resonant Dual-Spin Satellite with an Energy Damper," *J. Spacecr. Rockets*, **13** (No. 7), 416–423 (1976).

Garber, T. B.: "Influence of Constant Disturbing Torques on the Motion of Gravity-Gradient Stabilized Satellites," *AIAA J.*, **1** (No. 4), 968–969 (1963).

Garg, S. C., and Hughes, P. C.: "Use of Fibers in Gravity-Gradient Stabilization Systems," *J. Spacecr. Rockets*, **8** (No. 10), 1085–1087 (1971).

Gebman, J. R., and Mingori, D. L.: "Perturbation Solution for the Flat Spin Recovery of a Dual-Spin Spacecraft," *AIAA J.*, **14** (No. 7), 859–867 (1976).

Gelman, H. [1]: "The Second Orthogonality Condition in the Theory of Proper and Improper Rotations. I. Derivation of the Conditions and of Their Main Consequences," *J. Res. Natl. Bur. Stand.—B. Math. Sci.*, **72B** (No. 3), 229–237 (1968).

Gelman, H. [2]: "The Second Orthogonality Condition in the Theory of Proper and Improper Rotations. II. The Intrinsic Vector," *J. Res. Natl. Bur. Stand.—B. Math. Sci.*, **73B** (No. 2), 125–138 (1969).

Gelman, H. [3]: "A Note on the Time Dependence of the Effective Axis and Angle of a Rotation," *J. Res. Natl. Bur. Stand.—B. Math. Sci.*, **75B** (Nos. 3 & 4), (1971).

Gibbs, J. W.: *Vector Analysis*, Dover, New York, 1960.

Goldman, R. L.: "Influence of Thermal Distortion on Gravity Gradient Stabilization," *J. Spacecr. Rockets*, **12** (No. 7), 406–413 (1975).

Goldstein, H.: *Classical Mechanics*, Addison-Wesley, Reading, Mass., 1950.

Graham, J. D.: "Dynamics of Satellites Having Long Flexible Extendible Members," in *Proc. of the XVII International Astronautical Congress*, Madrid, October 1966.

Grasshoff, L. H. [1]: "Influence of Gravity on Satellite Spin Axis Attitude," *ARS J.*, **30** (No. 12), 1174–1175 (1960).

Grasshoff, L. H. [2]: "An Onboard, Closed-Loop, Nutation Control System for a Spin-Stabilized Spacecraft," *J. Spacecr. Rockets*, **5** (No. 5), 530–535 (1968).

Gray, A.: *A Treatise on Gyrostatics and Rotational Motion*, Dover, New York, 1959.

Greenwood, D. T.: *Principles of Dynamics*, Prentice-Hall, Englewood Cliffs, N.J., 1965.

Greub, W.: *Linear Algebra* (4th edition), Springer-Verlag, New York, 1975.

Grubin, C. [1]: "Dynamics of a Vehicle Containing Moving Parts," *J. Appl. Mech.*, **29**, 486–488 (1962).

Grubin, C. [2]: "Derivation of the Quaternion Scheme Via the Euler Axis and Angle," *J. Spacecr. Rockets*, **7** (No. 10), 1261–1263 (1970).

Grubin, C. [3]: "Attitude Determination for a Strapdown Inertial System Using the Euler Axis/Angle and Quaternion Parameters," Paper No. 73-900, in *Proc. of the AIAA Guidance and Control Conference*, Key Biscayne, August 1973.

Grubin, C. [4]: "Quaternion Singularity Revisited," *J. Guid. Control.* **2** (No. 3), 255–256 (1979).

Hadley, G.: *Linear Algebra*, Addison-Wesley, Reading, Mass., 1961.

Hagedorn, P. [1]: *Non-Linear Oscillations*, Clarendon, Oxford, 1981.

Hagadorn, P. [2]: "Über die Instabilität konservativer Systeme mit gyroskopischen Kräften," *Arch. Rat. Mech. Anal.*, **58**, 1–9 (1975).

Hahn, W. [1]: *Theory and Application of Liapunov's Direct Method*, Prentice-Hall, Englewood Cliffs, N.J., 1963.

Hahn, W. [2]: *Stability of Motion*, Springer-Verlag, New York, 1967.

Haines, G. A., and Leondes, C. T.: "Eddy Current Nutation Dampers for Dual-Spin Satellites," *J. Astronaut. Sci.*, **21** (No. 1), 1–25 (1973).

Hale, J. H.: *Oscillations in Nonlinear Systems*, McGraw-Hill, New York, 1963.

Halfman, R. L.: *Dynamics* (in 2 volumes), Addison-Wesley, Reading, Mass., 1962.

Hara, M.: "Effects of Magnetic and Gravitational Torques on Spinning Satellite Attitude," *AIAA J.*, **11** (No. 12), 1737–1742 (1973).

Hartbaum et al.: "Configuration Selection for Passive Gravity-Gradient Satellites," NASA SP-107, 1966.

Heiskanen, W. A., and Moritz, H.: *Physical Geodesy*, W. H. Freeman, San Francisco, 1967.

Hitzl, D. L. [1]: "Nonlinear Attitude Motion Near Resonance," AAS/AIAA Astrodynamics Specialist Conference, AAS Paper No. 68-124, Jackson, Wy., September 1968.

Hitzl, D. L. [2]: "Resonant Attitude Instabilities for a Symmetric Satellite in a Circular Orbit," *Celestial Mech.*, **5**, 433–450 (1972).

Ho, J. Y. L.: "Direct Path Method for Flexible Multibody Spacecraft Dynamics," *J. Spacecr. Rockets*, **14**, 102–110 (1977).

Hodge, W. F.: "Effect of Environmental Torques on Short-Term Attitude Prediction for a Rolling-Wheel Spacecraft in a Sun-Synchronous Orbit," NASA TN D-6583, Washington, D.C., January 1972.

Hooker, W. W. [1]: "A Set of r Dynamical Attitude Equations for an Arbitrary n-Body Satellite Having r Rotational Degrees of Freedom," *AIAA J.*, **8** (No. 7), 1205–1207 (1970).

Hooker, W. W. [2]: "Equations of Motion for Interconnected Rigid and Elastic Bodies: A Derivation Independent of Angular Momentum," *Celestial Mech.*, **11**, 337–359 (1975).

Hooker, W. W., and Margulies, G.: "The Dynamical Attitude Equations for an n-Body Satellite," *J. Astronaut. Sci.*, **12** (No. 4), 123–128 (1965).

Hrastar, J. A.: "Testing a Satellite Automatic Nutation Control System," NASA TN D-7596, Washington, D.C., May 1974.

Hubert, C.: "Spacecraft Attitude Acquisition from an Arbitrary Spinning or Tumbling State," *J. Guid. Control*, **4** (No. 2), 164–170 (1981).

Huckins, E. K., Breedlove, W. J., and Heinbockel, J. D.: "Passive Three-Axis Stabilization of the Long Duration Exposure Facility," Paper No. AAS 75-030, AAS/AIAA Astrodynamics Specialist Conference, Nassau, July 1975.

Hughes, P. C. [1]: "Geometrical Properties in State Space of Linear Differential Equations with Periodic Coefficients," University of Toronto Inst. for Aerospace Studies Tech. Note No. 127, Toronto, Canada, July 1968.

Hughes, P. C. [2]: "Attitude Stability for the Yaw-Wheel Class of Orbiting Gyrostats," *CASI J.*, **29** (No. 3), 268–284 (1983).

Hughes, P. C. [3]: "Attitude Stability of an Orbiting Gyrostat in Conical Equilibrium," *J. Guidance, Control, & Dynamics*, **8**, (No. 4), 417–425 (1985).

Hughes, P. C., and Cherchas, D. B. [1]: "Influence of Solar Radiation on the Spin Behaviour of Satellites with Long Flexible Antennae," *CASI Trans.*, **2** (No. 2), 53–57 (1969).

Hughes, P. C., and Cherchas, D. B. [2]: "Spin Decay of Explorer XX," *J. Spacecr. Rockets*, **7** (No. 1), 92–93 (1970).

Hughes, P. C., and Gardner, L. T.: "Asymptotic Stability of Linear Stationary Mechanical Systems," *J. Appl. Mech.*, **41** (No. 7), 228–229 (1975).

Hughes, P. C., and Golla, D. F.: "Stability Diagrams for a Rigid Gyrostat in a Circular Orbit," University of Toronto Inst. for Aerospace Studies Tech. Note No. 243, Toronto, Canada, May 1984.

Hultquist, P. F.: "Gravitational Torque Impulse on a Stabilized Satellite," *ARS J.*, **31** (No. 11), 1506–1509 (1961).

Huseyin, K.: *Vibrations and Stability of Multiple Parameter Systems*, Sijthoff and Noordhoff, Rockville, Md. 1978.

Ickes, B. F.: "A New Method for Performing Digital Control Systems Attitude Computations Using Quaternions," *AIAA J.*, **8** (No. 1), 13–17 (1970).

INTELSAT/ESA: *Proc. of the INTELSAT/ESA Colloquium on Dynamic Effects of Liquids on Spacecraft Attitude Control*, Washington, D.C., April 1984.

Iorillo, A. J. [1]: "Nutation Damping," Hughes Aircraft Co. Rept. IDC 2230.14/69, Sunnyvale, Ca. October 1964.

Iorillo, A. J. [2]: "Nutation Damping Dynamics of Axisymmetric Rotor Stabilized Satellites," ASME Annual Winter Meeting, Chicago, 1965.

Jankovic, M.: "Deployment Dynamics of Flexible Spacecraft," Ph.D. dissertation, University of Toronto Institute for Aerospace Studies, 1979.

Jerkovsky, W.: "The Structure of Multibody Dynamics Equations," *J. Guid. Control*, **1** (No. 3), 173–182 (1978).

Johnson, C. R.: "TACSAT I Nutation Dynamics," AIAA Third Communication Satellite Systems Conference, AIAA Paper 70-455, Los Angeles, Calif., April 1970.

Johnson, K. R.: "Effect of Dissipative Aerodynamic Torque on Satellite Rotation," *J. Spacecr. Rockets*, **5** (No. 4), 408–413 (1968).

Junkins, J. L., Jacobson, I. D., and Blanton, J. N.: "A Nonlinear Analog of Rigid Body Dynamics," *Celestial Mech.*, **7**, 398–407 (1973).

Kalman, R. E., and Bertram, J. E.: "Control System Analysis and Design via the Second Method of Liapunov: Continuous Time Systems," *J. Basic Eng.*, 371–393 (1960).

Kamm, L. J.: "'Vertistat': An Improved Satellite Orientation Device," *ARS J.*, **32** (No. 6), 911–913 (1962).

Kane, T. R. [1]: "Attitude Stability of Earth-Pointing Satellites," *AIAA J.*, **3** (No. 4), 726–731 (1965).

Kane, T. R. [2]: *Dynamics*, Holt, Rinehart and Winston, New York, 1968.

Kane, T. R. [3]: "Solution of Kinematical Differential Equations for a Rigid Body," *J. Appl. Mech.*, 109–113 (1973).

Kane, T. R., and Barba, P. M.: "Attitude Stability of a Spinning Satellite in an Elliptic Orbit," *J. Appl. Mech.*, 402–405 (1966).

Kane, T. R., and Fowler, R. C.: "Equivalence of Two Gyrostatic Stability Problems," *J. Appl. Mech.*, 1146–1147 (1970).

Kane, T. R., and Likins, P. W.: "Gravitational Forces and Moments on Spacecraft," NASA CR-2618, Washington, D.C., October 1975.

Kane, T. R., Likins, P. W., and Levinson, D. A.: *Spacecraft Dynamics*, McGraw-Hill, New York, 1983.

Kane, T. R., and Marsh, E. L.: "Attitude Stability of a Symmetric Satellite at the Equilibrium Points in the Restricted Three-Body Problems," *Celestial Mech.*, **4**, 78–90 (1971).

Kane, T. R., Marsh, E. L., and Wilson, W. G.: Letter to the Editor, *J. Astronaut. Sci.*, **9**, 108–109 (1962).

Kane, T. R., and Mingori, D. L.: "Effect of a Rotor on the Attitude Stability of a Satellite in a Circular Orbit," *AIAA J.*, **3** (No. 5), 936–940 (1965).

Kane, T. R., and Shippy, D. J.: "Attitude Stability of a Spinning Asymmetrical Satellite in a Circular Orbit," *J. Astronaut. Sci.*, **10** (No. 4), 114–119 (1963).

Kane, T. R., and Wang, C. F.: "On the Derivation of Equations of Motion," *SIAM J.*, **13**, 487–492 (1965).

Kanning, G.: "The Influence of Thermal Distortion on the Performance of Gravity-Stabilized Satellites," NASA TN D-5435, Washington, D. C., November 1969.

Kaplan, M. H., and Patterson, T. C.: "Attitude Acquisition Maneuver for Bias Momentum Satellites," *COMSAT Tech. Rev.*, **6** (No. 1), 1–23 (1976).

Katucki, R. J., and Moyer, R. G.: "Systems Analysis and Design of a Class of Gravity Gradient Satellites Utilizing Viscous Coupling Between Earth's Magnetic and Gravity Fields," in NASA SP-107, Washington, D.C., 1966.

Kaula, W. M.: *Theory of Satellite Geodesy, Applications of Satellites to Geodesy*, Blaisdell, New York 1966.

King-Hele, D.: *Theory of Satellite Orbits in an Atmosphere*, Butterworth, London, 1964.

Klemperer, W. B.: "Satellite Librations of Large Amplitude," *ARS J.*, **30** (No. 1), 123–124 (1960).

Klumpp, A. R. [1]: "Singularity-Free Extraction of a Quaternion from a Direction-Cosine Matrix," *J. Spacecr. Rockets*, **13** (No. 12), 754–755 (1976).

Klumpp, A. R. [2]: "Reply by Author to R. H. Spurrier," *J. Spacecr. Rockets*, **15** (No. 4), 256 (1978).

Korn, G. A., and Korn, T. M.: *Mathematical Handbook for Scientists and Engineers*, McGraw-Hill, New York, 1961.

Kutzer, A.: "HELIOS," Paper 77-266, 13th AIAA Annual Meeting, Washington, D.C., January 10–14, 1977.

Lanczos, C.: *The Variational Principles of Mechanics* (2nd edition), University of Toronto Press, Toronto, Canada, 1962.

Landon, V. D.: "Early Evidence of Stabilization of a Vehicle Spinning about Its Axis of Least Inertia," in *Proc. of the Symposium on Attitude Stabilization and Control of Dual-Spin Spacecraft*, USAF Rept. No. SAMSO-TR-68-191, 1968.

Landon, V. D., and Stewart, B.: "Nutational Stability of an Axisymmetric Body Containing a Rotor," *J. Spacecr. Rockets*, **1** (No. 6), 682–684 (1964).

Lange, B.: "Linear Coupling Between Orbital and Attitude Motions of a Rigid Body," *J. Astronaut. Sci.*, **28** (No. 3), 150–167 (1970).

LaSalle, J., and Lefschetz, S.: *Stability by Liapunov's Direct Method*, Academic, New York, 1961.

Leimanis, E.: *The General Problem of the Motion of Coupled Rigid Bodies about a Fixed Point*, Springer-Verlag, New York, 1965.

Leipholz, H.: *Stability Theory*, Academic, New York, 1970.

Lerner, G. M., and Coriell, K. P.: "Attitude Capture for the Geodynamics Experimental Ocean Satellite GEOS-3," AAS/AIAA Astrodynamics Specialist Conference, Nassau, Bahamas, July 1975.

Li, T., and Longman, R. W.: "Stability Relationships between Gyrostats with Free, Constant-Speed, and Speed-Controlled Rotors," *J. Guid. Control*, **5** (No. 6), 545–552 (1982).

Likins, P. W. [1]: "Stability of a Symmetrical Satellite in Attitudes Fixed in an Orbiting Reference Frame," *J. Astronaut. Sci.*, **12** (No. 1), 18–24 (1965).

Likins, P. W. [2]: "Attitude Stability Criteria for Dual-Spin Spacecraft," *J. Spacecr. Rockets*, **4** (No. 12), 1638–1643 (1967).

Likins, P. W. [3]: *Elements of Engineering Mechanics*, McGraw-Hill, New York, 1973.

Likins, P. W. [4]: "Analytical Dynamics and Nonrigid Spacecraft Simulation," JPL Tech. Rept. 32-1593, July 1974.

Likins, P. W. [5]: "Point-Connected Rigid Bodies in a Topological Tree," *Celestial Mech.*, **11** (No. 3), 301–317 (1975).

Likins, P. W., and Columbus, R. L.: "Effects of Eddy-Current Damping on Satellite Attitude Stability," *AIAA J.*, **4** (No. 6), 1123–1125 (1966).

Likins, P. W., and Fleischer, G. E.: "Large-Deformation Modal Coordinates for Nonrigid Vehicle Dynamics," JPL Tech. Rept. 32-1565, November 1972.

Likins, P. W., and Roberson, R. E.: "Uniqueness of Equilibrium Attitudes for Earth-Pointing Satellites," *J. Astronaut. Sci.*, **13**, 87–88 (1966).

Likins, P. W., Tseng, G. T., and Mingori, D. L.: "Stable Limit Cycles Due to Nonlinear Damping in Dual-Spin Spacecraft," *J. Spacecr. Rockets*, **8** (No. 6), 568–574 (1971).

Likins, P. W., and Wrout, G. M.: "Bounds on the Librations of Parametrically Resonant Satellites," *AIAA J.*, **7** (No. 6), 1134–1139 (1969).

Longman, R. W. [1]: "Gravity-Gradient Stabilization of Gyrostat Satellites with Rotor Axes in Principal Planes," *Celestial Mech.*, **3**, 169–188 (1971).

Longman, R. W. [2]: "Stability Analysis of All Possible Equilibria for Gyrostat Satellites under Gravitational Torques," *AIAA J.*, **10** (No. 6), 800–806 (1972).

Longman, R. W. [3]: "Stable Tumbling Motions of a Dual-Spin Satellite Subject to Gravitational Torques," *AIAA J.*, **11** (No. 7), (1973).

Longman, R., Hagedorn, P., and Beck, A.: "Stabilization Due to Gyroscopic Coupling in Dual-Spin Satellites Subject to Gravitational Torques," *Celestial Mech.*, **25**, 353–373 (1981).

Longman, R. W., and Roberson, R. E.: "General Solution for the Equilibria of Orbiting Gyrostats Subject to Gravitational Torques," *J. Astronaut. Sci.*, **16** (No. 2), 49–58 (1969).

MacMillan, W. D.: *Dynamics of Rigid Bodies*, Dover, New York, 1936.

Martin, E. R.: "Experimental Investigations on the Fuel Slosh of Dual-Spin Spacecraft," *COMSAT Tech. Rev.*, **1** (No. 1), 1–20 (Fall 1971).

Mayer, A.: "Rotations and Their Algebra," *SIAM Rev.*, **2** (No. 2), 77–122 (April 1960).

McIntyre, J. E., and Gianelli, M. J. [1]: "Bearing Axis Wobble for a Dual Spin Vehicle," *J. Spacecr. Rockets*, 945–951 (1971).

McIntyre, J. E., and Gianelli, M. J. [2]: "The Effect of an Autotracking Antenna on the Nutational Stability of a Dual Spin Spacecraft," AIAA Fourth Communications Satellite Systems Conference, AIAA Paper 72-571, Washington, D.C., April 1972.

Meirovitch, L. [1]: "On the Effects of Higher-Order Inertia Integrals on the Attitude Stability of Earth Pointing Satellites," *J. Astronaut. Sci.*, **15** (No. 1), 14–18 (1968).

Meirovitch, L. [2]: *Methods of Analytical Dynamics*, McGraw-Hill, New York, 1970.

Meirovitch, L., and Wallace, F. B., Jr.: "Attitude Instability Regions of a Spinning Unsymmetrical Satellite in a Circular Orbit," *J. Astronaut. Sci.*, **14** (No. 3), 123–133 (1967).

Meyer, G.: "Design and Global Analysis of Spacecraft Attitude Control Systems," NASA TR R-361, Washington, D.C., March 1971.

Mickens, R. E.: *Nonlinear Oscillations*, Cambridge University Press, Cambridge, 1981.

Mingori, D. L.: "Effects of Energy Dissipation on the Attitude Stability of Dual-Spin Satellites," *AIAA J.*, **7** (No. 1) 20–27 (1969).

Mingori, D. L., and Kane, T. R.: "The Attitude Stabilization of Rotating Satellites by Means of Gyroscopic Devices," *J. Astronaut. Sci.*, **14** (No. 4), 158–166 (1967).

Mingori, D. L., Tseng, G. T., and Likins, P. W.: "Constant and Variable Amplitude Limit Cycles in Dual-Spin Spacecraft," *J. Spacecr. Rockets*, **9** (No. 11), 825–830 (1972).

Mobley, F. F., and Fischell, R. E.: "Orbital Results from Gravity-Gradient-Stabilized Satellites," in NASA SP-107, 1966.

Modi, V. J., and Brereton, R. C. [1]: "Libration Analysis of a Dumbbell Satellite Using the WKBJ Method," *J. Appl. Mech.*, **33** (No. 3), 676–678 (1966).

Modi, V. J., and Brereton, R. C. [2]: "The Stability Analysis of Coupled Librational Motion of an Axi-Symmetric Satellite in a Circular Orbit," in *Proc. of the XVIII International Astronautical Congress*, Belgrade, pp 109–120 (1967).

Modi, V. J., and Brereton, R. C. [3]: "Periodic Solutions Associated with the Gravity-Gradient-Oriented System," Part I. "Analytical and Numerical Determination," *AIAA J.*, **7** (No. 7), July 1969, pp. 1217–1225; Part II. "Stability Analysis," *AIAA J.*, **7** (No. 8), 1465–1468 (1969).

Modi, V. J., and Flanagan, R. C.: "Effect of Environmental Forces on the Attitude of Gravity Orientated Satellites. Part 1: High Altitude Orbits," *Aeronaut. J.*, **75**, 783–793 (1971).

Modi, V. J., and Neilson, J. E. [1]: "Attitude Dynamics of Slowly Spinning Axi-Symmetric Satellites under the Influence of Gravity Gradient Torques," in *Proc. of the XX International Astronautical Congress*, Mar del Plata, 1969, Pergamon, New York, 1972.

Modi, V. J., and Neilson, J. E. [2]: "On the Periodic Solutions of Slowly Spinning Gravity Gradient System," *Celestial Mech.*, **5**, 126–143 (1972).

Modi, V. J., and Pande, K. C.: "Solar Pressure Induced Librations of Spinning Axisymmetric Satellites," *J. Spacecr. Rockets*, **10** (No. 9), 615–617 (1973).

Modi, V. J., and Shrivastava, S. K. [1]: "Coupled Librational Motion of an Axi-Symmetric Satellite in a Circular Orbit," *Aeronaut. J.*, **73** (No. 704), 674–680 (1969).

Modi, V. J., and Shrivastava, S. K. [2]: "Approximate Solution for Coupled Librations of an Axisymmetric Satellite in Circular Orbit," *AIAA J.*, **9** (No. 6) 1212–1214 (1971).

Modi, V. J., and Shrivastava, S. K. [3]: "Librations of Gravity-Oriented Satellites in Elliptical Orbits through Atmosphere," *AIAA J.*, **9** (No. 11), 2208–2215 (1971).

Mohan, S. N., Breakwell, J. V., and Lange, B. O.: "Interaction Between Attitude Libration and Orbital Motion of a Rigid Body in a Near Keplerian Orbit of Low Eccentricity," *Celestial Mech.*, **5**, 157–173 (1972).

Moran, J. P.: "Effects of Plane Librations on the Orbital Motion of a Dumbbell Satellite," *ARS J.*, **31**, 1089–1096 (1961).

Morgan, S. P., and Yu, E. Y.: "Meteoritic Disturbances of Gravitationally Oriented Satellites," *J. Spacecr. Rockets*, **2** (No. 6), 857–862 (1965).

Mortensen, R. E.: "Strapdown Guidance Error Analysis," *IEEE Trans. Aero. Electron. Syst.*, **AES-10** (No. 4), 451–457 (1974).

Morton, H. S., Junkins, J. L., and Blanton, J. N.: "Analytical Solutions for Euler Parameters," *Celestial Mech.*, **10**, 287–301 (1974).

Musen, P.: *Astronom. J.*, **59**, 262 (1954).

NASA [1]: "Satellite Meteorology 1958–1964," NASA SP-96, Washington, D.C., 1966.

NASA [2]: "Models of Mars' Atmosphere (1967)," NASA SP-8010, May 1968.

NASA [3]: "Propellant Slosh Loads," NASA SP-8009, August 1968.

NASA [4]: "Models of Venus' Atmosphere (1968)," NASA SP-8011, December 1968.

NASA [5]: "Meteoroid Environment Model—1969 (Near Earth to Lunar Surface)," NASA SP-8013, March 1969.

NASA [6]: "Magnetic Fields—Earth and Extraterrestrial," NASA SP-8017, March 1969.

NASA [7]: "Spacecraft Magnetic Torques," NASA SP-8018, March 1969.

NASA [8]: "Models of Earth's Atmosphere (120 to 1000 km)," NASA SP-8021, May 1969.

NASA [9]: "Spacecraft Gravitational Torques," NASA SP-8024, May 1969.

NASA [10]: "Spacecraft Radiation Torques," NASA SP-8027, October 1969.

NASA [11]: "Spacecraft Mass Expulsion Torques," NASA SP-8034, December 1969.

NASA [12]: "Spacecraft Aerodynamic Torques," NASA SP-8058, January 1971.

NASA [13]: "Turbular Spacecraft Booms (Extendible, Reel Stored)," NASA SP-8065, February 1971.

NASA [14]: "Passive Gravity-Gradient Libration Dampers," NASA SP-8071, February 1971.

NASA [15]: "Solar Electromagnetic Radiation," NASA SP-8005, May 1971.

NASA [16]: "Earth Albedo and Emitted Radiation," NASA SP-8067, July 1971.

Nayfeh, A. H., and Mook, D. T.: *Nonlinear Oscillations*, Wiley-Interscience, New York, 1979.

Neer, J. T. [1]: "Hughes Gyrostat Stabilization Concept," *CASI J.*, 103–107 (1970).

Neer, J. T. [2]: "INTELSAT IV Nutation Dynamics," AIAA Fourth Communications Satellite Systems Conference, AIAA Paper 72-537, Washington, D.C., April 1972.

Nidey, R. A.: "Gravitational Torque on a Satellite of Arbitrary Shape," *ARS J.*, **30** (No. 2), 203–204 (1960).

Nurre, G. S.: "Effects of Aerodynamic Torque on an Asymmetric, Gravity-Stabilized Satellite," *J. Spacecr. Rockets*, **5** (No. 9), 1046–1050 (1968).

Ohkami, Y.: "Computer Algorithms for Computation of Kinematical Relations for Three Attitude Angle Systems," *AIAA J.*, **14**, 1136–1137 (1976).

Palmer, J. L., Blackiston, H. S., and Farrenkopf, R. L.: "Advanced Techniques for Analysing and Improving Gravity Gradient Satellite Pointing Accuracies at High Orbital Altitudes," AFFDL-TR-66-206, June 1967.

Parlett, B. N.: *The Symmetric Eigenvalue Problem*, Prentice-Hall, Englewood Cliffs, N.J., 1980.

Patterson, G. N.: *Molecular Flow of Gases*, Wiley, New York, 1956.

Patterson, W. B., and Kissell, K. E.: "Comparison of the Theoretical Solar Radiation Effects and the Observed Accelerations of the PAGEOS Satellite," *J. Spacecr. Rockets*, **12** (No. 9), 539–543 (1975).

Paul, B.: "Planar Libration of an Extensible Dumbbell Satellite," *AIAA J.*, **1** (No. 2), 411–418 (1963). See also comment by F. W. Niedenfuhr, *AIAA J.*, **1** (No. 7), 1713–1714 (1963).

Paul, B., West J. W., and Yu, E. Y.: "A Passive Gravitational Attitude Control System for Satellites," *Bell Systems Tech. J.*, 2195–2238 (1963).

Pengelley, C. D.: "Gravitational Torque on a Small Rigid Body in an Arbitrary Field," *ARS J.*, 420–422 (1962).

Perry, G. E., and Slater, J. D.: "Measurements of Spin Rates of the Prospero Satellite," *J. Br. Interplanet. Soc.*, **26**, 167–169 (1973).

Phillips, K.: "Active Nutation Damping Utilizing Spacecraft Mass Properties, *IEEE Trans. Aerospace. Electron. Syst.*, **AES-9** (No. 5), 688–693 (1973).

Pisacane, V. L., Pardoe, P. P., and Hook, B. J., "Stabilization System Analysis and Performance of the GEOS-A Gravity-Gradient Satellite (Explorer XXIX)," *J. Spacecr. Rockets*, **4**, (No. 12), 1623–1630 (1967).

Pringle, R., Jr. [1]: "Bounds on the Librations of a Symmetrical Satellite," *AIAA J.*, **2** (No. 5), 908–912 (1964).

Pringle, R., Jr. [2]: "Stability of the Force-Free Motions of a Dual-Spin Spacecraft," *AIAA J.*, **7** (No. 6), 1054–1063 (1969).

Pringle, R., Jr. [3]: "Effect of Perturbed Orbital Motion on a Spinning Symmetrical Satellite," *J. Spacecr. Rockets*, **11** (No. 7), 451–455 (1974).

Puri, N. N., and Gido, J. F.: "Nutational Stability Criteria for a Dual Spin Spacecraft—the Damper Reaction Torque and Quadratic Function Method," in *Proc. of the XIX International Astronautical Congress*, New York, 1968, Vol. II, Pergamon, New York, 1970, pp. 527–548.

Roberson, R. E. [1]: "Attitude Control of a Satellite Vehicle—An Outline of the Problems," in *Proc. of the VIII International Astronautical Congress*, Barcelona, October 1957.

Roberson, R. E. [2]: "Gravitational Torque on a Satellite Vehicle," *J. Franklin Inst.*, **265**, 13–22 (1958).

Roberson, R. E. [3]: "Torques on a Satellite from Internal Moving Parts," *J. Appl. Mech.*, **25**, 196–200 (1958).

Roberson, R. E. [4]: "Stability and Control of Satellite Vehicles," in *Handbook of Astronautical Engineering* (Ed.: H. H. Koelle), McGraw-Hill, New York, 1961.

Roberson, R. E. [5]: "Generalized Gravity-Gradient Torques," Chapter 4 in *Torques and Sensing in Earth Satellites* (Ed.: S. F. Singer), Academic, New York, 1964.

Roberson, R. E. [6]: "Equilibria of Orbiting Gyrostats," *J. Astronaut. Sci.*, **15** (No. 5), 242–248 (1968).

Roberson, R. E. [7]: "Kinematical Equations for Bodies Whose Rotation is Described by the Euler-Rodrigues Parameters," *AIAA J.*, **6** (No. 5), 916–917 (1968).

Roberson, R. E. [8]: "Euler's Problem on Permanent Axes of Rotation, Extended to the Spinning Gyrostat," *J. Appl. Mech.*, 1154–1156 (1970).

Roberson, R. E. [9]: "The Equivalence of Two Classical Problems of Free Spinning Gyrostats," *J. Appl. Mech.*, 707–708 (1971).

Roberson, R. E. [10]: "A Form for the Translational Dynamics Equations for Relative Motion in Systems of Many Non-Rigid Bodies," *Acta Mech.*, **14**, 297–308 (1972).

Roberson, R. E., and Hooker, W. W.: "Gravitational Equilibria of a Rigid Body Containing Symmetric Rotors," *Proc. of the XVII International Astronautical Congress*, Madrid, 1966.

Roberson, R. E., and Tatistcheff, D.: "The Potential Energy of a Small Rigid Body in the Gravitational Field of an Oblate Spheroid," *J. Franklin Inst.*, 262 (No. 3), 209–214 (1956).

Roberson, R. E., and Wittenburg, J.: "A Dynamical Formalism for an Arbitrary Number of Rigid Bodies, with Reference to the Problem of Satellite Attitude Control," in *Proc. of the IFAC Congress London*, 1966, Butterworth, London, 1968.

Robinson, A. C.: "On the Use of Quaternions in Simulation of Rigid-Body Motion," WADC Tech. Rept. 58-17, December 1958.

Robinson, W. R.: "Attitude Stability of a Rigid Body Placed at an Equilibrium Point in the Restricted Problem of Three Bodies," *Celestial Mech.*, **10**, 17–33 (1974).

Rumyentsev, V. V.: "Dynamics and Stability of Rigid Bodies," Centro Internazionale Matematico Estivo, Rome, 1972.

Russell, W. J.: "On the Formulation of Equations of Rotational Motion for an *N*-Body Spacecraft," TR-0200 (4133)-2, Aerospace Corp., El Segundo, Calif., 1969.

Sabroff, A. E. [1]: "Two-Damper Passive Gravity-Gradient Stabilization System," NASA SP-107, Washington, D.C., 1966.

Sabroff, A. E. [2]: "Advanced Spacecraft Stabilization and Control Techniques," *J. Spacecr. Rockets*, **5** (No. 12), 1377–1393 (1968).

Sabroff, A., Farrenhopf, R., Frew, A., and Gran, M.: "Investigation of the Acquisition Problem in Satellite Attitude Control," AFFDL Tech. Rept. TR-65-115, December 1965.

Sarychev, V. A.: "Effect of Earth's Oblateness on the Rotational Motion of an Artificial Earth Satellite," *ARS J.*, 834–838 (1962).

Sarychev, V. A., and Mirer, S. A.: "The Use of Gyrodampers in Passive Satellite Attitude Control Systems," *J. Astronaut. Sci.*, **25** (No. 3), 179–205 (1977).

Sarychev, V. A., Mirer, S. A., and Sazonov, V. V.: "Plane Oscillations of a Gravitational System Satellite-Stabilizer with Maximal Speed of Response," *Acta Astronaut.*, **3**, 651–669 (1976).

Sarychev, V. A., and Sazonov, V. V. [1]: "Spin-Stabilized Satellites," *J. Astronaut. Sci.*, **24** (No. 4), 291–310 (1976).

Sarychev, V. A., and Sazonov, V. V. [2]: "Gravity Gradient Stabilization of Large Space Stations," *Acta Astronaut.*, **8** (Nos. 5–6), 549–573 (1981).

Schaaf, S. A., and Talbot, L.: "Mechanics of Rarefied Gases," Section 16 in *Handbook of Supersonic Aerodynamics*, NAVORD Rept. 1488, U.S. Navy Ordinance Bureau, Washington, D.C., Vol. 5, February 1959.

Schecter, H. B.: "Dumbbell Librations in Elliptic Orbits," *AIAA J.*, **2** (No. 6), 1000–1003 (1964).

Scher, M. P.: "Effects of Flexibility in the Bearing Assemblies of Dual-Spin Spacecraft," *AIAA J.*, **9** (No. 5), 900–905 (1971).

Scher, M. P., and Farrenkopf, R. L.: "Dynamic Trap States of Dual-Spin Spacecraft," *AIAA J.*, **12** (No. 12), 1721–1725 (1974).

Schindler, G. M.: "On Satellite Librations," *ARS J.*, **29** (No. 5), 368–370 (1959).

Schlengel, L. B.: "Contribution of Earth's Oblateness to Gravity Torque on a Satellite," *AIAA J.*, **4** (No. 11), 2075–2077 (1966).

Schlichting, A.: *Boundary Layer Theory*, McGraw-Hill, New York, 1960.

Schnapf, A.: "The TIROS IX Wheel Satellite," *RCA Review*, 3–40 (March 1966).

Schneider, C. C., and Likins, P. W.: "Nutation Dampers vs. Precession Dampers for Asymmetric Spacecraft," *J. Spacecr. Rockets*, **10**, 218–222 (1973).

Sen, S., and Bainum, P. M.: "Motion of a Dual-Spin Satellite During Momentum Wheel Spinup," *J. Spacecr. Rockets*, **10** (No. 12), 760–766 (1973).

Shepperd, S. W.: "Quaternion from Rotation Matrix," *J. Guid. Control*, **1** (No. 3), 223–224 (1978).

Shrivastava, S. K., and Modi, V. J.: "Effect of Atmosphere on Attitude Dynamics of Axi-Symmetric Satellites," in *Proc. of the XX International Astronautical Congress*, 1969, Pergamon, New York, 1972.

Sincarsin, G. B., and Hughes, P. C. [1]: "Gravitational Orbit–Attitude Coupling for Very Large Spacecraft," *Celestial Mech.*, **31**, 143–161 (1983).

Sincarsin, G. B., and Hughes, P. C. [2]: "Torques from Solar Radiation Pressure Gradient During Eclipse," *J. Guid. Control Dynam.*, **6** (No. 6), 511–517 (1983).

Slafer, L. I.: "Dual-Spin Spacecraft Nutation Control Using Articulated Payloads," *IEEE Trans. Aerospace. Electron. Syst.*, **AES-16** (No. 1), 74–82 (1980).

Slafer, L. I., and Marbach, H.: "Active Control of the Dynamics of a Dual-Spin Spacecraft," *J. Spacecr. Rockets*, **12** (No. 5), 287–293 (1975).

Slafer, L. I., and Smay, J. W.: "COMSTAR Spacecraft Product-of-Inertia Coupled Electronic Nutation Damper," *J Guid. Control*, **1** (No. 4), 273–278 (1978).

Smay, J. W., and Slafer, L. I.: "Dual-Spin Spacecraft Stabilization Using Nutation Feedback and Inertia Coupling," *J. Spacecr. Rockets*, **13** (No. 11), 650–659 (1976).

Spencer, T. M. [1]: "Attitude Dynamics of a 'Nearly-Spherical' Dual-Spin Satellite and Orbital Results for OSO-7," in *Proc. of the 13th IUTAM Congress*, Moscow, August 1972.

Spencer, T. M. [2]: "Attitude Dynamics of a 'Nearly-Spherical' Dual-Spin Satellite and Orbital Results for OSO-7," *J. Spacecr. Rockets*, **10** (No. 5), 289–290 (1973).

Spencer, T. M. [3]: "Energy-Sink Analysis for Asymmetric Dual-Spin Spacecraft," *J. Spacecr. Rockets*, **11** (No. 7) 463–468 (1974).

Spencer, T. M. [4]: "Cantilevered-Mass Nutation Damper for Dual-Spin Spacecraft," in *Proc. of the Symposium on Attitude Stabilization and Control of Dual-Spin Spacecraft*, USAF Rept. No. SAMSO-TR-68-191, 1968.

Sperling, H. J.: "The General Equilibria of a Spinning Satellite in a Circular Orbit," *Celestial Mech.*, **6**, 278–293 (1972).

Spurrier, R. A.: "Comment on 'Singularity-Free Extraction of a Quaternion from a Direction-Cosine Matrix'," *J. Spacecr. Rockets*, **15** (No. 4), 255 (1978).

Staugaitis, C. L. and Predmore, R. E.: "Mechanical Properties of Advanced Gravity Gradient Booms," NASA TN D-5997, Washington, D.C., October 1970.

Stern, T. E.: *Theory of Nonlinear Networks and Systems*, Addison-Wesley, Reading, Mass., 1965.

Stoker, J. J.: *Nonlinear Vibrations in Mechanical and Electrical Systems*, Wiley-Interscience, New York, 1950.

Stuelpnagel, J.: "On the Parameterization of the Three-Dimensional Rotation Group," *SIAM Rev.* **6** (No. 4), 422–430 (1964).

Symon, K. R.: *Mechanics* (3rd edition), Addison-Wesley, Reading, Mass., 1971.

Synge, J. L., and Griffith, B. H.: *Principles of Mechanics*, McGraw-Hill, New York, 1959.

Szebehely, V.: *Theory of Orbits*, Academic, New York, 1967.

Taylor, R. S., and Conway, J. J.: "Viscous Ring Precession Damper for Dual-Spin Spacecraft," in *Proc. of the Symposium on Attitude Stabilization and Control of Dual-Spin Spacecraft*, USAF Rept. No. SAMSO-TR-68-191, 1968.

Thomas, L. C., and Cappellari, J. O.: "Attitude Determination and Prediction of Spin-Stabilized Satellites," *Bell System Tech. J.*, 1657–1726 (1964).

Thomson, W. T. [1]: *Introduction to Space Dynamics*, Wiley, New York, 1961.

Thomson, W. T. [2]: "Spin Stabilization of Attitude Against Gravity Torque," *J. Astronaut. Sci.*, **9**, 31–33 (1962).

Thomson, W. T. [3]: "Passive Attitude Control of Space Vehicles," Chapter 7 in *Guidance and Control of Aerospace Vehicles* (Ed.: C. T. Leondes), McGraw-Hill, New York, 1963.

Thompson, W. T., and Reiter, G. S.: "Motion of an Asymmetric Spinning Body with Internal Dissipation," *AIAA J.*, **1** (No. 6), 1429–1430 (1963).

Tinling, B. E., and Merrick, V. K.: "Exploitation of Inertial Coupling in Passive Gravity-Gradient-Stabilized Satellites," *J. Spacecr. Rockets*, **1**, 381–387 (1964).

Tinling, B. E., Merrick, V. K., and Watson, D. M.: "A Split-Damper Inertially Coupled Passive Gravity-Gradient Attitude Control System," *J. Spacecr. Rockets*, **4** (No. 11), 1437–1442 (1967).

Tonkin, S. W.: "A Basic Attitude Instability of Spacecraft with Imperfect Momentum Wheels," *Automatica*, **16**, 415–418 (1980).

Tossman, B. E., et al.: "MAGSAT Attitude Control System Design and Performance," AIAA Paper 80-1730, 1980.

Truesdell, C.: "Whence the Law of Moment of Momentum?," in *Essays in the History of Mechanics*, Springer-Verlag, New York, 1968.

Tseng, G. T.: "Nutational Stability of an Asymmetric Dual-Spin Spacecraft," *J. Spacecr. Rockets*, **13** (No. 8), 488–493 (1976).

Tsuchiya, K.: "Attitude Behavior of a Dual-Spin Spacecraft Composed of Asymmetric Bodies," *J. Guid. Control.* **2** (No. 4), 328–333 (1979).

Turner, R. E.: "A Matrix Method for Studying Motions of Spacecraft Consisting of Interconnected Rigid Bodies," NASA TN D-4749, Washington, D.C., September 1968.

van den Broek, I. P.: "Attitude Perturbations of a Spin Satellite Due to Gravity-Gradient Torques in a Regressing Orbit," Delft University of Technology, Dept. of Aero. Eng. Report VTH-145, August 1967.

Velman, J. R.: "Simulation Results for a Dual-Spin Spacecraft," TR-0158 (3307-01)-16, Aerospace Corp., El Segundo, Calif., 1967.

Vigneron, F. R. [1]: "Motion of Freely Spinning Gyrostat Satellites with Energy Dissipation," *Astronaut. Acta*, **16**, 373–380 (1971).

Vigneron, F. R. [2]: "Stability of a Dual-Spin Satellite with Two Dampers," *J. Spacecr. Rockets*, **8** (No. 4), 386–389 (1971).

Vigneron, F. R. [3]: "Dynamics of Alouette-ISIS Satellites," *Astronaut. Acta*, **18**, 201–213 (1973).

Vigneron, F. R., and Garrett, T. W.: "Solar Induced Distortion Atmospheric Drag Coupling in Alouette Satellites," Canadian Defence Research Telecommunications Establishment Report 1171, February 1967.

Vigneron, F. R., and Krag, W. E.: "Optical Measurements and Attitude Motion of Hermes after Loss of Stabilization," *J. Guid. Control*, **5** (No. 5), 539–541 (1982).

Vigneron, F. R., and Millar, R. A.: "Flight Performance of the Stabilization System of the Communications Technology Satellite," *J. Guid. Control*, **1** (No. 6), 404–412 (1978).

Vigneron, F. R., and Staley, D. A.: "Satellite Attitude Acquisition by Momentum Transfer—The Controlled Wheel Speed Method," *Celestial Mech.*, **27**, 111–130 (1982).

Volosov, V. M.: "Averaging in Systems of Ordinary Differential Equations," *Russ. Math Surv.*, **17**, 1–126 (1962).

Walker, J. A., and Schmitendorf, W. E.: "A Simple Test for Asymptotic Stability in Partially Dissipative Symmetric Systems," *J. Appl. Mech.*, **40** (No. 4), 1120–1121 (1973).

Wallace, F. B., Jr., and Meirovitch, L.: "Attitude Instability Regions of a Spinning Symmetrical Satellite in an Elliptic Orbit," *AIAA J.*, **5** (No. 9), 1642–1650 (1967).

Weiss, R., et al.: "SEASAT-A Attitude Control System," *J. Guid. Control*, **1** (No. 1), 6–13 (1978).

Wertz, J. R. (Editor): *Spacecraft Attitude Determination and Control*, D. Reidel, Dordrecht, Holland, 1978.

Whisnant, J. M., Anand, D. K., Pisacane, V. L., and Sturmanis, M.: "Dynamic Modeling of Magnetic Hysteresis," *J. Spacecr. Rockets*, **7** (No. 6), 697–701 (1970).

Whisnant, J. M., Waszkiewicz, P. R., and Pisacane, V. L.: "Attitude Performance of the GEOS-II Gravity-Gradient Spacecraft," *J. Spacecr. Rockets*, **6** (No. 12), 1379–1384 (1969).

White, E. W., and Likins, P. W.: "The Influence of Gravity Torque on Dual-Spin Satellite Attitude Stability," *J. Astronaut. Sci.*, **16** (No. 1), 32–37 (1969).

Whittaker, E. T.: *A Treatise on the Analytical Dynamics of Particles and Rigid Bodies*, Cambridge University Press, Cambridge, 1970 (1st edition: 1904).

Wiggins, L. E. [1]: "Theory and Application of Radiation Forces," Report LMSC-4384055, Lockheed Missiles and Space Co., Sunnyvale, Calif., September 1963.

Wiggins, L. E. [2]: "Relative Magnitudes of the Space Environment Torques on a Satellite," *AIAA J.*, **2** (No. 4), 770–771 (1964).

Wilcox, J. C.: "A New Algorithm for Strapped-Down Inertial Navigation," *IEEE Trans. Aerospace. Electron. Syst.*, **AES-3** (No. 5), 796–802 (1967).

Wilkes, J. M.: "General Expression for a Three-Angle Rotation Matrix," *J. Guid. Control*, **2** (No. 2), 156–157 (1979).

Willems, P. Y. (Ed.): *Gyrodynamics*, Springer-Verlag, New York, 1974.

Williams, C. J. H.: "Dynamics Modelling and Formulation Techniques for Non-Rigid Spacecraft," in *Symposium on the Dynamics and Control of Non-Rigid Spacecraft*, Frascati, Italy, 1976, pp. 53–70.

Wilson, R. A.: "Rotational Magnetodynamics and Steering of Space Vehicles," NASA TN D-566, 1961. (See also Paper 65, *Proc. of the XI International Astronautical Congress Stockholm*, Springer-Verlag, New York, 1960.)

Wittenburg, J.: *Dynamics of Systems of Rigid Bodies*, B. G. Teubner, Stuttgart, 1977.

Yu, E. Y.: "Optimum Design of a Gravitationally Oriented Two-Body Satellite," *Bell System Tech. J.*, 49–76 (1965).

Zago, H.-D., and Leiss, F.: "Attitude Measurement and Control System of the Aeronomy Satellite AEROS," Paper No. 73-856, AIAA Guidance & Control Conference, Key Biscayne, Fl, August 1973.

Zajac, E. E. [1]: "Capture Problem in Gravitational Attitude Control of Satellites," *ARS J.*, **31** (No. 10), 1464–1466 (1961).

Zajac, E. E. [2]: "Limits on the Damping of Two-Body Gravitationally Oriented Satellites," *AIAA J.*, **1** (No. 2), 498–500 (1963).

Zajac, E. E. [3]: "Comments on 'Stability of Damped Mechanical Systems' and a Further Extension," *AIAA J.*, **3** (No. 9), 1749–1750 (1965).

Zajac, E. E. [4]: "The Kelvin–Tait–Chetayev Theorem and Extensions," *J. Astronaut. Sci.*, **11** (No. 2), 46–49 (1964).

INDEX

Absolute velocity, 40
Accommodation coefficients:
  aerodynamic, 250, 254–255, 258, 275
  radiative, 261–263
Aerodynamic torque, 248–260, 275–278, 314, 337, 356, 415
  on flat plate, 275–276
  on sphere, 277–278
AEROS satellite, 397–398
Alouette I satellite, 411,414
Alouette II satellite, 413
Angle of attack, 251
Angular momentum:
  of continuum, 54
  of gyrostat, 67, 160
  of N-wheel system, 180–183
  of point mass system, 45
  of rigid body, 58, 87, 95, 105
    with damper, 62
    in orbit, 347
    with wheel, 67
Angular velocity, 22–29, 530–532
  law of addition, 35
  linearization about simple spin, 29
  relative, 67
  second-order expansion for, 38
  of symmetrical rigid body, 96–98
  of tri-inertial rigid body, 105–111
Anik C satellite, 423–424, 455–456
Apparent gyrostat, 158–159, 178, 201, 216
Atmosphere, 259
ATS satellite, 333–334
Attitude acquisition, 305, 451–452, 467
Attitude-orbit coupling, 2, 284–293
Axis/angle variables, *see* Euler axis/angle variables

Bias momentum satellites, 423–444, 455–470

Capture, *see* Attitude acquisition
Cayley-Klein parameters, 18
Center of pressure, 252
Characteristic equation, 495
  for gravity gradient satellite, 296, 322–323, 352–353
  for gyrostat, 172, 188
    with damper, 221
    with nonspinning carrier, 161, 184
    in orbit, 436–437
  for rigid body with damper, 150
  for spinning rigid body, 118, 120
  for symmetrical gyrostat, 190
  for Thomson equilibrium, 360
  for zero momentum gyrostat, 164, 189
Conical equilibrium, 363–371, 416, 430–433, 474–475, 478
Continuum, 50, 52–55, 249
Control, 4
  active, 281–282, 424, 446, 453–455, 462–464
  passive, 281–282
Control moment gyro, 160
Crew motion, 270, 334

Dampers, 323–335, 391–400
  tuning, 152, 391, 420, 421
Damping, 61, 139–140, 308, 519–520
  complete, 514
  eddy current, 266
  for gravity gradient satellites, 315–346
  in gyrostats, 192–231
  pervasive, 152, 323, 331, 514
  in spinners, 139–155, 383–400
Damping matrix, 510
  for gravity gradient satellite, 317, 320, 331
  for gyrostat:
    with damper, 221
    with gimballed momentum wheel, 477
  for quasi-rigid body, 145
  for rigid body with damper, 149
  for Thomson equilibrium, 383
DeBra-Delp region, 296–298, 301, 316–318, 320–323
Density (mass), 53
Deployment, 70
Diffuse reflection, 250, 276
Direction cosines, 7, 29, 30–31
  differential equations for, 24
  sphere, 113–114, 122–124, 174, 303–304
Direction cosine matrix, *see* Rotation matrix
Directional stability, 121–124, 134, 143–146, 149, 153–154, 161, 165, 172–175, 190, 197–198, 230, 315, 317, 360–362, 383, 418, 520–521
Direct-path method, 81
Dissipation, *see* Damping
DODGE satellite, 339–342, 345
Dual-spin satellite, *see* Bias momentum satellites
Dynamics formulations, 3–4

Eddy current damping, 266, 280, 327, 383
Eddy current torque, 406–407
Ellipsoids, 108–112, 114, 186
Elliptic functions, 105, 107, 132, 134, 299–300, 350, 374–375, 417
Energy sink hypothesis, 3, 140–141, 193–217, 384
Equations of motion, *see* Motion equations
Error function, 256, 277
ESRO IV satellite, 399
Euler angles, 18–20, 29, 30–31
  differential equations for, 26–27
  linearization, 22, 27
  rotation matrices for, 20–21
  for symmetrical rigid body, 98–100
Euler axis/angle variables, 17, 30–31
  differential equations for, 24–25
  linearization, 22, 27
  from rotation matrix, 13–14
Euler parameters, 17–18, 29, 30–31
  differential equations for, 26
  linearization, 22, 27
  matrix operations with, 34
  for symmetrical rigid body, 103–104
  for tri-inertial rigid body, 132–133
Euler-Rodrigues variables, 30–31, 33, 35
Euler's theorem, 10–13, 32
Explorer I satellite, 3, 409
Explorer IV satellite, 413
Explorer VII satellite, 413
Explorer VIII satellite, 413
Explorer XI satellite, 413
Explorer XVIII satellite, 394
Explorer XX satellite, 412
Explorer XXXXV satellite, 413

Floquet theory, 224, 307, 374–377, 380, 386, 418, 499–502
Fluids interface, 3
Free-molecular flow, 249–260, 275–278
  over flat plate, 275–277
  over sphere, 277–278
Fuel sloshing, 3, 233, 270

Gas jets, 2, 129, 136, 233, 269
GEOS I satellite, 337–338
GEOS II satellite, 328, 337
GEOS C satellite, 457–458
GGTS satellite, 326, 338–339
Gibbs parameters, 33
Gimbals, 30
Gravitational potential, 2, 234–235
Gravitational stabilization, 281–353
  design considerations, 313–353
  in elliptic orbit, 308–313
  flight experience, 335–346
  librations of symmetrical satellite, 301–305
  librations of tri-intertial satellite, 305–308, 352–353
  pitch motion, 298–301
Gravitational torque, 233–248
Gravity gradient force field, 286, 346
Gravity gradient potential, 236–238, 272–273
Gravity gradient torque, 236–238, 272–273, 283
  average, 401–405
  generalized, 244, 273–274
  multibody spacecraft, 246–248
  with oblateness, 240–244, 273
  several primaries, 239
Gyricity, 511

Gyricity matrix, 510
for gravity gradient satellite, 295, 317, 320, 331
for gyrostat, 172
with damper, 221
with gimballed momentum wheel, 477
with nonspinning carrier, 161
in orbit, 435–436
for rigid body with damper, 149
for spinning rigid body, 117
for Thomson equilibrium, 359, 383
for zero momentum gyrostat, 164
Gyric stabilization, 117, 295
Gyrostats, 65–70, 423–455
apparent, 158–159, 178, 201, 216
conical equilibrium, 430, 441–444
cylindrical equilibrium, 430, 437–440
with dampers, 218–225, 226, 230
hyperbolic equilibrium, 430, 440–442
Kelvin's, 158–159, 178, 201, 216
motion equations, 157–158
nonspinning carrier, 161–164, 198–200
in orbit, 424–444
quasi-rigid, 193–211, 226–228, 230
simple spins, 160–161
symmetrical, 189, 203–205, 427–428
zero-momentum, 164–165

Hamiltonian, 84, 294, 297, 302, 316, 367–368, 379, 472, 473
Heat pipe, 225
HELIOS satellite, 399–400
Hermes satellite, 42, 468–470
Herpolhode, 113
Hill method, 302
Hurwitz, *see* Routh-Hurwitz criteria
Hyperbolic equilibrium, 363–371, 430–433, 473–475

Impulse function, 127
Impulsive approximation, 127–129, 137
Inertia matrix (system), 510
for bias momentum system, 69
for gyrostat, 69, 172
with damper, 219, 221
with gimballed momentum wheel, 477
with nonspinning carrier, 161
in orbit, 435–436
for gravity gradient satellite, 295, 317, 320, 331
for point mass system, 50
for rigid body, 60, 88
with damper, 64, 149, 154
with wheel, 69
for spinning rigid body, 117
for Thomson equilibrium, 359, 383
for zero momentum gyrostat, 164
Inertia ratios, 118, 134
Infinitesimal approximation, *see* Linear approximation
Instantaneous axis, 36, 101
INTELSAT IV satellite, 446, 449–451
Invariable plane, 112–113
ISIS satellite, 395, 407, 413

Jupiter, 40

Kelvin's gyrostat, 158–159, 178, 201, 216
Kelvin-Tait-Chetayev theorem, 149, 317
Kinematics, rotational, 6–38, 530, 32
Kinetic energy:
of bias momentum system, 68
of continuum, 55
of gyrostat, 68
of N-wheel system, 184
of point mass, 41
of point mass system, 46–47, 51, 84
of rigid body, 58, 87, 95, 105
with damper, 62
in orbit, 347
with wheel, 68

Lagrange region, 296, 316–318, 320–323
Lagrange's planetary equations, 289
Landon's rule, 146, 204, 207, 209, 215–216, 230
LDEF, 334–335, 346
Legendre functions, 241
Legendre polynomials, 241
Liapunov functions, 115, 122–123, 142–144, 187–188, 196, 197, 297, 302–305, 368–371, 383, 473, 505–510, 513
Liapunov's method, 114, 504–510
Lie series, 381
Likins-Pringle equilibria, 362–372, 383
Limit cycle, 225, 490–491
Linear approximation, 493–502
Linear periodic systems, 499–502
Linear stationary systems, 494–497
Linear time-variable systems, 497–502

Magnetic field, 264–265
Magnetic-hysteresis torque, 266, 324
Magnetic moment, 264, 280
Magnetic torque, 264–266, 280, 326, 405–406, 415
MAGSAT satellite, 464–465

Major-axis rule, 140–146
  for gyrostats, 195, 198, 217
  modified by damper, 151–152
Major-or-minor-axis rule, 3, 120–121, 134, 315
Mass matrix, *see* Inertia matrix (system)
Mathieu equation, 312
Maxwell distribution function, 255, 277
  drifting, 256–257
Mean free path, 248
Method of averaging, 225, 289, 292, 372, 380, 403
Mingori's gyrostat, 223
Moments of inertia:
  of bias momentum system, 66
  first, 43
  of gyrostat, 66
  mixed, 71
  parallel-axis theorem, 87
  principal, 86, 132
    with damper, 62
    with wheel, 66
  second, 43, 85
  of symmetrical rigid body, 85, 89
Momentum:
  of bias momentum system, 67
  of continuum, 54
  of gyrostat, 67
  of point mass, 41
  of point mass system, 50
  of rigid body, 58
    with damper, 62
    with wheel, 67
Momentum wheel, 459–462
Motion equations, 39–59
  for bias momentum system, 70
  for gyrostat, 70
  for point mass, 42
  for point mass system, 50
  quasi-Lagrangian, 42, 52, 60
  for rigid body, 55–61, 88, 93–138
    with damper, 62
    with wheel, 70
  for symmetrical rigid body, 96–104
Multi-rigid-body dynamics, 70–83, 90–92
  augmented bodies, 79
  connection barycenter, 79
  direct-path method, 81
  formulations, 78
  modeling, 76
Negative-definite function, 505
Newton-Euler formulation, *see* Vectorial mechanics
Norms, 482–483
NRL satellites, 324–327, 344–345
Nutation angle, 99, 130, 152, 205–206, 213, 392–393

Orlando's formula, 497
Orbital stability, 138, 489–490
Orbit interface, 1–2
  orbit-attitude coupling, 2, 284–293
  osculating ellipse, 285, 289
Orthonormal matrices, 8–9, 37–38
OSO I satellite, 447, 467
OSO VIII satellite, 455
OV1-10 satellite, 332, 334, 338

PAGEOS satellite, 412, 413
Parallel-axis theorem, 87
Parametric excitation, 259, 312
Pegasus, 266, 381
Penumbra, 279
Permanent rotations, *see* Simple spins
Pioneer satellite, 398
Poinsot's visualization, 112
Point mass, 40–42, 83
  system of, 42–52, 84
Polhode, 109–112, 133, 142, 153, 165–171
Positive-definite function, 505
Potential energy, 87, 472
  due to gravity, 234–235, 347
  of rigid body in orbit, 348
Poynting vector, 260
Precession, 100–103, 138, 163, 191, 205, 460
  angle, 100, 103
  frame, 180, 191, 207
  frequency, 100, 130, 162, 176, 190, 216
  orbit, 372, 381
  prograde, 101
  retrograde, 101
Principal axes of inertia, 85
Principal moments of inertia, 86
Principal rotations, 15, 25
Prospero satellite, 413
Pure spins, *see* Simple spins

Quasi-rigid body, 140, 205–208
  gravity gradient satellite, 315–318
  gyrostat, 193–211
  stability, 143–146
Quaternions, *see* Euler parameters

RAE satellite, 341–343
Reaction wheel, 159–160, 463–464
Reference frames, 6–38
  body-fixed, 57, 74
  magnetic, 265, 279
  orbiting, 236, 282–283, 348

principal axis, 85
reference spin, 358
stroboscopic, 359
Reflection matrix, 32
Relative spin rate, 97, 130, 138, 182, 206
Resonances, 292
nonlinear, 307–308, 379–380, 421, 453
pitch-orbit, 311
Reynolds' transport theorem, 53
Rigid body, 3, 55–61, 93–138
angular momentum, 95
attitude stability, 114–124
Euler motion equations for, 95
kinetic energy, 95
motion equations for, 55–61
88, 95
in orbit, 284–313, 293–313
orientation, 6, 55
spinning in orbit, 356–381
stability, 114–124
symmetrical, 96–104, 181
with torque, 124–129
torque-free motion, 96–124
Roberson equilibria, 428–444
Rotation matrix, 9–10
differential equation for, 22–23
eigenvalues of, 31
Euler axis and angle, in terms of, 12
geometrical interpretation, 13
product of, 31
Taylor expansion for, 21
Routh-Hurwitz criteria, 117, 151, 221, 322, 389, 495–497

Salyut satellite, 334, 346
SAS satellite, 465
Satellites:
dumbell, 292, 348
1963-22A, 323–325, 336–337
1963—38B, 323
1963–49B, 323
AEROS, 397–398
Alouette I, 411, 414
Alouette II, 413
Anik C, 423–424, 455
ATS, 333–334
DODGE, 339–342, 345
ESRO IV, 399
Explorer I, 3, 409
Explorer IV, 413
Explorer VII, 413
Explorer VIII, 413
Explorer XI, 413
Explorer XVIII, 394
Explorer XX, 412
Explorer XXXXV, 413
GEOS-I, 337–338
GEOS-II, 328, 337
GEOS C, 457–458
GGTS, 326, 338–339
HELIOS, 399–400
HERMES, 423–424, 468–470
INTELSAT IV, 446, 449–451
ISIS, 395, 407, 413
LDEF, 334–335, 346
NRL satellites, 324–327, 344–345
OSO I, 447, 467
OSO VIII, 455
OV1-10, 332, 334, 338
PAGEOS, 412, 413
Pegasus, 266, 381
Pioneer, 398
Prospero, 413
RAE, 341–343
Salyut, 334, 346
SAS, 465
SEASAT, 464–466
Skylab, 29
SKYNET, 413
solar power satellite, 279
Space shuttle, 30
Space station, 346
Sputnik I, 411
Sputnik III, 413
SYNCOM, 393–394, 410
TACSAT, 448–449
Telstars, 413
TIROS, 412–414
Vanguards, 413
SEASAT satellite, 464–466
Self-excited motion, 124–127, 136
Simple spins, 114, 153
for gyrostat, 160–161, 165–178, 187
Simulation, 4–5
Solar-motoring torque, 407–408
Solar-pressure constant, 263
Solar-pressure torque, 260–264, 292, 314, 337, 356
on flat plate, 278
penumbral, 263, 279
on sphere, 279
Skylab, 29
SKYNET satellite, 413
Solar sailing, 2
Space shuttle, 30
Space station, 346
Specular reflection, 250, 276
Spin stabilization, 354–422

Sputnik I satellite, 411
Sputnik III satellite, 413
Stability boundaries, 497–498
Stability definitions:
  asymptotic stability, 488, 490
  attitude stability, 116
  directional stability, 116, 121–124
  global stability, 488
  gyric stability, 515–519
  infinitesimal stability, 502
  instability, 487
  Liapunov stability, 482–487
  local stability, 488
  orbital stability, 489, 490
  of origin, 492–493
  robustness, 439
  static stability, 513
  total stability, 488
  uniform stability, 487
  w-stability, 114
Stability theory, 4, 480–521
  linear periodic systems, 499–502
  linear stationary systems, 494, 497
  linear systems, 493–502, 510–520
  linear time-variable systems, 497–502
Stiffness matrix, 510
  for gravity gradient satellite, 295, 317, 320, 331
  for gyrostat, 172
    with damper, 221
    with gimballed momentum wheel, 477
    in orbit, 435–436
  for rigid body with damper, 149, 154
  for spinning rigid body, 117
  for Thomson equilibrium, 359, 383
  for zero momentum gyrostat, 164
Structures interface, 3
Sturm's theorem, 517
SYNCOM satellite, 393–394, 403, 410

TACSAT satellite, 448–449
Telstar satellites, 413
Thermal flutter, 269, 337
Thermodynamics, second law, 3
Thomson equilibrium, 357–362, 369, 372, 385–386, 401, 418, 420
Thrusters, *see* Gas jets
TIROS satellites, 412–414
Topological tree, 77
Torques, 232–280
  aerodynamic, 248–260
  disturbing, 2
  eddy current, 266, 280
  Earth radiation, 263, 264
  gravitational, 233–248
  magnetic, 264–266, 280
  magnetic hysteresis, 266
  meteoroidal, 266–268
  solar pressure, 260–264
  vs. altitude, 272
TRANSIT satellite, 336–337
Trap states, 453
Tri-inertial rigid body, 104

Vanguard satellites, 413
Variational approximation, *see* Linear approximation
Vectorial mechanics, 39–40, 51–52
Vectrix (pl. vectrices), 8, 522–534
Venus,260